Dictionary *of* THEORIES

Area Editors

The area editors involved in the preparation of the headword list and vetting of the entries are as follows:

Astronomy/Engineering/Physics	Michael Sprackling Department of Physics King's College London
Business/Economics	Donald Rutherford Department of Economics University of Edinburgh
Chemistry	Michael Freemantle IUPAC Oxford
Life sciences/Medicine	Diane and Mark Dittrick Freelance editors USA
Linguistics/Literary theory/Stylistics	Roger Fowler School of English and American Studies University of East Anglia
Mathematics/Computer science	Jonathen Borwein now Shrum Professor of Science Simon Fraser University Vancouver
Natural & Earth sciences	Michael Allaby Freelance editor UK
Philosophy	A R Lacey Department of Philosophy King's College London
Politics/History	Rodney ~~Ronald~~ Barker Department of Government London Schoool of Economics and Political Science
Psychology/Sociology	Noel Sheehy School of Psychology Queens University Belfast

Dictionary *of* THEORIES

Jennifer Bothamley

Gale Research International Ltd

LONDON • DETROIT • WASHINGTON D.C.

ISBN 1-873477-5-8

Published in the United Kingdom by Gale Research International Limited
Published simultaneously in the United States by Gale Research Inc.
(An affiliated company of Gale Research International Limited)

A CIP catalogue record for this book is available from the British Library.

Typeset by Florencetype Ltd, Kewstoke, Avon
Printed in the United Kingdom by BPCC

Contents

Introduction

The idea for this dictionary first came from Mark Annis, Senior Assistant Librarian for Northumberland County Library Service. He wanted a book that covered 'theories' in all subject areas and which gave the librarian a brief explanation of, for example, the **butterfly effect**, its subject area – Physics not Biology – and a suggestion as to where to look for a fuller explanation.

The first problem when compiling a dictionary of this type is what constitutes a theory. The *Shorter Oxford English Dictionary* defines a theory as:

a) A scheme or system of ideas or statements held as an explanation or account of a group of facts or phenomena; a hypothesis that has been confirmed or established by observation or experiment and is propounded or accepted as accounting for the known facts; a statement of what are held to be the general laws, principles or causes of something known or observed.
b) that department of an art or technical subject which consists in the knowledge (or theory of knowledge), statement of the facts on which it depends, or its principles or methods, as distinguished from the practice of it
c) systematic statement of the general principles or laws of some branch of mathematics; a set of theorems forming a connected system

Hence not only theories, but also principles, hypotheses, rules, paradoxes, laws, principles and various 'ism's, 'ologies and 'sis's have been included. Coverage is thus not only of the sciences, with its very clearly defined idea of what is a theory, but also the arts, where the definition becomes more hazy.

Separate entries are given where a theory is covered by one or more subject areas, but with a different emphasis; where the definition covers a variety of areas the entry shows the multiple areas covered and gives an explanation for any slight variations in the perception of the theory. For example, **field theory** is covered separately by Mathematics, Physics and Psychology, as each area has a different definition of what field theory means to it. On the other hand, **futurism**, a term used both in Art and Literary theory and developed from the same movement, has only one entry for the two areas.

Each entry includes not only a field label, but also, where appropriate: an indication of when the theory was first proposed and by whom; a concise description of the theory and a comment about its validity; and a one item bibliography giving details of where a fuller explanation can be found. This is either a basic textbook or a general text dedicated to the subject (see Peter Principle).

There are two indexes. One covers all the people mentioned in the text and the entries in which they appear; the other lists all entries under their appropriate subject area, with entries which cover more than one area appearing under both – **Hering theory of colour vision** is found under both the Physics and Psychology headings.

As always with a first edition, as many theories have been left out as have been included; any suggestions for what should be included in the next edition will be gratefully received and should be sent to the publishers address.

Acknowledgements

The editors would like to thank the following for their help, guidance, expertise and hardwork: Dr Sophie Botros, Sarah Hall, Jim Hopkins, Dr Keith Hossack, Dr Chris Hughes, Alan McIntosh, David Pickering, Dr Gabriel Segnal, Dr Lucas Siorvanes.

The Philosophy area editor would also like to acknowledge the following titles: A Flew, ed., *Dictionary of Philosophy* (Macmillan, 1984); A Bullock, O Stallybrass and S Trombley, eds., *Fontana Dictionary of Modern Thought*, 2nd edn (Fontana, 1988); G Vesey and N A Foulkes, eds., *Dictionary of Philosophy* (Collins, 1990); P A Angeles *Dictionary of Philosophy* (Harper and Row, 1981); P Edwards, ed., *Encyclopedia of Philosophy* (Collier Macmillan, 1973).

Contributors

AB	Antonia Boström, freelance editor and writer, London
ABA	Alan Barnard, Department of Social Anthropology, University of Edinburgh
AL	Adrian Lewis, Department of Combinatorics and Optimization, University of Waterloo
AM	Alan McIntosh, freelance editor and writer, Edinburgh
ARL	A R Lacey, Department of Philosophy, King's College London
AV	Aruna Vasudevan, editor, London
DP	David Pickering, freelance editor and writer, Buckingham
DT	D H Tarling, Department of Geological Sciences, University of Plymouth
GD	Gerard Delaney, freelance editor and writer, London
HR	Helen Roberts, Co-ordinator of Research, Barnardo's
JB	John Bowers, School of Mathematics, University of Leeds
KH	Karla Harby, freelance editor and writer, New York
LL	Lucinda Lubbock, freelance editor and writer, London
MB	Michael Bean, Department of Pure Mathematics, University of Waterloo
MF	Michael Freemantle, IUPAC, Oxford
ML	Martha Limber, Department of Applied Mathematics, University of Waterloo
MLD	Madeleine Ladell, freelance editor and writer, London
MS	Michael Sprackling, Department of Physics, King's College London
NS	Noel Sheehy, School of Psychology, Queens University Belfast
PH	Paul Hannon, freelance editor and writer, London
RA	Robin Allaby, Department of Biochemistry, UMIST
RB	~~Ronald~~ Rodney Barker, Department of Government, London Schoool of Economics and Political Science
RF	Roger Fowler, School of English and American Studies, University of East Anglia

A

***a priori* theories** *Accountancy* Popular in the 1960s, this approach to accounting relied heavily on deductive reason for the development and improvement of accounting systems. Proponents of the method used assumed axioms instead of experience to develop cause and effect principles.

PH

abandonment (20th century) *Philosophy* Concept central to atheistic EXISTENTIALISM. According to such Existentialists as Sartre, God does not exist and life therefore has no intrinsic purpose or meaning. Man has been 'abandoned' in the universe and must create his own morality and code of values without the assistance of any divine being. *See also* ATHEISM.

W Kaufmann, ed, *Existentialism from Dostoevsky to Sartre* (New York, 1956; London, 1957)

DP

abc conjecture (1985) *Mathematics* Due to the mathematician J Oesterlé and generalized by D W Masser, this is the conjecture that there exists a positive number k such that for all positive integers a, b and c with

$$a + b = c$$

and

$$(a,b,c) = 1$$

we have

$$z < G^k$$

where G is the greatest square-free factor of abc. This conjecture has profound implications in the study of DIOPHANTINE EQUATIONS; for example, that the Fermat equation has at most a finite number of solutions. *See* FERMAT'S LAST THEOREM.

C L Stewart and Kunrui Yu, 'On the ABC Conjecture', *Math. Ann.*, 291 (1991), 225–30

MB

ABC model *Psychology* A model of behaviour based on the operant conditioning paradigm of American psychologist Burrhus Frederic Skinner (1904–90).

The ABC model states that behaviour may be changed by manipulating either the conditions preceding the behaviour or the consequences following the behaviour. Behaviour that is rewarded will occur more often while behaviour that is not, or punished, will occur less frequently. This is a basic assumption of clinical and educational psychologists' operating behaviour modification programmes. *See* BEHAVIOUR THERAPY.

B F Skinner, *About Behaviourism* (New York, 1974)

NS

Abegg's rule *Chemistry* Empirical rule named after its discoverer, German chemist Richard Abegg (1896–1910).

Solubilities in water of salts of alkali metals with strong acids (such as hydrochloric, nitric and sulphuric acids) decrease on descending the alkali metal group in the periodic table; that is, from lithium to caesium. The solubilities of alkali metal salts of weak acids (such as ethanoic acid) increase on descending the group. The salt sodium chloride is an exception, being less soluble in water than potassium chloride.

MF

Abegg's rule of eight (1904) *Chemistry* Named after German chemist Richard Abegg (1869–1910), noted for his theory of valency. The sum of the negative and positive valencies of an element is eight. For example, phosphorus has a valency of -3 in PH_3 and $+5$ in PCl_5, respectively. The rule is limited to non-metallic elements such as phosphorus and sulphur.

J R Partington, *General and Inorganic Chemistry* (London, 1949)

MF

Abel's limit theorem (19th century) *Mathematics* Named after the Norwegian mathematician Niels Henrick Abel (1802–29). The result that the limit assigned to a convergent series by ABEL SUMMATION exists and agrees with the sum.

M Hazewinkel, ed., *Encyclopaedia of Mathematics* (Dordecht, 1988)

ML

Abel's summation (19th century) *Mathematics* Although named after the Norwegian mathematician Neils Henrick ABEL (1802–29), this method of SUMMABILITY THEORY can be found in works by Euler and Leibniz.

It is used to compute a possibly divergent infinite series of complex numbers. The series

$$\sum_{k=0}^{\infty} a_k$$

can be summed by the Abel method to S if the series

$$\sum_{k=0}^{\infty} a_k x^k$$

is convergent for any $0 < x < 1$ and

$$\lim_{x \to 1^-} \sum_{k=0}^{\infty} a_k x^k = S.$$

K Knopp, *Theory and Application of Infinite Series* (New York, 1990)

ML

Abelian theorems *Mathematics See* TAUBERIAN THEOREMS.

ability-to-pay principle (16th century–) *Economics* Taxation should be levied according to an individual's ability to pay; that is, the more prosperous will meet a proportionately higher percentage of the national tax demand. This principle was extended by the Swiss philosopher Jean Jacques Rousseau (1712–78), the French political economist Jean-Baptiste Say (1767–1832) and the English economist John Stuart Mill (1806–73).

The principle (in contrast to the BENEFIT APPROACH PRINCIPLE) is based on the notion of equal sacrifice, is generally regarded as the most equitable form of taxation, and is used in most industrialized economies; but equality of sacrifice is open to interpretation as it can be measured in absolute, proportional or marginal terms. *See* EQUAL SACRIFICE THEORY and TAX INCIDENCE.

R A Musgrave and P B Musgrave, *Public Finance in Theory and Practice* (New York, 1975)

PH

abiogenesis *Biology See* SPONTANEOUS GENERATION.

Abney's effect *Psychology* Named after the English chemist and physicist Sir William Abney (1843–1920), this is the phenomenon whereby when a large area is suddenly illuminated, the light seems to appear first in the centre of the patch (rather than appearing all at once) and then spreads outwards towards the edges. When the light is extinguished, the edges disappear first and the centre last.

L Kaufman, *Perception* (New York, 1979)

NS

Abney's law (19th century) *Physics* Named after its English discoverer, the chemist Sir William de Wivelslie Abney (1844–1920), this states that every beam of white light may be regarded as a mixture of a certain amount of white light plus light of a given wavelength. This wavelength defines the hue.

When the white light content is zero the light is said to be saturated. When a pure spectral colour is desaturated by adding white light to it, its hue shifts towards the red end of the spectrum if the wavelength of the spectral colour is less than 570 nm; and towards the blue end of the spectrum if the wavelength of the spectral colour is greater than 570 nm.

J Thewlis, ed., *Encyclopaedic Dictionary of Physics* (New York, Oxford and London, 1962)

MS

absence (1966) *Literary Theory* Key idea of French Marxist critic Pierre Macherey (1938–).

Literary texts are not simply to be 'interpreted', in a process of extracting what they say, since they manifest significant 'absences' (what is not said) related to Freud's unconscious. These absences point to the text's contradictory relationships to ideology.

P Macherey, *A Theory of Literary Production*, G Wall, trans. (London, 1978 [1966])

RF

absence paradox (19th century) *Philosophy* A source of humour used in music halls, but possibly ancient.

No person is ever present, because he is either not in Vladivostok or alternatively is not in Patagonia, so he must be somewhere else. If he is somewhere else, he certainly is not here. This argument understands a relative adverb as absolute.

JB

absolute *Philosophy* 19th century idealists from Hegel (1770–1831) to Bradley (1846–1924) – unlike the earlier 18th century idealist Berkeley (1685–1753) – were heavily influenced by Immanuel Kant (1724–1804). Kant insisted that there are limits to what we can, in principle, know about reality, and that we necessarily look at the world in certain ways only; for example in terms of substance and cause.

These idealists concluded that the world as we see it is somehow derivative, apparent, relative, incomplete, and even contradictory (Bradley). They used 'the absolute' as a term for reality as it really is, free from these limitations but unknowable by us. Spinoza (1632–77) and the Greek philosopher Parmenides (early 5th century BC: *see* ELEATICISM) have been seen as forerunners of this doctrine. For absolute idealism *see* OBJECTIVE IDEALISM.

A Quinton, *Absolute Idealism* (1972); reprint of Dawes Hicks lecture at British Academy in 1971

ARL

absolute advantage theory (18th century) *Economics* This theory of Scottish economist Adam Smith (1723–90) was an international trade theory that asserted individuals or nations trade because they have superior productivity in particular industries.

Governments may attempt to counter absolute advantage by erecting trade barriers, allowing young, uncompetitive, industries enough time to become established. Smith's theory was superseded by the comparative advantage approach. *See also* COMPARATIVE COSTS, HECKSCHER-OHLIN TRADE THEORY and TECHNOLOGICAL GAP THEORY.

C P Kindleberger and P H Lindert, *International Economics* (Homewood, Ill., 1978)

PH

absolute income hypothesis (1936) *Economics* Proposed by English economist John Maynard Keynes (1883–1946) as part of his work on the relationship between income and consumption, this theory was much refined during the 1960s and 1970s, notably by American economist James Tobin (1918–).

Consumption is a non-linear function of income. As income rises, consumption will rise but not necessarily at the same rate. In its developed form, this hypothesis is still generally accepted. *See* RELATIVE INCOME HYPOTHESIS and PERMANENT INCOME HYPOTHESIS. *See also* LIFE-CYCLE HYPOTHESIS.

J M Keynes, *The General Theory of Employment, Interest and Money* (New York, 1936)

PH

absolute pitch theory *Physics* Absolute pitch is the name given to the ability to recognize and define the pitch of a note without the use of a reference tone.

Theories relating to this ability can be divided into four groups, for none of which is there conclusive evidence. (1) Hereditary theory: a child so gifted is assumed to learn pitch names in early life, just as it learns the names of colours. (2) Learning theory: it is assumed that anyone can acquire absolute pitch by diligent and constant practice. (3) Unlearning theory: this assumes that the ability to develop absolute pitch is nearly universal, but is trained out of most people at an early age. (4) Imprinting theory: this theory assumes that absolute pitch is the result of rapid, irreversible learning at a specific developmental stage.

J Thewlis, ed., *Encyclopaedic Dictionary of Physics* (New York, Oxford and London, 1962)

MS

absolute reaction rate, theory of (1935) *Chemistry* Theory of reaction rates based on statistical thermodynamics. Also known as the transition state theory, it was developed by M Polanyi and particularly H Eyring following earlier work of Richard Chace Tolman (1881–1948) in 1927 and H Pelzer and the Hungarian physicist Eugene Paul Wigner (1902–) in 1932.

As the reactant molecules change towards product molecules there is an increase in potential energy until a maximum is reached. At this maximum, molecules have a critical configuration known as the transition state or activated complex. Only molecules with adequate total energy can attain this critical state. As the transition state changes to the product molecules there is a decrease in potential energy. The absolute rate theory assumes an equilibrium between reactant molecules and the activated complex. The equilibrium constant is determined by statistical thermodynamics. It is assumed that the activated complex differs from the reactant molecules in only one loose vibrational mode. On the basis of this theory it is possible to calculate the rates of chemical reactions.

M Freemantle, *Chemistry in Action* (Basingstoke, 1987)

MF

absolute space and time (17th century) *Physics* The concept of force as presented in Newtonian mechanics gives rise to the need to specify the velocity of a body and its rate of change. For this purpose some reference system is needed.

Newton postulated the existence of absolute space and absolute time with respect to which the forces and movements in nature should be determined. Absolute space is the reference frame against which the motion of the body in space is to be measured. Absolute time, according to Newton, 'flows uniformly on, without regard to anything external' and enables a unique date to be assigned to every event. However, as many commentators have pointed out, there is nothing to distinguish one interval of time from any other, except the different events that occur in them. Similarly, there is no way of distinguishing one part of space from another, except through the relation of the position of material bodies in space. 'All knowledge of space and time is essentially relative' (James Clerk Maxwell). Absolute space and absolute time are misconceived notions.

Newton, in fact, assumed that the distant galaxies are not moving and provide a reference frame embedded in (absolute) space. This is not true, but the relative positions of the galaxies and of the so-called fixed stars, as seen from Earth, change so very slowly that they provide an adequate reference frame in space, while the period of any oscillator provides the necessary measure of time interval. *See* NEWTONIAN MECHANICS.

J Thewlis, ed., *Encyclopaedic Dictionary of Physics* (New York, Oxford and London, 1962)

MS

absolute zero (17th century–) *Physics* Regardless of the temperature scale and unit used, there is a least possible temperature that is the same for all substances. The concept of an 'ultimate degree of cold' was described by the English philosopher John Locke (1632–1704), but more carefully formulated by the French theorist Guillaume Amontons (1663–1705).

On the most commonly used temperature scale (the thermodynamic scale), this lowest possible temperature is given the value zero and the temperature is referred to as the absolute zero of temperature. In macroscopic terms, absolute zero is the temperature at which a system would undergo a reversible isothermal process without heat transfer. In microscopic terms, the thermal energy of the atoms of a system is zero at absolute zero, except for the zero-point energy attributed to the atoms by quantum mechanics. According to the third law of thermodynamics, absolute zero is a well-defined temperature but is not attainable except as the limit to an infinite series of processes.

J Thewlis, ed., *Encyclopaedic Dictionary of Physics* (New York, Oxford and London, 1962)

MS

absolutism *Philosophy/Politics* In philosophy, a contrast to RELATIVISM in any of its senses. In its political sense, a description (more frequently than justification) of government without constitutional restrictions. The authority to govern cannot be qualified or restricted, because if it is, whatever restricts it is itself the final power.

Historically, one form it has taken has been the doctrine of the 'divine right of Kings', which was popular in Stuart England. Though this doctrine was expressed in the course of 16th and 17th century debates over sovereignty, the term absolutism became current only in the 19th century to describe regimes having either such a character or such an aspiration. Notable supporters of political absolutism have included Hegel and Mussolini. *See also* DIVINE RIGHT.

David Miller *et al*, eds *The Blackwell Encyclopaedia of Political Thought* (Oxford, 1987)

ARL, RB

abstinence rule *Psychology* A rule for clients in therapy, derived from the psychoanalytic theory of Austrian Sigmund Freud (1856–1939).

Anxiety and frustration are considered to be driving forces in the therapeutic process. In order to preserve such 'energy' for the therapeutic process, clients are encouraged not to engage in 'gratifications' (for example, smoking, idle conversation, high levels of sexual activity) outside the therapy sessions.

R J Corsini, *Current Psychotherapies* (Illinois, 1984)

NS

abstract art (20th century) *Art* Synonymous with 'abstraction', this term has been employed since the first decades of the 20th century to describe non-representational painting; a style which persists today.

Although elaborated upon and refined in its precise manifestations, abstract art essentially comprises non-figurative art, as first exemplified by the colour abstracts of Russian painter Wassily Kandinsky (1866–1944) around 1911. It then developed in CUBISM and later, on an international scale, in ABSTRACT EXPRESSIONISM.

H Osborne, *Abstraction and Artifice in Twentieth Century Art* (1979)

AB

abstract expressionism (1946) *Art* An artistic movement whose title was first used by the American critic Robert Myron Coates (1897–1973) in the *New Yorker* magazine to describe contemporary painting in New York.

Its roots lie in SURREALISM and AUTOMATISM, from which it adopted and developed theories of improvisation, spontaneity and the importance of the process of artistic creation. Stylistically, the painters included in this movement are not homogeneous; however, the 'drip paintings' of Jackson Pollock (1912–56) and the large monochromatic COLOUR FIELD PAINTING canvases of Mark Rothko (1903–70) represent just two approaches to the challenge of non-figurative, self-expressive painting. *See* ACTION PAINTING.

I Sandler, *Abstract Expressionism. The Triumph of American Painting* (1973)

AB

abstractionism *Philosophy* View that the mind gets some or all of its concepts by abstracting them from concepts it already has or from experience.

For example, one might abstract 'red' from a set of experiences, each involving red along with other properties; or (differently) abstract the generic concept 'animal' from the already possessed concepts of its species (cow, horse, and so on); or (differently again) abstract a determinable concept like 'colour' from the concepts of its determinate forms (red, blue, and so on). Abstractionism has been criticized on such grounds as that the resulting concepts or ideas would be impossibly empty or indeterminate (Berkeley (1685–1753)), or that it could not produce concepts that play the roles our concepts actually do play (Geach).

P T Geach, *Mental Acts* (1957), 18–44

ARL

absurd, theatre of the (1962) *Literary Theory* Phrase coined by the British critic Martin Esslin (1918–), but the relevant sense of 'absurd' was used in 1942 by French existentialist philosopher and writer Albert Camus (1913–60). The word was also employed in 1950 reviews of *The Bald Prima Donna* by Eugene Ionesco (1912–).

Anti-realistic theatre of the 1950s, typically the work of E Ionesco and Samuel Beckett (1906–89), celebrating the meaninglessness of life and the impotence of man in an uncaring universe. Theatrical techniques such as nonsensical dialogue, inconsistency of place and of character, absence of a named 'hero', grotesque and negative behaviour, make pointlessness palpable; these techniques have a history much longer than the absurd, going back to the turn of the 20th century.

M Esslin, *The Theatre of the Absurd* [1962], revised and enlarged edn (London, 1968)

RF

acceleration principle (1917) *Economics* Formulated by American economist John Clark (1884–1963), this principle is one of the basic pillars of modern macroeconomic theory and is an important tool in understanding the development of business cycles.

The acceleration principle links investment to output, each level of which needs a specific amount of capital. If output (and the amount of machinery required to make it) is expected to rise, the amount of capital within an economy must be increased. The accelerator equation is

$$I = \alpha \Delta t$$

where I is net investment in year t, α is the accelerator coefficient and Δt is the annual change in income. *See also* BUSINESS CYCLE, HARROD-DOMAR GROWTH MODEL, and TRADE CYCLE.

J M Clark, 'Business Acceleration and the Law of Demand', *Journal of Political Economy*, vol. xxv, 3 (March, 1917), 217–35; J R Hicks, *A Contribution to the Theory of the Trade Cycle* (Oxford, 1950)

PH

acceptability *Linguistics See* GRAMMATICALITY.

accident proneness (20th century) *Psychology* Hypothesis developed by the English theorists Major Greenwood and M Woods.

Early epidemiological studies of industrial accidents seemed to indicate that a small group of individuals have a disproportionately high level of accidents while the majority have very few. This was ascribed to a set of personality traits (for example, impulsiveness, emotional instability) that predispose the individual to repeated accidents. The evidence suggesting the existence of an accident prone psychological profile has since been discredited.

N P Sheehy and A J Chapman, 'Industrial Accidents', *International Review of Industrial and Organisational Psychology*, C L Cooper and I T Robertson, eds (Chichester, 1987)

NS

accommodation (1973) *Linguistics* Developed by British social psychologists, especially Howard Giles.

An individual may accommodate her or his speech to an addressee in order to be like the addressee and receive positive approbation (CONVERGENCE), or to affirm an identity separate from the addressee (DIVERGENCE). Accent, speech tempo, vocabulary, even language choice in multilingualism, may be adjusted to conform to or diverge from the addressee's norms.

H Giles and N Coupland, *Language: Contexts and Consequences* (Milton Keynes, 1991)

RF

accretion hypothesis (1866) *Geology* Initiated by the American geologist James Dwight Dana (1813–95) and mainly documented by the American A E J Engel in 1963, this theory proposes that continents have grown by the addition (accretion) of newly formed continental rocks to their edges.

The type area for such growth is North America where the older rocks are generally surrounded by wide strips of rocks of younger and younger age. This pattern of growth is much more difficult to see (or does not exist) in other continents, such as Africa. Nonetheless, new continental rocks are now considered to be generated in narrow zones bordering the oceanic deeps where oceanic rocks descend into the Earth's interior, but are mainly converted into true crustal materials when the zone is the site of continental collision.

A distinct accretion theory proposes that, at its beginning, the Earth was formed by the accumulation (accretion) of meteorite-like bodies which are termed planetisimals.

B F Windley, *The Evolving Continents* (New York, 1977)

DT

acid-growth hypothesis (mid-20th century) *Biology* Advocated by numerous botanists since the 1960s.

A disputed theory to explain the wall loosening which occurs when plant cell walls are made more plastic and extensible by treatment with auxins, a class of plant hormones. This theory holds that auxins cause certain cells in the stem or elsewhere to secrete H^+ into their surrounding walls, and that these ions lower the pH (that is, increase acidity) so that wall loosening and rapid growth can occur. The theory also seeks to explain the action of the Venus's flytrap plant in catching insects. Research reported since 1980 has seriously undermined this theory.

F Salisbury, *Plant Physiology* (Belmont, Calif., 1992)

KH

Ackeret's theory (1925) *Physics* Named after Jakob Ackeret who introduced it, this theory relates to a disturbance produced in a supersonic air-stream and its resemblance to sound waves.

When introduced into a supersonic air-stream, an aerofoil with a sharp leading edge and its whole surface at a small angle to the direction of air flow creates a disturbance. This comprises two plane waves occurring at the leading and trailing edges of the aerofoil, respectively, which are propagated outwards like sound waves.

J Thewlis, ed., *Encyclopaedic Dictionary of Physics* (New York, Oxford and London, 1962)

MS

Ackermann function *Mathematics* Attributed to the American philosopher and logician Robert John Ackerman (1933–).

The rapidly growing function defined recursively by

$$A(0,y) = 1 \quad \text{for } y \geqslant 0$$
$$A(1,0) = 2$$
$$A(x,0) = x + 2 \quad \text{for } x \geqslant 2$$
$$A(x,y) = A(A(x-1,y),y-1) \quad \text{for } x,y \geqslant 1.$$

Thus

$$A(x,1) = 2x$$
$$A(x,2) = 2^x$$
$$A(x,3) = 2^{2^{\cdot^{\cdot^{\cdot^{2}}}}}$$

and $A(x,4)$ is so rapidly growing that there is no accepted mathematical notation for such a function. Moreover, such a function is not effectively computable.

A V Aho, J E Hopcroft and J D Ullman, *Data Structures and Algorithms* (Reading, Mass., 1983)

AL

acoustic phonetics (1940s–) *Linguistics* Facilitated by electronic developments in the USA, which were initially aimed at providing 'visible speech' for the deaf.

The analysis (using instruments) of the physical properties of the sounds of speech, in terms of such analytic concepts as frequency, amplitude, complexity, and noise. The findings of acoustic phonetics have many practical applications, and are also an important input to the theory of PHONOLOGY. *See also* ARTICULATORY PHONETICS and AUDITORY PHONETICS.

P Ladefoged, *Elements of Acoustic Phonetics* (Chicago, 1962)

RF

acquired characteristics, inheritance of (19th century) *Biology* Also called use-inheritance, this is the notion that changes occur in an organism as a consequence of environmental influences and not genetic factors, and that these changes may be passed on to the organism's offspring. The theory is associated with French biologist Jean Baptiste Pierre Antoine de Monet, Chevalier de Lamarck (1744–1829), although the idea was widely believed before his time.

An example would be Lamarck's belief that giraffes have long necks because of generations of stretching upward to feed upon leaves in trees. This theory has been abandoned in the 20th century, as it cannot be experimentally confirmed and other theories better explain such observations. *See also* ADAPTATION, EVOLUTION, LAMARCKISM and LYSENKOISM.

J B Lamarck, *Zoological Philosophy* (1809; Chicago, 1984)

KH

act of identity (1970s) *Linguistics* A sociolinguistic concept developed by the British linguist R B LePage; similar to ACCOMMODATION.

The varieties of language, dialect, accent, speech style and so on, which are available in a community express a network of idealized social values. An individual performs a subjective and motivated ACT OF IDENTITY in choosing how to speak; expresses a social identity.

R B LePage, 'Projection, Focusing, Diffusion', *York Papers in Linguistics*, 9 (1980), 1–33

RF

act utilitarianism *Philosophy* Also called 'extreme' or direct UTILITARIANISM. Original, and 'official' form of utilitarianism which says that our duty on any occasion is to act in the way which will produce actual overall consequences better than (or at least as good as) those that any other act open to us would produce.

Difficulties in predicting consequences, including difficulties in principle where self-prediction is concerned, mean that as a practical prescription utilitarianism can only tell us to aim for the best probable outcome, and act utilitarianism has often been superseded by RULE UTILITARIANISM.

J J C Smart, 'Extreme and Restricted Utilitarianism', *Theories of Ethics*, P Foot, ed. (1967); defends the former

ARL

actantial theory (1966) *Stylistics* Developed by French structuralist Algirdas Julien Greimas.

Development of NARRATIVE GRAMMAR based on Vladimir Propp's *Morphology of the Folktale*. Greimas analyzed the roles of characters in narrative from a functional point of view, reducing Propp's 32 'functions of the *dramatis personae*' to just six which he called *actants*: sender/receiver, subject/object, helper/opponent. This approach to narrative grammar is analogous to CASE GRAMMAR.

A J Greimas, trans. D McDowell, R Schleifer and A Velie, *Structural Semantics* (Lincoln, 1983 [1966])

RF

action painting (1950s) *Art* Term invented by the American art critic Harold Rosenberg, with reference to ABSTRACT EXPRESSIONISM. It was used to describe the approach of American artists such as Willem de Koonig (1904–) and Jackson Pollock (1912–56), whose works emphasized the actions of artistic creation.

H Rosenberg, *The Tradition of the New* (1959)

AB

action theory (20th century) *Politics/Sociology* Theory of social investigation.

The social investigator cannot and should not be indifferent to what s/he studies. People are always both subjects and objects of research, and thus investigation should involve principled stands on the problems studied, and principled intentions of changing them.

Maggie Humm, *The Dictionary of Feminist Theory* (Hemel Hempstead, 1989)

HR

action theory *Psychology* Term coined by the American psychologist and philosopher William James (1842–1910).

Any psychological theory can be described as an action theory if it is concerned with the study of human goal-directed behaviour and the social basis for that behaviour. The three main components of any action theory are action, act and actor. An action can be defined as planned and goal-directed behaviour; an act as a unit of action occurring in a socially-defined situation and which is characterized by a goal; an actor as a person with the capacity for self-reflection.

A J Chapman and D M Jones, eds, *Models of Man* (Leicester, 1980)

NS

activated complex theory *Chemistry See* ABSOLUTE REACTION RATE, THEORY OF.

activation-synthesis hypothesis (20th century) *Psychology* Name given to a theory of American psychologists J Allan Hobson and R W McCarley.

A theory of dreaming which draws together the evidence indicating that the brain is active during sleep (activation), and postulates that dreams are a conscious interpretation (synthesis) of all this activity. Since many dreams do not reflect either the brain's or body's state they are considered sleeping interpretations of physical states. The theory continues to attract scientific interest.

J A Hobson and R W McCarley, 'The Brain as a Dream State Generator: An Activation-Synthesis Hypothesis of the Dream State', *American Journal of Psychiatry* vol. CXXXIV (1977), 121

NS

activation theory of emotion *Psychology* A generalization made by the American psychologist and philosopher William James (1842–1910).

Represented in many different theoretical conceptions of emotion, the assumption is that emotions lie collectively at the extreme pole of a continuum of arousal or activation, while the states of coma and deep sleep lie at the opposite pole. The arousal or activation refers to increased neurophysiological arousal. The primacy given to neurophysiological arousal over cognitive or 'mental' states varies with different theories of emotion. *See* CANNON-BARD THEORY OF EMOTION, COGNITIVE APPRAISAL THEORY, JAMES-LANGE THEORY, and THALMIC THEORY.

R Plutchik and H Kellerman, eds, *Emotion: Theory, Research and Experience* (New York, 1980)

NS

activism (20th century) *Philosophy/Psychology* Theory of the German-American psychoanalyst Erich Fromm (1900–80). Rarely used in psychology, this is the doctrine that any relationship between thought and reality is characterized by continuous activity on the part of the mind rather than passive sensory receptivity.

A J Chapman and D M Jones, eds, *Models of Man* (Leicester, 1980)

NS

activism (20th century) *Politics* Theory of political tactics. Argument and persuasion are limited by existing circumstances. In order to change those circumstances action is necessary which either achieves real change or demonstrates possible changes. At its best, activism can avoid pessimism and resignation; at its worst, it can encourage the belief that action is an alternative to thought.

Alan Bullock, Oliver Stallybrass and Stephen Trombley, eds, *The Fontana Dictionary of Modern Thought*, 2nd edn (London, 1988)

RB

activity theory of ageing (20th century) *Psychology* Theory developed by the American psychologist Robert J Havighurst.

The natural inclination of most elderly people to associate with others and to participate in group and community affairs is often blocked and disrupted by contemporary retirement practices. Disengagement and re-engagement are counterbalancing tendencies. Through the former, older people relinquish certain social roles; while the latter prevents the consequences of disengagement from going too far in the direction of social isolation, loneliness and so forth. The theory has been extensively criticized and has undergone several revisions. *See* DISENGAGEMENT THEORY OF AGEING.

D B Bromley, *The Psychology of Human Ageing* (Harmondsworth, 1966)

NS

actor-observer bias *Psychology* Theory of the American psychologist Harold H Kelley (1921–).

The term identifies a judgmental bias whereby individuals place grater emphasis on the role of situational factors when explaining their own behaviour, and stronger emphasis on dispositional qualities when accounting for similar behaviour in others. For instance, when someone trips we tend to think the fault may lie in their carelessness, whereas when we trip we tend to attribute it to external factors (such as a slippery floor). *See* FUNDAMENTAL ATTRIBUTION ERROR and SELF-SERVING BIAS.

NS

H H Kelley and J L Michela, 'Attribution Theory and Research', *Annual Review of Psychology*, vol. XXXI (1980), 457–501

NS

actualism (1968) *Art* Term first employed by French writer Alain Jouffrey to account for the effect of revolutionary situtations on art, for instance during the Paris riots in May 1968.

During such events the division between art and social reality ceases to exist, and its significance or irrelevance is made manifest.

A Jouffrey, 'What's to be Done about Art', *Art and Confrontation: France and the Arts in an Age of Change* (London, 1970)

AB

actualism (1785) *Geology* Observable present-day processes are considered to have existed during the geological past; hence all past geological features are interpretable in terms of processes acting on the Earth today. The theory was originated by the uniformitarian Scottish geologist James Hutton (1726–97), proselytized by the mathematician John Playfair (1748–1819), and developed by another Scottish geologist, Sir Charles Lyell (1797–1875).

While this approach now seems self-evident, it contrasts with the unnatural events invoked by adherents of CATASTROPHISM. There is a distinct link between actualism and PLUTONISM. *See also* UNIFORMITARIANISM.

C Lyell, *Principles of Geology* (1830); A Hallam, *Great Geological Controversies* (Oxford, 1983)

DT

actualization theory *Psychology* Proposed by the American psychologist Abraham Harold Maslow (1908–70).

The theory holds that individuals will work towards becoming everything that they possibly can; owing to the human need or psychological imperative for self-fulfilment and not merely a personal preference or inclination. Maslow postulated a hierarchy of needs: basic physiological needs, safety needs, belongingness needs, esteem needs and self-actualization needs. The final level of psychological development is achieved when all lower order needs are fulfilled and the actualization of the full personal potential takes place. The theory led to the development of actualization therapy.

A Maslow, *Motivation and Personality* (New York, 1970)

NS

acuity theory *Psychology* Developed by the German-French psychologist K R König (1832–1901).

A general body of theory relating to the acuity of the visual system. Visual acuity arises through an interaction of several factors including the physical spread of light, retinal processes and higher order neural processes. Many of these aspects of vision have been treated by numerous researchers, Konig being one of the earliest workers in the field.

R S Woodworth and H Schlosberg, *Experimental Psychology* (London, 1966)

NS

acupuncture (pre–2500 BC) *Biology/Medicine* Founded in the philosophy of ancient China, this is the theory that disease is caused by an imbalance between the *Yin* (passive, dark, female) and the *Yang* (active, light, male) aspects of the spirit, leading to the obstruction of the vital force (*ch'i*).

The goal of treatment is restoration of the Yin-Yang balance and hence health. To explain the well-established anaesthetic powers of acupuncture, scientists theorize that acupuncture needles induce natural opiates or that they interfere with the nerves' transmission of painful stimuli. Western physicians reject the notion that acupuncture can cure disease, however.

H Kruger, *Other Healers, Other Cures: A Guide to Alternative Medicine* (1974)

KH

adaptation (17th century–) *Biology* Described by the British natural theologians John Ray (1627–1705) and William Paley (1743–1805), the theory was later refined by Charles Darwin (1809–82) in his *Origin of Species by Means of Natural Selection: Or, the Preservation of the Favoured Races in the Struggle for Life* (1859).

This term refers to a process allowing organisms to change to become better suited for survival and reproduction in their given habitat. It is also used to describe a change in nervous response after exposure to a constant stimulus; for example, the decline in sensitivity to a particular smell experienced in humans after prolonged exposure.

P B Medawar and J S Medawar, *Aristotle to Zoos: A Philosophical Dictionary of Biology* (Oxford, 1983)

KH

adaptation level theory *Psychology* Theory of the American psychologist H Helson (1898–1977).

Posits that the neutral, adapted background provides a standard against which new stimuli are perceived. Adaptation level refers to a level on a sensory continuum to which the sense organ has adapted. For example, originally cool water may be made to feel warm if the person first adapts to cold water. Helson's theory refers primarily to sensory processes though it has since been generalized and applied to other fields such as that of attitude change.

H Helson, *Adaptation Level Theory: An Experimental and Systematic Approach to Behavior* (New York, 1964)

NS

adaptive evolution *Biology See* ADAPTATION.

adaptive expectations (1911) *Economics* Originally used by American economist Irving Fisher (1867–1947), the principle was popularized and given its current name during the 1950s in the study of hyper-inflation. It posits that future values may be calculated on the basis of previous values and their margin of error.

For example, next year's inflation rate can be estimated by judging how inaccurate this year's rate is compared with last year's forecast. The principle, by its nature, underestimates or overestimates constantly changing variables. Also called the error learning hypothesis, it was finally abandoned as inadequate in the early 1970s because forecasters frequently take into account information other than the past behaviour of the variable under study. *See* RATIONAL EXPECTATIONS and RANDOM WALK HYPOTHESIS.

I Fisher, *The Purchasing Power of Money* (New York, 1911); P Cagan, 'The Monetary Dynamics of Hyperinflation', *Studies in the Quantity Theory of Money*, M Friedman, ed. (Chicago, 1956)

PH

adaptive landscape *Biology See* SHIFTING BALANCE THEORY.

adaptive radiation (1898) *Biology* Also known as Hawaiian radiation, this is the rapid evolution of distinct species from a relatively recent common ancestor, with each species adapted to survive under specialized conditions. The idea is attributed to Henry Fairfield Osborn (1857–1935), paleontologist and influential president of the American Museum of Natural History.

On the Galapagos Islands, 14 species of Darwin's finches, each occupying different niches in the environment, are believed to have evolved from the same ancestral finch that colonized the islands. Hawaiian examples of adaptive radiation include hundreds of fruit fly species and more than 20 distinct kinds of honeycreeper birds.

S Carlquist, *Island Biology* (New York, 1974)

KH

adding-up problem (1894) *Economics* Identified by English economist Philip Wicksteed (1844–1927) among others, this refers to the difficulty of ensuring that the income earned by different factors of production when added together equals NATIONAL INCOME.

If each factor of production is paid the value of its respective Marginal Product, *ceteris paribus*, a situation will eventually arise in which total product (total income) will be exhausted. Wicksteed studied the problem in the model of a long-run economy in PERFECT COMPETITION. The Swiss mathematician Leonhard Euler (1707–83) argued that certain assumptions had to be made about the production function (that is, how factors of production are combined to produce output) to make national income equal the sum of income earned by individual factors of production. *See* EULER'S THEOREM.

P H Wicksteed, *An Essay on the Co-ordination of the Laws of Distribution* (London, 1932)

PH, AV

adequacy, levels of (1962) *Linguistics* Hierarchy of standards for grammars proposed by the American linguist Avram Noam Chomsky (1928–).

A GENERATIVE GRAMMAR is 'observationally' adequate if it identifies those sentences in a corpus of data that are grammatical sentences of the language; 'descriptively', if it analyzes the structure of sentences in relation to speakers' intuitions about them; it achieves 'explanatory' adequacy if the description accords with a set of psychologically valid LINGUISTIC UNIVERSALS.

N Chomsky, *Current Issues in Linguistic Theory* (The Hague, 1964), ch. 2

RF

adjacency pair *Linguistics See* CONVERSATIONAL ANALYSIS.

Adlerian theory *Psychology* Named after its originator, the Austrian psychiatrist Alfred Adler (1870–1937).

Also known as 'individual psychology', the theory asserts that all behaviour must take account of the social context in which it occurs. It emphasizes concepts of choice, individual responsibility and meaning of life. Central is the notion of inferiority and the struggle of the person against such feelings. A healthy personality is proactive rather than reactive: people should not merely adapt to environments or events but create and modify them; they should make choices for the future rather than see themselves as determined by their past.

A Adler, *The Theory and Practice of Individual Psychology* (New York, 1929)

NS

administered pricing (1930s) *Economics* Sometimes known as inflexible pricing, this term refers to the setting of prices by individual or public bodies rather than by the interaction of market forces of demand and supply.

In general, prices are set by adding a profit margin to the average cost. Seen in the pricing policies of monopolies, cartels, oligopolies or government organizations, the producer in the market dictates the market price, which is often held constant for a period of time. The fixed exchange-rate system, in which member countries established fixed values for the exchange rates of their currencies, is an example of administered pricing. Some economists argue that inflexible pricing has been a causal factor in COST-PUSH INFLATION. *See* MARK-UP PRICING.

G C Means, 'Industrial Prices and their Relative Inflexibility', *Senate Documents*, vol. XIII (Washington, DC, 1935); D Jackson, *Introduction to Economics: Theory to Economics* (London, 1982)

PH

administrative theory (20th century) *Politics/Management* Theory of public administration.

There is no single administrative theory, but rather a body of work which seeks to discover the general constraints, incentives, and structures within which administration, particularly administration in governmental or public institutions, is conducted.

Christopher Hood, *Administrative Analysis* (Brighton, 1986)

RB

adsorption theory of heterogeneous catalysis (1960s–70s) *Chemistry* Theory of adsorption developed to explain heterogeneous catalysis.

A catalyst increases the rate of a chemical reaction. Heterogeneous catalysis occurs when the catalyst and reacting system are in different phases; for example, when the catalyst is a solid and the reactants and products are gases. The reactant gas molecules diffuse to the surface of the catalyst where they accumulate in a process known as adsorption. Molecules next to one another on the surface react to form products. The products then break away from the solid in a process known as desorption. Finally, the product molecules diffuse away from the surface into the gas. The reaction between the gases ethene and hydrogen to form the gas ethane using a solid nickel catalyst is an example.

M Freemantle, *Chemistry in Action* (Basingstoke, 1987)

MF

advantage, principle of *Psychology* A theory proposed by the American psychologist Burrhus Frederic Skinner (1904–90).

Of two or more incompatible or inconsistent responses, one has the advantage of being more reliable or of occurring more frequently. The principle is sometimes used in behaviour modification programmes where, for example, clients can be taught to control panic responses by performing frequently practised relaxation exercises which are inconsistent with the physiological responses associated with panic.

G H Bower and E R Hilgard, *Theories of Learning* (Englewood Cliffs, N.J., 1981)

NS

adverse selection (1970) *Economics* American economist George Arthur Akerlof

(1940–) first noted this problem (sometimes referred to as the lemon problem), which arises from the inability of traders/buyers to differentiate between the quality of certain products.

The most cited example is the second-hand car industry, in which a trader dealing in, for example, Minis possesses product information that the other buyers/sellers in that market lack. He thus operates at a comparative advantage as the other people in the market cannot tell if he is selling a 'lemon' (poor-quality car). Consequently, there is risk involved in purchasing the good and while the lower price buyers are willing to take this risk, traders selling quality cars are not willing to sell at such a low price. There are three components to this theory: (1) there is a random variation in product quality in the market; (2) an asymmetry of information exists about the product quality; (3) there is a greater willingness for poor-quality car sellers to trade at low prices than higher-quality owners. Insurance and credit markets are areas in which adverse selection is important. *See* ASYMMETRICAL INFORMATION.

G A Akerlof, 'The Market for "Lemons": Quality Uncertainty and the Market Mechanism', *Quarterly Journal of Economics*, vol. LXXXIV (August, 1970), 3

PH, AV

AEF (Arthur's Education Fund) (1938) *Feminism* A theory of sexual inequality described by Virginia Woolf (1882–1941).

In any family the distribution of resources amongst the children will favour male children ('Arthur') and penalize female children. Priority will be given to the education and career promotion of males, and females will, by being deprived of an equal or in some cases of any share in such services, subsidize their more fortunate brothers.

Virginia Woolf, *Three Guineas* (London, 1938)

HR

aerial perspective *Art* The illusion of recession through the depiction of atmospheric effects.

In painting, distant landscapes or objects appear progressively fainter, cooler in hue and less distinct in outline. The technique is recognizable in the landscape backgrounds of Leonardo da Vinci (1452–1519), and in Chinese landscape painting.

AB

aesthetic distance *Literary Theory See* PSYCHICAL DISTANCE.

aestheticism (1870s–*c*.1900) *Art* An awareness of aesthetics in British artistic and literary society in the late 19th century.

Although it had no manifesto, the Aesthetic Movement, led by figures such as American artist James Whistler (1834–1903) and Irish writer Oscar Wilde (1854–1900), was important for its cult of 'art for art's sake' which pervaded all forms of art and literature.

AB

aether (ether) *Physics* First conceived of by the ancient Greeks, aether has several times been revived as a hypothetical medium supposed to fill all space.

It was introduced theoretically as the medium to carry light waves (the luminiferous aether), particularly when it was realized that light takes a finite time to travel in interplanetary space where there is not enough matter to support the propagation of a wave. Therefore, the luminiferous aether must possess elasticity and have a finite density. Subsequently, the aether was invoked to dispense with the idea of action-at-a-distance between electric charges, and was used to transmit the forces between electric charges by becoming stressed (the electromagnetic aether). The existence of the aether is now considered to be an unnecessary assumption.

J Thewlis, ed., *Encyclopaedic Dictionary of Physics* (New York, Oxford and London, 1962)

MS

affections, doctrine of the (15th century) *Music* The notion that music can and should be constructed so as to evoke in the listener certain idealized emotions, or 'affections'. Derived from Classical doctrine, this concept can be inferred from the writings of 15th-century theorists. It is dealt with more explicitly by the French philosopher René Descartes (1596–1650) in *Les passions de l'acme* (1649).

The desired affection (usually one per work, or section of a work) is animated through the composer's choice of large-scale elements, such as key, tempo and metre. The concept flourished during the Baroque period, along with the related DOCTRINE OF MUSICAL FIGURES, but later both fell from favour.

MLD

affective arousal theory *Psychology* Originated by the American psychologist D C McClelland (1917–).

Motives are said to derive from changes in affective or emotional states. McClelland placed particular emphasis on the need for achievement (abbreviated as N-ach) as an acquired motivating force and used it to make predictions about academic and occupational achievement. The theory has undergone extensive criticism and development but remains influential.

D C McClelland, *The Achieving Society* (Princeton, 1961)

NS

affective fallacy (1949) *Literary Theory* Identified by the American critic William Kurtz Wimsatt (jnr) (1907–75) and aesthetician Monroe C Beardsley (1915–).

'A confusion between the poem and its *results*' (Wimsatt and Beardsley), condemned for impressionism and RELATIVISM. In 'affective criticism' – common in the 19th century and rationalized in the 1920s by Ivor Armstrong Richards's pseudo-scientific emphasis on the emotive role of literature (*see* HARMONIZATION OF IMPULSES) – the critic erroneously gives an account of the psychological effects the poem has on him, instead of an objective account of the poem itself. *See also* INTENTIONAL FALLACY.

W K Wimsatt, *The Verbal Icon* (Kentucky, 1954)

RF

affine field theories, relativistic *Physics* Relativistic affine field theories apply the special properties associated with *n*-dimensional affine space in attempting to extend the general theory of relativity to explain the origin of the electromagnetic and meson fields.

Paul Davies, ed., *The New Physics* (Cambridge, 1989)

GD

agapism *Psychology* Unattributable to any one originator, the term is rarely used within psychology.

Derived from the Greek, agapism or agape refers to a complex form of love. It is sometimes used for the unselfish love taught by such figures as Christ and Buddha, but more often it describes erotic and sensual love accompanied by feelings of tenderness, protectiveness, self-denial and a preference for the features, gestures, speech and so on of the other person.

E Berscheid and E H Walster, *Interpersonal Attraction* (Reading, Mass., 1978)

NS

age-area hypothesis (1923) *Anthropology/Archaeology* Proposed by American anthropologist Clark Wissler (1870–1947).

Cultural changes typically take place at the centre of a culture area and spread outwards. Therefore older aspects of culture will be found on the outskirts and newer ones in the centre of the area. This hypothesis was a useful tool for the reconstruction of the history of culture areas, but its implication for dating archaeological sites was superseded by the development of radio-carbon dating in the second half of the 20th century. *See also* DIFFUSIONISM.

C Wissler, *Man and Culture* (New York, 1923)

ABA

ageing clock *Biology* *See* PROGRAMMED AGEING.

ageing hormone *Biology/Medicine* Espoused by W D Denckla of Harvard University, this is the theory that ageing and death are regulated by a special hormone secreted by the brain or pituitary gland that has yet to be discovered. This theory is an example of PROGRAMMED AGEING. *See also* AGEING, THEORIES OF.

J Behnke, C Finch and G Moment, *The Biology of Aging* (New York, 1978)

KH

ageing, theories of *Biology/Medicine* These are variations of the general idea that ageing (also termed senescence) can be explained, and are usually accompanied by the hope that the process can be reversed or stopped.

Such concepts are widespread and date from antiquity.

While there are dozens of theories of ageing, nearly all can be placed into one of two camps: the idea that ageing results from some form of wear and tear, or the idea that ageing is genetically programmed in the body. Some theories include elements of both ideas. While most theorists can cite supportive evidence, no one theory has been universally accepted by biologists. *See also* AGEING HORMONE, CODON-RESTRICTION THEORY, EPIPHENOMENALIST THEORY OF AGEING, ERROR CATASTROPHE THEORY OF AGEING, EXHAUSTION THEORY OF AGEING, FREE RADICAL THEORY OF AGEING, HAYFLICK LIMIT, IMMUNOLOGIC THEORY OF AGEING, PROGRAMMED AGEING, RANDOM HITS THEORY OF AGEING, RATE OF LIVING THEORY, SHANGRI-LA PHENOMENON, SOMATIC MUTATION THEORY OF AGEING and TESTICULAR EXTRACT THEORY.

KH

agency theory (1970s) *Economics* A theory developed in the 1970s, this refers to the variety of ways in which agents, linked by contractual arrangements with a firm, influence its behaviour.

These may include organizational and capital structure, remuneration policies, accounting techniques and attitudes toward risk-taking. Agency costs are deemed the total cost of administering and enforcing these arrangements. *See* THEORY OF THE GROWTH OF THE FIRM, THEORY OF THE FIRM, ORGANIZATION THEORY, MANAGERIAL THEORIES OF THE FIRM and THEORY OF BUREAUCRACY.

M C Jensen and W H Meckling, 'Theory of the Firm: Managerial Behavior, Agency Costs and Ownership Structure', *Journal of Financial Economics*, vol. III (1976), 305–60; R Watts and J Zimmerman, *Positive Accounting Theory* (Englewood Cliffs, N.J., 1986)

PH

agglutinating or agglutivative language (19th century) *Linguistics* Category of language in linguistic typology.

A type of language (exemplified by Turkish, Japanese and Hungarian) which adds one or a series of clearly distinct suffixes to words to make grammatical distinctions.

B Comrie, *Language Universals and Linguistic Typology*, 2nd edn (Oxford, 1989)

RF

aggregate demand theory (20th century) *Economics* Championed by English economist John Maynard Keynes (1883–1946), this asserts that the total demand within an economy helps to determine the level of output and growth.

If demand and consumption are lowered (for example, through wage cuts), overall economic activity will be reduced. Aggregate demand theory has been overshadowed in recent years by SUPPLY-SIDE ECONOMICS.

J M Keynes, *The General Theory of Employment, Interest and Money* (New York, 1936)

PH

agnosticism *Philosophy* Any view claiming that knowledge is unobtainable in a given area, whether merely in practice or (and more usually) in principle too. Term invented by Thomas Henry Huxley (1825–95) in 1869.

In religious matters, agnosticism should be distinguished from atheism (the view that there is no god), though it sometimes includes the view that sentences like 'There is a god' are meaningless rather than true or false. *See also* SCEPTICISM.

ARL

Airy hypothesis (1855) *Geology* A theory proposed by the British astronomer Sir George Biddell Airy (1801–92) to account for discrepancies between the calculated mass of continental or oceanic areas and their actual gravitational attraction. Airy concluded that the upper layers of the Earth must all have the same average density, so high areas must be underlain by roots of low density crustal rocks which project down into the uniformly denser mantle rocks beneath.

This ISOSTATIC THEORY is now known to be only a partial explanation, although it is applicable when comparing oceanic and continental areas. There is also evidence for some roots beneath most mountain belts. *See also* ISOSTATIC THEORY and PRATT HYPOTHESIS.

N-A Mörner, *Earth Rheology, Isostasy and Eustasy* (New York, 1980)

DT

akaryotic theory (1970) *Biology* First described by Roger Stanier, this term was coined by Marcello Barbieri of the University of Turin, Italy; this is the idea that the evolutionary ancestors of plant and animal cells were non-nucleated cells resembling bacteria (prokaryotes), but with a cytoplasm and biochemistry more like that of modern animal and plant cells (eukaryotes), which have nuclei.

Refinements of the Stanier model have been suggested by several scientists. *Compare with* EUKARYOTIC THEORY, PROKARYOTIC THEORY, SERIAL SYMBIOSIS THEORY and SYMBIOTIC THEORY OF CELLULAR EVOLUTION.

M Barbieri, *The Semantic Theory of Evolution* (Chur, Switzerland, 1985)

KH

alchemy (*c.*500 BC–16th century) *Chemistry* Alchemy was the predecessor of chemistry. The two main aims of alchemy were the transmutation of other metals into gold and the discovery of an elixir or universal remedy which would promote everlasting life.

The Macmillan Family Encyclopedia (London, 1980)

MF

aleatoric theory *Psychology* Unattributable to any one originator, this theory is rarely used within psychology.

Aleatoric theory sees human activity as largely embedded within historically contingent circumstances. Because the configuration of prevailing circumstances is not fundamentally systematic, cross-time trajectories of human activity may be viewed as dependent on chance. Aleatoric theory attempts to understand developmental trajectories in terms of the contemporary context and in so doing to address contemporary concerns.

A J Chapman and D M Jones, *Models of Man* (Leicester, 1980)

NS

aleatory music *Music See* INDETERMINACY.

Alexander's sub-base theorem (20th century) *Mathematics* Named after the American mathematician James Waddell Alexander (1888–1971), this is the result in TOPOLOGY whereby if every open covering of a given topological space by elements of a sub-base has a finite sub-covering, then the space is compact. The usual definition of compactness for a topological space is that every covering by open sets has a finite sub-covering. Hence, this theorem provides a condition for compactness which is often technically simpler to verify.

J R Munkres, *Topology: A First Course* (Prentice-Hall, 1975)

MB

Alexandroff compactification (1924) *Mathematics* Named after the Russian mathematician Pavel Sergeevich Aleksandrov (1896–1982), the Alexandroff (or one-point) compactification of a given topological space is essentially an extension of the space to a compact space with one additional point.

For example, the one-point compactification of the complex plane consists of all complex numbers together with the point at infinity. (This is best seen by considering complex numbers as points on the Riemann sphere.) Alexandroff gave conditions under which such a compactification is possible. *See also* TOPOLOGY and STONE-ČECH COMPACTIFICATION.

J R Munkres, *Topology: A First Course* (Prentice-Hall, 1975)

MB

Alfvén's theory of planetary formation (1952) *Astronomy* Named after its originator, the Swedish theoretical physicist Hans Olof Gösta Alfvén (1908–), this theory assumes that the formation of the planets was due to a cloud of gas, having a composition the same as the average composition of matter in the universe, falling towards the Sun, perhaps in the last stages of the formation of the Sun itself.

The Sun is assumed to have had a magnetic field strength above a certain value. Initially, the gas cloud is assumed to be ionized by electromagnetic action. Gravity causes the ionized gas to fall towards the Sun while the effect of the magnetic field opposes this motion. During the process the gas gradually loses its ionization. A transfer of angular momentum from the Sun to the gas takes place through the electromagnetic forces. The ionized gas condenses into small grains which move in ellipses and the planets are formed by the agglomeration of these grains. At a late stage of the formation of the larger planets, a similar process occurs around them, giving rise to the satellites. A good point of the theory is that the planets and satellites are formed by similar mechanisms, but the value of the Sun's magnetic dipole moment that must be assumed is very high.

S Mitton, ed., *The Cambridge Encyclopaedia of Astronomy* (London, 1973)

MS

algebra *Mathematics* The branch of mathematics whose principal concern is the study of formal systems, such as groups and fields, which generalize various aspects of ordinary arithmetic. A characteristic feature of 20th-century algebra is the use of highly sophisticated symbolic notation. In the broadest sense, algebra can be considered as the manipulation of any such formalized symbolism.

I N Herstein, *Topics in Algebra* (Wiley, 1975)

MB

algebraic addition theorem *Mathematics* A theorem or identity for any given function f which states that the values $f(x)$, $f(y)$, and $f(x + y)$ satisfy a polynomial relation for all possible arguments x and y.

For example,

$$\exp(x + y) = \exp(x) \exp(y)$$

is an algebraic addition theorem for the exponential function. It is an important result in COMPLEX FUNCTION THEORY that a meromorphic function has such an addition theorem if and only if it is rational, trigonometric, or elliptic.

Encyclopedic Dictionary of Mathematics (MIT Press, 1987)

MB

algebraic geometry *Mathematics* The branch of geometry which relies on algebraic methods.

Originally the term referred to all the work from the time of the French mathematicians Pierre de Fermat (1601–65) and René Descartes (1596–1650) in which ALGEBRA had been applied to geometry. Much of this work was related to the study of plane algebraic curves. In the latter half of the 19th century it referred to the study of algebraic invariants and birational transformations. In the 20th century, it refers to the study of birational transformations and algebraic varieties. In the last 50 years, algebraic geometry has become increasingly abstract and in many ways bears little resemblance to what was studied earlier. At the same time, results in this area have had profound implications in subjects as diverse as NUMBER THEORY, COMPLEX FUNCTION THEORY and CATASTROPHE THEORY. *See also* HILBERT'S PROBLEMS.

S S Abhyankar, *Algebraic Geometry for Scientists and Engineers*, Mathematical Surveys and Monographs 35 (AMS, 1990)

MB

algebraic number theory *Mathematics* The branch of NUMBER THEORY which employs algebraic methods.

A Frohlich and M J Taylor, *Algebraic Number Theory* (Cambridge, 1991)

MB

algebraic topology (1895) *Mathematics* Started by Henri Poincaré (1854–1912) when he defined the fundamental group of a surface in Euclidean geometry.

The ultimate geometric objective of TOPOLOGY is to find which pairs of geometric figures are homeomorphic; that is, can be transformed into each other by functions which change little for small changes of the variables. This investigation is assisted by associating various groups with each figure for comparison. This method has had many successes but in 1958 A A Markov proved by using the fundamental group that the ultimate objective is unattainable. *See* GROUP THEORY, HOMOLOGY THEORY and K-THEORY.

A H Wallace, *An Introduction to Algebraic Topology* (London, 1957)

JB

algorithmic-heuristic theory *Psychology* Originated by the American L Landa (1925–61).

A theory of performance, learning and instruction formulated to perform a number of tasks ranging from identifying quasi-algorithmic processes in topics other than mathematics to designing new instructional methodology and materials, based on the knowledge of algorithmic-heuristic processes. The theory is concerned with increasing the effectiveness and efficiency of performance, learning and instruction.

L N Landa, *Algorithmization in Learning and Instruction* (Englewood Cliffs, N.J., 1966)

NS

alienation (19th century–) *Politics/Psychology* Variously defined, but principally the positive theory of the German philosopher G W F Hegel (1770–1831) and the critical theory of the German social theorist Karl Marx (1818–83) of consciousness in history.

Hegel: human history consists of the progressive divisions or alienations of consciousness, whereby complexity and identity are recognized in growing division. Spirit creates its word in the act of knowing it. We know things only when they become separate from us. Marx: capitalism is characterized by the separation or alienation of the workers from what they make, from the productive process, from their true nature or 'species being' and from other people. *See also* MARXISM.

R Schacht, *Alienation* (New York, 1970)

RB

alienation effect (1930s) *Literary Theory* From the German *Verfremdungseffekt*, a theory developed by playwright Bertolt Brecht (1898–1956).

Use of estranging theatrical devices such as narration, unrealistic dialogue, anachronistic costume, and so on to unsettle the audience's reception of a play, forcing them into an attitude of critique. *See also* EPIC THEATRE, DEAUTOMATIZATION and ESTRANGEMENT.

J Willett, ed., *Brecht on Theatre* (London, 1974)

RF

all-or-none principle (1871, 1917) *Biology/Medicine* The idea that once a certain threshold of neuron stimulation has been exceeded a nerve will fire, but that until the threshold is passed, the nerve will not fire. First proposed for the contraction of cardiac muscle by American physiologist Henry Pickering Bowditch (1840–1911) in 1871, and for skeletal muscle by American physiologist Frederick Haven Pratt (1873–1958) in 1917.

The principle also states that once the threshold is passed, the strength of the stimulation has no effect on the strength of the nerve's reaction; that is, the nerve's response is the same for any impulse beyond the threshold amount. While the all-or-none principle holds for many neurons, detailed studies of some isolated nerves and receptors suggest that the intensity of sensation depends on the number of nerve impulses travelling along a single fibre. This idea is also applied to non-biological situations.

J Fulton, *Selected Readings in the History of Physiology* (Springfield, Ill., 1966)

KH

all-over painting (1950s) *Art* Term used for a non-formal style of American painting in which no differentiation is made between areas of the painted surface of a canvas; as exemplified by the COLOUR FIELD painters or the 'drip paintings' of American artist Jackson Pollock (1912–56).

AB

Allais paradox (1952) *Economics* Formulated by French economist Maurice Allais (1911–) to counter principles of GAME THEORY propounded by other economists and to establish a definition of rationality in risky situations.

Earlier analysis of game theory had suggested that in the following example, the preference of A over B should entail a preference of C over D. Allais stated, however, that many sensible people who wanted A over B would choose D or C, suggesting a paradox.

For example, do you prefer A over B, where A is the certainty of winning 100, and B is a 10 per cent chance of winning 500; and do you prefer C over D, where C is an 89 per cent chance of winning 100, and D a 1 per cent chance of winning nothing.

Equilibrium is such that an individual's marginal rate of substitution equals the marginal rate of transformation; that is C is an 11 per cent chance of winning 100 (89 per cent chance of winning nothing), or D is a 10 per cent chance of winning 500 (90 per cent chance of winning nothing).

See CO-OPERATIVE GAMES THEORY, COLLUSION THEORY and OLIGOPOLY THEORY.

PH

allegory (traditional) *Literary Theory* Western allegorical literature and interpretation are based on techniques and interpretations of the Old and New Testament.

Narrative fiction in which events, their sequence, and characters (by 'personification') stand for a more abstract development of ideas, usually moral or political. Exemplary works in the genre include Dante's *Divine Comedy* (*c.*1307), Langland's *Piers Plowman* (*c.*1377), Spenser's *Faerie Queene* (1590–96), Bunyan's *Pilgrim's Progress* (1678), and – an example of the sub-genre of 'beast fable' – Orwell's *Animal Farm* (1945).

A Fletcher, *Allegory: The Theory of a Symbolic Mode* (Ithaca, 1964)

RF

Allen's rule (1877) *Biology* Named after naturalist Joel Asaph Allen (1838–1921), this states that within a given species or genus, appendages such as tails, ears and limbs tend to be relatively smaller in animals that inhabit cooler climates than in those living in warmer climates. This tendency is correlated with the need to conserve heat in colder environments, and to radiate heat in warmer ones.

V Grant, *The Evolutionary Process: A Critical Study of Evolutionary Theory*, 2nd edn (New York, 1991)

KH

alliance theory (1949) *Anthropology* The name given to the kinship theory developed by French anthropologist Claude Lévi-Strauss (1908–).

The approach within kinship studies which emphasizes relations between groups by marriage, rather than the structure of kinship groups themselves. Of particular concern are those societies in which there exist formal rules for marriage between cousins. The theory is widely accepted as useful, though Lévi-Strauss's original interest in ideal rather than actual cases is now regarded with suspicion. *See also* KINSHIP and LINEAGE THEORY.

C Lévi-Strauss, *The Elementary Stuctures of Kinship* (London, 1969 [French edn, 1949])

ABA

alloparapatric speciation (1942) *Biology* The gradual emergence of new species from populations that started out geographically isolated (allopatric) but later become parapatric, meaning that they share a limited overlapping region where hybridization occurs. *See also* SPECIATION, THEORY OF and SPECIES, THEORIES OF. *Compare with* STASIPATRIC SPECIATION and SYMPATRIC SPECIATION.

C Barigozzi, *Mechanisms of Speciation* (New York, 1981)

KH

allopatric speciation (1942) *Biology* Coined by the German evolutionary biologist Ernst Mayr (1904–), this term describes the process whereby distinct species that evolve separately as a consequence of genetic isolation, perhaps caused by geological change that divides a once-unified population into two or more breeding units. The speciation begins with the emergence of subspecies, and then over a long period unique species may appear. Allopatric means 'different fatherland.' *Compare with* SYMPATRIC SPECIATION and ALLOTROPIC SPECIATION. *See also* FOUNDERS EFFECT.

E Mayr, *Systematics and the Origin of Species* (New York, 1942)

KH

alloy theory *Physics* An alloy is a macroscopically homogeneous mixture of metals. The term alloy theory is used to cover a group of theories attempting to explain the details of alloy phase diagrams, and the nature of the equilibrium phases that occur, in terms of the atomic properties of the elements making up the alloy.

J Thewlis, ed., *Encyclopaedic Dictionary of Physics* (New York, Oxford and London, 1962)

MS

allusion (traditional) *Literary Theory* An indirect reference in a text which activates a second text, the relationship between the two enriching the interpretation of the alluding text. The mechanics of cueing are various: partial or amended quotation, paraphrase, and so on. The importance of allusion as a critical concept has increased from the 1970s with the popularity of INTERTEXTUALITY.

Z Ben-Porat, 'The Poetics of Literary Allusion', *Poetics and the Theory of Literature*, 1 (1973), 105–28

RF

alpha, beta, gamma hypothesis *Psychology* Theories of the American psychologist Burrhus Frederic Skinner (1904–90).

The term refers to three divergent hypotheses relating to learning. The alpha hypothesis states that the frequency with which a behaviour is performed enhances learning. The beta hypothesis states that repetition frequency has no effect on learning. The gamma hypothesis states that repetition frequency hinders learning. All three hypotheses have been supported under different experimental circumstances.

G H Bower and E R Hilgard, *Theories of Learning* (Englewood Cliffs, N.J., 1981)

NS

alpha-beta theorem *Mathematics* *See* SCHNIRELMANN DENSITY THEORY.

als ob (1876) *Literary Theory* Philosophy of fiction propounded by the German philosopher Hans Vaihinger (1852–1933).

This is the philosophy of 'as if', an idealistic attitude basic to human motives: for the sake of some desired goal, we accept as if true a proposition known to be false. In responding to the arts, readers and viewers accept a fiction as if it were a representation of the real. *See* WILLING SUSPENSION OF DISBELIEF.

H Vaihinger, *The Philosophy of 'As If'* [1911], C K Ogden, trans. (New York, 1924)

RF

alternating series test, Leibniz's (18th century) *Mathematics* Named after the German philosopher, logician, mathematician and scientist Gottfried Willhelm Leibniz (1646–1716), this establishes that an alternating series (signs of successive items alternate between positive and negative) is convergent if the absolute value of the terms decrease monotonically to zero.

For example, the alternating series

$$\sum_{n=0}^{\infty}(-1)^n\frac{1}{n} = 1 - \frac{1}{3} + \frac{1}{5} - \frac{1}{7}\cdots$$

is convergent, since

$$|a_n| = \left|\frac{(-1)^n}{n}\right| = \frac{1}{n}$$

decreases monotonically to zero as n tends to infinity. Its sum is $\pi/4$.

H Anton, *Calculus with Analytic Geometry* (New York, 1980)

ML

alternation of generations (19th century) *Biology* Also called metagenesis. The Norwegian zoologist Johannes Iapetus Smith Steenstrup (1813–97) coined the term in his book, *On the Alternation of Generations* (English translation 1845). It refers to the alternation of reproductive forms in an organism's life-cycle, especially alternation of sexual and asexual reproduction.

True alternation of generations occurs in plants with a haploid, gamete-producing generation alternating with a diploid, spore-producing generation. The term is also applied to the life-cycles of animals (such as jellyfish) in which a free-swimming sexual stage alternates with an asexual, sedentary stage (but in jellyfish both generations are diploid).

KH

alternation theorem *Mathematics* A polynomial P of degree $n-1$ is a *best uniform approximant* to a continuous function f on the interval $[a,b]$ if $||r|| = \max_{[a,b]}|r(x)|$ is as small as possible, where $r = f - P$ is the *error function*. Best approximants are characterized by having at least $n + 1$ *alternations*: points x_i in the interval where $r(x_i) = -r(x_{i+1}) = \pm||r||$. This result, due to E Borel in 1905, can be extended to *generalized polynomials*

$$\textstyle\sum_1^n \lambda_i g_i$$

where the continuous functions $g_1,\ldots,g_n$ satisfy the *Haar condition* of being linearly independent on any n distinct points in the interval. It is fundamental in the design of numerical methods for computing best approximants.

AL

alternative hypothesis *Mathematics* *See* HYPOTHESIS TESTING.

alternative theorem *Mathematics* Any result presenting two systems of equations or inequalities and stating that exactly one system has a solution.

One case is the FREDHOLM ALTERNATIVE in FUNCTIONAL ANALYSIS, named after Swedish mathematician and physicist Erik Ivar Fredholm (1866–1927). For example, if A is a continuous linear operator with closed range, having adjoint A^* (for instance, a matrix and its transpose) then either the system $Ax = b$ is solvable for any b or the equation $A^*y = 0$ has a non-zero solution. Analogous results are fundamental in the study of integral equations. *See also* FARKA'S LEMMA.

AL

altruism (1850–55) *Biology/Sociology* Also called Hamilton's genetical theory of social behaviour, altruism is any behaviour of an animal that may be disadvantageous for the individual, but that benefits others of its species. The theory was espoused (as the 'law of mutual aid') by Prince Peter Kropotkin (1842–1921), a Russian philosopher; but the term was coined by Auguste Comte (1798–1857), French philosopher and social reformer.

A bird that warns the flock of impending danger by calling out, thereby making itself obvious to the predator, is nonetheless increasing the likelihood that its own kin (and their shared gene pool) will survive. Kropotkin believed that both animal and human survival depended on 'mutual aid', and recorded many examples from the animal kingdom. *Compare with* SURVIVAL OF THE FITTEST. *See also* HAPLODIPLOID HYPOTHESIS, INCLUSIVE FITNESS and KIN-SELECTION THEORY.

P Appleman, *Darwin: A Norton Critical Edition* (New York, 1970)

KH

altruism *Philosophy* In popular speech, a willingness to sacrifice one's own interests for those of others. It is this sense that is relevant to discussions of, for example, the evolutionary origins and role of altruism in animals. Philosophically, altruism is rather a view about what one ought to do, and contrasts with EGOISM and UNIVERSALISM.

It is a form of CONSEQUENTIALISM and prescribes that one should act so as to maximize the happiness or welfare of people (or possibly living creatures) in general, oneself alone excluded. Only these last three words distinguish altruism from universalism or UTILITARIANISM, and in principle it has varieties corresponding to the varieties of these, and is open to many of the same objections. It also faces objections which mirror some of those facing egoism; for example, to what extent it is really practicable.

ARL

Alvarez theory (1980s) *Biology/Geology* Proposed by American father and son geologists, Luis Alvarez (1911–) and Walter Alvarez (1940–), this asserts that showers of meteors and the resulting dust clouds caused the mass extinctions at the end of the Cretaceous epoch (145–65 million years ago).

The concept remains controversial, although geological evidence supporting it continues to accumulate. *See also* MASS EXTINCTION.

R Muller, *Nemesis, the Death Star* (New York, 1988)

KH

ambiguity (traditional; 1957; 1930) *Linguistics/Literary Theory* A potential for more than one meaning in an utterance. Ambiguity is usually resolved by context.

Traditional SEMANTICS concentrated on individual words like *trunk* and *bank* which are called 'homonyms'. In TRANSFORMATIONAL-GENERATIVE GRAMMAR, Avram Noam Chomsky (1928–) stipulated that speakers' INTUITIONS would include recognition of syntactic ambiguity, as in 'The old men and women boarded the bus'.

In poetry, the English poet and critic William Empson (1906–84) detected several kinds of multiple meaning, or types of ambiguity, used to convey richness, a complicated state of mind, creative contradiction, and so on. His exposition of ambiguity in 1930 was one of the major sources for the ideas of NEW CRITICISM.

J G Kooij, *Ambiguity in Natural Language* (Amsterdam, 1971); W Empson, *Seven Types of Ambiguity* (New York, 1966)

RF

Amonton's laws of friction (1699) *Physics* Named after the French theorist who formulated them, Guillaume Amontons (1663–1705). When one solid body slides over another, there is a resistance to the relative motion of the surfaces in contact; this is called friction.

Two basic laws describing the behaviour of two plane solid surfaces in contact were discovered by Amontons, though they were known earlier to Leonardo da Vinci (1452–1519). (1) The friction force between two solids with plane surfaces is independent of the area of contact of the solids. (2) The friction force between the solids is proportional to the load between the surfaces (the coefficient of friction), and is a constant for a given pair of materials. *See* COULOMB'S LAW OF FRICTION.

J Thewlis, ed., *Encyclopaedic Dictionary of Physics* (New York, Oxford and London, 1962)

MS

Ampère's circuital theorem (19th century) *Physics* Named after its discoverer, the French physicist André Marie Ampère (1775–1836).

This states that the magnetic circulation along all concentric paths around a straight wire carrying a current I is the same and equal to $\mu_0 I$, where μ_0 is the permittivity of free space. The circulation is defined as the line integral of the magnetic induction B around any closed path. Therefore, the theorem may be written

$$\oint B.\mathrm{d}L = \mu_0 I$$

where $\mathrm{d}L$ is an element of the path taken.

J Thewlis, ed., *Encyclopaedic Dictionary of Physics* (New York, Oxford and London, 1962)

MS

Ampère's laws of electromagnetism (19th century) *Physics* Any conductor carrying an electric current produces a magnetic field around it. These laws relate this magnetic field to the current producing it; and are named after their discoverer, the French physicist André Marie Ampère (1775–1836).

Law 1: A closed loop of a conducting wire carrying a current I produces the same magnetic field configuration as would a shell of magnetic material bounded by the loop and having a magnetization numerically equal to I.

Law 2: When a conducting wire carrying a current I is placed in a uniform magnetic field of induction B, the force on a length L of the conductor has a magnitude IBL and is in a direction at right angles to both the length of the conductor and the direction of the magnetic field.

J Thewlis, ed., *Encyclopaedic Dictionary of Physics* (New York, Oxford and London, 1962)

MS

Ampère's rule (19th century) *Physics* Named after its proposer, the French physicist André Marie Ampère (1775–1836). In any plane normal to a straight conductor carrying a current, the magnetic field lines form a series of circles concentric with the wire. Ampère's rule relates the direction of the magnetic field to the direction of the current flow.

Suppose a man is swimming in the conductor with the current, and that he turns to face a small compass needle; the north-seeking pole of the needle will be deflected towards his left hand. *See* CORKSCREW RULE.

J Thewlis, ed., *Encyclopaedic Dictionary of Physics* (New York, Oxford and London, 1962)

MS

Ampère's theory of magnetism (19th century) *Physics* The French physicist André Marie Ampère (1775–1836) found that solenoids produce magnetic field configurations identical to those produced by permanent magnets of the same dimensions. On this basis, he suggested that all magnets are simply collections of electric currents and around each individual molecule of a magnet an electric current is circulating.

J Thewlis, ed., *Encyclopaedic Dictionary of Physics* (New York, Oxford and London, 1962)

MS

anagenesis (1889) *Biology* Also called phyletic evolution, single-line evolution and virtual evolution, the term anagenesis was used in 1889 in a title by Alpheus Hyatt

(1838–1902), American evolutionist and co-founder of neo-Lamarckism.

It describes upward evolution; the progressive evolution of species into new species. The term is also used for evolution following a single line of development; that is, the gradual transformation of one species into another. *See also* EVOLUTION. *Compare with* CLADISTICS and CLADOGENESIS.

J Brown, *Biogeography* (St Louis, Mo., 1983)

KH

analogy (1974) *Biology* Associated with Konrad Zacharias Lorenz (1903–89), the Austrian zoologist and ethologist (animal behaviourist) who won the Nobel prize for Physiology or Medicine in 1973.

This is the concept that a particular structure or behaviour shared by two different species exists because both species were subjected to the same evolutionary pressures favouring the characteristic. The wings of birds and bats are analogous as wings (but not as forelimbs). *Compare with* CONVERGENT EVOLUTION and HOMOLOGY.

D Dewsbury, *Comparative Animal Behaviour* (New York, 1978)

KH

analysis-by-synthesis (1950s) *Linguistics* Named by the American linguists Morris Halle and K N Stevens.

Theory of speech perception in which the signal is recognized as a particular speech sequence by an active process in which hearers project their knowledge of possible linguistic structure onto the sounds they hear.

M Halle and K N Stevens, 'Speech Recognition: a Model and a Program for Research', *The Structure of Language*, J A Fodor and J J Katz, eds (Englewood Cliffs, 1964), 605–12

RF

analytic continuation *Mathematics See* IDENTITY THEOREM.

analytic/synthetic (1783) *Philosophy/Linguistics* Distinction first formulated by the German philosopher Immanuel Kant (1724–1804), adopted as a fundamental principle in linguistic SEMANTICS.

An analytic or necessary truth ('sentence' in linguistics) is true by virtue of its meaning: 'All bachelors are unmarried men'. A synthetic or contingent truth is true by virtue of empirical fact: 'Grass is green' is not necessarily true, but only if grass is green.

T M Olshewsky, ed., *Problems in the Philosophy of Language* (New York, 1969), ch. 5

RF

anarchism (19th century–) *Politics* Theory advocating society without coercive government.

Coercion and authority, of which government is the principal expression, are rejected as incompatible with freedom and autonomy. Society should be held together by voluntary co-operation. Human nature is naturally collaborative and social, and with the removal of all forms of power and authority a natural harmony can emerge or be cultivated. Anarchists have normally been suspicious of large scale organization, and large scale industrial organization in particular, and have envisaged social life being conducted in small communities in which a variety of skills are cultivated both by individuals and within the community as a whole. *See also* ANARCHO-CAPITALISM, ANARCHO-FEMINISM, ANARCHO-SYNDICALISM and BAKUNINISM.

David Miller, *Anarchism* (London, 1984)

RB

anarcho-capitalism (20th century) *Politics* Theory of unrestrained pursuit of economic self-interest.

Anarcho-capitalism is the advocacy of an extreme blend of ANARCHISM and free-market economics. Unlike anarchism, anarcho-capitalism assumes that individuals will naturally pursue their own self-interest. Unlike more mainstream advocacies of capitalist economics, anarcho-capitalism proposes doing away with government entirely and leaving all the functions normally carried out by the state, including law enforcement, to either individual or group voluntary action.

Murray Rothbard, *Ethics of Liberty* (Atlantic Highland, NJ, 1982)

RB

anarcho-feminism (20th century) *Politics/Feminism* ANARCHISM extended to take account of the oppression of women.

Of all forms of coercion and authority, that based on sex is the most fundamental and the most oppressive. The liberation of women requires the ending of all forms of oppression, but must strike at the root of male oppression or PATRIARCHY, which is both a form of exploitation in its own right, and one which pervades most other coercive and authority relationships.

H J Ehrlich, ed., *Reinventing Anarchy: What Are Anarchists Thinking These Days?* (London, 1979)

HR

anarcho-syndicalism (20th century) *Politics/ Sociology/Economics* Industrial version of ANARCHISM, and anarchist version of SYNDICALISM. In industrial societies, the appropriate body to conduct revolutionary or transformative politics consists of workers organized around their industry, rather than their trade. Since social power is at root economic rather than political, power can therefore be seized at its root.

David Miller, *Anarchism* (London, 1984)

RB

androcentrism (20th century) *Politics/ Sociology/Feminism* Theory attributed by feminists to male scholarship.

Androcentrism, or male centredness, is attributed to rather than claimed by theories and thinkers. Androcentrism takes male values or practices as the norm, and then explains female values or practices as deviations from, or unsuccessful aspirations towards, male ways of doing things.

Charlotte Perkins Gilman, *The Man Made World or Our Androcentric Culture* (London, 1911)

HR

anima mundi *Biology See* ANIMISM.

animal electricity (1791) *Biology* Proposed by Italian physiologist Luigi Galvani (1737–98) in *De viribus electricitatis in motu musculari commentarius* (1791), based on his experiments with frogs; this is the idea that electricity can cause muscle contraction, both when applied externally and when generated by animal tissues.

Modern biologists now know that electrical synapses are prominent in the neurons of cold-blooded animals, such as fish and amphibians, and that they operate more efficiently at low temperatures than chemical synapses do.

J Fulton, *Selected Readings in the History of Physiology* (Springfield, Ill., 1966)

KH

animal heat (*c*.350 BC) *Biology* Originating in ancient Greece, where it was promoted by Galen (129–*c*.AD 200), this is the notion that vital, innate heat resides in the bodies of humans and animals.

Animal heat was believed to be centred in the heart (especially the left ventricle), while the lungs were charged with the removal of gaseous wastes and cooling the body to prevent overheating. This model dominated until the 17th century, when other explanations for bodily warmth were proposed. For example, the Flemish biochemist Jean Baptiste van Helmont (1579–1644) and the English physician Thomas Willis (1621–73) attributed animal heat to fermentation.

E Mendelsohn, *Heat and Life* (Cambridge, Mass., 1964)

KH

animal language hypothesis (19th century) *Biology* The view that at least some non-human species can communicate with humans using a true language, even if a non-vocal one. An early experiment was performed by Sir John Lubbock (a neighbour of Charles Darwin), who attempted to teach his dog Sign Language for the Deaf in 1882.

Most experiments in this field feature attempts to teach language, including American Sign Language, to chimpanzees; but in 1987 Irene Pepperberg reported that an African grey parrot named Alex could communicate simple ideas in English. Many biologists remain sceptical of such reports, however. *See also* BEE LANGUAGE HYPOTHESIS.

R Milner, *Encyclopedia of Evolution: Humanity's Search for Its Origins*, (New York, 1990)

KH

animal magnetism (late 18th century) *Psychology* A theory by the Viennese physician Franz Mesmer (1743–1815).

A progenital theory of what later became known and understood as hypnosis. Mesmer believed that the observed hypnotic trance he induced in his patients was the result of an occult force (which he termed animal magnetism) which flowed from the hypnotist to the subject. The theory was soon discredited, although the term mesmerism remained and continued to interest the medical profession. The actual term 'neuro-hypnotism' (later shortened to 'hypnotism') was not coined until the mid-19th century by the English surgeon James Braid (1795–1860). *See* BRAID'S THEORY OF HYPNOSIS.

NS

animism (1720) *Biology/Medicine* Also called *anima mundi*, this is the doctrine that life is produced by an immaterial soul, or *anima*, that is distinct from the matter composing the body itself. It is associated with the German chemist and physician Georg Ernest Stahl (1660–1734).

Applied to medicine, animism held that all the body's workings were directed by *anima*, which could also prevent and fight disease. *See also* HEALING POWER OF NATURE.

W McDougall, *Body and Mind: A History and a Defense of Animism* (Westport, Conn., 1974)

KH

animism *Philosophy* A term variously used, in particular for the view that apparently inanimate parts of the universe (rivers, mountains, stars, and so on, as well as plants) are in fact animated and activated by souls or spirits; for example, Naiads (springs), Dryads (oak-trees) and so on.

Usually the term is applied to primitive beliefs of this nature rather than to philosophical views claiming to see life in the apparently lifeless (for which *see also* HYLOZOISM, PANPSYCHISM and VITALISM). 'Teleological animism' has been used for the view that some wholes organize themselves and their parts so as to fulfil certain aims which they themselves originate; *see also* ORGANICISM.

ARL

anomalous monism *Philosophy* View associated especially with the American philosopher Donald Davidson (1920–), saying that mental events are identical with certain physical events (hence the 'MONISM'), but that there are no laws which are purely mental, or which connect mental events with physical ones (hence the 'anomalous'; that there are no strict deterministic laws for predicting or explaining mental events is called the 'anomalism of the mental').

This is because whether two events are connected by a law depends on how they are described. Two mental events will also be two physical events, and described in physical terms may be connected by a law; but if one or both are described in mental terms (for example, as a decision rather than as a neurone-firing), no law will connect them. This is because, though any mental event is identical with some physical event, there is no reason to think that all mental events of a certain kind (for example, all decisions) are also physical events of one and the same kind (for example, a certain kind of neurone-firing). Davidson's view is a token IDENTITY THEORY OF MIND AND BODY.

D Davidson, 'Mental Events', *Experience and Theory*, L Foster and J W Swanson, eds (1970); reprinted in D Davidson, *Essays on Actions and Events* (1980)

ARL

anomie (19th century–) *Politics/Sociology* Characteristic of modern society described by French sociologist Emile Durkheim (1858–1917).

In modern societies, an absence of effective moral or cultural rules and restraints to individual wills leads to both social and individual breakdown, since frameworks are necessary for stable social life.

Emile Durkheim, *Suicide* (trans. Glencoe, Illinois, 1957)

RB

anterior-posterior development theory *Psychology* Not attributable to any one originator, the term refers to the anatomical observation that there is rapid growth of the head region as contrasted with lower areas of the body during foetal development. The anterior-posterior development gradient describes the progression of anatomical and motor development from 'head to tail'. The head and its movements develop first, then the upper trunk, arms and hands followed by the lower trunk and leg, foot and toe movements.

P H Mussen, J J Conger and J Kagan, *Readings in Child Development and Personality* (New York, 1975)

NS

anthropic principle *Astronomy* An idea that the universe possesses many of its extraordinary properties because they are necessary for the existence of life and observers.

S Mitton, ed., *The Cambridge Encyclopaedia of Astronomy* (London, 1973)

MS

anthropological linguistics (*c*.1900) *Linguistics* Pioneers included the German-American anthropologist Franz Boas (1858–1942) and the Polish-British anthropologist Bronislaw Malinowski (1884–1942). (*See also* CONTEXT OF SITUATION and PHATIC COMMUNION.)

Boas's study of the indigenous 'Indian' languages of America was one of the cornerstones of American STRUCTURAL LINGUISTICS. But the term means not only 'the study of language by anthropologists' but more broadly the study of language in relation to cultural contexts, yielding a rich range of topics. *See also* ETHNOGRAPHY OF COMMUNICATION.

D H Hymes, ed., *Language in Culture and Society* (New York, 1966)

RF

anthropomorphism (1858) *Biology/Philosophy* A term first used in biology by the encyclopaedist George Henry Lewes (1817–78) in *Seaside Studies*, this is the interpretation of animal behaviours in terms of human motivation; for example, the notion that a mother dog cares for her puppies because she loves them. Difficult or impossible to prove or disprove, anthropomorphism is viewed dubiously by most modern biologists.

W Beck, *Life: An Introduction to Biology* (New York, 1991)

KH

anthroposophy (19th–20th centuries) *Philosophy* The teachings of the German occult philosopher Rudolf Steiner (1861–1925), derived from an ancient Greek phrase meaning 'wisdom about man'.

Steiner held that the development of man's spiritual awareness was of paramount importance. He attempted to treat the investigation of spirituality as a 'scientific' study, and based much of his research upon his central contention that man's intelligence was derived from a more spiritually perceptive form of consciousness which required revival. Christ's self-sacrifice was, he argued, a catalyst for such a realization of man's spiritual potential. *See also* THEOSOPHY.

R Steiner, *Knowledge of the Higher Worlds*, G Metaxa, trans. (London and New York, 1923)

DP

antibody, theories of *Biology* A reference to the many theories seeking to explain how the body identifies foreign substances and responds to them.

Major theoretical models include those that are mathematical and population based, as well as those that are mostly non-mathematical. Despite increased understanding of immunology over the last few years, the field remains rich with uncertainties and theoretical controversies. *See also* CLONAL SELECTION THEORY, GERMLINE THEORY OF ANTIBODY DIVERSITY, INSTRUCTIVE THEORY, SELECTIVE THEORY OF ANTIBODY DIVERSITY, SIDE-CHAIN THEORY, SOMATIC MUTATION THEORY OF ANTIBODY DIVERSITY, and TWO-GENE HYPOTHESIS. *Compare with* ONE GENE-ONE POLYPEPTIDE.

T Kindt and J Carpa, *The Antibody Enigma* (New York, 1984)

KH

antigen template theory *Biology See* INSTRUCTIVE THEORY.

antigenic drift *Biology* The tendency for new strains of a virus to appear over time as a consequence of NATURAL SELECTION.

As the population develops immunity to existing strains of a virus (for example, an influenza virus), slightly different versions of the virus that are better equipped to evade the body's existing immune responses may become established.

H Zinsser, *Rats, Lice and History* (London, 1963)

KH

anti-hero (early 17th century–) *Literary Theory* The term occurs in *Notes from Underground* (1864) by the Russian novelist Fyodor Mikhailovich Dostoevsky (1821–81).

A fictional character embodying the reversal of the qualities of the 'noble' hero: witless, clumsy, incapable of coping or of keeping quiet, his actions give rise to a story of usually comic mishaps. The classic example is the hero of Cervantes' *Don Quixote* (1605, 1615). Such figures were much in vogue in the 1950s and 1960s; for example, Jim Dixon in Kingsley Amis's *Lucky Jim* (1957).

F M Dostoevsky, *Notes from Underground* [1864], S Shishkoff, trans. (Washington, D.C., 1969)

RF

antilanguage (1976) *Linguistics* A term coined by the British linguist Michael Alexander Kirkwood Halliday (1925–).

A variety of a language used by a social group which is in an antagonistic relationship to official society and authority, such as thieves and prison inmates, for secrecy, solidarity and verbal play. Characterized by deformations of words, substitution of new words (RELEXICALIZATION), proliferation of words for key or problematic concepts within the group (OVERLEXICALIZATION).

M A K Halliday, *Language as Social Semiotic* (London, 1978)

RF

anti-literature (1935) *Literary Theory* Term coined by the British poet David Gascoyne (1916–) to describe writing which reverses and transgresses the conventions of traditional literary forms. The concept reflects Gascoyne's support for SURREALISM.

D Gascoyne, *A Short Survey of Surrealism* (1935)

RF

antimatter (1928) *Physics* A theory based on the relativistic mathematical description of Schrödinger's wave equation by Paul Adrien Maurice Dirac (1902–84), joint winner of the 1933 Nobel prize for Physics.

Antimatter is matter composed entirely of antiparticles. These have the property that they correspond with other particles of identical mass and spin; but their charge, baryon number, strangeness, charm and isospin quantum numbers, though equal in magnitude, are opposite in sign (for example, the electron/positron and up quark/up antiquark). A particle and its antiparticle interact by annihilation, producing photons or other elementary particles or antiparticles in their place and conserving energy and momentum in the process. The existence of antimatter in the universe, however, has so far not been detected.

Paul Davies, ed., *The New Physics* (Cambridge, 1989)

GD

antinomianism *Philosophy* View of certain Christians that the duties of a Christian are not to be circumscribed by obedience to a moral law or set of laws.

More widely, the view that justification is by faith rather than by obedience to such laws. More widely still, any view that seeks to justify the actions of certain agents as superior to – and properly to be exempt from – the requirements of law, whether moral or legal. (Not connected in sense with 'antinomy', meaning 'contradiction.'). *See also* SITUATIONISM.

ARL

anti-novel (20th century with earlier antecedents) *Literary Theory* A form of fictional writing which ostentatiously deviates from the technical conventions of the novel. The classic example is *Tristram Shandy* (1759–67) by English writer Laurence Sterne (1713–68), but this 'genre' is in general a 20th century practice.

Many different kinds of departures from the norms of the novel may be employed: deviations in narrative technique such as lack of a story, or a labyrinthine one, distortions of time, over-long, highly specific descriptions; all kinds of verbal dexterity (as used by Joyce and Nabokov); use of another genre's form (Nabokov's *Pale Fire* (1962) appears to be a scholarly edition of a poem); tricks of physical format such as loose pages (B S Johnson), blank or decorated pages (Sterne), and so on. *See* 'NOUVEAU ROMAN'.

G M Hyde, *Vladimir Nabokov: America's Russian Novelist* (London, 1977)

RF

antirealism *Philosophy* A view, primarily associated with the Oxford logician M A E

Dummett (1925–), which insists that we can only understand a statement if we understand under what circumstances someone who asserted it would say something true, and that we can only understand this if we could manifest our understanding, at least in principle, by asserting it in the relevant circumstances.

It follows that we could not understand any alleged truths that transcend all possibility, even in principle, of being verified. The view gains plausibility when we ask what sense it makes to talk of understanding something when we could never in any circumstances manifest a knowledge of it. But realists (of the relevant kind) insist on the contrary that truth must be prior to, and independent of, our means of ascertaining it. Antirealism has its roots in LOGICAL POSITIVISM, and EMPIRICISM generally, as well as in the later philosophy of Wittgenstein (1889–1951), and in mathematical INTUITIONISM.

M A E Dummett, *Truth and Other Enigmas* (1978)

ARL

Antonoff's rule (1907) *Chemistry* Named after G N Antonoff who suggested it. When two liquid phases are in equilibrium, their interfacial tension equals the difference between the surface tensions of the two liquids.

S Glasstone, *Textbook of Physical Chemistry* (London, 1948)

MF

apartheid (20th century) *Politics/Sociology* South African/Afrikaans theory of legally enforced racial separation.

Apartheid rested on the argument that racial distinctions were fundamental to the character of human societies, and that they were biologically based. Human beings should thus be categorized by race, and this categorization should be legally enforced in social, economic, and political affairs. Races should not intermarry, be educated together, or live in the same areas. In practice apartheid was not a policy of racial segregation between blacks and whites, but a justification for a STATE enforced programme of white supremacy. It was in effect, therefore, the application of theories of RACISM.

Graham Evans and Jeffrey Newham, *The Dictionary of World Politics* (Hemel Hempstead, 1990)

RB

apathy, constructive (20th century) *Politics* Theory of the political inaction of the majority in democracies, developed by American political scientists in the second half of the 20th century.

Only a small number, an élite, are politically active in democracies. Any great degree of ACTIVISM on the part of the masses would be destabilizing, and subversive of democracy. Therefore the political apathy of the mass of the population, far from being regrettable, is a necessary condition of political stability.

Thomas Dye and Harmon Zeigler, *The Irony of Democracy* (Belmont, California, 1970)

RB

Apéry's theorem (1978) *Mathematics* The result, recently proved by French mathematician Roger Apéry (1916–), that the value of the ZETA FUNCTION at three is irrational.

A van der Poorten, 'A proof that Euler missed . . . Apéry's proof of the irrationality of $\zeta(3)$', *The Mathematical Intelligencer* (Springer, 1979)

MB

aphasia (19th century) *Linguistics* A clinical term (relevant to linguistics from the 1940s onwards), referring to a breakdown of or disturbance to language, of pathological origin.

Two arguments connect aphasia to linguistic theory. (1) In 1941, the Russian linguist Roman Jakobson (1896–1982) argued that loss of language mirrored in reverse the stages of child language acquisition; the stages in each case gave clues to LINGUISTIC UNIVERSALS. (2) More recently, types of aphasia linked to damage in specific areas of the brain have been cited as evidence of CEREBRAL LOCALIZATION, part of the argument that humans are biologically adapted to language. *See also* BROCA'S AREA and WERNICKE'S AREA.

R Jakobson, *Child Language, Aphasia, and Phonological Universals* (The Hague, 1968 [1941]); E H Lenneberg, *Biological Foundations of Language* (New York, 1967)

RF

apical dominance *Biology* The tendency for the bud at the tip of the stem of most plants to inhibit the growth of lateral (or axillary) buds.

Apical dominance aids plant survival because if the apical bud is damaged or removed, a lateral bud will grow to take its place. (Consequently, gardeners often pinch off apical buds to encourage fuller lateral growth.) Several hypotheses have been put forward to explain apical dominance, including those based on the action of plant hormones (especially cytokinins); but currently no hypothesis enjoys wide acceptance.

F Salisbury, *Plant Physiology* (Belmont, Calif., 1992)

KH

Apollonian/Dionysian (1872) *Literary Theory* Distinction proposed by the German philosopher Friedrich Wilhelm Nietzsche (1844–1900).

The Apollonian impulse in the poet strives toward rationality and perfection of form; the Dionysian toward music and passion. The ideal tragic art balances the two impulses. *See also* NAIVE/SENTIMENTAL.

F Nietzsche, *The Birth of Tragedy*, excerpted in H Adams, ed., *Critical Theory since Plato* (New York, 1971), 636–41

RF

Apollonius's theorem (*c.*200 BC) *Mathematics* Discovered by Apollonius of Perga (*c.*255–170 BC).

Let A and B be two points in the Euclidean plane and let r be a positive number other than 1. Then the locus of a point P such that the ratio of the lengths $AP/BP = r$ is a circle. A generalization of this result provides a definition of a conic in projective geometry.

JB

apoplast-symplast concept (1930s) *Biology* Introduced by the German E Münch, and developed by Alden S Crafts and Theodore C Broyer. This is an explanation of root pressure (that is, the hydrostatic pressure sometimes produced in the roots of vascular plants by the osmosis of ground water); a process which causes fluid to flow up the xylem.

According to this theory, the interconnecting walls and water-filled xylem parts of a plant (the 'dead' parts) are considered a single system called the apoplast, while the cytoplasm (the 'living' part) is the symplast. Ions diffuse into the roots from the soil via the apoplast into the symplast, where they increase inside the cells to levels higher than those outside, in the apoplast. Differences in ion concentrations are thought to create an osmotic system in the roots.

F Salisbury, *Plant Physiology* (Belmont, Calif., 1992)

KH

apprenticeship novel *Literary Theory See* BILDUNGSROMAN.

approximation theorem (20th century) *Mathematics* Folklore. Any two numbers are approximately equal. This holds whenever the range of approximation is greater than the modulus (positive value) of the difference of the numbers.

John Bowers, 'A Drop of the Hard Stuff', *Invitation to Mathematics* (Oxford, 1988)

JB

apriorism *Philosophy* Apart from its popular meaning of dogmatism, apriorism is an alternative – though less common – term for 'rationalism' in its philosophical senses; that is the views that there are *a priori* concepts, or substantive *a priori* truths, or both.

ARL

aquatic theory of human evolution (20th century) *Biology* The controversial idea that humans evolved through a semi-aquatic phase which explains their relative hairlessness, their relatively high amount of body fat, and the instinct for swimming they exhibit at birth. Originally proposed by German biologist Max Estenhofer in 1912, and later by the British biologist Sir Alister Clavering Hardy (1896–), the theory was popularized in the 1980s by Elaine Morgan, a Welsh writer. Mainstream anthropologists view this theory with scepticism.

E Morgan, *The Aquatic Ape* (New York, 1982)

KH

arbitrage pricing policy *Economics* This term refers to the practice of simultaneously buying an asset (currency or commodity) while trading in a low-price market, and selling it in a higher price market with the intention of making a profit from the transaction.

It is argued that the effect of the arbitrageur's action is eventually to eliminate any existing price differentials between the two (or more) markets. Arbitrage pricing is practised by dealers in the commodity and currency markets, where it serves to iron out price disparities across markets.

S A Ross, 'The Arbitrage Pricing of Capital Asset Pricing', *Journal of Economic Theory*, vol. XIII (1976), 341–60

PH, AV

arbitrariness of sign *Linguistics See* SIGN.

archetype (1920s–) *Literary Theory* Sources in anthropology (*see also* MYTH CRITICISM) and JUNGIAN THEORY (psychology).

A 'primordial image' residing in the collective imagination of a people, expressed in myths and in the figurative dimension of literature: exile, rebirth, earth goddess, and so on.

M Bodkin, *Archetypal Patterns in Poetry* (Oxford, 1934)

RF

Archimedean property *Mathematics* Named after Greek mathematician, physicist, and inventor Archimedes (*c.*287–212 BC), who is generally regarded as the greatest mathematician of antiquity.

It is the order axiom for the real number system which states that if $x > 0$ and if y is an arbitrary real number, then there is a positive integer n such that $nx > y$. Geometrically, it means that any line segment can be covered by a finite number of line segments of a given positive length.

T M Apostol, *Mathematical Analysis* (Addison-Wesley, 1974)

MB

Archimedes principle (3rd century BC) *Physics* Named after its discoverer, the Ancient Greek philosopher Archimedes (*c.*287–212 BC), this states that when a body is completely or partially immersed in a fluid (a liquid or gas), the fluid exerts an upward force on the body (the upthrust or buoyancy force) equal in magnitude to the weight of fluid displaced by the body. The upthrust acts through the centre of mass of the displaced fluid.

J Thewlis, ed., *Encyclopaedic Dictionary of Physics* (New York, Oxford and London, 1962)

MS

area rule (20th century) *Physics* A term used in high-speed hydrodynamics in either of two ways: (1) to describe certain results relating the drag on bodies travelling at supersonic speeds to the cross-sectional area; (2) to describe formulae giving the wave drag of an aircraft's wing-fuselage combination, particularly with regard to reducing the drag by suitable shaping of the fuselage.

J Thewlis, ed., *Encyclopaedic Dictionary of Physics* (New York, Oxford and London, 1962)

MS

areal linguistics (1900s) *Linguistics* Also known as areal typology, this is the study of the relationships between geographically adjacent languages (such as those of the Balkans) which have influenced each other and therefore share features in common even though they may be genetically unrelated. Such a group of languages is called a *Sprachbund* (language union).

B Comrie, *Language Universals and Linguistic Typology*, 2nd edn (Oxford, 1989)

RF

aristocracy *Politics/Sociology* Theory of hierarchy. Originally from the Greek 'rule by the best', but now attribution of quality to hereditary élites.

In any society there will be a minority of people of superior skill, wisdom, experience and moral fibre. These qualities are likely to be transmitted through families and are thus to all intents and purposes hereditary. It is appropriate that such people should take a leading role in social and political affairs. They have a duty to do so – *noblesse oblige* – as do others to defer to them.

Roger Scruton, *A Dictionary of Political Thought* (London, 1982)

RB

aristogenesis *Biology See* NEO-DARWINIAN EVOLUTION.

Aristotelian critics *Literary Theory See* CHICAGO CRITICS.

Aristotelian logic (3rd century BC) *Mathematics* Named after Greek philosopher and scientist Aristotle (384–322 BC), this term refers to those of his logical theories based on a deductive inference known as a syllogism (which consists of two premises and a conclusion). Aristotelian logic continued to develop and remain influential throughout the Middle Ages. *See also* 'MODUS PONENS' and 'MODUS TOLLENS'.

W Gellert, S Gottwald, M Hellwich, H Kästner and H Küstner, eds, *The VNR Concise Encyclopedia of Mathematics* (New York, 1975)

ML

Aristotelianism (4th century BC–) *Politics/ Philosophy* Theory of politics derived from the work of the Greek philosopher Aristotle (384–322 BC).

Humans are naturally political, and the political life of the free citizen in the self-governing STATE or '*polis*' is the highest form of life, the essence of the 'good life'. Inequalities breed envy which is disruptive of political stability. The best form of constitution is a mixture of leadership, aristocracy, and citizen participation. The Aristotelian ideal of political life as an end in itself, not a means to other ends, has underlain much liberal thinking about politics, the vote, and CITIZENSHIP. *See also* ARISTOTLE'S FOUR CAUSES.

J Barnes, *Aristotle* (Oxford, 1982)

RB

Aristotle's four causes *Philosophy* Theory derived from the work of Greek philosopher Aristotle (384–322 BC). 'Cause' is a misleading, but traditional, translation of a word meaning 'factor responsible', or perhaps 'explanatory factor'.

The 'four causes' provide answers to four questions one might ask about something, for example, a man: 'What is it made from?' 'Flesh and so on' (*material cause*); 'What *is* its form or essence?' 'A two-legged creature capable of reason (say)', (*formal cause*; 'What produced it?' 'The father (on Aristotle's biology)' (*efficient cause*); 'For what purpose?' 'To fulfil the function of a man (roughly meaning 'to live a life in accordance with reason') (*final cause*).

The doctrine, or parts of it, can then be extended in various ways (in particular to cover events and states as well as objects), and undergoes various complications in the process; but its primary application is to objects, especially biological objects and artifacts. The four causes, especially the first two, are closely linked to Aristotle's important dichotomy between matter and form (HYLOMORPHISM).

Aristotle, *Physics*, book 2

ARL

arousal theory *Psychology See* ACTIVATION THEORY OF EMOTION.

Arrhenius' dissociation theory (1883–87) *Chemistry* Theory, also known as the theory of electrolytic dissociation, developed by Swedish physical chemist Svante August Arrhenius (1859–1927). Arrhenius received the Nobel prize for Chemistry in 1903 for his theory. Electrolytes are compounds which break apart (dissociate) spontaneously to form ions when they dissolve in water.

The ions are the agents of electrolysis. Strong electrolytes dissociate completely whereas weak electrolytes dissociate partially. The degree of dissociation depends on the concentration of the electrolyte in water.

M Freemantle, *Chemistry in Action* (Basingstoke, 1987)

MF

Arrow-Debreu model (1954) *Economics* Named after American economist Kenneth Arrow (1921–) and French-born economist Gerard Debreu (1921–), who examined the dynamics of the whole economic system and were able to prove the existence of a multi-market equilibrium in which no excess demand or supply exists.

The theory is based on two assumptions: first, that a competitive equilibrium exists if each person in the economy possesses some quantity of every good available for sale in that market; second, that exploitable labour resources exist which are capable of being used in the production of desired goods/ services. This model is considered to have been a major advance in economic theory.

See GENERAL EQUILIBRIUM, and WALRAS'S STABILITY.

K J Arrow and G Debreu, 'The Existence of an Equilibrium for a Competitive Economy' *Econometrica*, vol. XXII (1954), 265–90; Y Balasko, *Foundations of the Theory of General Equilibrium* (New York, 1986)

PH

Arrow impossibility theorem *Mathematics* In modelling any democratic process one would like a fixed procedure for *aggregating* the preferences of a group of individuals into an overall ordering. Arrow's result shows that this is not possible in general for a group larger than two if the procedure is required to satisfy four natural conditions: (1) the procedure always produces an ordering; (2) any universally shared preference should be reflected; (3) the outcome should not depend on preferences for irrelevant options; (4) each individual can influence the outcome.

AL

art as device (1917) *Stylistics* Literary theory of Russian formalists (*see* FORMALISM, RUSSIAN), defined by Viktor Shklovsky (1893–1984).

Objecting to late 19th/early 20th century theories which saw poetry in terms of symbols and images, the formalists argued for the centrality of rhetorical 'devices' (*priem*) such as PARALLELISM, archaism, METAPHOR, which cast the text into a 'poetic language' distinct from 'ordinary language', causing DEAUTOMATIZATION.

V Shklovsky, 'Art as Technique', *Russian Formalist Criticism*, L T Lemon and M J Reis, trans. and eds (Lincoln, Nebraska, 1965), 3–24

RF

art informel (1950) *Art* A French term meaning 'art without form', this was adopted by the critic Michel Tapie to describe abstract painting (similar to American ABSTRACT EXPRESSIONISM) opposed to the rigour of CUBISM or the geometrical abstraction of 'DE STIJL' and SUPREMATISM, where the artist's emotions and subconscious fantasies are expressed.

Pioneers of this movement were the German painter Wols (1913–51) and French artist Hans Hartung (1904–89), Jean Fautrier (1898–1964) and Jean Dubuffet (1901–85). After 1954, TACHISM was introduced as the term to describe all non-geometrical abstraction.

LL

art of the Third Reich (1933–45) *Art* Following the rise to power of Nazi leader Adolf Hitler (1889–1945), German art came under the tight control of Paul Joseph Goebbels (1897–1945). As propaganda minister, he brought artists under the control of the *Reichskulturkammer*, which body decreed that all forms of art should be racially conscious.

The art which was subsequently produced until the fall of National Socialism is typified by cleanly-realist works which are traditionalist and aim to promote Nazi ideology. All other works were branded degenerate and removed or destroyed. Meanwhile, however, much valuable non-Nazi art was acquired (as spoils of war) from across Europe for private consumption or gain.

H Osborne, *Oxford Companion to 20th-century Art* (Oxford, 1981)

LL

arte mat (1917) *Art* The name given by Italian artists Giorgio de Chirico (1888–1978) and Carlo Carra (1881–1966) to the style of painting that resulted from their encounter at the military hospital in Ferrara in 1917. In English it is often referred to as 'metaphysical painting'.

Although short-lived and not strictly a school, according to de Chirico's brother Alberto Savinio (1891–1952) its significance was: 'the total representation of spiritual necessities within plastic limits – power to express the spectral side of things – irony'. Metaphysical painting is imbued with an air of mystery, ambiguity and incongruity, achieved through unreal perspective and striking lighting.

U Apollonio, *Pittura metafisica* (Milan, 1945); G de Chirico, *Valori Plastici* (1918–21)

AB

arte povera (1970) *Art* An Italian art movement which began after a similarly named exhibition (Turin, Mus. Civ.) organized by critic Germano Celant.

Relations with CONCEPTUAL ART and

MINIMALISM are evident, but its most typical expression is the HAPPENING. Practitioners refuse to recognize the 'product' or 'work', concentrating on the process rather than the result. This term also refers to the base materials used by such artists as Pistoletto, Mario Ceroli and Mario Merz.

LL

articulatory phonetics (19th century) *Linguistics* Phonetics was part of both the Indian and Greek grammatical traditions; influential modern English phoneticians included Henry Sweet (1845–1912) and Daniel Jones (1881–1967).

Study of the sounds of speech in terms of the structure and actions of the vocal organs (diaphragm, lungs, glottis, pharyngeal, nasal and oral cavities, tongue, teeth, lips) which are involved in producing them. *See also* ACOUSTIC PHONETICS, AUDITORY PHONETICS, and CARDINAL VOWELS.

P Ladefoged, *A Course in Phonetics*, 2nd edn (New York, 1982)

RF

artificial intelligence *Psychology* Pioneered by the English mathematician Alan Mathison Turing (1912–54).

A relatively new discipline which draws principally from computer science and psychology, especially cognitive psychology. It is concerned with investigating existing intelligent systems, exploring the potential for developing new systems and applying this knowledge in the pursuit of technological solutions to applied problems (for example, medical diagnosis). Cognitive science has emerged within psychology as a sub-discipline concerned with the application of formal techniques to evaluate and develop psychological theory.

M Sharples *et al.*, *Computers and Thought: A Practical Introduction to Artificial Intelligence* (Cambridge, 1989)

NS

artificial intelligence and language (1950s) *Linguistics* In this area, there was early unsuccessful work in MACHINE TRANSLATION; the field flourished only from the 1970s, with acknowledgement of the linguistic complexity involved.

Use of computational procedures, with programs based on appropriate grammars of natural language, to understand, by machine simulation, the linguistic abilities of speakers. Major areas include PARSING, and the representation of and access to meaning. Engineering applications include the development of intelligent knowledge-based systems for translation and for voice recognition.

B J Grosz and K Sparck Jones and B L Webber, *Readings in Natural Language Processing* (Los Altos, 1986)

RF

artificial languages (17th century, late 19th century) *Linguistics* There was discussion of 'universal languages' in 17th century rationalist philosophy; the International Language Movement flourished from the late 19th century: pioneers included the German priest J M Schleyer (1832–1912), inventor of Volapük, and the Polish doctor Lazarus Ludwig Zamenhof (1859–1917), who created Esperanto.

Constructed languages devised with the intention of facilitating international communication; usually assembled from common roots and grammatical markers taken from English and major European languages.

The notation of formal logic, and computer programming languages such as BASIC and PROLOG, are also artificial languages, but with different purposes.

A Large, *The Artificial Language Movement* (Oxford, 1985)

RF

artificial selection (1859) *Biology* Contrasted at length with NATURAL SELECTION by English biologist Charles Darwin (1809–82), this is the deliberate breeding of animals, cross-pollination of plants, or development of particular microbial strains to suit human needs. More formally, the term refers to the choosing of genotypes that will contribute to the gene pool for succeeding generations.

C Darwin, *On the Origin of Species by Means of Natural Selection, or the Preservation of Favoured Races in the Struggle for Life* (London, 1859); *The Variation of Animals and Plants Under Domestication* (London, 1868)

KH

Artin's conjecture on primitive roots (20th century) *Mathematics* Named after the German-born American mathematician Emil Artin (1898–1962).

This is a quantitative form of the conjecture that every integer which is not a perfect square is a primitive root of infinitely many primes; that is, for every non-square integer a, there are infinitely many primes p for which a^{p-1} is congruent to 1 mod p.

Encyclopedic Dictionary of Mathematics (MIT Press, 1987)

MB

Arts and Crafts Movement (*c.*1886) *Art* A name coined by the English printer and bookbinder Thomas Cobden-Sanderson (1840–1922), a member of the Arts and Crafts Exhibition Society, for a philosophy which proved significant in the development of British design at the end of the 19th century.

Espousing an idealistic approach to design and the arts, the movement, whose champions included the English designer William Morris (1834–96), rejected the overly ornate style of the Victorians. Instead they favoured a simpler, decorative repertoire, drawn largely from medieval and Celtic art.

AB

as if hypothesis (1911) *Psychology* Term coined by the German philosopher Hans Vaihinger (1852–1933) to describe how thinking and acting proceed by unproven or contradictory assumptions which are treated as if they were true.

Vaihinger considered knowledge as largely a network of such 'as if' strategies. It is a variety of thinking in which one formulates a coherent conjecture and proceeds with further investigation on the presumption that it is probably true. Many principles of science are in fact hypothetical.

J S Bruner, *Beyond the Information Given* (New York, 1973)

NS

as if, philosophy of *Literary Theory See* 'ALS OB'.

Ascoli's theorem (19th century) *Mathematics* Named after the Italian analyst Giulio Ascoli (1843–96), this result specifies conditions for the limit of a sequence of continuous functions to be a continuous function.

More precisely, if a family of functions is equicontinuous and pointwise bounded, then it is totally bounded in the uniform norm. Each sequence in such a family has a norm convergent subsequence on compact sets. The complex case is known as the *Arzela-Ascoli theorem*.

G B Folland, *Real Analysis* (New York, 1984)

ML

asset valuation theory (20th century) *Accountancy* There are two principal methods of asset valuation. Historical cost accounting values assets at the production cost or purchase price, less depreciation. Fixed assets (land and buildings) are valued at net historical costs, and current assets at cost or net realizable value, whichever is the lower. The second method, CURRENT COST ACCOUNTING (also known as inflation accounting), is used in times of rising prices and values assets on the basis of net realizable value or current replacement cost.

Asset valuation policy may change during the life of a firm. It also varies widely from company to company and country to country.

PH

assimilation-contrast theory *Psychology* Proposed by the Austrian-American F Heider (1896–).

A theory of information processing which states that information which is broadly concordant with one's beliefs and attitudes is likely to be accepted and processed by reciprocal approximation of divergent viewpoints. Discordant information is likely to be rejected. For instance, where a listener is presented with discordant information he or she may respond by stating his or her own position in an uncomprising fashion. Hence the theory is sometimes labelled 'the boomerang effect'. *See* ASSIMILATION THEORY.

G Lindzey and E Aronson, eds, *The Handbook of Social Psychology* (London, 1969)

NS

assimilation, law of *Psychology* Proposed by the Swiss psychologist Jean Piaget (1896–1980).

The process of assimilation of some aspect of an organism's environment into its behaviour. In a novel situation an organism will react in much the same way as it did in other similar situations encountered previously. For instance, as applied to memory the law states that novel objects or events must be assimilated into the existing cognitive structure before they can be remembered. The law appears in different psychological domains. *See* ASSIMILATION THEORY and ASSIMILATION-CONTRAST THEORY.

A Baddeley, *The Psychology of Human Memory* (New York, 1976)

NS

assimilation theory *Psychology* Proposed by the Turkish psychologist M Sherif (1906–).

A theory of attitude change based on the supposition that attitudes are altered by changes in the relationship between the originally-held position and the reliability of the source of this new attitude. An individual can either assimilate the new attitude or create or maintain a contrasting position. The theory has been criticized for its failure to predict accurately specific instances in which an individual's attitudes will change and when they will not. *See* ASSIMILATION CONTRAST THEORY.

M Billig, *Social Psychology and Intergroup Relations* (London, 1976)

NS

association, laws of *Psychology* Proposed by the Scottish metaphysician Thomas Brown (1778–1820).

A number of empirical and theoretical generalizations about the manner in which associations are made. Aristotle proposed relations between elements which lead to associations and these 'relations' serve as an associationistic model of human cognition. Subsequently, the following laws have been forwarded: similarity, where two similar memory contents are linked; contrast, where two contrasted elements are linked; and contiguity in space and time, where two simultaneous or immediately successive elements are linked together. *See* ASSOCIATIONISM, ASSOCIATIVE LEARNING and CONNECTIONISM.

G H Bower and E R Hilgard, *Theories of Learning* (Englewood Cliffs, N.J., 1981)

NS

associationism *Philosophy* A doctrine developed primarily by the Scottish philosopher David Hume (1711–76) and the English psychologist David Hartley (1705–57). (Psychology and philosophy of mind were not then distinguished, but Hartley offered a physiological basis for what in Hume was a purely mentalistic doctrine.)

Ideas, regarded rather as sensations or as mental images, were associated in the mind according to certain laws, mainly concerning contiguity and resemblance, and thereby led to further ideas, and to the functioning of mental life in general. The doctrine had analogies to physical theories whereby physical atoms moved under the influence of physical laws like that of GRAVITATION. Eventually it came to be rejected – for example, by F H Bradley (1846–1924) and H L Bergson (1859–1941) – because such psychological 'atoms' are not in fact to be found and the proposed mechanism was far too simplistic; also, it was not always clear whether contiguity and resemblance apply to the ideas themselves or to the things they are of.

Associationism is also called 'association of ideas', a phrase apparently due to the English philosopher John Locke (1632–1704), in his *Essay Concerning Human Understanding*. The doctrine, however, has far older roots, going back to Greek thought and even to primitive 'sympathetic' magic. An allied modern doctrine is that of the conditioned reflex.

ARL

associationism (4th century BC–) *Psychology* Proposed by the Scottish philosopher James Mill (1773–1836).

The theory that higher-order mental or behavioural processes result from the association of simpler mental or behavioural elements. Complex mental processes such as thinking, learning and memory comprise associative links formed between ideas according to specific laws and principles. This general hypothesis was first described by Aristotle and came to fruition in the 17th and 18th centuries. In recent years it has lost much of its explanatory power since most cognitive processes are too complex to be solely formed by associative connections. *See* ASSOCIATION, LAWS OF, ASSOCIATIVE LEARNING and CONNECTIONISM.

G H Bower and E R Hilgard, *Theories of Learning* (Englewood Cliffs, N.J., 1981)

NS

associative chain theory *Psychology* Proposed by the American psychologist Edward Lee Thorndike (1874–1949).

One of the first theories of complex behaviour arguing that each of the components of a sequential action is linked associatively to the antecedent component of that action. The whole act is therefore a chain of elementary acts. For example, a sentence may be considered to comprise no more than a chain of words. The theory is regarded as simplistic and untenable. *See* ASSOCIATIONISM.

G H Bower and E R Hilgard, *Theories of Learning* (Englewood Cliffs, N.J., 1981)

NS

associative law *Mathematics* The theorem or axiom of any mathematical system that a given binary operation $*$ has the property that the bracketing of its arguments may be disregarded. In particular,

$$(a*b)*c = a*(b*c).$$

Any binary operation possessing this property is said to be associative. For example, conjunction is associative but the vector product is not. *See also* COMMUTATIVE LAW and DISTRIBUTIVE LAW.

MB

associative learning (17th century–) *Psychology* Proposed in its modern form by the American psychologist John Broadus Watson (1878–1958).

The theory describes learning as the linking or binding of mental elements. Early forms of the theory are associated with the early empiricist philosophers Thomas Hobbes (1588–1679), John Locke (1632–1704), Bishop George Berkeley (1685–1753) and David Hume (1711–76). Later notions of associations between stimulus-response units are linked with behaviourists such as J B Watson and Burrhus Frederic Skinner (1904–80); and with contemporary work by propositional and imaginal cognitive psychologists such as the Canadian psychologist Allan Uhro Paivio (1925–) and Gordon Howard Bower (1932–). *See* ASSOCIATION, LAWS OF and CONNECTIONISM.

G H Bower and E R Hilgard, *Theories of Learning* (Englewood Cliffs, N.J., 1981)

NS

associative shifting, principle of *Psychology* Proposed by the American psychologist Edward Lee Thorndike (1874–1949).

Responses to one set of stimuli can be evoked by other stimuli in a similar situation. It is a principle of learning which holds that if a response can be maintained while the stimulus environment is gradually altered or 'shifted' (by adding or subtracting elements), the same response will eventually occur in a completely new situation.

G H Bower and E R Hilgard, *Theories of Learning* (Englewood Cliffs, N.J., 1981)

NS

asymmetrical information *Economics* This situation exists when one side of the market possesses information lacked by others in that market. Employers in labour markets often possess more information about the current/future status of their industry than trade unions or workers, and can use this as a basis of negotiation. However, it can be seen as an imperfection in the working of the market mechanism and may lead to economic inefficiency. *See* ADVERSE SELECTION.

G Akerlof, 'The Market for"Lemons": Quality Uncertainty and the Market Mechanism', *Quarterly Journal of Economics*, vol. LXXXIV (August 1970), 3

PH, AV

asymmetry *Art* A term referring to the absence of symmetry in a painting or object. In order to achieve a realistic effect, a certain degree of asymmetry is necessary; for instance, in the representation of the human face, which exemplifies the lack of absolute symmetry in nature. This lack of symmetry was exploited in Baroque and Rococo art in order to give movement and excitement to compositions.

AB

atavism (1825–35) *Biology/Politics/Sociology* Meaning the reversion to an earlier type (a throwback) that has been absent during intervening generations, the concept of atavism dates from antiquity. In modernity, the idea was applied by Cesare Lombroso

(1835–1909), an Italian criminologist who believed that 'the criminal mind' was caused by a primitive human state. American sociologist Thorstein Veblen (1857–1929) applied atavism to predatory industrialists in his *The Theory of the Leisure Class* (1899).

While the idea has often been used to explain 'undesirable' people, in biology it describes the appearance of structures or colours in individuals that are not evident in their parents or grandparents. It is believed that this is caused by the inheritance of a recessive or complementary gene. The word 'atavisim' is rarely used by modern biologists. *See also* MENDEL'S LAWS and HEREDITARIANISM

P B Medawar and J S Medawar, *Aristotle to Zoos: A Philosophical Dictionary of Biology* (Oxford, 1983)

RB, KH

atheism *Philosophy* The contention that there is no God and that religious faith in such an entity is a consequence of man's imagination or gullibility.

It has been argued that virtually everyone is an atheist as few people profess to believe in all the gods or other divine personalities devised by man. More conventionally, the atheist denies the existence of God as perceived by Christians and adherents of other major world religions, pointing out the lack of material evidence for such a being and the dilemma posed by the idea of a loving God permitting suffering and other evils to continue in the world. Atheists frequently point to the apparently ever-widening gulf between modern science and religious faith, and sometimes claim that God has become 'the god of the spaces in between' known scientific facts. As science advances, so many atheists argue that God's natural territory is decreasing and that 'blind' faith is increasingly untenable and even immoral when the acceptance of such ideas depends increasingly upon the denial of reason. *See also* AGNOSTICISM, HUMANISM and THEISM.

J Monod, *Chance and Necessity* (Oxford, 1982)

DP

atomic theory *Chemistry See* BOHR'S THEORY OF THE HYDROGEN ATOM, DALTON'S ATOMIC THEORY and RUTHERFORD'S ATOMIC THEORY.

atomic uniformity, principle of *Philosophy* Principle used by the English economist John Maynard Keynes (1883–1946) in trying to justify induction.

It said that, if induction is to work, a complex change must be resolvable into a set of component changes each of which is separately attributable to some distinct feature of the preceding state of affairs. *See also* INDUCTIVE PRINCIPLE.

J M Keynes, *A Treatise on Probability* (1921), 249

ARL

atomism *Philosophy* As a physical theory atomism was invented by Leucippus and Democritus in the 5th century BC, developed by Epicurus a century or so later, and revived in the 17th century.

In it matter consisted of tiny indivisible, indestructible and unchanging bits of solid stuff, differing in shape and size, and jostling each other in the void to constitute the material world. They were responsible for colours, smells, tastes, and so on (the 'secondary' qualities; *see also* EFFLUXES), but did not themselves have them. In the earliest phase physical divisibility may not have been distinguished from conceptual or mathematical divisibility, despite the atoms' having shapes. The rival 'continuous' theory of matter, with no void, was held by Aristotle (384-322 BC) and the Stoics, but atomism has proved more fruitful in the development of modern physical theory, despite its enormous differences.

More generally, any theory can be called atomism which analyzes a certain set of phenomena in terms of a set of (not necessarily physical) building blocks, each having a narrowly circumscribed set of properties (for example, sensations or ideas in the case of SENSATIONALISM or ASSOCIATIONISM). *See also* LOGICAL ATOMISM and SEMANTIC ATOMISM.

ARL

atonality (early 20th century) *Music* Most frequently associated with the pre-serial music of the Second Viennese School (and also called pantonality), this term refers to music which is not ordered according to a tonal system, nor organized by serial principles.

Though atonal music may use a hierarchic ordering of pitch, it is characterized by its absence of tonal centres, triads or modes.

Thus it is defined more by what it lacks (notably a concept of harmonic dissonance and resolution) than by what it retains. *See also* CHROMATICISM.

R Reti, *Tonality, Atonality, Pantonality: A Study of Some Trends in Twentieth Century Music* (London, 1958)

MLD

attitude theories *Philosophy* In effect another name for SPEECH ACT THEORIES; though, strictly speaking, attitude theories analyze the meaning of certain words or sentences in terms of the expression of attitudes rather than the performing of various other acts that one can perform by speaking, such as prescribing or denying. The name therefore applies to EMOTIVISM more happily than to, say, PRESCRIPTIVISM or the speech act theory of NEGATION.

ARL

attraction, law of (20th century) *Psychology* Proposed by the American psychologist Ellen Berscheid.

Not a law as such but a term referring to the empirical evidence indicating that people who like each other will position themselves closer together than those who have no particular liking for each other. Attraction itself is conceived as the act of adjusting proximity relationships between individuals depending upon their liking for each other. *See* GAIN LOSS THEORY OF ATTRACTION.

E Berscheid and E H Walster, *Interpersonal Attraction*, 2nd edn (Reading, Mass., 1978)

NS

attribution theory *Psychology* Proposed by the Austrian-American F Heider (1896–).

A theory primarily concerned with social perception: how one perceives oneself and others. Attribution involves ascribing a characteristic to oneself or another person. Attribution theory holds that in social situations an individual observes another performing some behaviour and, based on these behavioural data, proceeds to infer something of the other's intentions and motivational disposition. This inference is then used to account for the observed conduct. *See* ATTRIBUTION THEORY OF EMOTION and NON-COMMON EFFECTS PRINCIPLE.

F Heider, *The Psychology of Interpersonal Relations* (New York, 1958)

NS

attribution theory of emotion *Psychology* Proposed by the American Stanley Schachter (1922–).

The name commonly given to Schachter's theory of emotion, also referred to as 'cognitive arousal theory', 'cognitive evaluation theory' and 'Schachter and Singer's theory'. The physiological activations asssociated with various and even opposite emotions are similar, and it is the cognitive attribution or evaluation concerning the causes of physiological changes that determine to a great extent what emotion is experienced. A considerable amount of Schachter's work has been verified empirically. *See* ATTRIBUTION THEORY.

H Gleitman, *Psychology* (London, 1991)

NS

auditory phonetics (19th and 20th centuries) *Linguistics* Research into hearing has been continuous in the modern period; from the 1960s it has been linked to PSYCHOLINGUISTICS, and more recently ARTIFICIAL INTELLIGENCE.

Study of the physiology, neurology and psychology of speech perception is a complex and interdisciplinary field. Hearing is not simply a physical process, but is guided by linguistic and other knowledge; *see* ANALYSIS-BY-SYNTHESIS, ACOUSTIC PHONETICS and ARTICULATORY PHONETICS.

P Lieberman and S E Blumstein, *Speech Physiology, Speech Perception, and Acoustic Phonetics* (Cambridge, 1988)

RF

***aufbau* principle** *Chemistry* From the German word *aufbau*, meaning 'building-up' (this hypothesis is also known as the building-up principle), this principle determines the electronic configurations of atoms or molecules.

Electrons in their ground states occupy atomic or molecular orbitals in order of increasing orbital energy levels. The lowest energy orbitals are always filled first. The periodic classification of elements is based on the *aufbau* principle. *See also* HUND'S RULE and PAULI EXCLUSION PRINCIPLE.

M Freemantle, *Chemistry in Action* (Basingstoke, 1987)

MF

Auger effect (20th century) *Physics* Named after the French physicist Pierre Auger (1899–), who pioneered its study. It is possible for an atomic nucleus to interact directly with, and absorb, an electron in one of the innermost shells (internal conversion). This vacant place must be filled by an electron from a higher shell and this transition may be accompanied by the emission of an X-ray. The Auger effect is a competing process in which the energy is transferred to another atomic electron which is then emitted (the Auger electron).

Whichever process occurs, a vacancy is left in a higher electron shell and the process continues until the whole electron distribution in the atom has been re-established.

J Thewlis, ed., *Encyclopaedic Dictionary of Physics* (New York, Oxford and London, 1962)

MS

augmented transition network grammar (1970) *Linguistics* Developed by W A Woods and used in ARTIFICIAL INTELLIGENCE and PSYCHOLINGUISTIC research.

It is a more sophisticated GRAMMAR than a FINITE STATE GRAMMAR, and was used computationally to determine the syntax of a sentence. It has also been invoked as a model for how humans employ syntax in understanding sentences, but is little used now.

A Garnham, *Psycholinguistics* (London, 1985)

RF

Austinianism (19th century) *Politics/Law* Theory of law expounded by English lawyer John Austin (1790–1859).

Law consists of commands issued by a sovereign. It is judged law not because we ought to obey it or because it fulfils any moral criteria, but because it is habitually obeyed. A sovereign, which can be a person or a group, is the holder of final and unlimited power, and will as a matter of observable fact be found in any society. This theory of law is the most simple kind of LEGAL POSITIVISM.

H A L Hart, *The Concept of Law* (Oxford, 1961)

RB

author, death of the (1968) *Literary Theory* Proclaimed by French critic Roland Barthes (1915–80), revised and refined by French philosopher Michel Foucault (1926–84).

The traditional notion of an author pre-existing the text and inscribing a fixed set of meanings in it is challenged (*see also* INTENTIONAL FALLACY); 'author' is an illusion produced by the reader's constructive act of reading (*see also* 'LISIBLE' AND 'SCRIPTIBLE'). For Foucault, this author-in-the-text is employed by readers as a control against a chaos of subjective meanings.

R Barthes 'The Death of the Author' [1968], *Image–Music–Text*, S Heath, ed., and trans. (London, 1977), 142–48; M Foucault, 'What is an Author?' [1969], *Textual Strategies*, J V Harari, ed. (London, 1979), 141–60

RF

authoritarianism (20th century) *Politics* A critical term for despotic regimes.

Authoritarianism is a manner rather than a style of governing, which neither takes seriously nor tolerates the expression of dissenting opinion nor the pursuit of contrary policies. Sometimes used as an alternative to TOTALITARIANISM by conservatives who do not wish to use that designation for military or despotic regimes which sustain private industrial property.

David Robertson, *The Penguin Dictionary of Politics* (London, 1985)

RB

authority *Politics* A theory justifying special functions or powers.

Persons or institutions have authority if they possess qualities of insight, or experience, or skill which mark them off from others, and which entitle them to make decisions which those others are unqualified to make. Authority may thus justify governing others, or making pronouncements about morals, or scientific or technical matters. The concept of authority is regarded with deep suspicion by subscribers to ANARCHISM.

R Flathman, *The Practice of Political Authority: Authority and the Authoritative* (Chicago, 1980)

RB

autocracy *Politics* Absolute and arbitrary rule.

An autocrat is not constrained by laws or coventions, and autocracy is essentially unpredictable as a form of government. Unlike authority, which implies a right or qualification to command, autocracy indicates no more than untrammelled power.

Roger Scruton, *A Dictionary of Political Thought* (London, 1982)

RB

autocrine hypothesis (1980) *Biology/Medicine* Postulated by American research physicians Michael Benjamin Sporn (1933–) and George Joseph Todaro (1937–), this is the theory that cancerous cells both produce growth factors and respond to them, unlike most normal cells (which are regulated only by the growth factors produced by other cells).

R Weinberg, *Oncogenes and the Molecular Origins of Cancer* (Cold Spring Harbor Laboratory, 1989)

KH

autogenesis *Biology See* LAMARCKISM.

autokinetic effect *Psychology* Named by the German psychologist Kurt Koffka (1886–1941).

Also called the autokinetic illusion/phenomenon, this is a kind of apparent motion in which a small, static spot of light viewed in a dark room appears to move. The movement is apparently not due to eye movements and may cover as much as 20° of the visual field. The effect can be reliably produced and consequently has been employed in psychological investigations of suggestibility (and in the establishment of group norms under such influences).

E G Boring, *A History of Experimental Psychology* (New York, 1957)

NS

automata theory (mid-20th century) *Mathematics* Originating with the study of mathematical models of nervous systems and electronic computers, this theory concerns mathematical models of transformers of discrete information (abstract machines) and their capacity to solve various types of problems by means of the algorithms available to them.

There are several types of automata, including TURING MACHINES, linear bounded automata, pushdown automata, and finite automata. These models are also used in the theory of complexity of computation of algorithms. *See also* CELLULAR AUTOMATON and COMPLEXITY THEORY.

I Peterson, *The Mathematical Tourist* (New York, 1988)

ML

automatic stabilization *Economics* A term first used by the American economist Albert Hart (1909–), this refers to elements built-in to fiscal policy which serve to decrease automatically the impact of fluctuations on economic activity. (Such elements are sometimes called *built-in stabilizers*.)

In general terms, built-in stabilizers can be anything which serves to reduce the value of the MULTIPLIER; this dampens fluctuations in NATIONAL INCOME caused by shifts in autonomous spending. The two principal stabilizers are: taxes, and government transfer payments (chiefly, unemployment and supplementary benefits). Although stabilizers may reduce fluctuations in national income, they will not totally eliminate them. *See* FINE-TUNING.

A G Hart, '"Model Building" and Fiscal Policy', *American Economic Review*, vol. XXXV (September, 1945), 531–58

PH

automatism (1940s) *Art* This term describes both the aims and methods of SURREALISM in allowing unconscious images and chance effects to form a work; for example, the use of ink blots by French painter Francis Picabia (1879–1953).

The principle was adopted by the New York Surrealists of the 1940s and provided the basis for ACTION PAINTING, ART INFORMEL; and, particularly, a group of seven Canadian artists active in the 1940s who called themselves *Les Automatistes*.

LL

automimicry (1960s) *Biology* Proposed by biologists Lincoln and Jane van Zandt Brower. This term refers to MIMICRY (as seen in the Monarch butterfly) in which the minority of individuals who lack a particular characteristic (in this case, unpalatabilty to birds) are indistinguishable from all others of the species, most of which possess the characteristic.

Monarch butterflies become unpalatable from eating poisonous plants while in the larvae stage; those few larvae that grow up eating non-poisonous plants lack the protective characteristic, but not the protective coloration.

R Owen, *Camouflage and Mimicry* (Chicago, 1982)

KH

autonomy of literary text (*c*.1930–60) *Literary Theory* For sources and context, *see* NEW CRITICISM.

A literary work is an independent aesthetic object, autonomous in the sense of being free of its author's determination, independent of social and historical context, 'self-reflexive' in the sense of referring only to itself and not to the phenomenal world.

C Brooks, *The Well Wrought Urn* (London, 1968 [1947])

RF

autonomy of syntax (1960s) *Linguistics* Tenet of TRANSFORMATIONAL-GENERATIVE GRAMMAR.

A GRAMMAR has three components: SYNTAX, SEMANTICS, and PHONOLOGY. A set of syntactic rules can be constructed without reference to the other two components; semantics and phonology both need to make reference to syntax, and are therefore not autonomous.

A Radford, *Transformational Grammar* (Cambridge, 1988)

RF

autosuggestion *Psychology* Pioneered by the French pharmacist Emile Coue (1857–1926).

Also termed self-suggestion, this began as a crude technique for self-improvement; the crux of which was the phrase 'every day in every way I am getting better and better', to be repeated by the client 20 or 30 times a day. It involves the process of giving suggestions to oneself and has evolved into the modern use of autohypnosis as part of hypnotherapy and relaxation techniques.

J R Hilgard, *Personality and Hypnosis: A Study of Imaginative Involvement* (Chicago, 1979)

NS

Auwers-Skita rule *Chemistry* The rule originally suggested that the *cis* isomer of *cis-trans* isomeric hydroaromatic compounds had the lower molecular refractivity but the higher refractive index and density. The rule, after several modifications, now states that among alicyclic epimers with the same dipole moment the isomer with the highest enthalpy has the highest density, refractive index and boiling point.

MF

available space theory *Biology* The theory that a new leaf bud only arises when a certain minimum of space has become available for it on the stem. *Compare with* REPULSION THEORY.

F Salisbury, *Plant Physiology* (Belmont, Calif., 1992)

KH

average cost pricing (1939) *Economics* Research by Robert Ernest Hall (1943–) and Charles J Hitch showed that, whilst prices represent the average cost of production and distribution, firms typically set their pricing policies by determining the average cost of production, and adding a profit margin which does not appear to vary with market demand. *See* MARK-UP PRICING and MARGINAL COST PRICING.

R E Hall and C J Hitch, 'Price Theory and Business Behaviour', *Oxford Economic Papers* vol. II (May, 1939), 12–45

PH

average reader (AR) (1959) *Stylistics* Also called '*Lecteur moyen*'. A technique formulated by the stylistician Michael Riffaterre.

Group of cultivated readers used as informants to identify where stylistically significant points occur in a literary text; preferred to formal linguistic analysis, which produces arbitrary and imperceptible patterns. In 1966 replaced by 'superreader', *archilecteur*, analytic ideal reader equipped with relevant linguistic, literary and historical competence. *See also* CONVERGENCE.

M Riffaterre, 'Describing Poetic Structures', *Structuralism*, in J Ehrmann, ed. (New York, 1966), 188–230

RF

Avogadro's law (1811) *Chemistry* Named after the Italian physicist and chemist Amedeo Avogadro (1776–1856) who proposed it. (Also known as Avogadro's hypothesis, Avogadro's principle and Avogadro's theory.) The law states that equal volumes of all gases at the same temperature and pressure contain equal numbers of molecules.

The number of molecules in one mole of gas is always 6.022×10^{23}. This quantity is known as Avogadro's constant. It follows from Avogadro's law that 1 mole of any gas always occupies the same volume at constant temperature and pressure. The law is only valid for an ideal gas. *See also* GAS LAWS.

M Freemantle, *Chemistry in Action* (Basingstoke, 1987)

MF

axiom of choice *Mathematics* If S is any non-vacuous class of non-empty sets, there is a function f defined on S such that for each A in S, $f(A)$ lies in A. It is equivalent to the HAUSDORFF MAXIMALITY THEOREM. *See also* ZORN'S LEMMA and the WELL-ORDERING PRINCIPLE.

JB

axiomatic set theory (20th century) *Mathematics* The presentation of SET THEORY as a formal set of uninterpreted axioms and rules of inference, rather than as the formalization of a given body of knowledge. The most standard axiomatization is now known as ZERMELO-FRAENKEL SET THEORY. *See also* NAIVE SET THEORY.

MB

axiomatic theories *Economics* These are theories relating to consumer behaviour and rationality, and are an essential part of CONSUMER DEMAND THEORY and indifference curve analysis.

The axioms of *rationality* are: *completeness* (the ability to order every available combination of goods according to preference); *transitivity* (relationship between different combination preferences); and *selection* (the consumer will aim for the most desired combination). The axioms of *behaviour* comprise the axioms of *dominance* (also known as the axioms of greed), which are: *continuity* (relating to indifference curve analysis) and *convexity* (the assumption that the indifference curve will be convex to the origin). *See* CONSUMER DEMAND THEORY.

PH, AV

B

Babinet's principle (19th century) *Physics* Named after its discoverer, the French physicist Jacques Babinet (1794–1872), this principle relates to the effects of complementary diffracting screens.

Two screens S_1 and S_2 are said to be complementary if S_2 is the screen obtained by making the opaque parts of screen S_1 transparent and the transparent parts opaque. Babinet's principle states that, except in regions that are illuminated when no diffracting screens are present, complementary diffracting screens produce identical intensity distributions.

J Thewlis, ed., *Encyclopaedic Dictionary of Physics* (New York, Oxford and London, 1962)

MS

background knowledge *Linguistics* *See* MUTUAL KNOWLEDGE.

backlash theory (20th century) *Politics/Feminism* Theory of male counterattack against FEMINISM.

Women have made slow and steady, if relatively insubstantial, advances. But in response to this a semi-consciously articulated campaign, or 'backlash' has developed in the last quarter of the 20th century to discredit women's aspirations and to remove their gains, and to do so by any means from propaganda to violence.

Susan Faludi, *Backlash* (London, 1992)

HR

Baeyer's strain theory (1885) *Chemistry* Named after German chemist Johann Friedrich Wilhelm Adolf von Baeyer (1835–1917) who pioneered synthesis of organic compounds. Baeyer was Professor of Chemistry at Strasburg and Munich and won the Nobel prize for Chemistry in 1905. Also known simply as strain theory, this was formulated to explain the relative stability of certain types of organic compounds.

When carbon is bonded to four other atoms, the angle between any pair of bonds is the tetrahedral angle 109.5°. Deviations from this angle in cyclic (ring) compounds such as cyclopropane cause molecules to be strained and therefore relatively unstable. The greater the deviation from this angle the more unstable a molecule is and thus the more prone it is to ring opening reactions. Parts of Baeyer's theory have been discarded as they are based on false assumptions. The theory does not apply to rings with more than four carbon atoms.

R T Morrison and R N Boyd, *Organic Chemistry* (Boston, 1987)

MF

Baire's category theorem (1899) *Mathematics* Named after the French mathematician René Baire (1874–1932), this is the result in TOPOLOGY (and of fundamental importance in classical FUNCTIONAL ANALYSIS) whereby the intersection of any countable collection of dense open subsets of a complete metric space is itself dense in the space. The term 'category' was used by Baire to classify certain topological spaces and bears no relationship to the subject in abstract ALGEBRA known as CATEGORY THEORY.

Béla Bollobás, *Linear Algebra* (Cambridge, 1990)

MB

Baker-Nathan effect *Chemistry* This effect was originally observed for reactions such as that of *para*-substituted benzyl bromides with pyridine. The observed reaction rates are opposite to those predicted for alkyl groups which release electrons by the inductive effect. The Baker-Nathan effect is explained by considering delocalization of the σ-electrons. When a hydrogen atom (H) in an organic molecule is β to an unsaturated carbon atom (H—C—C=), the hydrogen–carbon σ-bond (H—C) becomes less localized by partial σ,π-conjugation.

MF

Baker's theorem (1966) *Mathematics* Named after the British mathematician Alan Baker (1939–), this is the result in NUMBER THEORY whereby if $\alpha_1,\ldots,\alpha n$ are non-zero algebraic numbers such that $\log\alpha_1,\ldots,\log\alpha n$ are linearly independent over the rationals, then $1, \log\alpha_1,\ldots,\log\alpha n$ are linearly independent over the field of all algebraic numbers. This result, which generalizes the GELFOND-SCHNEIDER THEOREM, furnishes the transcendence of $e^{\beta_0}\alpha_1^{\beta_1}\ldots\alpha_n^{\beta_n}$ and indeed of any non-vanishing linear form

$$\beta_1 \log \alpha_1 + \cdots + \beta_n \log \alpha_n$$

in which the αs and βs are non-zero algebraic numbers. Quantitative versions of this theorem have also played a crucial role in the effective solution of a wide variety of DIOPHANTINE EQUATIONS. For his significant contributions, Baker was awarded a Fields Medal.

A Baker, *Transcendental Number Theory* (Cambridge, 1979)

MB

Bakuninism (19th century) *Politics* Theory of REVOLUTION attributed to and derived from Russian revolutionary exponent of ANARCHISM Michael Bakunin (1814–76).

Human nature is essentially sociable and co-operative. Religion, government and capitalism all distort human nature, and in order to allow it to develop freely and properly, the institutions that distort it must be removed by revolutionary action. Hence 'the urge to destroy is also a creative urge'.

A Kelly, *Michael Bakunin* (Oxford, 1982)

RB

balance hypothesis (1953) *Biology* Proposed by Lewis, this is the theory that parasitism is based on a balance between the host's ability to provide substances that favour the growth of the parasite and those which inhibit it. Substances of both kinds have been observed in many different species.

KH

balance of power *Politics* Theory of international relations.

Equality of power between nations is conducive to peace, since it constrains any one nation from engaging in war with another. Thus weak nations may be promoting general stability by arming, but strong nations may destabilize balance by continuing to arm. The theory has in practice justified deeply complex calculations of comparative military advantage, particularly in relation to nuclear weapons. *See also* COLLECTIVE SECURITY.

David Robertson, *The Penguin Dictionary of Politics* (London, 1986)

RB

balance of terror (20th century) *Politics* An aggressive version of BALANCE OF POWER.

When nations possess massive destructive weaponry then they are deterred from attacking each other not by the likelihood of effective defence, but by the probability of their own destruction.

Roger Scruton, *A Dictionary of Political Thought* (London, 1983)

RB

balance theory *Psychology* Proposed by the Austrian-American F Heider (1896–).

Those theories which explain many aspects of human behaviour as an attempt to re-establish some form of psychological balance. The balance theory put forward by Heider is concerned with attitudes and attitude change. It holds that people tend to retain attitudes and opinions which they deem to be compatible and reject those which are less agreeable. A person will attempt to resolve attitudes and opinions which are out of balance with each other. *See* COGNITIVE CONSISTENCY THEORY.

F Heider, *The Psychology of Interpersonal Relations* (New York, 1958)

NS

balanced budget multiplier (1940s) *Economics* Attributed to J Gelting in 1941 and Norwegian economist Trygve Haavelmo (1911–) in 1945, this is the effect on NATIONAL INCOME of equal changes in government expenditure and revenues.

The multiplier effect on income of an increase in government expenditure exactly matched by an increase in taxation will lead to a situation where the balance of the government's budget will remain unchanged. The balanced budget multiplier is important in understanding government management of the economy. *See* MULTIPLIER, MULTIPLIER-ACCELERATOR, EQUILIBRIUM, GENERAL EQUILIBRIUM THEORY, and PARTIAL EQUILIBRIUM THEORY.

T Haavelmo, 'Multiplier Effects of a Balanced Budget', *Econometria*, vol. XIII (October 1945), 311-18

PH, AV

Baldwin effect (19th–20th century) *Biology* Named after the American psychologist James Mark Baldwin (1861–1934), this is a theory explaining how a particular trait (phenotype) becomes genetically fixed in a population, and is also known as genetic assimilation.

When an advantageous trait appears in a few individuals in response to environmental cues, an individual could, by chance, become endowed with the advantageous trait even in the absence of environmental input. In time the gene responsible would spread through the population because of the survival advantage conferred to those possessing the trait. This theory is not universally accepted by evolutionists. *Compare with* LAMARCKISM.

F Hitching, *The Neck of the Giraffe: Where Darwin Went Wrong* (New Haven, Conn., 1987)

KH

Balkanization (20th century) *Politics* A theory of the fragmentation of states, its name derived from the area of south-eastern Europe.

States are broken up into smaller and more numerous units, formally autonomous but so small and so taken up with conflict amongst themselves that they present no threat to their more powerful neighbours.

RB

Banach-Alaoglu theorem (1940) *Mathematics* Named after Polish mathematician Stefan Banach (1892–1945) and Leonidas Alaoglu (1914–), this is the result in FUNCTIONAL ANALYSIS whereby the unit ball of a normed linear space is compact in the weak-star topology.

Béla Bollobás, *Linear Analysis* (Cambridge, 1990)

MB

Banach-Steinhaus theorem (1927) *Mathematics* Named after Polish mathematician Stefan Banach (1892–1945) and Hugo Steinhaus (1887–1972), this theorem is also known as the principle of uniform boundedness. It is the result in FUNCTIONAL ANALYSIS whereby a pointwise-bounded family of continuous linear operators from a Banach space to a normed linear space is uniformly bounded.

Béla Bollobás, *Linear Analysis* (Cambridge, 1990)

MB

Banach-Tarski theorem *Mathematics* A famous counter-intuitive result often used to emphasize the fragility of our ideas of volume. In particular it demonstrates that a solid sphere in three-dimensional Euclidean space can be decomposed into a finite number of pieces which may then be rigidly translated to construct two spheres of the same size as the original.

AL

band theory of ferromagnetism *Physics* A theory which explains ferromagnetism in terms of the behaviour of electrons in the unfilled bands of solids.

The number of electrons, the magnitude of the exchange interaction, the form of the band, and the temperature all determine the degree of magnetization, which is governed by the FERMI-DIRAC STATISTICAL AND DISTRIBUTIVE LAW. *See also* BAND THEORY OF SOLIDS and MAGNETISM, THEORIES OF.

Robert M Besançon, ed., *The Encyclopedia of Physics* (New York, 1985)

GD

band theory of solids *Physics* The band theory is a description of the behaviour, quantum states and energy values of electrons under the influence of the positively charged nuclei of crystalline solids as opposed to their behaviour in isolated atoms and molecules.

The discrete atomic energy levels combine to form the valence and conduction bands, and these are separated by bands of forbidden energy. Conducting materials have overlapping valence and conduction bands with available free states where the electrons are free to move. Insulators have a wide forbidden band separating the filled valence band and empty conduction band. Semiconductors have a narrow forbidden band across which electrons can move to the conduction band through thermal fluctuations.

Robert M Besançon, ed., *The Encyclopedia of Physics* (New York, 1985)

GD

bang-bang principle *Mathematics* The idea in CONTROL THEORY that for a linear dynamical system $\dot{x} = ax + bu$, if a state is reachable in a fixed time by choices of the control variable $u(t)$ lying between 0 and 1, then it is reachable using only $u(t) = 0$ or 1 at any time t. If u represents the activity of an engine, we thus never need to run the engine at intermediate powers. The result extends to higher dimensions, and is a consequence of the LIAPUNOV CONVEXITY THEOREM.

AL

Barba's law (1880) *Physics* This law is named after its discoverer and states that, in a tensile test in which samples are tested to fracture, geometrically similar test pieces of a given material deform in a similar manner.

If, for example, the test pieces are right circular cylinders, the same value for the elongation at fracture is obtained for samples of a given material provided that the ratio of the length to the diameter has the same value.

J Thewlis, ed., *Encyclopaedic Dictionary of Physics* (New York, Oxford and London, 1962)

MS

bargaining theory (20th century) *Politics* A theory of politics as negotiation.

Politics – particularly the internal politics of government, legislatures, and parties – is usefully seen as a series of negotiations or bargains between groups with differing but not fundamentally incompatible INTERESTS. Such a view discounts the likelihood of significant ZERO-SUM situations, and is opposed to CONFLICT THEORY.

Geoffrey Roberts and Alistair Edwards, *A New Dictionary of Political Analysis* (London, 1991)

RB

Barkhausen effect (1919) *Physics* Named after the German physicist Heinrich Georg Barkhausen (1881–1956) who first studied the effect. When the magnetization of a ferromagnetic material is changed by altering the strength of the applied magnetic field, the change in the magnetization takes place in a series of discontinuous steps, even when the rate of change of the applied magnetic field strength is extremely slow.

The curve of magnetization against magnetizing field is made up of a series of small steps. These discontinuities result from irreversible changes in the domain structure of the material.

J Thewlis, ed., *Encyclopaedic Dictionary of Physics* (New York, Oxford and London, 1962)

MS

Barlow's rule (1906) *Chemistry* Rule for valency introduced by English chemists William Barlow (1845–1934) and William Jackson Pope (1870–1939). The volume occupied by the atoms in a given molecule is proportional to the valencies of the atoms. The lowest values for the valencies are used.

Sir William A Tilden, *Chemical Discovery and Invention in the Twentieth Century* (London, 1919)

MF

Barnum effect *Psychology* Named after the American showman P T Barnum (1810–91), to whom is accredited the aphorism 'There's a sucker born every minute.'

The term refers to the widespread predisposition to believe that general and vague personality descriptions or predictions have specific relevance to certain individuals. This effect has frequently contaminated research on personality assessment, and is often the principle behind predictions by astrologers and so on.

H Gleitman, *Psychology* (London, 1991)

NS

baroque (late 19th century) *Literary Theory* Traditionally derogatory term rehabilitated in 1888 by the German art historian Heinrich Wölfflin (1864–1945).

In the late Renaissance (roughly 1580–1680), an elaborate, highly decorative and expansive style of poetry and prose deploying much figurative language, including CONCEITS, sounds-patterning, and grand narrative designs. The Italian poet Torquato Tasso's *Gerusalemme Liberata* (1584) is regarded as the prototypical Baroque work; in English, the METAPHYSICAL POETS of the early 17th century, and much ornate prose of the period.

R Wellek, 'The Concept of Baroque in Literary Scholarship' [1946], and 'Postscript 1962', *Concepts of Criticism*, S G Nichols (jnr), ed. (New Haven, 1963), 69–127

RF

baryon conservation, law of *Physics* A law relating to interactions between subatomic particles.

Baryon is a collective term for the classes of particles known as nucleons and hyperons. For any imaginable process, the number of baryons minus the number of antibaryons in a system is conserved, allowing only for the creation or annihilation of baryon-antibaryon pairs. However, theories such as GRAND UNIFIED THEORIES (GUTs) point to occasions when it may not always be conserved; for example, at the very high energies achieved in the early universe.

Paul Davies, ed., *The New Physics* (Cambridge, 1989)

GD

base-pairing rules *Biology See* CHARGAFF'S RULES.

basement effect *Psychology See* CEILING EFFECT.

basic colour terms (1969) *Linguistics* Pioneering study by the American psychologists B Berlin and P Kay.

Contrary to earlier assumptions that a language could cut up the colour spectrum in unpredictably various ways, Berlin and Kay demonstrated the existence of eleven hierarchically ordered basic colour terms which named 'focal colours': colours which are made perceptually salient by the human visual apparatus. Not all languages possess all terms; a language which possesses six (say) will have the first six in the hierarchy, and invented terms for the other five will be readily learned. *See also* BASIC LEVEL TERMS, NATURAL CATEGORY, and PROTOTYPE.

B Berlin and P Kay, *Basic Color Terms* (Berkeley, 1969)

RF

basic level terms (1970s) *Linguistics* Concept developed by the American cognitive psychologists B Berlin and Eleanor Rosch.

A central part of the mechanism by which language categorizes experience. Words such as 'car', 'chair', 'rose' refer to an object as that cluster of attributes which are most perceptually salient, and most useful, for speakers. Superordinate, more general terms ('plant', 'flower') invoke fewer shared attributes; more specific terms ('floribunda', 'hybrid tea') give more information than is generally required for communication. *See also* BASIC COLOUR TERMS, NATURAL CATEGORY and PROTOTYPE.

E Rosch and B B Lloyd, eds, *Cognition and Categorization* (Hillsdale, NJ, 1978)

RF

basis theorem *Mathematics* The result in LINEAR ALGEBRA whereby every vector space has a basis; that is, there is a set of linearly independent vectors such that every vector in the space can be expressed as a linear combination of vectors in this set.

Hence, a basis can be thought of as a collection of atoms from which a vector space is built. While a space can have many different bases, the cardinality of any two is the same. *See also* ZORN'S LEMMA and HILBERT'S THEOREM.

T W Hungerford, *Algebra* (Springer, 1974)

MB

Bateman's principle (1948) *Biology* The theory that in most polygamous species, some males will produce disproportionately large numbers of offspring, and that consequently the variance in reproductive success will be greater among males than among females. That is, most females will reproduce, and the number of offspring they produce will vary little from female to female, while some males will produce many more offspring than other males. Bateman supported his hypothesis with studies of fruit flies (*Drosophila melanogaster*). *See also* SEXUAL SELECTION.

KH

Batesian mimicry (1862) *Biology* Named after the British naturalist and explorer Henry Walter Bates (1825–92), this term refers to the resemblance of a harmless organism to some other poisonous, dangerous or unpalatable species which may be conspicuous in appearance.

This mimicry confers on the harmless species the same protection from predators enjoyed by the species which has actually evolved the defence mechanism. It is also called pseudaposematic colouration. *Compare with* MÜLLERIAN MIMICRY.

J Sternburg, 'Batesian mimicry: Selective Advantage of Color Pattern', *Science* vol. CXCV (1977), 681–83

KH

bathos (1728) *Literary Theory* Established by the English poet and critic Alexander Pope (1688–1744), parodying Longinus's *On the Sublime* (1st century AD).

A mishandled attempt at an elevated style, flopping from a rhetorical height to a ludicrous anticlimax.

A Pope, *Peri Bathous: Or, Martinus Scriblerus, His Treatise Of the Art of Sinking in Poetry* [1728], *The Prose Works of Alexander Pope*, R Cowler, ed., vol II (Oxford, 1986), 171–276

RF

Bauhaus (1919) *Art* A school of architecture and industrial arts, formed in Weimar by the German architect Walter Gropius (1883–1969).

Its first manifesto, which owed something to the British ARTS AND CRAFTS MOVEMENT, emphasized the artist as craftsman; and insisted on the unity of the arts in a building, the elimination of the division between monumental and decorative elements of a building, and the importance of design in industrial mass-production. Following the closure of the school under the Nazi regime in 1933, its style was disseminated internationally by its former associates.

G Naylor, *Bauhaus* (1968)

AB

bay region theory (1972) *Biology* Named by K D Bartle and D W Jones after the appearance of the chemical structure of polycyclic aromatic hydrocarbons, which possess a 'hollow' reminiscent of an aquatic bay; this theory postulates that chemical carcinogenesis involves a specific structural element (a diol epoxide) of certain polynuclear aromatic hydrocarbons.

H Hiatt, *Origins of Human Cancer: Book B* (Cold Spring Harbor Laboratory, 1977)

KH

Bayes's theorem *Mathematics* Named after British probability theorist and theologian Thomas Bayes (1702–61), this is the fundamental result of statistics whereby the *conditional probability* of an event E given an event A as

$$P(E|A) = P(A|E)\frac{P(E)}{P(A)};$$

more generally, when E_n is an element of a set E_i that constitutes a partition of the sample space,

$$P(E_n|A) = \frac{P(A|E_n)P(E_n)}{\sum_i [P(A|E_i)P(E_i)]}.$$

Thus prior estimates of probability can be revised in the light of observations.

JB

Bayes's theory of statistical sampling (18th century) *Mathematics* Named after British probability theorist and theologian Thomas Bayes (1702–61), this is a method of statistical estimation based on the assumption that to any unknown parameter in a statistical problem, there can be assigned a probability distribution. The optimal rule is called a *Bayesian decision function* which assigns a decision to each result of an observation. Bayesians differ from adherents of FREQUENCY THEORY in believing that individual events have probabilities.

R V Hogg and A T Craig, *Introduction to Mathematical Statistics* (New York, 1967)

ML

Bayesianism *Philosophy* Belief that the use of induction in science can be rationally justified by appeal to Bayes's Theorem (Thomas Bayes, 1702–61)).

This says that the probability of one proposition, given another, equals the probability of the second, given the first, multiplied by the prior probability of the first (that is the probability it already has, irrespective of the second) and divided by that of the second. The first proposition will be the one we are interested in, while the second will be some piece of evidence; and according to the theorem this will support the first proposition in proportion as its own prior probability is low, which suits our intuitions. However, though the theorem itself is undisputed, its usefulness depends on our assigning suitable prior probabilities, with numerical values, to the two propositions. This raises problems.

'Bayesianism' is also sometimes used of any conception of rationality based on maximizing expected utilities, which links it to SUBJECTIVIST THEORIES OF PROBABILITY.

ARL

BCS theory (1957) *Physics* Named after the American physicist and double Nobel prizewinner (1956 and 1972) John Bardeen (1908–91); and Leon Neil Cooper (1930–) and John Robert Schrieffer (1931–), both of whom also shared the 1972 Nobel prize.

The theory relates to the phenomenon of superconductivity shown by many metals, alloys and intermetallic compounds whose electrical resistance vanishes when cooled below a certain transition temperature. In the superconducting state, electrons are not free to move independently, and form dynamic pairs which interact through lattice vibrations. These Cooper pairs, as they are known, obey the BOSE-EINSTEIN STATISTICAL AND DISTRIBUTIVE LAW and are the basis of the BCS theory. The theory is less successful, however, in explaining the properties of the recently-discovered high-temperature (100 K) superconductors which rely on heavy-fermion systems.

Paul Davies, ed., *The New Physics* (Cambridge, 1989)

GD

beanbag genetics *Biology See* PARTICULATE GENETICS.

Beaufort wind scale (1805) *Meteorology* The British Admiral Sir Francis Beaufort (1774–1857) defined 13 scales (0–12) for estimating the force of wind on a sailing ship, and this was adopted by the Royal Navy in 1838.

Initially based on practical observations (particularly of the amount and type of sail used in calm to gale force winds and the ability of ships to withstand more violent storms and hurricanes), the scale was later modified to refer more specifically to sea states (for example, the formation of white horses, the amount of spray), and, on land, the effect on trees, dust, and so on. This numerical scale is difficult to quantify in terms of wind speeds, but is still widely used as a practical guide to likely effects.

H U Roll, *Physics of the Marine Atmosphere* (New York, 1969)

DT

bee language hypothesis (1945) *Biology* The hypothesis that foraging honey bees use a sophisticated 'waggle dance' to communicate accurately the distance and direction of food sources to other bees in the hive. Although the phenomenon had been recognized for centuries, it was first described in detail by Austrian zoologist Karl von Frisch (1886–1982), co-winner of the 1973 Nobel prize for Physiology or Medicine.

Although disputed for many years by researchers who believed that the bees were actually responding to olfactory cues, von Frisch's work was vindicated by a series of experiments designed by J L Gould. *See also* ANIMAL LANGUAGE HYPOTHESIS.

J Gould, *Ethology: the Mechanisms and Evolution of Behavior* (New York, 1982)

KH

Beer-Lambert law *Chemistry* A combination of BEER'S LAW and LAMBERT'S LAW OF ABSORPTION, this states that when light is transmitted through a solution or gas, its intensity after transmission (I_f) is related to the intensity before transmission (I_i) by

$$\log_{10}(I_f/I_i) = -eCl$$

where e is the extinction coefficient of the sample, c is its concentration and l is the length of the cell containing the solution or gas through which the light passes. The product *eCl* is the optical density or absorbance of the sample. *See* LAMBERT'S LAW OF ABSORPTION which refers specifically to cell length (that is, absorbing medium thickness) of pure liquids and not concentration.

P W Atkins, *Physical Chemistry* (Oxford, 1978)

MF

Beer's law *Chemistry* Absorption is proportional to the number of absorbing molecules, that is the concentration of the absorbing medium. *See also* BEER-LAMBERT LAW.

MF

behaviour constraint theory *Psychology* Proposed by the American psychologist Martin E P Seligman (1942–).

The normal pattern of coping with environmental threats is one whereby the person perceives a loss of control, reacts by attempting to regain control and, if these attempts fail, perhaps experiences a sense of helplessness. The theory has undergone several revisions and led to the formulation of a range of therapeutic programmes designed to allow clients to regain a sense of control and influence over their own lives. *See* FRUSTRATION-FIXATION HYPOTHESIS.

M E P Seligman, *Helplessness: On Depression, Development and Death* (San Francisco, 1975)

NS

behaviour therapy *Psychology* Developed by the South African-born American psychologist Joseph Wolpe (1915–).

Also known as behavioural psychotherapy, this approaches the problems of the client by concentrating on changing certain behaviour patterns which are leading to those problems. The therapist attempts to modify the ineffective or maladaptive patterns of behaviour by applying learning techniques. The approach contrasts with others which look to the unconscious and probe thoughts and feelings in an attempt to change certain aspects of personality. *See* BEHAVIOURISM.

A E Kazdin, *Behavior Modification in Applied Settings* (Homewood, Ill., 1980)

NS

behaviouralism (20th century) *Politics* Theory of politics, particularly prevalent in North America in the third quarter of the 20th century.

Social and particularly political life are best studied in a manner analogous to that used in the natural sciences: observation, measurement, calculation. When this is done, regularities or laws of human political behaviour will be discovered, which will hold good for all times and all places. The methods of investigation implied or required are thus mass surveys of voting, participation, and other forms of observable political action.

Bernard Crick, *The American Science of Politics* (Berkeley, 1959)

RB

behaviourism (1913) *Biology/Philosophy/Psychology* Also referred to as the stimulus-response model, this term was coined by the American psychologist John Broadus Watson (1878–1958) in his paper, 'Psychology as the Behaviorist Sees It'. It is a theory of animal and human behaviour holding that actions can be explained entirely as responses to stimuli; and asserting that observable and measurable behaviour is the only suitable material for scientific psychological investigation.

Behaviourism rejects genetically based explanations and (subjective, therefore unreliable) mentalistic concepts such as introspection. One objection to it is that it is not always clear what counts as behaviour, and how far one can give a pure description of behaviour without bringing in interpretation. However, while much of the theory has fallen out of popularity, it continues to have a profound influence in a variety of effectiveness programmes in clinical and educational psychology. *See* BEHAVIOUR THERAPY, EFFECT, LAW OF, EPIPHENOMENALISM and LEARNING THEORY. *Compare with* ETHOLOGY.

J B Watson, *Behaviorism* (1925); D Dewsbury, *Comparative Animal Behaviour* (New York, 1978); B F Skinner, *About Behaviourism* (New York, 1974)

KH, ARL, NS

behaviourism in linguistics (*c.*1920–late 1950s) *Linguistics* American STRUCTURAL LINGUISTICS was dominated by the psychological model of John Broadus Watson (*see*

BEHAVIOURISM). Language was regarded by such pioneers as Leonard Bloomfield as a set of habits or behaviour patterns operating under stimulus-response mechanisms; no internal faculty of language was tolerated.

In the late 1950s Avram Noam Chomsky (1928–) rejected this position, attacking its proponent Burrhus Frederic Skinner (1940–). *See* MENTALISM IN LINGUISTICS.

L Bloomfield, *Language* (New York, 1933); B F Skinner, *Verbal Behavior* (New York, 1957)

RF

Bell-Magendie law (1811) *Psychology* The body of anatomical evidence first disclosed by Scottish anatomist and surgeon Sir Charles Bell (1774–1842), and (11 years later) by French physiologist François Magendie (1783–1855).

The Bell-Magendie law states that the ventral roots of the spinal nerves are motor in function, and the dorsal roots are sensory.

R Thomson, *The Pelican History of Psychology* (Harmondsworth, 1968)

NS

Bell's theorem (1964) *Mathematics* Reality as described by the theory of quantum mechanics appears to involve random elements. It is natural to try to explain this inherently probabilistic behaviour by invoking hidden variables, which are deterministic but cannot be observed. Bell's result shows that no such explanation is possible in standard interpretations of quantum mechanics.

AL

Bellman principle of optimality (1957) *Mathematics* The key idea of the area of OPTIMIZATION THEORY known as *dynamic programming*, where problems evolving over several (n) time periods are considered.

The Bellman principle asserts that an optimal solution of the n-period problem can be constructed recursively from a solution to the $(n-1)$-period problem: one optimizes first over the first period and then over the remaining $n-1$ periods using the previous solution.

AL

belongingness, law of *Psychology* Proposed by the American psychologist Edward Lee Thorndike (1874–1949).

Stimuli are more likely to become associated if they are related to each other in some way so that one may elicit another. This law may also be described as the generalization that an array of stimuli will be more likely to be perceived or reacted to as a whole if the elements belong to each other in some way.

G H Bower and E R Hilgard, *Theories of Learning* (Englewood Cliffs, N.J., 1981)

NS

benefit approach principle (17th century–) *Economics* A traditional principle of taxation expounded by English philosophers Thomas Hobbes (1588–1679) and John Locke (1632–1704), and by Dutch jurist Hugo Grotius (1583–1645).

Taxation is levied broadly in relation to the benefits that people receive in public services. Since taxes are paid individually and public services are provided collectively, a taxation system based on this principle is open to criticism. (For example, it would be difficult to allocate the costs for national health, defence or foreign policy to individuals in proportion to their consumption.) The benefit approach principle received modern refinement by Swedish economist Erik Lindahl (1891–1960). *See* ABILITY-TO-PAY PRINCIPLE and EQUAL SACRIFICE THEORY.

R A Musgrave and A T Peacock, eds, *Classics in the Theory of Public Finance* (London, 1958)

PH

Bergen frontal theory (1920s) *Meteorology* Named after Bergen in Norway, where a group of meteorologists recognized that huge air masses of different temperature and humidity tend to move as coherent units over considerable distances, separated by relatively narrow zones within which they interact. The relatively narrow zones between these air masses are termed fronts.

These concepts are now fundamental to weather forecasting, and the factors controlling the movement and interaction of such masses are crucial to understanding climatic changes over timescales of a few years to decades.

DT

Bergeron-Findeism theory *Meteorology* *See* BERGERON'S THEORY.

Bergeron's theory (1930s) *Meteorology* Originally proposed by the German climatologist Alfred Wegener (1880–1930)), this hypothesis was mainly developed by the Swedish meteorologist Tor Bergeron and subsequently extended by the German meteorologist Walter Findeism. The theory provided an explanation for the formation of raindrops in an ice/water cloud.

At temperatures of −12 to −30°C, ice crystals can grow by sublimation while supercooled water droplets still evaporate. The ice grains can then fall into warmer air where they melt to form water drops. This theory does not explain all water droplet formation as, in tropical latitudes, the temperatures within a rising cloud may not reach 0°C and rain formation is then thought to be explicable by the COALESCENCE THEORY.

C D Ahrens, *Meteorology Today: An Introduction to Weather, Climate and the Environment* (West Publ. Co., 1991)

DT

Bergmann's rule (1847) *Biology* Named after biologist C Bergmann, this states that within a given species or genus the average size of individuals tends to be larger in cooler climates than in warmer ones.

This rule is usually explained by noting that the volume of any solid increases faster than its surface area as the solid grows larger. Thus, larger bodies have a smaller surface-to-volume ratio than smaller bodies. In cold climates, a relatively smaller surface area is advantageous because metabolism rates are more nearly proportional to body surface area than to body volume. The rule holds true in 75–90 per cent of bird species and 60–80 per cent of mammal species.

V Grant, *The Evolutionary Process: A Critical Study of Evolutionary Theory*, 2nd edn (New York, 1991)

KH

***bergschrund* hypothesis** *Geology* A theory to explain the formation of crevasses.

A crevasse usually occurs between the top of a glacier and its surrounding rocks. The theory proposes this to be due to the gravitational slide downhill of the glacier, and erosion of the rocks which occurs when the glacial ice plucks off rock fragments as it separates from the wall. Water falling into the crevasse then refreezes against and within cracks in the rock wall.

DT

Bernal model of a monatomic liquid (1950s) *Physics* This is a simulation of the atomic arrangement in a monatomic liquid, first made by John Desmond Bernal (1901–71).

Using the result from X-ray work that simple liquids show short-range order but not long-range order, Bernal assumed that the instantaneous atomic arrangement in a simple liquid would be that of an assembly of spheres showing random packing. To study the properties of his model he allowed Plasticene spheres to accumulate in a container with no attempt to produce an orderly arrangement (random packing). By squeezing the spheres together and examining the shapes produced Bernal was able to deduce a value for the first co-ordination number (the number of nearest neighbours) in a random stacking arrangement, and obtained the value of 9.3 ± 0.8.

J Thewlis, ed., *Encyclopaedic Dictionary of Physics* (New York, Oxford and London, 1962)

MS

Bernoulli-Euler law (1744) *Physics* This law is named after the Swiss mathematicians Daniel Bernoulli (1700–82) (who suggested it in 1742) and Leonhard Euler (1707–83) (who derived it, two years later).

The law states that, for an elastic beam of thickness a bent to a radius of curvature R, such that $R \gg a$, the bending moment M is given by

$$M = EI/R$$

where E is Young's modulus for the material of the beam and I is the second moment of area of the cross-section of the beam about an axis which is normal to the plane of bending and passes through the neutral plane. For a beam of thickness a and width b, the expression for I is

$$I = ba^3/12.$$

J Thewlis, ed., *Encyclopaedic Dictionary of Physics* (New York, Oxford and London, 1962)

MS

Bernoulli trials (17th century) *Mathematics* Named after Swiss analyst, probability theorist and physicist Jakob Bernoulli (1654–1705), this is a scheme of PROBABILITY THEORY based on independent trials, each of which can have one of two results, success or failure. The probability of an event is determined by the *binomial distribution*. *See* BINOMIAL THEOREM.

S Ross, *Probability Theory* (New York, 1976)

JB

Bernoulli's hypothesis (18th century) *Economics* Proposed by Swiss mathematician Daniel Bernoulli (1700–82), this theory suggests added dimensions to the evaluation of risk.

Acceptance of a risk depends not only on the nominal value of what may be lost but also on the intrinsic value, or utility, of it to the person accepting the risk. *See* BOUNDED RATIONALITY, ST PETERSBURG PARADOX and UNCERTAINTY.

K Pearson, *The History of Statistics in the 17th and 18th Centuries* (New York, 1978)

PH

Bernoulli's theorem *Mathematics See* LARGE NUMBERS, LAW OF.

Bernoulli's theorem (18th century) *Physics* This theorem is named after its discoverer, the Swiss mathematician Daniel Bernoulli (1700–82); and relates to the steady flow of incompressible fluids that have negligible viscosity.

In its simplest form it relates to the flow along a streamline when conditions are steady. When conditions are steady, a streamline coincides with a line of motion of a particle of the fluid. Under these conditions the theorem states that, at any point along a particular streamline, the sum of the pressure energy, the potential energy and the kinetic energy of the incompressible, non-viscous fluid is constant. If at any point on the streamline, ρ is the density of the fluid, v is its speed and p is its pressure

$$\frac{1}{2}\rho v^2 + p + \rho U = \text{constant}$$

where U is the potential energy of unit mass of fluid due to an external force such as gravity. The theorem may be applied to compressible fluids such as gases as a reasonable approximation provided that v is much less than the speed of sound in the gas.

J Thewlis, ed., *Encyclopaedic Dictionary of Physics* (New York, Oxford and London, 1962)

MS

Berthelot-Thomsen principle (19th century) *Chemistry* Named after French physical chemist and statesman Marcellin Pierre Eugène Berthelot (1827–1907), and Danish chemist Hans Peter Jürgen Julius Thomsen (1826–1909) who was noted for his work on thermochemistry. This principle states that of all possible chemical reactions that may take place, the one that leads to the greatest release of energy will occur. Exceptions include processes involving changes of state.

MF

Bertrand duopoly model (1883) *Economics* Developed by French mathematician Joseph Bertrand (1822–1903), the model is a variant of the standard duopoly (a market characterized by two suppliers).

A supplier in the Bertrand duopoly assumes his competitor will not change prices in response to his price cuts. If each follows this logic, an equilibrium will be established and neither firm will benefit from charging a different price, thereby making price equal to marginal cost. The model has been criticized because it ignores production costs and entry by new firms. *See* COURNOT DUOPOLY MODEL and DUOPOLY THEORY.

M Shubik, *Strategy and Market Structure* (New York, 1959)

PH

Bertrand's postulate (19th century) *Mathematics* Named after French mathematician Joseph Louis Bertrand (1822–1903), this is the result (proved by Chebyshev) whereby for every integer n greater than one, there is always a prime between n and $2n$. This result essentially says that the primes are well distributed, and can be regarded as the forerunner of extensive modern researches on the difference between consecutive primes.

A Baker, *A Concise Introduction to the Theory of Numbers* (Cambridge, 1984)

MB

Bethe–Heitler theory of energy loss (1934) *Physics* Named after its originators, the German-born American physicist Hans Albrecht Bethe (1906–) and Walter Heitler

(1904–81); this is a theory of the energy loss of high-energy electrons in passing through matter, using Dirac's relativistic wave equation.

When a charge is accelerated, it emits electromagnetic radiation (*bremsstrahlung*) so that a charged particle deflected by an atomic nucleus or electron of an absorbing medium may lose energy and make a transition to a lower energy state. Inelastic collisions of this type are an important means of energy loss for high-energy electrons. Bethe and Heitler calculated the *bremsstrahlung* emission of an electron in a screened Coulomb field.

J Thewlis, ed., *Encyclopaedic Dictionary of Physics* (New York, Oxford and London, 1962)

MS

Betti's reciprocal theorem (1872) *Physics* This theorem is named after its discoverer, the Italian physicist Enrico Betti (1823–92), and is a special case of a more general theorem given by John William Strutt, 3rd Baron Rayleigh (1842–1919).

Betti's theorem states that when an elastic body that obeys HOOKE'S LAW is subjected to two sets of body and surface forces, then the work that would be done by the forces of the first set acting over the displacements produced by the forces of the second set acting alone, is equal to the work that would be done by the forces of the second set acting over the displacements produced by the forces of the first set acting alone.

J Thewlis, ed., *Encyclopaedic Dictionary of Physics* (New York, Oxford and London, 1962)

MS

Bezold-Brücke effect *Psychology* Named after the German scientist Johannes Friedrich Wilhelm von Bezold (1837–1907) and the Austrian physiologist Ernst Wilhelm Ritter von Brücke (1819–92).

The visual phenomenon whereby changes in perceived hue are associated with changes in luminance. More precisely, yellowish reds and yellowish greens are perceived as yellower with increases in illumination; and bluish reds and bluish greens appear more blue. Purer reds, yellows, greens and blues do not show this effect.

L Kaufman, *Perception* (New York, 1979)

NS

Bezout's lemma (18th century) *Mathematics* Named after French mathematician Etienne Bezout (1730–83), this is the result whereby if f and g are polynomials with greatest common divisor d, then there exist polynomials a and b such that $d = af + bg$. This result, which generalizes a similar result for the integers, follows directly from the EUCLIDEAN ALGORITHM for polynomials.

I N Herstein, *Topics in Algebra* (Wiley, 1975)

MB

Bezout's theorem (1770) *Mathematics* Named after French mathematician Etienne Bezout (1730–83), this is one of the oldest theorems in ALGEBRAIC GEOMETRY.

The result essentially is that two plane algebraic curves with degrees m and n respectively and with no common component have exactly mn points of intersection, counting multiplicity and points at infinity.

S S Abhyankar, *Algebraic Geometry for Scientists and Engineers*, Mathematical Surveys and Monographs 35 (AMS, 1990)

MB

bibliographical criticism (1980s) *Literary Theory* Scholarly criticism of individual works which pays close attention to the textual format and circumstances of publication of a work, on the grounds that these factors affect reading and meaning.

J J McGann, *The Beauty of Inflections: Literary Investigations in Historical Method and Theory* (Oxford, 1985)

RF

bidialectalism (1960s) *Linguistics* A term coined by analogy to 'bilingualism', this refers to the possession of two or more mutually intelligible dialects, or the relationship between dialects in contact. The study of bidialectalism contributes to an understanding of linguistic change. *See also* ACCOMMODATION.

P Trudgill, *Dialects in Contact* (Oxford, 1986)

RF

Bieberbach conjecture *Mathematics* A celebrated open problem of complex analysis, finally proved by L de Branges in 1985. The result states that if the function f on the unit disc of the complex plane is *holomorphic* (that is, complex-differentiable) and *one-to-one* (which is to say $f(z_1) \neq f(z_2)$ when $z_1 \neq z_2$), and has the power series inside the disc $f(z) = z + a_2z^2 + a_3z^3 + \ldots$, then the coefficients satisfy the growth condition $|a_n| \leqslant n$ for all n.

AL

big bang theory (1960s) *Physics* A cosmological theory that attempts to explain the origin of matter and radiation in the universe in terms of a cataclysmic explosion supposed to have occurred 10–20 billion years ago.

Elementary particles and antiparticles were created within a fraction of a second after the big bang, followed by photons of radiation. In the following minutes, deuterium and helium nuclei formed, followed by neutral hydrogen atoms when the temperature had dropped sufficiently. At this point in the expansion process, matter became decoupled from radiation, and interacted to form stars and galaxies resulting in a further cooling to the present observed temperature of the microwave background. The theory has been successful in explaining the expansion of the universe, the measured cosmic abundance of helium and the microwave background. *See also* FRIEDMANN UNIVERSES, NUCLEOSYNTHESIS and STEADY-STATE THEORY OF THE UNIVERSE.

Paul Davies, ed., *The New Physics* (Cambridge, 1989)

GD

bilateral monopoly *Economics* A market characterized by a single seller and a single buyer, otherwise known as a monopoly and a monopsony.

Examples of a bilateral monopoly frequently occur in the public sector where a government education employer negotiates with a single teachers' union on pay and conditions. *See* DUOPOLY THEORY, MONOPOLISTIC COMPETITION and MONOPOLY THEORY.

PH

Bildungsroman (from *c.*1750) *Literary Theory* This term emerges in German criticism in the second half of the 18th century, and is synonymous with the 'novel of development' or 'novel of education' or 'apprenticeship novel'. It refers to a variety of the novel in which the plot traces the development of a young person's ideas and sentiments towards maturity in relation to a series of obstacle and opportunities.

Goethe's *Die Lieden des jungen Werthers* (1774), *The Sorrows of Young Werther*, is taken to be the archetypal example. In English, Fielding's *Tom Jones* (1749) and Dickens's *David Copperfield* (1849–50) are well-known examples.

M Beddow, *The Fiction of Humanity: Studies in the Bildungsroman from Wieland to Thomas Mann* (Cambridge, 1982)

RF

billiard ball model (20th century) *Politics* Theory of international relations.

It is not necessary to study the internal politics of nations or governments, since their interactions on the international scene can be understood in terms of the pressures they exercise upon each other, and the responses to those pressures. An application to international politics of BEHAVIOURALISM.

Graham Evans and Jeffrey Newnham, *The Dictionary of World Politics* (Hemel Hempstead, 1990)

RB

bimetallism (18th century–) *Economics* A system operating in the USA and in the Latin Union (1865–73) in Europe during the 19th century, this used two metals (most commonly gold and silver) to serve as an international standard of value and means of payment.

The precursor of the Gold Standard, bimetallism was preferred by such as Scottish economist Adam Smith (1723–90) because of the relative stability of having two precious metals connected to each other in a fixed ratio.

A Smith, *An Inquiry into the Nature and Causes of the Wealth of Nations* (London, 1776); M Friedman and A J Schwartz, *A Monetary History of the United States* (Princeton, N.J., 1963); L Walras, *Théorie mathématique du bimetallisme* (1881)

PH

binary opposition (1940s) *Linguistics* Principle affirmed by the Russian linguist Roman Jakobson (1896–1982).

A DISTINCTIVE FEATURE in PHONOLOGY has only two values, '+' and '−'. This dichotomous principle is fundamental to STRUCTURALISM, going far beyond phonology.

R Jakobson and C G M Fant and M Halle, *Preliminaries to Speech Analysis. The Distinctive Features and their Correlates* (Cambridge, Mass., 1951)

RF

binding *Linguistics* *See* GOVERNMENT AND BINDING.

binomial nomenclature (1749) *Biology* A system of classification first proposed by the Swedish botanist Carolus Linnaeus (1707–78).

Linnaeus introduced the system of classifying an organism using a Latin generic noun, for the genus, and a specific adjective, for the species. Until that time organisms had been labelled with brief Latin descriptions (polynomial). This change greatly facilitated EVOLUTION THEORY, to which, ironically, Linnaeus was opposed. The system was first published in 1753 in the *Species Plantarum* which is still considered to be the starting point of modern botanical nomenclature.

RA

binomial theorem (17th century) *Mathematics* Derived independently by Sir Isaac Newton in 1665 and by James Gregory in 1670, this result gives the expansion of the nth power of a binomial as a polynomial with $n + 1$ terms. To be specific,

$$(x+a)^n = \sum_{k=0}^{n} \binom{n}{k} x^{n-k} a^k$$

where n is a positive integer and

$$\binom{n}{k} = \frac{n!}{(n-k)!k!}.$$

Thus,

$$(x+1)^4 = 1 + 4x + 6x^2 + 4x^3 + x^4.$$

More generally, if α is a real number and z is a complex number whose modulus is strictly less than one, then

$$(1+z)^\alpha = \sum_{k=0}^{\infty} \binom{\alpha}{k} z^k$$

where

$$\binom{\alpha}{k} = \frac{(\alpha-1)(\alpha-2)\cdots(\alpha-k+1)}{k!}.$$

See also MULTINOMIAL THEOREM, LARGE NUMBERS, LAW OF and PASCAL'S TRIANGLE.

H Anton, *Calculus with Analytic Geometry* (New York, 1980)

MB

bioeconomics (1970s) *Economics* Developed by American economist Gary Becker (1930–), this term refers to an area of economics in which sociobiology is applied to explain human behaviour in a capitalist economic system.

Becker asserted that the combined assumptions of maximizing behaviour, market equilibrium and stable preferences were at the heart of the economic approach. Seeing the economic approach as comprehensive, he decided that it could be applied to all types of human behaviour; for example, racial discrimination, fertility, education, crime, marriage, and political problems. Becker's argument assumes that human behaviour can be seen as involving participants who maximize their utility from a constant set of preferences, and accumulate an optimal amount of information and other factors in a variety of markets.

G S Becker, *The Economic Approach to Human Behaviour* (Chicago, Ill., 1977); C W Clark, *Mathematical Bioeconomics: the Optimal Management of Renewable Resources* (New York, 1976)

PH

bioenergetics (1910–15) *Biology* The principles and theories underlying how energy reaches the Earth and is used by its inhabitants; living systems as energy converters. More specifically, the principle that the Sun's energy is converted to energy in living things in ways consistent with the LAWS OF THERMODYNAMICS.

KH

biogenesis (1870) *Biology* A term coined by the distinguished British biologist Thomas Henry Huxley (1825–95), this refers to the production of living organisms from other living organisms.

Biogenesis refers especially to the fundamental concept of modern biology that life comes only from the reproduction of living things, and that species can produce offspring only of the same species. *Compare with* SPONTANEOUS GENERATION.

KH

biogenetic law *Biology See* RECAPITULATION.

biogenetics *Biology See* GENETIC ENGINEERING.

biogeny *Biology See* BIOGENESIS.

biological clock (1950–55) *Biology* The internal sense of timing responsible for triggering certain behaviours in animals and changes in plants, despite the absence of external environmental cues.

Substantial experimental evidence for the existence of biological clocks has been reported in a wide variety of organisms. Cell membranes probably play a role in the clock's mechanism, but its physiological basis in both animals and plants is poorly understood. *Compare with* BIORHYTHMS. *See also* CIRCADIAN RHYTHMS and CIRCANNUAL RHYTHMS.

J Brady, *Biological Clocks* (Baltimore, Md, 1979)

KH

biological death clock *Biology See* PROGRAMMED AGEING.

biological determinism *Politics/Sociology/Biology* Theory of human character.

Normally attributed to thinkers rather than claimed by them. Human character is determined by physical, biological characteristics, which are inherited. RACISM and SEXISM both frequently employ the assumptions of biological determinism to divide people into groups which are alleged to differ in ability and inclination.

Maggie Humm, *The Dictionary of Feminist Theory* (Hemel Hempstead, 1989)

RB

biological species concept *Biology See* SPECIES CONCEPT.

biological theory of evolution *Biology See* NEO-DARWINISM.

bioprogram hypothesis (1981) *Linguistics* Proposed by the British linguist Derek Bickerton.

The creolization of a pidgin (*see* CREOLE AND PIDGINS) by the first generation of native speakers involves the realization of the same innate semantic and syntactic features which characterize all first-language acquisition. *See also* INNATENESS HYPOTHESIS.

D Bickerton, *Roots of Language* (Ann Arbor, 1981)

RF

biorhythms (1897–1932) *Biology/Medicine* Proposed by physicians in Vienna, Berlin, Philadelphia and elsewhere, biorhythms are claimed to be three precise cycles (set into action at birth) which control human behaviour. These are: a physical cycle of 23 days; an emotional cycle of 28 days; and an intellectual cycle of 33 days.

The first half of each cycle is considered its optimum time, while hazardous 'critical days' occur when two or more cycles intersect. There is little, if any, scientific support for this theory. *Compare with* BIOLOGICAL CLOCK.

KH

Biot-Savart law (19th century) *Physics* Named after its discoverers, the French physicists Jean-Baptiste Biot (1774–1862) and Felix Savart (1791–1841), this law relates the magnetic flux density B of the magnetic field produced by a current I flowing through a straight conductor in free space to the distance R from the wire.

The magnetic flux density is given by

$$B = \mu_0 I/2\pi R$$

where μ_0 is the permittivity of free space and the lines of magnetic induction are circles concentric with the axis of the conductor and perpendicular to it.

This term is also applied to the more general expression for the magnetic flux density produced by any current element.

J Thewlis, ed., *Encyclopaedic Dictionary of Physics* (New York, Oxford and London, 1962)

MS

Biot's law (1818) *Physics* Named after its discoverer, the French physicist and astronomer Jean-Baptiste Biot (1774–1862). The rotation produced by an optically active medium is proportional to the length of the light path in the medium and, for solutions, to the concentration. The rotation is approximately proportional to the inverse square of the wavelength of the light.

J Thewlis, ed., *Encyclopaedic Dictionary of Physics* (New York, Oxford and London, 1962)

MS

Birkhoff's theorem (1946) *Mathematics* A result first discovered by D König in 1936 and sometimes also attributed to American mathematician John von Neumann (1953). Like KÖNIG'S THEOREMS, it can be translated via a network flow problem into a linear programming formulation (*see* DUALITY THEORY OF LINEAR PROGRAMMING).

A *doubly stochastic matrix* is one whose entries are all non-negative and whose rows and columns each sum to 1; while a *permutation matrix* has exactly one 1 in each row and column, the other entries being 0. The result states that any doubly stochastic matrix is a convex combination of permutation matrices (*see also* CARATHÉODORY THEOREM).

AL

Birkhoff's ergodic theorem (1931) *Mathematics* A famous convergence result for a system evolving step by step in time, also known as the *strong* or *pointwise* ergodic theorem (to distinguish it from the *mean ergodic theorem* of von Neumann).

Suppose T is a measure-preserving transformation (in the natural sense) on any measure space (so, for example, T might preserve the total 'length' of set in [0,1]). Beginning with any initial integrable function $f(x)$, the Cesaro means of the iterates $f(T^n x)$,

$$\frac{1}{n}\left(f(x) + f(Tx) + \ldots f(T^{n-1}x)\right)$$

(which corresponds in practice to the 'time mean' observed over a long period) converge almost everywhere to a limit function f^* which is *invariant*: $f^*(x) = f^*(Tx)$ almost everywhere. This result is very important in COMMUNICATION THEORY and INFORMATION THEORY.

AL

birth order theory *Psychology* Proposed by the Austrian psychiatrist Alfred Adler (1870–1937).

Part of Adler's theory of personality which emphasized the interaction between factors of heredity, environment and individual creativity. Since each child has a different genetic composition and is born into a somewhat different social setting, each interprets its environment differently. Thus, it is important to investigate similarities among people born in the same ordinal position and differences in the ways they interpret their experiences. There is an extensive and controversial literature on the topic. *See* ADLERIAN THEORY.

C S Hall and G Lindzey, *Introduction to Theories of Personality* (Chichester, 1985)

NS

Bishop-Phelps theorem (1961) *Mathematics* Two results in FUNCTIONAL ANALYSIS establishing the density of support points and support functionals for a closed convex subset C of a real Banach space X.

A point x in C is a support point if there is a closed supporting hyperplane for C at x (geometrically a tangent plane). In other words, there is a continuous linear functional which attains its maximum over C at x (*see* SUPPORT THEOREM). Support points are dense in the boundary, and such functionals (called support functionals) are dense (in norm) in the dual space X^* (*see also* the JAMES THEOREM). These results are the precursors of some important variational principles in OPTIMIZATION THEORY.

AL

bitonality (19th century) *Music* Most often found in music of the first half of the 20th century, the best-known exponents of this concept include composers Igor Stravinsky (1882–1971), Darius Milhaud (1892–1974) and Charles Ives (1874–1954). Bitonality is the theory that two distinct keys or tonalities may be traced as simultaneously present in a piece of music.

The effect may be achieved either by sounding together two separate musical strands, each in a single key different from the other; or, more subtly, by alternating between notes from both keys.

MLD

bivalence, law or principle of *Philosophy* Theory that every proposition is either true or false.

Possible objections are of two kinds. First, can we decide what counts as a proposition in the relevant sense? Second, might not the principle fail for some presumably genuine propositions; for example, 'Jones was brave' (where Jones died peacefully after a life entirely devoid of danger)? The law of bivalence is not necessarily the same as that of EXCLUDED MIDDLE.

M A E Dummett, 'Truth', *Proceedings of the Aristotelian Society* (1958–59)

ARL

black body law (19th century) *Physics* An ideal or 'black' body is the ideal absorber of radiation: it absorbs all electromagnetic radiation of all wavelengths that falls on it at all temperatures. It is also the ideal radiator of electromagnetic radiation and the radiation that it emits per unit area per unit time is a function of temperature only.

The equation giving the energy u emitted by a black surface of area A at a temperature T per unit time is

$$u = \sigma A T^4$$

where σ is a constant known as the Stefan-Boltzmann constant. The equation, known as the Stefan-Boltzmann equation, was deduced in 1879 by the Austrian physicist Joseph Stefan (1835–93) on the basis of measurements of the emission from a platinum surface made by the Irish physicist John Tyndall (1820–93); and was, in 1884, shown theoretically by the Austrian, Ludwig Boltzmann (1844–1906) to be valid for black surfaces only. The value of σ is 5.6705×10^{-8} W m^{-2} K^{-4}.

J Thewlis, ed., *Encyclopaedic Dictionary of Physics* (New York, Oxford and London, 1962)

MS

black box model (20th century) *Politics* Theory of national and international politics.

The actions of parties, pressure groups and states, and the relations between them, are fruitfully studied by looking at 'inputs' and 'outputs' or pressures and actions on the one hand, and the resultant policies on the other. The internal responses and calculations – or the ideological or pragmatic considerations of the institutions involved – are treated as irrelevant or invisible; as if they took place within a 'black box', and did not need to be considered when examining what goes into the box, or what comes out of it. Akin to BILLIARD BALL MODEL. *See also* AXIOMATIC THEORY and SYSTEMS THEORY.

Alan Bullock, Oliver Stallybrass, and Stephen Trombley, eds, *The Fontana Dictionary of Modern Thought*, 2nd edn (London, 1988)

RB

black box theory *Mathematics* Any theory that gives a formal set of rules for calculating the output of a system from the input, regarding the system as a black box; that is, without specifying how the calculation is performed.

For example, in many algorithms (*see also* COMPUTABILITY THEORY) in combinatorial OPTIMIZATION THEORY (*see also* COMBINATORICS), one often assumes the existence of a suitable oracle or black box for verifying a particular property of the current solution, such as whether it belongs to a given polyhedron. The exact details of the oracle are not required.

AL

black hole (1969) *Physics* A name suggested by John Archibald Wheeler (1911–) in 1969.

A black hole is an astronomical body whose gravitational field is sufficiently strong to cause relativistic curving of space-time around it. This results in gravitational self-closure to create a region from which particles and photons cannot escape (the event horizon). Black holes are thought to form from the gravitationally unstable cores of massive stars which explode as supernovae. They may also be the unseen components of certain binary systems. The centres of some galaxies may contain supermassive black holes of 10^6 to 10^9 solar masses.

Paul Davies, ed., *The New Physics* (Cambridge, 1989)

GD

Blackman curve (1905) *Biology* Named after the British biologist F F Blackman, this graphic representation is also known as the limiting factor. It is an ideal curve illustrating how insufficient quantities of some essential nutrient will retard plant growth even if all other nutrients are available in abundance.

The Blackman model is rarely observed except under highly controlled conditions. Essentially the same theory was proposed earlier by the German chemist Justus, Freiherr von Liebig (1803–73) in his 'law of the minimum'. *See also* LIEBIG'S LAW. *Compare with* TOLERANCE, LAW OF.

F Salisbury, *Plant Physiology* (Belmont, Calif., 1992)

KH

Blackwell-Rao theorem *Mathematics See* RAO-BLACKWELL THEOREM.

Blagden's law (18th century) *Physics* Named after English physical chemist Charles Blagden (1748–1820), who stated that the freezing-point of water is lowered by dissolved salts in the simple inverse ratio of the proportion that the water bears to it in the solution. The law was, in fact, first discovered by Richard Watson (1737–1816) in 1771. *See also* COPPET'S LAW and RAOULT'S LAW.

J Thewlis, ed., *Encyclopaedic Dictionary of Physics* (New York, Oxford and London, 1962)

MS

Blanc's rule *Chemistry* On pyrolysis, butanedioic (succinic) and pentanedioic (glutaric) acids yield cyclic anhydrides whereas hexanedioic (adipic) and heptanedioic (pimelic) acids yield cyclic ketones. There are some exceptions.

R J W Cremlyn, *A College Organic Chemistry* (London, 1970)

MF

Blanquism (19th century) *Politics* Theory of politics ascribed to and derived from the French socialist Louis-Auguste Blanqui (1805–81).

Capitalist and reactionary regimes are to be overthrown by a revolutionary coup d'état, carried out by an élite of dedicated revolutionaries. They will then use the power of the STATE to introduce a regime based on equality. Critics of LENINISM often portray it as a species of Blanquism.

S Bernstein, *August Blanqui and the Art of Insurrection* (London, 1971)

RB

blastema theory (early 19th century) *Medicine* Based on a theory by the Scottish physician John Hunter (1728–93), this asserts that solid body parts develop from plastic lymph, a fluid circulating in blood and providing the material for tissue maintenance, growth and repair.

The development of microscopy, which provided evidence for cell theory, helped discredit the blastema notion by the end of the 1860s.

KH

Blaue Reiter (1911) *Art* Meaning 'The Blue Rider' (after an almanac of that title), this was the name of an artistic group in Munich headed by Russian artist Vasily Kandinsky (1866–1944), and Germans August Macke (1887–1914) and Franz Marc (1880–1916) who broke away from other Expressionists in the *Neue Künstlervereinung*.

The group had no precise artistic programme, although Kandinsky and Macke outlined their aims in the catalogue to the group's first exhibition as being the expression of the artist's feelings/desires in 'multiple fashion'. At further exhibitions the group encompassed French artist Robert Delaunay (1885–1941) and the Austrian composer Arnold Schönberg. After Macke's and Marc's deaths in World War I the group disbanded.

LL

blending inheritance (19th century) *Biology* Also discussed as the 'paint pot problem' by Henry Fleeming Jenkin (1833–85), a Scottish engineer, this is the belief that distinct characteristics of the father and mother always take on an intermediate, or averaged, expression in the offspring.

For example, a tall mother and a short father would be expected to produce offspring of medium height. The paint pot theory held that with each succeeding generation, distinctive traits would tend to become completely diluted, just as a single drop of black paint ultimately disappears when stirred into a bucket of white paint. Such metaphors were abandoned once the field of genetics advanced, and the intermediate traits sometimes observed in offspring could be explained by recessive genes. *See also* MENDELISM. *Compare with* PARTICULATE INHERITANCE.

D Hull, *Darwin and His Critics* (Chicago, 1973)

KH

Bloch's theory *Physics* Named after the Swiss-born American physicist Felix Bloch (1905–83), joint winner of the 1952 Nobel prize for Physics.

A theory giving a mathematical formulation of the BAND THEORY OF SOLIDS and treating the effect of the periodic field of a crystal lattice on the motion of free electrons. The theory uses a wave function containing the periodicity of the lattice by modulating the original plane wave function.

Robert M Besançon, ed., *The Encyclopedia of Physics* (New York, 1985)

GD

Blondel–Rey law *Physics* Named after André Eugène Blondel (1863–1938), who asserted that the process of vision is not instantaneous: it takes an observer some finite time to perceive that a stimulus has reached the eye.

If a steady source of luminance L_a is adjusted to appear of the same brightness as a flash of a source of luminance L exposed for a time t, L_a is said to be the effective luminance of the flash of luminance L. The Blondel–Rey law states that

$$L_a = (Lt)/(t + t_0)$$

where t_0 is a constant that depends upon the subject and the visual conditions.

J Thewlis, ed., *Encyclopaedic Dictionary of Physics* (New York, Oxford and London, 1962)

MS

Bode's law (1772) *Astronomy* Sometimes called the Titius-Bode law, this was first stated by German astronomers Johann Daniel Titius (1729–96) in 1766 and (independently) Johann Ehlert Bode (1747–1826) six years later. This is a simple numerical relationship connecting the distances at which the various planets move around the Sun.

To obtain the relationship, write down the sequence:

0 1 2 4 8 16 32 64 128 256

multiply by 3 to give:

0 3 6 12 24 48 96 192 384 768

add 4 to give:

4 7 10 16 28 52 100 196 388 772

These numbers are approximately equal to the distances of the planets from the Sun, taking the Earth's distance as 10, the actual values being: Mercury 4, Venus 7.2, Earth 10, Mars 15.2, asteroids 26.5, Jupiter 52.0, Saturn 95.4, Uranus 191.6, Neptune 300.7, Pluto 395. The law was stated before the discovery of Uranus, the asteroids, Neptune and Pluto. Uranus and the asteroids fit the law well, but Neptune and Pluto fail.

S Mitton, ed., *The Cambridge Encyclopaedia of Astronomy* (London, 1973)

MS

body art *Art* Use of the human body as an artistic medium. This method was inspired by exponents of the HAPPENING during the late 1950s and early 1960s, and by French artist Yves Klein (1928–62) who employed 'imprints' of the female body on canvases. During the 1970s Bruce McLean and Vito Acconci pioneered the idea of ordinary facial or body movements as art. Others, including Gina Pane, have used self-abuse as a medium; and Keith Arnatt self-burial, filmed as a work.

LL

body reaction theory of emotion *Psychology* *See* JAMES-LANGE THEORY OF EMOTION.

body type *Psychology* *See* CONSTITUTIONAL THEORY.

Bohr-Sommerfeld theory of the atom (1915) *Physics* Named after the Danish physicist Niels Henrik David Bohr (1885–1962) and the German, Arnold Sommerfeld (1868–1951), this theory suggested the possiblity of splitting a magnetic field's energy levels.

Sommerfeld investigated the Bohr model of the atom but considered the possibility of new energy levels arising from the existence of elliptical orbits in addition to the circular orbits studied by Bohr. This necessitated two quantum conditions, one specifying the radial component of the angular momentum of the electron about the nucleus; and the other the azimuthal component. When non-relativistic mechanics are used, the theory shows no new energy levels for the hydrogen atom; but introduces the concept of degeneracy: orbits of differing eccentricity in which the electron has the same energy. This gives the possibility of splitting the energy levels in a magnetic field.

J Thewlis, ed., *Encyclopaedic Dictionary of Physics* (New York, Oxford and London, 1962)

MS

Bohr-Wheeler theory of nuclear fission (1930) *Physics* This theory, introduced by the Danish physicist Niels Henrik David Bohr (1885–1962) and John Archibald Wheeler (1911–), describes the process of low-energy induced nuclear fusion in terms of compound nucleus formation and the LIQUID DROP MODEL OF THE NUCLEUS.

Using the liquid drop model, an unexcited nucleus is represented by a spherical liquid drop in which the surface tension effect of the nuclear forces balances the repulsive Coulomb forces of the nuclear charges. If the drop is distorted (by the absorption of a low-energy neutron, for example), the surface area increases, which corresponds to a weakening of the nuclear binding. For small distortions, this is counterbalanced by the loss of Coulomb energy as the two parts of the nucleus are now farther away from each other. However, if the distortion becomes sufficiently great, the long-range Coulomb repulsion overcomes the short-range nuclear binding force and fission occurs.

J Thewlis, ed., *Encyclopaedic Dictionary of Physics* (New York, Oxford and London, 1962)

MS

Bohr's theory of the hydrogen atom (1913) *Physics* Named after its Danish originator, Niels Henrik David Bohr (1885–1962), this is a theory to explain the line spectra observed for atomic hydrogen.

The model of the hydrogen atom developed by the British chemist Ernest Rutherford (1871–1937), using only 'classical' physics, predicts that the electron should radiate electromagnetic radiation continuously and spiral into the nucleus. To give the atom stability and to explain the observed spectral lines, Bohr postulated several new features to be incorporated into the model.

(1) The electron can rotate in certain stable circular orbits around the nucleus without radiating electromagnetic radiation (stationary orbits). (2) In these stable orbits, the angular momentum of the electron about the nucleus must be an integral multiple of $h/2\pi$, where h is Planck's constant. (3) An atom only radiates or absorbs electromagnetic radiation when the electron makes a transition from one stable orbit to another. The frequency ν of the radiation is given by

$$h\nu = E_1 - E_2$$

where E_1 is the energy of the electron in the initial stationary orbit and E_2 is that in the final stationary orbit. This model, which is essentially an *ad hoc* patching up of Rutherford's model, has been superseded by quantum mechanics. *See* RUTHERFORD'S MODEL OF THE ATOM.

J Thewlis, ed., *Encyclopaedic Dictionary of Physics* (New York, Oxford and London, 1962)

MS

bolshevism (1903–) *Politics* Theory of revolutionary action associated with the Bolshevik Party in Russia under the leadership of V I Lenin (1870–1924).

In order to move from capitalism to SOCIALISM and eventually to COMMUNISM, a 'vanguard party' is necessary to lead the working class against the capitalist state. This party will hold opinions in advance of the class which it leads, and in order to succeed must be organized along military lines of loyalty and command, rather than in a democratic or open manner.

Neil Harding, *Lenin's Political Thought*, 2 vols (London, 1977, 1981)

RB

Boltzmann's distribution law (1868) *Physics* Named after the Austrian physicist Ludwig Boltzmann (1844–1906) who did important work on the kinetic theory of gases.

The law describes the statistical distribution of velocities and energies of gas molecules at thermal equilibrium and obeying classical (non-quantum) mechanics. For an assembly of weakly-interacting particles that are distinguishable, the number n_j of particles having an enery E_j is related to the number n_k having an energy E_k by the equation

$$\frac{n_j}{n_k} = \exp \frac{-(E_j - E_k)}{kT}$$

where T is the temperature and k is Boltzmann's constant. In the case of indistinguishable particles, Boltzmann statistics reduce to Bose-Einstein statistics for bosons (photons, α-particles and all nuclei with an even mass number) and Fermi-Dirac statistics for fermions (electrons, protons and neutrons). *See also* BOSE-EINSTEIN STATISTICAL AND DISTRIBUTION LAW and FERMI-DIRAC STATISTICAL AND DISTRIBUTION LAW.

Robert M Besançon, ed., *The Encyclopedia of Physics* (New York, 1985)

GD, MS

Boltzmann's entropy theory (1877) *Physics* Named after the Austrian physicist Ludwig Boltzmann (1844–1906).

This theory relates the entropy S, or degree of disorder of a system, to the thermodynamic probability W, where W is the number of microscopically distinct states of a system. The relationship has the form $S = k \ln W$, where k is the Boltzmann constant. W may also be regarded as the number of solutions of the Schrödinger equation for the system giving the same energy distribution. A perfect crystal lattice at the absolute zero of temperature has $W = 1$, and so S is zero ($\ln 1 = 0$). In practice, solids have a certain amount of configurational entropy due to disorder.

Robert M Besançon, ed., *The Encyclopedia of Physics* (New York, 1985)

GD

Boltzmann's superposition principle *Physics* Named after the Austrian physicist Ludwig Boltzmann (1844–1906).

This principle expresses the fact that the behaviour of viscoelastic materials can be described by linear differential equations with constant coefficients and time as variable. At any instant of time, the stress (or strain) under an arbitrary strain (or stress) history is a linear superposition of all strains (or stresses) applied at previous times multiplied by the values of a weighting function corresponding to the time intervals having elapsed since imposition of the respective strains (or stresses).

Robert M Besançon, ed., *The Encyclopedia of Physics* (New York, 1985)

GD

Bolzano's theorem *Mathematics See* INTERMEDIATE VALUE THEOREM.

Bolzano-Weierstrass theorem (19th century) *Mathematics* Proved by Czech analyst Bernhard Bolzano (1781–1848) and later independently by German mathematician Wilhelm Weierstrass (1815–97), this is the result whereby every bounded infinite subset of n-dimensional Euclidean space has a cluster point. In particular, every bounded infinite sequence has a convergent subsequence. *See also* HEINE-BOREL COVERING THEOREM.

G B Folland, *Real Analysis* (New York, 1984)

ML

Boo Hurrah theory *Philosophy* Slightly disrespectful title for EMOTIVISM as a theory of ethics, because it analyzes moral judgments as expressions of unfavourable or favourable emotion.

ARL

boomerang effect *Psychology See* ASSIMILATION-CONTRAST THEORY.

bootstrap theory *Physics* A theory that is concerned with the self-consistency of a more general theory. A bootstrap is any means by which a system brings itself into some desired state; for example, positive feedback of the output of an electronic circuit used to control conditions in the input circuit.

GD

Borel-Cantelli lemma (20th century) *Mathematics* Named after French mathematician Emile Borel (1871–1956) and Italian mathematician Francesco Cantelli.

This is the frequently used lemma of PROBABILITY THEORY on infinite sequences of random events, whereby if the sum of the individual probabilities of events from a probability space is finite, then the probability that infinitely many events occur is zero. If the events are mutually independent, then the probability that infinitely many events occur is one. This is called the Borel criterion for the ZERO-ONE LAW.

M Hazewinkel, ed., *Encyclopaedia of Mathematics* (Dordecht, 1988)

ML

Born-Haber cycle (1919) *Chemistry* Named after German theoretical physicist Max Born (1882–1970), and German chemist Fritz Haber (1868–1934) who is famous for his discovery of the synthesis of ammonia. Born received the Nobel prize for Physics in 1954 and Haber the Nobel prize for Chemistry in 1918. The cycle, which is based on HESS'S LAW, is a graphical method of displaying the formation of an ionic compound in a series of simple hypothetical steps.

It is used to calculate the lattice enthalpy of an ionic compound. For example, the lattice enthalpy (ΔH_{latt}) of sodium chloride (NaCl) can be calculated from its enthalpy of formation (ΔH_{form}), the ionization energy (I) and enthalpy of sublimation (ΔH_{subl}) of sodium and the electron affinity (E) and enthalpy of atomization (ΔH_{atom}) of chlorine:

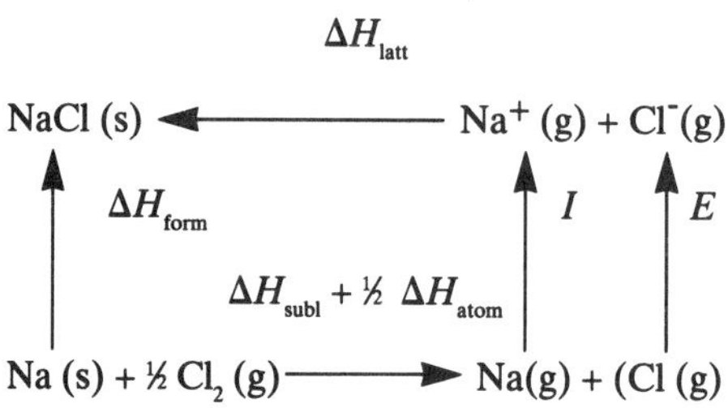

The equation is:

$$\Delta H_{latt} = \Delta H_{form} - (\Delta H_{subl} + \frac{1}{2}\Delta H_{atom} + I + E)$$

M Freemantle, *Chemistry in Action* (Basingstoke, 1987)

MF

Born's theory of melting (1939) *Physics* Named after its originator, the German physicist Max Born (1882–1970), this states that the rigidity modulus of a crystal decreases with temperature.

Born's theory is based on the assumption that the melting-point corresponds to the temperature at which the rigidity modulus of the material becomes zero; that is, at the melting-point the crystal structure becomes unstable with respect to variations in shape. However, because of thermal fluctuations, the crystal structure should collapse before this stability limit is reached, and this is borne out by experiment.

J Thewlis, ed., *Encyclopaedic Dictionary of Physics* (New York, Oxford and London, 1962)

MS

Borsuk-Ulam theorem (1933) *Mathematics* Named after its independent discoverers Karol Borsuk and Stanislaw M Ulam (1909–84).

There is no function which changes sign when all variables change sign, which has values which change by a small amount when all variables change by a small amount and which maps an n-dimensional analogue of a sphere with unit radius into an $(n-1)$-dimensional analogue of a sphere with unit radius. For example, there is no such function which maps a sphere of radius 1 into a circle of radius 1.

JB

Bose-Einstein statistical and distributive law (1924) *Physics* Named after the Indian physicist Satyendra Nath Bose (1894–1974) and the German-Swiss-American mathematical physicist Albert Einstein (1879–1955).

Bose-Einstein statistics apply to particles (bosons) which do not obey the PAULI EXCLUSION PRINCIPLE; for example, photons. The Bose-Einstein distribution law gives n, the average number of identical bosons in a state of energy E as,

$$n = \frac{1}{\exp(\alpha + E/kT) - 1},$$

where k is the Boltzmann constant, T is the thermodynamic temperature, and α is a quantity depending on temperature and the concentration of the particles. At high temperatures and low concentrations, the distribution law tends to the classical (Boltzmann) distribution, $n = A \exp(-E/kT)$. *See also* BOLTZMANN DISTRIBUTION LAW and FERMI-DIRAC STATISTICAL AND DISTRIBUTION LAW.

Robert M Besançon, ed., *The Encyclopedia of Physics* (New York, 1985)

GD

Bouguer-Lambert law *Chemistry See* LAMBERT'S LAW OF ABSORPTION.

Bouguer's law *Chemistry See* LAMBERT'S LAW OF ABSORPTION.

bounded rationality (1980s) *Economics* Developed by American behaviourist

Herbert Simon (1916–), this analysis of decision-making accepts that there are cognitive limits to an individual's knowledge and capacity to act rationally. *See* UNCERTAINTY and BERNOULLI'S HYPOTHESIS.

R M Cyert and J G March, *A Behavioral Theory of the Firm* (Englewood Cliffs, N.J., 1975); H A Simon, *Models of Bounded Rationality* (Cambridge, Mass., 1982)

PH

Bowen's reaction principle (1928) *Geology* American geologist Norman Levi Bowen (1887–1956) proposed that solid mineral crystals within a molten rock would continue to react with the molten fraction. On this basis, he elucidated a sequence of chemical changes whereby rocks of different composition could be obtained from the same parent material, depending (in part) on how quickly the mineral grains were removed from the melt.

In particular, Bowen proposed the geochemical principles behind these interactions. Whilst still considered to govern the general trends in a series of very complex interactions, these principles cannot be applied in detail to real rocks.

N L Bowen, *The Evolution of Igneous Rocks* (New Jersey, 1928)

DT

Bowman-Heidenhain hypothesis (late 19th century) *Biology/Medicine* Named after the English oculist Sir William Bowman (1816–92), who first described the anatomy of the kidney in 1842; and Rudolf Peter Heinrich Heidenhain (1834–97), the German physician who espoused the hypothesis later in the century. It postulates that only water is separated at the Malpighian bodies of the kidney, while the dissolved constituents of the urine are secreted by the epithelium of the urinary tubules.

Heidenhain believed that the secretion of urea by the tubule epithelium constituted a 'vitalistic' process, meaning that it relied upon a life-force that defies analysis. This hypothesis dominated the field for many years, bringing kidney research to a standstill. *See also* REDUCTIONISM and VITALISM. *Compare with* LUDWIG'S THEORY and CUSHNY'S THEORY.

J Fulton, *Selected Readings in the History of Physiology* (Springfield, Ill., 1966)

KH

Boyle's law (1662) *Chemistry* Named after its discoverer, the Irish chemist Robert Boyle (1627–91). In continental Europe it is often known as Mariotte's law, after the French physicist, Edme Mariotte (1620–84), who discovered it independently in 1676.

The volume (V) of a given mass of gas is inversely proportional to its pressure (P) at constant temperature; that is

$$V = \frac{C}{P}$$

where C is a proportionality constant. The law is valid only for an ideal gas. *See also* GAS LAWS.

M Freemantle, *Chemistry in Action* (Basingstoke, 1987)

MF

brachistochrone problem (1696) *Mathematics* The classical and motivating problem of the CALCULUS OF VARIATIONS that sought the path which would allow a constrained weighted particle under the force of gravity to move from one particular point to another (not in the same vertical line) in the least possible time. The Swiss mathematician Jakob Bernoulli (1667–1748) first discovered that the required curve called a brachistochrone is a cycloid.

Encyclopedic Dictionary of Mathematics (MIT Press, 1987)

MB

bracketing (1933/1957) *Linguistics* A notation for CONSTITUENT STRUCTURE: each constituent is enclosed in a pair of brackets which may be labelled with a symbol for the category of constituent: S(NP(the child) VP(V is NP(father PrepP(of NP(the man))))). Equivalent to TREE DIAGRAM.

K Brown and J Miller, *Syntax*, 2nd edn (London, 1991)

RF

Bragg's law (1913) *Physics* Named after its discoverer, the British chemist William Lawrence Bragg (1890–1971). Bragg realized that the three conditions which must be satisfied to obtain a diffracted beam when a monochromatic beam of X-rays falls on a crystal could be replaced by a single equation if the diffraction was treated as a reflection from a particular set of crystal planes.

When a monochromatic beam of X-rays of wavelength λ strikes a crystal, the directions in which reflected rays are observed (that is, in which constructive interference of the X-rays occurs) are given by the equation

$$n\lambda = 2\,d\sin\theta$$

where θ is the grazing angle of incidence of the X-rays on the crystal planes, *n* is an integer and *d* is the spacing of the crystal planes that are effectively reflecting the X-rays.

J Thewlis, ed., *Encyclopaedic Dictionary of Physics* (New York, Oxford and London, 1962)

MS

Braid's theory of hypnosis *Psychology* Named after the English surgeon James Braid (1795–1860).

Braid used the term 'nervous sleep' to describe hypnosis, and saw it as allied to but not identical with natural sleep. This 'nervous sleep' (neurhypnotism, later shortened to hypnotism) is supposed to be caused by protracted ocular fixation which brings about a state of fatigue in the relevant brain centres. In later years Braid placed greater emphasis on the role of psychological factors such as sustained mental concentration and the role of suggestibility. *See also* ANIMAL MAGNETISM.

J R Hilgard, *Personality and Hypnosis: A Study of Imaginative Involvement* (Chicago, 1979)

NS

brain field theory *Psychology* Proposed by the German-American psychologist Wolfgang Köhler (1887–1967).

Also referred to as isomorphism, literally meaning 'equal form'. The theory is associated with the Gestalt school of psychology, especially the work of the German-American psychologist Kurt Lewin (1890–1947), which holds that there is a relationship between excitatory fields in brain cortex and conscious experience. Isomorphism refers to supposed electrical brain states or 'fields' of the same shape as perceived objects. The correspondence is not presumed to be between the physical stimulus and the brain but between the perception of the stimulus and the brain. *See* GESTALT THEORY.

A D Ellis, ed., *A Sourcebook of Gestalt Psychology* (London, 1938)

NS

branching evolution *Biology See* CLADISTICS and CLADOGENESIS.

branching rule *Physics* This is a rule used in the construction of energy level diagrams for electrons in atoms, and gives the multiplicity of terms. It is concerned with the addition of additional electrons to given states. *See* COMPOUND NUCLEUS THEORY OF BOHR AND BREIT AND WIGNER.

J Thewlis, ed., *Encyclopaedic Dictionary of Physics* (New York, Oxford and London, 1962)

MS

Brans-Dicke theory (1961) *Physics* Named after the American physicists Carl Henry Brans (1935–) and Robert Henry Dicke (1916–).

An alternative theory of gravity to general relativity in which a scalar gravitational field is postulated in addition to the curved space-time metric. Brans and Dicke attempted to apply MACH'S PRINCIPLE by looking for a framework in which the gravitational constant *G* arises from the structure of the universe, and would have a value which decreased slowly with time. The theory differs from general relativity in its predictions for observable effects. *See also* RELATIVITY.

Jayant V Narlikar, *Introduction to Cosmology* (Cambridge, 1993)

GD

break and exchange model *Biology See* BREAKAGE AND REUNION MODEL.

breakage and reunion model *Biology* Also called the break and exchange model, this is the theory that during genetic recombination, parts of parental chromosomes are traded as a result of the physical breakage of the chromosomes and the recombination (reunion) of the broken pieces.

KH

Bredt's rule (1924) *Chemistry* Qualitative rule formulated by J Bredt. A double bond cannot be placed with one terminus at the bridgehead of a bridged ring system unless the rings are large enough to accommodate the double bond without excessive strain.

The rule has been quantified by the proposal that the sum of the number of atoms in the three bridges between the two bridgeheads of a bicyclic system determines whether that system can accommodate a bridgehead double bond.

Compendium of Chemical Terminology: IUPAC Recommendations (Oxford, 1987)

MF

Breit-Wigner equation (1936) *Physics* Named after the Russian-American Gregory Breit (1899–1981) and Hungarian-born American physicist Eugene Paul Wigner (1902–) who introduced it, this equation describes the properties of discrete energy levels (resonances) within the range of energies in which the compound nucleus model is appropriate.

It was obtained by the application of quantum mechanics to the compound nucleus model of nuclear reactions, and expresses reaction rates as cross-sections for the reaction. For the special case where the projectile is a low-energy neutron and the compound nucleus de-excites by the emission of a gamma ray, the equation leads to

$$\sigma = \text{constant} \times \frac{1}{v}$$

where σ is the cross-section for the reaction and v is the neutron speed, a result known as the $(1/v)$ law.

J Thewlis, ed., *Encyclopaedic Dictionary of Physics* (New York, Oxford and London, 1962)

MS

Brewster's law (1812) *Physics* Named after its discoverer, the British scientist Sir David Brewster (1781–1868).

When unpolarized light strikes a plane dielectric (non-conducting) surface, the reflected light is partially plane-polarized. Brewster noticed that, when this light's angle of incidence (measured from the normal to the surface) is such that it is at right angles to the direction of the refracted ray, the reflected light is completely plane-polarized. The light's angle of incidence at which this occurs is i_p, and is given by Brewster's law as

$$i_p = \tan^{-1} n$$

where n is the refractive index of the dielectric for light travelling from vacuum (or air) to the dielectric.

J Thewlis, ed., *Encyclopaedic Dictionary of Physics* (New York, Oxford and London, 1962)

MS

Brianchon's theorem (1806) *Mathematics* Although Charles-Julien Brianchon (1785–1864) proved the theorem by a different method, it can be deduced from PASCAL'S MYSTIC HEXAGRAM THEOREM by exchanging the words 'point' and 'line' and making linguistic adjustments. The theorem is: if a hexagon is circumscribed about a conic, its three diagonals are concurrent.

JB

bricolage (1962) *Stylistics* Term coined by French structural anthropologist Claude Lévi-Strauss (1908–) for processes of myth-formation in pre-literate, pre-scientific cultures.

The construction of something else out of materials to hand, for example patchwork. Myths are models of social life which make analogical use of natural materials and processes: animals, weather, and so on. *Bricolage* is basically a metaphor-making process and has been extended to language as such; also applied to Modernist texts which are 'patchworks' of other writings and styles (for example, T S Eliot's *The Waste Land*), the juxtapositions between *ad hoc* materials producing new significances. *See also* INTERTEXTUALITY.

T Hawkes, *Structuralism and Semiotics* (London, 1977)

RF

British empiricists *Philosophy* Name applied primarily to John Locke (1632–1704), George Berkeley (1685–1753), and David Hume (1711–76), with lesser figures such as Francis Bacon (1561–1626) and Thomas Reid (1710–96). *See also* EMPIRICISM, SUBJECTIVE IDEALISM, REGULARITY THEORY OF CAUSATION, BUNDLE THEORIES and CONTINENTAL RATIONALISTS.

ARL

Britten-Davidson model (1969) *Biology* Proposed by American geneticists Roy John Britten (1919–) and Eric Harris Davidson (1937–), this is a theory describing the regulation of gene expression in cells with a nucleus (eukaryotes).

According to the model, chromosomes in the nucleus contain 'sensor genes' that recognize certain substances found in the cell. When an inducer (a small molecule that encourages the production of larger amounts of enzymes for its metabolism) enters the nucleus, it binds to the sensor gene and promotes production of an 'integrator gene'. This gene then produces a specific activator ribonucleic acid (RNA), which goes on to bind to the receptor site of a structural gene. This model is analogous to the JACOB-MONOD MODEL of gene regulation in prokaryotes (simpler cells without nuclei).

P Sheeler, *Cell and Molecular Biology* (New York, 1987)

KH

Broca's area (1861) *Linguistics* Named after the French neurologist Paul Broca (1824–80).

The area of the inferior frontal gyrus and the precentral gyrus in the left cerebral hemisphere; damage to this area of the brain may result in 'Broca's aphasia', restriction of speech output to short, ungrammatical sentences. Broca's area/aphasia and WERNICKE'S AREA are classic reference-points in clinical neurolinguistics and aphasiology.

M Garman, *Psycholinguistics* (Cambridge, 1990)

RF

Brønsted catalysis law *Chemistry See* BRØNSTED RELATION.

Brønsted-Lowry theory (1923) *Chemistry* Theory of acids and bases named after Danish chemist Johannes Nicolaus Brønsted (1879–1947) and English chemist Thomas Martin Lowry.

An acid is a substance consisting of molecules or ions which donates protons (H^+), and a base is a substance consisting of molecules or ions which accepts protons. In the following example, hydrogen chloride gas (HCl) is an acid and water (H_2O) is an acid:

$$HCl(g) + H_2O(l) \rightarrow H_3O^+(aq) + Cl^-(aq)$$

See also LEWIS'S THEORY OF ACIDS AND BASES.

M Freemantle, *Chemistry in Action* (Basingstoke, 1987)

MF

Brønsted relation *Chemistry* Named after Danish chemist Johannes Nicolaus Brønsted (1879–1947). The name applies to either of the equations:

$$k_{HA}/p = G(K_{HA}q/p)^{\alpha}$$

and

$$k_{A}/q = G(K_{HA}q/p)^{-\beta}$$

where α, β and G are constants for a given reaction series, k_{HA} and K_A are catalytic coefficients (or rate coefficients of reactions whose rates depend on the concentrations of HA and/or of A^-), K_{HA} is the acid dissociation constant of the acid HA, p is the number of equivalent acidic protons in the acid HA, and q is the number of equivalent basic sites in the conjugate base A^-.

The Brønsted relation is often termed the Brønsted catalysis law, or catalysis law. Although justifiable on historical grounds, this name is not recommended, since Brønsted relations are now known to apply to many uncatalyzed reactions.

Compendium of Chemical Terminology: IUPAC Recommendations (Oxford, 1987)

MF

Brouwer's theorem (1912) *Mathematics* Named after Dutch logician Luitzen Egbertus Jan Brouwer (1881–1966), this FIXED POINT THEOREM states that a compact convex set mapped into itself has a fixed point.

For example, in the complex plane any continuous mapping of the unit disc into itself has a fixed point. This theorem was extended by Schauder and Tychonoff to be true for a normed space or a locally convex space.

J B Conway, *A Course in Functional Analysis* (New York, 1985)

ML

Brouwer's theorem on domain invariance *Mathematics See* INVARIANCE OF DOMAIN THEOREM.

Brownian system (1780) *Biology/Medicine* Also called 'Brunonianism', this was created by the Scottish physician John Brown (1735–88) and is a comprehensive medical theory postulating that life and its vital functions are the sum of the constant responses of an organism to its environment.

Health represented a balance between stimuli and 'excitability', defined as the ability to receive impressions from outside the body and respond to them. When stimuli were excessive or inadequate, disease occurred, and physicians tried to restore the balance by prescribing musk, camphor, rich soups, opium, spicy foods and (especially) brandy.

G Risse, 'Schelling's *Naturephilosophie* and John Brown's *Elements of Medicine*', *Bulletin of the History of Medicine*, vol. L (Fall, 1976) 321–34

KH

Bruck-Ryser-Chowla theorem (1950) *Mathematics* A result of COMBINATORICS concerning the existence of symmetric block designs – collections of v *blocks*, each consisting of k of v possible points. The result states that if each pair of points is contained in λ blocks, then v is even implies $v - \lambda$ is a perfect square, while v is odd implies the equation

$$x^2 = (k - \lambda)y^2 + (-1)^{(v-1)/2}\lambda z^2$$

has a non-trivial integer solution. The converse remains an open problem, with the exception of the projective plane of order 10 ($v = 111$, $k = 11$, $\lambda = 1$), which was proved not to exist with computer assistance by C Lamb *et al.* (1990).

AL

Brueckner's theory of nuclear matter (20th century) *Physics* Named after its originator, Keith Allan Brueckner (1924–), this uses a field theory technique and is used to give a formal expression which represents the effect of inter-nucleon interactions in the nucleus.

The effective mass of a nucleon is found to depend on the momentum of the nucleon inside the nucleus, and the effective interaction is given by an integral equation involving the original interaction.

J Thewlis, ed., *Encyclopaedic Dictionary of Physics* (New York, Oxford and London, 1962)

MS

Brunonianism *Biology/Medicine See* BROWNIAN SYSTEM.

budget-maximization theory (20th century) *Politics/Management* Theory of bureaucracy.

Bureaucrats will seek to expand the budgets of their departments, irrespective of what their formal political commitments or personal ambitions may be. This will lead them both to expand the numbers of their staff and the volume of their funds and facilities, and to extend or develop the areas over which they have responsibility. *See also* PARKINSON'S LAW.

Patrick Dunleavy, *Democracy, Bureaucracy and Public Choice* (London, 1991)

RB

Buffon's needle problem (1777) *Mathematics* First posed by French naturalist and mathematician Georges Louis Leclerc de Buffon (1707–88), this is the problem of determining the probability that a needle which is thrown at random will land on one of a set of parallel lines in a plane.

When the needle has length l and the lines are a (where $a > l$) units apart, the probability is $2l/\pi a$. This classical problem was responsible for the development of the theory of geometric probabilities. It was an experimental check on the BERNOULLI THEOREM and has also been used as a Monte Carlo method, with limited success, for computing π.

M Hazewinkel, ed., *Encyclopaedia of Mathematics* (Dordecht, 1988)

ML

building-up principle *Chemistry See* AUFBAU PRINCIPLE.

bundle hypothesis *Psychology* Proposed by the German-American psychologist Wolfgang Köhler (1887–1967).

Associated with, and criticized by, the Gestalt school of psychology, the theory holds that 'a whole is no more than a sum of its parts'. Gestalt psychologists believe that a complex precept is more than the sum or bundle of its several stimulus elements. For instance, a musical tune can be described as a bundle of notes but Gestalt Psychology would argue that the tune comprises an organizational entity which is more than the sum of its constituent elements. *See* GESTALT THEORY, DERIVED PROPERTIES, POSTULATE OF.

A D Ellis, ed., *A Sourcebook of Gestalt Psychology* (London, 1938)

NS

bundle theories *Philosophy* Theories that analyze a given item as a mere bundle of items of some other kind; where the first item would normally be thought of as something substantive and independent, the other items being somehow related to it, or dependent on it and owing their existence to it.

There are two main examples. The first treats ordinary objects as mere bundles of properties, so that my chair is simply a set of properties (blackness, four-leggedness, and so on) somehow put together, there being no independent chair that has these properties. The second example treats the mind or self as simply a set of experiences and not as something which has these experiences. *See also* NO-OWNERSHIP THEORY and REDUCTIONISM.

David Hume, *A Treatise of Human Nature* (1739–40); see book 1, part 1, §6 for bundle theory of the self (which, however, Hume withdraws in the Appendix)

ARL

Bunsen-Kirchoff law (1859) *Chemistry* Named after German-born chemist Robert Wilhelm Bunsen (1811–99), famous for his laboratory burner, and German physical chemist Gustav Robert Kirchoff (1824–87), noted for the development of the spectroscope. This law states that every element has a characteristic emission spectrum of bright lines, and a characteristic absorption spectrum of dark lines.

A F Holleman and H C Cooper, *A Text-Book of Inorganic Chemistry* (London, 1916)

MF

Bunsen-Roscoe effect *Psychology* Named after the German physicist and chemist Robert Wilhelm Bunsen (1811–99) and the English chemist Sir Henry Enfield Roscoe (1833–1915).

The principle that the absolute threshold for vision is a mathematical function expressed as the product of the duration and intensity of the stimulus. It is a keystone concept in photochemistry theory and had an important impact in early work on human visual perception. The law holds moderately well for short durations of up to approximately 50 milliseconds.

L Kaufman, *Perception* (New York, 1979)

NS

bureaucracy (20th century) *Economics/Politics/Sociology* Theory of organization in modern society, developed in particular by the German sociologist Max Weber (1864–1920).

Bureaucracy is the characteristic form of organization in modern society, not only in its government but in commerce and social institutions. Responsibility is vested in full-time officials whose livelihood is derived from their salaries and who are appointed on merit. Bureaucracy works with written records, regular procedures, and accumulated precedent. It involves clear hierarchies of responsibility and command which enable the resources of an institution to be applied with maximum effect. *See* THEORY OF THE FIRM, THEORY OF THE GROWTH OF THE FIRM and ORGANIZATION THEORY.

David Beetham, *Bureaucracy* (Milton Keynes, 1987)

RB

Buridan's ass (14th century) *Philosophy* A theoretical illustration of the dilemma posed by the need to make a decision between two equally attractive proposals. The concept itself dates from much earlier times, being discussed first by Aristotle (384–322 BC).

In Buridan's example, an ass faces starvation when it is unable to choose between two equally appetizing piles of hay.

DP

Burnet's theory *Biology* *See* CLONAL SECTIONAL THEORY.

Busch theory (1962) *Biology/Medicine* Named after Harris Busch (1923–), physician and biochemist at Baylor College of Medicine (Houston, Texas), this is a theory of cancer causation.

It holds that there are three stages of carcinogenesis: initiation, promotion and acceleration. Part of the cell's normal deoxyribonucleic acid (DNA) is believed to aid carcinogenesis once it becomes freed from combining with the suppressor protein that normally inhibits malignancy. A cancer-causing agent binds to the DNA-suppressor combination during initiation, frees the DNA from its suppressor protein during promotion, and allows it to form cancerous ribonucleic acid (RNA) during acceleration. *See also* CANCER, THEORIES OF.

H Busch, *Biochemistry of the Cancer Cell* (New York, 1962)

KH

business cycle *Economics* Fluctuations in economic activity usually follow a pattern of depression, recovery, boom and recession. The cycles vary in time from more than one year to 12 years, although long wave cycles may extend to 60 years.

Causes for the cycles are varied but include credit changes, over-investment, under-consumption, fluctuations in agricultural output, the interaction of the multiplier and accelerator, war, and international politics. The cyclical nature of elections in many Western industrialized democracies has also had an impact on the general level of economic activity of a country. *See* KONDRATIEFF CYCLES, SUNSPOT THEORY, PRODUCT LIFE-CYCLE THEORY, ACCELERATION PRINCIPLE, FINE TUNING, MULTIPLIER-ACCELERATOR MODEL, POLITICAL BUSINESS CYCLE and TRADE CYCLE.

A F Burns and W C Mitchell, *Measuring Business Cycles* (New York, 1946); G Haberler, *Prosperity and Depression. A Theoretical Analysis of Cyclical Movements* (London, 1958); C P Kindleberger and P H Lindert, *International Economics* (Homewood, Ill., 1978); D Ricardo, *On the Principles of Political Economy and Taxation* (London, 1817)

PH

butterfly effect (1963) *Mathematics* Discovered by American meteorologist Edward Norton Lorenz (1917–), this term refers to a STRANGE ATTRACTOR (resembling the flapping wings of a butterfly) which is produced by a set of DIFFERENTIAL EQUATIONS describing air flows in the atmosphere known as the Lorenz equations. In meteorological terms, this models the unpredictability of local weather patterns. *See* CHAOS THEORY.

I Peterson, *The Mathematical Tourist* (New York, 1988)

ML

Buys-Ballot's law *Meteorology* Formulated by the Dutch meteorologist Christoph Buys-Ballot (1817–90), this law states that, in the northern hemisphere, winds blow anticlockwise around low pressure areas and clockwise in the southern hemisphere.

Winds do not blow directly from high to low pressure areas owing to the influence of the Earth's rotation. This gives rise to the Coriolis force which deflects the direction of flow.

DT

C

Cailletet and Mathias law (1886) *Chemistry* Named after L Cailletet and E Mathias, and subsequently verified by Sydney Young (1857–1937) in 1900 and others. Also known as the law of the rectilinear diameter, this states that the relationship between the mean density of a liquid and its saturated vapour at a specified temperature is a linear function of the temperature.

S Glasstone, *Textbook of Physical Chemistry* (London, 1948)

MF

calculus of variations, optimal control *Mathematics* The calculus of variations is a classical area of OPTIMIZATION THEORY (developed initially by L Euler around 1744) concerned with choosing a function $x(t)$ to minimize or maximize an objective expression like $\int f(x,\dot{x},t),dt$ subject to certain constraints.

Such problems arise, for example, when trying to find the shortest path between two given points on a surface. The area evolved into OPTIMAL CONTROL, where we consider the Cantor-Bendixson theorem. Any set of real numbers can be decomposed into at most a countable set $\{x_1,x_2,\dots\}$ and a set P which is *perfect*, which means that the *derived set* of all *cluster points* of P (limits of non-constant sequences from P) is equal to P. This fact considerably simplifies the topological structure of the real line.

AL

caloric theory (18th–mid-19th century) *Physics* This theory regarded heat as an imponderable, indestructible fluid of zero density, surrounding the ultimate particles of matter, and called caloric.

The Scottish physicist Joseph Black (1728–99) believed the temperature of a body was determined by the amount of caloric that it contained. Caloric was considered to be a highly elastic fluid, the self-repulsion of which was responsible for thermal expansion. Whilst indestructible, it could pass from one body to another until equilibrium was achieved. The conservation of heat in calorimetry experiments was assured on this model. Caloric was also considered able to enter into chemical combination with the atoms of a substance (producing a phase change) when the heat became latent and did not affect a thermometer. Caloric theory was abandoned largely as a result of the work of the English physicist James Prescott Joule (1818–89). *See also* PHLOGISTON THEORY.

J Thewlis, ed., *Encyclopaedic Dictionary of Physics* (New York, Oxford and London, 1962)

MS

Cambridge capital controversies (1930s–) *Economics* This term refers to the debate between British and American economists concerning the neoclassical approach to economics. They were based at Cambridge University (England) and the Massachusetts Institute of Technology (Cambridge, USA) respectively.

The Modern School, particularly influenced by Alfred Marshall (1842–1924) and headed by the English economists Arthur Cecil Pigou (1877–1959) and John Maynard Keynes (1883-1946), refuted the microeconomic ideas of neo-classical economics; especially as seen in the work of the Americans Paul Samuelson (1915–) and Robert Solow (1924–).

Emphasizing the macroeconomic approach, the English denied (among other things) the existence of a functional relationship between the rate of profit and the capital intensity of an economy, and demonstrated the possibility of capital re-switching. (That is, if it is possible when the rate of return falls for firms to switch to more capital intensive methods of production (thus increasing the rate of investment), it is also possible that under certain circumstances the rate of return will reach such a level that firms switch from more to less capital intensive methods of production.) The Modern School also questioned the existence of an aggregate production function and the determination of savings and interest levels. *See* CAPITAL THEORY.

G H C Harcourt, *Some Cambridge Controversies in the Theory of Capital* (Cambridge, 1972)

PH

camera lucida (1674) *Art* An optical instrument invented by English physicist and inventor Robert Hooke (1635–1703) in about 1674.

In 1807 the English chemist and natural philosopher William Hyde Wollaston (1766–1828) developed one of benefit to artists, comprising a four-sided prism of glass with one angle of 90° opposite another of 135° and the remaining two at 67.5°. Such instruments were chiefly used for copying or reducing drawings, particularly by 19th-century landscape artists.

AB

camera obscura (15th century) *Art* A mechanical aid for drawing from nature, consisting of lenses and mirrors arranged in a darkened tent. A mirror fixed at an angle of 45° reflects the view through a double convex lens onto a sheet of paper placed at the focus of the lens.

Its invention is associated with the Italian architect and theorist Leon Battista Alberti (1404–72). It was taken up by 18th-century *vedutisti* ('view painters') such as Antonio Canaletto (1697–1768), and was employed by Thomas Wedgwood *c.*1794 in early photographic processes.

AB

cancer, theories of *Biology/Medicine* A reference to the well-accepted concept that cancer has a variety of causes depending on the type of cancer, and that more than one cause may be involved in a particular instance of the disease.

An individual's genetic endowment plus his or her exposure to substances or experiences known to encourage cancer appear to interact, making some people more susceptible to the disease than others. Some cancers are always associated with genetic abnormalities, while others continue to defy attempts to understand their aetiology. *See also* BUSCH THEORY, CATABOLIC DELETION HYPOTHESIS, DELETION HYPOTHESIS, FEEDBACK DELETION HYPOTHESIS and WARBURG THEORY.

KH

Cannon-Bard theory of emotion *Psychology* Named after the American physiologist Walter Bradford Cannon (1871–1945), and the French physician Louis Bard (1857–1930).

The integration of emotional expression is controlled by the thalamus, in the brain, sending relevant excitation patterns to the brain cortex at the same time that the hypothalamus controls the behaviour. The value of the theory was in the question it raised for theorists: do we experience an emotion because we perceive our bodies in a particular way or are there specific emotional neural patterns which respond to environmental events and then release bodily and visceral expressions? *See* ACTIVATION THEORY OF EMOTION, COGNITIVE APPRAISAL THEORY and JAMES-LANGE THEORY.

H Gleitman, *Psychology* (New York, 1991)

NS

canon (*c.*1860–) *Literary Theory* From the Greek word *kańon*, meaning 'rule'. Various ecclesiastical usages, the most relevant (dating from the 4th century AD) denoting the set of acknowledged biblical scriptures. In modern English literary theory, the idea is most associated with critics Matthew Arnold (1822–88) and Frank Raymond Leavis (1895–1978), particularly the latter's *The Great Tradition* (London, 1948).

A set of literary works felt to be quintessential for a national literary culture, identified by qualities such as 'high seriousness', 'creativity', and 'authenticity'. In the Arnold-Leavis tradition, the moral and spiritual value of the literature is affirmed as against philistinism. A canon has, needless to say, never been agreed, but certain 'classic' books (and exclusions of books) tend to be maintained by publishers and examination syllabuses.

J Guillory, 'Canon', *Critical Terms for Literary Study*, F Lentricchia and T McLaughlin, eds (Chicago, 1990), 233–49

RF

Cantor set (1883) *Mathematics* Named after German mathematician Georg Ferdinand Ludwig Philip Cantor (1845–1918) and also known as Cantor's ternary set, this is the subset of the interval [0,1] which is formed by successively removing the open middle third, then the open middle third of each of the remaining intervals, and so on. The representation of these numbers in base 3 contains no ones. The set plays a fundamental role in modern theories of integration dynamics and chaos.

Encyclopedic Dictionary of Mathematics (MIT Press, 1987)

MB

Cantor's diagonal theorem (19th century) *Mathematics* Named after German mathematician Georg Ferdinand Ludwig Philip Cantor (1845–1918), this is the result in SET THEORY whereby the power set of a given set X (that is, the set of all subsets of X, denoted $\wp(X)$) has cardinality which is strictly greater than the cardinality of the set X itself. In particular, members of X and $\wp(X)$ cannot be put into one-to-one correspondence and so there are 'more' real numbers than natural numbers. The name of the theorem is related to the fact that the proof uses a diagonal process.

P R Halmos, *Naive Set Theory* (Springer, 1974)

MB

Cantor's intersection theorem (19th century) *Mathematics* Named after German mathematician Georg Ferdinand Ludwig Philip Cantor (1845–1918), this is the result in TOPOLOGY whereby in a complete metric space, any nested sequence of closed sets whose diameters approach zero contains a unique intersection point.

J R Munkres, *Topology: A First Course* (Prentice-Hall, 1975)

MB

Cantor's paradox (19th century) *Mathematics* Named after German mathematician Georg Ferdinand Ludwig Philip Cantor (1845–1918), this is the paradox which arises in Cantor's formulation of SET THEORY from the assumption that there is an all-inclusive infinite set (that is, a set which contains all sets).

If there were such a set, then every subset would be a member of it; however, by CANTOR'S DIAGONAL THEOREM, every set has strictly more subsets than it has members. In modern set theory this is interpreted to mean that there is no largest cardinal number.

P R Halmos, *Naive Set Theory* (Springer, 1974)

MB

capital logic (20th century) *Politics/Economics* A theory within MARXISM.

The form and character of political institutions are determined by those of economic ones. The 'logic' of capital thus shapes that of government. The theory has been criticized as mechanical and inflexible.

Bob Jessop, *The Capitalist State* (Oxford, 1982)

RB

capital theory (18th century–) *Economics* Developed over a period of 250 years by many economists ranging from Adam Smith (1723–90) to Karl Marx (1818–83), this analyzes links between the theories of production, growth, value and distribution to explain why capital produces a return that keeps capital intact yet yields interest or a profit which is permanent.

Classical economists interpreted capital as raw materials and the wages fund; Marxist economists saw capital as a social mode of production; the Austrian school maintained that time was crucial to the concept of capital. The Cambridge view is given under CAMBRIDGE CAPITAL CONTROVERSIES.

C J Bliss, *Capital Theory and the Distribution of Income* (New York, 1975); H G Harcourt, *Some Cambridge Controversies in the Theory of Capital* (Cambridge, 1972)

PH

capital asset pricing model (1960s) *Economics* Developed by American economist James Tobin (1918–) among others, this studies the relationship between risk and return by suggesting that asset prices will adjust to ensure that the return on an asset compensates for the risk undertaken by the investor, when held with a perfectly diversified portfolio. The model has dominated economic understanding of the financial sector. *See* PORTFOLIO SELECTION THEORY, MARGINAL EFFICIENCY OF CAPITAL, ARBITRAGE PRICING THEORY, and UNCERTAINTY.

J Tobin, 'Liquidity Preference as Behaviour Towards Risk', *Review of Economic Studies*, vol. xxv (February, 1958), 65–86; H Levy and M Sarnat, eds, *Financial Decision Making Under Uncertainty* (New York, 1977)

PH

Carathéodory's extension theorem *Mathematics* A result derived by C Carathéodory (1873–1950) which allows the construction of measures from the simpler idea of an outer measure. For example, beginning with the primitive notion of the volume of a finite box in Euclidean space, we can try to minimize the volume of a covering of a given set by a union of boxes, giving Lebesgue outer measure. Carathéodory's result then extends this definition to Lebesgue measure.

AL

Carathéodory's principle (1909) *Physics* This is an axiomatic statement of the second law of thermodynamics formulated by German mathematician Constantin Carathéodory (1873–1950).

In the neighbourhood of any arbitrary equilibrium state J of a thermally isolated system, there are other equilibrium states J' that are inaccessible from the state J by means of a quasi-static process. This axiom allows the existence of an integrating factor for the heat transfer in an infinitesimal reversible process for a physical system of any number of degrees of freedom.

J Thewlis, ed., *Encyclopaedic Dictionary of Physics* (New York, Oxford and London, 1962)

MS

Carathéodory theorem (1907) *Mathematics* Named after C Carathéodory (1873–1950), this states that if a point x in n-dimensional Euclidean space can be written as a *convex combination* of points $y_1,\ldots,y_m$ (meaning

$$x = \sum_1^m \lambda_i y_i$$

where the numbers λ_i are non-negative and sum to 1), then x can be written as a convex combination of at most $n+1$ of the points. This result is important in restricting the number of variables which can be non-zero at an optimal solution of a linear program. For example, it gives an approach to the ALTERNATION THEOREM via the DUALITY THEORY OF LINEAR PROGRAMMING.

AL

Caravaggism *Art* A term describing the style and technique of the Italian artist Michelangelo da Caravaggio (1573–1610), adopted by artists both in Italy (such as B Manfredi, Orazio Gentileschi (1563–1647) and G B Caracciolo) and abroad (for instance, Jusepe de Ribera (1588–1656), Gerard van Honthorst (1590–1656) and the 'Northern *Caraviggisti*').

These artists were influenced by Caravaggio's harsh effects of light and shade; his unconventional, often brutal, portrayal of religious subjects; and bold, realistic handling of genre and still life. All this was in constrast to the more classical and harmonious style of his leading contemporary, Annibale Carracci (1540–1609).

LL

carbon-ratio theory *Geology* This states that in any region, the density of the hydrocarbons varies inversely with the carbon ratio in the coals; that is, the heaviness of crude oil is greater if there is a low percentage of volatile carbon in the coals in the same region.

DT

carcinogenesis *Biology/Medicine See* CANCER, THEORIES OF.

Cardano's formula (1545) *Mathematics* Named after Italian mathematician and physician Girolamo Cardano (1501–76) – also known as Jerome Cardan – who published it in his *Ars magna*. The original discoverer was probably fellow Italian mathematician Nicolo Tartaglia (*c.*1500–57).

The formula gives the solutions of the general (normalized) cubic equation

$$x^3 + bx^2 + cx + d = 0$$

In particular, they are

$$x_1 = -\frac{b}{3} + \alpha - \frac{p}{3\alpha}$$

$$x_2 = -\frac{b}{3} + \omega\alpha - \frac{\omega^2 p}{3\alpha}$$

$$x_3 = -\frac{b}{3} + \omega^2\alpha - \frac{\omega p}{3\alpha}$$

where ω is either root of the equation $x^2 + x + 1 = 0$, is any value of

$$\left(-\frac{q}{2} \pm \sqrt{\frac{q^2}{4} + \frac{p^3}{27}}\right)^{1/3}$$

and

$$p = c - \frac{b^2}{3}$$

$$q = \frac{2b^3}{27} - \frac{bc}{3} + d\,.$$

The quantity

$$\Delta^2 = \frac{q^2}{4} + \frac{p^3}{27}\,,$$

called the discriminant of the cubic, determines the nature of the roots. In particular, the equation has three real roots if Δ^2 is negative and exactly one real root if Δ^2 is positive. There is a similar formula for quartic equations.

Encylopedic Dictionary of Mathematics (MIT Press, 1987)

MB

cardinal vowels (1844) *Linguistics* The idea was proposed by the British linguist Alexander John Ellis (1814–90); the term 'cardinal vowel positions' was used by Henry Sweet (1845–1912) in 1877; the system was fully developed by the British phonetician Daniel Jones (1881–1957).

A reference system of ideal vowel sounds, mapping a space within the mouth bounded by the extreme positions which the tongue can assume while still sounding a vowel.

D Jones, *English Pronouncing Dictionary* (London, 1917)

RF

care-taker language *Linguistics* *See* MOTHERESE.

Carleson's theorem (1966) *Mathematics* Named after its discoverer, Lennart Axel Edvard Carleson (1928–).

The FOURIER SERIES of real multiples of $\sin nx$ and $\cos nx$ for a real function $f(x)$ of which the square is integrable converges almost everywhere to the sum $f(x)$. The result generalizes to the case where $[f(x)]^p$ is integrable for $p > 1$. Andrei Nikolaevich Kolmogorov (1903–87), however. proved that the result is false if $f(x)$ is merely integrable.

JB

Carlsbad law *Chemistry* The mineral orthoclase feldspar forms crystals which may twin according to several laws. The Carlsbad law is the most common of these. It applies to crystals with a composition plane on 010 and with the vertical *c* (or *z*) axis as twin axis. It is commonly an interpenetrant (penetration) twin.

MF

carnivalization (1965) *Literary Theory* Theory of Russian Mikhail M Bakhtin (1895–1975).

Development of DIALOGISM in which the metaphor of carnival is applied to the structure of those narratives which invert conventional relationships, subvert power, and celebrate a 'grotesque canon of the body': paradigmatically, F Rabelais's *Gargantua and Pantagruel* (1532–48), L Sterne's *Tristram Shandy* (1760–61), W S Burroughs's *Naked Lunch* (1959).

M M Bakhtin, *Rabelais and his World*, trans. H Iswolsky (Bloomington, 1984 [1965])

RF

Carnot cycle (1824) *Physics* Named after its originator, the French scientist Nicholas Léonard Sadi Carnot (1796–1832). The

Carnot cycle is a set of four thermodynamically reversible processes that may be performed by any thermodynamic system. It is often used as a model for the working substance of a heat engine that is operating between two heat reservoirs at different temperatures.

The system is imagined to start in thermal equilibrium with a hot reservoir at a temperature T_h. The sequence of processes forming the cycle is then as follows. (1) The system undergoes an isothermal expansion at a temperature T_h and extracts a quantity of heat Q_h from the high-temperature reservoir. (2) The system is then isolated from the reservoir and undergoes an adiabatic expansion until its temperature falls to T_c, that of the cold reservoir. (3) The system is placed in thermal contact with the reservoir at temperature T_c and undergoes an isothermal compression. In this process the system rejects heat Q_c to the reservoir. (4) The isothermal compression is stopped with the system in such a state that, when it undergoes an adiabatic compression, its temperature reaches T_h when it returns to its original volume, thus closing the cycle. An engine operating in a Carnot cycle is called a Carnot engine.

J Thewlis, ed., *Encyclopaedic Dictionary of Physics* (New York, Oxford and London, 1962)

MS

Carnot's theorem (1824) *Physics* In his discussion of the efficiency of heat engines, the French physicist Nicholas Léonard Sadi Carnot (1796–1832) made use of the concept of an ideal heat engine (a Carnot engine) operating between two heat reservoirs at different temperatures. The engine operated in such a way that no finite temperature gradients were set up in its working substance or between the working substance and either reservoir. No frictional processes were present.

On the basis of the non-occurrence of perpetual motion, Carnot was able to demonstrate that: (1) no heat engine operating between two reservoirs can be more efficient than a Carnot engine operating between the same reservoirs (Carnot's theorem); (2) the efficiency of a Carnot engine depends only on the temperatures of the reservoirs and not on the nature of the working substance (the corollary to Carnot's theorem).

J Thewlis, ed., *Encyclopaedic Dictionary of Physics* (New York, Oxford and London, 1962)

MS

case grammar (1966) *Linguistics* Proposed by the American linguist Charles J Fillmore (1929–) as a modification of A N Chomsky's STANDARD THEORY.

The deep structure of sentences is analyzed as a verb accompanied by a set of nouns each in a covert 'case': the cases allow distinctions in the functions of nouns such as 'agent', 'beneficiary', 'location'. (Differs from 'case' in Latin grammar, which is a surface structure inflection on the ends of words.) Fillmore's case grammar as such was never completely developed; but the concept of 'case' remains in recent linguistics; *see also* THETA THEORY.

C J Fillmore, 'The Case for Case', *Universals in Linguistic Theory*, E Bach and R T Harms, eds (New York, 1968), 1–90

RF

Casorati-Weierstrass theorem (19th century) *Mathematics* Named after the Italian mathematician Felice Casorati (1835–90) and German mathematician Karl Theodor Wilhelm Weierstrass (1815–97).

It is the seminal result in COMPLEX FUNCTION THEORY that the set of values of an analytic function arising from any neighbourhood of an essential singularity of the function is dense in the complex numbers.

S G Krantz, 'Functions of One Complex Variable', *Encyclopedia of Physical Science and Technology* (Academic Press, 1987)

MB

Castle-Hardy-Weinberg law *Biology See* HARDY-WEINBERG PRINCIPLE.

catabolic deletion hypothesis *Biology/Medicine* Also called the catabolic enzyme deletion hypothesis, this is the theory that cancer is caused by the loss of at least one catabolic enzyme because of a deletion mutation in deoxyribonucleic acid (DNA). This mutation was thought to increase the availability of polymers, permitting continued cell growth. *See also* CANCER, THEORIES OF and DELETION HYPOTHESIS.

KH

Catalan's constant (19th century) *Mathematics* Named after Belgian mathematician Eugène Charles Catalan (1814–94).

The sum of the series

$$\sum_{n=0}^{\infty}(-1)^n(2n+1)^{-2} = 1 - \frac{1}{9} + \frac{1}{25} - \frac{1}{49} + \ldots$$

which is equal to approximately 0.915965. It is unknown whether this constant is irrational or rational.

ML

catallaxy (20th century) *Politics/Economics* A theory of a desirable society, set out by the Austrian economist and political theorist F A Hayek (1899–1992).

A 'catallaxy' is a market order without planned ends, characterized by the 'spontaneous order' which emerges when individuals pursue their own ends within a framework set by law and tradition. The function of government is to maintain the RULE OF LAW which guarantees fair and equal procedures, but is neutral as to goals. *See also* UTOPIANISM.

F A Hayek, *Law, Legislation and Liberty* (London, 1982)

RB

catalysis law *Chemistry See* BRØNSTED RELATION.

catalyzer *Stylistics See* KERNEL, CATALYZER, INDEX.

catastrophe theory (20th century) *Politics/Sociology/Economics/Mathematics* (1) In politics, a species of theory, rather than a particular theory, derived from the work of applied mathematician René Thom (1923–).

Political or other systems develop up to a point where internal contradictions lead to disruption or 'catastrophe'. Beyond that point the theories which explained the previous order of things will no longer hold. MARXISM applies such a theory to capitalism.

(2) In economics, the theory has its roots in the work of French mathematician Jules Henri Poincairé (1854–1912). A mathematical study of the transition from one state of equilibrium to another and the ensuing instability. The analysis shows how many stable equilibria exist given a choice of control variables but does not show which of them will be in a particular system.

(1) Roger Scruton, *A Dictionary of Political Thought* (London, 1982); (2) P T Saunders, *An Introduction to Catastrophe Theory* (Cambridge, 1980)

RB, PH

catastrophe theory (1966) *Mathematics* Discovered by René Thom (1923–), this theory classifies the classes of surfaces which can be transformed into each other by functions which have regular tangents at each point. The applications use this classification to predict sections of the surfaces where sudden changes occur and suggest methods for avoiding them, as one may sometimes avoid jolting down a staircase by using a ramp instead.

T Poston and I Stewart, *Catastrophe Theory and its Applications* (London, 1978)

JB

catastrophic event theories of the solar system (18th–20th centuries) *Physics*

Theories proposing the formation of the planets during a catastrophic encounter between the Sun and another object. The first such theory by de Buffon in 1745 postulated that a comet collided with the Sun. Most later theories, such as the planetesimal theory (1901–05) of Chamberlain and Moulton and the tidal theories (1916–18) of Jeffreys and Jeans, assumed an approach or collision involving another star. Woolfson's capture theory attempted to overcome objections that ejected hot material would have dissipated into space before it could form planets by proposing that material was drawn off a relatively cool tenuous protostar moving past the Sun. More recently, the contracting nebula hypothesis, originally proposed in the 19th century, has begun to be revived.

Valerie Illingworth, ed., *Macmillan Dictionary of Astronomy* (London, 1985)

GD

catastrophism (late 18th century) *Biology/Geology* Developed by the Swiss zoologist George Cuvier (1769–1832) and English geologist William Whewell (1794–1866), this theory posited Earth history as a succession of worlds, each separated by periods when sudden catastrophic and supernatural events (usually violent floods) were dominant and, by definition, inexplicable.

The theory accounted for mass extinctions of species (evidenced by fossil remains), and suggested that the present world was established after the biblical Flood. It was directly opposed by adherents of UNIFORMITARIANISM. Whilst the geological record bears witness to sudden catastrophies in Earth history, it is now recognized that sudden results can also result from a series of natural, gradual progressions. *See also* DILUVIANISM.

L Eisely, *Darwin's Century* (New York, 1958); R Porter, *The Making of Geology: Earth Science in Britain, 1660–1815* (Cambridge, 1977)

KH, DT

catecholamine hypothesis *Psychology* Not attributable to any one originator, the term describes that body of evidence which suggests a biochemical basis for schizophrenia: the symptoms of the psychosis are the result of a build-up of catecholamines in particular synaptic clefts in the brain.

The hypothesis is supported by evidence that chronic use of amphetamines produces a 'model' paranoid schizophrenic psychosis and that substances that function to mildly increase the level of the amines have antidepressant effects. Therapy is directed towards restoring a normal balance of catecholamines.

H Kaplan, I Freedman and B H Sadock, *Comprehensive Textbook of Psychiatry* (Baltimore, 1980)

NS

categorial grammar (1930s) *Linguistics* Developed by the Polish logicians Kazimierz Ajdukiewicz and Stanislaw Leśniewski (1886–1936), so named by the Israeli linguist Yehoshua Bar-Hillel in 1970.

System of grammatical analysis with two fundamental categories Sentence and Name from which Functors or functions are derived. A sentence can be analyzed as a combination of a predicate function and the arguments which complete it. There was renewed interest in categorial grammar in the 1980s, but disagreement on what counts as a basic category. *See also* MEREOLOGY.

R T Oerle and E Bach, and D Wheeler, eds, *Categorial Grammars and Natural Language Structures* (Dordrecht, 1988)

RF

categorical imperative *Philosophy* Term from German philosopher Immanuel Kant (1724–1804), who claimed to derive morality – in the form of an imperative valid for all rational beings – from reason.

The general idea was that I may not act in ways that I cannot, without inconsistency, will that everyone else should act in too. Suppose that to gain some advantage I make a promise, intending not to keep it: were I to will that everyone may break their promises, I would in effect be willing for the institution of promising to break down; in which case I could not use it as I originally tried to. The imperative is categorical because not conditional on one's own desires, like 'If you want money, work hard' (an imperative which Kant would call hypothetical). Problems arise in deciding when the categorical imperative does indeed apply, and whether hypothetical imperatives are really imperatives at all. *See also* DEONTOLOGY.

H J Paton, *The Moral Law* (1948); standard translation of Kant's main relevant work

ARL

category mistake *Philosophy* Term introduced by English philosopher Gilbert Ryle (1900–76) for cases where we talk of something in terms appropriate only to something of a radically different kind.

For example, 'The Prime Minister is in London, and the Foreign Secretary is in Paris, and the Home Secretary is in Bristol, but where is the Government?' The Government is not another person alongside its members. Ryle used the notion primarily to claim that mind and body cannot be spoken of in parallel ways, but are in different 'categories'. One problem is to say when things are indeed in different categories.

G Ryle, *The Problem of Mind* (1949), ch. 1

ARL

category theory (20th century) *Mathematics* An abstract theory which arose out of the work of American mathematicians Samuel Eilenberg (1913–) and Saunders MacLane (1909–) in algebraic TOPOLOGY during the

1940s. It provides a convenient language of unification for much of 20th-century ALGEBRA. *See also* BAIRE CATEGORY THEOREM.

T W Hungerford, *Algebra* (Springer, 1974)

MB

catharsis (*c*.330 BC) *Literary Theory* Due to Greek philosopher Aristotle (384–322 BC).

'Purging' of the audience's emotions of pity and fear induced by a tragic drama, through controlled exposure to those same emotions. *See also* TRAGEDY.

Aristotle, *Poetics*, S H Butcher, trans., *Critical Theory since Plato*, H Adams, ed. (New York, 1971), 48–66

RF

Cauchy condensation test (19th century) *Mathematics* Named after French mathematician and physicist Baron Augustin Louis Cauchy (1789–1857), this is a test for the convergence of an infinite series.

If $\{p_n\}$ is a sequence of decreasing positive terms, then the series Σp_n and $\Sigma 2^n p_{2^n}$ either both converge or both diverge. The second series is viewed as a condensation of the first.

K Knopp, *Theory and Application of Infinite Series* (New York, 1990)

ML

Cauchy criterion (19th century) *Mathematics* Named after French mathematician and physicist Baron Augustin Louis Cauchy (1789–1857).

(1) The necessary and sufficient condition for an infinite sequence of numbers to converge is that the absolute value of the difference between two terms with sufficiently large indices tends to zero. Such a sequence is called a Cauchy sequence.

(2) More generally, conditions for the convergence of a series, improper integrals, a family of functions, and so on, which are established by verifying that a Cauchy sequence is obtained and then utilizing the completeness of the underlying metric space.

M Hazewinkel, ed., *Encyclopaedia of Mathematics* (Dordrecht, 1988)

ML

Cauchy-Hadamard theorem (1821) *Mathematics* Named after French mathematicians Baron Augustin Louis Cauchy (1789–1857), who stated the theorem in his lectures, and Jacques Hadamard (1865–1963) who made the formulation and proof explicit.

The theorem states that the complex power series

$$\sum_{k=0}^{\infty} a_k(z-c)^k$$

has radius of convergence

$$\frac{1}{\limsup_{n\to\infty} |a_n|^{1/n}} .$$

J B Conway, *Functions of One Complex Variable* (New York, 1973)

ML

Cauchy integral theorem (19th century) *Mathematics* Named after French mathematician and physicist Baron Augustin Louis Cauchy (1789–1857), this is the result (of fundamental importance in COMPLEX FUNCTION THEORY) whereby the complex line integral of an analytic function around a simple closed curve is zero. *See also* RESIDUE THEOREM OF CAUCHY.

S G Krantz, 'Functions of One Complex Variable', *Encyclopedia of Physical Science and Technology* (Academic Press, 1987)

MB

Cauchy-Kowalewska theorem (*c*.1850) *Mathematics* Named after mathematicians Baron Augustin Louis Cauchy (1789–1857), who proved a version for one equation, and Sonja Kowalewska (1850–91), who proved the general result by a different method.

Let S be a system of r partial DIFFERENTIAL EQUATIONS in r unknown functions $v_1, v_2, \ldots, v_k$ which are functions of the $p+1$ variables $x_1, x_2, \ldots, x_{p+1}$ in the form

$$D_{p+1}v_j = h_j(x_1, x_2, \ldots, x_{p+1}, v_1, v_2, \ldots, v_r, D_1v_1, D_2v_1, \ldots, D_iv_k, \ldots, D_pv_r),$$

where D_i represents partial differentiation with respect to x_i and the functions h_j can be expressed as a Taylor's series in its variables, which do not include partial differential coefficients with respect to x_{p+1}. Then S has a unique solution in which $v_1, v_2, \ldots, v_r$ have

Taylor's series in the variables x_1, $x_2,\ldots,x_{p+1}$. This is the only general result on systems of partial differential equations, but the systems that occur in applications very rarely satisfy the conditions of the theorem.

JB

Cauchy lemma (19th century) *Mathematics* Named after French mathematician and physicist Baron Augustin Louis Cauchy (1789–1857), this is the result in GROUP THEORY whereby if G is a finite group and p is a prime number which divides the order of G, then G contains an element of order p. *See also* SYLOW'S THEOREMS.

I N Herstein, *Topics in Algebra* (Wiley, 1975)

MB

Cauchy mean value theorem (19th century) *Mathematics* Named after French mathematician and physicist Baron Augustin Louis Cauchy (1789–1857), but also known as the generalized mean value theorem. This is the result whereby if f and g are functions which are differentiable in an interval (a,b) and continuous on $[a,b]$, then

‹P/S›

for some point c in the open interval (a,b).

M Spivak, *Calculus* (1980)

MB

Cauchy-Schwartz inequality (19th century) *Mathematics* Named after French mathematician and physicist Baron Augustin Louis Cauchy (1789–1857) and German analyst and complex function theorist Hermann Amandus Schwartz (1843–1921).

The inequality

‹P/S›

which is true for any inner product. In Euclidean space it is known as the Cauchy inequality. The complex version is known as Schwartz's inequality and the integral version is known as the Bunyakovsii inequality.

G B Folland, *Real Analysis* (New York, 1984)

ML

Cauchy's stress theorem (19th century) *Physics* Named after its discoverer, French mathematician Baron Augustin Louis Cauchy (1789–1857), this states that the force at a point on a surface of a body of fluid is given by sn where s is a symmetric second-order Cartesian tensor and n is a unit vector normal to the surface. This is the fundamental hypothesis of fluid mechanics.

JB

causal principle *Philosophy* Name for a variety of principles, such as that every event has a cause, that the same cause must have the same effect, or that the cause must have at least as much reality as the effect.

This last principle (somewhat akin to the principle of sufficient reason) usually says that what causes something to be of a certain sort must itself be of that sort to at least the same degree; for example, what makes something hot must itself be hot. This goes back to Aristotle's principle that actuality is prior to potentiality: that is, what is potentially so-and-so can only be made actually so by something that is itself actually so.

ARL

causal realism *Philosophy* The view that substantive causal connections exist in reality, as opposed to the reductionist approach of the REGULARITY THEORY OF CAUSATION.

ARL

causal theories *Philosophy* Any theory which analyzes a concept in terms of causation can be called a causal theory of that concept. In particular, causal theories have been offered of knowledge, meaning, memory, perception, and reference. All such theories can of course exist in different versions.

ARL

causal theories of meaning *Philosophy* Theories which explain the meaning of a word or sentence in terms of its effect on the hearer, or in terms of the cause of its utterance by the speaker.

Such theories are also sometimes called 'stimulus/response theories', and they have some kinship with BEHAVIOURISM. An objection is that most such views ignore the roles of intention and insight in genuine, as against merely apparent, communication.

H P Grice, 'Meaning', *Philosophical Review* (1957); reprinted in P F Strawson, ed., *Philosophical logic* (1967)

ARL

causal theories of perception *Philosophy* Any theory which says that the object of perception plays a causal role in the perception itself.

The object may cause us to have a certain experience without itself being perceived (we may have to infer its existence, or 'construct' it from experiences rather as we 'construct' the average man from real men: *see also* PHENOMENALISM). Or we may perceive the object but our experience in doing so only counts as perceiving it if it is itself caused by the object. Or it may simply be that whenever we do perceive an object it has a causal role to play in our doing so (without that role forming part of the analysis of perception). Objections to the first two views include that of ensuring that the causal chain is of the right kind.

A J Ayer and L J Cohen, 'The Causal Theory of Perception' (symposium), *Proceedings of the Aristotelian Society*, supplementary volume (1977)

ARL

causal theories of reference *Philosophy* Any theory saying that if we are to refer to an object we must be in some relevant causal contact with it. We cannot therefore refer to fictitious objects, but must be using their names in some other way.

Suppose that I try talking about one 'Ebenezer Pilkington, who is "F"' (where 'F' is some elaborate description, satisfied by one person only, who happens also to be the only person called 'Ebenezer Pilkington'); and suppose that all this is pure coincidence and I have never heard of him or been in any causal contact with him. On the views in question I am not referring to him, whatever else I am doing. These views are akin to the CAUSAL THEORY OF NAMES, though not the same since only people refer.

S Kripke, *Naming and Necessity* (1980)

ARL

causal theory of knowledge *Philosophy* Any theory which says that to know a truth one must believe it and one's belief must stand in a certain causal relation to the truth itself.

For example, I know that Caesar crossed the Rubicon if his doing so caused some historian to write a book saying so, which caused my local library to buy it, which caused me to read and believe it. The causal connection might be more complex than a simple chain, and the knower might have to make some inferences. Objections include the case of timeless truths like those of mathematics, which do not seem to cause anything; and the possibility that the causal chain might be of the wrong sort, so that intuitively one would not say that here was a case of knowledge. (This entry ignores the distinction between facts and events).

A I Goldman, 'A Causal Theory of Knowing', *Journal of Philosophy* (1967)

ARL

causal theory of memory *Philosophy* Any theory holding that for me to remember something some present mental experience of mine (or perhaps some present piece of behaviour of mine) is causally related to something relevant in the past.

This 'something relevant' may be what is remembered, but may also be something merely connected with that: I remember to put the cat out if I am caused to do so by some previous intention of mine to do so, but this intention is not what I am remembering. Difficulties arise over ensuring that the causal chain is of the right kind, and this leads causal theories to be specified in terms of the TRACE-THEORY OF MEMORY.

C B Martin and M Deutscher, 'Remembering', *Philosophical Review* (1966); critical

ARL

causal theory of names *Philosophy* Theory advanced especially by American philosophers Saul Kripke (1940–) and Hilary Putnam (1926–) that whether a currently used name names a certain object depends on whether current use of the name causally depends on its use by people who originally dubbed the object with that name.

'Homer' names whatever person the Greeks used it (or a Greek variant of it) to address (even if that person was not a poet at all). 'Homer' does not mean (as the rival DESCRIPTIVE THEORY OF NAMES holds) 'whoever wrote the *Iliad* and *Odyssey*'. Kripke and Putnam also extend the theory to cover words for 'natural kinds', like 'tiger' or 'water'. 'Water' names whatever stuff it was first applied to, and does not mean 'H_2O' or 'colourless tasteless liquid', and so on. *See also* CAUSAL THEORIES OF REFERENCE.

S P Schwartz, ed., *Naming, Necessity and Natural Kinds* (1977)

ARL

Cavalieri's principle (1635) *Mathematics* Named after Italian mathematician and physicist Francesco Bonaventura Cavalieri (1598–1647), this statement was familiar to the ancient Greeks.

The principle states that the volumes of two solids are equal if they have the same height and if the cross-sections have the same area.

M Hazewinkel, ed., *Encyclopaedia of Mathematics* (Dordrecht, 1988)

ML

Cayley representation theorem (19th century) *Mathematics* Named after British mathematician Arthur Cayley (1821–95), this is the result whereby every group is isomorphic to a subgroup of the symmetric group. (The symmetric group on *n* letters can be considered as the set of permutations of the numbers 1,2,...,*n*; that is, the set of bijective functions on this set.) The result says that the study of groups can be essentially reduced to the study of permutation groups. *See also* ISOMORPHISM THEOREM and GROUP THEORY.

I N Herstein, *Topics in Algebra* (Wiley, 1975)

MB

Cayley-Hamilton theorem (19th century) *Mathematics* Named after British mathematician Arthur Cayley (1821–95) and Irish mathematician William Rowan Hamilton (1805–65), this is the result in LINEAR ALGEBRA whereby every square matrix A satisfies its own characteristic equation (that is, the polynomial in x which is defined as the determinant of the matrix $xI - A$). This result is often useful when calculating inverse matrices.

I N Herstein, *Topics in Algebra* (Wiley, 1975)

MB

ceiling effect *Psychology* Term coined by the American psychologist Lee Joseph Cronbach (1916–).

This phrase is used in a variety of ways but usually to indicate the maximal score on a test or a limit on the performance of a task. 'Ceiling effect' refers to: (1) the absence of improvement in performance due to the fact that the individual is already functioning at maximum capability (the reverse being termed a 'basement effect'); (2) the restriction on a subject's score as he/she approaches this theoretical ceiling; (3) in pharmacology, the maximum dose of a drug that will produce the desired effect.

A S Reber, *Dictionary of Psychology* (Harmondsworth, 1986)

NS

cell theory (1838–39) *Biology* Advanced by German physiologist Theodor Schwann (1810–82), based on work by Matthias Jakob Schleiden (1804–81), this theory holds that cells are the basic building blocks of all living things. Reproduction and growth occur through cell division. Cell theory is a fundamental tenet of modern biology.

W F Bynum, E J Browne, R Porter, *Dictionary of the History of Science* (Princeton, NJ, 1981); T Schwann, *Mikroskopische Untersuchungen* (Microscopical Investigations) (1839)

KH

cell theory of ageing *Biology See* HAYFLICK LIMIT.

cellular automaton (1950s) *Mathematics* Introduced by American mathematician John von Neumann and Polish-born American mathematician Stanislaw M Ulam (1909–84), this is a mathematical model of physical phenomena which is based on the idea of an array of cells, each of which represents an automaton or abstract machine which can take on a finite number of states. It is the mathematical equivalent of an array of robots, each of which is programmed to perform specific tasks. Each cell is simple, but together the system is capable of complex behaviour. *See also* AUTOMATA THEORY.

I Peterson, *The Mathematical Tourist* (New York, 1988)

JB

cellular origin theory *Biology* The idea that viruses composed of deoxyribonucleic acid (DNA) were once the components of living cells that have broken away to exist independently. This theory represents the favoured explanation for the emergence of

DNA viruses. *See also* RETROGRADE EVOLUTION. *Compare with* PROTOVIRUS HYPOTHESIS.

H Fraenkel-Conrat, *Virology* (Englewood Cliffs, NJ, 1988)

KH

Celsius temperature (1740) *Physics* This refers to temperature measured on a scale devised by Anders Celsius (1701–44). The unit on this scale is the degree celsius, symbol °C. On this scale the melting-point of ice is given the value zero degrees celsius (0°C), and the boiling-point of water under a pressure of one atmosphere the value of 100°C. (In his original scale, Celsius had the values inverted.) The scale has been superseded by the International Temperature Scale and now a temperature t°C is defined by the equation

$$t/°\mathrm{C} = T/K - 273.15$$

where T is the thermodynamic temperature, measured in kelvins, symbol K. *See* THERMODYNAMIC SCALE OF TEMPERATURE.

J Thewlis, ed., *Encyclopaedic Dictionary of Physics* (New York, Oxford and London, 1962)

MS

central dogma (20th century) *Biology* Formulated after the British scientist Francis Harry Compton Crick (1916–) and the American James Dewey Watson (1928–) explained the structure and function of deoxyribonucleic acid (DNA), this essential concept of modern genetics states that genetic information passes in only one direction, from deoxyribonucleic acid (DNA) to the protein.

More formally, it is the idea that DNA serves as a template both for its own duplication and for the synthesis of ribonucleic acid (RNA), which in turn serves as the template for protein synthesis. Thus, a change in a protein sequence cannot cause a subsequent change in the nucleic acid sequence.

J Watson, *The Double Helix: A Personal Account of the Discovery of the Structure of DNA* (London, 1981)

KH

central limit theorem (1812) *Mathematics* The first version is called the Laplace theorem and was proved by French mathematician Pierre Simon, Marquis de Laplace (1749–1827). A special case was studied by French mathematician Abraham de Moivre (1667–1754) in 1730.

A fundamental result of PROBABILITY THEORY, this states that the sum or mean of a sufficiently large sequence of independent identically distributed *random variables* having finite expectations and variances has a probability distribution that is approximately normal. Therefore, in particular, if a large enough sample is drawn from a population the sum or mean of the sample values can be treated as if they came from a normally distributed random variable.

S Ross, *Probability Theory* (New York, 1976)

ML

central neural theory of emotion *Psychology* *See* CANNON-BARD THEORY OF EMOTION.

central place theory (1933) *Economics* Developed by the German economic geographer Walter Christaller (1894–1975), this theory studies how cities and towns develop hierarchies of economic activity from the population size and the distance inhabitants are prepared to travel for goods and services.

The central place is frequently located at the geographical periphery of a region, and its location is determined by applying the laws of marketing, distribution and traffic. *See* GRAVITY MODEL, LEAST COST LOCATION THEORY and WEBER'S THEORY OF THE LOCATION OF THE FIRM.

W Christaller, *The Central Places of Southern Germany* (Englewood Cliffs, N.J., 1966); A Losch, *The Economics of Location* (Newhaven, Conn., 1954)

PH

centralization *Politics* A belief in the efficiency of the concentration of powers.

Organizations will run more effectively, and policies be formed and implemented more efficiently, if power is exercized centrally. This is believed to facilitate coherent action, and the most appropriate mobilization and application of resources. It is contrasted with SUBSIDIARITY.

RB

centres of origin (1926) *Biology* A theory proposed by the Russian plant geneticist Nikolai Ivanovich Vavilov (1887–1943).

Vavilov argued that the region in the world where the greatest genetic diversity of a single species occurs (the centre of diversity) is also the region where this species evolved: the centre of origin. Assuming natural mutations (which give rise to genetic diversity) occur at a constant rate, then the largest number of mutations should occur where the species has existed longest and therefore where it evolved. Although the two can often broadly correlate, the relationship does not hold as a rule and does not allow for species which do not have a continuous geographical range.

N I Vavilov, 'Studies in the origins of cultivated plants', *Bulletin of Applied Botany and Plant Breeding*, vol. XVI (1926), 1–245

RA

centrifugal speciation (1957) *Biology* A theory proposed by the American biologist W L Brown Jr.

The most likely place for a new, or derived, species to evolve is towards the centre of the progenitor species' geographical range. This is because the genetic diversity of any species is greatest in the centre of its range, where because of this the greatest number of genetically deviant organisms will occur. In an uneven population density, new species will become established in the spaces, and with a higher frequency in the central region. This is an accepted mechanism of speciation.

W L Brown, 'Centrifugal speciation', *Quarterly Review of Biology*, vol. XXXII (1957), 247–77

RA

cerebral localization (1836) *Linguistics* Observation of the 19th century Frenchman M Dax, developed by modern neurolinguistics from *c.*1940.

The language faculty seems to be usually located in the left cerebral hemisphere, the dominant hemisphere for most people (determining handedness also), though linguistic functions of the right hemisphere should not be discounted. There is some evidence that particular aspects of language are controlled by more precise locations in the brain: *see also* BROCA'S AREA and WERNICKE'S AREA.

M Garman, *Psycholinguistics* (Cambridge, 1990)

RF

CES production function (1961) *Economics* Proposed by the American economist Kenneth Arrow (1921–), Hollis Chenery (1918–), B Minhas, and Robert Solow (1924–), this is also known as the constant elasticity of substitution function.

This is a linearly homogenous production function with a constant elasticity of input substitution, which takes on forms other than unity. It replaced the COBB-DOUGLAS PRODUCTION FUNCTION model which looked at physical output as a product of labour and capital inputs. The equation for the CES model is:

$$Q = A(ak - b^{-b} + (1 - c)L - b^{-b}) - 1/b$$

where Q is output, K capital and L labour and a, b, c are constants.

K Arrow, H Chenery, B Minhas and R Solow, 'Capital-Labour Substitution and Economic Efficiency', *Review of Economics and Statistics*, 43, 3 (August, 1961), 225–50

PH

Cesaro summation (1890) *Mathematics* Named after Italian analyst and geometer Ernesto Cesaro (1859–1906), this is a method of SUMMABILITY THEORY to compute the limit of a possibly divergent sequence of numbers as the limit, as n approaches infinity, of the means of the first n numbers. If the sequence is convergent, this limit agrees with the original limit. For example, this method gives a limit of 1/2 to the sequence 1,0,1,0,....

K Knopp, *Theory and Application of Infinite Series* (New York, 1990)

ML

Ceva's theorem (1678) *Mathematics* Named after Giovanni Ceva (1647–1734), its discoverer. In the same book he proved a theorem which had been stated by Menelaus in about 98 AD and which is the dual of this, in that the statement is obtained by exchanging the words 'line' and 'point' and making consequent linguistic changes.

Let the sides BC, CA, AB of the triangle ABC be divided at R, S, T in the ratios r, s, t

to *1*. Then the three lines *AR*, *BS*, *CT* are concurrent if and only if $rst = 1$. *See also* MENELAUS'S THEOREM.

JB

chain of being (4th century BC–18th century) *Biology/Philosophy* Also called the great chain of being and *scala natura*. Based on ideas of Plato (*c*.427–*c*.347 BC) and Aristotle (384–322 BC), but popularized in biology in the writings of German philosopher Gottfried Leibniz (1646–1716), French naturalist Georges Louis Leclerc, Comte de Buffon (1720–88), and Swiss philosopher Charles Bonnet (1720–93). This is the influential concept that all of nature – from non-living matter to sophisticated organisms to spiritual beings – forms an unbroken physical and metaphysical series.

The theory's biological significance developed in the 18th century, when species were arranged in a graded series or hierarchy. Certain species, such as the green hydra, were thought to be crucial links in the series. The notion that particular metals, animals or classes of humans outrank others is no longer accepted as part of mainstream science. *See also* ORTHOGENESIS and SPECIESISM.

A Lovejoy, *The Great Chain of Being: A Study of the History of an Idea* (Cambridge, Mass., 1936)

KH

chain rule (17th century) *Mathematics* Attributed to British physicist, astronomer, and mathematician Sir Isaac Newton (1642–1727), this theorem of calculus is used in differentiation of a function which states that

$$\frac{dy}{dx} = \frac{dy}{dt} \cdot \frac{dt}{dx}$$

where y is a differentiable function of t and t is a differentiable function of x. Similarly, for partial differentiation of a function the theorem states that

$$\frac{\partial f}{\partial x} = \left(\frac{\partial f}{\partial u} \cdot \frac{\partial u}{\partial x}\right) + \left(\frac{\partial f}{\partial v} \cdot \frac{\partial v}{\partial x}\right)$$

where f is a function of u and v, each of which is a function of x.

H Anton, *Calculus with Analytic Geometry* (New York, 1980)

ML

Chamberlin-Moulton hypothesis *Physics* *See* PLANETESIMAL THEORY OF PLANETARY FORMATION.

Chandrasekhar's limit (1931) *Astronomy* Named after the American astronomer Subrahmanyan Chandrasekhar (1910–).

A star is prevented from collapsing if its kinetic energy (radiation pressure) dominates its gravitational potential energy. However, the gravitational potential energy increases as the mass of the star increases until a critical mass is reached at which the star collapses. For a star consisting of N particles, gravitational collapse is inevitable when

$$N = (hc/2\pi Gm^2)^{3/2} \approx 2 \times 10^{57}$$

where h is Planck's constant, c is the speed of light in a vacuum, G is the universal gravitational constant and m is the particle mass. The number N is known as the Chandrasekhar limit and the corresponding mass ($=mN$) as the Chandrasekhar mass.

S Mitton, ed., *The Cambridge Encyclopaedia of Astronomy* (London, 1973)

MS

chaos theory (1960s) *Mathematics* The study of phenomena which appear random, but in fact have an element of regularity which can be described mathematically.

Chaotic behaviour has been found to exist in a wide range of applications, such as seizures or epilepsies, atmospheric prediction models and fibrillation of the heart. *See also* KNEADING THEORY, STRANGE ATTRACTOR, FEIGENBAUM PERIOD-DOUBLING CASCADE and RUELLE-TAKENS SCENARIO.

R L Devaney, *An Introduction to Chaotic Dynamical Systems* (New York, 1989)

ML

character (*c*.330 BC) *Literary Theory* Originated with the Greek philosopher Aristotle (384–322 BC).

Fictional illusion of the existence of a person participating in a narrative. In classical times characters were treated as 'types'; modern humanism in, typically, the 19th and 20th century novel has valued the illusion of individuality, detail of personality. The idea is disfavoured under POSTSTRUCTURALISM. *See also* FLAT AND ROUND CHARACTERS.

W J Harvey, *Character and the Novel* (Ithaca, 1965)

RF

Chargaff's rules (1950) *Biology* Stated by Austrian-American biochemist Erwin Chargaff (1905–), these are generalizations about the relationship among the various components of deoxyribonucleic acid (DNA).

More formally, in a DNA sample of duplex molecules from any source, the number of adenine residues will (approximately) equal the number of thymines, while the number of guanines will equal the cytosines. Thus the ratio of purines (adenine and guanine) to pyrimidines (thymine and cytosine) will always equal approximately 1. By contrast, Chargaff observed that the (A + T)/(G + C) ratio varied considerably. Also, he noted that the nitrogenous base composition is constant for cells from different organs of the same species, and is characteristic of the DNA of that species.

H Judson, *The Eighth Day of Creation* (New York, 1979)

KH

charity, principle of *Philosophy* Principle named by N L Wilson – in *Review of Metaphysics* (1958–59), page 532 – that when interpreting another speaker, especially of an unknown language, we should make those assumptions about his intelligence, knowledge, sense of relevance and so on, that will make most of what he says come out true. *See also* HUMANITY, PRINCIPLE OF.

G Sundholm, 'Brouwer's Anticipation of the Principle of Charity', *Proceedings of the Aristotelian Society* (1984–85), 264–68

ARL

Charles's law (1787 and 1802) *Chemistry* Law proposed by French chemist Jacques Alexandre César Charles (1746–1823) in 1787 and developed and published by French scientist Joseph Gay-Lussac (1778–1850) in 1802. Also known as the Gay-Lassac law, this states that at constant pressure, gas volume (V) of a given mass of gas is proportional to its absolute temperature (T); that is,

$$V \propto T$$

Charles's law was not originally published. It was developed by Gay-Lassac who, in 1802, stated that the volume of a gas changes by $1/273$ of its volume at 0°C for every 1°C change in temperature. The law is valid only for an ideal gas. *See also* GAS LAWS.

M Freemantle, *Chemistry in Action* (Basingstoke, 1987)

MF

charm theory *Physics* Charm is a quantum number used in the theory of quarks and hadrons. The charm quark has charm +1 and the charm antiquark has charm −1. All other quark flavours have charm 0. The charm of a particle is the number of charmed quarks minus the number of anticharmed quarks.

Paul Davies, ed., *The New Physics* (Cambridge, 1989)

GD

Charpentier's law *Psychology* Named after the Frenchman, P M A Charpentier (1852–1916).

Also referred to as Charpentier's illusion, the law states that the product of an object's light intensity and of the stimulus or image area (size of the retina, dependent on the angle of vision) is a constant for the magnitude of the threshold value of light sensitivity (optical threshold stimulus). It is a visual-perception rule that the product of the foveal image area and the intensity of light is constant for threshold stimuli. (AI = K where A is the area of the image and I is the intensity of the stimulus and K is a constant.)

L Kaufman, *Perception* (New York, 1979)

NS

Chebyshev's theorem *Mathematics* A form of the WEAK LAW OF LARGE NUMBERS.

chemical combination laws *Chemistry* Also known as the chemical laws. The composition of chemical compounds and the changes in composition that take place during chemical reactions are governed by four important laws. These are called the laws of chemical combination and are the: (1) LAW OF CONSERVATION OF MATTER; (2) LAW OF CONSTANT COMPOSITION; (3) LAW OF MULTIPLE PROPORTIONS; (4) LAW OF RECIPROCAL PROPORTIONS.

M Freemantle, *Chemistry in Action* (Basingstoke, 1987)

MF

chemical equilibrium, law of *Chemistry See* EQUILIBRIUM LAW.

chemical laws *Chemistry See* CHEMICAL COMBINATION LAWS.

chemiosmotic coupling hypothesis *Biology See* CHEMIOSMOTIC THEORY.

chemiosmotic theory (1961) *Biology* Also known as the chemiosmotic coupling hypothesis, this was proposed by British biochemist Peter Dennis Mitchell (1920–), who received the Nobel prize for Chemistry in 1978.

It posits that hydrogen ions are pumped across the inner membrane of the mitochondria (that part of an animal cell responsible for energy) or across the thylakoid membrane of chloroplasts (which serve the same function in plants) as a consequence of electrons passing through the electron transport chain. The flow of protons that results supplies the energy required to form adenosine triphosphate (ATP), the nucleotide that stores energy and provides the mechanism for energy exchange in virtually all living cells. Although rejected for many years, Mitchell's theory now enjoys wide acceptance.

W Beck, K Liem and G Simpson, *Life: An Introduction to Biology* (New York, 1991)

KH

chiaroscuro (1681) *Art* Meaning 'bright/dark', this term was first used by the Italian art historian Filippo Baldinucci (*c.*1624–96) to describe paintings which rely on the dramatic effects of contrasting light and shadow in their composition. The foremost exponents of this style were Michel Angelo Merisi Caravaggio (1569–1609) and Rembrandt (1606–69).

AB

chiasmatype hypothesis (1909) *Biology* Proposed by F A Janssens. This is the theory in genetics that crossing over between nonsister chromatids results in the formation of chiasmata, the familiar cross-shaped joints between nonsister chromatids. This theory provided an important foundation for later research in genetics.

A Sturtevant, *A History of Genetics* (New York, 1965)

KH

Chicago critics (1930s–) *Literary Theory* Group of critics practising at the University of Chicago, notably Ronald Salmon Crane (1886–1967), Elder Olson (1909–), Wayne C Booth (1921–).

Otherwise called 'Chicago Neo-Aristoteleans' for their adherence to the poetic theory of Aristotle (384–322 BC). Opposed to NEW CRITICISM (whose proponents they regarded as a-theoretical, subjective, and over-preoccupied with language), they based their work on Aristotle's idea of the literary work as a concrete artistic whole, an organic union of identifiable components, an example of a specific GENRE, and an IMITATION of significant human actions.

R S Crane, ed., *Critics and Criticism* (Chicago, 1952)

RF

chicken (20th century) *Politics* An application of GAME THEORY.

Political conflicts arise which are analogous to the game of chicken in which two contestants will, for instance, drive towards each other in motor cars. If both hold their course, both will crash. The one who saves herself, and so her opponent, is the 'chicken'. As in other 'games', the action of each participant is seen to be dependent in part on their estimate of the likely behaviour of the other. The Cuban missile crisis of 1962 is often described in these terms.

Robert Abrams, *Foundations of Political Analysis* (New York, 1979)

RB

Child's law (1911) *Physics* Also known as the Child-Langmuir law (named after C D Child and American chemist Irving Langmuir (1881–1957), its formulators) this law relates to the current flowing in a thermionic diode.

When the current is controlled by the space charge (that is, the repulsions between the negatively-charged electrons limit the current), the current I flowing in a diode with a potential difference V between the cathode and anode is given by

$$I = CV^{3/2}$$

where C is a constant. The law is only a good approximation as its theoretical derivation neglects the Maxwellian distribution of the initial speeds of the electrons. *See* MAXWELL'S EQUATIONS.

J Thewlis, ed., *Encyclopaedic Dictionary of Physics* (New York, Oxford and London, 1962)

MS

chilling injury, theory of *Medicine See* LIPID-MEMBRANE HYPOTHESIS OF CHILLING INJURY.

Chinese remainder theorem *Mathematics* Apparently known to the Chinese at least 1500 years ago, this is the result whereby if $n_1, n_2, \ldots, n_k$ are natural numbers which are coprime in pairs (that is, the greatest common divisor of n_i and n_j is 1, for $i \neq j$), then for any integers $c_1, \ldots, c_k$ there is an integer x such that n_j divides $x - c_j$ for all j with $1 \leqslant j \leqslant k$. Equivalently, the system of linear congruences $x \equiv c_j \pmod{n_j}$, $1 \leqslant j \leqslant k$ has a solution; in fact, there is a unique solution modulo $n = n_1, \ldots, n_k$.

A Baker, *A Concise Introduction to the Theory of Numbers* (Cambridge, 1984)

ML

Cholodny-Went model (1926) *Biology* Suggested by N Cholodny and demonstrated by the Dutch-American botanist Frits W Went (*ob.*1990).

The theory that light on one side of a plant causes auxin, a plant hormone, to be transported towards the shaded side where it induces growth. While evidence for the model is quite strong, some research reports since the 1970s have cast doubt on the theory.

F Salisbury, *Plant Physiology* (Belmont, Calif., 1992)

KH

chosisme (1950s) *Literary Theory* French for 'thingism', this is a descriptive technique favoured by practitioners of the 'NOUVEAU ROMAN'.

A kind of hyper-realism which employs very detailed decriptions of the physical appearance of inanimate objects, this is a style of writing practised particularly by the French novelists Alain Robbe-Grillet (1922 –) and Michel Butor (1926–). *See* TROPISM.

M Butor, *La modification* (Paris, 1957)

RF

Christian socialism (19th century) *Politics/Religion* Theory of the social responsibility of the Christian Church, developed within the Anglican Church in Britain in the 19th century.

Christian teaching has social and economic implications, and the Church should take a lead in both improving the conditions of the working class and providing moral and cultural leadership to allow its members a fuller participation in the life of society.

E Norman, *The Victorian Christian Socialists* (Cambridge, 1987)

RB

chromaticism (16th century–) *Music* Derived from the Greek word for colour, this is the notion that in tonal music certain notes foreign to those belonging to a key may be present without undermining the key of the piece.

Chromatic notes are the five pitches that, together with the seven pitches of a major or minor scale ('diatonic'), make up a complete chromatic scale (12 pitches to an octave, each a semitone apart). In music notated using a key signature, chromatic notes are indicated by an 'accidental' placed in front of a diatonic note. Extreme chromaticism ultimately leads to ATONALITY.

MLD

chromosome theory of cancer (1912) *Biology* Also called the mutation theory of cancer, this is attributed to German biologist Theodor Boveri (1862–1915), although mitotic irregularities in cancerous tissues had been observed in 1890 by David von Hansemann. This is the theory that cancer is caused by abnormal chromosomes in cells as a result of irregularities in mitosis, the usual method of cell division.

The first experimental support for this theory in humans occurred in 1960, with the discovery of the Philadelphia chromosome in patients with chronic myeloid leukemia. *Compare with* SOMATIC MUTATION THEORY.

T Bovari, *Zur Frage der Entstehung Maligner Tumoren* (The Origin of Malignant Tumours) (1914); J Rowley, *Chromosomes and Cancer: From Molecules to Man* (New York, 1983)

KH

chromosome theory of heredity (1902) *Biology* A unifying theory of the American geneticist W S Sutton (1877–1916).

Mendel's Laws of Inheritance can be explained if the factors to which Mendel referred (which were in fact the fundamental units of inheritance, or genes) occur at specific sites on chromosomes. This remains unchallenged.

RA

chronaxie (1903–09) *Biology* A term coined by French physiologists Louis Lapicque (1866–1952) and his wife, Marcelle Lapicque, this is the theory that the ability of a tissue (for example, a nerve fibre) to excite another tissue (for example, a skeletal muscle fibre) depends on the two tissues having similar excitatory time factors; the Lapicques called this time factor the chronaxie. Biologists now consider chronaxie to be the minimum time that an electric current of twice the threshold strength must flow for a tissue to be excited.

J Fulton, *Selected Readings in the History of Physiology* (Springfield, Ill., 1966).

KH

chronistics *Biology* The study of the time it takes for evolutionary events to occur. Organisms believed to have evolved rapidly are called tachytelic, while those believed to have evolved slowly are bradytelic. Intermediate time frames are termed horotelic.

KH

chunk *Linguistics* *See* MAGICAL NUMBER SEVEN.

circadian rhythms (1950s) *Biology* Term coined by Franz Halberg at the University of Minnesota. From the Latin *circa*, meaning 'around', and *dies*, meaning 'day'. These comprise a cyclic pattern of behaviour or physiology operating over a period of approximately (but not exactly) 24 hours.

Circadian rhythms have been observed in fungi, plants, animals and single-celled organisms. Most human circadian rhythms vary from 25 to 27 hours. *See also* BIOLOGICAL CLOCKS and CIRCANNUAL RHYTHMS.

J Brady, *Biological Clocks* (Baltimore, Md, 1979)

KH

circannual rhythms (1970–75) *Biology* From the Latin *circa*, meaning 'around', and *annus*, meaning 'year'. These comprise a cyclic pattern of behaviour or physiology observed over a period of approximately one year.

For example, the optimum germination of many seeds coincides with certain times of year, even when those seeds have been stored under conditions of unvarying light, temperature and humidity. *See also* BIOLOGICAL CLOCK and CIRCADIAN RHYTHMS.

F Salisbury, *Plant Physiology* (Belmont, Calif., 1992)

KH

circular reaction *Psychology* Discovered by the Swiss psychologist Jean Piaget (1896–1980).

A concept for which use is still found in recording early infant development, 'circular reaction' describes a voluntary act or reflex that generates its own repetition, sometimes without apparent motive or reward. According to Piaget, the reactions are of three kinds: primary, secondary and tertiary. Primary circular reactions are centred on the infant's own body; the secondary involve stereotyped manipulations of objects; and the tertiary are concerned with repetition with variations from cycle to cycle.

J H Flavell, *Cognitive Development* (Englewood Cliffs, N.J., 1977)

NS

circulation of élites (19th century–) *Politics* Theory of political change described by Italian social scientist Vilfredo Pareto (1848–1923).

Changes of regime, revolutions, and so on occur not when rulers are overthrown from below, but when one élite replaces another. The role of ordinary people in such transformation is not that of initiators or principal actors, but as followers and supporters of one élite or another.

Geraint Parry, *Political Elites* (London, 1969)

RB

citizenship *Politics* Theory of the value of political participation.

People are citizens when they participate to the fullest possible extent in the politics and government of their societies. Ideally all citizens are available for public office, though this is not feasible in large modern states. Citizens engage in political discussion and participation not as means to some other end such as the protection of their INTERESTS, but as a form of intelligent social activity worthwhile for its own sake. Arguments for citizenship frequently draw on ARISTOTELIANISM.

Geoff Andrews, ed., *Citizenship* (London, 1991)

RB

civil disobedience *Politics/Law* Theory of principled law breaking. The expression was used by the American writer Henry David Thoreau (1817–62) in his essay 'On the Duty of Civil Disobedience', and put into action by Mahatma (Mohandas) Gandhi (1869–1948).

There is a conflict between a general obligation to obey the law in free and constitutional societies, and the obligation to follow one's conscience. If a major clash occurs, the citizens may express their disagreement by openly breaking the law to demonstrate profound dissent from a particular policy. In so doing they accept the penalties which follow such action. *See also* SATYAGRAHA.

H A Bedau, ed., *Civil Disobedience: Theory and Practice* (Indianapolis, 1979)

RB

civil society (19th century–) *Politics* Theory of non-governmental aspects of modern society, developed by German philosopher G W F Hegel (1770–1831) and German social theorist Karl Marx (1818–83).

Hegel: in modern societies there is a civil or non-political sphere which, whilst it might be affected by government, is separate from it. Thus whilst politics takes place in political society, religion, economics, and voluntary association take place in civil society. Marx: the distinction between STATE and civil society obscures the dependence of the one on the other, and the function of the state in maintaining a particular form of economic power.

David Miller, ed., *The Blackwell Encyclopaedia of Political Thought* (Oxford, 1987)

RB

cladistics (1960s) *Biology* Also called branching evolution, and phylogenetic systematics, this is a system for classifying organisms based on shared characteristics derived from a common ancestor. It is most closely associated with W Hennig's *Phylogenetic Cladistics* (1966), but the idea probably dates from the Italian zoologist Daniele Rosa (1857–1944).

More formally, cladistics may be understood as the reconstruction of phylogenies based on characteristics unique to a particular taxonomic group (apomorphic). Modern cladistics relies heavily on analysis of genetic material and serum proteins (rather than on similarities in outward appearance) to classify species. *See also* CLADOGENESIS and CLADOGRAM. *Compare* ANAGENESIS.

T Steussy and T Duncan, *Cladistic Perspectives on the Reconstruction of Evolutionary History* (New York, 1984)

KH

cladogenesis (1953) *Biology* First used in an English language publication by the British evolutionary biologist Sir Julian Sorell Huxley (1887–1975), this term refers to 'branching' evolution which occurs when populations diverge and evolve new species sharing common ancestral types. It is also a form of adaptive evolution that is conducive to developing a greater variety of organisms. *See also* CLADISTICS, CLADOGRAM and EVOLUTION. *Compare with* ANAGENESIS.

E Delson, *Ancestors: The Hard Evidence* (New York, 1985)

KH

cladogram (1960s) *Biology* Most closely associated with W Hennig's *Phylogenetic Cladistics* (1966), this is a branching line diagram using CLADISTICS to describe the theoretical evolution of a species from ancestral stock, based on shared, unique structures suggesting a common ancestor. *See also* CLADOGENESIS.

J Brown, *Biogeography* (St Louis, Mo., 1983)

KH

class dealignment (20th century) *Politics* Theory of voting.

Whereas previously there was a significant relationship between voting and social class, this link is now being broken and other considerations – of policy, moral issues, religious alliance, and so on – are taking its place.

Henry Drucker *et al.*, eds, *Developments in British Politics 2* (London, 1986)

RB

class struggle (19th century–) *Politics/Economics* Account of historical change given by German social theorist Karl Marx (1818–83).

The principal division within society is between classes. Classes are groups defined in terms of their relationship to the means of production. They have conflicting INTERESTS, and this conflict is the basis of political competition. Class struggle therefore – rather than pursuit of the national interest, or ideological difference – is the motor of history. *See also* MARXISM.

David McLellan, *The Thought of Karl Marx* (London, 1981)

RB

class voting (20th century) *Politics* Explanation of voting.

At some point in the development of industrialism and of democracy, economic class becomes the single most important factor in voting. Members of the working class tend to vote for left wing parties, those of the middle classes for right wing parties. Were this theory an adequate account of voting, all governments would have been socialist.

Patrick Dunleavy *et al.*, eds, *Developments in British Politics 3* (London, 1991)

RB

classical macroeconomic model (18th–19th century) *Economics* Developed by French economist Jean-Baptiste Say (1767–1832) and later revised by other classical economists.

Through the assumptions of factor and product price flexibility, and in the absence of regulations which prevent the market adjustment of demand and supply, full employment equilibrium will be reached in the classical macroeconomic model. *See* NEW CLASSICAL MACROECONOMICS, SAY'S LAW, STATIONARY STATE, and WAGES FUND DOCTRINE.

A Smith, *An Inquiry into the Nature and Causes of the Wealth of Nations* (London, 1776); J B Say, *Traité d'Economie Politique (1803)* (London, 1817)

PH

classical theory of money (17th–19th century) *Economics* Money was considered a commodity, the price of which was determined by the amount of time needed to produce it (that is, gold and silver mining).

This approach was revised by English economist David Ricardo (1772–1823) and the ensuing Bullionist Controversy which paved the way for the Currency School and Banking School of economics which, in turn, provided the framework of the modern British banking system. *See* QUANTITY THEORY OF MONEY.

F W Fetter, *Development of British Monetary Orthodoxy 1719–1875* (Cambridge, Mass., 1965)

PH

classical theory of probability *Philosophy/Mathematics* Theory generally attributed to French mathematician and astronomer Pierre-Simon, Marquis de Laplace (1749–1827) in his *Essai philosophique sur les probabilités* (1820). It says that the probability of an occurrence in a given situation is the proportion, among all possible outcomes, of those outcomes that include the given occurrence.

The main difficulty lies in dividing up the alternatives so as to ensure that they are equiprobable, for which purpose Laplace appealed to the controversial principle of INDIFFERENCE. A related difficulty is that the theory seems to apply to at best a limited range of rather artificial cases, such as those involving throws of dice.

H E Kyburg, *Probability and Inductive Logic* (1970), ch. 3

ARL

classicism *Art* A term describing adherence to classical ideals of beauty, proportion and symmetry as exemplified in Ancient Greek and Roman architecture and sculpture, most notable during the Renaissance and neoclassical period.

The term also refers to any other period or work of art where artist(s) display a preference for order and objectivity over formlessness and subjectivity.

LL

classicism/Romanticism (1798/1808) Antithesis first publicly formulated by the German critic Karl Wilhelm Friedrich von Schlegel (1772–1829) and developed by his brother August Wilhelm von Schlegel (1767–1845).

An opposition of two kinds of aesthetic ideal and of artistic practice. Classicism disciplines the poet's imagination with strict rules of form, emphasizing craftsmanship and technical precision. ROMANTICISM is devoted to the free expression of the poet's feelings and views. The German writer Johann Goethe (1749–1832) called the classic 'healthy' and the romantic 'sickly'. *See also* APOLLONIAN/DIONYSIAN.

R Wellek, 'The Concept of Romanticism', *Concepts of Criticism* (New Haven, 1963)

RF

Clausius' theorem and inequality (19th century) *Physics* Named after the German physicist Rudolf Julius Emanuel Clausius (1822–88) who derived this result which relates to any system undergoing a cyclic process.

In an elementary part of a cyclic process, let the system absorb a quantity of heat $\mathrm{d}Q$ in a reversible manner from a source at a temperature T. If the cyclic process is irreversible

$$\oint \mathrm{d}Q/T < 0.$$

This is the inequality of Clausius. If the cyclic process is reversible

$$\oint \mathrm{d}Q/T = 0.$$

This is the theorem of Clausius.

J Thewlis, ed., *Encyclopaedic Dictionary of Physics* (New York, Oxford and London, 1962)

MS

Clausius' virial theorem (1870) *Physics* Named after the German physicist Rudolf Clausius (1822–88), who was responsible with the Scottish physicist William Thomson Kelvin (1824–1907) for formulating the second law of thermodynamics.

This theorem makes it possible to determine the equation of state of a liquid or gas when the force F_{ij} between two atoms i and j is known as a function of their separation r_{ij}. The product $F_{ij}r_{ij}$ is summed over all pairs of atoms in the fluid and the average taken, that is

$$\sum F_{ij}\, r_{ij}\, .$$

Then the molar volume V_m of the fluid at a pressure p and temperature T is given by

$$pV_m = RT + \frac{1}{3}\sum F_{ij}\, r_{ij}\, ,$$

where R is the molar gas constant.

Robert M Besançon, ed., *The Encyclopedia of Physics* (New York, 1985)

GD

clay theory of evolution *Biology/Geology* *See* MINERAL THEORY.

Cleopatra's nose *History* Anonymous theory of the role of chance in history.

History is as much influenced by unpredictable and unique factors as it is by general processes, movements, or trends. The shape of Cleopatra's nose, which was allegedly one of her more alluring features, beguiled Anthony and thus caused havoc amongst the military government of Rome. *See also* COCK-UP THEORY and CONTINGENCY THEORY.

RB

clerisy (19th century–) *Politics* Advocacy of cultural ÉLITISM, particularly by English poet Samuel Taylor Coleridge (1772–1834) and Anglo-American poet T S Eliot (1888-1965).

In any society, there will be a minority who are distinguished by exceptional intellect, refinement, or understanding. They should, without necessarily ruling directly, set the tone and the agenda for the life of the majority. Clerisy is thus a formalized recommendation of élitism.

T S Eliot, *The Idea of a Christian Society* (London, 1939)

RB

climax theory (1904) *Biology* A theory of ecological succession, first proposed by the American ecologist Frederic Edward Clements (1874–1945).

If left to run its course, in any one climatic region of the world, vegetation change over time from bare soil to a complex community (succession) will develop only one type of community (for example, an oak woodland) regardless of the starting conditions. This theory is considered rather extreme and has been subsequently modified by Tansley (1939) and Whittaker (1953) to allow a variety of climax communities, depending on many factors, which are viewed as continuous rather than a definite end point.

M Begon, J L Harper and C R Townsend, *Ecology: Individuals, Populations and Communities* (London, 1987)

RA

clinical linguistics (1941; 1970s) *Linguistics* A seminal paper by the Russian linguist Roman Jakobson (1896–1982) was published in 1941 (*see* APHASIA), but consolidated research took place much later.

The description and classification of speech and language disabilities in terms of the categories of descriptive linguistics, in order to identify systematic patterns of impairment in individuals and to facilitate clinical remediation. 'Clinical linguistics is still at a primitive stage of development' (Crystal). *See also* LANGUAGE PATHOLOGY and NEUROLINGUISTICS.

D Crystal, *Clinical Linguistics*, 2nd edn, (1986)

RF

cloaca theory *Psychology* Its name derived from the Latin word for sewer or drain, a theory proposed by the Austrian founder of psychoanalysis Sigmund Freud (1856–1939).

Young children often hold the notion that birth takes place through the anus and is a form of defecation. Psychoanalytic theory makes much of this fantasy and Freud considered it to be a possible source of bisexuality. Other theorists consider that it reflects an understandable confusion in the minds of young children. *See* INFANTILE BIRTH THEORIES.

E Jones, *The Life and Works of Sigmund Freud* (New York, 1953)

NS

clock theory *Biology See* PROGRAMMED AGEING.

clonal selection theory (1957) *Biology* Idea originated by N K Jerne and refined by Sir Frank Macfarlane Burnet (1899–), the Australian immunologist who won the Nobel prize for Physiology or Medicine in 1960. This controversial theory asserts that antibodies form because of continual mutations in the antibody producing part of the genes, creating a diverse assortment of antibody variants.

Normally each variant is produced by only a few cells, but if a particular antibody is needed for immune response to an antigen (that is, 'selected'), those particular cells undergo rapid clonal proliferation to meet the demand while the less useful antibody variants die out. *See also* ANTIBODY, THEORIES OF and INSTRUCTIVE THEORY.

F Burnet, *The Clonal Selection Theory of Acquired Immunity* (1959)

KH

closed graph theorem (20th century) *Mathematics* The useful result in TOPOLOGY that a linear function f on a topological space X is continuous if and only if its graph $\{(x,f(x)):x\mp X\}$ is closed (in the appropriate product topology).

Béla Bollobás, *Linear Analysis* (Cambridge, 1990)

MB

closed texts *Stylistics See* OPEN AND CLOSED TEXTS.

closing lemma (1967) *Mathematics* A general form of the theorem was proved by C Pugh. It states that given any diffeomorphism f of the unit sphere with an irrational rotation number, there is another diffeomorphism of the unit sphere which is arbitrarily close to f and has rational rotation number. More general theorems exist for the C^1- topology.

R L Devaney, *An Introduction to Chaotic Dynamical Systems* (New York, 1989)

ML

closure (1970s) *Literary Theory* Evaluative as well as theoretical term, used either positively or negatively.

An aesthetically satisfying sense of formal completeness given by, for instance, rhyme in poetry or a clear outcome in narrative.

Closure is disvalued in the sceptical atmosphere of DECONSTRUCTION and POSTSTRUCTURALISM, suggesting authoritarian singleness of meaning, MONOLOGISM. *See also* OPEN AND CLOSED TEXTS.

B H Smith, *Poetic Closure* (Chicago, 1968)

RF

closure, law of *Psychology* Proposed by the German psychologist Kurt Koffka (1886–1941).

Associated with the Gestalt school of psychology, the law posits that we have a tendency to perceive incomplete objects as complete by filling in or closing up gaps in sensory inputs and viewing asymmetric and unbalanced stimuli as symmetric and balanced. The brain synthesizes the missing parts of a perceived image thereby closing the gap between reality and the desired 'picture'. For example, if an occasional note is deleted from a song the listener may unconsciously supply the missing notes. *See* GESTALT THEORY.

A D Ellis, ed., *A Sourcebook of Gestalt Psychology* (London, 1938)

NS

clubs, theory of (1956) *Economics* Based on work by American economists Charles Tiebout (1924–68) and James Buchanan (1919–), this studies the optimal size of groups of people with a shared consumption (pools, clubs, museums), and the optimal provision of the goods or services.

A club good is excludable in that it is possible to prevent its consumption by entire groups of people, but it is also a non-rival good in that its consumption by one individual does not curb the consumption of another individual.

C Tiebout, 'A Pure Theory of Local Expenditures', *Journal of Political Economy*, vol. LXIV (1954), 416–24

PH

co-evolution (1960s) *Biology* Coined by American population biologist Paul Ralph Ehrlich (1932–) and botanist Peter Hamilton Raven (1936–), this term refers to evolutionary changes that occur in genetically unrelated species as they interact with each other in their environment.

For instance, as cheetahs developed greater speed the better to prey upon antelopes, the antelopes evolved greater speed to escape the cheetahs. Predation, competition for food or shelter, mutally beneficial arrangements and symbiosis are examples of co-evolutionary forces.

D Futuyma, *Coevolution* (Sunderland, Mass, 1983)

KH

co-operative games theory (1940s) *Economics* A branch of GAMES THEORY dealing with co-operative rather than simply competitive players.

Games theory attempts to study the interaction of individual decisions (given specific assumptions about decisions made under risk), the general environment and individual behaviour patterns. Co-operative games theory is used in the analysis of cartels and other forms of market collusion. *See* COLLUSION THEORY, OLIGOPOLY THEORY and ALLAIS PARADOX.

J von Neumann and O Morgenstern, *The Theory of Games and Economic Behaviour* (Princeton, N.J., 1944)

PH

co-operative principle of conversation (1967) *Linguistics* Proposed by the American ordinary language philosopher Herbert Paul Grice (1913–).

Grice makes a fundamental assumption that participants in a conversation behave co-operatively. This 'co-operative principle' is maintained by speakers observing four 'maxims': *quantity* (appropriate amount of information); *quality* (truthfulness); *relation* (relevance); *manner* (clarity). *See also* IMPLICATURE and RELEVANCE.

H P Grice, 'Logic and Conversation' [1967], P Cole and J L Morgan, eds *Syntax and Semantics*, 3: *Speech Acts* (New York, 1975), 41–58

RF

coalescence theory *Meteorology* In a rising air mass within a cloud in which the air is warmer than some −12°C, the larger raindrops will rise at a slower rate than small water droplets, and therefore collide with them, growing by collision until they are large enough to fall under their own weight (*c*19m). In colder clouds, rain formation is mostly explicable by the BERGERON THEORY.

R R Rogers, *A Short Course in Cloud Physics* (New York, 1989)

DT

Coase theorem *Economics* Named after the English-born American economist Ronald Harry Coase (1910–).

This hypothesis asserts that as long as there are well-defined property rights (and no transaction costs), externalities will not cause a breakdown in the allocation of resources. Externalities being defined as the benefits or costs to a society of the process of consumption or production; for example, pollution, disease and spill-overs.

R Coase, 'The Problem of Social Cost', *Journal of Law and Economics* 3, 1 (1960), 1–44

PH

Cobb-Douglas production function (1928) *Economics* Developed by American economist Paul Douglas (1892–1976) and mathematician Charles Cobb.

A production function which shows physical output as the Douglas labour and capital inputs; that is

$$Q = AL^aK^b$$

where Q is output, A, a, b are constants, and L and K are labour and capital, respectively.

Capital can be interchanged with labour without affecting output. Cobb and Douglas also suggested that the share of labour and of capital within an economy are relatively constant over time, although this production function was largely abandoned after 1961. *See* CES PRODUCTION FUNCTION, INPUT-OUTPUT ANALYSIS and THEORY OF PRODUCTION.

P H Douglas, 'Are There Laws of Production?', *American Economic Review*, vol. XXXVIII (March, 1948), 1–41

PH

COBRA (1948) *Art* An artistic movement formed out of the Danish *Spiralen* group (*Co*penhagen), the Belgian *Bureau Internationale de Surréalisme Révolutionnaire* (*Br*ussels), and the Dutch Experimental Group (*A*msterdam).

Its adherents had affinities with American ACTION PAINTING in their emphasis of the unconscious and spontaneous. Their works are distinctive for their abstracted compositions, violent brushwork and saturated colour.

Cobra 1948–51, exh. cat. Stedlijk Museum (Amsterdam, 1962)

AB

cobweb theory (1934) *Economics* Named by Hungarian-born economist Nicholas Kaldor (1908–86), this stems from a simple dynamic model of cyclical demand which involves time lags between the response of production and a change in price (most often seen in agricultural sectors).

The theory focuses on the process of adjustment in markets by tracing the path of prices and outputs in different equilibrium situations. It is so named because its graphic representation resembles a cobweb with the equilibrium point at the centre of the cobweb. It is sometimes referred to as the hog-cycle (after the phenomenon observed in American pig prices during the 1930s). *See* ADAPTIVE EXPECTATIONS.

N Kaldor, 'A Classificatory Note on the Determination of Equilibrium', *Review of Economic Studies*, vol I (February, 1934), 122–36; M Nerlove, 'Adaptive Expectations and Cobweb Phenomena', *Quarterly Journal of Economics*, vol. lxxii (1958), 227–40

PH

cock-up theory (20th century) *Politics/History* Anonymous sceptical view of history.

History is frequently to be explained by the ordinary errors and inadequacies of people, especially powerful people, rather than by any GRAND THEORY. The theory expresses a common scepticism towards both intellectual and personal pretension, and the view that behind most great men there is a far more mundane mixture of idleness and incompetence. A developed application of this theory towards warfare suggests that there are three kinds of military events: cock-ups, routs, and national disasters. *See also* CONTINGENCY THEORY.

Norman F Dixon, *On the Psychology of Military Incompetence* (London, 1976)

RB

code (1970) *Stylistics* Fundamental concept in SEMIOTICS, given an important application by the French structuralist Roland Barthes (1915–80).

A text is a weaving of a number of cultural codes, systems of signs which give significances over and above the linguistic meanings of the words and sentences. Barthes distinguishes five, having to do with narrative structure, character type, social values, and so on. The primary activity of the reader is bestowing the significances of the codes onto the text.

R Barthes, *S/Z*, trans. R Miller (London, 1975 [1970])

RF

code-switching (*c*.1964) *Linguistics* Sociolinguistic phenomenon studied by the American ethnographer of communication John J Gumperz (1922–).

Depending on the nature of the communicative situation, speakers within a particular SPEECH COMMUNITY may choose between two varieties of their language (for example dialects, or standard and regional) to signify different social values (for example formality, local pride) or to redefine the situation.

J-P Blom and J J Gumperz, 'Social Meaning in Linguistic Structures; Code-switching in Norway', *Directions in Sociolinguistics: The Ethnography of Communication*, J J Gumperz and D H Hymes, eds (New York, 1972), 409–34

RF

codon-restriction theory (1971) *Biology/Medicine* Proposed by B L Strehler and colleagues, this theory proposes that as cells age they lose the ability to translate genetic information from messenger ribonucleic acids (mRNAs), resulting in errors during protein synthesis. Why this should happen is not explained by the theory. *See also* AGEING, THEORIES OF. *Compare with* ERROR CATASTROPHE THEORY OF AGEING.

A Roy and B Chatterjee, *Molecular Basis of Aging* (Orlando, Fla, 1984)

KH

cognitive appraisal theory *Psychology See* ATTRIBUTION THEORY OF EMOTION.

cognitive arousal theory of emotion *Psychology See* ATTRIBUTION THEORY OF EMOTION.

cognitive conditioning *Psychology* Developed by the American psychologist Edward Chase Tolman (1886–1959).

A type of BEHAVIOUR THERAPY based on the notion of the stimulus-response unit. Its purpose is to modify behaviour that is considered to be faulty. The thoughts associated with the behaviour to be modified are paired with an aversive stimulus, so that they become associated with aversion, thereby discouraging such behaviour. The procedure is repeated until the behaviour is modified.

A E Kazdin, *Behavior Modification in Applied Settings* (Homewood, Ill., 1980)

NS

cognitive consistency theory *Psychology* Proposed by the American psychologist L Festinger (1919–).

Imbalanced, inconsistent or contradictory opinions, beliefs or attitudes give rise to tension or 'energy' which serves as a force motivating behaviour. Festinger believed that the relations between cognitions may be irrelevant, consonant or dissonant. Dissonance exists between two elements when those two elements are opposing. It is an unpleasant state which the individual tries to resolve by modifying cognitions or introducing new cognitions to his or her system: a cognitive restructuring. The theory is often criticized for being impossible to disprove. *See* DISSONANCE THEORY and BALANCE THEORY.

L Festinger, *A Theory of Cognitive Dissonance* (Stanford, 1957)

NS

cognitive dissonance theory *Psychology See* COGNITIVE CONSISTENCY THEORY and DISSONANCE THEORY.

cognitive evaluation theory of emotion *Psychology See* ATTRIBUTION THEORY OF EMOTION.

cognitive labelling theory of emotion *Psychology See* ATTRIBUTION THEORY OF EMOTION.

cognitive learning theory *Psychology* Proposed by the American psychologist Edward Chase Tolman (1886–1959).

Learning involves central constructs and new ways of perceiving events. It is a basic assumption in cognitive therapy which holds that emotional problems are the result of faulty ways of thinking and distorted attitudes and perceptions towards oneself and others. The cognitive therapist directs and guides the client in his/her cognitive restructuring, a process which involves correcting and revising those faulty perceptions and attitudes. The therapist assists this process by eliciting evidence to the contrary from the client, or citing such evidence himself.

R D Gross, *Psychology: The Science of Mind and Behaviour* (London, 1992)

NS

cognitive map *Psychology* Proposed by the American psychologist Edward Chase Tolman (1886–1959).

The picture built up by an organism on the basis of experiences. The individual is not a passive receptor of information about the environment but seeks and collects clues about features and relationships, constructing an internal map. For example, an individual will have a cognitive map of the building in which he works, and a rat will have a cognitive map of the maze which it has run.

R D Gross, *Psychology: The Science of Mind and Behaviour* (London, 1992)

NS

cognitive personality theory *Psychology* Proposed by the German- American psychologist Kurt Lewin (1890–1947).

Concerned with the development of individual differences in the process of thinking as they affect the perceptions, attitudes and behaviour of individuals. Personality is centred on the person's thinking processes. The popularity of the cognitive approach to personality has been on the increase, and has led to greater interest in the cognitive styles of individuals. More recently, contributions from the field of artificial intelligence have been generating research interest.

R D Gross, *Psychology: The Science of Mind and Behaviour* (London, 1992)

NS

cognitive psychology *Psychology* *See* COGNITIVE THEORY.

cognitive theory *Psychology* While the term cannot be attributed to any specific individual, it may be linked historically with the work of the German scientist Wilhelm Wundt (1832–1920), regarded as the first experimental psychologist.

Cognitive theory is a collective term for psychological theories seeking to explain thought processes with reference to the relationship between subject and object, thought and world. Cognitive theory is principally concerned with investigating the conditions for cognition: the structural and functional architecture of the knowing or cognizing organism or system.

R G Gross, *Psychology: The Science of Mind and Behaviour* (London, 1992)

NS

coherence (1970s) *Stylistics* Traditional concept clarified by recent text-linguistic theories. *See also* COHESION.

The unity and intelligibility of (literary) texts. On the basis of linguistic and semiotic cues, readers naturalize and render coherent textual meanings, building a 'textual world' which orders and makes sense of details in terms of existing conventional knowledge. *See also* COHESION and MENTAL MODEL.

R de Beaugrande and W Dressler, *Introduction to Text Linguistics* (London, 1981)

RF

coherence theory of truth *Philosophy* A theory maintaining that a proposition will be true if it forms part of a system of mutually coherent propositions which is wider than any rival system. The coherence or consistency in question must of course be definable independently of truth, which may be difficult.

The theory is favoured especially by OBJECTIVE IDEALISM, which rejects the sharp distinction between what is known and the knowing of it that the CORRESPONDENCE THEORY OF TRUTH seems to require. Idealists tend to add, however, that only the system as a whole is fully true, the individual component propositions being only partly true or true to some degree. (*See also* DEGREES OF TRUTH.) Weaker versions of the theory say that such coherence provides a criterion of truth in some or all cases, truth itself being defined – if at all – in some other way.

H Joachim, *The Nature of Truth* (1906)

ARL

cohesion (1976) *Linguistics* System of TEXT LINGUISTICS outlined by the British linguists Michael Alexander Kirkwood Halliday (1925–) and Ruqaiya Hasan.

A text is not a random, unstructured sequence of sentences; its parts are held together by a mass of 'cohesive ties' (pronouns, ellipses, vocabulary choices, and so on) which give it semantic coherence and consistency of reference to its topic; cohesion is part of 'textual function' in FUNCTIONAL GRAMMAR.

M A K Halliday and R Hasan, *Cohesion in English* (London, 1976)

RF

cohesion, law of *Psychology* Proposed by the American psychologist Edward Chase Tolman (1886–1959).

States that behavioural acts which are spatially and temporally proximate tend to become integrated into more complex acts. The term can also be used in the context of group dynamics to refer to the forces which hold a group together. Cohesion is dependent upon: (1) the extent to which interaction within a group has positive qualities for the group's members; (2) the extent to which a group's activities are rewarding for its members; (3) the usefulness of group membership for achieving an individual's objectives.

R D Gross, *Psychology: The Science of Mind and Behaviour* (London, 1992)

NS

cohesion species concept (20th century) *Biology* Proposed by American biologist Alan Templeton of Washington University (St Louis, Missouri).

One of several models for the SPECIES CONCEPT holding that a species can be defined without resorting to interbreeding as a defining characteristic, making the model fully applicable to species which reproduce asexually. Templeton defines a species as the 'most inclusive population of individuals having the potential for phenotypic cohesion through intrinsic cohesion mechanisms'. These mechanisms include factors that define the limits of spread of new genetic variations via gene flow (for example, successful fertilization systems), and factors that define the environmental niche and limit the spread of genetic variation via NATURAL SELECTION and GENETIC DRIFT (for example, ecological constraints). *See also* SPECIATION, THEORY OF.

M Ereshefsky, *The Units of Evolution: Essays on the Nature of Species* (Cambridge, Mass., 1992)

KH

cohesion theory (1914) *Biology* Outlined in book form with supporting data by Irish plant physiologist Henry H Dixon, but discussed as early as 1894 by Dixon and Irish physicist John Joly (1857–1933); this is the theory that the sun supplies the energy for lifting water to the upper reaches of a plant (even very tall trees) through the evaporation of water from the plant's leaves.

More formally, the concept that the loss of water vapour through the stomata from the surfaces of the leaf parenchyma cells causes tension (and ultimately cohesion) in the water column of the xylem. This tension is transmitted through the water column of the plant to the absorbing cells of the roots, which respond to the increased diffusion pressure deficit by increasing absorption of water from the soil. Originally controversial, this theory is now the most widely accepted explanation for the ascent of sap in plants. It is also called Dixon cohesion theory, Dixon-Askenazy cohesion theory, and transpiration-cohesion-tension theory. *Compare with* APOPLAST-SYMPLAST CONCEPT.

F Salisbury, *Plant Physiology* (Belmont, Calif. 1992)

KH

cold art (20th century) *Art* From the German '*Kalte Kunst*', this is the name given to a branch of CONSTRUCTIVISM based on mathematical principles and formulae. It is related to CONCRETE ART.

Cold art's leading practitioner, the Swiss artist Max Bill (1908–), saw rationality as the most essential form of thinking and aimed at producing works which were as self-contained and intrinsically balanced as music. His paintings typically comprise geometric forms in primary colours. Other exponents include the American artists Robert Morris (1931–) and Sol Lewitt (1928–).

C H Waddington, *Behind Appearance* (1960)

LL

Collatz sequence *Mathematics* The unsolved conjecture due to Lothar Collatz (1910–91) that every sequence determined by starting at any natural number n, replacing it by $3n + 1$ and then dividing out powers of two, and repeating the process finitely returns to 1.

For example, $3 \to 10 \to 5 \to 16 \to 1$ and $7 \to 22 \to 11 \to 34 \to 17 \to 52 \to 13 \to 40 \to 5 \to 16 \to 1$. This is illustrative of a large class of seemingly innocent modern mathematical problems related to the RECURSIVE PRINCIPLE.

JB

collective action, logic of (20th century) *Politics* Theory of the operation of political groups described by American political scientist Mancur Olson (1932–).

Rational calculation of self interest will lead people to 'free ride'. Workers will enjoy a pay raise negotiated by a union whether they are members of it or not, and so need extra incentives if they are to be active supporters of the union. Rational self interest alone will not therefore explain why organizations successfully promote 'public goods' which people can enjoy whether or not they worked to secure them.

Mancur Olson, *The Logic of Collective Action: Public Goods and the Theory of Groups* (Cambridge, Mass., 1965)

RB

collective bargaining theory (20th century) *Economics* This term refers to studies carried out by UK political economist Alfred Marshall (1842–1924) into the negotiation of wage rates and conditions of employment by representatives of the labour force (usually trade union officials) and management. Collective bargaining generally involves the joint regulation of agreed procedures. *See* THEORY OF COUNTERVAILING POWER and SYNDICALISM.

S Webb and B Webb, *Industrial Democracy* (London, 1897); H A Clegg, *Trade Unionism under Collective Bargaining. A Theory Based on Comparisons of Six Countries* (Oxford, 1976)

PH

collective electron theory *Physics* The theory is concerned with the problem of finding appropriate wave functions for individual electrons moving in the periodic potential of a crystal lattice by assuming the electrons to be shared by the whole crystal. It may be contrasted with the Heisenberg or localized electron theory. *See also* BAND THEORY OF SOLIDS and FERROMAGNETISM, HEISENBERG THEORY OF.

GD

collective memory (20th century) *Psychology* Proposed by the Swiss psychiatrist Carl Gustav Jung (1875–1961).

Also called the 'collective unconscious' or the 'racial unconscious'. Jung believed in an aspect of the unconscious shared by all people across races. This collective unconscious he held to be inherited, transpersonal and consisting of the residue of the evolution of the species. A central weakness of the proposition lies in the difficulty of testing it fairly; however, it stimulated much interest in mythology and its possible archetypal content, and in literature particularly this analytic technique was accepted and explored.

C G Jung, *Memories, Dreams, Reflections* (London, 1963); J F Jacobi, *The Psychology of Jung* (Yale, 1943)

RA, NS

collective model of the atomic nucleus (1950) *Physics* Originated by American physicist Leo James Rainwater (1917–), and developed three years later by the Danish physicist Niels Henrik David Bohr (1885–1962) and B R Mottleson. In this model, a nucleus that is away from a closed shell configuration takes on an ellipsoidal shape and shows rotational energy levels comparable with the lowest observed energy levels of nuclei.

This model combines features from the LIQUID DROP MODEL OF THE ATOMIC NUCLEUS and the SHELL MODEL OF THE ATOMIC NUCLEUS, both of which give rise to awkward anomalies. For example, the energy levels for surface vibrations in the liquid drop model are much higher than those observed. As in the shell model, the collective model assumes that the nucleons do not interact with each other to a first approximation, but that they move in a distorted potential field which arises from the collective motions of the liquid drop model.

J Thewlis, ed., *Encyclopaedic Dictionary of Physics* (New York, Oxford and London, 1962)

MS

collective security (20th century) *Politics* Theory in international relations.

Nations achieve security from attack by collective military agreements of all nations with each other to join forces against any aggressor. An alternative to BALANCE OF POWER.

Graham Evans and Jeffrey Newnham, *The Dictionary of World Politics* (Hemel Hempstead, 1990)

collectivism (20th century) *Politics/Sociology* Theory of social and political organization.

The major problems facing individuals are ones which they share in common with others. The solutions, correspondingly, are most effective when pursued in common or collectively. The provision of welfare and the cultivation of the economy, in particular, are best conducted through collective rather than individual action. The highest and most general form of collective action is the state, and collectivism is frequently a synonym for the advocacy of governmental solutions to social problems. Collectivism differs from COMMUNALISM and is frequently contrasted with INDIVIDUALISM.

A V Dicey, *Law and Public Opinion in England during the Nineteenth Century*, 2nd edn (London, 1914, reprinted London 1962)

RB

collegialism *Politics/Management* Non-managerial theory of organizational power.

Organizations and enterprises are most appropriately run by co-operative agreement amongst those directly involved. Collegialism is particulary associated with the professions, and is in that sense a form of middle class GUILD SOCIALISM. It is contrasted with MANAGERIALISM, which recommends placing control in a core of specialists who are not directly involved in whatever task the organization or profession performs.

Roger Scruton, *A Dictionary of Political Thought* (London, 1982)

RB

collision theory (1918–35) *Chemistry* Theory of chemical reaction rates developed by W C McC Lewis (1918), C N Hinshelwood (1923), and particularly H Eyring (1935). Theory of the rates of bimolecular gas-phase reactions developed from the KINETIC THEORY OF GASES to account for the influence of concentration and temperature on reactions.

The rate of reaction is the frequency of collisions multiplied by the fraction of collisions with an energy of at least the activation energy. The theory only applies to a specific step in a reaction and not necessarily the reactant species represented by the stoichiometric equation. Experimental values for rates of reaction are generally lower than those calculated from the theory.

M Freemantle, *Chemistry in Action* (Basingstoke, 1987)

MF

collision theories of planetary formation (18th century–) *Astronomy* An early collision theory was that proposed by the French naturalist Georges Louis Leclerc, Comte de Buffon (1707–88), who surmised that the planets were formed as a result of a head-on collision between a comet and the Sun. Chunks of matter were splashed out of the Sun during the collision and these chunks then condensed to form the planets.

In 1880, a similar theory was proposed by A W Bickerton, in whose model the comet was replaced by a second star, which was assumed to have collided with the Sun at a grazing incidence. Bickerton concluded that the debris of the collision would form a third, nebulous body; condensations in which would form the planets.

These two theories suffer from a serious objection. The two large planets Jupiter and Saturn are each surrounded by a number of satellites, forming a miniature solar system. It seems unlikely that these two satellite systems and the solar system had different origins. However, it is very unlikely that three grazing collisions can have occurred, one for the formation of each system.

S Mitton, ed., *The Cambridge Encyclopaedia of Astronomy* (London, 1973)

MS

collocation (1951) *Linguistics* Idea proposed by the British linguist John Rupert Firth (1890–1960).

The distributional tendency of certain words to appear close together – 'collocate' – in texts: 'dark' and 'night', 'time' and 'save', 'spend' and 'waste', and so on. Firth claims that the range of a term's collocates is part of its meaning.

J R Firth, 'Modes of Meaning', *Papers in Linguistics 1934–1951* (London, 1957)

RF

collusion theory *Economics* Co-operation between two or more companies producing similar goods may entail similar pricing or output levels. These conditions of competition may be more akin to a monopoly market. Collusion is outlawed in most capitalist economies. *See* MONOPOLY, OLIGOPOLY THEORY, MONOPOLISTIC COMPETITION, MONOPOLY CAPITALISM and ADMINISTERED PRICING.

G J Stigler, 'A Theory of Oligopoly', *Journal of Political Economy*, 72, 1 (February, 1964), 44–61

PH

colonialism *Politics/Economics* Theory of the territorial extension of national power.

Nations and economies will seek to extend their influence by colonizing weaker or less economically developed areas. Theories of colonialism differ as to whether the principal motive in this expansion is military or economic. *See also* IMPERIALISM.

Geoffrey Roberts and Alistair Edwards, *A New Dictionary of Political Analysis* (London, 1991)

RBN

colour field painting (1950s–60s) *Art* A style which developed in the USA, rejecting the gestural and tactile qualities of ABSTRACT EXPRESSIONISM.

American painters such as Ellsworth Kelly (1923–) and Barnett Newman (1905–70) produced large-scale monochromatic canvases whose saturation of colour does not suggest form or representation, and which tend to overwhelm the spectator.

AB

combination principle *Psychology* Proposed by the American psychologist Burrhus Frederic Skinner (1904–90).

Simultaneously occurring stimuli or stimuli in close temporal and physical proximity may elicit a combined response; or, responses linked to separate stimuli will occur together when a stimulus eliciting either response is presented. The principle is associated with theory and research in BEHAVIOURISM.

B F Skinner, *The Behaviour of Organisms* (New York, 1938)

NS

combination principle of Ritz (1908) *Physics* This principle, also known as the Ritz series formula after its originator, the German theoretical physicist Walter Ritz (1878–1909). Although preceding the Bohr model of the atom, this principle was used to solve atomic problems.

In modern terms, the principle may be written as follows. The wave numbers N of the lines in a given atomic spectral series may be written in the form

$$N = R[(1/A^2) - (1/B^2)]$$

where R is the Rydberg constant, A is a constant for the particular series and B has various integer values for the different lines in the series.

J Thewlis, ed., *Encyclopaedic Dictionary of Physics* (New York, Oxford and London, 1962)

MS

combinatorics *Mathematics* The area of mathematics which grew out of classical questions dating back at least to French mathematician Blaise Pascal (1623–62) counting permutations, combinations and subdivisions of finite sets of objects.

It now encompasses a wide variety of areas using algebraic, asymptotic and probabilistic methods and techniques to study the existence, enumeration and structural characterization of such discrete mathematical objects as ordered sets, graphs (*see also* GRAPH THEORY), codes (see COMMUNICATION THEORY) and designs (*see also* OFFICER PROBLEM and BRUCK-RYSER-CHOWLA THEOREM). When an objective function is involved we speak of problems in combinatorial OPTIMIZATION THEORY, such as for example the *travelling salesman problem*.

AL

combined gas law *Chemistry See* GENERAL GAS LAW.

combining volumes law *Chemistry See* GAY-LUSSAC'S LAW OF COMBINING VOLUMES.

comedy (traditional) *Literary Theory* No single authoritative theory.

Comedy has been an important and prolific form of drama from the plays of the Greek Aristophanes (*c*.448–*c*.380 BC), and is a mode which finds expression in non-dramatic genres too. Aristotle's *Poetics* (4th century BC) has little to say about comedy, but seems to regard it as an antithesis to TRAGEDY. Theories of comedy which attempt to reconstruct this antithesis oversimplify comedy's diversity.

Clearly, comedy seeks to engage and amuse its audience, and laughter is an outcome; but it can lead the audience to a perspective on realistic and serious problems. Wiliness and mischief, and ability to extricate oneself from difficult predicaments, are features of the comic character, as are fluency and verbal play. Comedy has many sub-genres – comedy of manners, romantic comedy, farce, comedy of humours, of intrigue, and so on – and any overall theory would need to isolate their common elements.

P Lauter, ed., *Theories of Comedy* (New York, 1964)

RF

commodity theory of money *Economics* This term refers to a system of money based on a specific commodity; that is, any good suitable for exchange or consumption.

The system is usually linked to a specific quantity of the commodity whose value is determined by its price in the marketplace. The Gold Standard was a commodity money system. *See* BIMETALLISM, CLASSICAL THEORY OF MONEY, CURRENCY PRINCIPLE and SPECIE FLOW MECHANISM.

M Friedman, *Essays in Positive Economics* (Chicago 1953), 204–50; P Vilar, *A History of Gold and Money, 1450–1920* (London, 1976)

PH

common descent (19th century) *Biology* Attributed to English biologist Charles Darwin (1809–82), this is the concept that related species share a common evolutionary ancestor. This theory explains the similarities of structures observed in the families and *genera* of living things. *See also* DARWINISM.

D Kohn, *The Darwinian Heritage* (Princeton, NJ, 1985)

KH

common fate, law of *Psychology* Proposed by the German-American psychologist Wolfgang Köhler (1887–1967).

Also termed the factor of uniform density, this is a principle of the Gestalt school of psychology. It summarizes the considerable volume of empirical evidence indicating that aspects of a perceptual field which function, operate or move in a similar manner tend to be perceived together. *See* GESTALT THEORY.

A D Ellis, ed., *A Sourcebook of Gestalt Psychology* (London, 1938)

NS

common good *Politics* Theory of shared INTERESTS. There exists a desirable end for governmental or public policy which is good for the whole society. This 'common good' can be discovered by informed and reasoned thought, and though it may overlap with the good of particular groups or individuals, is different from and greater than the interest of any one of them.

Roger Scruton, *A Dictionary of Political Thought* (London, 1982)

RB

common ion effect *Chemistry* A solubility effect which follows from the MASS ACTION LAW.

If an electrolyte with an ion in common with a sparingly soluble solute is added to a solution containing that solute then the solubility of the solute decreases, resulting in precipitation of the solute. For example, when the electrolyte sodium chloride is added to an aqueous silver chloride solution, silver chloride is precipitated.

M Freemantle, *Chemistry in Action* (Basingstoke, 1987)

MF

communalism *Politics* Advocacy of communes.

There is a positive value in individuals co-operating with each other in small groups, rather than relying on individual effort. Communalism differs from COLLECTIVISM in that it values groups small enough for all members to be familiar with one another; and in seeing the benefits of co-operation as consisting as much in social and cultural satisfaction as in material efficiency. Communalism provides the ethical aspiration of which COMMUNISM was, in its 19th century version, the strategic or political application.

Roger Scruton, *A Dictionary of Political Thought* (London, 1982)

RB

communication theory *Mathematics* The study of the effectiveness of communication channels in transmitting information, pioneered by C E Shannon in 1948.

The information is encoded, possibly using an error-correcting code (*see also* COMBINATORICS), transmitted through a channel (which may be subject to certain probabilities of error), and the message is then decoded. The average effectiveness of this procedure can be measured in terms of the *Shannon capacity* of the channel (*see also* INFORMATION THEORY). This area of probabilistic analysis is becoming increasingly relevant in high-speed telephone and computer networks.

AL

communications theory (20th century) *Politics/Sociology* Explanation of human society in terms of communication of meanings.

The existence of social relations and of human society depends on the creation of shared meanings and understandings. These in turn depend on the communication of such meanings amongst people, and in this process language plays a central role. There are many differing accounts of these processes. *See also* DIFFUSION THEORY.

Allan Bullock, Oliver Stallybrass, and Stephen Trombley, eds, *The Fontana Dictionary of Modern Thought*, 2nd edn (London, 1988)

RB

communicative competence (1966) *Linguistics* A term coined by the American ethnographer of communication Dell H Hymes.

Hymes criticized the limitations of A N Chomsky's theory of LINGUISTIC COMPETENCE: the 'ideal speaker-listener' does not correspond to real speakers' abilities; performance in Chomsky is negative, but could be reformulated as an account of language variation and appropriateness in contexts of use. Communicative competence would include the range of abilities requisite for a speaker to communicate effectively in real situations. *See also* ETHNOGRAPHY OF COMMUNICATION.

D H Hymes, 'On Communicative Competence' [1971], *Sociolinguistics*, J B Pride and J Holmes, eds (London, 1972)

RF

communicative dynamism (1960s) *Linguistics* Part of FUNCTIONAL SENTENCE PERSPECTIVE. The relative degrees to which the elements of a sentence 'contribute toward the further development of the communication' (Firbas). *See also* PRAGUE SCHOOL.

J Firbas, 'On the Dynamics of Written Communication in the Light of the Theory of Functional Sentence Perspective', *Studying Writing: Linguistic Approaches*, C R Cooper and S Greenbaum, eds (Beverly Hills, 1986), 40–71

RF

communism *Politics* Theory of ideal society.

The management of material resources should be a common enterprise of the whole society. A communist society would be characterized by the absence of private property, classes based on economic possession, or any form of state. The term was loosely and strictly nonsensically applied to the state socialist regimes of Eastern Europe between 1917 and 1989. Communism is thus a form of COMMUNALISM.

David Miller *et al.*, *The Blackwell Encyclopaedia of Political Thought* (Oxford, 1987)

RB

communitarianism (20th century) *Politics* Theory of democracy.

Because of the large number of voters in modern democracies, elections provide a greatly diluted form of political participation. Representative, indirect politics – with voters meeting and discussing and choosing at local level – achieves some of the involvement in a political community which otherwise would be lost.

Anne Phillips, *Engendering Democracy* (Cambridge, 1991)

RB

commutative law *Mathematics* The theorem or axiom of any mathematical system that a given binary operation $*$ has the property that the order of its arguments may be disregarded. In particular,

$$a*b = b*a\ .$$

Any binary operation possessing this property is said to be commutative. For example, ordinary addition of real numbers is commutative but subtraction is not. *See also* ASSOCIATIVE LAW and DISTRIBUTIVE LAW.

MB

compactness theorem (20th century) *Mathematics* The result in logic whereby a set of formulae has a model if and only if every finite subset has a model. An equivalent formulation is that a set of formulae is consistent if and only if every finite subset is consistent.

This result, which follows almost immediately from GÖDEL'S COMPLETENESS THEOREM, can be used to construct, among other things, non-standard models of arithmetic that allow for constructing models of the real numbers violating the ARCHIMEDEAN PROPERTY.

J E Rubin, *Mathematical Logic: Applications and Theory* (Saunders, 1990)

MB

comparative and historical linguistics (19th century) *Linguistics* The predominant linguistic approach of the 19th century, stimulated by the discovery of the ancient Indian language Sanskrit, and of pre-Christian Indian linguistics, at the end of the 18th century. *See also* INDO-EUROPEAN.

Study of the historical development of languages, especially the nature of sound change, comparison of the structures of languages, and reconstruction of 'family relationships' between them. *See* GRIMM'S LAW and VERNER'S LAW.

H Pedersen, *The Discovery of Language: Linguistic Science in the Nineteenth Century*, J W Spargo, trans. (Bloomington, 1962), on the linguists and their work; R Anttila, *An Introduction to Historical and Comparative Linguistics* (New York, 1972), on method

RF

comparative costs (18th–19th century) *Economics* A feature of the comparative advantage principle developed by English economists Robert Torrens (1780–1864) and David Ricardo (1772–1823). Postulates that trade is beneficial between countries even if one country is more efficient in every sector because of the difference in internal production costs. Ricardo used the example of English and Portuguese products.

PH

comparative judgment, law of *Psychology* Proposed by the German experimental psychologist Gustav Theodor Fechner (1801–87).

This holds that a stimulus will be judged relative to or in comparison with some other stimulus. It is a law commonly used in psychological investigations where a standard or anchor stimulus is employed to allow people to make other judgments for alternative stimuli. For example, in studies on colour discrimination people may be asked to compare the lightness or darkness of a sample of colours against a fixed standard colour. *See* DISCRIMINATION.

R S Woodworth and H Schlosberg, *Experimental Psychology* (London, 1966)

NS

comparison test (17th century) *Mathematics* Attributed to French mathematician and physicist Baron Augustin Louis Cauchy (1789–1857), this test establishes convergence of an infinite series with non-negative terms, by verifying that each term of the series is bounded above by the corresponding term of a convergent series.

For example, the series

$$\sum \frac{1}{2^k + k}$$

converges since

$$\tfrac{1}{2^k+k} < \tfrac{1}{2^k} \text{ and } \sum \left(\tfrac{1}{2}\right)^k$$

is a convergent geometric series.

H Anton, *Calculus with Analytic Geometry* (New York, 1980)

ML

compatibilism *Philosophy* View that freewill and DETERMINISM are compatible. Even though all our actions are caused, it is held, we can still be free in the only senses that are desirable or possible. (Indeed, it is sometimes added, we would not be free at all if our actions were uncaused, since they would then be arbitrary and unpredictable, and not really actions at all.) Compatibilists could in principle be indeterminists but in fact are nearly always (soft) determinists. Incompatibilists, who think that freewill and determinism are incompatible, may accept either of these and reject the other, though they mostly tend in practice to accept freewill and reject determinism.

R E Hobart, 'Free Will as Involving Determinism and Inconceivable Without it', *Mind* (1934); reprinted in B Berofsky, ed., *Free Will and Determinism* (1966)

ARL

compensation principle *Economics* This principle has its roots in the work of French engineer and economist Jules Dupuit (1804–66), English political economist Alfred Marshall (1842–1924) and Italian sociologist and economist Vilfredo Pareto (1848–1923). It refers to a transfer mechanism by which total economic welfare is maximized when individuals who gain from a change in the economy compensate those who have suffered because of the change.

The principle was subsequently developed into an important feature of welfare economics through the work of Hungarian-born economist Nicholas Kaldor (1908–86) and English economist John Hicks (1904–89). Since it does not rely on the physical transfer of money, critics maintain that the compensation principle lacks a quantifiable verification of the relative gains and losses. *See* COST BENEFIT ANALYSIS, PARETO OPTIMALITY, SCITOVSKY PARADOX and SOCIAL WELFARE FUNCTION.

V Pareto, 'Il massimo di utilita dato dalla libera concorrenza', *Giornale degli Economisti Series*, 2, 9 (July, 1894), 48–66

PH

competence, linguistic *Linguistics* *See* LINGUISTIC COMPETENCE.

competitive-exclusion principle (Gause principle) (1934) *Biology* Named after the Russian ecologist G F Gause.

If two species have identical resource requirements then they cannot coexist in the same environment, unless resources are unlimited. This is accepted in ecological thinking and agrees with the concept of the ecological niche.

M Begon, J L Harper and C R Townsend, *Ecology: Individuals, Populations and Communities* (London, 1987)

RA

complementarity theory (1927) *Physics* A theory attributable to the Danish physicist Niels Bohr (1885–1962).

This theory proposes that the wave and particle aspects of nature complement rather than contradict one another, since no experiment can ever reveal both. For example, an experiment that demonstrates the particle-like nature of electrons (for example, the photoelectric effect) will not show their wave-like nature (for example, as in electron diffraction), and vice versa. *See also* DE BROGLIE'S RELATION.

GD

complementary distribution *Linguistics* *See* DISTRIBUTION and PHONEME.

complementary phenomena *Psychology* Proposed by the German- American psychologist Wolfgang Köhler (1887–1967).

The term refers to the processes used by the visual system to structure incomplete stimuli or groups of stimuli into totalities. These processes include 'filling in' and enlarging in order to complete the visual field; for example, the completion of the blind spot. The phenomenon of perceptual reliability is explained by Gestalt laws of pragnanz, compactness, similarity, proximity and so on. Familiarity of experience is also influential.

L Kaufman, *Perception* (New York, 1979)

NS

complex function theory (19th century) *Mathematics* The study of functions whose value and argument are complex numbers. In particular, it is the study of analytic (also known as holomorphic) functions; that is, functions which possess a power series expansion in a neighbourhood of each point of the domain of definition. Such functions arise in all branches of mathematics, including Fourier analysis, partial DIFFERENTIAL EQUATIONS, ALGEBRAIC GEOMETRY, and NUMBER THEORY. These functions are also used to model important physical phenomena in fields such as aerospace engineering, geophysics, and hydrodynamics. The special feature of analytic functions is that properties on a small open set determine the behaviour of the function on its entire domain. *See also* CAUCHY'S INTEGRAL THEOREM.

S G Krantz, 'Functions of One Complex Variable', *Encyclopedia of Physical Science and Technology* (Academic Press, 1987)

MB

complex interdependence (20th century) *Politics* Theory of international politics. Nations are involved in a complex network of interdependence such that states are not necessarily the appropriate unit of analysis.

Graham Evans and Jeffrey Newnham, *The Dictionary of World Politics* (Hemel Hempstead, 1990)

RB

complexity theory *Mathematics* The attempted classification according to difficulty of *algorithms* (*see also* COMPUTABILITY THEORY) and computational methods, often for problems in COMBINATORICS or OPTIMIZATION THEORY.

Common measures count the number of operations (floating or fixed point, or binary for example) required for solution, with particular concern for its eventual growth as a function of problem size. The method is polynomial time (or in class *P*) if the growth is bounded by a polynomial (*Compare with* KHATCHIAN ALGORITHM), while it is in class *NP* if a solution can be verified in polynomial time. The celebrated open conjecture $P = NP$ (generally believed false) asserts that these two classes are identical. *See also* NP-COMPLETE PROBLEM and P=NP PROBLEM.

AL

componential analysis (1950s) *Linguistics* First proposed in ANTHROPOLOGICAL LINGUISTICS for analysis of kinship terminology; draws on concepts of DISTINCTIVE FEATURE analysis in PHONOLOGY; adapted for TRANSFORMATIONAL-GENERATIVE GRAMMAR by the American linguists Jerrold Jacob Katz (1932–) and Jerry Alan Fodor (*see* INTERPRETIVE SEMANTICS).

Meanings of words can be represented as ordered bundles of semantic features or markers, perhaps universal, and usually regarded as binary in form: ±HUMAN, and so on. Thus, *child* would be +HUMAN, −ADULT, 0MALE. Componential semantic analysis has been applied with interesting results to distinct and closed areas of the vocabulary such as kinship terminology and classifications of animals, but becomes very difficult to formalize for the general vocabulary. *See also* PROTOTYPE.

W H Goodenough, 'Componential Analysis and the Study of Meaning', *Language*, 32 (1956), 195–216; G Leech, *Semantics* (Harmondsworth, 1974)

RF

composite commodity (1939) *Economics* Developed by English economist John Hicks (1909–89), this term describes a group of goods whose relative prices do not vary and can thus be treated as one commodity. *See* OWN RATE OF INTEREST.

J R Hicks, *Value and Capital* (Oxford, 1939)

PH

compound nucleus theory of Bohr and Breit and Wigner (1936) *Physics* This theory provides a useful model for the description of low-energy nuclear reactions, and is named after the Dane Niels Henrik David Bohr (1885–1962), Russian-American Gregory Breit (1899–1981) and the Hungarian-born American Eugene Paul Wigner (1902–) who introduced it.

On this model a low-energy nuclear reaction is considered to take place in two stages. In the first stage the bombarding particle is absorbed by the target nucleus to form an excited unstable compound nucleus in which the excitation energy is rapidly shared among all the nucleons. In the second stage, the compound nucleus emits particles or radiation to leave a lower energy nucleus, known as a recoil nucleus.

J Thewlis, ed., *Encyclopaedic Dictionary of Physics* (New York, Oxford and London, 1962)

MS

comprehension axiom *Mathematics* Also known as the axiom of specification, this principle of SET THEORY states that for every set *A* and every condition *S*, there is a set *B* whose elements are exactly those elements of *A* for which *S* holds. *See also* CANTOR'S PARADOX.

P R Halmos, *Naive Set Theory* (Springer, 1974)

MB

Compton's rule *Chemistry* Named after Arthur H Compton (1892–1962), this states that the melting point of an element on the absolute scale of temperature (in kelvins) equals half the product of the relative atomic mass of the element and its specific latent heat of fusion. The rule is empirical.

MF

computability theory *Mathematics* The study of the capabilities, limitations and complexity (*see also* COMPLEXITY THEORY) of automatic procedures or *algorithms* for solving problems, which originated in D Hilbert's aims to formalize mathematics (1901).

Such procedures are often modelled on a simple idealized computer called a TURING MACHINE. Hilbert's original aim was thrown into doubt after the exhibition by Kurt Gödel (1906–78) of unsolvable problems: an example is the HALTING PROBLEM for Turing machines.

AL

computational linguistics (1950s) *Linguistics* This is a general term covering a variety of applications for computational procedures in the area of natural language, the most important being ARTIFICIAL INTELLIGENCE AND LANGUAGE, computational analysis of a CORPUS, and MACHINE TRANSLATION. Also included are such topics as the use of the computer in literary studies, STYLISTICS, authorship studies, and so on.

D G Hays, *Introduction to Computational Linguistics* (London, 1967)

RF

computational psychology *Philosophy* An approach to learning which postulates events in the brain which 'represent' inferences and so on. It mediates between methodological BEHAVIOURISM and a purely introspective approach. It broadens out into cognitive science when it studies artificial intelligence and areas bordering on CYBERNETICS and so on. *See also* CONNECTIONISM.

M A Boden and D H Mellor, 'What is Computational Psychology?' (symposium), *Proceedings of the Aristotelian Society*, supplementary volume (1984)

ARL

conceit (early 20th century) *Literary Theory* Poetic device which became popular with the rediscovery of the METAPHYSICAL POETS.

In the 16th century, the 'Petrarchan conceit' was a hyperbolic comparison of the attributes of one thing (for example the poet's mistress) with those of another, somewhat distant (for example a landscape). In the early 17th century, the English metaphysical poets specialized in conceits based on elaborate and far-fetched comparisons (*see* John Donne's 'The Flea'). The style was reintroduced in the 1920s.

K K Ruthven, *The Conceit* (London, 1969)

RF

concentration theory of cutaneous cold *Psychology* Proposed by the American W L Jenkins.

Developed from the SPOT THEORY OF TOUCH, this theory states that the perception of cold intensity depends on the average concentration of active spots of encapsulated nerve endings beneath the area of the skin stimulated. There are insufficient encapsulated endings to satisfy the theory, but Jenkins's hypothesis stimulated considerable work on the primary functional sensory systems in the skin: warmth, cold, touch and pain.

R S Woodworth and H Schlosberg, *Experimental Psychology* (London, 1966)

NS

conceptual art (1967) *Art* A cerebral approach to art first championed in 1967 by the American sculptor Sol Lewitt (1928–) in the magazine *Art Forum* as a reaction against the FORMALISM of post-war art.

The idea or concept is the most important aspect in the artistic process. The planning and concept are decided beforehand, but the end result is intuitive and without recognizable purpose.

U Meyer, *Conceptual Art* (1972)

AB

conceptualism *Philosophy* Any view which emphasizes concepts when analyzing something.

Primarily, conceptualism is a view about universals (things normally denoted in English by words ending in '-hood', '-ness', or '-ty'). It says that these are concepts in the mind (though not necessarily confined to an individual mind), and neither non-material objects with a real existence independent of any mind (as realists hold) nor mere words (as nominalists hold). In connection with IDENTITY, conceptualism says a governing concept is always involved: A cannot just be the same thing as B; it must always be the same so-and-so as B.

H Stamland, *Universals* (1972)

ARL

concrete art (1930) *Art* Dutch artist Theo van Doesburg (1883–1931) coined this name in his manifesto (also signed by Carlsund, Helion, Tutundjian and Wantz) for a distillation of Constructivist ideas aiming to create self-sufficient art, using planes and colours, with no other significance than itself and using controlled and precise techniques.

In 1936 the Swiss artist Max Bill (1908–) and Frenchman Jean Arp (1887–1966) adopted this term, Bill organizing several exhibitions of '*konkrete Kunst*'. After World War II these ideas were adopted by groups in Sweden (*Concretists*) and Italy (*Movimento per l'arte concreta*).

T van Doesburg, *Art Concret* (1930)

LL

concrete poetry (early 20th century) *Literary Theory* Essentially due to MODERNISM, but there existed an earlier Renaissance tradition of 'emblems', for example poems of George Herbert (1593–1633) shaped like wings or an altar.

Poem, usually a single page for instantaneous apprehension, in which format and typography offer a visual meaning. Often 'iconic', the shape of the poem representing some non-linguistic object.

M E Solt, ed., Concrete Poetry: A World View (Bloomington, 1968)

RF

concrete universal (1947) *Literary Theory* Phrase coined by the American critic William Kurtz Wimsatt (jnr) (1907–75) for a traditional belief about poetry.

Poetry uniquely exhibits a linguistic paradox: it refers to, or creates the fiction of, highly specific, individual, concrete, entities; but treats them in such a way that the themes or implications of poems have universal application.

W K Wimsatt, *The Verbal Icon* (Kentucky, 1954)

RF

conditioned reflex *Biology See* PAVLOV'S DOG.

Condorcet's principle (1785) *Economics* Named after French mathematician and philosopher Marie Antoine Condorcet (1743–94), this principle deals with the complexity of voting and choices, by which the final choice is made by rejection of all other alternatives in a series of paired contests.

Condorcet observed that 'majority voting is the best voting rule when only two people can vote'. *See* PARADOX OF VOTING, COMPENSATION PRINCIPLE, SOCIAL WELFARE FUNCTION and IMPOSSIBILITY THEOREM.

M Condorcet, *Essai sur l'application de l'analyse à la probabilité des décisions rendues à la pluralité des voix* (Paris, 1785)

PH

confirmation principle *Philosophy* Alternative name for a weak version of the VERIFIABILITY PRINCIPLE, whereby in order to be meaningful a statement must, if not a tautology, be confirmable or disconfirmable by observation.

ARL

conflict theory *Politics/Sociology* Theory of politics as moderated antagonism.

Politics is best seen as competition or conflict over resources, power, or prestige. The term covers a wide range of theories, amongst which, appropriately, there is no agreement, but of which MARXISM is the most familiar. *See also* DIALECTICAL MATERIALISM.

Alan Bullock, Oliver Stallybrass, and Stephen Trombley, eds, *The Fontana Dictionary of Modern Thought*, 2nd edn (London, 1988)

RB

conflicting behaviour effect *Psychology* A theory proposed by the American psychologist Burrhus Frederic Skinner (1904–90).

When a stimulus produces two types of behaviour simultaneously, and a conflict exists between these two behaviours, a number of consequences may result: the behaviours may extinguish one another; they may alternate, causing ambivalent behaviour; they may repress or encourage each other; or they may completely reorientate behaviour. For example, if fighting cocks are made simultaneously to fight and to run away, they begin to pretend to peck at the ground. *See* ADVANTAGE, PRINCIPLE OF.

B F Skinner, *The Behaviour of Organisms* (New York, 1938)

NS

Confucianism (5th century BC) *Philosophy* Body of teaching associated with the Chinese philosopher Confucius (*c*.551–479 BC).

Confucianism was the traditional state religion of China until the Communists suppressed it after the Cultural Revolution in the 1960s. Confucian ideas are drawn from the five books of 'Analects' compiled from the sayings of Confucius himself and his disciples. Although followers acknowledge the existence of a 'divine will' (*ming*), they place great emphasis upon the moral responsibility of man for his deeds. Men are born with innate goodness and must develop their capacity for greater wisdom through benevolence, self-control, and obedience to their own true nature. Strict codes of social behaviour and ritual sacrifice to one's ancestors are among the features of Confucianism, which still exerts a strong influence upon contemporary Chinese society and culture.

H G Creel, *Confucius* (New York, 1949; London, 1951)

DP

congruity theory/principle *Psychology* Proposed by the American psychologist Charles Egerton Osgood (1916–).

Attitude change is alleged to depend upon the discrepancy between the initial attitudes of the receiver towards the source and the content of the message, taking into account the positive and negative nature of the communication. Predictions of actual attitude change are not particularly successful and various additional assumptions must be made: a weighted averaging model, in which the relative weight of the source depends on the attitude of the receiver to the source, describes attitude change more successfully. *See* COGNITIVE CONSISTENCY THEORY and DISSONANCE THEORY.

C E Osgood and P H Tannenbaum, 'The Principle of Congruity in the Prediction of Attitude Change', *Psychological Review*, vol. LXII (1955), 42–55

NS

connectionism *Psychology* Proposed by the American psychologist Edward Lee Thorndike (1874–1949).

Simpler mental or behavioural elements are combined, associated or connected to form higher-order mental or behavioural processes and these elements are defined as stimuli and responses. The theory fell into disuse, being regarded as too simplistic to explain complex cognitive processes. More recently there has been a resurgence of interest with the availability of sophisticated computers which can be programmed to perform extremely complex pattern-recognition tasks using the principles contained within connectionism. *See* ASSOCIATIVE LEARNING and ASSOCIATIONISM.

E L Thorndike, *Animal Intelligence* (New York, 1911)

NS

connectionism *Philosophy/Computing* A theory of the mind with many versions. They have in common that they set up models which employ simple interactions between the nodes in a computer network in such a way that sets of these interactions occur at the same time (or 'in parallel', hence 'parallel processing').

This uses the information processing in the brain or nervous system as a model, and dispenses with separate elements in the system to carry the separate pieces of information; for example, sentences in a code which represent memories, thoughts, and so on. (*See also* TRACE THEORY OF MEMORY, LANGUAGE OF THOUGHT.) It is disputed whether this whole approach represents a fundamentally new way of looking at thought, or whether it simply gives a microlevel analysis of an older or classical view such as that used in cognitive psychology.

W Bechtel and A Abrahamsen, *Connectionism and the Mind* (1991)

ARL

connexive implication *Philosophy* Term used in a kind of RELEVANCE LOGIC, existing in different versions but similarly motivated and using ideas from Aristotle (384–322 BC) and Boethius (*c.*AD 480–524).

The relevant kind of implication is defined as holding when the antecedent of a conditional proposition is incompatible with the negation of the consequent. This bans implications of the forms (where P and Q are propositions) 'If not P then P' and 'If P then Q, and if P then not Q', though both of these are valid in classical logic.

Connexive implication has been criticized for leading to the exclusion of logical principles it is in fact implausible to exclude.

R Routley and H Montgomery, 'On Systems Containing Aristotles' Thesis', *Journal of Symbolic Logic* (1968); critical, with references to expositions of the idea

ARL

consciousness, motor theory of *Psychology* Proposed by the American psychologist John Broadus Watson (1878–1958).

Consciousness is regarded as no more than an epiphenomenon; muscular and glandular action represent the true realities which are amenable to scientific investigation. The theory postulates that what one may actually experience as consciousness is a mere correlate of action, while what one senses or perceives is dependent on how one reacts to it. The theory is not accepted today. *See* BEHAVIOURISM.

B F Skinner, *About Behaviourism* (New York, 1974)

NS

consciousness raising (20th century) *Politics/Feminism* Feminist theory of understanding.

Women's understanding of their social and political situation is achieved by group discussion in which personal experience is related and re-assessed. Individual experience is thus shown to be both typical (rather than isolated or abnormal), and a basis for wider social and political action and understanding.

Maggie Humm, *The Dictionary of Feminist Theory* (London, 1989)

HR

consent *Politics/Law* Theory of basis of political obligation.

The sole justification for enforcing laws is if those against whom they are enforced have consented to the general arrangements in which the laws are drawn up. Thus the consent of the governed is the only thing that can legitimate a system of government. There is disagreement about how one can tell when that consent has been given, and in what ways it is given.

H Beran, *The Consent Theory of Political Obligation* (Oxford, 1987)

RB

consequentialism *Philosophy* Doctrine that the moral rightness of an act or policy depends entirely on its consequences; the moral goodness of the agent depending on the act's expected or intended consequences.

This is one form of TELEOLOGY. UTILITARIANISM is one form of consequentialism. Objections include the apparent moral counterintuitiveness of many consequentialist prescriptions, especially in connection with justice, reward and punishment, and with the difficulties in deciding what the consequences of an action are. *See also* DEONTOLOGY.

P Foot, 'Utilitarianism and the Virtues', *Mind* (1985); with discussion by S Scheffler

ARL

conservation laws *Physics See* CONSERVATION OF CHARGE, CONSERVATION OF ENERGY, CONSERVATION OF MASS-ENERGY, CONSERVATION OF MATTER and CONSERVATION OF MOMENTUM.

GD

conservation of charge *Physics* The principle states that the total net charge of any system remains constant.

GD

conservation of energy *Physics* *See* CONSERVATION OF MASS-ENERGY.

conservation of interfacial angle (1669) *Physics* Named after its discoverer, the Danish naturalist Nicolas Steno (1638–86).

From measurements of the interfacial angles of sections cut from differently distorted crystals of quartz, Steno showed that analogous angles in the different sections were always the same, whatever the actual size and shape of the sections themselves. The law is true for crystals generally, as was shown by Domenico Guglielmini (1655–1710) in the years 1688–1705. *See also* STENO'S LAWS.

J Thewlis, ed., *Encyclopaedic Dictionary of Physics* (New York, Oxford and London, 1962)

MS

conservation of mass-energy *Physics* The principles of the conservation of energy and of the conservation of mass state, respectively, that for an isolated system, the total energy and the total mass remain constant. In classical physics, these two laws were independent despite limited experimental evidence for the conservation of mass. In Einstein's theory of relativity, the conservation of energy is assumed, and the transfer of energy E in any process involves the transfer of mass $m = E/c^2$, where c is the speed of light in a vacuum. Hence the conservation of energy ensures the conservation of mass.

Robert M Besançon, ed., *The Encyclopedia of Physics* (New York, 1985)

GD

conservation of matter (19th century) *Physics*

This principle states that matter can neither be created nor destroyed. In ordinary chemical reactions, it is also known as the conservation of mass, by which the total mass of the reactants is always equal to the total mass of the products. *See also* CONSERVATION OF MASS-ENERGY.

conservation of momentum *Physics* In any system of interacting particles, the linear momentum in any fixed direction remains constant unless an external force acts in that direction. Similarly, for a system rotating about a fixed axis, the angular momentum remains constant unless an external torque is applied.

Robert M Besançon, ed., *The Encyclopedia of Physics* (New York, 1985)

GD

conservatism (19th century–) *Politics* Political theory based on tradition.

The institutions of human society and government have evolved slowly and survive because they have stood the test of time. Social life is thus formed by processes beyond the rational comprehension of any single person. The principle virtue is thus caution, in not seeking ambitiously to go beyond the limits of individual reason. Human abilities are not equally distributed, and it is for the good of all that those of superior ability lead in politics, economics and social affairs. Private PROPERTY is good for individuals and for society as a whole, and governments should preserve it but not interfere in its distribution.

Noel O'Sullivan, *Conservatism* (London, 1976)

RB

conservative paradox (1991) *Politics* Theory of government described by English political scientist Rodney Barker (1942–).

Conservatives, and others, assume that those whom they govern but who are in revolt or rebellion or are otherwise disaffected, share the same values and beliefs as their rulers. Therefore, like naughty children who accept the values of their parents, they will be not only subdued but chastened by the coercive use of state power. In fact, since the disaffected citizens hold values different from those of their rulers, they will regard state coercion against them as further proof of government's unfitness. Thus in seeking to sustain power, government undermines it.

Rodney Barker, *Political Legitimacy and the State* (Oxford, 1990)

RB

consistent empiricism *Philosophy* Name given by the German philosopher Moritz Schlick (1882–1936), a member of the VIENNA CIRCLE, to his own version of LOGICAL POSITIVISM.

M Schlick, 'Meaning and Verification', *Philosophical Review* (1936), 343

ARL

consolidation theory of learning *Psychology* Proposed by the Canadian psychologist Donald Olding Hebb (1904–).

That which is learned is hypothesized to undergo a process of consolidation whereby that material is retained. The process of consolidation is thought to involve long-term neurophysiological changes, although the exact operations of the nervous system which account for consolidation are not known. In general, consolidation involves some change in neurophysiology relating to an acquired behaviour. The term has also been used as a kind of neurological metaphor for the transition from short-term to long-term memory. *See* DECAY THEORY.

A D Baddeley, *Working Memory* (Oxford, 1986)

NS

conspiracy *Politics/History* Theory of political power.

In many or all social and political situations, there is a small group of people who deliberately, in concert, and in secret pursue an objective of their own. They do so under cover of the ostensible constitution or political arrangements. Conspiracy theory has variously identified communists, Jews, the secret service, business, Masons, and many others as the principal actors. Paradoxically, conspiracy theory which attributes overwhelming power to small groups has often been used to justify the power of other groups and the persecution of minorities.

Roger Scruton, *A Dictionary of Political Thought* (London, 1982)

RB

constancy, law/principle of *Psychology* First proposed by the Austrian founder of psychoanalysis Sigmund Freud (1856–1939).

This asserts that all mental processes tend towards the state of equilibrium and stability of the inorganic state. It is similar to the nirvana principle in psychoanalysis which holds that it is the tendency of all instincts and life processes to seek the stability and equilibrium of the inorganic state. This is the goal of the death instinct.

E Jones, *The Life and Works of Sigmund Freud* (New York, 1953)

NS

constancy hypothesis *Psychology* Proposed by the German-American psychologist Wolfgang Köhler (1887–1967).

Posits that strict ISOMORPHISM (a one-to-one correspondence) holds between the proximal stimulus and the sensory experience. The hypothesis assumes that the context within which the observation takes place will have absolutely no effect on perception: a given stimulus will always produce a given response. The absence of a treatment of contextual or environmental factors constrains the validity of the hypothesis. *See* BRAIN FIELD THEORY.

A D Ellis, ed., *A Sourcebook of Gestalt Psychology* (London, 1938)

NS

constant composition, law of (1797) *Chemistry* Law attributed to French chemist Louis Joseph Proust (1755–1826). Also known as the law of constant proportions, the law of definite proportions, and the law of definite composition.

A pure compound always consists of the same elements combined in the same proportion by weight. For example, the compound carbon dioxide (CO_2) invariably contains 27.37 per cent carbon and 72.73 per cent oxygen by mass. This is equally true whether the samples are taken, for example, from the North Pole, the South Pole, the Sahara Desert or from the Moon. The French chemist Claude Louis (Compte) Berthollet (1748–1822) had believed that compounds of varying compositions were possible. Proust's law helped to disprove this theory until the 1930s when intermetallic compounds and some sulphides with slightly variable compositions were discovered. Such compounds are now called berthollides to distinguish them from daltonides, or compounds conforming to this law.

M Freemantle, *Chemistry in Action* (Basingstoke, 1987)

MF

constant heat summation, law of *Chemistry* *See* HESS'S LAW.

constant proportions, law of *Chemistry* *See* CONSTANT COMPOSITION, LAW OF.

constituent, constituent structure (1933/1957) *Linguistics* Basic unit of analysis proposed by Leonard Bloomfield (1887–1949) for American STRUCTURAL LINGUISTICS, formalized by Avram Noam Chomsky (1928–) as PHRASE STRUCTURE GRAMMAR; based on PARSING in TRADITIONAL GRAMMAR.

A constituent is a well-formed syntactic unit which is a part of another unit. Each of the words in 'The child is father of the man' is a constituent, and also 'the child', 'is father of the man', 'father of the man', 'of the man' and 'the man'; other sequences such as 'child is' and 'of the' are not constituents. Constituents are included in one another hierarchically, as can be seen in a TREE DIAGRAM or BRACKETING representation.

K Brown and J Miller, *Syntax*, 2nd edn (London, 1991)

RF

constitutional theory (2nd century–) *Psychology* First proposed by the Greek physician Galen (AD 130–*c.*200).

A general term for theories postulating that specific psychological characteristics are associated with aspects of the physical constitution. There are three major constitutional theories. Galen's is based on four basic types: sanguine, melancholic, choleric and phlegmatic. The German psychiatrist Ernst Kretschmer (1888–1964) proposed three basic types: pyknic (stocky), aesthenic (slender), and athletic (muscular); and one mixed type, termed dysplastic (disproportioned). The American psychologist William Herbert Sheldon (1898–1970) hypothesized three fundamental constitutional types: ectomorphic (thin), mesomorphic (muscular) and endomorphic (fat). *See also* HUMORAL THEORY.

W H Sheldon and S S Stevens, *Varieties of Human Temperament: A Psychology of Constitutional Differences* (New York, 1942)

NS

constitutionalism *Politics* Theory of government limited by law.

Constitutions, either in the form of written law or established convention, can provide a framework within which government should operate, and by which it can be prevented from despotic conduct or the infringement of rights. The theory of constitutionalism has much in common with that of the RULE OF LAW. It is an optimistic theory, and has been criticized for failing to examine the realities of political power beneath legal or constitutional surfaces.

David Miller, ed., *The Blackwell Encyclopaedia of Political Thought* (Oxford, 1987)

RB

constructivism (1914) (1) *Art* Russian avant-garde movement pioneered in *c.*1914 by the artist Vladimir Tatlin (1885–1953) and current until *c.*1921.

Following the examples of collage in CUBISM and FUTURISM, Tatlin proposed a 'culture of materials' in which ILLUSIONISM and simulated effects in art were eschewed in favour of an art based on the construction of real materials. After 1917, artistic links with industry were emphasized together with the development of a non-objective 'production art' for the improvement of society. Outside the USSR, the Russian brothers Naum Gabo (1890–1977) and Antoine Pevsner (1886–1962) pursued a less austere interpretation of the movement, as set out in their *Realistic Manifesto* (1920).

AB

(2) *Literary Theory* Similarly inspired by modern engineering and technology, literary adherents in the 1920s demanded that poems should be integrated artefacts like machines, constructed with economy and achieving their effects swiftly. Prominent members were A N Chicherin (1889–1960) and I Selvinski (1899–1968).

G Janecek, ed., *The Look of Russian Literature: Avant-Garde Experiments, 1900–1930* (Princeton, 1984); G Rickey, *Constructivism. Origins and Evolution* (1968)

RF

constructivism *Philosophy/Mathematics* A view in the philosophy of mathematics which insists that mathematical entities (numbers, sets, proofs, and so on) can only be said to exist if they can be constructed; that is if some method can be specified for arriving at them on the basis of things we accept already.

One advantage of this is that various paradoxes can be excluded before they arise. A disadvantage may be that certain things are excluded that appear to be intuitively acceptable. Varieties of constructivism include INTUITIONISM, and (usually) FINITISM, while FORMALISM is sometimes included and sometimes contrasted with it.

ARL

consumer demand, theory of (20th century) *Economics* This is the analysis of demand with regard to consumer behaviour and rationale when changes occur in variable factors such as price, income, substitute goods. Choice and revealed preference are two important factors affecting consumer demand. *See* AXIOM THEORIES.

P Newman, *The Theory of Exchange* (Englewood Cliffs, N.J., 1965)

PH

consumer surplus (19th–20th century) *Economics* Identified by French engineer and economist Jules Dupuit (1804–66) and later developed by English economist Alfred Marshall (1842–1924), this theory assumes that the price paid by consumers for a good never exceeds and seldom equals the amount they are willing to pay rather than forgo the good.

The satisfaction derived from the good is greater than that derived from products given up in making the purchase; thus, the consumer receives a surplus in satisfaction in excess of the price paid for the good. *See* COST BENEFIT ANALYSIS, PRICE DISCRIMINATION, SOCIAL WELFARE FUNCTION and COMPENSATION PRINCIPLE.

A Marshal, *Principles of Economics* (London, 1890); A Bergson, 'A Note on Consumer's Surplus', *Journal of Economic Literature*, vol. XIII (1975), 38–44

PH

contemporaneity principle *Psychology* Term coined by the German-American psychologist Kurt Lewin (1890–1967).

Encompasses any approach to the investigation and explanation of behaviour which stresses the role of current circumstances, rather than previous circumstances and behaviours. No psychological theory can be completely ahistorical; behaviourist approaches, for example, would have minimal reference to past experience whereas classical psychoanalytic theories are strongly historical. *See* CONTEXT THEORY.

K Lewin, *Principles of Topological Psychological* (New York, 1938)

NS

contestable markets theory (1982) *Economics* Developed by American economist William Baumol (1922–), this theory defines contestability as the effectiveness of barriers to entry and exit in a market. Perfect competition, with complete freedom of movement, is perfectly contestable. By removing or reducing barriers, competition will be enhanced. *See* IMPERFECT COMPETITION, MONOPOLISTIC COMPETITION and STRUCTURE-CONDUCT-PERFORMANCE THEORY.

W J Baumol, J C Panzar and R D Willig, *Contestable Markets and the Theory of Industry Structure* (New York, 1982)

PH

context effect *Psychology* Proposed by the American educationist and philosopher John Dewey (1859–1952).

Nothing an individual can do can take place in a vacuum. Context effect is a general term for the behavioural effects resulting from the specific context within which a stimulus is presented or a response is made. Because context effects are omnipresent, the significance of a particular behaviour or response cannot be ascertained in isolation, but must be studied in context.

T R Sarben, 'Contextualism: A World View for Modern Psychology', *Nebraska Symposium on Motivation*, A W Landfield, ed. (Lincoln, 1977)

NS

context of situation (1923) *Linguistics* A term coined by the Polish anthropologist Bronislaw Malinowski (1884–1942) (*see also* PHATIC COMMUNION); the linguistic theory was formulated by the British linguist John Rupert Firth (1890–1960) in the 1930s; later, much developed by M A K Halliday (1925–).

Malinowski observed that utterances in unfamiliar languages only become intelligible when related to what is going on in the situation in which they are spoken. Firth formalized context of situation by specifying its components in such a way that typical situation types could be identified and described, thus clarifying the description of meaning in speech events.

J R Firth, *Papers in Linguistics 1934–1951* (London, 1957)

RF

contextualism *Philosophy* Any view that sees some phenomenon as relative to a context, or insists on the relevance of context for interpretation.

In aesthetics, the doctrine that works of art can be appreciated only by reference to their context, circumstances of production, artist's intuitions, and so on (*see also* ISOLATIONISM). In ethics, the view that values are instrumental (that is, relative to certain ends); and the view that moral problems only arise – and can only be solved – against a background of principles which themselves can only be assessed by taking some further principles for granted for the moment. In philosophy of science, the doctrine that theoretical terms can only have contextual meaning; that is meaning which they get by playing a role in a deductive system with empirically testable consequences.

ARL

contiguity theory *Psychology* Proposed by one of the most influential learning theorists, the American psychologist E R Guthrie (1886–1959).

All learning occurs in one trial or not at all: the last observed response to occur in the presence of a stimulus is the one which is assumed to have been learned. In order to account for gradual improvements in performance, the theory holds that most behaviours are made up of many complex responses. Most stimuli are also complex so that many stimulus-response connections are needed for effective behaviour. The theory was particularly popular when BEHAVIOURISM occupied a predominant position in learning theory, but it has since been abandoned. *See* CONTINUITY THEORY and DISCONTINUITY THEORY OF LEARNING.

E R Guthrie, *The Psychology of Learning* (New York, 1935)

NS

continental drift theory (1912) *Biology/Geology* The hypothesis that continents move independently over the Earth's surface in a similar way to icebergs. This theory was supported by evidence collated by the German meteorologist Alfred Wegener (1880–1930) during the early 20th century.

On the basis of palaeontological evidence, it was suspected in the 19th century that now widely separated continents were once interconnected. These super-continents were termed Laurasia (now North America, Greenland, Europe and Siberia) and Gondwanaland, possibly united as Pangaea some 300 million years ago. (In biology, continental drift helps explain the wide geographic distribution of species that are closely related genetically.) As the notion of drift came to be accepted, ocean basins were understood not to be sunken land masses (as had been posited), but geologically-recent phenomena formed as the continents moved apart. The theory has now been subsumed by PLATE TECTONICS. *See also* GONDWANALAND HYPOTHESIS, SEA-FLOOR SPREADING and VICARIANCE.

A Wegener, *The Origin of Continents and Oceans* (1915); W Glen, *Continental Drift and Plate Tectonics* (Columbus, Ohio, 1975); U B Marvin, *Continental Drift: The Evolution of a Concept* (Washington, 1973)

KH, DT

Continental rationalists *Philosophy* Name primarily applied to René Descartes (1596–1650), Baruch Spinoza (1632–77), and Gottfried Wilhelm Leibniz (1646–1716), together with various lesser figures including Cartesians (followers in a general sense of Descartes) like Arnold Geulincx (1625–69) and Nicolas Malebranche (1638–1715).

See also RATIONALISM, OCCASIONALISM, DOUBLE ASPECT THEORY OF MIND, PRE-ESTABLISHED HARMONY, BRITISH EMPIRICISTS.

ARL

contingency theory *Biology* *See* 'WONDERFUL LIFE' THEORY.

contingency theory *Politics/History* A sceptical theory of politics and history.

Any historical situation or set of political events is as likely to be shaped by particular circumstances – 'contingencies' – as by any general structures, rules, or frameworks.

Understanding thus depends as much on looking at the peculiarities of the case as at the general conditions in which it occurs. *See also* CLEOPATRA'S NOSE and COCK-UP THEORY.

RB

contingency theory of leadership *Psychology* Proposed by the Austrian psychologist Fred Edward Fiedler (1922–).

The theory posits two classifications of leaders: (1) those motivated by the need to accomplish assigned tasks (task-orientated); (2) those motivated by close and supportive relations with members of the group (people-orientated). The effectiveness of the leader is contingent upon both the leader's personality and the characteristics of the leadership situation. The model has generated considerable controversy and led to the development of more than 120 published tests, many of which are in commercial use. *See also* FIEDLER'S CONTINGENCY MODEL OF LEADERSHIP, LEADERSHIP THEORIES, PATH-GOAL THEORY and NORMATIVE DECISION THEORY.

F E Fiedler, *A Theory of Leadership Effectiveness* (New York, 1967)

NS

continued fraction algorithm *Mathematics* A procedure which can be used to express any real number as a continued fraction; that is, in the form

$$a_0 + \cfrac{1}{a_1 + \frac{1}{a_2 + \cdots}}.$$

For example,

$$\frac{1+\sqrt{5}}{2} = 1 + \cfrac{1}{1 + \frac{1}{1+\cdots}}.$$

For a given number α, this procedure generates a sequence of successively better rational approximations to α (called the convergents to α) in such a way that every rational number in the sequence is the best approximation among all rational numbers having the same denominator. For example, the first three convergents to π are

$$\frac{22}{7}, \frac{333}{106}, \frac{355}{113}.$$

This algorithm can be used to solve PELL'S EQUATION and is also related to the EUCLIDEAN ALGORITHM.

A Baker, *A Concise Introduction to the Theory of Numbers* (Cambridge, 1984)

MB

continuity equation *Physics* This is an expression of the law of conservation of mass. It implies that matter cannot be created or destroyed and states that the rate of increase of matter in a fixed volume is equal to the difference between the rate of flow of matter into the volume and the rate of flow of matter out of the volume.

J Thewlis, ed., *Encyclopaedic Dictionary of Physics* (New York, Oxford and London, 1962)

MS

continuity, law or principle of *Philosophy* Principle of Gottfried Wilhelm Leibniz (1646–1716) which can be roughly rendered as saying that when the difference between two causes is diminished indefinitely, so is the difference between their effects (though Leibniz would not put it in these causal terms, since for him God is the only true cause). 'Nature makes no leaps', as he says in the Preface to his *New Essays on the Human Understanding* (*c.*1704). *See also* PRE-ESTABLISHED HARMONY.

G H R Parkinson, ed., *Leibniz: Philosophical Writings* (1973), 158, with refs

ARL

continuity theory/hypothesis *Psychology* Proposed by the American psychologist Clark Leonard Hull (1884–1952).

A general theoretical principle of learning which has appeared in a variety of models. Learning results from a progressive, continuous process of trial and error. Unproductive responses are extinguished while those yielding results are strengthened. Problem-solving is seen as a step-by-step learning process in which the correct response is discovered, practised and reinforced. The theory does not account particularly well for 'sudden insight' and successful performance on the first trial. *See* CONTIGUITY THEORY.

C L Hull, *Principles of Behaviour* (New York, 1943)

NS

continuity thesis *Economics* An assertion that a continuum exists between the classical and neoclassical schools of economy which was not disrupted by the rise of the Marginalist movement of the 1870s.

The thesis maintains that English economist Alfred Marshall (1842–1924) did not overturn classical economics but simply used sharper mathematical tools to refine Ricardian economics. *See* NEOCLASSICAL THEORY.

G F Shove, 'The Place of Marshall's Principles in the Development of Economic Theory', *Economic Journal*, vol. LII (1942), 294–329

PH

continuum hypothesis (1878) *Mathematics* Due to the German mathematician Georg Ferdinand Ludwig Philip Cantor (1845–1918), this is the celebrated hypothesis that there is no set with cardinality strictly greater than the cardinality of the set of rational numbers, but strictly less than the cardinality of the continuum (that is, the set of real numbers).

The hypothesis is consistent with the axioms of SET THEORY as demonstrated by Kurt Gödel (1906–78) in 1940. In fact, it has been shown by Cohen in 1963 that the negation of the continuum hypothesis is also consistent with the axioms of set theory. Thus it is in some sense a matter of choice as to whether one accepts this hypothesis. *See also* HILBERT'S PROBLEMS and ZERMELO-FRAENKEL SET THEORY.

Encyclopedic Dictionary of Mathematics (MIT Press, 1987)

MB

contract theory *Economics* This is the study of the manner in which labour and capital agree the parameters of production, and the amount of risk and rewards each side will bear. *See* AGENCY THEORY, COLLECTIVE BARGAINING THEORY and THEORY OF THE FIRM.

A M Okun, *Prices and Quantities* (Washington, DC, 1981)

PH

contracting Earth hypothesis (17th century) *Geology* Probably originated by the French philosopher René Descartes (1596–1650), and more clearly formulated by Gottfried Wilhelm Leibnitz (1646–1716), this theory postulates that since the Earth's supposed formation from molten rocks, it has been cooling and therefore contracting.

Recalculations were necessitated by the discovery (in the late 19th and early 20th centuries) of radioactivity and the Earth's radiogenic heating. Contraction of less than 1 per cent is now thought to have occurred during the last 4 billion years, but this factor may become increasingly important over the next five billion years. *See also* EXPANDING EARTH HYPOTHESIS, PLUTONISM and NEPTUNISM.

DT

contraction mapping theorem *Mathematics* A theorem asserting the existence and uniqueness of a fixed point of a contraction mapping (a mapping which uniformly shrinks distances) of a complete metric space onto itself. *See also* BROUWNER'S THEOREM, TARSKI FIXED POINT THEOREM and KAKUTANI FIXED POINT THEOREM.

J B Conway, *A Course in Functional Analysis* (New York, 1985)

ML

contractualism *Philosophy* Any theory basing either moral obligation in general, or the duty of political obedience, or the justice of social institutions, on a contract, usually called a 'social contract'. The idea goes back at least as far as Plato's *Crito* (*c.*395 BC), and contractualists (or contractarians) have also included Thomas Hobbes (1588–1679), John Locke (1632–1704), Jean Jacques Rousseau (1712–78), and various modern writers.

The contract may be an allegedly historical one or a tacitly implied one, or an imaginary one. It may be between people who set up a sovereign, or between the people and the sovereign, or between the individual and society or the state, or between hypothetical beings in a setting making for impartiality.

J Rawls, *A Theory of Justice* (1972)

ARL

contradiction, law of *Philosophy* Also called the law (or principle) of non-contradiction. One of the traditional three laws of THOUGHT (the other two being the laws of IDENTITY and of EXCLUDED MIDDLE). Variously formulated as saying that no proposition can be both true and not true; or that nothing can be –

without qualification – the case and not the case at the same time; or that nothing can – without qualification – both have and lack a given property at the same time.

The law cannot be logically proved without begging the question, though arguments of a different kind (among those called transcendental arguments) have been offered in its defence since Aristotle (384–322 BC) in his *Metaphysics* (book 4, chapter 4). However, recently a notion of DIALETHEISM has been defended which allows breaches of the law in certain cases. *See also* PARACONSISTENCY.

G Priest, 'Contradiction, Belief and Rationality', *Proceedings of the Aristotelian Society* (1985–86)

ARL

contrast, law of *Psychology* First proposed by the Scottish metaphysician Thomas Brown (1778–1820).

An associationist principle which states that observing or thinking about a particular quality tends to produce a recall of its opposite. The closely related principle of maximum contrast refers to the tendency of the infant and young child to gravitate to perceptual opposites and striking contrasts rather than subtle contrasts. *See* ASSOCIATIONISM.

A Warren, *A History of the Association Psychology* (New York, 1921)

NS

control theory *Mathematics* The study of the evolution of a dynamical system governed by DIFFERENTIAL EQUATIONS which may be influenced by choice of control variables.

Applications include modelling engineering control systems or economies. Typical questions concern which states are reachable (*see also* BANG-BANG PRINCIPLE) or stable. When some objective or cost function is involved we speak of OPTIMAL CONTROL, one fundamental result of which is the PONTRYAGIN MAXIMUM PRINCIPLE. *See also* CALCULUS OF VARIATIONS.

AL

control theory (1940–) *Psychology* Most famously developed by the Austrian-Canadian biologist Ludwig von Bertalanffy (1901–72).

A developing body of theory based on a feedback-system paradigm, also called 'cybernetic psychology' or 'systems theory psychology'. In the 1940s, a number of scientists postulated the notion that the principles of homeostasis (self-regulation) may be applied to the mind as well as the body. However, an increasing belief that homeostatic principles could be seen to underlie all behaviour is contradicted by some evidence, particularly that pertaining to social interaction.

D J McFarland, *Feedback Mechanisms in Animal Behaviour* (London, 1971)

NS

Convention T *Philosophy* Also called 'Criterion T'. A device due to Polish logician Alfred Tarski (1901–83) and originally used in defining truth for a formal language, but later used (by American philosopher Donald Davidson (1930–)) to give an account of meaning in terms of truth.

The details are complex, but roughly: consider the sentence '*La neige est blanche*' is true (in French) and if and only if snow is white. This is called a T-sentence for French and is in fact true. A definition of truth for French will satisfy convention T if it entails all true T-sentences for French and no false ones. Truth here is taken to be relative to a language, as with Tarski's semantic theory of truth, and the 'definition' will be simply a conjunction of all the true T-sentences. Tarski simply assumed that 'snow is white' is a translation of the French sentence, but Davidson tried to dispense with this *assumption* about meanings and provide a theory of meaning (for French in this case). Davidson's initial proposal (modified later) was that a set of axioms might be constructed which would logically entail these T-sentences, and that such a set, with the rules of logic, would constitute a theory of meaning from French.

G Evans and J McDowell, eds, *Truth and Meaning* (1976)

ARL

conventionalism *Philosophy/Mathematics* Any theory appealing to convention to explain something which is not obviously of conventional origin (as, for example, the symbols chosen for some purpose are).

Among older writers, conventionalism is associated especially with Jules Henri Poincaré (1854–1912) and Pierre-Maurice Duhem (1861–1916); and among modern ones with Willard Van Orman Quine (1908–).

In logic and mathematics conventionalism says that *a priori* truths are true only because of linguistic convention (which raises the question whether the correct application of such conventions is itself conventional: if it is we are in danger of an infinite regress of conventions). In science, conventionalism says that it is conventional which of competing and inconsistent scientific systems we adopt (though it need not be arbitrary – one may be simpler than another, or more likely to lead to further ideas, and so on; *see also* INSTRUMENTALISM).

K Britton and J O Urmson and W C Kneale, 'Are Necessary Truths True by Convention?' (symposium), *Proceedings of the Aristotelian Society*, supplementary volume (1947)

ARL

convergence *Biology See* CONVERGENT EVOLUTION.

convergence *Linguistics See* ACCOMMODATION.

convergence (1959) *Stylistics* Term defined by the stylistician Michael Riffaterre, who cites a 1946 mention by J Marouzeau.

The co-occurrence at a point in a text of a cluster of several stylistic devices – for example, unusual word-order, archaic words, alliteration, repetition, all falling together at one point – will make the stylistic device more perceptible.

M Riffaterre, 'Criteria for Style Analysis' [1959], *Essays on the Language of Literature*, S Chatman and S R Levin, eds (Boston, 1967), 412–30

RF

convergence theory (1960s) *Politics/Economics* Theory of development of industrial nations, first proposed by Dutch economist Jan Tinbergen (1903–).

The division between capitalist democracies in Western Europe and North America, and communist states in Eastern Europe, is fading. The nature of industrial society leads both types to converge towards a common centre. Capitalist societies are becoming more organized and managerial, communist societies more open to market methods and to greater political and civil freedoms. Convergence theory received a violent acceleration after 1989.

J Tinbergen, 'On the Theory of Trend Movements', *Selected Papers* (1959); J K Galbraith, *The New Industrial State* (Boston, Mass., 1967); David Robertson, *The Penguin Dictionary of Politics* (London, 1985)

RB

convergent evolution (late 19th century) *Biology* The term 'convergence' was used by English biologist Charles Darwin (1809–82) in *Descent of Man* (1871), while 'convergent evolution' came into common use in the 1960s. This is the phenomenon whereby particular characteristics are shared by species which are otherwise quite unrelated.

For example, the ability to fly appears in some mammals and reptiles, as well as in birds. Adaptation explains how these similarities arise. *Compare with* ANALOGY and HOMOLOGY.

F Magill, *Magill's Survey of Science: Life Science Series* (Englewood Cliffs, NJ, 1991)

KH

conversational analysis (1968) *Linguistics* The main model was developed by the American ethnomethodologists E A Schegloff and H Sacks.

Study of the strategies speakers use to organize talk as significantly ordered sequences of meaningful actions, including the conventions for TURN-TAKING. Sequences of turns are two-part ADJACENCY PAIRS: in the second part of the pair, speaker B analyzes speaker A's first part as a particular kind of act (for example 'summons') and responds accordingly (for example 'answer'). *See also* DISCOURSE ANALYSIS.

M Coulthard, *An Introduction to Discourse Analysis: New Edition* (London, 1985)

RF

conversational maxims *Linguistics See* CO-OPERATIVE PRINCIPLE OF CONVERSATION.

Cook's constant (probably 19th century) *Mathematics* Unpublished, but well known among school children, this is the difference between the answer required and the answer obtained. Adding Cook's constant to any calculation therefore always produces the desired result. The constant can be used legitimately by proving that the expression for it is equal to 0. *See also* FIDDLER'S FACTOR.

JB

cook's law (probably 19th century) *Mathematics* Educational folklore. To prove that an equation holds, cook the more complicated side until it matches the simpler side. This works best when the simpler side is a single number or standard function.

JB

Coolidge effect *Psychology* Named in reference to a joke about Calvin Coolidge (1872–1933), 30th president of the USA.

The term is applied to a phenomenon whereby the males of many species will show high continuous sexual performance for extended periods of time with successive introductions of new receptive females.

G Bermant and J M Davidson, *Biological Basis of Sexual Behaviour* (New York, 1974)

NS

Cope's law or rule (19th century) *Biology* Named after the American palaeontologist Edward Drinker Cope (1840–97), this is the concept that phyletic lineages have a tendency to be founded by small-sized organisms, and that organisms tend to increase in size through evolutionary time. This may occur because larger organisms are better able to control resources, such as feeding territories, than smaller ones.

W Beck and K Liem and G Simpson, *Life: An Introduction to Biology*, 3rd edn (New York, 1991)

KH

Copenhagen interpretation (1927) *Physics* Named after the native city of the Danish physicist Niels Bohr (1885–1962) who originally proposed this interpretation of quantum phenomena.

This is the orthodox interpretation of the mathematical formalism of quantum mechanics put forward by Bohr and the German physicist Werner Karl Heisenberg (1901–76) (and vigorously opposed by Albert Einstein (1879–1955)) in which no reality can be ascribed to the microscopic world. The Copenhagen interpretation states that quantum mechanics is a complete theory in the sense that its basic hypotheses about the particle and wave parameters, and about uncertainty relations, are ultimate final reflections of the real world and not capable of further modification.

Paul Davies, ed., *The New Physics* (Cambridge, 1989)

GD

Coppet's law *Chemistry* Named after the French physicist Louis Cas de Coppet (1841–1911), this states that the lowering of the freezing point of a solution below 0°C is proportional to the amount of solute dissolved in the solution. *See also* BLAGDEN'S LAW and RAOULT'S LAW.

MF

copy-choice hypothesis (20th century) *Biology* An explanation of genetic recombination based on the idea that a new strand of deoxyribonucleic acid (DNA) alternates between strands from the mother and strands from the father during its replication.

KH

core grammar (1970s) *Linguistics* Theory and goal of the American linguist Noam Chomsky (1928–), core grammar is the set of universal principles or tendencies which underlie all human languages. Chomsky has worked towards specifying these for English and for natural language generally.

A Radford, *Transformational Syntax* (Cambridge, 1981)

RF

core, theory of the (1881) *Economics* Developed by Irish-born economist and statistician Francis Edgeworth (1845–1926), this theory analyzes those parts of the economy which cannot be improved upon by individual or concerted action.

The theory is a fundamental equilibrium aspect of modern macroeconomics, the core coinciding with a set of price equilibria under perfect competition.

F Y Edgeworth, *Mathematical Psychics* (1881); W Hildenbrand, *Core and Equilibrium of a Large Economy* (Princeton, N.J., 1974)

PH

Coriolis's theorem (1831) *Physics* Named after the French physicist Gaspard Gustave de Coriolis (1792–1843).

The theorem states that for a particle moving in a Newtonian frame of reference F_1 which is itself moving with respect to a second frame of reference F_2, the total acceleration of the particle is the vector sum of its acceleration with respect to F_1, the acceleration of F_1 relative to F_2, and the Coriolis acceleration, which equals twice the vector product of the angular velocity of F_2 with respect to F_1 and the linear velocity of the particle with respect to F_2.

Robert M Besançon, ed., *The Encyclopedia of Physics* (New York, 1985)

GD

corkscrew rule (19th century) *Physics* A rule suggested by the Scottish physicist James Clerk Maxwell (1831–79) for remembering the relation between the direction of the current flow in a linear conductor and the direction of the associated magnetic field.

If an observer imagines that a normal (right-handed) corkscrew is being driven in the direction of the current, the direction of the magnetic field lines is that in which the corkscrew is being turned. *See* AMPÈRE'S RULE.

J Thewlis, ed., *Encyclopaedic Dictionary of Physics* (New York, Oxford and London, 1962)

MS

corporatism (20th century) *Politics/Sociology/Economics* Descriptive and prescriptive theory of government's relations with society. It is sometimes termed neo-corporatism to distinguish it from the political theories of FASCISM.

In the prescriptive version, all those engaged in a common enterprise, particularly as a means of making a living, have a common interest and should deal with government through their leaders as, for example, educational workers, or workers in agriculture, rather than 'horizontally' as labourers, clerical workers, managers, and so on. This was being attempted by Italian FASCISM under Benito Mussolini (1883–1945). Descriptively, first in the work of the French sociologist Emile Durkheim (1858–1917) and later in a version fashionable in the 1970s, modern governments deal with organized INTERESTS by negotiating with the leaders of the great 'corporations' of labour and capital/management. Corporatism went into both theoretical and practical decline in the 1980s.

David Miller *et al.*, eds, *The Blackwell Encyclopaedia of Political Thought* (Oxford, 1987)

RB

corpus (pl. corpora) (1960s) *Linguistics* The idea of basing linguistic analysis on a corpus of observed sentences was problematized by Avram Noam Chomsky (1928–) in his attack on earlier American linguistics, but now corpora are regarded as an important resource for linguistic research.

A recorded body of linguistic data which have either survived by accident (for example, the corpus of known Anglo-Saxon texts) or been collected and organized on systematic principles; corpora of the latter type include a number of modern computer-readable resources compiled for various research operations.

R Garside and G Leech and G Sampson, *The Computational Analysis of English: A Corpus-Based Approach* (London, 1987)

RF

corpuscular theory of light (17th century) *Physics* A theory strongly supported by the English physicist Isaac Newton (1642–1727) that treats light as a stream of non-interacting particles or corpuscles.

The law of reflection is pictured in this theory as a form of elastic reflection of the corpuscles, in which the corpuscles behave as perfectly smooth and perfectly elastic spheres. To explain refraction, only a fraction of the corpuscles in a beam must be able to penetrate the surface separating the two media. Further, to give Snell's law, the speed of light must be higher in a medium of greater refractive index. This is not observed in practice. The theory, as a complete theory to explain the behaviour of light, was abandoned in the middle of the 19th century, but it is now realized that light does indeed show particle behaviour in its interaction with matter. *See also* REFRACTION OF LIGHT, LAWS OF.

J Thewlis, ed., *Encyclopaedic Dictionary of Physics* (New York, Oxford and London, 1962)

MS

correlational sociolinguistics (1960s) *Linguistics* The 'classical' sociolinguistic model was developed by the American linguist William Labov initially as a critique of the GENERATIVE GRAMMAR of Avram Noam Chomsky (1928–). Labov's researches have been replicated in an English context by Peter Trudgill. Also known as URBAN DIALECTOLOGY.

Chomsky regarded variation within a language as part of 'linguistic performance' and therefore outside the scope of linguistics proper (*see* LINGUISTIC COMPETENCE). Labov showed that linguistic variation was systematic and motivated when related to variation in social factors such as class, gender and age, and to the degree of formality/casualness in speech. He studied the speech of a representative population (characterized by sociological methods), concentrating on variables such as whether or not words like *card* were pronounced with an R. The incidence of *r*s was shown statistically to correlate with the social variables, so language offered a kind of map of the population's social structure. Labov's studies are a methodological model, and offer insights on linguistic change and on attitudes to language. *See also* HYPERCORRECTION and SOCIAL MARKERS IN SPEECH.

W Labov, *Sociolinguistic Patterns* (Philadelphia, 1972); P Trudgill, *The Social Differentiation of English in Norwich* (Cambridge, 1974)

RF

correspondence principle (1923) *Physics* A principle introduced by the Danish physicist Niels Henrik David Bohr (1885–1962) as a way of establishing relationships between classical and quantum physics.

The principle states that predictions of the quantum theory for the behaviour of an atomic system must correspond to the predictions of classical theory in the limit of large quantum numbers.

J Thewlis, ed., *Encyclopaedic Dictionary of Physics* (New York, Oxford and London, 1962)

MS

correspondence or relational theories of meaning *Philosophy* Theories which analyze the meaning of words in terms of things they stand for in some sense, be these objects of various kinds (*see also* NAMING THEORIES OF MEANING) or ideas and so on. (*See also* IDEATIONAL THEORIES OF MEANING; and PICTURE THEORY OF MEANING for such a theory concerning sentences.)

ARL

correspondence theory/law *Psychology* First applied in this field by the American psychologist Frank Anderson Logan (1924–).

Borrowed from the work of Danish physicist Niels Bohr (1885–1962), the theory holds that whatever is true of 'molecular' behaviour is also true of 'molar' behaviour, and that a unifying principle can be found. Although a general theory of behaviour, it is most frequently associated with learning theory. It is often used in an analogical sense, there being substantive difficulties distinguishing between 'molecular' and 'molar' behaviour. *See* MICROMOLAR THEORY.

A J Chapman and D M Jones, eds, *Models of Man* (Leicester, 1980)

NS

correspondence theory of truth *Philosophy* The strictest form of the theory defines truth as a structural correspondence between what is true (a belief, judgment, proposition, sentence, and so on) and what makes it true (an event, fact, state of affairs, and so on).

Because of difficulties in defining such a relation (difficulties also facing the PICTURE THEORY OF MEANING), the theory is often weakened to saying simply that what is true is so because there is a relevant fact, without any correspondence of structure. In an even weaker version (held by Aristotle) something is true if it simply 'says things as they are,' a view which approaches the REDUNDANCY THEORY OF TRUTH (*see also* SEMANTICS, TRUTH-CONDITIONAL). Like the COHERENCE THEORY, the correspondence theory may offer merely a criterion (rather than an analysis) of truth.

D W Hamlyn, 'The Correspondence Theory of Truth', *Philosophical Quarterly* (1962)

ARL

corresponding stages, law of *Biology See* VON BAER'S LAWS.

corresponding states, law of (1881) *Chemistry* Law stated by Dutch physicist Johannes Diderik van der Waals (1837–1923) who won the Nobel prize for Physics in 1910. This states that real gases behave similarly in terms of their reduced variables.

For example, real gases in the same state of reduced volume and temperature exert approximately the same reduced pressure. The reduced variables (pressure, volume and temperature) of a gas are the actual variables divided by the critical values. For example, the reduced pressure (p_r) of a gas equals its actual pressure (p) divided by its critical pressure (p_c)

$$p_r = \frac{p}{p_c}$$

This behaviour can be expressed in a general form of the van der Waals equation of state. The equation is approximately valid for gases composed of non-polar molecules, but does not hold for molecules that are non-spherical or polar.

P W Atkins, *Physical Chemistry* (Oxford, 1978)

MF

cosine law of emission *Chemistry See* LAMBERT'S LAW OF EMISSION.

cosmic censorship, principle of (20th century) *Astronomy* A hypothesis suggested by English theoretical physicist Roger Penrose (1931–), this postulates that singularities in space and time are always surrounded by event horizons that prevent them being observed and from influencing the outside world.

A singularity is a portion of the edge of the universe where the laws of physics break down, possibly where temperature and density are infinite. An event horizon is the boundary of a black hole. No signals or particles can travel from inside the event horizon to the outside.

S Mitton, ed., *The Cambridge Encyclopaedia of Astronomy* (London, 1973)

MS

cosmic insemination *Biology See* PANSPERMIA.

cosmological principle *Astronomy* The name was given by the British astrophysicist Edward Arthur Milne (1896–1950). From astronomical observations it has been concluded that, on a very large scale, matter in the universe is distributed uniformly and isotropically. Thus, at this scale, the universe will look the same, from no matter which point. There are fluctuations in density but, on average, the universe looks the same from all locations.

In their STEADY-STATE THEORY OF THE UNIVERSE, the British cosmologist Sir Hermann Bondi (1919–) and the Austrian-born astronomer Thomas Gold (1920–) made use of an extension of the cosmological principle, known as the perfect cosmological principle. This states that, apart from local irregularities, the universe presents the same aspect to observers at any place and at any time.

S Mitton, ed., *The Cambridge Encyclopaedia of Astronomy* (London, 1973)

MS

cost-benefit analysis (19th century–) *Economics* First examined by French engineer and economist Jules Dupuit (1804–66) and later developed by 20th century economists, this is the determination of the total value of a proposed investment's inputs and outputs.

Such analysis examines opportunity costs, EXTERNALITIES, shadow prices and estimates of future interest rates. The technique was first used in the assessment of projects under the US *Flood Control Act, 1936*, and received a firmer theoretical underpinning by English economist John Hicks (1904–89) in a 1943 paper on consumer surpluses. *See* CONSUMER SURPLUS, COMPENSATION PRINCIPLE and SOCIAL WELFARE FUNCTION.

PH

cost-push inflation (1950s) *Economics* Examined by American and European economists in response to high inflation levels, this term refers to a rise in prices triggered by an increase in the costs of production (such as wages or commodity prices) in the absence of an increase in demand.

The phenomenon was experienced in the 1950s as wage rates jumped, and in the

1970s as oil prices soared. *See* ADMINISTERED PRICING, DEMAND-PULL INFLATION, MARK-UP PRICING and ADAPTIVE EXPECTATIONS.

R J Gordon, 'Understanding Inflation in the 1980s', *Brookings Papers on Economic Activity*, vol. I (1985), 263–89; W Thorp, *The New Inflation* (New York, 1959)

PH

costruzione leggitima (15th century) *Art* An Italian term meaning 'legitimate construction', this refers to a scientific perspective developed by the Italian architect Filippo Brunelleschi (1377–1446) in order to depict space. It is also described in the treatise *Della Pittura* (1432) by Italian architect and theorist Leon Battista Alberti (1404–72). The area of the composition is divided into equal zones of recession, connected by orthogonals to the vanishing point. By the application of these rules, objects will appear parallel to the picture plane.

AB

Coulomb's law of electrostatic force (1785) *Physics* Named after its discoverer, the French physicist Charles Augustin de Coulomb (1736–1806). The magnitude of the force of interaction between two point charges at rest is directly proportional to the product of the charges and inversely proportional to the square of the distance between them and acts along the line joining the two charges.

J Thewlis, ed., *Encyclopaedic Dictionary of Physics* (New York, Oxford and London, 1962)

MS

Coulomb's law of friction (1781) *Physics* Named after its discoverer, the French physicist Charles Augustin de Coulomb (1736–1806), this states that, for two surfaces in relative motion, the force of kinetic friction is almost independent of the relative speed of the surfaces.

Coulomb's discovery was based on recognition of the distinction between static friction (the force to start two surfaces in relative motion) and kinetic friction (the force to maintain the two surfaces in relative motion). He showed that kinetic friction could be much lower than static friction. *See* AMONTON'S LAWS OF FRICTION.

J Thewlis, ed., *Encyclopaedic Dictionary of Physics* (New York, Oxford and London, 1962)

MS

counterpart theory *Philosophy* Term used in connection with the MODAL REALIST analysis of necessity, possibility, and counterfactual conditional statements (those where the antecedent is presented as being false).

Consider 'If Hitler had invaded England he would have won.' Assuming his invasion was a possibility, there will be possible worlds in which he does, and in some of these he wins and in some he loses (assuming neither of these is impossible). 'If Hitler . . .' will be true if (ignoring a minor subtlety) the nearest world in which he invaded is also one in which he won. One problem (apart from providing criteria of 'nearness') is of that transworld identity: in remote worlds, is Hitler still really Hitler? To answer this and associated problems he is given a counterpart in certain of the possible worlds. Counterpart theory then studies the conditions under which he has such a counterpart, and what is involved in his having one.

D K Lewis, *On the Plurality of Worlds* (1986)

ARL

countervailing power (1952–) *Politics/Economics* Theory of political modification of markets, formulated by American economist J K Galbraith (1908–).

In the classic liberal economy, goods and services are provided and prices set by free bargaining. Modern economies give massive powers to large business corporations to bias this process, and there arise 'countervailing' powers in the form of trade unions, citizens' organizations and so on, to offset business's excessive advantage. *See also* MONOPOLISTIC COMPETITION THEORY.

J K Galbraith, *The Anatomy of Power* (London, 1984)

RB

coupling (1962) *Stylistics* Term coined by the American stylistician Samuel R Levin.

In poetry, structures which are parallel both in word-order and in meaning, thus giving a double emphasis to their relationship, special power or wit. Much exploited by Dryden and Pope. Development of

Roman Jakobson's POETIC PRINCIPLE. *See also* PARALLELISM.

S R Levin, *Linguistic Structures in Poetry* (The Hague, 1962)

RF

Cournot duopoly model (1838) *Economics* Named after French economist Augustin Cournot (1801–77), this model shows two firms that react to one another's output changes until they eventually reach a position from which neither would wish to depart.

Both firms eventually expand to such a degree that they have equal shares in the market and secure only normal profits. *See* DUOPOLY THEORY, BERTRAND DUOPOLY MODEL and BILATERAL MONOPOLY.

A Cournot, *Récherches sur les principes mathématiques de la théorie des richesses* (Paris, 1838)

PH, AV

courtly love (11th century–) *Literary Theory* Developed by the French poet Chrétien de Troyes (d.*c*.1183), the Italians Dante (1265–1321) and Petrarch (1304–74), and the English poet Geoffrey Chaucer (*c*.1340–1400).

Conventional literary representation of sexual love in terms of the pursuit and idolization of a woman, abasement of the suitor, ceremonious courtly politeness; devotion to love as the highest emotion and philosophy. Courtly love was a source of many beliefs and motifs in the representation of sexual relations well into the modern period.

C S Lewis, *The Allegory of Love* (London, 1936)

RF

covariation principle *Psychology* Proposed by the American psychologist Harold H Kelley (1921–).

When required to make decisions under uncertainty, people infer causal connections on the basis of co-occurrence between events. Thus, people will attribute behaviour to a causal factor if that factor was present whenever the behaviour occurred and was absent whenever it did not occur. It is a principle defining a judgmental error which has been found to be important in explaining interpersonal judgments. *See* ATTRIBUTION THEORY and ACTOR-OBSERVER BIAS.

H H Kelley and J L Michela, 'Attributuon Theory and Research', *Annual Review of Psychology*, vol. XXXI (1980), 457–501

NS

covering law model *Philosophy* A model of explanation associated especially with German logician Carl G Hempel (1905–), who regarded it as adequate for all types of explanation. Basically a statement is explained if it is derived from a set of laws together with certain factual statements, as we might explain 'Fido barks' by saying 'All dogs bark and Fido is a dog'.

The laws must be true general statements, and subject to certain restrictions to exclude accidental 'explanations' like 'Fido barks because he is a pet of mine and all my pets (as it happens) bark'. The laws, however – though general (for example not mentioning particular objects) – need not be universal, and the derivation of the conclusion may be inductive and not deductive; explanations can be statistical or probabilistic as well as 'deductive-monological'. Problems concern the scope of the theory, what restrictions must be placed on the relevant general statements, and the relevance of background knowledge.

C G Hempel, *Aspects of Scientific Explanation* (1965), ch. 12

ARL

CPT theorem (1954–55) *Physics* A theorem developed by the physicist G Lüders and the Austrian-born American theoretical physicist Wolfgang Pauli (1900–58), winner of the 1945 Nobel prize for Physics. It is also known as the conjugation theorem.

The theorem states that the simultaneous operation of charge conjugation C, parity inversion P, and time reversal T, on a physical process is a fundamental symmetry of relativistic quantum field theory. The laws of relativistic quantum mechanics remain unchanged if in any process particles are swapped with antiparticles, the process is exchanged with its mirror image, and the direction of time is reversed. The effect of the threefold CPT operation produces another equally allowable process which is also described by the same theoretical framework. No experimental evidence exists for violation of CPT invariance.

Paul Davies, ed., *The New Physics* (Cambridge, 1989)

GD

Craig's theorem (1953) *Philosophy* Proof concerning the formal description of scientific theories, expounded by William Craig.

According to Craig, a formal expression of a scientific theory is divisible into 'theoretical' and 'observational' vocabularies, and (since the 'observational' terms are all deducible) it follows that a description can be produced that consists only of 'observational' terms and in which all 'theoretical' terms are omitted (although he did not suggest that these were in fact undesirable).

DP

Cramer's rule *Economics* A method for solving multivariate simultaneous linear equations.

$$a_{11}x_1 + a_{12}x_2 + \ldots a_{1n}x_n = b_1$$

$$\vdots$$

$$a_{n1}x_1 + a_{n2}x_2 + \ldots a_{nn}x_n = b_n$$

where x_i is the ith variable, a_{ij} is the constant coefficient on x_j in the ith equation, and b_i is the constant on the right-hand side of the ith equation. This can be written in matrix notation as AX = b, where A is the matrix containing the elements a_{ij}, **b** is the vector containing the elements b_i; **X** is the vector of values of the variables x_i. The value of x_k which satisfies the set of simultaneous equations is found by Cramer's rule by replacing the kth column of the matrix A by the vector **b**, forming a new matrix A_k. The value of $\mathbf{X}_k$ is then the determinant of A_k divided by the determinant of A, that is

$$X_k = |A_k| / |A| \; ; \quad k = 1, \ldots, n$$

K A Fox and T K Kaul, *Intermediate Economic Statistics* (Melbourne, Fla, 1980)

PH

creation of charge theory *Physics/Geology* *See* DYNAMO THEORY.

creationism *Biology See* SPECIAL CREATION.

creative evolution *Philosophy* The theory which French philosopher Henri Bergson (1859–1941) substituted for the Darwinian mechanism of his day. Bergson's theory mediated between the mechanism of natural selection and an outright teleological view, appealing to an *élan vital* ('vital impetus') which guided evolution in a certain direction; not in what he saw as the mechanistic non-explanatory fashion of current ORTHOGENESIS, nor under the influence of a pre-envisaged (and therefore in a sense pre-existing) end, which for him is a mere 'inverted mechanism' which excludes invention and the unforeseen.

Part of Bergson's popular appeal was his espousal of a theory of evolution that did not (as Darwin's seemed to) exclude religion. Bergson uses detailed scientific arguments as well as philosophical ones, but it is unclear that he can say why evolution takes the path that it does (or any coherent path) and is not merely chaotic.

H L Bergson, *L'évolution creatrice* (1907), trans. as *Creative Evolution* (1911)

ARL

creative resultants, principle of *Psychology* Proposed by the German physiologist and psychologist Wilhelm Max Wundt (1832–1920).

The philosophical principle that when several elemental components are organized or co-ordinated the resulting synthesis has properties and characteristics that are fundamentally different in kind from those of the separate components viewed independently. The combination of elements into a significant whole produces emergent properties exclusive to that form. A similar idea can be found in the principle of Gestalt Psychology which states that 'a whole is more than a sum of its parts'. *See* GESTALT THEORY.

W M Wundt, *Outlines of Psychology* (Leipzig and Heidelberg, 1907)

NS

creativity (18th century) *Literary Theory* For context *see* ORIGINALITY and ROMANTICISM.

The poet's ability to produce new, unprecedented writing; *see also* INVENTION. The artist's creativity, now a cliché, was shocking at the dawn of Romanticism: in orthodox Christianity, creation had been reserved for God. *See also* LITERARY MODE OF PRODUCTION.

R Williams, *Keywords* (Glasgow, 1976)

RF

creoles and pidgins (1880s) *Linguistics* The academic study of creoles and pidgins was founded by the German linguist Hugo Schuchardt (1842–1927); studies have flourished from the 1970s in relation to the Chomskyan theory of INNATENESS.

A pidgin is a simplified language, not native to any of its speakers, developed for communication between people who do not share a common language. A creole is a pidgin grammatically developed and elaborated (with inflections, and so on) as a native language. *See also* BIOPROGRAM HYPOTHESIS.

S Romaine, *Pidgin and Creole Languages* (London, 1988)

RF

Crespi effect *Psychology* Named after the American psychologist L Crespi (1916–).

Refers to a disproportionate increase in a response when an increase in incentive occurs. For example, rats learning a maze run to the goal faster when the incentive is a larger amount of food than if it is a smaller amount. Once performance levels are established, exchanging the amounts yields an immediate change in performance: rats switched from a large reward to a small one run more slowly than expected, while rats switched from small to large run faster.

L P Crespi, 'Amount of Reinforcement and Level of Performance', *Psychological Review*, vol. LI (1944), 341–57

NS

crisis management (20th century) *Politics* Attempted general account of strategy in international conflict. A middle way requires to be found between total intransigence and surrender.

Graham Evans and Jeffrey Newnham, *The Dictionary of World Politics* (Hemel Hempstead, 1990)

RB

crisis of capitalism (19th century–) *Politics/Economics* Prediction in MARXISM of the collapse of capitalism.

A species of CATASTROPHE THEORY, envisaging that the inherent contradictions of the capitalist system will lead, through political conflict, to the collapse or overthrow of capitalism. Twentieth century versions of this theory see the crisis of capitalism as being a LEGITIMACY CRISIS rather than a simple economic or political one.

Jon Elster, *An Introduction to Karl Marx* (Cambridge, 1986); P M Kenway, 'Marx, Keynes and the Possibility of Crisis', *Cambridge Journal of Economics*, 4, 1 (March, 1980), 23–26

RB

crisis theory (20th century) *Politics/History* Any theory which sees history or political events making their most significant advances through crises.

Crises may be of many kinds: CRISIS OF CAPITALISM, LEGITIMACY CRISIS, crisis of confidence amongst rulers, or of faith or support amongst subjects.

James O'Connor, *The Meaning of Crisis* (Oxford, 1987)

RB

critical linguistics (1979) *Linguistics* Proposed by a group of linguists at the University of East Anglia in the UK.

The structure of language in use encodes systems of belief or ideology which originate in social and political structures; in representing the world, linguistic choices signify values, usually unconsciously. Critical linguistics – which has been much applied to official, media, political and commercial texts – is a method of analysis for making explicit these immanent social meanings.

R Fowler and R Hodge and G Kress and T Trew, *Language and Control* (London, 1979)

RF

critical period (1967) *Linguistics* Proposed by the American neurolinguist Eric Heinz Lenneberg.

Claim that there is a critical period for language acquisition from two years up to early teens, related to maturation and lateralization of the brain; the evidence however is not clear.

E H Lenneberg, *Biological Foundations of Language* (New York, 1967)

RF

critical realism *Philosophy* Name introduced by American philosopher Roy Wood Sellars (1880–1973) in *Critical Realism* (1916) for his attempt to mediate between direct realism and IDEALISM by saying that the objects of perception are neither objects themselves nor ideas and so on in the mind but sets of properties of these objects.

D Drake *et al.*, eds, *Essays in Critical Realism* (1920)

ARL

critical theory (20th century) *Politics* An academic development of MARXISM.

Critical theory was developed in Germany and then America in the 1930s by the 'Frankfurt School'. It stressed cultural and intellectual developments in place of the more traditional Marxist concern with the economy and with class.

David Miller *et al.*, eds, *The Blackwell Encyclopaedia of Political Thought* (Oxford, 1987)

RB

criticism (17th century) *Literary Theory* From the Greek word *kritos*, meaning a 'judge'.

The default term for any discourse about literature. In relation to individual works of literature, it means any or all of the activities of evaluation, description, classification, and interpretation. Raymond Williams has shown (*Keywords*, Glasgow, 1976) that negative evaluation or fault-finding has been a persistent connotation. Since the 1970s, many writers have rephrased the goals of criticism or critique so that it concerns the conditions and constraints on the beliefs and assumptions found in texts: movements like CULTURAL MATERIALISM, LINGUISTIC CRITICISM and NEW HISTORICISM, as well as MARXIST CRITICISM, make this explicit. *See also* LITERARY THEORY, POETICS and PRACTICAL CRITICISM.

W K Wimsatt (jnr) and C Brooks, *Literary Criticism: A Short History* (New York, 1957); C Belsey, *Critical Practice* (London, 1980)

RF

cross-cutting relationships, law of (1669) *Geology* Implicit in the diagrams of the Danish naturalist Nicolas Steno (1638–86), this principle is used to determine the original positions and sequences of sedimentary deposits and rocks.

Sedimentary deposition in water tends to lie approximately horizontal. If deposition is interrupted by erosion, as by the cutting of a channel, subsequent deposits first fill the channel. When preserved, the channel cuts across the previous horizontal beds and can be used to determine the original 'way up' of the layers and therefore their sequence. Similarly, rocks cut by a geological fracture (fault) may be overlaid by rocks unaffected by the fault; hence, they are shown to be younger than the faulted beds. *See* STENO'S LAWS.

DT

crowding hypothesis *Economics* Occupations with few or no barriers to entry become crowded, thereby depressing wage levels. Some sectors dominated by women or immigrants display the phenomenon.

English economist John Stuart Mill (1806–73) and Irish-born economist Francis Edgeworth (1845–1926) both used this model of discrimination in their economic analyses. *See* DUAL LABOUR MARKET THEORY, LABOUR FORCE PARTICIPATION, OCCUPATION SEGREGATION and SEGMENTED LABOUR MARKET THEORY.

J S Mill, *Principles of Political Economy with Some of Their Applications to Social Philosophy* (London, 1848)

PH

crowding-out (1970s) *Economics* The displacement of private spending by government expenditure financed by borrowing. When a government borrows heavily, interest rates may be forced to rise, thereby curbing individual consumption.

B M Friedman, 'Crowding out or Crowding in? Economic Consequences of Financing Government Deficits', *Brookings Papers on Economic Activity*, vol. IX (1978), 593–641

PH

Crum-Brown's rule *Chemistry* Named after the Scottish chemist Alexander Crum-Brown (1869–1908), noted for his research into organic chemistry.

Substitution takes place in the *meta* position to group X on a benzene ring if the compound HX can be converted by direct oxidation to HOX; if this is not possible, a mixture of *ortho-* and *para-*isomers are formed. For example, the group $-NO_2$ on the benzene ring is *meta*-directing since HNO_2 can be readily oxidized to HNO_3.

The rule is empirical and not universally applicable.

MF

crystal field theory (1930s) *Chemistry* Developed by German-born American physicist Hans Albrecht Bethe (1903–), who won the Nobel prize for Physics in 1967.

Ligands of transition-metal complexes are represented by point charges which interact with the d-atomic orbitals of the central metal atom. The theory provides a general but approximate explanation of the structure of a transition-metal complex. LIGAND FIELD THEORY is an extension of crystal field theory.

P W Atkins, *Physical Chemistry* (Oxford, 1978)

MF

Cubism (1908) *Art* This term was first coined by the French critic Louis Vauxcelles in his 1908 review of an exhibition of paintings by Georges Braque (1882–1963). *Les Demoiselles d'Avignon* (1906–07) by Spanish artist Pablo Picasso (1881–1973) and Braque's *Nude* (1907–08) are now considered to represent the first paintings of Analytical Cubism (*c.*1906–09).

Continuing the experiments of French artist Paul Cézanne (1839–1906) in pictorial analysis, picture space is kept shallow and objects are represented simultaneously from all angles, producing a faceted, cubic appearance. The later phase of this movement, Synthetic Cubism (*c.*1912–14), made use of *papiers colles* and other collage techniques. The movement had far-reaching significance for 20th century art.

J Golding, *Cubism: A History and an Analysis, 1907–1914* (1959)

AB

cultural absolutism *Psychology* Not attributable to any one originator, this is a general term referring to the opinion that values, concepts and achievements of diverse cultures may be understood and judged according to a universal standard. Cultural absolutism in the context of psychology holds that a psychological theory (for example of intelligence) developed in one culture has equal validity in a different cultural setting. *See* CULTURAL RELATIVISM.

H Triandis, *Handbook of Cross-Cultural Psychology* (Boston, 1980)

NS

cultural bias hypothesis *Psychology* Proposed by the Anglo-American psychologist Raymond Bernard Cattell (1915–).

A general term referring to the bias whereby the 'typical' group represented through the test development process is favoured over those who are neglected and thereby less likely to achieve a high score. Psychometrically, the problem is one of determining objectively whether the cultural bias hypothesis is or is not a valid explanation of observed difference between any two specific subpopulations on any specific test.

A Anastasi, *Psychological Testing* (London, 1982)

NS

cultural determinism *Psychology* Postulated by the German-born anthropologist Franz Boas (1858–1942), this is the point of view that human behaviour is primarily shaped and controlled by cultural and social factors. Culture is regarded as supra-organic and to be approached as an object of scientific study independent of the individuals who make up the culture at any one time. *See* CULTURAL ABSOLUTISM and CULTURAL RELATIVISM.

H Triandis, *Handbook of Cross-Cultural Psychology* (Boston, 1980)

NS

cultural epoch theory *Psychology* Not attributable to any one originator, this term is rarely used within psychology.

It holds that all cultures evolve through a series of stages (or epochs). In addition, each person develops in a manner which mirrors this sequence. The development of an organism is seen as a microcosmic replaying of the evolution of its species. The theory does not have wide currency within psychology.

H Triandis, *Handbook of Cross-Cultural Psychology* (Boston, 1980)

NS

cultural feminism (20th century) *Feminism* Feminist theory and strategy.

The predominant culture is male or patriarchal, and women must therefore discover or cultivate their own cultural institutions. These would form an alternative, rather than a challenge, to patriarchal culture.

C Bunch, *Building Feminist Theory* (New York, 1981)

HR

cultural materialism (1970s) *Literary Theory* Primarily British critical approach. Major source was the Welsh Marxist critic Raymond Williams (1921–88), who developed the relevant analytic sense of 'culture' in a series of influential writings from the 1960s; other sources include Louis Althusser (1918–90), Pierre Macherey (1938–), Antonio Gramsci (1891–1937) and Michel Foucault (1926–84). The expression 'cultural materialism' also appears in anthropology.

Literary texts are treated as social and signifying practices within a historically specified culture, and analyzed in relation to such materially defined topics as power, patronage, and patriarchy. Criticism of this type refuses the notion of the transcendent literary text, and brings to bear strong historical scholarship. *See also* NEW HISTORICISM.

J Dollimore and A Sinfield, eds *Political Shakespeare* (Manchester, 1985); R Williams, *Problems in Materialism and Culture* (London, 1980)

RF

cultural relativism (*c.*1911) *Anthropology/Psychology* Generally attributed to German-American anthropologist Franz Boas (1858–1942) but better articulated by his American students, especially Melville J Herskovits (1895–1963).

All cultures are equally good. Cultures each have their own values and ways of understanding the world, and therefore each ethnic group needs to be understood in its own, culture-specific terms. Cultural relativism was the dominant theory in American anthropology at least until World War II. In its extreme form, however, it came under attack by philosophers who argued that it was untenable, since logically it should deny the possibility of absolute values of right and wrong. *See* CULTURAL ABSOLUTISM.

J Tennekes, *Anthropology, Relativism and Method* (Assen, Holland, 1971)

ABA, NS

culture of poverty (*c.*1961) *Anthropology* Coined by American anthropologist Oscar Lewis (1914–70).

The notion that poverty is not simply a matter of material deprivation, but develops its own values and social order. It is applicable to many cases, but critics maintain that it is politically unacceptable to suggest, as Lewis did, that poverty is self-perpetuating. *See also* CULTURAL RELATIVISM.

O Lewis, *The Children of Sanchez* (New York, 1961)

ABA

Curie's law *Physics* Sometimes called the Curie–Langevin law after its French discoverer Pierre Curie (1859–1906) and interpreter in atomic terms Paul Langevin (1872–1946). This states that for paramagnetic materials (solids, liquids and gases) in which the interaction between the dipoles is weak, the volume magnetic susceptibilty χ_m is inversely proportional to the temperature T; that is,

$$\chi_m = a/T$$

where a is a constant characteristic of the material, known as the Curie constant.

J Thewlis, ed., *Encyclopaedic Dictionary of Physics* (New York, Oxford and London, 1962)

MS

Curie-Weiss law *Physics* Named after the French physicists Pierre Curie (1859–1906) and Pierre Weiss (1865–1940).

For a material composed of strongly interacting dipoles, the behaviour is paramagnetic for temperatures above some critical temperature T_c, known as the Curie temperature, while at temperatures below T_c the behaviour is ferromagnetic. For such a material, the volume susceptibility χ_m at a temperature T above T_c is given by the Curie-Weiss law:

$$\chi_m = a/(T - T_c)$$

See CURIE'S LAW.

J Thewlis, ed., *Encyclopaedic Dictionary of Physics* (New York, Oxford and London, 1962)

MS

currency principle (mid-19th century) *Economics* The principle is concerned with the metallist belief that the money supply or currency in circulation should be strictly related to the amount of gold deposited with the Bank of England. If this money supply were controlled, monetary stability would follow.

The *Bank Acts* of 1844 and 1845 embodied this principle and helped establish the framework of modern British banking. *See* CLASSICAL THEORY OF MONEY, BIMETALLISM, FREE BANKING THEORY and REAL BILLS DOCTRINE.

F W Fetter, *Development of British Monetary Orthodoxy 1719–1875* (Cambridge, Mass., 1965)

PH

Cushny's theory (1917) *Biology/Medicine* Called the modern theory of urine secretion by its proposer, the Scottish pharmacologist Arthur Robertson Cushny (1866–1926), who extended ideas originally suggested by Carl Friedrich Wilhelm Ludwig (1816–95), eminent German physiologist and physician. This theory of kidney function held that mechanical filtration alone can explain glomerular fluid in the kidney, without recourse to vitalistic forces, even though the exact mechanism remained somewhat obscure.

'One part of the kidney filters off the plasma colloids, another absorbs a fluid of unchanging composition,' Cushny wrote. Cushny called the constituents of the plasma 'threshold bodies,' which he claimed were taken up by cells of the tubules in 'definite proportions' – determined by their normal values in the plasma – and then returned to the blood. The concept of theshold bodies forms the basis of modern renal function tests. *Compare with* BOWMAN-HEIDENHAIN HYPOTHESIS and LUDWIG'S THEORY.

A R Cushny, *The Secretion of Urine* (1917); J Fulton, *Selected Readings in the History of Physiology* (Springfield, Ill., 1966)

KH

customs union theory (18th century) *Economics* A customs union is a grouping of countries with a common external tariff, but with free trade, free movement of labour and capital among themselves. The theory examines the impact on trade in general following the removal of barriers (such as quotas and tariffs) between the countries and their establishment against other countries.

Dating back to the classical economic concept of free trade expounded by Scottish economist Adam Smith (1723–90), and English economists David Ricardo (1772–1823) and Robert Torrens (1780–1864), the theory has received modern updating by Canadian-born economist Jacob Viner (1892–1970). The European Community, COMECON and EFTA are modern examples of customs unions.

J Viner, *The Customs Union Issue* (New York and London, 1950)

PH

cybernetics (1948) *Physics* Cybernetics was defined by the American mathematician Norbert Wiener (1894–1964) in 1948 as 'control and communication in the animal and the machine'.

The term 'communication' implies the transfer of information while purposeful control requires information about the present situation and about the ultimate situation that is required. In this sense, cybernetics may be regarded as the study of information handling. The information may be used to modify the system and it may also be necessary to obtain information from other sources and process all the information before using it to effect control to bring about the desired outcome.

N Wiener, *Cybernetics* (New York, 1948)

MS

cybernetics *Politics* Application of mechanical and scientific models for the understanding of political life.

The manner in which political systems and institutions co-ordinate and control their own actions can be explained principally in terms of their internal 'mechanisms', and their actions understood principally in terms of these internal processes, rather than by reference to outside pressures as in BLACK BOX THEORY or BILLIARD BALL THEORY.

Roger Scruton, *A Dictionary of Political Thought* (London, 1982)

RB

cycle *Linguistics* *See* TRANSFORMATIONAL CYCLE.

cyclical theory *History/Politics/Economics* Theory of large scale historical or political change.

Events, economies, and political systems move through cycles similar to the natural life-cycles of living beings. These cycles can be observed, but there is no obvious explanation for them. The most familiar version is Spengler's DECLINE OF THE WEST.

RB

cynicism *Philosophy* Philosophy of the movement started by Diogenes of Sinope (4th century BC) and possibly influenced by Antisthenes, a contemporary and disciple of Socrates (469–399 BC). The movement lasted intermittently for some 800 years or more, flourishing mainly in its first two centuries and again under the early Roman Empire (first two centuries AD).

The Cynics were akin to and in some ways forerunners of the Stoics, but they confined themselves to ethics and never gained the respectability of the Stoics. Like the Stoics they advocated self-sufficiency, and the avoidance of emotional entanglements and slavery to desire, but unlike STOICISM they rejected social conventions and constraints as unnatural. The name 'Cynics' means 'doggy ones', perhaps because of their uninhibited habits (though other etymologies have been suggested).

D R Dudley, *A History of Cynicism* (1937)

ARL

D

d'Alembert's principle (18th century) *Physics* Named after its originator, the French philosopher and mathematician Jean le Rond d'Alembert (1717–83), this states that the condition for a mechanical or electromagnetic system's equilibrium is that the virtual work of the applied forces vanishes.

J Thewlis, ed., *Encyclopaedic Dictionary of Physics* (New York, Oxford and London, 1962)

MS

Dada (*c.*1915–16) *Art* With its name chosen at random (being the French word for 'hobby horse'), this movement first manifested itself in Zurich (*c.*1915–16) with the establishment of the Cabaret Voltaire by German poet and musician Hugo Ball (1886–1927).

The movement, which soon spread to other international centres, was a reaction to the horror of war, and reflected disillusionment with society. It aimed through iconoclastic and revolutionary means to subvert conventional taste. Adopting provocative and irrational approaches (reminiscent of FUTURISM), it attempted to undermine art itself. It ceased to be a significant artistic force after about 1922.

D Ades, *Dada and Surrealism* (1974)

AB

Dalton's atomic theory (1803) *Chemistry* The first modern atomic theory, named after the English chemist, mathematician and physicist John Dalton (1766–1844) who proposed it in a public lecture in 1803 and published it five years later.

Each element is composed of small particles of identical size and mass. These particles, which Dalton called atoms, are indivisible and remain unchanged during a chemical reaction. Dalton worked out the relative weights of the atoms of elements such as hydrogen, oxygen, nitrogen and sulphur. He gave a symbol to each element. Atoms are now known not to be indivisible particles. *See also* RUTHERFORD'S ATOMIC THEORY.

M Freemantle, *Chemistry in Action* (Basingstoke, 1987)

MF

Dalton's law *Chemistry* *See* MULTIPLE PROPORTIONS, LAW OF.

Dalton's law of partial pressures (1801) *Chemistry* Named after the English chemist, mathematician and physicist John Dalton (1766–1844) who proposed it.

The total pressure of a mixture of gases equals the sum of the partial pressures of the component gases at the same temperature. (The partial pressure of a component gas is the pressure the gas would exert if it alone occupied the total volume at the same temperature.) The law is only valid for mixtures of ideal gases.

M Freemantle, *Chemistry in Action* (Basingstoke, 1987)

MF

Danielli-Davson model (1934) *Biology* Proposed by James F Danielli and Hugh Davson, both of University College,

London. This early concept of cell membranes proposed two lipid layers with lipid material sandwiched between, and with protein material lying outside the lipid layers.

This model allowed for selective permeability of the membrane, meaning that certain molecules can pass through while others are barred. This model was superseded by the UNIT MEMBRANE HYPOTHESIS and, later, the FLUID-MOSAIC MODEL.

P Sheeler, *Cell and Molecular Biology* (New York, 1987)

KH

Danielli-Davson-Robertson model *Biology* *See* UNIT MEMBRANE HYPOTHESIS.

dark matter theory (20th century) *Astronomy* Dark matter is non-baryonic matter which does not interact with radiation and is therefore 'dark' for all astronomical purposes.It is thought to comprise about 90 per cent of the matter in the universe and can only be detected by its gravitational effects. The nature of this matter, if it exists, is at present a mystery. Some of it could come from massive neutrinos, and Doppler shifts from the opposite sides of rotating galaxies suggest that there is non-radiating matter present. It has been claimed that one huge concentration of dark matter, the Great Attractor, has been detected by its effect on the motion of many galaxies.

The universe is at present expanding, but whether it will continue to do so depends on the average density of matter in the universe.

If matter is relatively dense, gravitational attraction will slow down (and eventually stop) the expansion and make the universe contract. If matter is not sufficiently dense, the expansion will continue indefinitely. There is a critical density of matter that will just stop the expansion. Some theorists believe that the universe is closed and that the average density of matter is just equal to this critical density. However, for this to be true, there must be a large amount of unseen matter in the universe.

S Mitton, ed., *The Cambridge Encyclopaedia of Astronomy* (London, 1973)

GD, MS

Darwinian evolution *Biology* *See* DARWINISM.

Darwinian fitness (19th century) *Biology* Named after English biologist Charles Darwin (1809–82), and discussed by British naturalist Alfred Russel Wallace (1823–1913) in his book *Darwinism: An Exposition of the Theory of Natural Selection with Some of Its Applications* (1889). The idea that those individuals best adapted to their environment are the most likely to survive.

Darwin also used 'fitness' to mean robustness, and he applied the term to those possessing more advanced nervous systems. *Compare with* SURVIVAL OF THE FITTEST.

R Milner, *The Encyclopedia of Evolution: Humanity's Search for Its Origins* (New York, 1990)

KH

Darwinian selection *Biology* *See* NATURAL SELECTION.

Darwinism (19th century) *Biology* Term coined by the British naturalist Alfred Russel Wallace (1823–1913) after the eminent evolutionary theorist Charles Robert Darwin (1809–82).

Darwin's original hypothesis holds that individuals possessing characteristics enabling them to produce more viable offspring than less well-endowed individuals will tend to have those characteristics maintained and spread throughout the population (that is, via NATURAL SELECTION). Darwinism also encompasses other concepts, including GRADUALISM, COMMON DESCENT and SPECIATION. *See also* EVOLUTION. *Compare with* LAMARCKISM.

C R Darwin, *The Origin of Species by Means of Natural Selection, or the Preservation of Favoured Races in the Struggle for Life* (1859); A R Wallace, *Darwinism: An Exposition of the Theory of Natural Selection with Some of Its Applications* (1889); R Milner, *The Encyclopedia of Evolution: Humanity's Search for Its Origins* (New York, 1990)

KH

Davisian cycle (20th century) *Geology* The American geologist William Morris Davis (1850–1934) proposed that land-forms evolved through a steady progression of shapes following their initial uplift until the eventual levelling of the land (peneplanation). Repeated uplifts could be recognized from the identification of different stages of stream morphology. While still a useful concept, the practical application in inhomogeneous terrains is difficult.

DT

de Broglie's relation (1924) *Physics* Named after the French physicist Louis-Victor Pierre Raymond, 7th Duke de Broglie (1892–1987), winner of the 1929 Nobel prize for Physics.

The law states that any particle of mass m moving with a speed v will, in some experiments, display wavelike properties corresponding to a wavelength λ given by $\lambda = h/mv$, where h is the Planck constant. *See also* PLANCK'S QUANTUM THEORY.

Sybil B Parker, ed., *McGraw-Hill Encyclopedia of Science and Technology* (New York, 1987)

GD

de facto **and** ***de jure*** **theories of meaning** *Philosophy* A distinction associated with USE THEORIES OF MEANING. *De facto* theories give the meaning of a word in terms of how it is actually used; *de jure* theories give it in terms of how it should be used, or of rules for its use, claiming that actual usage may be incorrect.

L J Cohen, *The Diversity of Meaning* (1962, revised 1966), ch. 2

ARL

de Moivre's formula *Mathematics* Named after the French-born mathematician Abraham de Moivre (1667–1754), this is the result whereby

$$(\cos x + i \sin x)^n = \cos nx + i \sin nx$$

for all real numbers x and all positive integers n. Equating the real and imaginary parts of this expression provides a simple way of expressing cos nx and sin nx as polynomials in sin x and cos x. *See also* BINOMIAL THEOREM.

Encyclopedic Dictionary of Mathematics (MIT Press, 1987)

MB

De Morgan's laws *Mathematics* Named after Indian-born British mathematician and logician Augustus De Morgan (1806–71), this element of SET THEORY states that the complement of the intersection of two sets is the union of their complements, and that the complement of the union of two sets is the intersection of their complements. This is more compactly expressed as

$$(S \cap T)^c = S^c \cup T^c,$$

$$(S \cup T)^c = S^c \cap T^c.$$

Similar formulations of these laws may be found in other branches of mathematics including Boolean ALGEBRA and logic.

Encyclopedic Dictionary of Mathematics (MIT Press, 1987)

MB

De Stijl (1917) *Art* Name of a movement and periodical founded by Dutch artist Theo van Doesburg (1883–1931) in Holland; its other leading protagonist was fellow Dutch artist Piet Mondrian (1872–1944).

In the first edition, van Doesburg outlined the new aesthetic, or 'neo-Plasticism', as a rejection of expression and individuality in order to create work which is universal, abstract and mechanically composed. The favoured format was the line, right angle and three primary colours. These principles were modified after 1924 to allow other angles and planes; however, in the last issue (published in 1932 by van Doesburg's widow) the early rigour of the movement was restated. *De Stijl* was highly influential on both European and American artists and architects.

LL

deautomatization (1917) *Stylistics* Literary theory of the Russian formalists (*see* FORMALISM, RUSSIAN) defined by Viktor Shklovsky (1893–1984).

In daily life, perception becomes dulled or AUTOMATIZED through familiarity. Art prevents automatization by confronting the reader with the difficulty of poetic DEVICES, making perception an active process and restoring clarity of perception and offering a new, unfamiliar perspective. *See also* ART AS DEVICE, DEFAMILIARIZATION.

V Shklovsky, 'Art as Device', *Russian Formalist Criticism*, L T Lemon and M J Reis, trans. and eds (Lincoln, Nebraska, 1965), 3–24

RF

Debye-Hückel limiting law *Chemistry See* DEBYE-HÜCKEL THEORY.

Debye-Hückel theory (1923) *Chemistry* Theory of electrolytes, named after Dutch physicist Peter Debye (1884–1966) and E Hückel, which marked the start of a new era in electrochemistry. Debye won the Nobel prize for Chemistry in 1936.

The theory concerns the behaviour of ions in solution. Ionic solutions deviate from ideality because of long-range electrostatic interactions between ions. The theory assumes an ion has an actual position in solution and is surrounded by an atmosphere of oppositely charged ions. The theory provides an equation for the calculation of the activity coefficient of an ion in dilute solution. As the ionic strength approaches zero, the equation reduces to the limiting form (the Debye-Hückel limiting law). This yields activity coefficients in excellent agreement with experimental values for strong electrolytes such as sodium chloride. The extended Debye-Hückel theory takes into account the size of ions.

P W Atkins, *Physical Chemistry* (Oxford, 1978)

MF

Debye's T^3 law (1912) *Physics* Named after its discoverer, the Dutch-American physicist Peter Joseph Wilhelm Debye (1884–1966), this is a theory of specific heat capacity. It predicts that, at temperatures close to absolute zero, the specific heat capacity of a non-metallic solid should vary with temperature according to a T^3 law. *See* DEBYE'S THEORY OF HEAT CAPACITY.

J Thewlis, ed., *Encyclopaedic Dictionary of Physics* (New York, Oxford and London, 1962)

MS

Debye's theory of heat capacity (1912) *Physics* Named after its originator, the Dutch-American physicist Peter Joseph Wilhelm Debye (1884–1966).

For dielectric solids the heat capacity is associated with the vibrations of the atoms. In Debye's theory, the vibrational modes are modelled by sound waves in an elastic continuum. The boundaries of the solid determine the allowed wavelengths and the sound speed then determines the frequencies. The number of frequencies is limited to $3N$ (corresponding to the number of degrees of freedom of the N point atoms in the crystal) by cutting off the spectrum of frequencies. Each mode is allocated the energy of a quantum simple harmonic oscillator of that frequency. The model predicts that the molar heat capacity at constant volume should tend to the value $3R$ when the temperature T is very large and should vary as T^3 as T approaches absolute zero. R is the molar gas constant.

J Thewlis, ed., *Encyclopaedic Dictionary of Physics* (New York, Oxford and London, 1962)

MS

decay law for radioactivity (1902) *Physics* Formulated by the British physicists Ernest Rutherford (1871–1937) and Frederick Soddy (1877–1956).

If $N(0)$ is the number of active atoms of a given radioactive nuclide at a time taken as zero, then the number of active atoms $N(t)$ at a time t later, assuming no replenishment, is given by

$$N(t) = \mathrm{N}(0)\exp(-\lambda t)$$

where λ is a constant characteristic of the nuclide, known as the decay constant.

J Thewlis, ed., *Encyclopaedic Dictionary of Physics* (New York, Oxford and London, 1962)

MS

decay theory *Psychology* Proposed by the Canadian psychologist Donald Olding Hebb (1904–).

Also called trace decay theory, this is a concept of forgetting where the learned material leaves in the brain a trace or impression which gradually recedes and disappears unless practised and used. Decay is used as a biological metaphor to characterize the gradual degradation or disintegration of neural traces. This decay could be prevented by the process of consolidation. *See* CONSOLIDATION THEORY OF LEARNING.

A D Baddeley, *Working Memory* (Oxford, 1986)

NS

decentring (1960s) *Literary Theory* Aim/effect of French POSTSTRUCTURALISM and DECONSTRUCTION.

Attack on the dominance and centrality of certain authoritarian Western concepts (for example God, man, reason). In deconstructive criticism, texts are decentred by a process of analysis which denies a central meaning; meanings are disseminated and deferred. *See also* 'DIFFÉRANCE'.

C Norris, *Deconstruction* (rev. edn, London, 1991)

RF

decision making theory (20th century) *Politics* Theory of politics as rational.

Government and politics can be treated as a series of decisions taken by persons and institutions who make rational decisions, or who act as if they made rational decisions, in the light of their INTERESTS and the circumstances in which they operate.

Patrick Dunleavy, *Democracy, Bureaucracy and Public Choice* (Hemel Hempstead, 1991)

RB

decision theory *Economics* Decision theory is a fundamental aspect of economics in which individuals make rational choices from a range of alternatives that may be characterized by varying price. In determining a course of action, an individual must evaluate conditions of uncertainty and risk. *See* UNCERTAINTY, BOUNDED RATIONALITY, BERNOULLI'S HYPOTHESIS and CAPITAL ASSET PRICING MODEL.

A K Sen, 'Choice Functions and Revealed Preferences', *Review of Economic Studies*, 38, 2 (1971), 307–17

PH

decision theory (*c.*1960) *Mathematics* The determination of strategies to make decisions when, at best, the probabilities of events are known in order to maximize the sum of the products of the probabilities and the estimated utilities of the desired eventualities.

JB

decline of the West (1918) *Politics/History* Historical theory of German writer Oswald Spengler (1880–1936).

A version of CYCLICAL THEORY which presents civilizations as having 'seasons', and Western Europe as entering its 'winter' in the 20th century, with predictions of war and despotic leadership.

David Miller *et al.*, eds, *The Blackwell Encyclopaedia of Political Thought* (Oxford, 1987)

RB

deconstruction (1960s) *Literary Theory* Approach developed by the French philosopher Jacques Derrida (1930–) in the context of POSTSTRUCTURALISM. Coinage 'destruction' refers to the 'destructive analysis' of the German philosopher Martin Heidegger (1889–1976).

Deconstruction's philosophical basis is a questioning and strategic subversion of established Western authoritarian concepts (*see* DECENTRING). In literary (or other) criticism, analysis of texts is based on a profound scepticism concerning the stability of meaning in language (*see* 'DIFFÉRANCE'): the deconstructive critic does not seek a central interpretation, but lets the mind play with presuppositions and intertextualities. Deconstruction has been very influential in the USA. *See also* GRAMMATOLOGY, LOGOCENTRISM, PHONOCENTRISM and PUNCEPT.

J Derrida, *Of Grammatology*, trans. and introd. G C Spivak (Baltimore, 1977 [1967])

RF

decorum *Art* An aesthetic concept first developed in Antiquity and elaborated into a formal code during the Renaissance and Baroque periods. The constituent parts of a work of art should be mutually congruent and the style and execution appropriate to the subject. Breaches of such rules constituted social neglect. The works of Italian artist Raphael (1483–1520) represented a model of decorum.

AB

decorum (*c.*20 BC) *Literary Theory* The idea is in Aristotle's *Poetics* (*c.*330 BC) as *to prepon*, but the authority most cited is *Art of Poetry* by Horace (65–8 BC).

Literary propriety: style and action must be appropriate to subject and character. A hierarchy of styles (high, middle and low) was assumed, suitable for different ranks of character. Decorum was an important criterion (in theory) until ROMANTICISM insisted on individual expressivity.

Horace, *Art of Poetry*, E C Wickham, trans., *Critical Theory since Plato*, H Adams, ed. (New York, 1971)

RF

decrementless conduction, theory of (1923) *Biology/Medicine* Proposed by the noted Japanese physiologist Genichi Kato (1890–), this was the surprising hypothesis that the physical length of a nerve subjected to a narcotic (such as chloroform, cocaine or alcohol) has no effect whatsoever on the depth of narcosis required to abolish conduction of impulses in the nerve.

Kato used 10 cm nerves from the *Bufo vulgaris Japonicus* – a huge (350 g) toad native to Japan – as his experimental model.

J Fulton, *Selected Readings in the History of Physiology* (Springfield, Ill., 1966)

KH

Dedekind cut (19th century) *Mathematics* Named after the German mathematician Julius Wilhelm Richard Dedekind (1831–1916), this finding arose from the latter's efforts to put the theory of irrational numbers on a firm foundation.

It is a partition of the set of rational numbers into two disjoint subsets such that all the members of one of the subsets are less than all the members of the other subset. For example, the pair

$$(\{x \in Q : x^2 > 2\}, \{x \in Q : x^2 < 2\})$$

is defined to be the irrational number $\sqrt{2}$.

W Rudin, *Principles of Mathematical Analysis* (McGraw-Hill, 1976)

MB

deduction theorem *Philosophy* Let 'P' and 'Q' stand for (simple or compound) propositions. The deduction theorem says that: if Q can be logically inferred from P, then 'If P then Q' can be proved as a theorem in the logical system in question. This gives a method for dispensing with rules of inference in favour of axioms and theorems; but it does not hold for all logical systems, and in any case not all rules of inference can be dispensed with, for reasons due to Lewis Carroll.

L Carroll, 'What the Tortoise Said to Achilles', *Mind* (1895)

ARL

deductivism *Philosophy* Name sometimes applied to the claim, especially associated with Austrian philosopher Karl Raimund Popper (1902–), that since induction is logically invalid, science should dispense with it in favour of deduction. *See also* INDUCTIVISM, FALSIFICATIONISM, HYPOTHETICO-DEDUCTIVE METHOD and VIENNA CIRCLE.

ARL

deep and surface structure *Linguistics See* STANDARD THEORY.

defamiliarization *Stylistics See* DEAUTOMATIZATION.

defect theory *Psychology* Proposed by the English scientist Sir Francis Galton (1822–1911).

This theory postulates that the cognitive processes of the mentally handicapped are qualitatively different from those of normal individuals. Contemporary theorists are more likely to maintain that mental retardation is the result of slower development of these processes rather than of any qualitative difference. These two positions have important implications for the design of intervention programmes to facilitate the development of those disadvantaged through mental handicap.

P Mittler, *The Psychological Assessment of Mental and Physical Handicap* (London, 1970)

NS

deference (20th century) *Politics* Theory of political order, and of voting.

In obeying laws and rulers, people are moved by 'deference' to those whom they regard as in some often unspecified sense their superiors. Similar attitudes explain the support given to conservative and right wing parties by voters who it might be argued would benefit from redistributive or socialist policies. *See also* ÉLITISM.

Robert McKenzie and Alan Silver, *Angels in Marble* (London, 1968)

RB

definite composition, law of *Chemistry See* CONSTANT COMPOSITION, LAW OF.

definite descriptions, theory of (20th century) *Mathematics* Proposed by Bertrand Arthur William Russell (1872–1970).

A definite description is an expression which is capable of determining just one item, like 'the Tower of London'. The key to the theory is that the sentence 'The British mathematician with a Nobel Prize is Bertrand Russell' should be rendered by 'There is only one British mathematician who has a Nobel Prize and he is Bertrand Russell.'

KD

definite proportions, law of *Chemistry See* CONSTANT COMPOSITION, LAW OF.

degeneration theory (1870s) *Biology/Medicine* Ancient idea promoted by German zoologist Anton Dohrn and British naturalist E Ray Lankester (1847–1929), this is the theory that species can become simpler over time, rather than evolving into more complex species or remaining the same.

Parasites and whales are considered examples of degeneration because they have lost structures (organs and legs) during evolution. This theory was also applied to human history and called upon to explain the mental illnesses and the 'uncivilized' behaviour of 'primitive' peoples, applications that are now discredited.

E Lankester, *Degeneration: A Chapter in Darwinism* (London, 1880)

KH

degrees of truth *Philosophy* There are two main sources of the idea that truth has degrees. One is objective idealism, as explained under COHERENCE THEORY OF TRUTH. The other arises because many predicates are essentially vague. When does a heap of sand become large? It seems plausible that a 100-grain heap is not large, and that a heap which is not large can never become large by acquiring just one extra grain. It seems to follow that a small heap which has grains added one at a time can never become large. (This is one version of the paradox known as that of the heap, the *sorites*, or the bald man; being invented in ancient Greece and only revived quite recently.) The paradox (like many) is harder to answer than appears at first, and one answer takes the form of letting truth have degrees.

R M Sainsbury, *Paradoxes* (1988), ch. 2

ARL

deism (18th century) *Philosophy* The doctrine that belief in a passive creator-God is entirely consistent with reason, without recourse to established religion or the supernatural. In its 18th-century heyday, deism was particularly associated with such philosophical writers as François Marie Arouet de Voltaire (1694–1778) and Jean Jacques Rousseau (1712–78) (although similar thoughts had been expressed by other thinkers in earlier centuries).

In an age when scientific discoveries were encroaching further and further upon long-accepted biblical ideas of God's intervention in the world, deist philosophers argued that man was master of his own fate, having been given responsibility for his own deeds by God after the supreme being had brought the universe into existence. Since the creation of the world, the deists held, God had taken no direct active part in the world's affairs. To fulfil his obligations to God man would have to rely upon the exercise of his own intellect. However, such writers as Rousseau still referred to an 'unknowable' inner faith in God that defied scientific analysis. This 'irrational' faith has since been much mauled by philosophers who find it inconsistent with the deists' desire to base their beliefs upon reason alone.

M Wiles, *God's Action in the World* (London, 1986)

DP

deixis (3rd to 2nd centuries BC) *Linguistics* Part of this concept was recognized by the ancient Greek grammarians; *deixis* was the Greek word for 'pointing', used for demonstratives.

Now used in a broader sense for the orientation of what is being talked about in relation to speaker, hearer and spatiotemporal context. Aspects of language involved in *deixis* include demonstrative and personal pronouns, and tense.

J Lyons, *Semantics* (Cambridge, 1977), ch. 15

RF

deletion hypothesis (1957) *Biology* Attributed to American Van Rensselaer Potter (1911–), this is the hypothesis that cancer is caused by the loss of one or more of the cell's constituents of deoxyribonucleic acid (DNA), probably enzymes or proteins. *See also* CANCER, THEORIES OF.

KH

DeLorean theory (1990) *Art/Literary Theory/Economics* Theory of academic work described by English political scientist Rodney Barker (1942–).

Academic and artistic activities are measured according to the amount of funding they attract, rather than by the quality or quantity of their output. Named after the unsuccessful motor car manufacturer John DeLorean (1925–), who attracted large amounts of money for his plant in Northern Ireland but produced very few cars.

demand for money theory (20th century) *Economics* Also known as liquidity preference, this theory deals with the desire to hold money rather than other forms of wealth (for example stocks and shares). It is particularly associated with the work of English economist John Maynard Keynes (1882–1946).

Keynes distinguished three motives for holding money: the transaction motive (to meet day-to-day needs); the speculative motive (in anticipation of a fall in the price of assets); and the precautionary motive (to meet unexpected future outlays). The amount of money held is determined by the interest rate and the level of NATIONAL INCOME. Monetarists argue that the demand for money is no longer a function of the interest rate and income but that the RATE OF RETURN on a wider spectrum of physical and financial assets influences demand. *See* QUANTITY THEORY OF MONEY.

J M Keynes, *The General Theory of Employment, Interest and Money* (New York, 1936); D Fisher, *Money, Demand and Monetary Policy* (Hemel Hempstead 1989)

PH

demand-pull inflation (1940) *Economics* Outlined by English economist John Maynard Keynes (1883–1946), this term describes a rise in prices triggered by an excess of demand for the available supply in the economy.

Keynes raised an important concept of an inflationary gap, which replaced the notion that inflation was caused by a rise in the money supply. *See* COST-PUSH INFLATION and QUANTITY THEORY OF MONEY.

J M Keynes, *How to Pay for the War* (London, 1940)

PH

demand theory (19th century–) *Economics* First raised as a fundamental principle of microeconomics by French economist Léon Walras (1834–1910), this is the analysis of the relationship between the demand for goods or services and prices or incomes. The theory examines purchasing decisions of consumers and the subsequent impact on prices.

The theory was subsequently developed by English economist Alfred Marshall (1842–1924), Italian Vilfredo Pareto (1848–1923), Soviet Eugen Slutsky (1880–1948), American Kenneth Arrow (1921–) and the French-born Gerard Debreu (1921–). *See* AGGREGATE DEMAND THEORY, CONSUMER DEMAND THEORY and SLUTSKY THEOREM.

L Walras, *Eléments d'économie politique pure* (Lausanne, 1876); A Marshall, *Principles of Economics* (London, 1890)

PH

democracy *Politics* Government by the whole people. Democratic theory begins with the justification of government by the people, usually in terms either of the RIGHTS of individual citizens, or the need to protect their INTERESTS effectively. It then proceeds to the two questions of what government by the people means, and how, if at all, it can be implemented.

David Held, *Models of Democracy* (Oxford, 1987)

RB

democratic centralism (20th century) *Politics* Theory of Communist Party organization.

Power in Communist Parties was derived from the rank and file, but it could and should then be used to preserve disciplined orthodoxy. In practice this meant that central despotism over the membership could be justified by reference to the CONSENT of those thus subjected. It is now of largely historical interest, though survives as a term of political abuse. *See also* BOLSHEVISM.

Tom Bottomore, *A Dictionary of Marxist Thought*, 2nd edn (Oxford, 1991)

RB

democratic élitism (20th century) *Politics* Theory of democracy set out by the American economist and political scientist Joseph Schumpeter (1883–1946).

The theory of DEMOCRACY had attributed power to the people; that of ÉLITISM had attributed it to minorities. In democracies this polarity is resolved, by élites competing for popular electoral support to enable them to exercise power subject to its being renewed at the next election.

Geraint Parry, *Political Elites* (London, 1969)

RB

demographic transition (20th century) *Economics* This is a process by which underdeveloped countries experience a change in their birth and mortality rates because of a change in the economic development of the state.

Over time and with rising influence, high birth and death rates are replaced by slower or declining birth rates. *See* MALTHUSIAN POPULATION THEORY, SECULAR STAGNATION THEORY.

F Notestein, 'Population: the Long View', *Food for the World*, T W Schultz, ed. (Chicago, 1945)

PH

denotation and connotation (1843) *Philosophy/Linguistics* Distinction drawn by the English philosopher John Stuart Mill (1806–73).

A word denotes the class of entities to which it may be used to refer; it connotes the qualities usually associated with those entities. *See also* the non-equivalent distinction SENSE AND REFERENCE.

J Lyons, *Semantics* (Cambridge, 1977) ch. 7

RF

density wave theory (20th century) *Astronomy* This postulates that the spiral arms in galaxies cannot be permanent condensations of matter.

Permanent condensations can only survive as a pattern if a whole galaxy rotates as a rigid body, but observation shows that differential rotation prevails. Therefore any material in a galactic disc is wound up and features should be obliterated in a few galactic years. This suggests that the spiral arms must be a wave phenomenon. What are seen as spiral patterns are not permanent ridges of matter, but the locus of the crests of a density wave.

S Mitton, ed., *The Cambridge Encyclopedia of Astronomy* (London, 1973)

MS

deontology *Philosophy* Strictly, the study of duty, but in practice a particular view that duty is the primary moral notion, and that at least some of our duties (for example, keeping promises) do not depend on any value that may result from fulfilling them. Immanuel Kant (1724–1804) is probably the most famous deontologist. *See also* CATEGORICAL IMPERATIVE.

Consequentialists (or other teleologists), by contrast, base all duties on the production of value; they treat all duties as 'forward-looking'. Deontologists normally treat at least some duties as 'backward-looking', in that their justification depends on a past event or situation, and some as neither 'forward' nor 'backward-looking', in that their justification appeals to some absolute rule (for example against taking life).

ARL

dependency theory (1957) *Economics* First formulated by American economist Paul Baran (1910–64), this theory proposes that, where a developing country for the most part specializes in producing one good (usually agricultural) for export, an exploitative relationship develops in which its financial and economic resources are controlled by the local élite and the international economy. *See* DEMOGRAPHIC TRANSITION.

P Baran, *The Political Economy of Growth* (New York, 1957); A G Frank, *Dependent Accumulation and Underdevelopment* (London, 1978)

PH

dependency theory (20th century) *Politics* Theory of international and domestic politics.

The third world, or the poor, or the working class, or women, have to choose within structures or societies where they are dependent on the industrial world, or the rich, or men, for the means of existence. Their 'choice' of subordinate roles is thus determined from the start.

J Nash and K Fernandez, *Women, Men and the International Division of Labour* (Albany, NY, 1983)

RB

depth theory *Psychology* Proposed by the Austrian founder of psychoanalysis Sigmund Freud (1856–1939).

Any psychology that postulates dynamic psychic (mental) activities that are unconscious. It embraces all schools deriving from Freud, including many that depart widely from his teaching.

J A C Brown, *Freud and Post Freudians* (Harmondsworth, 1975)

NS

derivational theory of complexity (1960s) *Linguistics* Psycholinguistic hypothesis of the American psychologist G A Miller and his co-workers.

The more transformations are involved in generating a sentence, the more difficult it will be to produce and comprehend. Modern psycholinguists are sceptical about the validity of the original formulation of this theory.

J Aitchison, *The Articulate Mammal*, 3rd edn (London, 1989)

RF

derivative tests *Mathematics* Attributed to the principal founders of calculus: the French mathematician Pierre de Fermat (1601–65), the British mathematician Sir Isaac Newton (1642–1727) and the German mathematician Gottfried Wilhelm Leibniz (1646–1716).

(1) First derivative test: a test for optimality of a critical point of a given function f which uses the first derivative in a neighbourhood of the point. If the first derivative of f is strictly positive in a neighbourhood to the left of the critical point and is strictly negative to the right, then the critical point is a local maximum. If the first derivative of f is strictly negative to the left and strictly positive to the right of a critical point, then that point is a local minimum.

(2) Second derivative test: a test for optimality of a critical point that uses second derivative information at the point. If the second derivative is strictly positive at the critical point and continuous in a neighbourhood of the critical point then it is a local maximum. If the second derivative is strictly negative at the critical point, then the point is a local minimum and if the second derivative is zero at the critical point, then the test is indeterminate and the first derivative test should be tried.

H Anton, *Calculus with Analytic Geometry* (New York, 1980)

ML

derived properties, postulate of *Psychology* Proposed by the German-American psychologist Wolfgang Köhler (1887–1967).

A term from Gestalt Psychology referring to parts of a perception derived from the whole. A derived property is a property of a stimulus due to the characteristics of the whole situation of which it is a part. The postulate emphasizes the fact that, in perception, parts of a stimulus cannot easily be disjointed from other parts. *See* BUNDLE HYPOTHESIS and GESTALT THEORY.

A D Ellis, ed., *A Sourcebook of Gestalt Psychology* (London, 1938)

NS

Deryagin-Landau and Verwey-Overbeek theory *Chemistry See* DLVO THEORY.

Desargues's theorem (1639) *Mathematics* Named after Girard Desargues (1591–1661) who discovered it, although his work was lost and rediscovered in the 19th century.

If the three lines joining corresponding vertices of two triangles in a plane are concurrent then the three points of intersection of the corresponding sides are collinear. The theorem does not hold in this simple form in Euclidean geometry, but it holds if the geometry is augmented by points at infinity; a construction that was proposed independently by Desargues and Johannes Kepler (1571–1630). Because the theorem does not mention length or angle, it is also a theorem of projective geometry, being deduced from PAPPUS'S THEOREM.

JB

Descartes's rule of signs (17th century) *Mathematics* Named after French scientist, philosopher and mathematician René Descartes (1596–1650), this rule states that the number of positive roots, counting multiplicity, of a polynomial with real coefficients is less than or equal to the number of changes in sign in the sequence of coefficients.

If it is known that all the roots of a given polynomial are real, then Descartes's rule gives the exact number of roots. This rule can be used to find the number of negative roots of a polynomial $f(x)$ by considering $f(-x)$. *See also* STURM'S THEOREM.

M Hazewinkel, ed., *Encyclopaedia of Mathematics* (Dordecht, 1988)

ML

descriptions, theory of *Philosophy* Theory invented by English philosopher and mathematician Bertrand Russell (1872–1970) in 1905 to show how denoting phrases like 'the present king of France' could still have meaning though there is nothing for them to denote.

Russell claimed that the grammatical form of 'The present king of France is bald' is misleading as to its logical form, which involves an existential claim; the sentence really says: 'There exists exactly one person reigning over France and there exists no person reigning over France who is not bald', which is false (because of its first clause) but uncontroversially meaningful. Because the analysis of sentences should not depend on what happens to exist, Russell applied his analysis to all denoting phrases, calling most of them definite or indefinite descriptions (according to whether they involved the definite or indefinite articles (or their equivalents)). He also applied it to ordering proper names, which he treated as disguised descriptions.

The theory has been criticized as unnatural and unnecessary, but it and its ramifications have been immensely influential and are still current issues.

B Russell, 'On Denoting', *Mind* (1905); often reprinted

ARL

descriptive/prescriptive linguistics *Linguistics See* RULE.

descriptive theory of names *Philosophy* Theory that proper names, or some of them, and words for natural kinds, like 'tiger' or 'water', have meaning by specifying a description that the object or stuff concerned must satisfy for the name to apply to it.

For example 'tiger' means 'fierce animal with stripes . . .', 'water' means 'colourless tasteless liquid suitable for drinking . . .', 'Homer' means 'poet who wrote the *Iliad* and *Odyssey*'. Recently the theory has been attacked by advocates of the CAUSAL THEORY OF NAMES.

ARL

descriptivism *Philosophy* Any theory claiming that certain utterances have meaning by describing (or purporting to describe) some aspect of reality rather than in various other ways (for example, PRESCRIPTIVISM and EMOTIVISM). In practice the term is confined to ethical utterances. 'Lying is wrong' and 'You ought not to lie' both purport to state moral facts, though descriptivism leaves it open whether they are reducible to facts of another kind (NATURALISM, for example).

R M Hare, 'Descriptivism', *Proceedings of the British Academy* (1963); critical

ARL

detailed balancing principle *Physics* This principle states that, for a system in steady state equilibrium, every process of transformation or exchange of energy which occurs is accompanied by an analogous reverse process. The two processes occur with equal frequency. The principle is concerned with direct transfers by sudden processes, which may be collisions between molecules or other particles or with photons of radiation.

J Thewlis, ed., *Encyclopaedic Dictionary of Physics* (New York, Oxford and London, 1962)

MS

determinism *Philosophy/History* The general course of events is determined by structures deemed to be fundamental. These may be the economic system, the system of religious belief, the state of technology, and so on. Determinism is normally attributed to thinkers in a critical spirit, rather than claimed by them to describe their own views. Not the same as FATALISM (*see also* NECESSITARIANISM). For linguistic determinism, *see* SAPIR-WHORF HYPOTHESIS.

Hard determinists say freewill is incompatible with determinism and is therefore illusory. *Soft* determinists (including most determinists in practice) are compatibilists and can therefore accept freewill. Important issues here are: whether it makes sense for actions to be caused; how actions are related to the corresponding desires, intentions, and so on; and what the conditions are under which responsibility arises. A major problem for determinists is to say just what counts as 'every event being caused'. *Logical* determinism says that statements about tomorrow's events must be true or false today, and therefore tomorrow's events must be already fixed. This raises questions about the nature of truth, and about the law of EXCLUDED MIDDLE. Whether something is caused is independent of whether it is predictable. Evidence for the prediction might be in principle unobtainable (since it might have to include the effects of the prediction itself), and even an indeterminist might feel insulted if his own virtuous action were called unpredictable. *See also* HISTORICAL MATERIALISM.

Aristotle, *De Interpretatione* (*On Interpretation*), ch. 9; classic discussion of logical determinism; Alan Bullock, Oliver Stallybrass, and Stephen Trombley, eds, *The Fontana Dictionary of Modern Thought*, 2nd edn (London, 1988)

RB, ARL

determinism *Psychology* Proposed by the French mathematician and astronomer Pierre-Simon, Marquis de Laplace (1749–1827).

The belief that specific causal factors lie at the root of all events both physical and mental, and thereby all forms of behaviour. Some forms of determinism are more extreme than others: for example, in psychoanalysis the concept of freedom of choice is minimized; EXISTENTIALISM, on the other hand, claims it is possible to determine one's own goals, type of life and future within the limitations set by one's constitution and past experience.

A J Chapman and D M Jones, eds *Models of Man* (Leicester, 1980)

NS

deterrence (20th century) *Politics* Theory of nuclear strategy.

Nations can deter other nations from attacking them by the knowledge of certain retaliation. The most familiar form of such retaliation is possession of (sufficient) nuclear weapons. *See also* BALANCE OF TERROR and MUTUALLY ASSURED DESTRUCTION.

Graham Evans and Geoffrey Newnham, *The Dictionary of World Politics* (Hemel Hempstead, 1990)

RB

detonator theory (20th century) *Politics* Theory of insurrection.

Discontented populations, particularly under foreign occupation, are likened to explosive material which requires detonation by some additional action. This can be provided by trained élites, if necessary entering from abroad. The theory informed some of the activities of the Allies during World War II. *See also* FOCO THEORY.

David Stafford, *Britain and European Resistance 1940–1945* (London, 1979)

RB

deus ex machina (classical) *Literary Theory* Dramatic device particularly associated with the Greek dramatist Euripedes (*c*.484–406 BC).

Literally the Latin for 'a god from the machine', the phrase derives from the practice in Greek drama whereby a 'god' was lowered to the stage by a crane to rescue the hero or sort out the plot. This was condemned by Aristotle for unnaturalness. In modern usage, the term refers to any unmotivated, implausible trick used to resolve problems in the plot.

Aristotle, 'Poetics', *Critical Theory since Plato*, H Adams, ed., S H Butcher, trans. (New York, 1971)

RF

Deutsch's feedback model (1960) *Biology* Proposed by J Anthony Deutsch (1927–), this early model of animal behaviour states that an animal's body detects a deficit (for example, hunger) or an environmental stimulus (the presence of a predator) and excites a 'central structure' or 'link' in the nervous system.

The greater the deficit, the stronger the excitation. The excited link then activates the appropriate parts of the body to take corrective action (for example, feeding or fleeing). The resulting change in the deficit is fed back to an 'analyzer', which then inhibits the link so that it is no longer responsive. Over time this inhibition weakens, until the link is again sensitive to internal or external stimuli. This model, which has experimental support, constitutes an early application of the terminology of CONTROL THEORY of mathematics to problems in animal behaviour. *See also* LORENZ'S HYDRAULIC MODEL.

J Deutsch, *The Structural Basis of Behaviour* (Cambridge, 1960)

KH

development theory (20th century) *Politics* The industrialized nations represent the most advanced form of society, and other nations are categorized in terms of their approximation to this model. They can thus be judged more or less developed, and their efforts should be directed towards approaching as close as possible to the 'developed' model in their political, social, and economic institutions.

Roger Scruton, *A Dictionary of Political Thought* (London, 1982)

RB

deviation (1960s) *Stylistics* Prominent in the (independent) work of stylisticians Samuel R Levin and Geoffrey N Leech (1936–).

Style as departure from a linguistic norm. Quantitative (statistical) deviation characterizes a style which has an unusual frequency of some linguistic structure; the definition of a norm, however, is problematic. Qualitative (determinate) deviation is a creative bending of the rules of the language (*see also* GRAMMATICALITY and ACCEPTABILITY and FOREGROUNDING).

G N Leech, 'Stylistics', *Discourse and Literature*, T A van Dijk, ed. (Amsterdam, 1985), 37–57

RF

diachronic and synchronic (1916) *Linguistics* One of the foundational distinctions laid down by the French linguist Ferdinand de Saussure (1857–1913).

Distinction between studying language historically (which had been the dominant method in the 19th century), and describing it as a system existing at a particular point of time; the latter, the synchronic, was advocated by Saussure, and became the chief procedure of 20th century linguistics.

F de Saussure, *Course in General Linguistics*, W Baskin, trans. (Glasgow, 1974 [1916])

RF

dialectic *Philosophy* A term with various meanings for different philosophers, notably Plato, Aristotle, the Stoics, Kant, Hegel, and Marx.

For the ancients, dialectic was largely a matter of philosophical method. As embodying a theory about reality the term belongs primarily to Hegel, who thought that both reality itself and our thought about it (which were ultimately the same thing) developed by certain processes occurring (the thesis) which contained within themselves the seeds of their own frustration and destruction (the antithesis). This antithesis, however, contained its own antithesis (the negation of the negation) which emerged in the form of a synthesis of the original thesis and antithesis. This synthesis then formed the thesis for the next stage in the overall process.

Marx took over the general ideas but replaced Hegel's idealist equation of thought and reality by his own DIALECTICAL MATERIALISM, applying the above process to the material world. This becomes HISTORICAL MATERIALISM when applied to the development of societies and to human motivation, both of these resting ultimately on economic considerations.

Within psychology, dialectic may be described as any complex process of conceptual conflict or dialogue in which the generation, interpretation and clash of oppositions leads to a fuller mode of thought. It is not a prominent theory within the discipline, except in developmental psychology of the lifespan.

M P Honzik, 'Life-span Development', *Annual Review of Psychology*, vol. xxxv (1984), 309–31

NS, ARL

dialectic theory (5th century BC–) *Psychology* Dialectic may be described as any complex process of conceptual conflict or dialogue in which the generation, interpretation and clash of opposition leads to a fuller mode of thought. It is not a prominent theory within psychology except in development psychology of the life span.

M P Honzik, 'Life-span development', *Annual Review of Psychology*, 35, 309–31

NS

dialectical materialism (19th century–) *Politics* Term coined by German philosopher Josef Dietzgen (1820–88) and developed by Marx's successors.

History moved forward as a result of 'argument' between different material features of human society. Dialectical materialism was an ambitious and almost abstract attempt by Marx's successors to give an account of all human and natural events, and was subsequently largely discarded. *See also* HISTORICAL MATERIALISM.

Tom Bottomore, *A Dictionary of Marxist Thought*, 2nd edn (Oxford, 1991)

RB

dialectology (later 19th century) *Linguistics* The methodological pioneer of dialectology was J Gilliéron (1854–1926), a Swiss who worked in France.

Study of variations within a language primarily related to geographical area, for example differences of phonology and of vocabulary; CORRELATIONAL SOCIOLINGUISTICS, sometimes referred to as 'urban dialectology' (thus allowing 'dialect' to be construed as variation linked to social factors too), has much refined methods of dialectology since the 1960s.

W N Francis, *Dialectology* (London, 1983)

RF

dialetheism *Philosophy* A name for the view that the law of contradiction can on occasion and within certain limits be violated without irrationality; '. . . the view that some contradictions are true, or that some things are both true and false' (Priest, page 99). *See also* PARACONSISTENCY.

G Priest, 'Contradiction, Belief and Rationality', *Proceedings of the Aristotelian Society* (1985–86)

ARL

dialogism (1920s) *Literary Theory* Modern term for the fundamental principle underlying the work of the Russian critic Mikhail M Bakhtin (1895–1975) (pseud. V N Voloshinov, P N Medvedev); extremely influential since its discovery and translation in the 1960s.

All utterance is dialogic in that it anticipates the response of an addressee, is 'accented' towards another. In speech we define ourselves by our perceived relationship to others. *See also* ACCOMMODATION, CARNIVALIZATION, HETEROGLOSSIA, MONOLOGISM, MULTIACCENTUALITY and POLYPHONY.

M Holquist, *Dialogism* (London, 1990)

RF

dictatorship of the proletariat (19th century–) *Politics* Theory of course of revolution, coined by German social theorist Karl Marx (1818–83) but used most frequently by Russian revolutionary V I Lenin (1870–1924).

The capitalist STATE would be destroyed and replaced by a temporary dictatorship of the whole proletariat which, in abolishing capitalism, would itself then wither away and be replaced by COMMUNISM in which there would be neither coercive state nor class division. *See also* WITHERING AWAY OF THE STATE.

David Miller *et al.*, eds, *The Blackwell Encyclopaedia of Political Thought* (Oxford, 1987)

RB

Dido's problem *Mathematics See* ISOPERIMETRIC PROBLEM.

Die Brücke (1905–13) *Art* A name (German for 'the Bridge') adopted by a group of artists in Dresden with links to the Munich group BLAUE REITER. Members included Ernst Ludwig Kirchner (1880–1938), Erich Heckel (1883–1970), Karl Schmidt-Rottluff (1884–1976) and F Bleyl, joined in 1906 by Max Pechstein (1881–1951) and Emil Nolde (1867–1956).

Their art was analagous to that of the Fauves in France, based on nature rather than abstraction and using pure colour and bold draughtsmanship. However, the German group was Expressionist, being more concerned with spontaneity and emotion. Their subjects were often crudely and harshly portrayed but with great impact, especially through the medium of woodcut. *See* EXPRESSIONISM and FAUVISM.

LL

diegesis (*c*.373 BC) *Literary Theory* Theory postulated by Greek philosopher Plato (*c*.427–*c*.347 BC).

Plato distinguished diegesis (narration) from mimesis (representation of speech);

but *see* MIMESIS for broader modern meaning. The theory of diegesis has been much developed in structuralist theories of narrative. *See also* NARRATOR.

S Rimmon-Kenan, *Narrative Fiction: Contemporary Poetics* (London, 1983)

RF

Dieterici's rule (1899) *Chemistry* Equation proposed by C Dieterici. The rule is an equation of state for real gases

$$p(V - b)\, e^{a/RTV} = RT$$

where p is pressure, V volume, R the gas constant, T temperature and a and b constants. This is a modified form of the van der Waals equation. The rule allows for the influence of molecules near the boundary of the gas.

S Glasstone, *Textbook of Physical Chemistry* (London, 1948)

MF

difference principle (1971) *Politics* Theory for applying equality, proposed by the American political theorist John Rawls (1921–). Treating people unequally is only justified if by so doing the least advantaged member of society is made better off.

David Miller *et al.*, eds, *The Blackwell Encyclopaedia of Political Thought* (Oxford, 1987)

RB

differential equations (1676) *Mathematics* The term was first proposed by German philosopher and mathematician Gottfried Wilhelm Leibniz (1646–1716), who succeeded in solving many simple differential equations.

The study of equations which contain derivatives or differentials of a function. An ordinary differential equation (ODE) contains derivatives of a function of only one variable, while a partial differential equation (PDE) contains partial derivatives of a function of more than one variable.

M Hazewinkel, ed., *Encyclopaedia of Mathematics* (Dordecht, 1988)

ML

differential rent theory (19th century) *Economics* Raised as an issue by Scottish economist James Anderson (1739–1808) and English economist David Ricardo (1772–1823), this theory asserts that rent arises because of the differences in the fertility or location of agricultural land. No rent is paid on the worst land and the total amount of rent increases as the margin of cultivation is extended.

D Ricardo, *On the Principles of Political Economy and Taxation* (London, 1817)

PH

differentiated marketing *Marketing* The principle of providing products to satisfy specific needs of various consumer groups. This is one of the three basic marketing strategies, the others being undifferentiated and concentrated marketing.

A manufacturer begins with a basic product and then makes minor changes to meet the needs of each subgroup. The practice encourages product innovation and test marketing before these innovations are introduced to the broader market. Size of the firm is important, for it will need to achieve economies of scale and win substantial market share in each of the sub-markets. *See* HOTELLING'S LAW, MONOPOLISTIC COMPETITION, IMPERFECT COMPETITION and PRICE DISCRIMINATION.

M J Baker, *Marketing: An Introductory Text* (London, 1971)

AV, PH

différance (1968) *Literary Theory* Neologism formed by the French philosopher Jacques Derrida (1930–).

'Neither a word nor a concept', according to Derrida, it is in fact a central strategy in DECONSTRUCTION. The aberrant spelling (which cannot be heard in speech; *see also* PHONOCENTRISM) signals a term with a meaning unstably situated between 'differ' and 'defer', the two meanings of the French verb *différer*. Meanings, in Saussurean linguistics, are not positive entities but only the product of differences; and in communication there is no transcendant meaning, only an infinite deferral of meanings.

J Derrida, '*Différance*', A Bass, trans., repr. in H Adams and L Searle, eds *Critical Theory since 1965* (Tallahassee, 1986), 120–36

RF

diffusion laws *Chemistry See* GRAHAM'S LAW OF DIFFUSION.

diffusion theory *Physics* At the macroscopic level, diffusion is a universal process that leads to the elimination of concentration gradients in solids, liquids and gases. At the microscopic level, diffusion arises from the random motion of the atoms or molecules, activated by local fluctuations in the internal energy. The macroscopic laws that describe diffusion are known as Fick's laws, after their formulator, the German physiologist Adolf Eugen Fick (1829–1901).

First law: Consider a plane of unit area of cross-section drawn in the medium, normal to the x-direction, across which a concentration gradient $\partial c/\partial x$ exists. The net flux of matter in the x-direction, J_x, through the plane, from regions of high concentration to regions of low concentration, is given by

$$J_x = -D\,(\partial c/\partial x)\ .$$

This equation defines a quantity D, known as the diffusion coefficient, which turns out to be characteristic of the system and independent of $\partial c/\partial x$ and area.

Second law: If D is a constant

$$\partial c/\partial t = D(\partial^2 c/\partial x^2)\ .$$

This is the fundamental law upon which all experimental studies of diffusion depend.

J Thewlis, ed., *Encyclopaedic Dictionary of Physics* (New York, Oxford and London, 1962)

MS

diffusionism (1882) *Anthropology* A term coined by German geographer Friedrich Ratzel (1844–1904).

The historical approach which emphasizes the transmission of items of culture from one place to another rather than the development of culture through time. Its opposite is EVOLUTIONISM. Diffusionism was the main approach of German-Austrian anthropology prior to World War II, and in the early 20th century it also had adherents in the UK and the USA. It is no longer a major theoretical force.

R H Lowie, *The History of Ethnological Theory* (New York, 1937)

ABA

diglossia (1959) *Linguistics* Modelled on the French *diglossie*; coined by the American linguist Charles Albert Ferguson (1921–).

Some communities such as Greece, German-speaking Switzerland, and Arab countries deploy two varieties of a language: a local, vernacular, spoken version associated with informal, solidary communication; and a more prestigious, superposed version associated with education, government, religion and literature. Ferguson calls these 'Low' and 'High' respectively. Speakers and writers choose between L and H according to the situation, and thus signify different cultural values.

C A Ferguson, 'Diglossia', *Word*, 15 (1959), 325–40, repr. in P P Giglioli, ed., *Language and Social Context* (London, 1972), 232–51

RF

diluvianism (18th–19th centuries) *Geology* A theory which posited certain geological phenomena as evidence of the biblical Flood, it was particularly advanced by the scientific English clerics William Buckland (1784–1856), William Daniel Conybeare (1785–1873) and Adam Sedgwick (1787–1857).

In the valley bottoms of northern Europe and around the Alps, fine to coarse sediments were found to overlie chaotic mixtures of sands, gravels, clays and often large boulders. Adherents of the theory associated these 'diluvial' beds with deposition, disturbance by waters up to 2.7 km deep, and violent currents. Diluvianism gave way to UNIFORMITARIANISM by about 1870. *See also* CATASTROPHISM and DRIFT THEORY.

A Hallam, *Great Geological Controversies* (Oxford, 1983)

DT

dimension-psychological-personality theory *Psychology* Proposed by the German psychologist W Stern (1871–1938).

Part of Stern's personality theory, the concept is defined as the directions and time processes within which a person realizes himself/herself. The theory distinguishes two groups of personal dimensions: individual and world dimensions. The basic dimension of any person is considered to be the polarity within/without.

H Murray, *Explorations in Personality* (New York, 1938)

NS

diminishing returns, law of *Economics.* Sometimes referred to as variable factor proportions, this law states that as equal quantities of one variable factor are increased, while other factor inputs remain constant, *ceteris paribus*, a point is reached beyond which the addition of one more unit of the variable factor will result in a diminishing rate of return and the marginal physical product will fall.

PH

Dini's theorem (19th century) *Mathematics* Named after Italian geometer and analyst Ulisse Dini (1845–1918), this is the result whereby if a monotone decreasing sequence of continuous functions defined on a compact set converges pointwise to a continuous limit, then that convergence is uniform.

K Knopp, *Theory and Application of Infinite Series* (New York, 1990)

ML

Diophantine analysis *Mathematics* Named after Greek mathematician Diophantus of Alexandria (*fl.* AD 250), this technique is also known as Diophantine approximation. It is the area of NUMBER THEORY whose principal concern is the solubility of inequalities in integers.

Results in this area have important consequences in the study of DIOPHANTINE EQUATIONS and are closely related to the study of transcendental numbers. *See also* DIRICHLET'S THEOREM ON DIOPHANTINE APPROXIMATION, FERMAT'S LAST THEOREM, MINKOWSKI'S THEOREM ON CONVEX BODIES, CONTINUED FRACTION ALGORITHM, THUE-SIEGEL-ROTH THEOREM and GEOMETRY OF NUMBERS.

A Baker, *A Concise Introduction to the Theory of Numbers* (Cambridge, 1984)

MB

Diophantine approximation *Mathematics* *See* DIOPHANTINE ANALYSIS.

Diophantine equations *Mathematics* Named after the Greek mathematician Diophantus of Alexandria (*fl.* AD 250), these are any equations whose solutions are required to be integers.

The most common type is a polynomial equation in several unknowns with integer coefficients. In 1970, Matiyasevich, developing ideas of Davis, Robinson and Putnam, showed that there is no algorithm for determining whether a given Diophantine equation of polynomial type is solvable, thereby answering Hilbert's tenth problem. *See also* PELL'S EQUATION, HILBERT'S PROBLEMS, FERMAT'S LAST THEOREM, DIOPHANTINE ANALYSIS, BAKER'S THEOREM and THUE-SIEGEL-ROTH THEOREM.

A Baker, *A Concise Introduction to the Theory of Numbers* (Cambridge, 1984)

MB

Dirac's quantum theory (1928) *Physics* Named after the English mathematical physicist Paul Adrien Maurice Dirac (1902–84), joint winner of the 1933 Nobel prize for Physics.

Dirac's theory takes account of relativity in applying a mathematical formalism to the Schrödinger wave equation to deal with the wave mechanics of fermions. The theory shows that fermions must have spin ½, and it also predicts the existence of antiparticles. *See also* ANTIMATTER.

GD

direct action *Politics* Theory of politics derived from ANARCHISM.

Rather than accept the delays of representative politics, people with objectives or grievances in common should act directly to achieve their aims, either by demonstration or by such means as sit-ins, occupations, or the take-over of industrial premises.

April Carter, *Direct Action and Liberal Democracy* (London, 1973)

RB

direct democracy *Politics* Theory of direct self-government.

Democracy is rule by all the people, and thus requires direct participation in the making of decisions and the passing of laws. Such a form of government is only possible in small communities such as the city states of classical Greece, though it is aspired to in the arguments of Swiss philosopher Jean-

Jacques Rousseau (1712–78). Direct democracy is contrasted with representative democracy.

David Held, *Models of Democracy* (Oxford, 1987)

RB

directed democracy (20th century) *Politics* In states which are formally democratic, but where an élite or oligarchy governs in practice, democracy is claimed to exist because it is 'directed' or 'guided' rather than superseded. The term (which was sometimes applied to the Soviet Union) combines euphemism and criticism and is now used only ironically.

Roger Scruton, *A Dictionary of Political Thought* (London, 1982)

RB

directed panspermia (1981) *Biology* Proposed by British scientist Francis Harry Compton Crick (1916–), who shared the 1962 Nobel prize for Physiology or Medicine for his work in modelling deoxyribonucleic acid (DNA), this is the hypothesis that life on earth was started by bacteria sent via spaceship nearly 4 billion years ago from an advanced civilisation located somewhere in the universe.

Crick admits that no evidence exists for his theory; critics say that directed panspermia falls short as a theory of origins, because it fails to explain how life began on the originating planet. *See also* PANSPERMIA and 'INFINITE WORLDS THEORY'.

F Crick, *Life Itself, its Origin and Nature* (1981)

KH

Dirichlet's problem (19th century) *Mathematics* Named after French-born German mathematician Peter Gustav Lejeune Dirichlet (1805–59), this is a very important problem in COMPLEX FUNCTION THEORY.

Given an open bounded region A and a function ϕ which is continuous on the boundary of A, it is required to find a function u, continuous on the closure of A such that u is harmonic on A; that is, it satisfies Laplace's equation

$$\frac{\partial^2 u}{\partial x^2} + \frac{\partial^2 u}{\partial y^2} = 0,$$

and equals ϕ on the boundary of A. This situation arises in many physical problems in electrodynamics, heat, and POTENTIAL THEORY. A classical method due to Perron gives a general technique for seeking solutions to this problem. *See also* NEUMANN PROBLEM.

S G Krantz, 'Functions of One Complex Variable', *Encyclopedia of Physical Science and Technology* (Academic Press, 1987)

MB

Dirichlet's theorem (19th century) *Mathematics* Named after French-born German mathematician Peter Gustav Lejeune Dirichlet (1805–59), this posits that if f is a bounded periodic function which has at most a finite number of maxima and minima and a finite number of discontinuities in each period, then where f is continuous the FOURIER SERIES for f converges to f and where f is discontinuous it converges to the average of the right and left limits of f at the discontinuity.

K Knopp, *Theory and Application of Infinite Series* (New York, 1990)

ML

Dirichlet's theorem on arithmetical progressions (1837) *Mathematics* Named after French-born German mathematician Peter Gustav Lejeune Dirichlet (1805–59), this is the result in NUMBER THEORY whereby if a and b are positive integers with no prime factor in common, then the arithmetical progression

$$an + b \quad (n = 1,2,\ldots)$$

contains infinitely many primes. Equivalently, there are infinitely many primes of the form $an + b$. *See also* PRIME NUMBER THEOREM.

T M Apostol, *Introduction to Analytic Number Theory* (Springer, 1976)

MB

Dirichlet's theorem on Diophantine approximation (1842) *Mathematics* Named after French-born German mathematician Peter Gustav Lejeune Dirichlet (1805–59), this is the result whereby for any real number θ and any real number $Q > 1$, there exist integers p and q with $0 < q < Q$ such that

$$|q\theta - p| \leqslant \frac{1}{Q}$$

This assertion is a consequence of the PIGEON-HOLE PRINCIPLE. An important corollary is that if θ is irrational, then there exist infinitely many rationals

$$\frac{p}{q}$$

such that

$$\left|\theta - \frac{p}{q}\right| < \frac{1}{q^2}.$$

See also CONTINUED FRACTION ALGORITHM, HURWITZ'S THEOREM and THUE-SIEGEL-ROTH THEOREM.

A Baker, *A Concise Introduction to the Theory of Numbers* (Cambridge, 1984)

MB

disbelief *Literary Theory See* WILLING SUSPENSION OF DISBELIEF.

discontinuity theory of learning *Psychology* Proposed by one of the most influential learning theorists, the American psychologist E R Guthrie (1886–1959).

This is a theory of discrimination learning, holding that the organism must focus on those aspects of the stimulus which are critical to the required discrimination in order for learning to take place. It is regarded as belonging to the same general class as the ALL OR NONE PRINCIPLE. *See also* CONTIGUITY THEORY.

E R Guthrie, *The Psychology of Learning* (New York, 1935)

NS

discourse analysis (1970s) *Linguistics* Used either as a general term; or for a specific model, developed by British linguists at the University of Birmingham, studying classroom discourse.

(1) Umbrella for a number of approaches to language as continuous text or speech, emphasizing action and interaction. (2) The Birmingham model. SCALE AND CATEGORY grammar is extended 'above' the sentence to analyze the lesson genre as a set of 'transactions' consisting of 'exchanges' between speakers; exchanges are structured on 'moves' which in turn consist of 'acts'. *See also* CONVERSATIONAL ANALYSIS.

G Brown and G Yule, *Discourse Analysis* (Cambridge, 1983), on general use; J McH Sinclair and R M Coulthard, *Towards an Analysis of Discourse* (London, 1975), on Birmingham model

RF

discovery procedure (1930s–50s) *Linguistics* According to the American linguist Noam Chomsky (1928–), this is a major and false methodological assumption of American STRUCTURAL LINGUISTICS.

The school of linguistics which developed the ideas of the American Leonard Bloomfield (1887–1949) attempted to devise analytic procedures which – applied 'bottom-up' to recorded samples of speech, and without knowledge of meaning – could reveal the phonemic, morphological and syntactic systems of the language. This idea of discovery procedure was strongly denied by Chomsky in his formulation of GENERATIVE GRAMMAR.

L Bloomfield, *Language* (London, 1935 [1933]); N Chomsky, *Syntactic Structures* (The Hague, 1957), 55

RF

discrimination *Psychology* Particularly associated with the American psychologist Edward Lee Thorndike (1874–1949).

Discrimination in this sense is the ability to distinguish between types of stimuli, be they similar or different. In the context of learning through conditioning, it is the act of identifying and responding to relevant stimuli while suppressing a response to irrelevant stimuli. A different meaning is that found in social psychology, where discrimination is regarded as an act of exclusion or hostility directed against members of 'other' groups. *See* COMPARATIVE JUDGEMENT, LAW OF.

R Borger and A E M Seaborne, *The Psychology of Learning* (Harmondsworth, 1973)

NS

disengagement theory of ageing *Psychology* Proposed by the American psychologists E Cumming and R E Henry.

This concerns the psychological processes at work in reduced involvement in the social environment, contending that retirement involves a sharp decline in social interaction, a reduced life space and loss of social esteem and morale. This contrasts with the view that the individual shifts from one set of activities and social roles to others that may be equally or even more satisfying (re-engagement theory). The role of cultural values, attitudes, mode of life, economic and social factors were relatively neglected in the earliest formulations of the theory. *See* ACTIVITY THEORY OF AGEING.

D Bromley, *The Psychology of Human Ageing* (Harmondsworth, 1966)

NS

disequilibrium theory (20th century) *Economics* Found in the work of English economist John Maynard Keynes (1883–1946), this theory refers to a situation in which market equilibrium has not been reached or where there is a tendency for variable factors to change.

Keynes believed that an economy has a natural inclination towards disequilibrium. *See* EQUILIBRIUM THEORY, GENERAL EQUILIBRIUM THEORY, PARTIAL EQUILIBRIUM THEORY, FUNDAMENTAL DISEQUILIBRIUM, CLASSICAL MACROECONOMIC MODEL, LAW OF MARKETS, and COBWEB THEOREM.

J M Keynes, *The General Theory of Employment, Interest and Money* (New York, 1936); J D Hey, *Economics in Disequilibrium* (Oxford, 1981)

PH

disintegration (or transformation) theory (1902) *Physics*

The view that some atomic species are subject to spontaneous disintegration or transformation was first put forward by the British physicists Ernest Rutherford (1871–1937) and Frederick Soddy (1877–1956).

The main feature of the theory is that radioactive bodies contain unstable atoms, a fixed fraction of which decays in equal intervals of time. The residue of the decayed atom is a new element, which may also decay, and so on, until a stable element is reached.

J Thewlis, ed., *Encyclopaedic Dictionary of Physics* (New York, Oxford and London, 1962)

MS

disinterestedness (1790) *Literary Theory* Aesthetic experience discussed by the German philosopher Immanuel Kant (1724–1804); attitude of critic stipulated by the English critic Matthew Arnold (1822–88).

Aesthetic judgment is 'disinterested' in taking no account of the purpose or practical use of a work; Arnold's ideal critic is disinterested in avoidance of personal involvement in the arguments of feelings promoted by a work. *See also* AESTHETIC DISTANCE.

I Kant, *Critique of Judgment* (1790); M Arnold, 'The Function of Criticism at the Present Time' [1865], *Critical Theory since Plato*, H Adams, ed. (New York, 1971), 379–99, 583–95

RF

dislocation model of melting (20th century) *Physics* A model of the melting process, based on the idea that the melting transition results from a catastrophic proliferation of dislocations in the material.

A liquid at its melting-point is then pictured as a crystal that is saturated with dislocation lines. The energy of a dislocation line introduced into a crystal depends on the density of dislocation lines existing in the crystal; and the greater the density of dislocation lines, the lower the energy needed to create still more. To avoid the production of long-range stresses, the dislocations generated in the crystal should be dipoles; that is, close pairs of opposite sign.

J Thewlis, ed., *Encyclopaedic Dictionary of Physics* (New York, Oxford and London, 1962)

MS

displacement, displaced speech (1933) 'Displaced speech' coined by the American linguist Leonard Bloomfield (1887–1949); 'displacement' popularized by the American linguist Charles Francis Hockett (1916–) in the 1960s.

The use of language to refer to objects not present to the speakers in time and/or space; claimed to be a 'design feature' of human language.

C F Hockett, 'The Problem of Universals in Language', *Universals of Language*, J H Greenberg, ed. (Cambridge, Mass., 1966), 1–29

RF

displacement law in complex spectra *Physics* The arc spectrum of an element is similar to the first spark spectrum (the singly ionized atom spectrum) of the element one place higher in the periodic table, or to the second spark spectrum of the element two places higher.

J Thewlis, ed., *Encyclopaedic Dictionary of Physics* (New York, Oxford and London, 1962)

MS

displacement law in radioactive decay (1914) *Physics* A set of rules set out by Frederick Soddy (1877–1956) connecting the type of decay with the displacement in the periodic table of the daughter element relative to the parent.

When a parent atom emits an alpha particle, the daughter is displaced two columns to the left in the periodic table; that is, the atomic number of the daughter is two less than that of the parent. When a parent atom emits a negative beta particle, the daughter is displaced one column to the right in the periodic table; that is, the atomic number of the daughter is that of the parent plus one.

J Thewlis, ed., *Encyclopaedic Dictionary of Physics* (New York, Oxford and London, 1962)

MS

displacement law of Wien *Physics See* WIEN'S DISPLACEMENT LAW.

dissociation of ideas (1901) *Literary Theory* Stipulation of the French critic Rémy de Gourmont (1858–1915). The avoidance and questioning of familiar associations and, in writing, a technique of following disconnected or marginal lines of thought.

R de Gourmont, *La culture des idées* [1901] (Paris, 1983)

RF

dissociation of sensibility (1921) *Literary Theory* Phrase coined by the American poet Thomas Stearns Eliot (1888–1965).

Contentious theory of a failing in 17th century English poetry after John Donne (1572–1631). Donne, and the Elizabethan and Jacobean dramatists, enjoyed 'a direct sensuous apprehension of thought': intellection (for example, scientific theory) 'modified [Donne's] sensibility'. Dissociation of sensibility set in with John Milton (1608–74) and John Dryden (1631–1700), for whom thinking and emotion were separate experiences. *See also* METAPHYSICAL POETS.

T S Eliot, 'The Metaphysical Poets', *Selected Essays* (London, 1951)

RF

dissociation theory *Chemistry See* ARRHENIUS DISSOCIATION THEORY.

dissonance theory *Psychology* Proposed by the American L Festinger (1919–).

Dissonance refers to emotional state, tension or energy which results when two simultaneously held ideas, attitudes or opinions are inconsistent or conflicting. Dissonance may also result from a discrepancy between belief and overt behaviour. Individuals modify their cognitions or behaviours in order to achieve dissonance reduction. The theory accurately predicts that individuals will reduce dissonance, but cannot predict precisely how this will be achieved. *See* COGNITIVE CONSISTENCY THEORY.

L Festinger, *A Theory of Cognitive Dissonance* (Stanford, 1957)

NS

distance paradox (5th century BC) *Mathematics* Also known as the dichotomy paradox, this is one of the paradoxes of Zeno of Elea (*c.*490–435 BC).

Nobody can actually walk because before he can complete his first step he must cover half the distance, and before that a quarter of the distance and so on infinitely. Therefore, before a person can move he must carry out an infinity of operations, though they are to be carried out simultaneously so the supposed contradiction is not obvious. (Perhaps one has no difficulty if one does not think about the task.)

JB

distinctive feature (1930s, 1940s) *Linguistics* Phonological theory developed by PRAGUE SCHOOL linguists, especially Nikolay S Trubetzkoy (1890–1938) and Roman Jakobson (1896–1982).

Minimal acoustic properties such as 'tense' and 'strident' which distinguish words and meanings; a 'bundle' of distinctive features makes a recognisable sound or PHONEME.

C A M Baltaxe, *Foundations of Distinctive Feature Theory* (Baltimore, 1978)

RF

distribution (1930s) *Linguistics* Major principle of American STRUCTURAL LINGUISTICS, also compatible with the STRUCTURALISM of French linguist Ferdinand de Saussure (1857–1913). The terms 'complementary distribution' and 'free variation' originated in an article by American linguist Morris Swadesh (1909–67) in 1934.

The set of positions in the chain of speech in which a linguistic item may occur; for example, a verb can occur after an auxiliary but not after a definite article. Two items are said to be in 'complementary distribution' if they do not occur in the same environments (for example, the allophones of a phoneme), or in 'free variation' if they are interchangeable. *See* PHONEME.

Z S Harris, *Structural Linguistics* (Chicago, 1960 [1951])

RF

distribution law (1872) *Chemistry* Law attributed to French physical chemist and statesman Marcellin Pierre Eugène Berthelot (1827–1907). Also known as the partition law.

When a solute is soluble in two immiscible liquids, the solute distributes itself between the two liquids. The ratio in which it is distributed is governed by the distribution law. This states that a solute distributes itself between two immiscible liquids in a constant ration of concentrations irrespective of the amount of solute added.

M Freemantle, *Chemistry in Action* (Basingstoke, 1987)

MF

distribution theory (18th century–) *Economics* One of the fundamental components of modern economic theory, first examined by French economist Anne-Robert Jacques Turgot (1727–81), this is an explanation of how NATIONAL INCOME is distributed between different groups involved in the production process.

Functional distribution of income is that earned by the owners of the various factors of production. The income earned by individual factors is determined by the demand and supply for them. Distribution theory (which also examines the stage at which goods and services reach consumers) was developed subsequently by Scottish economist Adam Smith (1723–90), English economist David Ricardo (1772–1823), German political economist Karl Marx (1818–83) and others. *See* INCOME DISTRIBUTION THEORY, MARGINAL PRODUCTIVITY THEORY OF DISTRIBUTION, EULER'S THEORY, LABOUR THEORY OF VALUE and CAPITAL THEORY.

A J Turgot, *Réflections sur la formation et la distribution des richesses* (Paris, 1766)

PH

distributism (20th century) *Politics/Economics* Theory of politics and economics of Anglo-French writer Hilaire Belloc (1870–1953).

Since PROPERTY is desirable, it should be distributed to all households, each of which should have enough on which to make a living. The theory had more application to agricultural society, and was ill-adapted to an industrial one.

Rodney Barker, *Political Ideas in Modern Britain* (London, 1978)

RB

distributive law *Mathematics* The theorem or axiom of any mathematical system which states that a given pair of binary operations $\star$ and $\bullet$ has the property that one of the operations, say $\star$, distributes over the other; that is,

$$a \star (b \bullet c) = (a \star b) \bullet (a \star c) .$$

For example,

$$a(b + c) = ab + ac$$

is the distributive law for multiplication over addition. On the other hand, notice that

$$a + bc \neq (a + b)(a + c) .$$

See also ASSOCIATIVE LAW and COMMUTATIVE LAW.

MB

disuse, principle of *Psychology* Proposed by the American psychologist Edward Lee Thorndike (1874–1949).

One of several laws of conditioning, this states that responses and learned associations will weaken and disappear with disuse. With the passage of time, events of the past will appear dim and faded, being less well recalled. Practice is essential for retention.

E L Thorndike, *Animal Intelligence* (New York, 1911)

NS

divergence *Linguistics See* ACCOMMODATION.

divergence theorem (19th century) *Mathematics* Also known as 'Gauss's theorem', after German mathematician and astronomer Carl Friedrich Gauss (1777–1855), this theorem concerns vector analysis.

The triple integral of the divergence of a function over a region is the surface integral of the normal component of the function taken over the boundary of the region. That is, for a smooth function **F**

$$\int\int_{\partial R} \mathbf{F} \cdot \mathbf{n}\, \mathbf{ds} = \int\int\int_{\mathbf{R}} \mathbf{div\, F\, dV}$$

where **n** is the external unit normal to the surface. *See also* STOKES'S THEOREM.

H Anton, *Calculus with Analytic Geometry* (New York, 1980)

ML

divine right *Politics* Doctrine of royal authority reintroduced by Augustus (63 BC–AD 14) and stemming from the dual role of Roman emperors as both gods and rulers.

Kings derive their power from God. Their authority is thus unlimited except by divine law itself, and subjects are obliged to obey as they are obliged to obey God himself.

J W Allen, *A History of Political Thought in the Sixteenth Century* (London, 1960)

RB

divisionism (1899) *Art* A name first used by French artist Paul Signac (1863–1935) for a method of painting with pure colour which later developed into scientific POINTILLISM.

Signac proposed that to achieve the brightest and purest colour, all those of the spectrum should be used, applied in small daubs or dots of unmixed pigment which vary according to the size of the painting. At a distance these fuse in the spectator's eye.

P Signac, *De Delacroix au Néo-Impressionisme* (1899)

AB

Dixon-Askenazy cohesion theory *Biology See* COHESION THEORY.

Dixon cohesion theory *Biology See* COHESION THEORY.

DLVO theory (1941 and 1948) *Chemistry* Theory developed by B V Deryagin and L Landau (1941), and independently by E J W Verwey and J Th G Overbeek (1948).

The theory relates lyophobic colloid stability to energy of interaction between particles. At a given interparticle distance, the interaction energy (V) consists of two interparticle terms: the attractive potential (V_A) and the repulsive potential (V_R). All three terms (V, V_A and V_R) vary with distance between particles. Variation of V with interparticle distance leads to either a maximum value for V, in which case the colloid is stable, or no maximum, in which case coagulation occurs.

W J Popiel, *Introduction to Colloid Science* (New York, 1978)

MF

DNA (deoxyribonucleic acid) *Biology See* WATSON-CRICK MODEL.

Dollo's principle, law or rule (19th–20th century) *Biology* Named after Belgian biologist Louis Antoine Marie Joseph Dollo (1857–1931), this is the principle that evolution is irreversible, that structures and functions lost are not regained.

Modern thinking has refined this principle to account for cases in which some organisms 're-evolve' structures similar to those of ancestors, but lost in intervening species. For example, the evolution of whales shows ancestral species with simple, peg-like teeth, followed by species with cusped teeth, followed by modern whales, which have conical teeth without cusps. *See also* EVOLUTION.

R Milner, *The Encyclopedia of Evolution: Humanity's Search for Its Origins* (New York, 1990)

KH

dominant (1920s–1930s) *Stylistics* The idea originated with the Russian formalists (*see*

FORMALISM, RUSSIAN) and the PRAGUE SCHOOL; it was defined in 1935 by Roman Jakobson (1896–1982).

The major, FOREGROUNDED, stylistic patterning which identifies a literary type: for example unrhymed iambic pentameter for blank verse; metaphors animating the inanimate in GOTHIC.

R Jakobson, 'The Dominant', *Readings in Russian Poetics*, L Matejka and K Pomorska, trans. and eds (Cambridge, Mass., 1971), 82–7

RF

dominant ideology *Politics* Marxist account of control through ideas.

In any society the dominant ideas are those of the dominant class. In capitalist societies, dominant ideology is thus a powerful means of maintaining the system. *See also* ANDROCENTRISM.

Nicholas Abercrombie, Stephen Hill and Bryan S Turner, *The Dominant Ideology Thesis* (London, 1980)

RB

dominated convergence theorem (early 20th century) *Mathematics* Also known as 'Lebesgue's theorem', after French analyst Henri Léon Lebesgue (1875–1941), this gives conditions for the limit of the integral of a sequence of functions to be the integral of the limit of the sequence.

If $\{f_n\}$ is a sequence of integrable functions convergent almost everywhere to f and there exists a non-negative integrable function g such that $|f_n| \leqslant g$ for all n, then f is integrable and

$$\lim_{n\to\infty} \int f_n \, d\mu = \int f \, d\mu.$$

This theorem can be generalized to an arbitrary complete measure space. *See also* MONOTONE CONVERGENCE THEOREM and FATOU'S LEMMA.

G B Folland, *Real Analysis* (New York, 1984)

ML

dominator modulator theory *Psychology* Not attributable to any one originator, this theory refers to a general body of neuropsychological evidence indicating that chromatic vision depends upon a modulator which mediates the dominant receptor for brightness. A dominator is a retinal cell which responds to light over the entire visible spectrum. Modulators respond selectively to the blue, green, yellow and red spectral ranges. The 'theory' (the body of evidence) occupies a central position in investigations of visual perception.

J E Hochberg, *Perception* (Englewood Cliffs, N.J., 1978)

NS

domino theory (20th century) *Politics* American theory, during the Cold War, of defence against Soviet influence.

Nations over which the USA and the Soviet Union competed for influence were like a line of dominoes, and if one fell to 'communism', a process would be started which would 'lose' all of them.

Graham Evan and Jeffrey Newnham, *The Dictionary of World Politics* (Hemel Hempstead, 1990)

RB

Donders method *Psychology* Formulated by the Dutch oculist and physiologist Franciscus Cornelis Donders (1818–89).

Also referred to as the subtraction method, this emerged from Donders's study of three types of reaction time: simple reaction time, discrimination reaction time, and choice reaction time. In order to obtain an estimate of discrimination time, Donders subtracted simple reaction time from discrimination reaction time. Subtraction of discrimination reaction time from choice reaction time yields an estimate of how long it takes to make a choice from fixed alternatives. *See also* LADD-FRANKLIN THEORY OF COLOUR VISION.

R S Woodworth and H Schlosberg, *Experimental Psychology* (London, 1966)

NS

Doomsday theory (20th century) *Politics* Extension of the theory of MUTUALLY ASSURED DESTRUCTION.

If a nation possessed a Doomsday machine, or system of nuclear weapons so formidable that any attack on it would lead to total global destruction, aggressors would be deterred.

David Robertson, *The Penguin Dictionary of Politics* (London, 1986)

RB

Doppler effect (1842) *Physics* Named after the Austrian physicist Christian J Doppler (1803–53) who discovered the effect in sound waves. The first correct application of the principle to light waves was probably made by the French physicist Armand Hippolyte Louis Fizeau (1819–96).

When the source of a wave train and the observer are in relative motion with respect to a medium in which the waves propagate, the observed frequency of the waves is different from the frequency of the source. If, for example, an observer moves with a speed v_0 towards a stationary source of sound waves of frequency f_0, the frequency f heard by the observer is given by

$$f = [(v + v_0)/v]\, f_0$$

where v is the speed of the sound waves in the medium.

J Thewlis, ed., *Encyclopaedic Dictionary of Physics* (New York, Oxford and London, 1962)

MS

dose-response curve *Biology See* TOLERANCE, LAW OF.

double aspect theory of mind *Philosophy/Psychology* Theory that mind and body, or mental events and some cerebral events, are two aspects of a single thing. The theory resembles NEUTRAL MONISM but is more limited, applying only to certain cerebral (or perhaps neural) events. It is often attributed to Baruch Spinoza (1632–77), but interpretations of him differ. When the body becomes privileged and usurps the function of the single underlying thing, the theory veers towards the identity theory of mind and body.

D M MacKay, *The Clockwork Image* (1974)

ARL, NS

double bind theory *Psychology* Proposed by the Anglo-American psychologist Gregory Bateson (1904–80).

The theory describes a situation faced by a person who is receiving conflicting signals from another person. An example is a child whose parent finds close relationships difficult. The parent reaches to the child with simulated love and yet withdraws and becomes cold when the child approaches. The child is in a double bind situation: whatever action s/he performs will not be suitable and any ideas of what s/he is supposed to do will be confused. Bateson believed this to be a crucial causal factor in cases of schizophrenia and autism.

H Kaplan, I Freedman and B Sadock, *Comprehensive Textbook of Psychiatry* (Baltimore, 1980)

NS

double consciousness (20th century) *Politics/Feminism* Theory of consciousness.

In a patriachal society women have two consciousnesses: one as members of a male dominated system, one as women oppressed by it. The theory has many precedents, amongst them the Christian injunction to be 'in the world but not of it'.

Maggie Humm, *The Dictionary of Feminist Theory* (London, 1989)

HR

double effect doctrine *Philosophy* Ethical doctrine, associated especially though not exclusively with Roman Catholicism. Though we may not intentionally produce evil, we may intentionally do (in pursuit of a suitably greater good) what we foresee will in fact produce evil, provided we regard this evil as an unwanted side-effect which we would avoid if possible. The occurrence of the evil must not be necessary as a means to our intended end. The doctrine might, for example, forbid us to torture an innocent person to gain vital information in a good cause, while allowing us to kill an innocent civilian while bombing a munitions depot, provided that the intended good greatly outweighs the expected evil.

J T Morgan, 'An Historical Analysis of the Principle of Double Effect', *Theological Studies* (1949)

ARL

double entry accounting (*c.*1300) *Accountancy* First used in northern Italian city states, this method of bookkeeping forms the foundation of modern accountancy. It was first described in book form by Italian friar and mathematician Luca Pacioli (1445–*c.*1517) in 1494.

Each business transaction is recorded twice, in the form of a debit and a credit. Total debits in the accounts will always equal, or cancel out, the total credits showing that the account books have balanced correctly.

PH

double helix *Biology See* WATSON-CRICK MODEL.

double negation principle *Philosophy* Principle that, for any proposition P, P logically implies not-not-P, and not-not-P logically implies P. Classical logic accepts both these halves of the principle, but intuitionist logic accepts only the first half, and not the second. This is because it accepts the law of contradiction (and so, given P, cannot allow not-P), but rejects the law of EXCLUDED MIDDLE (and so, given not-not-P, does not consider itself forced to accept P).

G T Kneebone, *Mathematical Logic and the Foundations of Mathematics* (1963), 243-50; elementary account of intuitionism

ARL

dramatic irony *Literary Theory See* IRONY.

dramatis personae *Literary Theory See* PERSONA

dream theory *Psychology* Proposed by the Austrian founder of psychoanalysis Sigmund Freud (1856–1939).

Developed within psychoanalysis, the contents of dreams are analyzed for underlying or disguised motivations, symbolic meanings or evidence of symbolic representations. Dreams were assumed by Freud to be expressions of wish fulfilment; but since most wishes had been repressed, the deep meaning of dreams had to be interpreted through a veil of censorship, disguise and symbolism. During dream analysis the individual relates the dream and then free associates in order to derive insight into underlying symbolics and dynamics. *See* ACTIVATION SYNTHESIS HYPOTHESIS.

S Freud, *The Interpretation of Dreams* (Harmondsworth, 1976)

NS

drift theory (18th–19th centuries) *Geology* A theory, particularly associated with adherents of DILUVIANISM, which claimed that the unconsolidated gravels, sands, and clays covering most of northern Europe and North America were deposited as a consequence of Noah's Flood.

Against this, the Scottish geologist Sir Charles Lyell (1797–1875) suggested these sediments might have been dropped (during an Ice Age) from drifting icebergs, which also carved the valleys. All such sediments are still termed 'drift' deposits, but this origin is only valid for certain deposits (particularly erratic blocks).

A Hallam, *Great Geological Controversies* (Oxford, 1983)

DT

drive reduction theory *Psychology* Proposed by the American psychologist Clark Leonard Hull (1884–1952).

The theory states that the goal of all behaviour is to reduce or alleviate a drive state. Hull regarded drive reduction as the theoretical mechanism through which reinforcement worked. Any event which served to reduce a drive state also reinforced the response that preceded it, increasing the likelihood of its reoccurrence. The theory occupied a central position in Drive Theories of learning and was superseded by more comprehensive and refined theories. *See* DRIVE STIMULUS THEORY.

C L Hull, *Principles of Behaviour* (New York, 1943)

NS

drive stimulus theory *Psychology* Proposed by the American psychologist Clark Leonard Hull (1884–1952).

This theory is concerned with the efferent neural impulses which occur in a drive state. Hull argued that the reduction of these stimuli plays an important role in the regulation of reinforcement. Each drive stimulus is supposed to be associated with a particular set of responses that served to reduce it. The theory has been superseded by more comprehensive accounts of the role of reinforcement in learning. *See* DRIVE REDUCTION THEORY.

C L Hull, *Principles of Behaviour* (New York, 1943)

NS

Drude's law *Physics* Named after its discoverer, the German physicist Paul Karl Ludwig Drude (1863–1906), this law relates the specific rotation of monochromatic plane-polarized light by an optically active substance to the wavelength of the light used:

$$\alpha = k\,(\lambda^2 - \lambda_0^2)$$

where α is the specific rotation, λ is the wavelength of the light used, λ_0 is the wavelength of the light of the nearest absorption band and k is a constant known as the rotation constant.

J Thewlis, ed., *Encyclopaedic Dictionary of Physics* (New York, Oxford and London, 1962)

MS

dual-coding theory *Psychology* Proposed by the Canadian psychologist Allan Urho Paivio (1925–).

The theory holds that human memory is composed of two coding systems: a visual imagery system and a verbal system. The theory is concerned with explaining how representations in visual memory (for example, an image of a pen) are related to and integrated with different representational forms in verbal memory (for example, the word 'pen'). The theory occupies a central position in contemporary COGNITIVE THEORY.

A Paivio, *Mental Representations: A Dual Coding Approach* (New York, 1986)

NS

dual decision hypothesis (1969) *Economics* A refinement by R W Clower of English economist John Maynard Keynes's (1883–1946) unemployment equilibrium, this theory relates to initial and revised demand and supply plans.

If planned demand (demand based on prices which reflect full employment) varies from actual demand (based on a fall in income due to unemployment), the consumer and trader will have to revise their expenditure/selling plans as the consumer is restricted in his/her demand choices due to income. *See* GENERAL THEORY OF EMPLOYMENT, INTEREST AND MONEY.

R Clower, ed. *Monetary Theory* (Harmondsworth, 1969)

PH

dual economy theory (1953) *Economics* Attributed to J H Boeke from study of post-war Indonesia, this theory refers to an economy in which rich, capital-intensive modern sectors exist in the same model as comparatively poor, traditional, labour intensive sectors.

Economists have deliberated over whether an economy should achieve economic growth through its technically advanced sectors or whether resources should be spread evenly across the whole economy to achieve a more balanced growth. *See* DEPENDENCY THEORY, and DEMOGRAPHIC TRANSITION.

J H Boeke, *Economics and Economic Policy of Dual Societies as Exemplified by Indonesia* (New York, 1953)

PH

dual labour market theory *Economics* Examined by American economist Gary Becker (1930–) among others, this theory asserts that the labour market is divided into two sectors: the primary sector (in which labour is skilled, well-paid, and comparatively well educated); and the secondary market (which comprises relatively unskilled, badly paid and less educated individuals).

Although some wage differentials exist because of differences in education, work experience and the degree of skill required by the job, differentials also exist because of discrimination. The secondary sector is made up of a disproportionately high level of ethnic minorities, women and unskilled poor white labour. Becker argued that the forces governing discrimination are money costs and ignorance; that movement from the primary to the secondary sector was difficult; and that discrimination can be detrimental both to those who suffer it and those who impose it. *See* CROWDING HYPOTHESIS, SEGMENTED LABOUR MARKET THEORY, LABOUR MARKET DISCRIMINATION, NON-COMPETING GROUPS, SEARCH THEORY and INSIDER-OUTSIDER WAGE DETERMINATION.

G S Becker, *The Economics of Discrimination* (Chicago, 1971)

PH

dual labour market theory (20th century) *Feminism* Feminist economic theory.

There are two distinct economic systems. The first is what is normally understood as the economic system, the second is the patriarchal system of producing goods and services in the household. The subordination of women is rooted in the latter, but will have consequences throughout the 'conventional' economic system as well.

Maggie Humm, *The Dictionary of Feminist Theory* (London, 1989)

HR

dual state theory (20th century) *Politics* Explanation of apparent ambivalence in actions of capitalist states.

Those areas of the state's work which are not important to the INTERESTS of capital are relatively open to democratic and public pressures; others which more closely touch the interests of capital are 'insulated'.

Henry Drucker *et al.*, eds, *Developments in British Politics 2* (London, 1986)

RB

dual voice *Stylistics See* FREE INDIRECT DISCOURSE.

dualism *Philosophy* Any view which analyzes a given subject-matter, be it the universe as a whole or merely some area of concern, in terms of exactly two fundamentally distinct and opposed ideas.

Reality may be divided into matter and spirit, a person into body and soul, propositions into analytic and synthetic, judgments into factual and evaluative, to name a few philosophically influential dualisms. Opponents may say a given dualism is not sharp, or not exhaustive, or even does not exist at all.

A O Lovejoy, *The Revolt against Dualism* (1930); defends one kind of dualism

ARL

duality of patterning (1960s) *Linguistics* The term comes from the American linguist Charles Francis Hockett (1916–).

Duality of patterning is regarded as a design feature of human language: language is structured on hierarchically ordered levels; a pattern of units on one level (for example phonemes) makes up a unit on the next level (for example morphemes).

C F Hockett, 'The Problem of Universals in Language', *Universals of Language*, J H Greenberg, ed. (Cambridge, Mass., 1966), 1–29

RF

duality theory of linear programming *Mathematics* Linear programming is that area of OPTIMIZATION THEORY concerned with optimizing a linear objective function subject to linear constraints.

It is widely applicable in problems of resource allocation. One form of the duality theory states that $\bar{x}$ in n-dimensional Euclidean space $\mathbb{R}^n$ is optimal for the linear program

$$\min \{\langle c,x\rangle \mid Ax = b, x \geqslant 0 \text{ coordinatewise}\}$$

(where A is an $m \times n$ matrix) if and only if it satisfies the constraints and there is a $\bar{y}$ in $\mathbb{R}^m$ satisfying the constraints for the dual problem

$$\max \{\langle y,b\rangle \mid yA \leqslant c \text{ coordinatewise}\}$$

with equal objective value $\langle c,\bar{x}\rangle = \langle \bar{y},b\rangle$. In this case the optimal values of the two problems are equal. This characterization (originating with G B Dantzig and J von Neumann in 1947) is fundamental for numerical techniques. *Compare with* KHATCHIAN ALGORITHM.

AL

Dühring's rule (1878) *Chemistry* Discovered by U Dühring, this rule relates the vapour pressures of similar substances at different temperatures.

When the temperature at which a liquid exerts a particular pressure is plotted on a graph against the temperature at which a similar reference liquid exerts the same vapour pressure, a straight or almost straight line is obtained. Water is most often used as a reference liquid since accurate water vapour pressure data at different temperatures are widely available.

S Glasstone, *Textbook of Physical Chemistry* (London, 1948)

MF

Dulong and Petit law (1819) *Physics* Named after its discoverers, the French physicists Pierre Louis Dulong (1785–1838) and Alexis-Thérèse Petit (1791–1820); this is an empirical rule which states that, for most monatomic solids, the molar heat capacity at constant pressure has the value $3R$ at room temperature, where R is the molar gas constant. It was later realized that this value is, in fact, the high-temperature limit of the molar heat capacity at constant volume.

J Thewlis, ed., *Encyclopaedic Dictionary of Physics* (New York, Oxford and London, 1962)

MS

duopoly theory (1838) *Economics* First posited by French economist Augustin Cournot (1801–77), this theory examines the interaction of two firms in a market: each firm's output and prices are determined by the decisions of the other.

Cournot's model examined the reactions of the firms based on their output decisions, concluding that if one firm altered its output the other firm would also change its output by the same quantity. Eventually both firms would expand to an equilibrium point at which they would share the market and only retain normal profits. A later model by Joseph Bertrand (1822–1900) concentrated on price changes and concluded that if one firm altered its price and the other firm followed, both firms would eventually reach a position from which neither would wish to depart (equilibrium). *See* BERTRAND DUOPOLY MODEL, COLLUSION THEORY, and COURNOT DUOPOLY MODEL.

A Cournot, *Récherches sur les principes mathématiques de la théorie des richesses* (Paris, 1838); M Shubik, *Strategy and Market Structure* (New York, 1959)

PH

duplex theory of memory *Psychology* Proposed by the American psychologist Richard Chatham Atkinson (1929–).

A general theory of memory which assumes a flow of information from temporary sensory memory to short-term memory and then to long-term memory. The theory has enjoyed considerable popularity and is taken by some as having the status of fact. However, in recent times it has come under considerable attack and may be superseded by more sophisticated accounts. *See* LEVELS OF INFORMATION PROCESSING THEORY.

R C Atkinson and R M Shiffrin, 'Human Memory: A Proposed System and its Control Processes', *The Psychology of Memory and Motivation: Advances in Research and Theory*, K W Spence and K T Spence, eds (New York, 1968)

NS

duplexity theory of vision *Psychology* Proposed by the German, von Kries.

Also termed duplicity theory, this holds that there are two separate receptor mechanisms in the retina and they are the rods and the cones. The rods are achromatic and used in low illumination, while the cones are colour sensitive and used in high illumination. The theory is widely accepted.

R D Gross, *Psychology: The Science of Mind and Behaviour* (London, 1992)

NS

dynamic effect law *Psychology* Proposed by the Anglo-American psychologist Raymond Bernard Cattell (1905–).

This law states that specific behaviours become habitualized in proportion as they facilitate the attainment of a goal. The law emphasizes the significance of planning and the attainment of sub-goals within a plan rather than reinforcement of specific behavioural units.

R B Cattell, *Personality and Motivation Structure and Measurement* (New York, 1957)

NS

dynamic situations principle *Psychology* Proposed by the American psychologist Edward Chase Tolman (1886–1959).

This states that any stimulus pattern undergoes continuous changes due to such factors as uncontrolled variables, visceral change and responses. It is a widely held principle intended to emphasize that a stimulus is not a unitary, invariant event and that the relationships between stimuli and responses are not one-to-one or linear.

G A Kimble, *Hilgard and Marquis' Conditioning and Learning* (New York, 1961)

NS

dynamic state theory *Biology* The theory that proteins are virtually alive, being capable both of continuous metabolism and of providing the hereditary information necessary for life. This theory was discarded when it was learned that it is nucleic acids, not proteins, that carry hereditary information. *See also* GENOTYPE THEORY, ORIGINS OF LIFE and PHENOTYPE THEORY.

M Barbieri, *The Semantic Theory of Evolution* (Chur, Switzerland, 1985)

KH

dynamic system theory *Psychology* Proposed by the Austrian-Canadian biologist Ludwig von Bertalanffy (1901–72).

This holds that any change in one part of a dynamic system influences all interrelated parts. It is not a specific theory within psychology but a general theoretical orientation which sees itself as different from simpler mechanistic, componential accounts of behaviour. The theory finds particular favour in the psychology of groups and family interaction, and has contributed to important developments in clinical intervention. *See* CONTROL THEORY.

D J McFarland, *Feedback Mechanisms in Animal Behaviour* (London, 1971)

NS

dynamic theory *Psychology* Associated with the Austrian founder of psychoanalysis Sigmund Freud (1856–1939), and the German psychologist Wilhelm Max Wundt (1832–1920).

This is a general term referring to any psychological theory which is concerned with motivational processes and the unconscious. The term 'dynamism' is used to refer to a stable manner of behaving, the purpose of which is to fulfil drives and motives and to protect the individual from stress. *See* MEDIATION THEORY.

E Jones, *The Life and Works of Sigmund Freud* (New York, 1953)

NS

dynamical similarity principle *Physics* Two geometrically similar fluid flows are said to be dynamically similar when the flow field of one may be transformed into the flow field of the other by the same changes of scale that are needed to make the boundary conditions identical. If the equations of motion of the fluid are made non-dimensional by expressing speeds and lengths as fractions of these scales, the equations contain non-dimensional coefficients that determine the character of the flow. The general condition for dynamic similarity is that a particular coefficient should have the same value for the two flows.

J Thewlis, ed., *Encyclopaedic Dictionary of Physics* (New York, Oxford and London, 1962)

MS

dynamical theory of electron and X-ray diffraction (20th century) *Physics* This is a theoretical treatment of the process of electron or X-ray diffraction which takes account of the state of dynamic equilibrium that must be set up between the incident and diffracted beams in a crystal.

J Thewlis, ed., *Encyclopaedic Dictionary of Physics* (New York, Oxford and London, 1962)

MS

dynamical theory of heat (17th century–) *Physics* A dynamical view of the nature of heat was early expressed by the English physicist Robert Hooke (1635–1703) thus: 'Heat is a property of a body arising from the motion or agitation of its parts.'

By the 18th century, CALORIC THEORY prevailed (at least in Britain). This was put in doubt by the work of the Anglo-American scientist Benjamin Thompson, Count Rumford (1753–1814) in 1798; and conclusively disproved by the English physicist James Prescott Joule (1818–89) who converted work to heat by a variety of methods (electromagnetically, solid and liquid friction, expansion of a gas) and showed that the conversion factor was always the same. This result could not be explained on any plausible caloric model.

The current view of heat is that it is energy, but the term is reserved for use when the energy is in transit under a temperature difference. This flow of heat goes to change the internal energy of the system to which the heat is flowing. This distinction is necessary because the internal energy of a system can also be changed by the performance of work on the system.

J Thewlis, ed., *Encyclopaedic Dictionary of Physics* (New York, Oxford and London, 1962)

MS

dynamics of an asteroid (1809) *Astronomy* Initiated by C F Gauss (1777–1855), but reputed to have had its outstanding exposition in an elusive textbook by James

Moriarty (*c.*1840–91), with later contributions by other mathematicians including K Weierstrass (1815–97) and J E Littlewood (1885–1977).

The motion of an asteroid, which is now generally understood as a minor planet, is that of a body of negligible gravitational attraction in the gravitational field of two massive bodies, just like a spacecraft under the influence of the Earth and the Moon.

J F Bowers, 'James Moriarty: A Forgotten Mathematician', *New Scientist*, 124 (1989), parts 1696–7, 17–19

JB

dynamo theory (1919) *Geology* Proposing that some planetary and solar magnetic fields may be caused by the dynamo effect of electrically-conductive material circulating within a pre-existing magnetic field, this convective model of geomagnetism figured in the work of the American geophysicist Louis Agricola Bauer (1865–1932) and the Irish physicist Sir Joseph Larmor (1857–1942).

According to the theory, changes in the linkage between convective currents and their magnetic fields within the Earth's core explain how the Earth's magnetic field can change both direction and intensity on annual or longer timescales (secular variation), and even change polarity (north magnetic pole becoming the south magnetic pole and vice versa). Modern developments were pioneered in the late 1940s by the German-American theoretical physicist William Maurice Elsasser (1904–), and British geophysicist Sir Edward Crisp Bullard (1907–80).

W D Parkinson, *Introduction to Geomagnetism* (Scotland, 1983)

DT

dynamogenesis, principle of (20th century) *Psychology* Proposed by the American psychologist N Triplett, this principle asserts that motor responses are proportional to sensory activities; that is to say, there is a direct relationship between the frequency and intensity of motor activity and that of sensory activity. This is essentially an energy theory which attempts to account for the relationship between sensory energy entering an organism and the motor energy it expends.

R B Zajonc, 'Social Facilitation', *Science*, vol. CXLIX (1965), 269–74

NS

dyslexia (1887) *Linguistics* Also called 'word blindness'. A language disorder, of no agreed single origin, most often identified in spelling and sequencing difficulties but also affecting reading, and sometimes auditory perception of words. The centrality of writing in Western education means that dyslectic subjects are greatly disadvantaged at school unless their condition is acknowledged and remediated.

M E Thompson, *Developmental Dyslexia* (London, 1984)

RF

E

EAN rule *Chemistry See* EIGHTEEN-ELECTRON RULE.

Earnshaw's theorem (1842) *Physics* Named after its discoverer, Samuel Earnshaw (1805–88), this states that a system of particles interacting through forces that vary as the inverse square of the distance cannot be in stable static equilibrium. In particular, no purely stationary system of electric charges can be in stable equilibrium under their own influence.

J Thewlis, ed., *Encyclopaedic Dictionary of Physics* (New York, Oxford and London, 1962)

MS

earth works (1968) Related to ENVIRONMENT ART and LAND ART, this concept was conceived in the exhibition 'Earth Art' at Cornell University (1968). Rejecting the over-sophistication of MINIMALISM, as well as technological culture, art is created using the land as the environment and conveyed to the gallery/museum space through photographic documentation.

AB

Ebbinghaus's law *Psychology* Propounded by the German psychologist Hermann Ebbinghaus (1850–1909).

An increase in the quantity of material to be learned requires a disproportionate increase in the time required to learn it. For instance, Ebbinghaus found that it can take about 17 seconds for a person to learn a list of 12 meaningless syllables; and about 30 seconds for them to learn a list of 15, and so on.

R S Woodworth and H Schlosberg, *Experimental Psychology* (London, 1966)

NS

Eberlein-Smulyan theorem (1940–47) *Mathematics* A basic result in FUNCTIONAL ANALYSIS.

The *weak topology* on a Banach space X generally gives the weakest useful notion of convergence (in the sense of leaving the dual space X^* unchanged). The Eberlein-Smulyan theorem asserts the equivalence of compactness and sequential compactness in this setting, thus simplifying many proofs by allowing sequence arguments. The JAMES THEOREM gives another characterization of weak compactness.

AL

echo principle *Psychology* Propounded by the American psychologist Orval Hobart Mowrer (1907–82).

The tendency for one animal to imitate another, particularly if each is performing the same behaviour at the same time. The principle captures a fundamental feature of theories of learning which emphasize the role of imitation and modelling in the acquisition of new behaviours. The principle cannot on its own account for the diversity of ways in which humans and other animals have been observed to learn through watching and listening to others.

O H Mowrer, *Learning Theory and the Symbolic Processes* (New York, 1960)

NS

eclecticism and syncretism *Philosophy* Periods of philosophical innovation are often followed by periods of consolidation (some would say, decline) when progress is sought by selecting features from different philosophers, regarded as opposed to one another, and combining them to form a unified whole.

Alternatively it may be claimed that the philosophers were not really opposed to each other in the first place, but when read properly may be seen to have been saying the same thing; a notable example of this is the neo-Platonist treatment of Plato and Aristotle. Properly speaking, *eclecticism* is the former of these processes and *syncretism* the latter, but in practice the terms are used variously or even indifferently. *See* NEO-PLATONISM.

R J Hankinson, 'Galen's Anatomy of the Soul', *Phronesis* (1991); see p. 198, note 5

ARL

ecological art (1968) *Art* The critic Herb Aach was probably the first to use this term in 1968. Its origins lying in the work of French artists Marcel Duchamp (1887–1968) and Yves Klein (1928–62), ecological art aims to harness the forces of nature (its physical, biological and chemical processes) with the intention of revealing these by the construction of microcosmic models of natural phenomena. It is also linked to alchemy and the occult.

J Benthall, 'Art and Ecology', *Science and Technology in Art Today* (London, 1972)

AB

ecological pyramid (1927) *Biology* A concept first recognized by the British ecologist Sir Charles Sutherland Elton (1900–91).

There is an architecture to food chains, such that each trophic level has a greater number of individuals than the trophic level above (there are fewer sheep than grass plants, and even fewer wolves). The resulting pyramid of numbers does not hold true for all situations: a tree as the first trophic level has only one member. Pyramids have since been modified to take into account the amount of energy per unit time in each trophic level which always forms a pyramid.

RA

ecology (1869) *Biology* The name of the discipline coined by the German biologist Ernst Heinrich Haeckel (1834–1919).

The scientific study of interrelationships between organisms and their environment. The term is derived from the Greek *oikos*, meaning house.

RA

economic determinism *Politics/History* Theory of historical and political events and structures.

Political or historical developments are determined by the underlying economic system, or by the distribution of power over economic resources. The most familiar theory to which this term is applied is MARXISM, though it is applied by critics (rather than claimed by supporters) who deny that the relationship between economic and other structures or events is so simple.

Jon Elster, *An Introduction to Karl Marx* (Cambridge, 1986)

RB

economic liberalism *Economics* This term refers to the maximum role of markets and competitive forces in an economy. The state's role is limited to the establishment of the necessary framework in which markets can operate and to the provision of services which private enterprise cannot provide. *See laissez-faire*.

PH

economic methodology *Economics* As exampled by the work of Nassau William (snr), and English economists John Stuart Mill (1806–73) and John Maynard Keynes (1883–1946), this is the process by which economics is explained: the philosophy of science applied to economics.

In the 19th century, British economists examined the premises of economic theory, arguing that the verification of economic theory was hazardous at best. Today, there is debate as to how useful methodology is in the training of modern economists. Mark Blaug has argued that economic methodology can set the criteria for the acceptance and rejection of research programmes. Empirical testing has become an integral part of 20th century economic study.

M Blaug, *The Methodology of Economics, or How Economists Explain* (Cambridge and New

York, 1980); Joseph A Schumpeter, *History of Economic Analysis* (New York, 1957)

PH

economic theory of politics (19th century–) *Economics* This is an analytical model which assumes that politicians are vote maximizers and voters are utility maximizers.

Each voter is assumed to support the political party that will provide him with the highest degree of utility when it is elected. Politicians, motivated by self-interest and the desire for public office, formulate policies which will attract the maximum number of votes. *See* PARADOX OF VOTING, and IMPOSSIBILITY THEOREM.

S de Brunhoff, *The State, Capital, and Economic Policy* (London, 1978)

PH

economic theory of the state (19th century–) *Economics* This is an analytical framework for examining the role of institutions, taxation and law in the creation of economic models.

S de Brunhoff, *The State, Capital, and Economic Policy* (London, 1978)

PH

economy principle *Psychology* Proposed by the Briton, Conwy Lloyd Morgan (1852–1936).

Also referred to as parsimony, this is the general heuristic stating that if two scientific theories or propositions are equally acceptable, the simpler is to be preferred. The principle is fine in theory but has restricted application because it is rare to find two postulates which can be regarded as equally acceptable. Moreover, in psychology and cognate disciplines there are no methodologies for measuring the complexity of theories. *See also* OCKHAM'S RAZOR.

A J Chapman and D M Jones, eds, *Models of Man* (Leicester, 1980); L S Hearnshaw, *A Short History of British Psychology* (London, 1964)

NS

ecosystem (1935) *Biology* The term coined by the British ecologist Sir Arthur George Tansley (1871–1955).

The term describes the interdependence of a group of organisms upon one another and their living (biotic) and non-living (abiotic) environment. The concept includes the energy-flow through food chains, and the cycling of nutrients through organisms, via the food chain, and through the abiotic environment in biogeochemical cycles.

RA

Eddington's theory (1923) *Physics* Named after the English astronomer Sir Arthur Stanley Eddington (1882–1944) who was professor and observatory director at Cambridge and originator of the phrase 'the arrow of time'.

The theory points out the need, in describing a physical system, to take account of the observer, the nature of the observations, and the measurement processes, as in relativity and quantum theory. Eddington claimed that the values of many important dimensionless constants in physics could be worked out from this theory.

Charles C Gillispie, ed., *Dictionary of Scientific Biography* (New York, 1981)

GD

effect, empirical law of *Psychology* Formulated by the American psychologist Edward Lee Thorndike (1874–1949).

Sometimes termed the 'weak law of effect', this states that a response which is reinforced is likely to recur. Stated as such, the law is inherently circular: a reinforcing or 'satisfying state of affairs' is anything which increases recurrence of a behaviour, and anything which increases recurrence is a 'satisfying state of affairs'. The law is better considered as an empirical generalization rather than as an explanatory theory. *See* EFFECT, LAW OF and EFFECT, NEGATIVE LAW OF.

E L Thorndike, *Animal Intelligence* (New York, 1911)

NS

effect hypothesis (1980) *Biology* Proposed by Elisabeth Vrba, this theory asserted that the speciation rate (that is, the birth rate of species) tends to vary from species to species, favouring some species more than others.

This rate might cause the trends of MACROEVOLUTION (since 'Trends may be the effects of any factors which incidentally result in differential speciation rates'), and of MICRO-EVOLUTION (for example, an ADAPTATION enabling members of a species to survive on the edges of the species range). In addition, this theory holds that those species which speciate rapidly will ultimately overwhelm those which speciate slowly.

H Genoways, *Current Mammalogy*, vol. I (New York, 1987)

KH

effect, law of (1904) *Psychology* Formulated by the American psychologist Edward Lee Thorndike (1874–1949).

A stimulus-response connection is strengthened when the response is followed by a satisfier. Elimination of incorrect responses is attributed to the occurrence of annoyers. Taking these two claims together, the law of effect postulates that one learns or retains responses which are followed by satisfiers, and refrains from responses followed by annoyers. *See* BEHAVIOURISM, EFFECT, NEGATIVE LAW OF and LEARNING THEORY.

D Dewsbury, *Comparative Animal Behaviour* (New York, 1978); E L Thorndike, *Animal Intelligence* (New York, 1911)

NS

effect, negative law of *Psychology* Formulated by the American psychologist Edward Lee Thorndike (1874–1949).

The negative or reciprocal of the LAW OF EFFECT, this states that behavioural responses which are followed by an 'annoying state of affairs' are less likely to recur. Thorndike later withdrew the law, pointing out that punishment does not lead to the unlearning of behaviour but rather to its suppression.

E L Thorndike, *Animal Intelligence* (New York, 1911)

NS

efficient market hypothesis (20th century) *Economics* Dating back to work on the RANDOM WALK HYPOTHESIS by French economist Louis Bachelier (1870–1946), this precept asserts that stock market prices are the best available estimates of the real value of shares since the market has taken account of all available information on an individual stock. *See* RATIONAL EXPECTATIONS and ADAPTIVE EXPECTATIONS.

H Roberts, 'Stock Market "Patterns" and Financial Analysis: Methodological Suggestions', *Journal of Finance*, 14, 1 (March, 1959), 1–10

PH

effluxes (or effluences), theory of *Philosophy* Theory associated with Greek atomism and its revival in the corpuscularian philosophy of the 17th century as well as by non-atomists like Empedocles (5th century BC).

It holds that objects continually emit films from their surfaces, which cause them to be perceived, much as we ourselves might explain smell. Lucretius (1st century BC) also uses the theory to explain dreams and imagination, and thought in general.

T Lucretius, *De Rerum Natura* (*On the Nature of Things*), book 4

ARL

egalitarianism *Politics* Principle for guiding treatment of individuals.

A preliminary assumption that all people should, in the absence of good reason to the contrary, be treated equally. Thus the onus is on those who argue that sex should be a disqualification to a career, or that poverty should be a barrier to the provision of education, to prove their case, rather than vice versa.

David Miller *et al.*, eds, *The Blackwell Dictionary of Political Thought* (Oxford, 1987)

RB

ego-alter theory *Psychology* Proposed by the Austrian founder of psychoanalysis Sigmund Freud (1856–1939).

A variant of personality theory referring to an individual who is said to be so close to himself that he appears to exhibit a second self. This term is not generally used in contemporary psychology, except in an analogical sense, and is regarded as being insufficiently defined to be amenable to scientific investigation.

E H Erickson, *Childhood and Society* (New York, 1950)

NS

egocentric predicament *Philosophy* Term coined by Ralph Barton Perry (1876–1957) for the idea that all our knowledge of the world must take the form of mental representations within our own minds (sensations, images, ideas, and so on), which the mind then operates upon in various ways.

Thus we can never have any direct contact with reality outside our minds, and so, it seems, may not be justified even in thinking it exists. The predicament faces various kinds of EMPIRICISM, which are in danger of slipping into SOLIPSISM.

R B Perry, 'The Egocentric Predicament', *Journal of Philosophy, Psychology and Scientific Method* (1910), later called *Journal of Philosophy*. Reprinted in W G Muelder and L Sears, eds, *The Development of American Philosophy* (1940)

ARL

egoism *Philosophy* In ordinary speech, selfishness (also called 'egotism', a word never used in philosophy). As a philosophical doctrine egoism is either psychological egoism (for which *see* HEDONISM), or ethical egoism, which contrasts with UNIVERSALISM and ALTRUISM. Like them it is a form of CONSEQUENTIALISM, and prescribes that everyone should always act so as to maximize his own happiness or welfare.

Egoism must be distinguished from the absurd doctrine, which has no name, that everyone should maximize the speaker's happiness – but this very distinction leads to an objection to egoism: can I consistently prescribe egoism to you, since I shall be encouraging you to act in your interest and not in mine? Also seeking my own happiness may not be the best way of achieving it. However, sometimes 'enlightened egoism' is advocated as a means rather than an end, on the grounds that for everyone to pursue their own interest will maximize the general prosperity.

H Sidgwick, *The Methods of Ethics*, book 2 (1874; 7th and final edn 1901)

ARL

Egoroff's theorem *Mathematics* Named after the Russian analyst D F Egoroff (1869–1931), this element of MEASURE THEORY states that if a sequence of integrable functions $\{f_n\}$ converges almost everywhere to an integrable function f on a finite measure space X, then f_n converges to f uniformly on the complement of a set of arbitrarily small measure in X.

G B Folland, *Real Analysis* (New York, 1984)

ML

Ehrenfest's theorem (20th century) *Physics* Proposed by Austrian theoretical physicist Paul Ehrenfest (1880–1933), this states that the motion of a quantum mechanical wave packet will be identical to that of the classical particle that it represents when any potential acting upon it is effectively constant over the dimensions of the wave packet.

J Thewlis, ed., *Encyclopaedic Dictionary of Physics* (New York, Oxford and London, 1962)

MS

eight-N rule *Chemistry See* HUME-ROTHERY'S RULE.

eighteen-electron rule *Chemistry* Rule that allows the systematization of much of the structural chemistry of transition metal organometallic compounds. Also known as the effective atomic number (EAN) rule, and as the sixteen and eighteen electron rule.

In a stable transition metal complex, the central atom has 18 valence electrons. The valence shell of the atom therefore has the same electronic configuration as the outer shell of the next highest noble gas. The number of electrons in this shell is known as the EFFECTIVE ATOMIC NUMBER. Molecules and ions with only 16 valence electrons about the central metal atom may be just as stable and sometimes more stable than those with 18 electrons about the same atom.

K F Purcell and J C Kotz, *Inorganic Chemistry* (Philadelphia, 1977)

MF

Einstein-de Sitter universe (1932) *Physics* Named after the German-Swiss-American mathematical physicist Albert Einstein (1879–1955), and the Dutch physicist Willem de Sitter (1872–1934).

De Sitter published a solution of Einstein's field equations which implied a universe which is empty in the dynamic sense, although expanding in the kinematic sense. The de Sitter universe showed, however, that empty space-times could be obtained as ten solutions of general relativity. Although the property of emptiness was embarrassing, and contrary to MACH'S PRINCIPLE, its property of expansion turned out to contain the germ of truth. The Einstein-de Sitter universe is the simplest solution of the Friedmann equations. *See also* FRIEDMANN UNIVERSES.

Jayant V Narlikar, *Introduction to Cosmology* (Cambridge, 1993)

GD

Einstein shift or displacement (1911–16) *Physics* A result calculated by the German-born mathematical physicist Albert Einstein (1879–1955).

Using his special theory of relativity and the principle of equivalence, Einstein predicted in 1911 that when a ray of light from a star passes close to the Sun's limb (as seen by an observer on Earth), it will be deflected through an angle D given by

$$D = 2GM/c^2r$$

where G is the gravitational constant, M is the mass of the Sun, c is the speed of light in a vacuum and r is the distance of closest approach of the path of the light ray to the centre of the Sun. Later, in 1916, Einstein used his general theory of relativity to establish a corrected formula for D, namely,

$$D = 4GM/c^2r$$

See RELATIVITY, THEORY OF GENERAL and RELATIVITY, THEORY OF SPECIAL.

J Thewlis, ed., *Encyclopaedic Dictionary of Physics* (New York, Oxford and London, 1962)

MS

Einstein-Stark law (1912) *Chemistry* One of two basic laws of photochemistry, named after German-born American physicist and mathematician Albert Einstein (1878–1955) and J Stark. Einstein won the Nobel prize for Physics in 1921. The law is also known as the Stark-Einstein law, and as the (Einstein) law of photochemical equivalence.

In a photochemical process, one quantum of energy is absorbed by the molecule responsible for the primary photochemical process. For example, in the stratosphere, the following photochemical process takes place:

O_3	+ photon →	O_2	+	O
ozone molecule		dioxygen molecule		oxygen radical

The photon is a quantum of energy. The law does not necessarily imply that only one product molecule is produced when one photon is absorbed. If the subsequent reaction is a chain reaction then absorption of one photon might lead to several product molecules. The other basic law of photochemistry is the GROTTHUSS-DRAPER LAW.

P W Atkins, *Physical Chemistry* (Oxford, 1978)

MF

Eisenstein's criterion (19th century) *Mathematics* Named after German mathematician Ferdinand Gotthold Max Eisenstein (1823–52), this states that it is a sufficient condition for a polynomial with integer coefficients to be irreducible.

Specifically, if there is a prime p which divides every coefficient of the polynomial except for the leading coefficient but is such that p^2 does not divide the constant coefficient, then the polynomial is irreducible over the rationals (that is, it cannot be factored as a product of polynomials of lower degree whose coefficients are rational numbers). For example, $x^2 + 9x + 3$ is seen to be irreducible, on using $p = 3$. *See also* GAUSS'S LEMMA.

I N Herstein, *Topics in Algebra* (Wiley, 1975)

MB

Einstein's theory of gravitation *Physics See* RELATIVITY, THEORY OF GENERAL.

Einstein's theory of heat capacity (1907) *Physics* Named after its originator, the German-born mathematical physicist Albert Einstein (1879–1955), this theory was the first application of quantum mechanics to solid state physics.

Einstein assumed that the atoms in a non-metallic solid are all independent, and that each atom acts as a simple harmonic oscillator with a common frequency f. The energy E_i of each oscillator is then given by

$$E_i = nhf$$

where h is Planck's constant and n can take integer values. Using a Maxwell-Boltzmann distribution for the probability that an oscillator has an energy E_i, the molar heat capacity at constant volume is given by

$$C_{V,m} = 3R\,F(x)$$

where

$$x = hf/kT$$

and

$$F(x) = x^2e^x/(e^x - 1)^2$$

and k is Boltzmann's constant. This equation agrees fairly well with experiment at high temperatures but, although it predicts that $C_{V,m}$ tends to zero as the temperature T tends to zero, the fall is much more rapid than that observed.

J Thewlis, ed., *Encyclopaedic Dictionary of Physics* (New York, Oxford and London, 1962)

MS

Einstein's theory of mass-energy equivalence (1905) *Physics* Named after the German-Swiss-American mathematical physicist Albert Einstein (1879–1955).

This theory proposes that mass and energy are related by the equation $E = mc^2$, where E is energy, m is mass and c is the speed of light in a vacuum. *See also* CONSERVATION OF MASS-ENERGY.

Paul Davies, ed., *The New Physics* (Cambridge, 1989)

GD

Einstein's theory of photochemical equivalence (1912) *Physics* Also known as the Stark-Einstein law of photochemical equivalence after the German physicist Johannes Stark (1874–1957), winner of the 1919 Nobel prize for Physics, and the German-Swiss-American mathematical physicist Albert Einstein (1879–1955), winner of the 1921 Nobel prize for Physics.

The law relates to the quantum yield or efficiency of a photochemical reaction, and states that one molecule reacts for each photon or quantum of light energy absorbed, according to the formula, $E = hNf$, where E is the energy per mole absorbed, f is the frequency of the absorbed radiation, h is the Planck constant (*see* PLANCK'S QUANTUM THEORY) and N is Avogadro's number. In practice, this theoretical efficiency is rarely achieved since the products formed through the primary process of light absorption usually undergo further reactions (secondary processes).

Robert M Besançon, ed., *The Encyclopedia of Physics* (New York, 1985)

GD

Einstein's theory of the photoelectric effect (1905) *Physics* Named after the German-Swiss-American mathematical physicist Albert Einstein (1879–1955), winner of the 1921 Nobel prize for Physics.

This theory explains the observation by Lenard in 1902 that when light is shone onto the surface of a solid, electrons are released whose maximum velocity is independent of the intensity and increases linearly with the frequency of the light. For a given frequency, the number of electrons is directly proportional to intensity. Einstein assumed that the energy of the incident radiation was transferred in discrete amounts (photons), each of magnitude $h\nu$, where h is the Planck constant (*see* PLANCK'S QUANTUM THEORY) and ν the frequency. Each photon absorbed will eject an electron with a maximum kinetic energy given by the equation

$$E = h\nu - \phi,$$

where ϕ is the work function of the material.

Paul Davies, ed., *The New Physics* (Cambridge, 1989)

GD

Eleaticism *Philosophy* A movement in 5th century BC Greek thought stemming from Parmenides of Elea (in southern Italy) and his two main disciples Zeno of Elea (not the Stoic) and Melissus of Samos. The main tenet was an insistence that any kind of change was impossible, and so (on the usual interpretation) was any kind of plurality. Reality was one and unchanging, and the changing multiplicity of things was an illusion.

The arguments used appealed to strict logic, and were the first to do so systematically; Zeno in particular produced a number of paradoxes which plurality and change involve. The Eleatics' main influence – especially on their contemporaries and on Plato a century later – took the form of an insistence that whatever is ultimately real must be permanent and unchanging except for moving in space, and this influence can still be felt in philosophers like Descartes (1596–1650), Leibniz (1646–1716) and Kant (1724–1804). *See also* MONISM.

W K C Guthrie, *A History of Greek Philosophy*, 2 (1965)

ARL

electoral competition (20th century) *Politics* Theory of party behaviour in democracies.

Political parties compete for votes, and their policies and conduct are explained by this more than by any other single factor. In so doing they frequently raise the expectations of voters and the expenses to which governments are then committed.

Patrick Dunleavy *et al.*, eds, *Developments in British Politics 3* (London, 1990)

RB

electrolytic dissociation, theory of *Chemistry* *See* ARRHENIUS DISSOCIATION THEORY.

electromagnetic theory of light (19th century) *Physics* This theory proposes that light is made up of both electric and magnetic field oscillations propagating through space. The fields are at right angles to each other and to the direction of propagation. The mathematical formalism of wave theory is sufficient to describe propagation through space, but quantum theory is required to describe its interaction with matter. *See also* EINSTEIN'S THEORY OF THE PHOTOELECTRIC EFFECT.

Robert M Besançon, ed., *The Encyclopedia of Physics* (New York, 1985)

GD

electron theory of metals (20th century) *Physics* This posits that metals are elements whose valence electrons are weakly bound and are set free by the energy released when the atoms bond together to form the crystal. Therefore, a metal has a regular periodic arrangement of spherically-symmetric positive ions surrounded by an electron gas formed of the released electrons. These electrons can move through the crystal (free electrons) and are responsible for the cohesion of the metal through their interaction with the positive ions, and for the good electrical and thermal conductivities of metals. When the electrons are treated as moving in a single potential well the theory is called the free electron theory. When the periodic nature of the potential produced by the positive ions is considered, the energies of the electrons are restricted to certain ranges or bands, giving the band theory.

J Thewlis, ed., *Encyclopaedic Dictionary of Physics* (New York, Oxford and London, 1962)

MS

electroneutrality principle *Chemistry* In a solution of an electrolyte, the concentrations of all the ionic species are such that the solution is electrically neutral. For example, in an aqueous solution of sodium chloride, the concentrations of ions are such that the positive charges of the sodium (Na^+) and hydroxonium ($H_{30}{}^+$) ions balance the negative charges of the chloride (Cl^-) and hydroxide (OH^-) ions.

MF

electronic art (1950s–) *Art* Related to KINETIC ART, this is a technique whereby the artist develops abstract patterns on screens using cathode rays or oscilloscopes.

The American artist Ben F Laposky has worked in this area since the 1950s, and the method has also been exploited by the Korean artist Nam June Paik in his use of several television sets and human performers, with extreme distortion of images.

B F Laposky, 'Electronic Abstractions', letter in *Studio International* (October 1976)

AB

electronic theory of valency (1916) *Chemistry* A theory of the nature of chemical bonds, developed by American chemist Gilbert Newton Lewis (1875–1946) and, independently, the German physicist Walther Kossel (1888–1956).

When atoms form bonds they try to achieve the most stable (that is, the lowest-energy) electronic configuration. They can do this in two ways. (1) By gaining or losing electrons to form ions. The ions have full (and thus stable) outer-shell electron configurations like those of the noble gases. The electrostatic force of attraction between oppositely charged ions is called an electrovalent bond, or (more commonly) an ionic bond. (2) By sharing electrons. This is called covalent bonding. A covalent bond consists of a shared pair of electrons. However, in some molecules and ions, both the shared electrons come from one atom. This is called co-ordinate bonding.

M Freemantle, *Chemistry in Action* (Basingstoke, 1987)

MF

electroweak theory (1967) *Physics* Also known as quantum flavourdynamics and GWS theory after the American physicists Sheldon Lee Glashow (1932–) and Steven Weinberg (1933–), and the Pakistani theoretical physicist Abdus Salam (1926–), who shared the 1979 Nobel prize for Physics.

Electroweak theory is a GAUGE THEORY giving a unified description of both the electromagnetic and weak interactions. The latter involve the exchange of photons and of massive $W^{\pm}$ and neutral Z^0 bosons between quarks and leptons. The predicted $W^{\pm}$ and Z^0 bosons were experimentally observed in 1983–84 in high-energy interactions. The GWS (standard) model also predicts the existence of a heavy spin 0 particle known as the Higgs boson.

Paul Davies, ed., *The New Physics* (Cambridge, 1989)

GD

élitism *Politics* Explanatory and normative theory of politics, originated by Italian social scientists Vilfredo Pareto (1848–1933) and Gaetano Mosca (1858–1941).

Power and initiative have always been and will always be exercised by a cohesive minority who are marked off from the mass of the population by some particular skill, or quality, or insight. One version of the theory does no more than observe this phenomenon, the other applauds it. *See also* CLERISY and IRON LAW OF OLIGARCHY.

Geraint Parry, *Political Elites* (London, 1969)

RB

elliptic curve (19th century) *Mathematics* An algebraic curve whose defining equation is of the form

$$y^2 = ax^3 + bx^2 + cx + d$$

where a, b, c, d are members of a specific field (usually the real or complex numbers, but sometimes a finite field). There is a very beautiful theory which connects elliptic curves, elliptic functions, and elliptic integrals. These subjects are beginning to enjoy a renaissance, in part because of their application to problems in areas such as communications security. The name 'elliptic' relates to the fact that certain elliptic integrals arise when calculating the circumference of an ellipse. *See also* MORDELL'S THEOREM.

M Reid, *Undergraduate Algebraic Geometry* (Cambridge, 1990)

MB

embedding (1960s) *Linguistics* Part of the early standard grammar outlined by the American linguist Avram Noam Chomsky (1928–). Not current in GENERATIVE GRAMMAR.

A transformational process which places ('embeds') one sentence within another: 'S(Mary believed S'(that John had arrived)). In this example the sentence 'John had arrived' is the object of the verb 'believed'. *See also* RECURSION.

K Brown and J Miller, *Syntax*, 2nd edn (London, 1991)

RF

embourgeoisement (20th century) *Politics* Alternative to theory of IMMISERATION.

Members of the working class under capitalism, far from becoming worse off and increasingly discontented, have become 'seduced' by bourgeois values. They are second class members of the system it is in their INTERESTS to overthrow.

RB

emergence theories *Philosophy* Theories of the development of some phenomenon (for example life consciousness) where something emerges out of a background from which it could not have been predicted and in terms of which it cannot be fully explained. What emerges may be a law of nature, or a science, though this would normally be because of emergent properties in the relevant sphere. The existence of emergence theories contrasts with REDUCTIONISM in the relevant sphere. *See also* HOLISM and VITALISM.

ARL

emic/etic (1930s–) *Linguistics* Distinction developed in American STRUCTURAL LINGUISTICS and further clarified in TAGMEMICS.

A linguistic unit may be characterized in terms of its place and function within a system (the 'emic' approach), or 'etically' in terms of its external, material form. The distinction is illustrated by the relationship between PHONETICS (study of the sounds of speech) and PHONEMICS (a more abstract description of the sound-systems of language).

V G Waterhouse, *The History and Development of Tagmemics* (The Hague, 1974)

RF

Emmert's law *Psychology* Named after its originator, Swiss scientist E Emmert (1844–1911), this states that visual after-images are perceived as smaller if they are projected nearer to the subject; that is, the perceived size of the after-image is directly related to the distance to which it is projected. The law is $l' = la'/a$; where l' is the size of the after-image; l is the size of the stimulus object; a' the distance of eye from after-image; and a the distance of eye from stimulus object.

S Coren, C Porac and L M Ward, *Sensation and Perception* (New York, 1984)

NS

emotion, theories of *Psychology* Particularly associated with American philosopher and psychologist William James (1842–1910), and Danish psychologist Carl Georg Lange (1834–1900).

Emotion is the conscious experience of a particular feeling or state which leads to both internal and external reactions. There is no unified theory of emotion. Many theories have been proposed and they differ in emphasis on the role of physiological and cognitive components, but all theories involve a treatment of both of these. *See* ATTRIBUTION THEORY OF EMOTION and JAMES-LANGE THEORY OF EMOTION.

S Schachter and J E Singer, 'Cognitive, Social and Physiology Determinants of Emotional Estates', *Psychological Review*, vol. LXIX (1962), 379–99

NS

emotive/referential language (1923) *Literary Theory* Distinction drawn by English theorists Charles Kay Ogden (1889–1957) and Ivor Armstrong Richards (1893–1979), in *The Meaning of Meaning*.

Statements in scientific language make references which are true or false; in poetry, truth is immaterial: poetic or emotive language is used 'for the sake of attitudes and emotions which ensue'. Distinction both defines literature as FICTION, and supports a psychological view of literary value (*see* HARMONIZATION OF IMPULSES). *See also* SENSE and FOUR KINDS OF MEANING. The literary theory was further elaborated in Richards' book cited below, and was later a fundamental influence on NEW CRITICISM.

I A Richards, *Principles of Literary Criticism* (London, 1924)

RF

emotive theory of truth *Philosophy* A theory developed to parallel EMOTIVISM in ethics, the point being that what criterion of truth we adopt (for example logical intuition, faith, workability, verifiability) depends upon our emotions or attitudes. *See also* PRAGMATIC THEORY OF TRUTH.

B Savery, 'The Emotive Theory of Truth', *Mind* (1955)

ARL

emotivism *Philosophy* Theory that value judgments, including moral judgments, do not state facts (though they appear to), but are expressions of emotions or attitudes. (*See also* BOO HURRAH THEORY.)

The theory, a form of SPEECH ACT THEORY, arose under the influence of LOGICAL POSITIVISM (though it had antecedents in the 18th century) because of the difficulties of finding verification conditions for such utterances, and the need to explain how they can affect behaviour. Emotivism is also a form of SUBJECTIVISM, but must be distinguished from the subjectivist view that while value judgements do describe something, what they describe are not special moral and such-like facts but human attitudes, whether the speaker's own or other people's. One obvious, though not necessarily insuperable, difficulty is that of accounting adequately for moral argument and moral disagreement.

C L Stevenson, *Ethics and Language* (1944)

ARL

empathy *Psychology* A term coined by the American psychiatrist Harry Stack Sullivan (1892–1949) as part of his theory of interpersonal psychiatry.

The theory posits the existence of communication channels permitting individuals covertly to indicate understanding, attitudes and judgments to others. The notion of sympathy is also included in Sullivan's use of the word 'empathy'. The concept has been taken from Sullivan's theory and applied and developed extensively within personality theory. *See* INTERPERSONAL THEORY.

H S Sullivan, *The Interpersonal Theory of Psychiatry* (New York, 1953)

NS

empathy, historical (20th century) *History* Theory of historical explanation set out by British philosopher and historian R G Collingwood (1889–1943).

Historical understanding involves re-thinking the thoughts of those whom we study, and thus re-creating the mind of the past.

Patrick Gardiner, *The Nature of Historical Explanation* (Oxford, 1961)

RB

empirical studies of literature (1980s) *Stylistics* Studies of the ways readers handle and respond to literary texts; differs from RECEPTION THEORY and READER RESPONSE CRITICISM in using research methods drawn from the social sciences.

S J Schmidt, *Foundations of the Empirical Study of Literature* (Hamburg, 1982)

RF

empiricism *Philosophy* Any theory emphasizing sense-experience (including introspection) rather than reason or intuition as the basis for either some or all of our knowledge; 'basis' referring usually to justification, though sometimes to psychological origin.

Empiricism can concern either propositions or concepts, rejecting (most) *a priori* ones; for John Locke (1632–1704) all concepts ('ideas') were empirical, but propositions connecting them could be known *a priori*. Extreme empiricists confine our knowledge to immediate experience (and introspection) and fall into the egocentric predicament. Less extreme ones allow other knowledge to be reached from such experience, or confine their empiricism to certain spheres. Empiricists usually allow propositions of mathematics and logic to be *a priori*, but insist that they are analytic (that is, roughly, reducible to tautologies) rather than synthetic (embodying substantive knowledge). *See also* RATIONALISM.

D Odegard, 'Locke as an Empiricist', *Philosophy* (1965)

ARL

empiricism in linguistics (1930s–1950s) *Linguistics* Leonard Bloomfield (1887–1949) stipulated that linguistics should proceed on the model of the natural sciences. This was understood by the American structuralists (*see* STRUCTURAL LINGUISTICS, 2) to be an instruction to work on the basis of data comprising a corpus of utterances from real speech, analyzing it with reference to agreed and mechanical procedures, and with no reference to 'unobservable entities'. This attitude was strongly challenged by TRANSFORMATIONAL-GENERATIVE GRAMMAR in the 1950s: *see* INTUITION and MENTALISM IN LINGUISTICS.

L Bloomfield, *Language* (London, 1935 [1933])

RF

empiriocriticism *Philosophy* A name for the version of POSITIVISM developed by Austrian physicist and philosopher Ernst Mach (1838–1916) and the German Richard Avenarius (1843–96), and coming between the original positivism of Auguste Comte (1798–1857) and the later LOGICAL POSITIVISM.

Science on this view aims at the most economical description of appearances, on which predictions of further appearances can be based; but all appeal to hidden entities or causes, not only metaphysical (as for Comte) but scientific too, is banned. The system was criticized by V I Lenin (1870–1924) in his *Materialism and Empirio-Criticism* (1908 translated by A Fineberg (1947)).

E Mach, *Popular Scientific Lectures* (1898, German original 1894 or 1896) 186ff

ARL

encoding specificity principle *Psychology* Formulated by the American psychologist Gordon Howard Bower (1932–).

If at the time information is being recalled a stimulus is present that was also present at the time of storing the information, the recall will be aided and therefore faster. The principle is widely accepted. *See* ENCODING VARIABILITY PRINCIPLE.

A J Flexser and E Tulving, 'Priming and Recognition Failure', *Journal of Verbal Learning and Verbal Behavior*, vol. XXI (1982), 237–48

NS

encoding variability principle *Psychology* Formulated by the American psychologist Gordon Howard Bower (1932–).

Encoding variability refers to the degree of variability in the environment and mood in which one learns the same material. The larger the encoding variability the better chance one has of doing well in an examination of that material. The effect is due to the increased likelihood that the examining situation resembles one of those in which the material was learned. *See* ENCODING SPECIFICITY PRINCIPLE.

J R Anderson, *Cognitive Psychology and its Implications* (New York, 1980)

NS

encroaching control (*c*.1920) *Politics* Theory of WORKERS' CONTROL, developed by British socialist, historian and political scientist G D H Cole (1889–1959).

Workers' control of industry is to be achieved by a slow extension of power from the bottom up, until the overall balance of power within an industry, factory, or firm tips decisively in favour of the workers within it.

Rodney Barker, *Political Ideas in Modern Britain* (London, 1978)

RB

end of history *History* After a series of conflicts, or strivings towards more and more complete forms of human expression, or a cumulative and finally successful process of humanity learning by its mistakes, a plateau is reached.

Thereafter, although economic, scientific, intellectual and artistic creativity can continue, all the 'grand issues' will have been resolved, and history as we have known it will have come to an end. Various thinkers from Hegel and Marx in the 19th century to Francis Fukuyama in the 20th have succeeded in giving such ideas brief attraction or notoriety.

Francis Fukuyama, *The End of History and the Last Man* (London, 1992)

RB

end of ideology (20th century) *Politics* Anglo-American liberal conservative theory of the 1950s and 1960s.

'Ideology' as a complete explanation of everything, and a blueprint for a utopian society, had been a delusion which had none the less justified tyranny in Nazi Germany and Stalinist Russia. But ideology had come to an end, and the future, at least in the Western capitalist democracies, promised the modest politics of small scale, 'housekeeping' reform. Critics pointed out that this view was itself 'ideological' in that it sought to rule out a range of principled arguments about politics.

Rodney Barker, *Political Ideas in Modern Britain* (London, 1978)

RB

endocentric/exocentric (1933) *Linguistics* A distinction in SYNTAX formulated by the American linguist Leonard Bloomfield (1887–1949).

An endocentric construction is one which is of the same class as its main or 'head' constituent; for example, *poor John*, which is a noun phrase based on a noun. An exophoric construction is of a different syntactic class from its head; for example, *on the table*, which is a prepositional phrase based on a noun phrase.

L Bloomfield, *Language*, (London, 1935 [1933])

RF

endosymbiosis *Biology See* SYMBIOTIC THEORY OF CELLULAR EVOLUTION.

Engel's law (1857) *Economics* Formulated by German-born statistician Ernst Engel (1821–96), this states that as incomes increase, the proportion of income spent on food falls. The law is accepted as a basic principle of income and consumption.

E Engel, *Die Productions und Consumptionsverhältnisse des Königsreiche Sachsen* (Berlin, 1877)

PH

enharmonic equivalence *Music* The notion that two pitches very close to each other may be considered as equivalent (and thus interchangeable) even though they retain their separate labels; for example G sharp and A flat. Aurally, the notes are represented by a mean pitch that corresponds to neither 'real' pitch.

MLD

énonciation (1966) *Stylistics* Concept developed by the French linguist Emile Benveniste.

The concrete act of utterance as opposed to *énoncé* (that which is spoken about, for example the past tense events of narrative). *See also* HISTOIRE AND DISCOURS.

E Benveniste, *Problems of General Linguistics* (Coral Gables, Fla., 1971 [1966])

RF

Enskog and Chapman theory of thermal diffusion (1912) *Physics* Named after its discoverers, David Enskog (1884–1947) and Sydney Chapman, this is a type of diffusion in gases not governed by Fick's laws.

Thermal diffusion arises even if the initial concentrations of the gases are uniform, when the temperature in the container varies from point to point. Then, the average speed of the molecules (which depends on the temperature) varies from point to point and the transport equation has to be modified appropriately. When the molecules are similar in mass, the larger molecules diffuse to cooler regions. When the molecules are not too similar in mass, the heavier molecules diffuse to cooler regions. *See* DIFFUSION THEORY.

J Thewlis, ed., *Encyclopaedic Dictionary of Physics* (New York, Oxford and London, 1962)

MS

entitlement theorem *Economics* This is a precept of distributive justice which maintains that individuals are entitled to their goods provided these were obtained by socially acceptable means such as sale, purchase, or as gifts. *See* RAWLS'S THEORY OF JUSTICE.

J Rawls, *A Theory of Justice* (Oxford, 1971)

PH

entropy, law of increase of (19th century) *Physics* A law derived by the German physicist Rudolf Julius Emanuel Clausius (1822–88), this states that the entropy of a thermally isolated closed system cannot decrease.

In any process that is thermodynamically reversible, the entropy of a thermally isolated closed system is unchanged. In any process that is thermodynamically irreversible (that is, in all natural processes, the entropy of a thermally isolated closed system increases).

J Thewlis, ed., *Encyclopaedic Dictionary of Physics* (New York, Oxford and London, 1962)

MS

envelope theorem (1931) *Economics* Originally proposed by Canadian economist Jacob Viner (1892–1970) and later developed by American economist Paul Samuelson (1915–), refers to a curve that encloses an entire family of curves, each of which contributes at least one point to the envelope.

The primary use of the envelope curve is in relating long-run to short-run cost curves. It has become a fundamental analytical tool in modern economics.

J Viner, 'Cost Curves and Supply Curves', *Readings in Price Theory* (Homewood, Ill., 1952)

PH

environment art (mid-1960s) *Art* The origins of this movement are attributed to American artist Allan Kaprow (1927–) who also championed the HAPPENING. Its roots lie in the *Merzbau* (1925), assemblages and kinetic lightshows of German poet and painter Kurt Schwitters (1887–1948). The environment is the three-dimensional context in which content, sensory sensations, kinetic stimuli and imagination are synthesized in an effort to involve, and sometimes overpower, the spectator.

A Kaprow, *Assemblage, Environments and Happenings* (1966)

AB

environmental stress theory *Psychology* Proposed by the American physiologist Walter Bradford Cannon (1871–1945).

Environmental stimuli cause organismic responses within an organism. Stressors can be either directly or indirectly environmentally orientated. Indirect stressors are termed as being intraorganismic. Intraorganismic stressors include cognitive and physiological reactions, which are originally derived from the environment. Responses to the environment may either give pain or pleasure to the organism. Prolonged exposure to certain environmental stimuli can lead to dysfunction within the organism.

H Selye, 'General Adaptation Syndrome and Diseases of Adaptation', *Journal of Clinical Endocrinology and Metabolism*, vol. VI (1946), 117

NS

environmentalism *Psychology* Originally associated with the English philosopher John Locke (1632–1704), and later the discoverer of natural selection Charles Robert Darwin (1809–80).

This term refers to the schools of philosophical and theoretical thought which are concerned with the influence and effects the environment has on determining an organism's behaviour. Environmental psychology is a specific branch of the discipline, studying the relationship between an organism's behaviour and the environmental context in which the organism exhibits that behaviour.

A Anastasi, *Fields of Applied Psychology* (New York, 1979)

NS

Eötvös' law (1886) *Physics* Named after the Hungarian physicist Roland Eötvös (1848–1919), this law relates the specific free surface energy of a liquid–vapour interface to properties of the liquid.

The law is expressed as

$$S(M_r/\rho)^{2/3} = K(\mathrm{T_c} - T)$$

where S is the specific surface free energy (surface tension) and ρ is the density of the liquid, both at a temperature T, T_c is the critical temperature, M_r is the relative molecular mass of the liquid and K is a constant.

J Thewlis, ed., *Encyclopaedic Dictionary of Physics* (New York, Oxford and London, 1962)

MS

epic theatre (1920s–30s) *Literary Theory* Technique of performance advocated by the German playwright Bertolt Brecht (1898–1956).

An anti-representational style of theatre in which narration, unfamiliar settings and other estranging techniques prevent the audience identifying with the subject, encouraging a rational critique of contemporary social, economic and poitical practices. *See also* ALIENATION EFFECT.

J Willett, ed., *Brecht on Theatre* (London, 1974)

RF

Epicureanism *Philosophy* Philosophy of Epicurus of Samos (342–271 BC) and his followers, notably the Roman poet Titus Lucretius (*c.*99–*c.*55 BC).

They developed the atomism of Leucippus and Democritus, and like their contemporary rivals the Stoics they were materialists, though their atomism involved the existence of empty space, which the Stoics rejected. Atoms fell downwards through space, but were subject to a random 'swerve' which served to explain both their agglomeration into objects and human freewill (the latter application being subject to the same objection – that free acts are random – as the modern use of Heisenberg's 'uncertainty principle' for that purpose).

The Epicureans had little interest in logic (unlike the Stoics) but their epistemology and ethics both gave primacy to sensations. In ethics these could lead to pleasures, which were 'static' (associated with states) or 'kinetic' (associated with processes), and were to be pursued; but the need to avoid pain led to heavy emphasis on 'static' pleasures of a calm and self-sufficient kind (though the value of friendship was stressed). Epicureanism was thus very far from the modern sense of 'self-indulgence'.

A A Long and D N Sedley, *The Hellenistic Philosophers* (1987); translations with commentary

ARL

epic (classical Greece) *Literary Theory* The epic was recognized as a distinct genre by the Greek philosophers Plato (*c.*427–*c.*347 BC) and Aristotle (384–322 BC).

A narrative poem on a grand scale and of heroic content and serious tone. Typically, the narrative concentrates on the superhuman endeavours of a single hero on behalf of a threatened nation, over a long period of time. Epic in pre-literate societies was an ORAL-FORMULAIC mode of composition and performance; it employed an elevated style adorned with characteristic stock phrases and epithets. The origins of the epic genre in Europe are the Greek *Iliad* and *Odyssey* of Homer (8th century BC). In English, the Anglo-Saxon poem *Beowulf* (8th century) and John Milton's *Paradise Lost* (1667) are regarded as epics.

M Bowra, *Heroic Poetry* (London, 1952)

RF

epigenesis (17th–18th century) *Biology* Coined by William Harvey (1578–1657) in his *On the Generation of Animals* (1651), this is the concept that all animals share a uniform pattern of development beginning with the egg.

Epigenetic development, a gradual process of increasing complexity, begins when the egg is acted upon by male semen. *Compare with* PREFORMATION.

KH

epiphenomenalism *Philosophy* Doctrine that some item under investigation is a mere by-product of some process and has no causal influence of its own. In particular, the claim that the mind is not a separate entity from the body, that conscious phenomena are mere by-products of cerebral or neural processes and have no causal effects on those processes.

One objection to this is the difficulty it has in explaining rational trains of thought: if a thought occurs only because a suitable stage has been reached in a purely mechanical process in the brain, why should the thought be logically related to one before it?

K Campbell, *Body and Mind* (1970); offers 'new' epiphenomenalism; compare also J Bricke's discussion in *Mind* (1975)

ARL

epiphenomenalism *Psychology* A term coined by the American psychologist Orval Hobart Mowrer (1907–82).

Defined by Mowrer as the philosophical argument that only physical material being is reality and that all other phenomena, such as the mind, are of no relevance. Epiphenomenalists believe that psychic experiences, conscious and unconscious thoughts are all by-products of processes within an individual's neurological system. Such a position came to be closely associated with BEHAVIOURISM.

O H Mowrer, *Learning Theory and Personality Dynamics* (New York, 1950)

NS

epiphenomenalist theory of ageing (1950s) *Biology* Also called extrinsic theory, epiphenomenalism was a term coined by British author Alex Comfort (1920–), and espoused by Hans Selye (1907–), a Canadian scientist known for his theories concerning stress. The theory that ageing is caused by general environmental insults which cannot be avoided or prevented. *See also* AGEING, THEORIES OF.

J Behnke and C Finch and G Moment, *The Biology of Aging* (New York, 1978)

KH

epistemic closure, principle of *Philosophy* Principle that, where P and Q are propositions, if we know that P, and know that P logically entails Q, we know that Q.

Sometimes said to support SCEPTICISM, because if I know that, for example, I am holding a pen, and know that if I am holding a pen I am not merely dreaming that I am doing so, then (by the principle) I know that I am not merely dreaming this: but, it is alleged, I cannot know that I am not dreaming because even if I were, things would still appear just as they do. Therefore, I cannot know that I am really holding a pen. *See also* the THEORY OF RELEVANT ALTERNATIVES which has been used to answer this.

J Dancy, *An Introduction to Contemporary Epistemology* (1985)

ARL

equal freedom (19th century) *Politics* Account of freedom given by the English social theorist Herbert Spencer (1820–1903).

All people have a right to freedom up to the point at which they infringe upon the equal right of others.

D Wiltshire, *The Social and Political Thought of Herbert Spencer* (Oxford, 1978)

RB

equal sacrifice theory *Economics* The surrender of equal measures of utility by taxpayers.

The theory has three sub-groups: (1) equal absolute sacrifice (where each taxpayer surrenders the same absolute degree of utility that he obtains from his income); (2) equal proportional sacrifice (where each sacrifices the same proportion of utility he receives from his income); (3) equal marginal sacrifice (where each gives up the same utility from the last unit of income). *See* ABILITY-TO-PAY PRINCIPLE, BENEFIT APPROACH PRINCIPLE, COMPENSATION PRINCIPLE, COST-BENEFIT ANALYSIS, SOCIAL WELFARE FUNCTION and TAX INCIDENCE.

PH

equality *Politics See* EGALITARIANISM.

equality, law of *Psychology* A GESTALT principle of perceptual organization, formulated by the German psychologist Max Wertheimer (1886–1943).

The law states that as the several components of a perceptual field become more similar, they will tend to be perceived as a unit. This is widely accepted as a fundamental feature of perception.

A D Ellis, ed., *A Sourcebook of Gestalt Psychology* (London, 1938)

NS

equality of states doctrine (20th century) *Politics* Theory of the equality in law of all sovereign states. The principle is clearer than is the descriptive accuracy. But belief in the principle itself can become a factor in international relations if expressed through institutions such as the United Nations.

Graham Evans and Jeffrey Newnham, *The Dictionary of World Politics* (Hemel Hempstead, 1990)

RB

equifinality *Geology* A theory stating that since different processes may result in the formation of identical land forms, it is difficult (or impossible) to determine a causative process from the shape of a land form alone.

DT

equilibrium *Art* The balance of a composition's components. This is achieved through the careful distribution of elements, the varied lines of the design and the balanced use of colour.

AB

equilibrium law *Chemistry* A law of chemical equilibrium derived from the LAW OF MASS ACTION, it may also be traced from the thermodynamic methods of German physical chemist August Friedrich Horstmann (1842–1929) in 1873, the American mathematical physicist Josiah Willard Gibbs (1839–1903) in 1876, and Dutch chemist Jacobus Henricus van't Hoff (1852–1911) in 1886.

This law states that for any chemical equilibrium at a given temperature, the product of the concentrations of the products, each raised to a power equal to its coefficent in the equation, divided by the product of the concentrations of the reactions, each raised to a power equal to its coefficient in the equation, is constant. For a general reversible chemical reaction,

$$aA + bB \rightleftharpoons cC + dD$$

the equilibrium law can be expressed mathematically as follows,

$$K_c = \left(\frac{[C]^c [D]^d}{[A]^a [B]^b} \right)_{eq}$$

where [ ] is concentration and K_c is the equilibrium constant. The law only applies to reactions in solution.

M Freemantle, *Chemistry in Action* (Basingstoke, 1987)

MF

equilibrium theory *Economics* Body of analysis examining the balance of interrelated variables of an economy and their tendency to resist change.

Classical economic theory, outlined by Scottish economist Adam Smith (1723–90) and English economist David Ricardo (1772–1823), regarded market prices as fluctuating around the natural price (which can be considered a central price towards which market prices tend). Another feature of the classical approach was the subsistence theory of wages which viewed expansions and contractions of a population as forces of equilibrium which make the subsistence wage rate the long-run equilibrium rate.

The partial equilibrium analysis outlined by English economist Alfred Marshall (1842–1924) is central to the neoclassical approach to equilibrium theory. It demonstrated the continuous nature of economic change. Marshall examined 'slices' of time with a range of new tools such as substitution, the elasticity coefficient, the representative firm, CONSUMER SURPLUS, quasi-rent, economies of scale, and the long and short run.

The French economist Léon Walras (1834–1910) developed a theory of general equilibrium, in which all the markets of an economy are studied, and in which all supplies, prices and outputs of goods and factors are determined simultaneously. *See* GENERAL EQUILIBRIUM THEORY, PARTIAL EQUILIBRIUM THEORY, WALRAS'S STABILITY, SAY'S LAW, INTERNAL AND EXTERNAL BALANCE, FUNDAMENTAL DISEQUILIBRIUM, LAW OF MARKETS, CLASSICAL MACROECONOMIC MODEL, ARROW-DEBREU MODEL, COBWEB THEORY, THEORY OF THE CORE and CATASTROPHE THEORY.

A Smith, *An Inquiry into the Nature and Causes of the Wealth of Nations* (London, 1776); A Marshall, *Principles of Economics* (London, 1890); L Walras, *Eléments d'économie politique pure* (Lausanne, 1876); P A Samuelson, *Foundations of Economic Analysis* (New York, 1965)

PH

equilibrium theory *Politics* Optimistic and consensual view of politics. Societies achieve stability through an equilibrium of competing but not conflicting forces, and social structures have a natural tendency to move towards such balance.

Roger Scruton, *A Dictionary of Political Thought* (London, 1982)

RB

equipartition of energy theorem (19th century) *Physics* In a system consisting of a large number of interacting classical particles in an equilibrium state at a temperature T, the mean kinetic energy per particle is the same for each degree of freedom and is equal to $\frac{1}{2}kT$, where k is Boltzmann's constant.

For example, the molecules in an ideal monatomic gas are assumed to possess only kinetic energy of translation and no potential energy. Each molecule then has three degrees of freedom corresponding to the three directions of translational motion in space and the average kinetic energy per molecule is $\frac{3}{2}kT$.

J Thewlis, ed., *Encyclopaedic Dictionary of Physics* (New York, Oxford and London, 1962)

MS

equipotential theory *Psychology* Proposed by the American psychologist Karl Spencer Lashley (1890–1958).

This theory assumes that all areas of the brain make an equal contribution to overall functioning. Hence, a particular area of brain injury is not important since only the amount of brain injury determines any behavioural deficit. Many equipotential theorists emphasize deficits in abstract or symbolic ability which are assumed to accompany all forms of brain damage. Opponents, who are in the majority, point to contradictory neuropsychological evidence. *See* LOCALIZATION THEORY and MULTIPLE CONTROL PRINCIPLE.

K S Lashley, *Brain Mechanisms and Intelligence: A Quantitative Study of Injuries to the Brain* (Chicago, 1929)

NS

equity theory *Psychology* Proposed by the American psychologists Harold H Kelley (1921–) and John W Thibaut (1917–).

Within psychology, this is a social theory which states that we try to create fairness in groups by rewarding individuals in proportion to the amount they contribute to the group as a whole. Proponents of this theory feel that it is the basis for many explanations of behaviour; of altruism, power and aggression, for example. The equity theory states that both costs and rewards are (or should be) distributed proportionately within the group. *See* SOCIAL EXCHANGE THEORY.

S Kiesler and R Baral, 'The Search for a Romantic Partner: The Effects of Self-Esteem and Physical Attractiveness on Romantic Behavior', *Personality and Social Behavior*, K Gergen and D Marlowe, eds (Reading, Mass., 1970)

NS

equivalence principle (1908) *Astronomy* A principle introduced by the German-born mathematical physicist Albert Einstein

(1879–1955), this forms the basis of his THEORY OF GENERAL RELATIVITY.

An observer has no means of distinguishing whether the laboratory is in a uniform gravitational field or in a uniformly accelerated frame of reference. A consequence of this principle is that the laws of nature must be written in such a way that it is impossible to distinguish between a uniform gravitational field and a uniformly accelerated frame of reference.

S Mitton, ed., *The Cambridge Encyclopaedia of Astronomy* (London, 1973)

MS

equivalent proportions, law of *Chemistry See* RECIPROCAL PROPORTIONS, LAW OF.

Erasistratus' theory of circulation (*c.*300 BC) *Biology/Medicine* Attributed to Greek physician Erasistratus of Julis (*c.*330 BC–*c.*250 BC) by Galen (*c.*130–*c.*200), influential physician of Rome. The theory that the right side of the heart receives blood and pumps it to the lungs (where it is consumed by the lungs), while the left side receives air (pneuma) from arteries in the lung and then pumps it throughout the body via the arterial system.

This mistaken belief that arteries (from *arteria*, meaning 'windpipe') carry air occurred because the thick, muscular walls of arteries stay open even after an animal is killed, suggesting a hollow air tube. This notion was proven incorrect by Galen.

J Fulton, *Selected Readings in the History of Physiology* (Springfield, Ill., 1966)

KH

Erastianism (16th century–) *Politics* Named after the Swiss theologian Erastus (1524–83). The STATE should exercise supremacy over the church.

Roger Scruton, *A Dictionary of Political Thought* (London, 1982)

RB

Eratosthenes' sieve for primes (2nd century BC) *Mathematics* Named after Greek mathematician Eratosthenes (*c.*276–195 BC), this is an algorithm for systematically finding all the prime numbers less than a given integer n, by sequentially eliminating all multiples of primes up to $\sqrt{n}$. For example, from the list of odd numbers up to 25, first multiples of 3 would be eliminated, then multiples of 5 and so on

3 5 7 ~~9~~ 11 13 ~~15~~ 17 19 ~~21~~ 23 ~~25~~

to determine all prime numbers less than 25. *See also* SIEVE.

I Peterson, *The Mathematical Tourist* (New York, 1988)

ML

ergodic hypothesis (20th century) *Physics* For a system in which the probability that a member lies in the phase space is constant, all accessible states have equal probability, so all are realized. This is the main hypothesis of statistical physics.

JB

Erlangen programme (1872) *Mathematics* Originated by Felix Klein (1849–1925), who proposed that 'Geometries can be classified by the groups of transformations of the points which preserve the properties of the geometries.'

For example, the group of transformations of Euclidean geometry must preserve the angle between two lines, but this restriction does not apply to projective geometry, in which angle is not defined. Although the programme has directed attention to valuable methods and results, it is essentially false because Euclidean geometry and augmented Euclidean geometry have the same group although the latter has points at infinity and complex points with properties totally unlike the points in the former.

JB

error analysis (1960s) *Linguistics* Part of second language learning theory, this was influenced by generative studies of first language acquisition.

Error analysis is the study of the errors made by learners in order to: (1) assess the learner's progress toward the target language; (2) identify characteristic difficulties for speakers of one specific language learning some other specific language; (3) investigate universals of language-learning. *See also* INTERLANGUAGE.

S Pit Corder, *Introducing Applied Linguistics* (Harmondsworth, 1973)

RF

error catastrophe theory of ageing (1963) *Biology/Medicine* Also called error theory, this theory was attributed to the American biologist Leslie K Orgel (1927–), who asserted that ageing occurs because of errors in protein synthesis which increase and accumulate over the lifespan of the organism.

Because protein synthesis is governed by other proteins, an error anywhere in the protein synthesizing machinery caused by any agent is thought to lead to a gradually accelerating cumulative increase in errors, until the 'error catastrophe' (death) occurs. *See also* AGEING, THEORIES OF. *Compare with* CODON-RESTRICTION THEORY and RANDOM HITS THEORY OF AGEING.

E Schneider, *The Genetics of Aging* (New York, 1978)

KH

error learning hypothesis *Economics See* ADAPTIVE EXPECTATIONS.

error theory *Biology/Medicine See* ERROR CATASTROPHE THEORY OF AGEING.

escalation (20th century) *Politics* A theory of war which asserts that in military encounters there is a tendency (which can be checked or avoided) for each action by one side to be met by a more violent response by the other. Hence a conventional war can escalate into a nuclear one.

David Robertson, *The Penguin Dictionary of Politics (London, 1986)*

RB

escape learning *Psychology* Theory advanced by the American psychologist Orval Hobart Mowrer (1907–82).

Learning whereby an organism learns to escape a harmful stimulus by using a response referred to as an instrumental or operant response. When talking of escape learning it is not uncommon to refer to avoidance learning. The result of avoidance learning is that an organism makes a response before the stimulus has occurred, therefore avoiding the consequences of the stimulus. *See also* OPERANT THEORY.

E R Hilgard and G H Bower, *Theories of Learning* (New York, 1975)

NS

essential contestability (20th century) *Politics* A number of concepts in political theory, such as POWER or 'freedom' are debated by people whose fundamental conceptions or values are so at odds that no resolution is possible. Thus there is no 'true' meaning to such terms.

Alan Bullock, Oliver Stallybrass, and Stephen Trombley, eds, *The Fontana Dictionary of Modern Thought*, 2nd edn (London, 1988)

RB

essentialism (*c*.400 BC) *Biology* From Plato and Aristotle; applied to biology by Swedish botanist Carolus Linnaeus (1707–78) and others in their search for the perfect 'type specimen' for each species. This is the notion that there exists an essential 'ideal type' for each species, and that any variations from that ideal type observed in individuals are inconsequential. This perspective supports that view that species are relatively fixed and can be categorized. *Compare with* POPULATION THINKING.

E Mayr, *The Growth of Biological Thought* (Cambridge, Mass., 1982)

KH

essentialism *Philosophy* Properly speaking, the doctrine that at least some objects have essences; that is, they have some of their properties essentially, not just because they are described in a certain way (a bishop is essentially in holy orders, yet could be defrocked without ceasing to be himself) but because they must have those properties to be themselves. You might perhaps have been of the opposite sex, while still being yourself, but might you have been a horse?

Under the influence of EMPIRICISM in general and LOGICAL POSITIVISM in particular, essentialism has been unpopular. Recently, though, it has enjoyed a revival especially led by S Kripke and H Putnam.

C Kirwan, 'How Strong are the Objections to Essence?', *Proceedings of the Aristotelian Society* (1970–71)

ARL

essentialism *Politics* A belief about HUMAN NATURE. There are 'essential' features of human character which constitute the timeless and universal foundations of human nature. Depending on the version of this theory, this essence may divide people into males and females, or leaders and led; or unite them as all equally aggressive, or self-seeking, or co-operative.

Maggie Humm, *The Dictionary of Feminist Theory* (London, 1989)

HR

estrangement (*c.*1915–) *Stylistics* Leading idea of formalist stylistics; ubiquitous. *See also* ALIENATION EFFECT, ART AS DEVICE and DEAUTOMATIZATION.

Verbal art is based on conventional deformations of language (*see also* DEVIATION), or rhetorical DEVICES, which make the world represented strange or unfamiliar: reality in a new light.

R Fowler, *Linguistic Criticism* (Oxford, 1986)

RF

ether *Physics See* AETHER.

ethnography of communication (1962) *Linguistics* Anthropologically based SOCIOLINGUISTIC model, the central concepts of which were developed by the American linguist Dell H Hymes.

Study of 'ways of speaking' as they differ from culture to culture. Speakers possess COMMUNICATIVE COMPETENCE; that is, knowledge of appropriate ways of speaking within their culture. The unit of analysis is the communicative event, a single verbal interaction (for example, a marriage service; seeking and receiving a telephone number) within a specific communicative situation as recognized by the culture (for example a wedding; directory enquiries). *See also* CODE-SWITCHING and CONTEXT OF SITUATION.

M Saville-Troike, *The Ethnography of Communication* (Oxford, 1982)

RF

ethnolinguistics *Linguistics See* ETHNOGRAPHY OF COMMUNICATION.

ethnomethodology *Linguistics See* CONVERSATIONAL ANALYSIS.

ethology (18th century) *Biology* A term coined by French zoologist Etienne Geoffroy Saint-Hilaire (1772–1844), although the idea predates him. In its classical form, this is the study of animal behaviour from the evolutionary perspective with emphasis on behaviour in the natural world, including such ideas as INSTINCT and LORENZ'S HYDRAULIC MODEL.

Once the dominant approach toward animal behaviour, ethology was challenged from within and without during the mid-1900s to such a degree that the term itself fell out of vogue. Currently ethology can be difficult to identify as a distinct paradigm within the field of animal behaviour. *Compare with* BEHAVIOURISM.

D Dewsbury, *Comparative Animal Behaviour* (New York, 1978)

KH

Euclidean algorithm *Mathematics* Named after Greek mathematician Euclid of Alexandria (*c.*300 BC), this is an iterative procedure for finding the greatest common divisor (gcd) of two integers or polynomials. It can also be used to solve the linear DIOPHANTINE EQUATION $ax + by = c$ and is related to the CONTINUED FRACTION ALGORITHM.

A Baker, *A Concise Introduction to the Theory of Numbers* (Cambridge, 1984)

MB

Euclid's axioms (3rd century BC) *Mathematics* Named after Greek mathematician Euclid (*c.*300 BC), these axioms were laid out in his *Elements*. The fifth postulate is called the axiom of parallels. Attempts to show that it is independent of the others led to the development of non-Euclidean geometry in the 19th century.

M Hazewinkel, ed., *Encyclopaedia of Mathematics* (Dordrecht, 1988)

ML

Eudoxus's axiom (method of exhaustion) (4th century BC) *Mathematics* Defined by Eudoxus of Cnidos (*c.*408–*c.*355 BC) as a measure of length to allow for the proven existence of irrational numbers.

If two lengths a and b are given with $a < b$ and if, starting with $b = b_0$, a length greater than $b_{i/2}$ is taken away from b_i to form b_{i+1}, then there is an integer k such that $b_k < a$. This axiom was used to calculate with irrational numbers until the decimal notation was introduced.

JB

eugenics (1883) *Biology/Politics/Psychology/Sociology* Coined by Sir Francis Galton (1822–1911), British scientist and cousin of evolutionary theorist Charles Darwin (1809–82), this is the doctrine that such human qualities as intelligence and character are inherited, and that humanity should take deliberate steps to produce fine offspring.

'Positive' eugenics refers to the reproduction of desirable types, whereas 'negative' eugenics refers to efforts to prevent undesirable persons from reproducing. Although of popular interest in the early 20th century, the rise of Nazism and the racism associated with eugenics caused it to fall into disrepute.

S Holmes, *The Eugenic Predicament* (New York, 1933); Pauline M H Mazumdar, *Eugenics, Human Genetics and Human Failings* (London, 1991)

KH

eukaryotic theory (20th century) *Biology* The theory that the very first living cells were those with nuclei (eukaryotes) and that non-nucleated cells such as bacteria (prokaryotes) evolved inside these eukaryotes, from which they were later expelled to become free-living organisms. Biologists who question this theory note that it assumes that more complex organisms predated simpler ones; and that eukaryotes require free oxygen for survival, which scientists believe was absent during the Earth's first 2,000 million years. *See also* ORIGINS OF LIFE. *Compare with* AKARYOTIC THEORY, PROKARYOTIC THEORY, SERIAL SYMBIOSIS THEORY and SYMBIOTIC THEORY OF CELLULAR EVOLUTION.

M Barbieri, *The Semantic Theory of Evolution* (Chur, Switzerland, 1985)

KH

Euler-Bernoulli law (18th century) *Physics* Named after Swiss mathematicians Leonhard Euler (1707–83) and Daniel Bernoulli (1700–82).

The torque on a thin elastic beam under gravity is EIk where E is Young's modulus (a measure of the elasticity of the material) and I is the moment of inertia of the cross-section about an axis through the centre of mass and perpendicular to the plane of the couple. The result is approximately true but only for small deformations.

JB

Euler's constant *Mathematics* Named after Swiss mathematician Leonhard Euler (1707–83).

It is the real number $\gamma \approx 0.57721566$ which is defined by the equation

$$\gamma = \lim_{n\to\infty} \left(1 + \frac{1}{2} + \frac{1}{3} + \cdots + \frac{1}{n} - \log n\right)$$

where $\log n$ denotes the natural logarithm of n. It is not known whether this number is algebraic or transcendental, or even if it is rational or irrational.

T M Apostol, *Introduction to Analytic Number Theory* (Springer, 1976)

MB

Euler's laws of motion (18th century) *Physics* Postulated by Swiss mathematician Leonhard Euler (1707–83), these state that the force acting on a small element of a fluid is equal to the rate of change of its momentum and its torque is equal to the rate of change of its angular momentum, provided suitable units are used. These axioms are deducible from Newton's laws of motion for finite sets of particles but their deducibility for continuous bodies is less clear.

JB

Euler's theorem (18th century) *Mathematics* Named after its discoverer, the Swiss mathematician Leonhard Euler (1707–83), this states that for all complex numbers z the following holds

$$\exp(iz) = \cos z + i \sin z \, .$$

This result ensures that all the functions of elementary analysis are either polynomials or aspects of the exponential function.

JB

Euler's theory (18th century) *Mathematics/Economics* Devised by Swiss mathematician Leonhard Euler (1707–83), this is a theory of distribution based on marginal productivity.

Euler showed that under constant returns to scale, if each factor of production is paid the value of its marginal product, total output (income) will be completely exhausted. *See* ADDING-UP PROBLEM, MARGINAL PRODUCTIVITY, THEORY OF DISTRIBUTION, RETURNS TO SCALE.

L C Young, *Lectures on the Calculus of Variations and Optimal Control Theory* (Philadelphia, 1969)

PH

evaluation measure or procedure (1960s) *Linguistics* The American linguist Noam Chomsky (1928–) stipulated that a proposed generative grammar of a language should be evaluated for its simplicity, explicitness and linguistic significance. This is a problematic notion. *See also* ADEQUACY, LEVELS OF.

N Smith and D Wilson, *Modern Linguistics; The Results of Chomsky's Revolution* (Harmondsworth, 1979)

RF

eventstructure (*c.*1945) *Art* Theory proposed by the British artist John Latham which stresses the importance of process over product: structures in events (through time) are more valid than structure in art objects (in space).

In 1967 Latham's ideas were adopted by Theo Botschiver, Jeffrey Shaw and Sean Wellesley-Miller who formed the Eventstructure Research Group (ERG) in Amsterdam. They proposed an alternative to 'museum art' by staging public events, encouraging physical participation, 'operational art' and 'art of real consequences'.

J Latham *'Eventstructure', Studio international*, vol. CLXXIV (September, 1967), 82

AB

evolution (19th century) *Biology* Although most closely identified with the evolutionary theorist Charles Darwin (1809–82), the term was first used in a biological context by British author Herbert Spencer (1820–1903). Evolution is the change in the gene pool of a population from generation to generation that may, in time, produce new species.

This theory offers an intellectual framework for understanding species change that is substantiated by an enormous volume of data. The key mechanism of evolution is NATURAL SELECTION, but other mechanisms involved include genetic mutations and the FOUNDERS EFFECT (genetic drift). *See also* ANAGENESIS, CLADOGENESIS, DOLLO'S LAW, DARWINISM, MACROEVOLUTION, MICROEVOLUTION and NATURAL SELECTION.

J Reader, *The Rise of Life* (New York, 1988)

KH

evolution in little bags *Biology See* 'LITTLE BAGS' THEORY OF EVOLUTION.

evolutionarily stable strategy (ESS) (1976) *Biology/Mathematics* Coined by mathematician John Maynard Smith, this is the somewhat controversial application of GAME THEORY to animal behaviour holding that the best strategy for solving some problem will evolve and persist in a population, unless environmental changes favour a new strategy.

An ESS is defined as the strategy which, if most individuals adopt it, cannot be improved upon by any other strategy. For example, many species engage in contests for food or mates, and yet refrain from killing or maiming the loser even though they could easily do so (thus ridding themselves of a competitor). Application of this model suggests that if all members of such a species adopt a restrained strategy, then the species is better off than if members fight more aggressively.

KH

evolutionary determinism (1930) *Biology* Proposed by the British geneticist Sir Ronald A Fisher (1890–1962).

Selection on a trait, such as size, in a population may favour the average individuals (stabilizing selection); or it may favour a particular direction, such as larger individuals, from deterministic processes, such as a farmer selecting the largest grains to replant each year. This is directional/deterministic evolution, and its relative importance as opposed to stabilizing selection is uncertain.

N Eldredge, *Unfinished Synthesis* (Oxford, 1985)

RA

evolutionism (*c.*1861) *Anthropology* Developed within social anthropology by British jurist Sir Henry Sumner Maine (1822–88) and others.

The idea that society gradually advances, often explained in terms of stages from 'primitive' to 'barbarian' to 'civilized'. Some societies are slower to advance than others, and these slower, 'primitive' societies are held to remain at lower stages of evolution. Evolutionism in social anthropology developed to some extent independently of evolutionary theory in biology. It was largely overthrown in the 1920s by the onset of FUNCTIONALIST THEORY. *See also* DIFFUSIONISM.

Adam Kuper, *The Invention of Primitive Society: Transformations of an Illusion* (London, 1988)

ABA

evolutionism (19th century–) *Politics* Optimistic theory of history.

Human society is steadily and slowly developing by successive improvements, either by reform and adaptation, or by the 'natural selection' of superior values and institutions. *See also* EVOLUTION, THEORY OF, DIALECTICAL MATERIALISM, and SOCIAL DARWINISM.

Roger Scruton, *A Dictionary of Political Thought* (London, 1982)

RB

Ewing's theory of ferromagnetism (1885) *Physics* Named after the Scottish engineer and physicist Sir (James) Alfred Ewing (1855–1935).

Ewing's theory is an elaboration of the 'molecular theory' of magnetism proposed by the German scientist Wilhelm Eduard Weber (1804–91), according to which each molecule of a ferromagnetic substance acts as a small magnet. In the unmagnetized state, these elementary magnets arrange themselves in closed chains so that the net external effect of their poles is zero. Realignment of their magnetic axes produces magnetization, leading to saturation when all are aligned in the direction of magnetization. Magnetic hysteresis results when the force necessary to break up the molecular chains prevents the substance from following the changes in the magnetizing field. The theory has been partially confirmed by experiment. *See also* MAGNETISM, THEORIES OF.

Robert M Besançon, ed., *The Encyclopedia of Physics* (New York, 1985)

GD

exchange theory *Psychology* *See* SOCIAL EXCHANGE THEORY.

excitation *Psychology* A term coined by the Russian physiologist Ivan Petrovich Pavlov (1849–1936).

Excitation is the process by which a stimulus sets up either a chain of reactions or a single change in a receptor. Examples include a state of tension or the drive state referred to in some theories of learning. *See* DRIVE REDUCTION THEORY.

N R Carlson, *Physiology of Behavior* (London, 1986)

NS

excluded middle, law of *Philosophy* One of the traditional THREE LAWS OF THOUGHT (along with the laws of IDENTITY and CONTRADICTION). Every proposition is either true or not true.

This is weaker than the law of bivalence (every proposition is true or false), since if there is a third truth value excluded middle can still hold, though bivalence will fail. (However, bivalence is sometimes treated as a version of excluded middle). For classical logic, excluded middle follows from the law of contradiction. Intuitionist logic accepts the latter but not excluded middle (*see also* DOUBLE NEGATION), for reasons connected with the 'Jones was brave' example (*see* BIVALENCE). *See also* DEGREES OF TRUTH.

P T Geach and W F Bednarowski, 'The Law of Excluded Middle' (symposium), *Proceedings of the Aristotelian Society*, supplementary volume (1956)

ARL

exclusion principle (1924) *Physics* A principle introduced by Austrian-Swiss theoretical physicist Wolfgang Pauli (1900–58) to describe the distribution of electrons in atoms.

An electron in orbit around the nucleus of a neutral atom is characterized by the values of four numbers, known as quantum numbers. The principle states that no two electrons (or any other particles with half-integral spins) can have the same set of quantum numbers. This principle explains the electronic structure of atoms since, in a many-electron atom, the electrons cannot all occupy the lowest energy state but must occupy levels of progressively increasing energy.

J Thewlis, ed., *Encyclopaedic Dictionary of Physics* (New York, Oxford and London, 1962)

MS

exclusion theory (20th century) *Politics/ Sociology* Also referred to as social closure theory.

Groups maximize their own benefits by excluding non-members. At the same time they establish their identity as much by excluding non-members as by defining the characteristics of membership. Identity thus depends on the identification of 'outsiders' or 'enemies'.

Frank Parkin, *Marxism and Class Theory: A Bourgeois Critique* (London, 1979)

RB

exercise, law of *Psychology* Proposed by the American psychologist Edward Lee Thorndike (1874–1949).

Thorndike proposed that 'all other things being equal', repeated performance of a task makes the task easier to complete and reduces the likelihood of error. The law is considered problematic because of uncertainty regarding Thorndike's expression 'all other things being equal'. *See* THORNDIKE'S LAW OF EFFECT and IDENTICAL ELEMENTS THEORY.

E L Thorndike, *The Fundamentals of Learning* (New York, 1932)

NS

exhaustion theory of ageing *Biology* The theory that ageing occurs when an essential nutrient is used up. This theory is one of several competing explanations for the ageing process. *See* AGEING, THEORIES OF.

KH

existentialism *Philosophy* Movement originating with Sören Kierkegaard (1813–55) and continuing later with Karl Jaspers (1883–1969), Gabriel Marcel (1889–1973), Martin Heidegger (1889–1976), Jean-Paul Sartre (1905–80), and various others, though it has had little influence in English-speaking philosophy. Fyodor Dostoevsky (1812–81) and Friedrich Nietzsche (1844–1900) are sometimes included.

The main idea is to distinguish the kind of being possessed by humans – called *Dasein* (Heidegger) or *être-pour-soi* (Sartre) – from that possessed by ordinary objects (*être-en-soi* for Sartre). The former, which is partly an actual condition of humans and partly something to be pursued, is essentially open-ended and free from determination by any already existing essence: 'existence precedes essence'. Hence the emphasis on freedom, choice and responsibility, evasion of which by relapsing into a 'thing-like' state (or trying to do so) is Sartrian 'bad faith'. Consciousness of this total open-endedness leads to dread (despair, anguish, *angst*, *angoisse*). In studying being, existentialists have been influenced by PHENOMENOLOGY.

N Langiulli, ed., *The Existentialist Tradition* (1971); selections, those from Abbagnano, Buber, Marcel, and Sartre being perhaps the most accessible

ARL

exit, voice, and loyalty (1970) *Politics* Theory of dissidence described by American political scientist Albert O Hirschman (1915–).

Those who are dissatisfied with the policies or general character of an organization have three choices: they can leave (exit); they can oppose (voice); or they can keep quiet (loyalty).

Albert O Hirschman, *Exit, Voice, and Loyalty* (Cambridge, Mass., 1970)

RB

exotic theory of fossil origins (early 18th century) *Biology* A theory of the German philosopher Gottfried Leibniz (1646–1716).

This theory tried to explain the existence of apparently foreign fossils, such as tropical plants in western Europe, by stating that these had been transported after death from their place of origin to their place of burial. This actually can happen (although not in the case of these European fossils) and so the theory was ahead of its time, giving way eventually to biblical explanations. However, such fossils can also occur because of climate change over geological periods of time.

D M Moore, *Green Planet: The Story of Plant Life on Earth* (Cambridge, 1982)

RA

expanding Earth hypothesis (late 19th century–) *Geology* A theory associated with the work of the Hungarian, L Egyed and the Australian geologist Warren Carey, this postulates that the Earth may expand if its internal radiogenic heat is not lost sufficiently fast by conduction and convection to the surface.

Egyed suggested that the universal gravitation constant (G) varies with time, causing corresponding changes in the Earth's volume. Such increases would cause splits in its outer shell, with remaining shell fragments moving apart; thus explaining the formation of the world's present ocean floors and CONTINENTAL DRIFT. There is in fact very little evidence for such variance with time, and none for significant expansion on other planets. If present at all, expansion is considered likely to be by less than 1 per cent; though the CONTRACTING EARTH HYPOTHESIS is generally preferred today.

H G Owen, 'Continental Displacement and Expansion of the Earth during the Mesozoic and Cenozoic', *Transactions of the Royal Society of London*, vol. CCLXXXIA (London, 1976), 284–313

DT

expansion of the universe (*c*.1930) *Astronomy* The wavelengths of spectral lines in the light from galaxies show a shift to longer wavelengths compared with the values for terrestrial laboratory sources. This was interpreted as a Doppler shift, showing that the distant galaxies are receding from the Earth and from each other; that is, that the universe is expanding.

Observation shows that the relative speed of recession is proportional to the present separation of the galaxies. The so-called characteristic expansion time is the reciprocal of the Hubble constant. The true age of the expanding universe is less than the characteristic expansion time because the galaxies have been slowing down under the effect of their mutual gravitational attraction. *See* HUBBLE'S LAW.

S Mitton, ed., *The Cambridge Encyclopaedia of Astronomy* (London, 1973)

MS

expectancy theory *Psychology* Proposed by the American psychologist Edward Chase Tolman (1886–1959).

Organisms acquire a disposition to behave in certain ways to certain stimuli. The theory assumes that the occurrence of a future or expected event is always dependent on an organism choosing and executing the correct behaviour. *See* LATENT LEARNING.

E C Tolman, 'Principles of Purposive Behaviorism', *Psychology: A Study of a Science*, S Koch, ed. (New York, 1959)

NS

exploitation *Politics/Economics* Exploitation occurs where a person or institution systematically gives to another rewards or services or goods disproportionately smaller than that other person's efforts or contribution would justify, and retains for themself a correspondingly greater amount.

David Miller *et al.*, eds, *The Blackwell Encyclopaedia of Political Thought* (Oxford, 1987)

RB

exploitive system (1960) *Biology* Proposed by British embryologist and geneticist Conrad Hal Waddington (1905–75), this is the theory that animals have the capacity to select, out of an array of options, the particular environments in which they live, and thus they have an influence on the pressures that NATURAL SELECTION will impose on them. Waddington's theory implies that an animal's capacity for choosing is a cause of evolution in the same way that natural selection is a cause of evolution.

H Plotkin, *The Role of Behavior in Evolution* (Cambridge, Mass., 1988)

KH

expressionism (20th century) *Art* Name applied to early 20th-century art (mainly northern European) where the artist's state of mind is paramount when depicting the real world, which is thus often depicted in distorted forms. More a tendency than a style, it was based on the increasingly acceptable belief by the 1900s that expression is of primary importance and that the formal structure of a work creates the emotional impact. It may be viewed as a reaction against IMPRESSIONISM and Jugendstil.

Definitions vary as to the leading artists/groups encompassed by this term; generally, the German artists Max Beckmann (1884–1950) and Emil Nolde (1867–1956) and Austrians Oskar Kokoschka (1886–1980) and Egon Schiele (1890–1918), as well as DIE BRÜCKE and BLAUE REITER artists form the core of Expressionism. To these can be added the French Georges Rouault (1871–1958), Jules Pascin (1885–1930),

André Denoyer de Segonzac (1884–1974) and the Ecole de Paris artists. The term is occasionally applied to Fauves André Derain (1880–1954) and Maurice Vlaminck (1876–1958). *See* FAUVISM.

L Richard, *Phaidon Dictionary of Expressionism* (1978)

LL

expressionism (early 20th century) *Music* A term borrowed from the visual arts, in its musical sense it is frequently associated with the music of the Austrian-born composer Arnold Schoenberg (1874–1951) and his pupils. Expressionism is the deliberate expression in music of the composer's innermost urges rather than a reaction to external stimuli.

Expressionist music distorts, exaggerates or rejects traditional techniques to produce intensity of feeling. It is typified by the atonal works of composers linked with the German *Blaue Reiter* group of artists; for example, Schoenberg's 1909 stage-work *Erwartung* ('Expectation/Suspense'). Expressionism may be interpreted as an extreme form of the Romantic aesthetic of 'feeling above form'.

J Willett, *Expressionism* (London, 1970)

MLD

extended standard theory (EST) (1970s) *Linguistics* Modified version of TRANSFORMATIONAL-GENERATIVE GRAMMAR developed by the American linguist Avram Noam Chomsky (1928–) and associates.

Set of modifications designed to overcome limitations of the STANDARD THEORY, placing deep structure and surface structure closer together, allowing semantic interpretation of the surface, and above all constraining the excessive power of transformations in the standard theory of 1965. A number of sub-theories were developed, including SUBJACENCY, TRACE THEORY, and X-BAR THEORY. Developed into the model known as GOVERNMENT AND BINDING *c*.1980.

A Radford, *Transformational Syntax* (Cambridge, 1981)

RF

extension of kinship terms (1964) *Anthropology* First proposed as a formal principle by American linguist and anthropologist Floyd G Lounsbury (1914–).

The practice of classifying distant relatives by the same kinship terms as close ones, and especially the theory developed by Lounsbury which emphasizes this practice. For example, whereas in many languages (though not in English) a person's father and father's brother are both called 'father', the latter is said to be a 'father' in an 'extended' sense. Opponents claim that rather than 'extension', one should talk about 'categories' which include both near and distant relatives. *See also* KINSHIP.

H W Scheffler and F G Lounsbury, *A Study in Structural Semantics: The Siriono Kinship System* (Englewood Cliffs, NJ, 1971)

ABA

extensionality thesis *Philosophy* Thesis beloved of logical atomists, logical positivists, and various kinds of nominalists and reductionists. It says that apparent exceptions to LEIBNIZ'S LAW can be dispensed with; that is, intensions can be reduced to extensions, or roughly, what holds true of objects does not depend on how they are described.

For logical atomism in particular the thesis says that all propositions are truth-functions of certain basic ones (that is, their truth or falsity follows, given the truth or falsity of the basic ones). An alternative version of the thesis (by Willard Van Orman Quine (1908–)) says that only if the above is true can a coherent system of logic be constructed; that is, there is no intensional logic.

W V O Quine, 'Reference and Modality', *From a Logical Point of View*, 2nd revised edn (1961); reprinted with discussions in L Linsky, ed., *Reference and Modality* (1971)

ARL

extermination theory of fossil origins (1539) *Biology* A theory of the German theologian Martin Luther (1483–1546).

Luther argued that fossils represented the creatures that had been destroyed by the biblical flood (diluvial theory). This was not widely accepted at the time. However, towards the end of the 18th century the theory was reissued (replacing the EXOTIC THEORY OF FOSSIL ORIGINS) to explain the existence of extinct plants in coal measures in Europe. This was developed into catastrophe theory, largely by the Frenchman Baron George Cuvier (1769–1832), proposing that a series of floods had occurred in the past extinguishing life each time, which was then recreated. This explained the existence of sedimentary deposits. The theory eventually gave way to the arguments of the British geologist Sir Charles Lyell (1797–1875) documented in the *Principles of Geology* (1830–1833); and the uniformitarian philosophy, which stated that the processes acting on organisms today are the same as those in the past.

D M Moore, *Green Planet: The Story of Plant Life on Earth* (Cambridge, 1982)

RA

externalism *Philosophy* Any view claiming that to analyze a certain phenomenon reference must be made to something outside a certain sphere within which the phenomenon might have been thought to be confined.

In particular, externalism appears in certain analyses of mental notions such as belief and knowledge. An externalist view of belief holds, for example, that one can only believe that King Arthur ruled Britain if King Arthur at least existed, and perhaps also was causally connected to the believer (*see* CAUSAL THEORIES OF REFERENCE). Otherwise, one's state of mind cannot be that belief but something else; for example, that the sentence 'King Arthur ruled Britain' (which does exist) says something true.

An externalist view of knowledge holds that to count as knowing something one must be suitably related (for example causally: *see also* CAUSAL THEORY OF KNOWLEDGE) to the thing or fact in question. Externalist theories are often subdivided into strong and weak versions, and so on, in various ways. *See also* INTERNALISM.

C McGinn, *Mental Content* (1989), especially part 1; with review by D Owens in *Mind* (1990), externalism about belief and thinking

ARL

externalities (20th century) *Economics* Arthur Cecil Pigou (1877–1959) developed earlier work by fellow English economists Henry Sidgwick (1838–1900) and Alfred Marshall (1842–1924) into an important feature of modern economic theory.

Economic activity generates costs and benefits, some of which are not incurred/enjoyed by the person performing the activity. When, for example, a company pollutes the environment, it may enjoy efficient production yet society is faced with the cost of the pollution. (Acid rain is a large-scale example of this.) Similarly, a firm that trains its workers well will lose some of this benefit when the worker moves to another firm, in which case society as a whole benefits. *See* COST-BENEFIT ANALYSIS, CONSUMER SURPLUS, COMPENSATION PRINCIPLE, COASE THEOREM and SOCIAL WELFARE FUNCTION.

A C Pigou, *Economics of Welfare* (London, 1920)

PH

extinction (18th century) *Biology* Associated with Baron George Léopold Chrétien Frédéric Dagobert Cuvier (1769–1832), eminent French anatomist, this term refers to the destruction of a species.

Early in the 18th century, extinction was thought impossible, so when fossils were first discovered naturalists believed that the organisms still lived somewhere on earth. Modern biologists assume that extinction is the likely fate of most organisms, and that once a species becomes extinct, it cannot re-evolve. *See also* MASS EXTINCTION.

R Milner, *The Encyclopedia of Evolution: Humanity's Search for Its Origins* (New York, 1990)

KH

extinction *Psychology* Concept advanced by the Russian physiologist Ivan Petrovich Pavlov (1849–1936).

The gradual diminution in the strength or rate of response when the unconditioned stimulus (US) or the reinforcement is withheld. Some aspects of a learning response need reinforcement for the establishment and subsistence of the response; when the reinforcement is removed the response will start to be extinguished. The rate of extinction is a function of the similarity between the learning situation and the conditions of extinction. It is also affected by the reinforcement schedule that operates during learning.

N J Mackintosh, *The Psychology of Animal Learning* (London, 1974)

NS

extrinsic vs intrinsic criticism (1940s) *Literary Theory* Fundamental distinction on which the American NEW CRITICISM movement is based.

Extrinsic study of literary works relates them to their contexts: biographical, historical, production, reception by readers. 'New Critics' argued against these approaches (*see also* AFFECTIVE FALLACY and INTENTIONAL FALLACY), claiming that 'the natural and sensible starting-point' is 'the works of literature themselves' (Wellek and Warren). In intrinsic criticism, the object of study is the text cut out of context, regarded as a verbal artefact. The distinction is in fact polemical rather than objective. *See also* LITERARY PRAGMATICS.

R Wellek and A Warren, *Theory of Literature* (London, 1963 [1949])

RF

extrinsic theory *Biology See* EPIPHENOMENALIST THEORY OF AGEING.

Eyring's theory of liquid viscosity (1937) *Physics* Named after its originator, the American physical chemist Henry Eyring (1901–81).

This theory takes the liquid to be essentially that described by the hole theory (a large number of small clusters of molecules) with the major contribution to the viscosity being the work necessary to drag molecules from one cluster over those of an adjacent cluster with a different drift velocity.

This entails the movement of molecules over an energy barrier which is a thermally-activated process. Movement of molecules over the energy barrier must occur in the absence of an applied shear stress, but the molecular motion is then random and there is no net flow. The shear stress tilts the intermolecular potential distribution, giving a net directional flow. A simple analysis of this model, assuming a Boltzmann distribution for the successful jumps by a molecule over the energy barrier, gives a temperature-dependence of the viscosity of the form

$$\ln \eta = A + B/T$$

and a pressure-dependence

$$\ln \eta = D + Ep$$

where A, B, D and E are constants and T is the temperature. *See* HOLE THEORY OF LIQUIDS.

J Thewlis, ed., *Encyclopaedic Dictionary of Physics* (New York, Oxford and London, 1962)

MS

F

Fabianism (19th century–) *Politics* British theory of SOCIALISM associated with the Fabian Society; particularly with its leading early members Beatrice and Sidney Webb (1859–1943 and 1858–1947). Named after Roman general Q Fabius Maximus Verrucosus (–203) – nicknamed 'Cunctator' meaning 'the delayer' – who won his battles by slow attrition of the enemy.

Capitalist societies are evolving peacefully towards the highest form of rational organization through a STATE organized both locally and centrally, which is socialism. The most effective means of assisting this development is thorough research and the dissemination of information.

David Miller *et al.*, eds, *The Blackwell Encyclopaedia of Political Thought* (Oxford, 1987)

RB

fabula* and *sjuzhet (1920s) *Stylistics* Russian Formalist (*see* FORMALISM, RUSSIAN) theory of fiction, most associated with Viktor Shklovsky (1893–1984).

Fabula is the underlying chronological structure of a story; *sjuzhet* the structure of its narration, with gaps, transpositions, flashbacks, shifts of tempo and perspective, and so on. *See also* STORY AND PLOT, 'HISTOIRE AND DISCOURS'.

V Shklovsky, 'Sterne's *Tristram Shandy*', *Russian Formalist Criticism*, L T Lemon and M J Reis, transs and eds (Lincoln, Nebraska, 1965), 25–57

RF

fabulation (1967) *Literary Theory* Term coined by the American critic Robert Scholes (1929–), but English editor and printer William Caxton (*c.*1422–*c.*1491) had used 'fabulator' in 1484.

Species of ANTI-NOVEL characterized by verbal play, black comedy, allegory and a disregard for the conventions of literary REALISM. The classic exemplar is held to be Cervantes's *Don Quixote* (1605, 1615), and a modern instance is John Barth's *Giles Goat-Boy* (1966). *See* METAFICTION.

R Scholes, *Fabulation and Metafiction* (Urbana, 1979)

RF

fact/value distinction *Philosophy* Claim that a sharp distinction can be drawn between factual statements and value judgments, only the former being regarded as *stating* anything and as being true or false; the latter having meaning by doing something other than stating (for example, expressing attitudes: *see* SPEECH ACT THEORIES).

The claim underlies EMOTIVIST and prescriptivist – as against descriptivist – views in ethics; and is related to (though not identical with) G E Moore's attack (in his *Principia Ethica* (1903), §§5–14) on the 'naturalistic fallacy' (*see also* NATURALISM). However, the sharpness of the distinction has proved to be hard to maintain. *See also* HUME'S LAW.

C Beck, 'Utterances which Incorporate a Value Statement', *American Philosophical Quarterly* (1967)

ARL

factor-price equalization theorem (20th century) *Economics* Formulated by American economist Paul Samuleson (1915–) on the basis of the HECKSHER-OHLIN TRADE THEORY, this postulates that free trade in commodities will eliminate price differentials, thereby effecting an equalization of factor prices; especially wages and interest rates. *See* COMPARATIVE COSTS, ABSOLUTE ADVANTAGE THEORY, CUSTOMS UNION THEORY and RYBCZYNSKI THEOREM.

A P Lerner, 'Factor Prices and International Trade', *Econometrica*, XIX (1952), 1–15

PH

factor theorem *Mathematics* The result that if c is a root of the polynomial $p(x)$ (that is, $p(c) = 0$), then $(x - c)$ divides $p(x)$, or equivalently, $(x - c)$ is a factor of $p(x)$. *See also* RATIONAL ROOT THEOREM and REMAINDER THEOREM.

MB

factor theory *Psychology* Associated with the work of the English scientist Sir Francis Galton (1822–1911).

Any theory or school of thought that analyzes behaviour and behavioural phenomena in terms of different factors. Tests are used to distinguish these factors as, for example, in the field of intelligence. This term is also used to describe theories based on two separately recognizable processes (for example, cognitive and physiological). *See* FACTOR THEORY OF LEARNING.

C E Spearman, *The Abilities of Man: Their Nature and Measurement* (New York, 1970)

NS

factor theory of learning *Psychology* Proposed by the American psychologist Clark Leonard Hull (1884–1952).

This theory characterizes the phenomenon of learning in terms of two factors, stating that because learning is such a complex process the classical and operant conditioning theories are needed to explain how it occurs. *See* OPERANT THEORY and PAVLOVIANISM.

R M Gagne, ed., *Learning and Individual Differences*, (Columbus, Ohio, 1967)

NS

faculty theory (4th century BC–) *Psychology* Faculty psychology can be traced back to Aristotle (384–322 BC), with more recent examples including the theory of craniology proposed by German anatomist Franz Joseph Gall (1758–1828).

The term refers to the mental processes, faculties or functions as distinct from the contents of experience. For instance, Jungian psychologists use the term to refer to the rational functions of thinking and feeling and the irrational functions of intuition and sensation. 'Faculty' in this sense is an inexact term and has fallen into disuse within the discipline. *See* MENTAL FACULTY THEORY.

C G Jung, *Collected Works*, H Read, M Fordham and G Adler, eds (Princeton, N.J., 1953–78)

NS

Fahrenheit temperature (1714) *Physics* This refers to temperature measured on a temperature scale named after Daniel Gabriel Fahrenheit (1686–1736), who introduced the scale and was the first person to make reliable mercury-in-glass thermometers.

The unit on this scale is the degree Fahrenheit, symbol °F. Fahrenheit's scale was based on the lowest temperature that could be produced with a freezing mixture of ice and common salt (assigned the value 0°F), and the temperature of the normal human body (assigned the value 96°F). Using these fixed points, Fahrenheit found that the freezing-point of water is 32°F and that the boiling-point under a pressure of one atmosphere is 212°F.

J Thewlis, ed., *Encyclopaedic Dictionary of Physics* (New York, Oxford and London, 1962)

MS

Fajans' rules (1923) *Chemistry* Rules of chemical bonding named after the Polish-American physical chemist Kasimir Fajans (1887–1975) who proposed them.

All compounds possess a degree of covalency. The polarization of a covalent bond and the degree of covalency are high if: (1) the charges on the ions are high; (2) the cation is small; (3) the anion is large. Ionic bonds are favoured by small ionic charges, large cations, small anions and cations possessing a noble gas electronic structure.

M Freemantle, *Chemistry in Action* (Basingstoke, 1987)

MF

Fajans-Soddy law *Chemistry* A law of radioactive displacement, also known as the Fajans-Soddy-Russell displacement law. This states that the atomic number of an element decreases by two when an α-particle is emitted, and increases by one when a β-particle is emitted.

M Freemantle, *Chemistry in Action* (Basingstoke, 1987)

MF

Falkentheorie (1870s) *Literary Theory* Theory of the novella or short tale propounded by the German writer Paul Heyse (1830–1914).

The beloved falcon occurs in the ninth tale of the fifth day in the *Decameron*, by Italian writer Giovanni Boccaccio (1313–75). It is the last sacrifice of a lover who has ruined himself wooing a mistress; and its bitter loss moves the mistress to submission. For Heyse the falcon is a symbol, characteristic of the clear silhouette which gives each story its individual and memorable identity.

E K Bennett, *The German Novelle* (1954)

RF

fallibilism *Philosophy* Doctrine that nothing can be known for certain; that is, there is no infallible knowledge, but there can still be knowledge. We need not have logically conclusive justifications for what we know.

This was particularly insisted on by the American pragmatist Charles Sanders Peirce (1839–1914) in his opposition to FOUNDATIONALISM. *See also* RELEVANT ALTERNATIVES THEORY and EPISTEMIC CLOSURE.

ARL

falling rate of profit (18th–19th century) *Economics* This view was held by all leading classical economists except the English writer Thomas De Quincey (1785–1859). Scottish economist Adam Smith (1723–90) regarded the decline in profit as a result of 'the competition of capitals' whereas English economist David Ricardo (1772–1823) noted that diminishing returns in agriculture raised the real wage and reduced profits with which wages were inversely related. *See* THEORIES OF PROFITS.

A Smith, *An Inquiry into the Nature and Causes of the Wealth of Nations* (London, 1776); D Ricardo, *On the Principles of Political Economy and Taxation* (London, 1817)

PH

false consciousness (19th century–) *Politics* Marxist theory of 'deception' of the working class.

Members of the working class in capitalist societies have INTERESTS which would rationally lead them to revolutionary action against the system. However IDEOLOGY creates a 'false consciousness' which hides this reality from them. They therefore require leading into 'correct' political activity by a 'vanguard party', a central concept of LENINISM. *See also* BOLSHEVISM.

L Kolakowski, *Main Currents of Marxism*, 3 vols (Oxford, 1978)

RB

falsificationism *Philosophy* Claim associated especially with Austrian philosopher Karl Raimund Popper (1902–) that science should aim not to verify or confirm hypotheses – as verificationists and inductivists in general claim – but to falsify them.

This is because science is interested (in Popper's view) in universal affirmative conclusions, of the form 'All *A*s are *B*s', and if the universe is infinite (or merely too vast to explore) such conclusions could never be verified. However, they could be falsified by the discovery of a counter-example. Popper also rejects the possibility of weak CONFIRMATION, replacing it by his own notion of corroboration – though how far this is really different might be disputed. However, while verificationists claim that verifiability is essential for meaningfulness, Popper claims only that falsifiability is essential for scientific – as against metaphysical – status (not that it is so for meaningfulness). *See also* HYPOTHETICO-DEDUCTIVE METHOD, INDUCTIVISM, DEDUCTIVISM and IMPROBABILISM.

A O'Hear, *Karl Popper* (1980)

ARL

fancy (1800s) *Literary Theory* Defined in contrast to IMAGINATION by the English poet Samuel Taylor Coleridge (1772–1834).

Fancy is a mechanical capability which, in composition, reassembles in a new order remembered perceptions. After Coleridge, the term was applied to lighter, unserious, writings.

S T Coleridge, *Biographia Literaria* [1815], ch. 13, repr. in H Adams, ed., *Literary Theory since Plato* (New York, 1971), 470–71

RF

fantastic (1970s) *Literary Theory* Modern idea of a genre, retrospectively applied to a variety of earlier writings. Tzvetlan Todorov's book *The Fantastic* (1973 [1970]) crystallized this interest, which also relates to recent developments in PSYCHOANALYTIC CRITICISM.

Fantastic works are concerned with the intrusion of the supernatural, uncanny or mysterious into real life; a dream world is created which is disturbing rather than gratifying, because the fantasy concerns what is normally repressed. English GOTHIC is a good example, or the stories of Edgar Allan Poe (1809–49), or the bizarre worlds of Franz Kafka (1883–1924) and Jorge Luis Borges (1899–1986).

R Jackson, *Fantasy: The Literature of Subversion* (London, 1981)

RF

Faraday's laws of electrolysis (1834) *Chemistry/Physics* Two laws formulated by English chemist and physicist Michael Faraday (1791–1867), renowned for his pioneering experiments in electricity and magnetism and considered by some the greatest experimentalist who ever lived.

(1) The mass of a substance produced at an electrode during electrolysis is proportional to the quantity of electricity passed. The quantity of electricity is often expressed in terms of the Faraday. One Faraday is the charge carried by one mole of electrons or one mole of singly charged ions.

(2) The number of Faradays required to discharge one mole of an ion at an electrode equals the number of charges on the ion.

M Freemantle, *Chemistry in Action* (Basingstoke, 1987)

MF

Faraday's law of electromagnetic induction (1831–32) *Physics* Named after its discoverer, the English chemist and natural philosopher Michael Faraday (1791–1867), this is usually understood to mean that when the flux ϕ of magnetic induction through a circuit is changing, an e.m.f. is set up in the circuit, of magnitude e proportional to the rate of change of flux

$$e \propto \mathrm{d}\phi/\mathrm{d}t\,.$$

This is, in fact, the form in which the law was expressed in 1845 by German physicist Franz Ernst Neumann (1798–1895). Faraday gave no explicit reference to the magnitude of the induced e.m.f. but stated that an induced e.m.f. is produced: (1) in a rigid, stationary circuit, across which there is a changing magnetic flux; (2) in a rigid circuit moving across a steady magnetic field in such a way that the flux across it changes; (3) in part of a circuit which cuts any magnetic flux. *See* FARADAY-NEUMANN LAW and LENZ'S LAW.

J Thewlis, ed., *Encyclopaedic Dictionary of Physics* (New York, Oxford and London, 1962)

MS

Faraday–Neumann law (1845) *Physics* Franz Ernst Neumann (1798–1895) expressly assumed the proportionality of induced e.m.f. in a circuit with a changing magnetic flux linkage and the rate of change of that flux linkage. *See* FARADAY'S LAW OF ELECTROMAGNETIC INDUCTION and LENZ'S LAW.

J Thewlis, ed., *Encyclopaedic Dictionary of Physics* (New York, Oxford and London, 1962)

MS

Farka's lemma (1902) *Mathematics* An ALTERNATIVE THEOREM stating that exactly one of the following systems has a solution (where A is an $m \times n$ matrix and b is a point in n-dimensional Euclidean space):

(1) $A \leqslant 0$ (coordinatewise) and $\langle b, x\rangle > 0$;

(2) $yA = b$ and $y \geqslant 0$ (coordinatewise)

This result may be regarded as the foundation of many characterization results in OPTIMIZATION THEORY including the DUALITY THEORY OF LINEAR PROGRAMMING and the KARUSH-KUHN-TUCKER THEOREM.

AL

Fasbender theorem *Mathematics* Perhaps the first duality result in OPTIMIZATION THEORY (*compare with* the DUALITY THEORY OF LINEAR PROGRAMMING and the FENCHEL DUALITY THEOREM). It states that if the solution of FERMAT'S PROBLEM of finding a point minimizing the sum of the distances to the vertices of a given triangle is not a vertex then the minimum sum is equal to the maximum altitude of an equilateral triangle circumscribing the given triangle.

AL

fascism (20th century) *Politics* Named after Italian Fascist Party of Benito Mussolini (1883–1945).

Advocacy of deferential national unity under a leader, without the liberal or democratic forms of politics and government which are regarded as an obstacle to the implementation of the national will. Ordinary people, rather than being democratic citizens, are to be organized in corporations according to the contribution they make to national well-being. The Latin 'fasces', a bound bundle of rods, symbolized strength through unity.

Walter Laqueur, ed., *Fascism: A Reader's Guide; Analysis, Interpretations, Bibliography* (Harmondsworth, 1979)

RB

fatalism *Philosophy* The view – beloved of Greek oracles and their adherents – that the future, or part of it, will be what it will be, irrespective of our desires and actions. If we try to evade what is destined, our actions will always be frustrated and somehow turned so as to bring about the fated result.

Probably never held as a serious philosophical doctrine, FATALISM should be distinguished from DETERMINISM and NECESSITARIANISM.

ARL

Fatou's lemma (1906) *Mathematics* Named after its discoverer, Pierre Fatou (1878–1929).

Let $f_n(x)$ be a sequence of non-negative measurable functions defined on a measurable set E. Then the integral over E of the lower limit of $f_n(x)$ is less than or equal to the lower limit of the integral over E of $f_n(x)$. *See also* DOMINATED CONVERGENCE THEOREM and MONOTONE CONVERGENCE THEOREM.

JB

faunal succession, law of (early 19th century) *Geology* Formulated by the British geologist William Smith (1769–1839), this states that each geological layer contains its own particular assemblage of fossils which can be recognized in the same bed when traced over large distances. Successive layers contain different assemblages, reflecting both evolutionary and environmental changes.

DT

Fauvism (1905) *Art* From the French word *fauves*, meaning 'wild beasts', this name refers to a small group of painters who in Paris exhibited works notable for the bold and expressive use of pure colour. In this they were influenced by the arbitrary application of colour for emotional effect by Dutch painter Vincent van Gogh (1853–90).

The most prominent Fauves were Henri Matisse (1869–1954) and André Derain (1880–1954). Others included Albert Marquet (1875–1947), Henri Charles Manguin (1874–1949) and Charles Camoin (1879–65), Raoul Dufy (1877–1953), Othon Friesz (1874–1949), Georges Braque (1882–1963), Maurice Vlaminck 1876-1958), Jean Puy (1876–1960) and Kees van Dongen (1877–1968). Despite their enormous importance for later colourists and the Expressionist movement, by 1909 most of the above had developed individual styles away from Fauvism.

G Diehl, *Les Fauves* (Paris, 1943); J Elderfield, *The 'Wild Beasts': Fauvism and its Affinities* (London, 1976)

LL

Fechner's law *Psychology* Formulated by the German psychologist and philosopher Gustav Theodor Fechner (1801–87).

As the strength of a physical stimulus increases geometrically so will psychological experiences increase arithmetically. Thus, the intensity of a subjective sensation (for example, loudness) is proportional to the logarithm of the physical stimulus (sound intensity). The law is written: $S = k \log R$, where S is sensation, k is a constant, and R is stimulus. The law is closely related to

WEBER'S LAW and is sometimes termed the 'Weber-Fechner law'. It is a better description of the relationship between the physical stimulus and the physiological receptor response than of that between the physical stimulus and subjective experience. *See* POWER LAW.

G T Fechner, *Elements of Psychophysics*, D M Mowes and E G Boring, eds, vol. I (New York, 1966); (German edn, 1860)

NS

Fechner's paradox *Psychology* Formulated by the German psychologist and philosopher Gustav Theodor Fechner (1801–87).

The finding that after viewing an object binocularly, the same object increases in brightness when viewed again monocularly.

G T Fechner, *Elements of Psychophysics*, D M Mowes and E G Boring, eds, vol. I (New York, 1966); (German edn, 1860)

NS

Fechner–Weber law *Physics/Psychology See* FECHNER'S LAW and WEBER'S LAW.

federalism *Politics* Theory of divided legislative power.

Government and legislation are best divided between a federal STATE and its constituent states, departments, or areas. Such division of power is normally regulated by a written constitution, with a supreme or constitutional court to adjudicate in disputes between the overall federal states and the separate component states.

K Sawer, *Modern Federalism* (London, 1969)

RB

feedback deletion hypothesis (1950s) *Biology* Attributed to American scientist Van Rensselaer Potter (1911–), this is the theory that cancer results from the loss of a repression mechanism that restrains the synthesis of deoxyribonucleic acid (DNA) in normal cells.

This theory, a modified version of the DELETION HYPOTHESIS, is one of many concerning carcinogenesis. *See* CANCER, THEORIES OF and FEEDBACK DELETION HYPOTHESIS.

KH

feeling *Literary Theory See* FOUR KINDS OF MEANING.

Feigenbaum's constant *Mathematics* Named after American physicist Mitchell Jay Feigenbaum (1944–) who discovered the constant numerically. Oscar Lanford later supplied a computer-assisted proof.

A constant equal to approximately 4.6692016 which occurs in the iteration of many one-dimensional maps. Systems in which the constant appears include the logistic map and the Hénon map. *See also* FEIGENBAUM'S PERIOD-DOUBLING CASCADE.

I Peterson, *The Mathematical Tourist* (New York, 1988)

ML

Feigenbaum's period-doubling cascade *Mathematics* Named after its discoverer, American physicist Mitchell Jay Feigenbaum (1944–), this is a theory about the onset of chaos.

It hypothesizes that when the forces acting on a physical dynamical system are changed, a periodic orbit is often replaced by another one close to it which makes two turns before returning to its starting point. This action is repeated resulting in periodic orbits of periods 2, 8, 16, 32 ... times as long as the original. The places where the period doublings occur have successive ratios which are approximately equal to FEIGENBAUM'S CONSTANT. The Feigenbaum cascade has been observed in hydrodynamic turbulence experiments. *See also* CHAOS THEORY, MANNEVILLE-POMMEAU PHENOMENON and RUELLE-TAKENS SCENARIO.

I Peterson, *The Mathematical Tourist* (New York, 1988)

ML

Feit-Thompson theorem (1963) *Mathematics* Named after Walter Feit (1930–) and John G Thompson who proved a conjecture by William Burnside (1852–1927). The order of any finite simple group is either a prime number or an even number. *See also* FINITE SIMPLE GROUPS.

W Feit and J G Thompson, 'Solvability of Groups of Odd Order', *Pacific Journal of Mathematics*, 13 (1963), 775–1029.

JB

Fejer's theorem (1904) *Mathematics* Named after Hungarian analyst Lipót Fejer (1880–1959), this states that the arithmetic means of the partial sums (Fejer sums) of a FOURIER SERIES for a continuous periodic function on $[-\pi,\pi]$ converge uniformly to that function. If the function f is only integrable then the Fejer sums converge to f almost everywhere and converge at every point where f is continuous.

M Hazewinkel, ed., *Encyclopaedia of Mathematics* (Dordrecht, 1988)

ML

felicity condition *Linguistics See* SPEECH ACT THEORY.

female choice (1871) *Biology* First recognized by British naturalist and evolutionary theorist Charles Robert Darwin (1809–82), but elaborated by others in the 1960s, this is a theory of reproductive strategy holding that females choose the mate who offers the best genetic endowment to pass on to her offspring and, in some species, that mate who is also best able to feed and protect her and her young.

Females may also seek males who have privileged access to resources crucial to survival of the female, the male and the offspring; territorial songbirds who mate for life (or for many years) are examples. *Compare with* MALE-MALE COMPETITION. *See also* SEXUAL SELECTION.

D Dewsbury, *Comparative Animal Behaviour* (New York, 1978)

KH

feminism *Politics/Feminism* Analysis of exploitation of women, and proposals for alternatives.

The varieties of feminism are very great. All involve the assumption that, as a matter of observable fact, women are a category oppressed and exploited by men as a category. Both the accounts of this division, and the proposals for reform, revolution, or transformation differ greatly.

Juliet Mitchell and Ann Oakley, eds, *What Is Feminism?* (Oxford, 1986)

HR

feminist aesthetics (1970s) *Art* Belief that oppressive social conditions have blurred the idea of 'women's art' throughout the history of European culture. Recent feminist art aims to put forward a specifically female consciousness. Favoured media are non-traditional, such as ready mades, collages and performances.

G Ecker, *Feminist Aesthetics* (London, 1985)

LL

feminist criticism (1960s–) *Literary Theory/Feminism* Emergence and development linked to modern political FEMINISM, but citing earlier sources, for example the English novelist Virginia Woolf (1882–1941).

Feminist criticism has many concerns and orientations, for example the analysis of repressive images of women in male fiction; the position of women writers in relation to CREATIVITY, linguistic varieties, literary production (called by E Showalter GYNOCRITICS); the recuperation of women writers by publication in new editions. *See also* ANDRO-CENTRISM, CULTURAL FEMINISM and GAZE.

M Jacobus, ed., *Women Writing and Writing about Women* (London, 1979)

RF

feminist linguistics (1960s–) *Feminism/Linguistics* Originally expressed in terms of political feminism, recently more technically articulated through the contribution of linguists such as J Coates and D Tannen.

Studies of women's language, of language and sexism, and of male bias in linguistic theory. *See also* GENDER AND LANGUAGE.

D Tannen, *Feminism and Linguistic Theory* (London, 1985)

RF

feminist methodology (20th century) *Politics/Sociology/Feminism* The application of feminist theory to methods and concepts of sociological investigation.

Feminist research practice requires a critical stance towards existing methodology in the social sciences. While an attention to the responsibilities, rights and particular knowledge of those studied, and a recognition of gendered power relationships in the conduct and process of research may not be unique to feminist methodology; they are an essential component of it. All ways of knowing are political. The use of feminist methodology implies a commitment to the empowerment of women.

Helen Roberts, ed., *Doing Feminist Research*, 2nd edn (London, 1990)

HR

Fenchel duality theorem (1949) *Mathematics* One of the central duality results in convex OPTIMIZATION THEORY (*compare with* the DUALITY THEORY OF LINEAR PROGRAMMING). The *Fenchel conjugate* of a convex function is

$$f^*(y) = \sup \{\langle x,y\rangle - f(x)\}$$

Under a regularity condition the *primal value* $\inf\{f(x) + g(x)\}$ is equal to the *dual value* $\sup\{-f^*(y) - g^*(-y)\}$. Solutions of the dual problem can be identified with LAGRANGE MULTIPLIERS. Other applications include von Neumann's MINIMAX THEOREM.

AL

Fermat primes (17th century) *Mathematics* Named after French lawyer and amateur mathematician Pierre de Fermat (1601–65) who is credited with founding modern NUMBER THEORY, these are prime numbers of the form

$$2^{2^n} + 1 \quad (n = 1,2,\ldots)$$

Fermat conjectured that all numbers of this form are prime. This is true for n = 1, 2, 3, 4, but fails for n = 5, as was proved by Euler.

A Baker, *A Concise Introduction to the Theory of Numbers* (Cambridge, 1984)

MB

Fermat's principle of least time (1662) *Physics* Named after the French mathematician Pierre de Fermat (1601–65).

The principle states that the path taken by a ray of light (or other waves) passing between two given points by reflection or refraction through any collection of media is the one in which the ray takes the least time compared with all other possible paths. It is now more usually expressed as the principle of stationary time. *See also* HAMILTON'S PRINCIPLE OF LEAST ACTION.

Robert M Besançon, ed., *The Encyclopedia of Physics* (New York, 1985)

GD

Fermat's last theorem (1637) *Mathematics* Named after French lawyer and amateur mathematician Pierre de Fermat (1601–65), this is the celebrated *conjecture* in NUMBER THEORY that the equation $x^n + y^n = z^n$ has no non-trivial solutions when n is greater than two. (If $n = 2$, the solutions are Pythagorean triples.)

Fermat made the claim, in his copy of a translation of Diophantus that 'I have assuredly found an admirable proof of this, but the margin is too narrow to contain it.' Today, Fermat's claim is not given much credence, but the result is known to hold for almost all exponents (that is, the set of exponents on which it fails has zero density). While many approaches have been attempted, the problem remains open. *See also* ABC CONJECTURE, DIOPHANTINE EQUATION.

A Baker, *A Concise Introduction to the Theory of Numbers* (Cambridge, 1984)

MB

Fermat's little theorem (1640) *Mathematics* Named after French lawyer and amateur mathematician Pierre de Fermat (1601–65) who stated it, but without proof. The Swiss-born mathematician Leonhard Euler (1707–83) gave the first demonstration about a century later.

It is the result that for any integer n and any prime p, p divides $n^p - n$. Equivalently, n^p is congruent to n modulo p.

A Baker, *A Concise Introduction to the Theory of Numbers* (Cambridge, 1984)

MB

Fermat's (Steiner's) problem *Mathematics* A classical question of OPTIMIZATION THEORY, due to French lawyer and amateur mathematician Pierre de Fermat (1601–65).

Given any triangle in the plane we wish to find the *Torricelli point*, the point which minimizes the sum of the distances to the vertices. The problem can be solved via the FASBENDER THEOREM, which considers a dual optimization problem (*compare with* DUALITY THEORY OF LINEAR PROGRAMMING and FENCHEL DUALITY THEOREM). The problem generalizes to location problems (A Weber, 1909) which seek to locate central facilities.

AL

Fermi's ageing theory (20th century) *Physics* Named after its originator, the Italian nuclear physicist Enrico Fermi (1901–54), this approximate theory gives an equation that relates the spatial distribution of neutrons in a medium to the energy of the neutrons.

The fundamental assumption made in deriving the age equation is that the slowing-down process can be assumed to be continuous; that is, a neutron loses energy continuously and not in discrete amounts. Phenomena due to the finite size of the individual energy losses are ignored. The theory is inaccurate for light elements.

J Thewlis, ed., *Encyclopaedic Dictionary of Physics* (New York, Oxford and London, 1962)

MS

Fermi-Dirac statistical and distributive law (1926) *Physics* Named after the Italian-born American nuclear physicist Enrico Fermi (1901–54), winner of the 1938 Nobel prize for Physics, and the English mathematical physicist Paul Adrien Maurice Dirac (1902–84).

Fermi-Dirac statistics apply to particles (fermions) for which the Pauli exclusion principle is obeyed (for example, electrons). The Fermi-Dirac distribution law gives n, the average number of identical fermions in a state of energy E as,

$$n = \frac{1}{\exp(\alpha + E/kT) + 1},$$

where k is the Boltzmann constant, T is the thermodynamic temperature, and α is a quantity depending on temperature and the concentration of the particles. At high temperatures and low concentrations, the distribution law tends to the classical BOLTZMANN DISTRIBUTION LAW,

$$n = A \exp(-E/kT) .$$

See also BOSE-EINSTEIN STATISTICAL AND DISTRIBUTION LAW.

Robert M Besançon, ed., *The Encyclopedia of Physics* (New York, 1985)

GD

Fermi's paradox (1950) *Astronomy* Most arguments for the existence of extraterrestrial intelligence (ETI) in the universe rest on the principle of mediocrity. This asserts that, on a cosmic scale, there is nothing special about the Earth or human beings and, consequently, intelligence should be common in the universe.

During a conversation at the Los Alamos Laboratories in 1950, Italian physicist Enrico Fermi (1901–54) responded to a claim that ETIs exist by asking 'Then, where are they?' So far, there is no hard evidence to support the hypothesis of the existence of ETI.

S Mitton, ed., *The Cambridge Encyclopaedia of Astronomy* (London, 1973)

MS

Fermi's theory of beta decay (1934) *Physics* Named after its originator, the Italian nuclear physicist Enrico Fermi (1901–54), this theory gives the probability of the ejection of a beta particle (electron) with a given energy per unit time from a radioactive nucleus.

The theory is based on an analogy with photon emission and asserts that electrons and neutrinos are created in the act of beta decay. The neutrino helps the electron to carry away the energy of the decay.

J Thewlis, ed., *Encyclopaedic Dictionary of Physics* (New York, Oxford and London, 1962)

MS

Ferry-Porter law *Psychology* Formulated by the American, E S Ferry (1868–1956); and the Englishman, T C Porter (*fl.* 1890s–1900s).

The principle in visual perception that critical flicker frequency increases with the logarithm of the brightness of the stimulus. The relationship is independent of the wavelength of the stimulus.

R S Woodworth and H Schlosberg, *Experimental Psychology* (London, 1966)

NS

fetishism *Politics/Anthropology* The attribution of qualities, often religious or magical, to material objects. A concept most familiar by its adaptation within MARXISM as 'commodity fetishism'.

Roger Scruton, *A Dictionary of Political Thought* (London, 1982)

RB

feudalism (Middle Ages) *Politics/History* A medieval theory in which society was deemed to be held together by ties of loyalty and oblgiation between persons in a hierarchy. The essential currency was land and military service, the former held in return for the latter. Feudalism was a hierachy of lords and vassals, and it was possible to be both in relation to different persons. Those at the bottom of the hierachy (serfs) had an excess of duties over rights.

David Miller, ed., *The Blackwell Encyclopaedia of Political Thought* (Oxford, 1987)

RB

Fibonacci numbers *Mathematics* Named after Italian mathematician Leonardo Fibonacci (*c.*1170–1250) – also known as Leonardo of Pisa – who introduced the Arabic number system to Europe.

It is the sequence of integers

$$0, 1, 1, 2, 3, 5, 8, 13, 21, 34 \ldots$$

with the property that each number in the sequence is the sum of the previous two. These numbers arise in many applications in mathematics and in the life sciences.

MB

Fick's laws of diffusion *Physics See* DIFFUSION THEORY.

fiction (traditional; 20th century) *Literary Theory* From the Latin word *fingere*, meaning 'to fashion', 'to shape'. Particularly deployed in discussion of the NOVEL after the usage of the American novelist Henry James (1843–1916); for example his essay 'The Art of Fiction' (1884).

An imaginary world (early derogatory usage attaches to this meaning); a text, for example a novel or short story, conveying such a world; the activity of shaping such a world through the practice of writing. The last sense is very prominent in the later 20th century, with METAFICTION and the 'NOUVEAU ROMAN' drawing attention to the artifice of NARRATIVE (and arguably of all language, through which we constructively make sense of the world). 'Fiction' is now often used interchangeably with 'novel-and-short-story'.

F Kermode, *The Sense of an Ending* (London, 1987); P Waugh, *Metafiction* (London, 1984)

RF

fiddler's factor (probably 19th century) *Mathematics* Unpublished, but well-known among schoolchildren.

The answer required divided by the answer obtained. The factor can be used legitimately by proving that the expression for it is equal to 1. *See also* COOK'S CONSTANT.

fideism *Philosophy* The idea that religious faith stands apart from orthodox reason and can never be reconciled with it.

Such religious thinkers as St Augustine (354–430) have argued that reason itself plays a subsidiary role to faith; while others (including Danish philosopher Sören Aaby Kierkegaard (1813–55)) have maintained that acceptance of various aspects of religious belief requires actual denial of certain rational truths.

DP

'Fido'-Fido theories *Philosophy* Theories which explain some concept in terms of a direct relation to an object. The 'Fido'-Fido theory of meaning says that the meaning of a word is an object it stands for (the name 'Fido' means the dog Fido); the term is thus a nickname for a naming theory of meaning. The 'Fido'-Fido theory of belief is an extension of this idea, and says that to have a belief is to stand in a direct relation to a proposition which one is believing.

S Schiffer, 'The "Fido"-Fido Theory of Belief', *Philosophical Perspectives* (1987)

ARL

Fiedler's contingency model of leadership *Psychology* Proposed by the Austrian psychologist Fred Edward Fiedler (1922–).

This model emphasizes the importance of both the leader's personality and the situation in which that leader operates. A leader is the individual who is given the task of directing and co-ordinating task-relevant activities, or the one who carries the responsibility for performing these functions when there is no appointed leader. Fiedler relates the effectiveness of the leader to aspects of the group situation. He also predicts that the effectiveness of the leader will depend on both the characteristics of the leader and the favourableness of the situation. *See* CONTINGENCY THEORY OF LEADERSHIP, LEADERSHIP

THEORIES, PATH-GOAL THEORY and NORMATIVE DECISION THEORY OF LEADERSHIP.

F E Fiedler and J E Garcia, *New Approaches to Leadership* (New York, 1987)

NS

field *Linguistics See* REGISTER.

field, semantic *Linguistics See* SEMANTIC FIELD.

field theory *Biology See* REPULSION THEORY.

field theory *Mathematics* The branch of ALGEBRA which is concerned with the study of fields.

A field is a formal system consisting of a set, two distinguished elements of the set (usually denoted 0 and 1) and two binary operations on the set (usually referred to as addition and multiplication) such that the set is a commutative group under addition, the set excluding the element 0 is a commutative group under multiplication, and the multiplication distributes over the addition. For example, the rational numbers form a field, but the integers do not. *See also* GROUP THEORY, GALOIS THEORY, COMMUTATIVE LAW and DISTRIBUTIVE LAW.

I N Herstein, *Topics in Algebra* (Wiley, 1975)

MB

field theory (19th–20th centuries) *Physics* A theory of the way in which the electromagnetic, gravitational or nuclear field potentials account for the propagation of a field and the consequent transfer of energy, momentum and so forth.

Douglas M Considine, ed., *Van Nostrand's Scientific Encyclopedia* (New York, 1989)

GD

field theory *Psychology* Developed from Gestalt Psychology and closely linked with the work of the German-American psychologist Kurt Lewin (1890–1947).

This term is applied to any theory which focuses on the object under condsideration and the environment in which it is placed. Field theorists therefore focus on the interaction of the object and other objects in the environment. *See* GESTALT THEORY.

K Lewin, *Field Theory in Social Science* (New York, 1951)

NS

fight or flight theory *Psychology See* ENVIRONMENTAL STRESS THEORY.

figure of speech (traditional) *Literary Theory* Part of RHETORIC.

Elocutio in traditional rhetoric was a set of stylistic conventions spelled out in handbooks by lists of figures of speech: stylistic devices for ornamentation such as alliteration, chiasmus, and METAPHOR. *See* SCHEME and TROPE.

W Nash, *Rhetoric* (Oxford, 1989)

RF

figured-bass harmony (17th century) *Music* Figured bass parts first appeared at the start of the 17th century but were referred to by name only in the writings of later theorists. In Baroque music, this term refers to the notion that harmony is implicitly present above a given bass line, and that if a particular harmonic structure is intended it may be indicated by a layer of numbers placed below the bass notes, thus preventing ambiguity.

The system of numbers was much used by 17th- and 18th-century keyboard players, who were required to provide an accompaniment but were supplied only with the music for the bass line. The player chooses how much of the implied harmony is sounded and how the chord may be spaced.

P Williams, *Figured Bass Accompaniment* (Edinburgh, 1970)

MLD

filial regression, law of (1889) *Psychology* Formulated by the English scientist Sir Francis Galton (1822–1911).

A genetic principle stating that regression towards the mean of traits will occur between members of filial generations; for example, the offspring of two people will tend to be larger than the average of their parents' size, but smaller than the population average. This law only applies when large numbers of the population are considered, and where environmental factors are not exerting a powerful influence on development *in utero*.

F Galton, *Natural Inheritance* (New York, 1973)

NS

filter theory *Psychology* Frequently associated with the work of English psychologist D Broadbent (1926–).

This term relates to any of the theories of perception or memory which postulate that material is absorbed through a process of matching. Neural processes accept stimuli which match and do not allow those which don't. Filter theory by itself is too simple to account for the complexity of perception and memory, but is regarded as central to many of the more complicated theories.

L Kaufman, *Perception* (New York, 1979)

NS

filtration theory of kidney function *Biology* *See* LUDWIG'S THEORY.

finalism *Philosophy* View that there are final causes (*see* ARISTOTLE'S FOUR CAUSES) in nature; that is, that at least some things other than the products of deliberate human activity can be explained in terms of their end or purpose.

The idea that the world, or certain features of it, were amenable to such explanation goes back at least to a century before Aristotle (384–322 BC) and probably much further, but Aristotle was the first to dispense with a conscious designer, be it God, the gods, or nature. Questions which arise include: What is really meant by assigning a final cause? How does assigning one relate to assigning a function (as in 'The heart's function is to pump blood')? Can one assign both a final cause and an ordinary mechanical cause without danger of overdetermination – do the notions co-operate or compete? What counts as good evidence for a final cause?

C Taylor, *The Explanation of Behaviour* (1964)

ARL

fine tuning (1960s) *Economics* This term is attributed to American economist Walter Heller (1915–87), and is used to refer to a short-run interventionist approach to the economy using monetary and fiscal policies to control fluctuations in demand.

A popular policy of British governments from 1945 to 1970, it involved altering fiscal policy continually to stabilize NATIONAL INCOME as close to its full-employment potential as possible. However, studies of the fine tuning policies in Britain during the 1950s and 1960s have shown little, if any, success. The main problem with fine tuning the economy is the difficulty of accurately gauging the magnitude and timing of fluctuations in demand. *See* DEMAND THEORY.

W W Heller, *New Dimensions to Political Economy* (New York, 1967)

PH

finite simple groups (1982) *Mathematics* The problem of classification was posed in 1870 by Camille Jordan (1838–1922).

A group G is finite if it is a set consisting of a finite number of elements and it is simple if the only subgroups N such that $xN = Nx$ for every element x in G are G and the subgroup consisting of the identity element. An unedited proof of over 10,000 pages compiled during a century by many mathematicians demonstrates that each finite simple group is a member of one of 18 infinite families of groups or is one of 26 exceptional groups which are unrelated to the others. This result is fundamental to the classification of all finite groups. *See also* GROUP THEORY, FEIT-THOMPSON THEOREM and JORDAN-HOLDER THEOREM.

D Gorenstein, *Finite Simple Groups* (New York, 1982)

finite state grammar/language (1955) *Linguistics*· A version of the finite-state Markov process in mathematics. A model of language proposed by the American linguist Charles Francis Hockett (1916–) in his *Manual of Phonology* (1955).

A finite state grammar models a sentence as a succession of 'states' progressing left-to-right. Each word chosen determines what can follow it. Chomsky proved that such a grammar could generate an infinite number of sentences, but could not cope with other aspects of language such as discontinuous structure.

N Chomsky, *Syntactic Structures* (The Hague, 1957)

RF

finite strain theory *Physics* A theory of elasticity which is applicable to large strains such as the compressive strains occurring in the interior of the Earth.

MS

finitism *Philosophy/Mathematics* Usually regarded as a form of CONSTRUCTIVISM in the philosophy of mathematics, emphasizing that the construction in question must be possible in finitely many steps with finitely many elements.

The views that the construction must be possible in practice (not just in principle), and that a mathematical statement only gets its sense from the way it is proved, are sometimes called 'strict finitism'. *See also* FORMALISM.

P Benacerraf and H Putnam, eds, *Philosophy of Mathematics* (1964), part 4; see especially p. 505

ARL

firm, theory of the (1838) *Economics* This is analysis of the behaviour of companies that examines inputs, production methods, output and prices. The first elementary examination of companies was made by French economist Antoine Cournot (1801–77) and later modified by (among others) English political economist Alfred Marshall (1842–1924).

The traditional theory assumes that profit maximization is the goal of the firm. More recent analyses suggest that sales maximization or market share, combined with satisfactory profits, may be the main purpose of large industrial corporations. *See* THEORY OF THE GROWTH OF THE FIRM, SATISFICING, ORGANIZATION theory, THEORY OF BUREAUCRACY and AGENCY THEORY.

O E Williamson, *The Economics of Discretionary Behaviour: Managerial Objectives in a Theory of the Firm* (Englewood Cliffs, N.J., 1964)

PH

first isomorphism theorem *Mathematics* Also called the homomorphism theorem, this is a result of fundamental importance in GROUP THEORY.

If G and H are groups and if θ is a homomorphism from G to H (that is, a map which preserves the group structure) then the group $G/\ker\theta$ is isomorphic to the image of G under θ (that is, there is a bijective homomorphism from $G/\ker\theta$ to the image of G under θ). Here $\ker\theta$ is the kernel of the homomorphism θ (the elements of G which are mapped to the identity element of H) and $G/\ker\theta$ denotes the quotient group formed by identifying all elements of $\ker\theta$ with the identity element in G. Similar isomorphism theorems arise in other areas of abstract ALGEBRA. *See also* CATEGORY THEORY and ISOMORPHISM THEOREM.

I N Herstein, *Topics in Algebra* (Wiley, 1975)

MB

first law of thermodynamics (19th century) *Physics* The first law of thermodynamics is an extension of the result obtained by English physicist James Prescott Joule (1818–89), and asserts that when the state of an otherwise isolated closed system is changed by the performance of work, the amount of work needed depends only on the change effected and not on the means by which the work is done nor on the stages through which the system passes.

The extension is that, when there is a temperature difference between the system and its surroundings, the change ΔU in the internal energy of the system in a given change is equal to the work W done on the system plus the heat Q absorbed by the system; that is

$$\Delta U = Q + W.$$

This is the statement of the first law, which may be regarded as an extension of the principle of the conservation of energy to situations in which heat transfer takes place.

J Thewlis, ed., *Encyclopaedic Dictionary of Physics* (New York, Oxford and London, 1962)

MS

Fisher-Behrens problem (early 20th century) *Mathematics* Named after the British statistician R A Fisher (1910–62), this is the problem in STATISTICS of finding a test for the equality of the means of two normally distributed populations with different variances given a sample from each. American statistician Henry Scheffé (1907–77) designed an exact test which only uses some of the information contained in the sample and is only unique when the sample sizes are equal. Behrens and Welsh devised a test which uses all the information but is only approximate.

ML

Fisher's fundamental theorem (1930) *Biology* Named after British biologist Sir Ronald A Fisher (1890–1962), this still-controversial theory asserts that ADAPTATION occurs through selection of alleles which (on average and in combination with others) increase the fitness of the organism, leading to a continuous improvement in the population's fitness for survival. *Compare with* SHIFTING BALANCE THEORY.

G Kelsoe and D Schulze, *Evolution and Vertebrate Immunity* (Austin, Tex., 1987)

KH

fission hypothesis (1879) *Geology* Propounded by the British scientists George Darwin and Osmond Fisher, this theorized that the Moon may have been spun off the Earth during its early formation. The Pacific basin was thought to be the area from which the Moon was drawn.

If correct, this hypothesis would mean that the Moon had a composition similar to the surface rocks of the Earth. This is now known not to be the case.

DT

Fitts's law *Psychology* Formulated by the American psychologist Paul Morris Fitts (1912–65).

Motor movement time is related to precision of movement time and in turn to the distance that has to be covered. The law is written as:

$$mt = a + (b \log_2 2D/W)$$

where a and b are constants, D is distance moved and W is the width of the target. The width of the target is used to measure precision of movement.

P M Fitts, 'Engineering Psychology', *Annual Reviews of Psychology*, vol. IX (1958), 261–95

NS

Fitzgerald-Lorentz contraction (1892) *Physics* Named after the Irish physicist George Francis Fitzgerald (1851–1901), and the Dutch physicist Hendrik Antoon Lorentz (1853–1928), joint winner of the 1902 Nobel prize for Physics. Also known as the Lorentz-Fitzgerald contraction.

A theory put forward to account for the failure of the Michelson-Morley experiment to detect the earth's motion through the electromagnetic aether. It states that a material body moving through the ether with a velocity v, contracts by a factor of $\sqrt{(1 - v^2/c^2)}$ in the direction of motion, where c is the speed of light in a vacuum. The result is incorporated in Einstein's special theory of relativity (*see* RELATIVITY, THEORY OF SPECIAL.

GD

five ways (13th century) *Philosophy* The five methods employed by St Thomas Aquinas (*c.*1225–74) in his attempt to prove the existence of God by reference to natural facts about the universe, in *Summa Theologiae*.

The five ways were: argument from design; the cosmological argument; the degrees of perfection argument; the First Cause theory; the First Mover theory; and natural theology.

DP

five-fourths power law (19th century) *Physics* Also known as the cooling law of Dulong and Petit, after the French physicists Pierre Louis Dulong (1785–1838) and Alexis Thérèse Petit (1791–1820) who carried out an extensive study of the cooling of bodies.

When a normal laboratory calorimeter or similar body is cooling in still air (so that natural convection controls the heat loss), the rate of loss of heat to the surroundings is proportional to $(T - T_s)^{5/4}$, where T is the temperature of the body and T_s is the temperature of the surroundings.

J Thewlis, ed., *Encyclopaedic Dictionary of Physics* (New York, Oxford and London, 1962)

MS

fixed point theorems (1910) *Economics* First used in economic analysis by Hungarian-born mathematical economist John von Neumann (1903–57) in 1937, as an extension of the mapping techniques developed earlier by German mathematician Luitzen Egbertus Jan Brouwer; these theorems represent a macroeconomic approach setting out specific targets with regard to the aims of full employment, price stability, economic growth and balance of payments equilibrium.

L Brouwer, 'Uber eineindeutige, stetige Transformationen von Flachen in sich', *Mathematische Annalen*, vol. LXIX (1910), 176–80; J von Neumann, 'A Model of General Economic Equilibrium', *Review of Economic Studies*, 13, 33 (1945–46), 1–9

PH

fixed point theorem *Mathematics* Any theorem giving conditions for a mapping to have a fixed point; that is, a point which is mapped into itself by a transformation.

Proofs for the existence of fixed points and methods for finding them are important since the solution of every equation $f(x) = 0$ can be reduced to finding fixed points of the equations $x \pm f(x) = x$. *See* BROUWER'S THEOREM, TARSKI FIXED POINT THEOREM and KAKUTANI FIXED POINT THEOREM.

M Hazewinkel, ed., *Encyclopaedia of Mathematics* (Dordrecht, 1988)

ML

flat and round characters (1927) *Literary Theory* Distinction drawn by the English novelist Edward Morgan Forster (1879–1970).

In fiction, flat characters are 'constructed around a single idea or quality'; they are 'types' like medieval 'humours' (*see* FOUR HUMOURS). Round characters are more complex, presumably more like life, and so preferred.

E M Forster, *Aspects of the Novel* (Harmondsworth, 1963 [1927])

RF

Fleming's rules *Physics* Named after the British physicist Sir John Ambrose Fleming (1849–1945), who devised them.

Fleming's left-hand rule (the motor rule) deals with the force experienced by a current-carrying conductor in a magnetic field. If the thumb and first two fingers of the left hand are extended so as to be mutually perpendicular, and if the first finger points in the direction of the magnetic field and the second in the direction of the (conventional) current, then the thumb points in the direction of the mechanical force on the conductor carrying the current.

Fleming's right-hand rule (the dynamo rule) deals with the e.m.f. set up when a conductor moves in a magnetic field. Let the thumb and first two fingers of the right hand be extended so as to be mutually perpendicular. If the first finger points in the direction of the magnetic field and the thumb indicates the direction of motion of the conductor, then the second finger indicates the direction of the e.m.f. and the direction that the current will flow when the circuit is completed.

J Thewlis, ed., *Encyclopaedic Dictionary of Physics* (New York, Oxford and London, 1962)

MS

Floquet theory (1883) *Mathematics* Named after G Floquet who proved the main theorem.

The theory investigates where y is a vector function of the variable t over the complex field and P is a matrix of which the elements are periodic functions of t and y satisfies $y' = Py$. The main theorem states that the equation has a solution $y = Z\exp(Rt)$, where Z has the same periods as P and R is a constant matrix.

JB

florigen (1930s) *Biology* Coined by Russian botanist Mikhail Chailakhyan. From the Latin *flora*, meaning 'flower' and *genno*, 'to beget'. A hypothetical substance or stimulus that seems to be passed from a grafted stem that is about to flower into the host plant, causing the host plant to flower.

Cross-species experiments suggest that this substance must be the same in all flowering plants. The inability to isolate florigen, however, has led some botanists to doubt whether the substance exists.

F Salisbury, *Plant Physiology* (Belmont, Calif., 1992)

KH

fluid-mosaic model (1972) *Biology* Introduced by S J Singer and G Nicholson of the University of California at San Diego, this is the idea that the cell membrane is composed of a double layer of lipid molecules with proteins interspersed along its surface.

Some proteins lie towards the outside of the layer (the extrinsic or peripheral proteins), while others are positioned closer to the interior (the intrinsic or integral proteins). This model provides the operating theory for most experimental work in cell biology today. *Compare with* DANIELLI-DAVSON MODEL and UNIT MEMBRANE HYPOTHESIS.

P Sheeler, *Cell and Molecular Biology* (New York, 1987)

KH

'flush toilet' model *Biology See* LORENZ'S HYDRAULIC MODEL.

focalization (1972) *Stylistics* Concept developed by the French narrative theoretician Gérard Genette (1931–).

The angle or perspective from which a story or part of it is focussed, which may or may not be the narrator's point of view; for example, a character's internal perspective.

G Genette, *Narrative Discourse* (Oxford, 1980)

RF

foco theory (20th century) *Politics* Theory of REVOLUTION associated with the Cuban guerrilla leader and politician Che Guevarra (1928–67).

It is not necessary to wait until revolutionary conditions have developed, since a dedicated small group can ignite a revolution, thus creating both the uprising and the conditions which make it possible. Foco theory thus goes further than either DETONATOR THEORY or LENINISM in its reliance on revolutionary ÉLITISM.

Roger Scruton, *A Dictionary of Political Thought* (London, 1982)

RB

folk psychology *Philosophy* Term used in recent philosophy of mind for the view that beliefs, desires and so on exist and operate much as common sense assumes they do; that is, the operations of the mind can be adequately explained in terms of such notions, or (more strongly) that they cannot be adequately explained without them.

Strictly the term refers to such explanations themselves rather than to claims about their adequacy, but it is given point by recent claims that beliefs, desires and so on not only play no essential part in explaining mental phenomena, but do not even exist; that is, no phenomenon, and no set of features of our mental life, corresponds to our use of words like 'belief' and 'desire'. Strictly, therefore, such use is, on this view, incoherent, though harmless for ordinary purposes. 'Folk psychology' is an analogue in philosophy of mind of NAIVE REALISM in epistemology.

ARL

forbidden clone hypothesis *Biology* The idea that the normal tolerance for self-antigens in an animal is created by the death, during fetal life, of the clones responsible for the synthesis of the corresponding antibodies. If mutation causes such normally forbidden clones to re-emerge during adult life, they lead to the synthesis of auto-antibodies and produce autoimmunity.

KH

forced movement theory *Psychology* Proposed by the German-American biologist Jacques Loeb (1859–1924).

In animal psychology, a widely accepted theory of orientation in annelids which attempts to explain how animals acquire the capacity to orient themselves. The theory is based on an analysis of the bilateral symmetrical distribution of receptor, conductor and contractile tissues. A symmetrical stimulation causes muscular contraction and 'forced movement'.

N R F Maier and T C Schneirla, *Principles of Animal Psychology* (New York, 1964)

NS

forced saving (1912) *Economics* Given its current name by Austrian economist Ludwig von Mises (1881–1973), this term refers to an an involuntary reduction in consumption which arises when an economy is in full employment and when it has an excess supply of loans.

The excess depresses the market rate of interest and stimulates demand for investment finance which precipitates general inflation. As prices rise, those with fixed incomes consume less and savings are 'forced' out of them. This additional saving finances the extra investment. English economist John Maynard Keynes (1883–1946) recommended compulsory saving as part of a tough fiscal policy for financing World War II and avoiding inflation. Forced saving was first outlined in 1804 by English economists Henry Thornton (1760–1815) and Jeremy Bentham (1748–1832) under the term 'forced frugality'. *See* PARADOX OF

THRIFT, LOANABLE FUNDS THEORY OF THE RATE OF INTEREST, LIFE-CYCLE HYPOTHESIS and RELATIVE INCOME HYPOTHESIS.

L von Mises, *Theory of Money and Credit* (London, 1934); J M Keynes, *How to Pay for the War* (London, 1940)

PH

foregrounding (1930s) *Stylistics* In Czech: *actualisace*. Concept particularly associated with Jan Mukarovsky (1891–1975), a member of the PRAGUE SCHOOL.

Poetic language is distinguished by the highlighting of verbal devices such as metaphor, alliteration, poetic diction, and so on, against the background of the norms of ordinary language. *See also* ART AS DEVICE.

J Mukarovsky, 'Standard Language and Poetic Language' [1932], *A Prague School Reader on Esthetics, Literary Structure, and Style*, P L Garvin, trans. and ed. (Washington, DC, 1964), 17–30

RF

forensic linguistics (1967) *Linguistics* First used by the Swedish linguist Jan Svartvik in *The Evans Statements: A Case for Forensic Linguistics*.

A form of stylistic analysis using standard techniques of lexical, grammatical, conversational and textual description to evaluate claims about the authenticity, accuracy or method of creation of texts (for example police records of interviews with, or statements by, an accused) which have evidential importance in a Court of Law.

J P French and R M Coulthard and J Baldwin, eds, *Forensic Linguistics* (London, 1993)

RF

formal language *Linguistics See* RESTRICTED AND ELABORATED CODES.

formalism (20th century) *Art* A concept associated with MODERNISM, especially by the British critics Clive Bell (1881–1964) and Roger Fry (1886–1934), and the American Clement Greenberg.

In response to a developing interest in non-European art in the first decades of the 20th century, Fry and Bell attempted to formulate a semi-scientific system in which visual analysis of the formal characteristics of art took precedence over the artist's intentions and its social function. Formalist approaches were also applied to post-World War II criticism, particularly by Greenberg, in relation to ABSTRACT EXPRESSIONISM. By the 1960s its importance as a critical concept gave way to alternative methodologies.

AB

formalism *Philosophy* Any doctrine emphasizing form as against matter or content, especially in aesthetics, ethics, and philosophy of mathematics. (The term is not, however, normally used of a metaphysical preoccupation with Platonic or Aristotelian forms.)

In ethics formalism sees the value or rightness of an action in what kind of action it is (what formal description it satisfies) rather than in its consequences (*see also* DEONTOLOGY, a commoner term). For formalists in the philosophy of mathematics – followers of the German David Hilbert (1862–1943) – the objects of mathematics are mere marks (for example of ink or chalk) which are subjected to certain rules and arranged into formal systems; there is a certain connection with FINITISM, at least as a method.

S Körner, The *Philosophy of Mathematics* (1960), chs 4, 5

ARL

formalism, Russian (1915–30) *Literary Theory/Stylistics* Theories emanating from groups of linguists and critics in Moscow and St Petersburg, notably Viktor Shklovsky (1893–1984), Vladimir Propp (1895–1970), Roman Jakobson (1896–1982).

The formalists produced a set of theories which have been enormously influential in literary theory and stylistics. The common element is the aim to find the essence of poetic art in linguistic elements of the structure of the text. *See also* ART AS DEVICE, DE-AUTOMATIZATION, DOMINANT, 'FABULA' AND 'SJUZHET', POETIC PRINCIPLE, 'SKAZ'.

V Erlich, *Russian Formalism: History, Doctrine* (The Hague, 1965)

RF

foundationalism *Philosophy* Doctrine that knowledge must have foundations; that is, if we are to know anything at all there must be some things that we can know incorrigibly, so that it is impossible – or perhaps does not even make sense – for us to be mistaken.

The usual candidates for such knowledge have been facts about the immediate data of the senses or of introspection, such as what colours, tastes, and so on, are currently present to us, or what state of feeling or state of mind we are in. But probably the most famous foundationalist statement has been that of René Descartes (1596–1650): '*Cogito ergo sum*' ('I think therefore I am'). One objection to foundationalism is that it is hard to make knowledge incorrigible without constricting it so far that it ceases to be knowledge at all (one can misclassify even one's own immediate experiences).

R Descartes, *Meditations on the First Philosophy* (1641), especially Meditations 1 and 2

ARL

founder effect (1954) *Biology* An effect first observed by the German biologist Ernst Walter Mayr (1904–).

Small populations of a species arriving on an island tend to diverge genetically as a group from the mainland population. Mayr argued that this was due to the fact that such a population would contain only a fraction of the genetic variety of that of the mainland population, and, as a result of this, would be subject to more intense selection pressure. This model is generally not accepted, but other models have since been proposed to explain this genetic divergence.

E Mayr, 'Change in genetic environment and evolution', *Evolution as a Process*, J Huxley, A G Harvey and E B Ford, eds (London, 1954), 157–80

RA

founder-flush hypothesis (1982) *Biology* Proposed by Carson, this is a model for the evolution of new species holding that the FOUNDERS EFFECT leads to a population increase and reduced pressure from NATURAL SELECTION, causing additive genetic variance to be preserved, or even to increase.

This theory developed from Carson's studies of ADAPTIVE RADIATION in Hawaii. *Compare with* GENETIC REVOLUTION. *See also* SPECIATION, THEORY OF and SPECIES, THEORY OF.

KH

founders effect or principle (mid-20th century) *Biology* Also called the Sewall Wright effect after Sewall Wright (1889–1988), the American population theorist who first described it, this is the notion that genetically distinct populations in isolated places may be established by a very small contingent from another island, continent or from some other source.

These 'founders' or ancestors, being few in number, usually carry only a sampling of the gene pool for the entire species. Consequently, members of the resulting new population are likely to be more similar to each other than are members of the same species in the parental population. Many biologists consider the founders effect to be a special case of GENETIC DRIFT. *See also* EVOLUTION.

F Magill, *Magill's Survey of Science: Life Science Series* (Englewood Cliffs, NJ, 1991)

KH

four humours *Philosophy* Four bodily juices (blood, phlegm, black bile, yellow bile) whose balance or imbalance in the body was commonly regarded in Ancient Greek medicine as the source of health or disease. The doctrine was made influential for later thought by Galen (129–*c.* AD 199), though its origins are much earlier in the Hippocratic tradition (see the treatise *On the Nature of Man* in, for example, volume 4 of the Loeb edition of the *Hippocratic Treatises*).

The four humours are linked to the four elements (fire, air, water, earth, or hot, cold, moist, dry – stuffs and qualities were not properly distinguished until the time of Aristotle in the 4th century BC), which first appear as such in the cosmology of Empedocles (5th century BC).

G Sarton, *A History of Greek Science*, 1 (1953); see index

ARL

four kinds of meaning (1929) *Literary Theory* Theory developed by English theorist Ivor Armstrong Richards (1893–1979).

As a preliminary theoretical basis for literary study, Richards distinguishes: (1) 'sense', items which are spoken about; (2) 'feeling', speaker's feelings or attitudes toward those items; (3) 'tone', attitude to the listener; (4) 'intention', speaker's purpose, conscious or unconscious.

I A Richards, *Practical Criticism* (London, 1929)

RF

Fourier heat conduction equation (1822) *Physics* This equation is named after Jean Baptiste Joseph Fourier (1768–1830) who introduced it.

Conduction is described by an equation known as Fourier's rate equation. Consider a small plane of area A drawn normal to the x-direction in the medium where heat conduction is taking place. The rate of conduction of heat in the x-direction through the area is dQ/dt, given by

$$\frac{dQ}{dt} = kA\frac{\partial T}{\partial x}$$

where $\partial T/\partial x$ is the temperature gradient normal to the area. This equation defines a quantity k, known as the thermal conductivity of the material. When the heat flow takes place along a narrow unlagged bar with its axis in the x-direction the rate equation leads to

$$\frac{dT}{dt} = a\frac{d^2T}{dx^2} - \mu(T - T_s)$$

where a is the thermal diffusivity of the material and μ is $hs/A\rho c_p$. Here T is the temperature of the rod at a distance x from the hotter end, T_s is the temperature of the surroundings, h is the surface emissivity of the bar, ρ is the density of the material of the bar, s is the circumferential distance of the bar and c_p is its specific heat capacity at constant pressure. This equation is known as Fourier's equation.

J Thewlis, ed., *Encyclopaedic Dictionary of Physics* (New York, Oxford and London, 1962)

MS

Fourier principle (1807 and 1810) *Physics* Named after the French mathematician Jean Baptiste Joseph, Baron de Fourier (1768–1830).

Fourier formulated and solved the differential equations for heat flow in a solid body for which he introduced the expansion of functions in trigonometric series. He showed that any single valued periodic function can be expressed as a summation (Fourier series) of sine and cosine terms, with frequencies that are multiples of the frequency of the function. The analysis of a periodic function into its simple harmonic components is called a Fourier analysis.

Robert M Besançon, ed., *The Encyclopedia of Physics* (New York, 1985)

GD

Fourier series (19th century) *Mathematics* Named after the French analyst and physicist Jean Baptist Joseph Fourier (1768–1830).

An infinite trigonometric series of the form

$$\frac{1}{2}a_0 + \sum_{n=1}^{\infty}[a_n\cos(nx) + b_n\sin(nx)]$$

where the a_n and b_n are the Fourier coefficients. The series is used to represent or approximate a single-valued periodic function by choosing appropriate coefficients. Tests such as DIRICHLET'S TEST and DINI'S TEST give conditions guaranteeing convergence of the series to the function.

M Hazewinkel, ed., *Encyclopaedia of Mathematics* (Dordrecht, 1988)

ML

Fourier transform (19th century) *Mathematics* Named after the French analyst and physicist Jean Baptiste Joseph Fourier (1768–1830).

The integral transform

$$F(x) = \int_{-\infty}^{\infty} f(x)\exp(-\imath yx)\,dx$$

which sends a function f to another function F. Expansion of the exponential into real and imaginary parts gives the Fourier sine and cosine transforms. Under certain conditions the Fourier transform is invertible and its inverse is given by the Cauchy principle value of

$$f(x) = \frac{1}{2\pi}\int_{-\infty}^{\infty} F(y)\exp(\imath yx)\,dy.$$

The RIEMANN-LEBESGUE LEMMA gives conditions for the Fourier coefficients to converge to zero.

M Hazewinkel, ed., *Encyclopaedia of Mathematics* (Dordrecht, 1988)

ML

fractal (1970s) *Mathematics* The term was coined by the Polish-born American mathematician Benoit Mandelbrot (1924–) from the Latin adjective *fractus* meaning 'broken'.

A set of points which is too irregular to be described by traditional geometric language. It has a detailed structure which is visible at arbitrarily small scales, and many fractals have a degree of self-similarity; that is, they are made of parts which resemble the whole, and which may be approximate or statistical. Applications of fractals are rapidly expanding and to date include statistical physics, natural sciences and computer graphics. For example, fractals are used in image processing to compress data and to depict apparently chaotic objects in nature such as mountains or coastlines. Examples of fractal objects include the KOCH CURVE, MENGER SPONGE and SIERPINSKI GASKET. *See also* CHAOS THEORY.

R L Devaney, *An Introduction to Chaotic Dynamical Systems* (New York, 1989)

ML

frame (1975) *Linguistics* Proposed by the American philosopher Marvin Minsky (1927–) and deployed by some psycholinguists and text linguists.

An important type of SCHEMA: a cluster of common ideas about some domain of experience – 'bathrooms', 'terrorism', and so on; activated in processing texts and utterances to form textual worlds; *see* SCRIPT.

M Minsky, 'A Framework for Representing Knowledge', *The Psychology of Computer Vision*, P H Winston, ed. (New York, 1975), 211–77

RF

Franck-Condon principle (1925) *Chemistry/Physics* Named after the German-born American physicist James Franck (1882–1964), joint winner of the 1925 Nobel prize for Physics, and the American theoretical physicist Edward Uhler Condon (1902–74).

The principle that in any molecular system, electronic redistribution from one energy state to another is rapid enough that the nuclei of the atoms involved can be considered to be stationary during the redistribution. The principle accounts for the vibrational structure of electronic spectra.

P W Atkins, *Physical Chemistry* (Oxford, 1990)

GD

Fraser-Darling effect (1938) *Biology* Named after the British ecologist and animal geneticist Sir Frank Fraser-Darling (1903–1979).

The effect is seen with large breeding colonies of birds. Visual and auditory stimuli cause a synchronization and acceleration of the breeding cycle of the whole colony. This reduces the risk of any single young falling foul to predation, since the entire population of eggs is produced in one go.

RA

Fredholm alternative (20th century) *Mathematics* Named after Swedish mathematician and physicist Erik Ivar Fredholm (1866–1927), this is the theorem in FUNCTIONAL ANALYSIS that if A is a continuous linear operator with closed range then either A is surjective or the adjoint A^* has a non-trivial kernel. Hence, if A is a matrix, then either the inhomogeneous equation $Ax = b$ is always solvable or else the transpose homogeneous equation $A^*y = 0$ has a non-zero solution.

Béla Bollobás, *Linear Analysis* (Cambridge, 1990)

MB

free banking theory (19th century) *Economics* A liberal approach to banking in the USA adopted by the New York legislature in 1838 and later extended to other states through the *National Bank Act, 1863*.

The theory maintained that the banking system should be open to all and not dependent on legislative approval, provided certain minimum deposit requirements are met. Banks would have the power to issue their own banknotes if backed by bonds deposited with a state auditor. *See* CLASSICAL THEORY OF MONEY, CURRENCY PRINCIPLE and REAL BILLS DOCTRINE.

H E Krooss, ed., *Documentary History of Banking and Currency in the United States* (New York, 1980)

PH

free indirect discourse (1912) *Stylistics* 'Free indirect style' discussed by the French stylistician Charles Bally (1865–1947). In German: *Erlebte Rede*.

In fiction, representation of a character's speech or thought which is not introduced by a reporting tag such as 'He said/thought that . . .'; hence 'free'. Often marked by a combination of present (character) and past (narrator) tenses; the narrator's and the character's FOCALIZATIONS are co-present, in dialogue. Important concept since the 1980s due to the popularity of M M Bakhtin's DIALOGISM.

B McHale, 'Free Indirect Discourse', *Poetics and the Theory of Literature*, 3 (1978), 249–87

RF

free radical theory of ageing (1955) *Biology/Medicine* Originally proposed by American physician and biochemist Denham Harman (1916–), this proposes that ageing occurs from the gradual deterioration of body cells and tissues caused by the damaging reactions of free radical compounds.

In mammals, oxygen (O_2) is the main source of these highly reactive compounds, which carry unpaired electrons called free radicals. Evidence for this controversial theory comes from studies of the effects of ionizing radiation on living systems; life span experiments in which diet is modified to alter free radical reaction levels; certain studies of evolution; and studies implicating free radicals as causative agents in disease. *See also* AGEING, THEORIES OF.

W Pryor, *Free Radicals in Biology* (London, 1984)

KH

free variation *Linguistics See* DISTRIBUTION and PHONEME.

Freese's theory of mutagenesis *Biology See* MUTAGENESIS, THEORY OF.

Frenkel-Andrade theory of liquid viscosity (1934) *Physics* Named after Russian physicist Yakov Ilyich Frenkel (1894–1954) and the English physicist Edward Neville da Costa Andrade (1887–1971) who, independently, established the model. This theory assumed the view that liquids are, in many respects, like broken-up solids and that some properties of solids survive into the liquid phase.

In particular, the theory assumed that a molecule in the liquid has a normal frequency of vibration about an equilibrium position, though this position is liable to move. The amplitude of vibration is taken to be larger than that in the solid phase and sufficient for molecules in adjacent layers of liquid to come into contact. When these layers have a non-zero relative speed, as they do in laminar viscous flow, they are able to share momenta, and this is taken to be the major contribution to the viscosity. As the temperature changes, the chief effect is on the probability that the contact between adjacent layers will be effective in momentum sharing. There is also a change in the molecular spacing and in the frequency of vibration. The simplest consideration of this theory suggests that the viscosity should depend on temperature T according to the equation

$$\eta = A \exp (c/T)$$

where A and c are constants for a particular liquid.

J Thewlis, ed., *Encyclopaedic Dictionary of Physics* (New York, Oxford and London, 1962)

MS

frequency, law of (1886) *Psychology* Formulated by the Scottish metaphysician Thomas Brown (1778–1820).

The more often an organism repeats the same response to a stimulus the stronger the response becomes and the smaller is the likelihood of that response diminishing. Also referred to as the law of repetition.

O M Mowrer, 'On the Dual Nature of Learning: A Re-interpretation of "Conditioning" and "Problem Solving"', *Harvard Eductional Review*, vol. XVII (1947) 102–48

NS

frequency theory of hearing *Psychology* Proposed by the Briton, W Rutherford (1839–99).

Also known as the telephone theory. Rutherford proposed that the action of a telephone diaphragm is representative of the action of the basilar membrane of the ear. The theory is no longer accepted.

W Rutherford, 'The Sense of Hearing', *Journal of Anatomy and Physiology*, vol. XXI (1886), 166–68

NS

frequency theory of probability *Mathematics/Philosophy* Theory due to German philosopher and mathematician Richard von Mises (1883–1954) in *Probability, Statistics and Truth* (1928, 2nd edition translated 1939). Here he defined the probability of something in terms of the relative frequency of its occurrence on occasions when it might occur.

Von Mises limits his definition to cases where we have a collective, or potentially infinite sequence of cases satisfying certain conditions of randomness (for example throws of a die). Since any frequency in a finite run is compatible with almost any frequency in a sufficiently long run, we must define the probability (for example, of a six on our die) as the limiting frequency of sixes as the number of throws extends indefinitely, assuming that there is such a limit.

Objections to the theory include the narrrowness of its scope (for example, it has no strict application to the probability of single events, still less to that of theories and so on); and the difficulties of ensuring, first, that the randomness conditions are satisfied; and second, that there *is* a limiting frequency. With regard to this latter objection, we seem in danger of being reduced to saying that *probably* there is a limiting frequency, with this 'probably' being unexplained IMPOSSIBILITY OF A GAMBLING SYSTEM.

H E Kyburg, *Probability and Inductive Logic* (1970), ch. 4

ARL

Fresnel-Arago laws (1819) *Physics* The conditions under which two beams of plane-polarized light produce interference fringes were determined by the French scientists Augustin Fresnel (1788–1827) and Dominique François Jean Arago (1786–1853), and are contained in the laws named after them.

(1) Two beams of light plane-polarized in mutually perpendicular planes do not produce interference fringes under any conditions. (2) Two beams of light polarized in the same plane interfere under the same conditions as two similar beams of unpolarized light, provided that they are derived from a single beam of light. (3) Two beams of plane-polarized light derived from perpendicular components of unpolarized light and afterwards rotated into the same plane do not produce interference fringes under any conditions.

J Thewlis, ed., *Encyclopaedic Dictionary of Physics* (New York, Oxford and London, 1962)

MS

Freudian aesthetics (20th century) *Art* Psychoanalytical methods and theories of the unconscious formulated by Sigmund Freud (1856–1939) were fundamental to the development of aesthetic theory in the early 20th century.

Freud's principle contribution was to highlight the importance of the unconscious in the production and appreciation of art; to relate daydreams and dreams to art and creativity; and to emphasize the correspondences between art and neurosis. The French psychoanalyst Jacques Lacan (1901–81) reinterpreted and developed these ideas. *See* FREUDIAN THEORY, DREAM THEORY and UNCONSCIOUS MEMORY.

J J Spector, *The Aesthetics of Freud: A Study in Psychoanalysis and Art* (London, 1973)

AB

Freudian slip *Psychology See* PARAPRAXIS.

Freudian theory *Psychology* Any theory or perspective which draws heavily on the work of the Austrian founder of psychoanalysis Sigmund Freud (1856–1939).

The theory is based on a dynamic model of psychic structures (id, ego, superego); the schema of psychic development of the person (oral, anal and genital stages); and a schema of psychopathology. While the theory is frequently criticized for presenting untestable postulates, it has provided a rich corpus of ideas which has influenced many parts of the discipline.

E Jones, *The Life and Works of Sigmund Freud* (New York, 1953)

NS

Freytag's pyramid (1863) *Literary Theory* Schematic account of the structure of plots in drama, proposed by the German novelist, dramatist and critic Gustav Freytag (1816–95).

Freytag analyzed the plot of a five-act play as a pyramidal, ascending-descending structure with the following sequence of elements: Introduction, Inciting moment, Rising action, Falling action, Catastrophe.

G Freytag, *Freytag's Technique of the Drama* [1863], E J MacEwan, trans. (New York, 1968)

RF

Friedel's law (1913) *Physics* Named after its originator, the French physicist Georges Friedel (1865–1933), this law applies to the diffraction of X-rays and neutrons by crystals.

The intensities of reflection for X-rays and neutrons from opposite sides of the same crystal planes are the same. Therefore, certain asymmetries, such as polarization and the presence or absence of a centre of inversion, cannot be detected. The law is not valid where anomalous dispersion occurs.

J Thewlis, ed., *Encyclopaedic Dictionary of Physics* (New York, Oxford and London, 1962)

MS

Friedmann universes (1922-24) *Physics* Named after the Russian physicist Alexander Friedmann (1888–1925).

Friedmann universes are cosmological models that combine de Sitter's notion of expansion with Einstein's notion of non-emptiness. The field equations of such models ignore any contributions from electromagnetic radiation, and they suppose that the matter in the universe can be approximated by dust. All Friedmann models have an epoch in the past which is referred to as the big bang epoch, and implies a breakdown of the concept of space-time geometry, itself an inevitable feature of Einstein's general relativity. *See also* EINSTEIN-DE SITTER UNIVERSE and STEADY-STATE HYPOTHESIS.

Jayant V Narlikar, *Introduction to Cosmology* (Cambridge, 1993)

GD

Fries's rule *Chemistry* A benzenoid is an organic compound, such as phenol, derived from or containing the benzene ring structure. A benzenoid ring contains three double bonds. Its electronic configuration is energetically favoured and thus relatively stable. Fries's rule states that the most stable form of a hydrocarbon containing a number of rings is that in which the maximum number of rings has the benzenoid structure.

MF

Fritz John conditions theorem (1948) *Mathematics* A weaker precursor in OPTIMIZATION THEORY of the KARUSH-KUHN-TUCKER THEOREM, asserting the existence of LAGRANGE MULTIPLIERS for constrained optimization problems. No regularity condition is required, but in consequence an additional non-negative multiplier λ_0 must be attached to the function to be minimized, $f(x)$. Thus, if λ_0 vanishes the resulting necessary optimality conditions are less useful.

AL

Frobenius theorem (1874) *Mathematics* Named after its discoverer, Georg Frobenius (1849–1917).

A total DIFFERENTIAL EQUATION $dy = H(y,z)dz$, where y and z are vector functions and H is a continuous matrix, is completely integrable if and only if there exists a continuous non-singular matrix which satisfies certain conditions concerning the exterior product.

P Hartman, *Ordinary Differential Equations* (New York, 1964)

JB

Fromm's theory of personality *Psychology* Originated by the German-American psychoanalyst Erich Fromm (1900–80).

This is an attempt to draw together the views of Freud and Marx. Fromm described his theory as 'dialectical humanism', focusing on human beings' unceasing struggle for dignity and freedom in the context of their need for connectedness with each other. The theory continues to attract interest, especially from therapists concerned with understanding the importance of contemporaneous factors in personal development. *See* CONTEMPORANEITY PRINCIPLE.

E Fromm, *To Have or to Be?* (New York, 1979)

NS

frontier molecular orbital theory *Chemistry* Frontier molecular orbitals are the lowest unoccupied molecular orbitals. The theory considers the relative energy and symmetry of these orbitals and the highest occupied molecular orbitals in order to describe bonding and reactions between chemical entities. *See also* MOLECULAR ORBITAL THEORY

MF

frozen accident theory *Biology* The theory that the genetic code evolved by chance until the current codon assignments (the triplet of nucleotides) were set. Thereafter, the code was unable to evolve further because it provided the organism with selective advantages that would have been lost or reduced if the code were changed.

KH

frustration aggression hypothesis *Psychology* Proposed by the Austrian founder of psychoanalysis Sigmund Freud (1856–1939).

This postulates that aggressive behaviour is a sign of some previous frustration and that frustration will lead to either covert or overt aggressive behaviour. The theory draws on psychoanalytic explanations of aggression and physiological theories of emotion. It does not adequately account for the range of aggressive behaviour experienced or observed.

R Ardens, *The Hunting Hypothesis* (New York, 1976)

NS

frustration fixation hypothesis *Psychology* Proposed by the American psychologist Clark Leonard Hull (1884–1952).

The experimental finding that animals may persist in performing inadequate and non-adaptive responses when continually frustrated or placed in a situation where they are confronted with a choice of two unpleasant outcomes. The behaviour represents an extreme stress reaction.

N R F Maier and T C Schneirla, *Principles of Animal Psychology* (New York, 1964)

NS

f-sum rule *Chemistry* Also known as the Thomas-Reiche-Kuhn sum rule, this states that the sum of the oscillator strengths (f values) of absorption transitions of an atom in a given state, less the sum of the oscillator strengths of the emission transitions of the atom in the same state, is equal to the number of electrons taking part in these transitions.

MF

Fubini's theorem (early 20th century) *Mathematics* Named after Italian analyst, algebraist and differential geometer Guido Fubini (1879–1943), this element of MEASURE THEORY gives conditions for the order of integration to be reversed for an iterated integral.

If $(X, \mathcal{M}, \mu)$ and $(Y, \mathcal{N}, \nu)$ are sigma finite measure spaces and f is an integrable functions with respect to $\mu \times \nu$, then

$$\int f \, d(\mu \times \nu) = \int \left[\int f(x,y) \, d\nu(y) \right] d\mu(x)$$
$$= \int \left[\int f(x,y) \, d\mu(x) \right] d\nu(y).$$

This is an extension of TONELLI'S THEOREM.

G B Folland, *Real Analysis* (New York, 1984)

ML

fugue (17th century) *Music* The term *fuga* was first used from the 14th century onwards to denote a composition with imitative entries. 'Fugue' in the strict sense dates from the 17th century and is best exemplified in *The Art of Fugue*, a series of didactic compositions by the German composer Johann Sebastian Bach (1685–1750).

It is a formal concept and compositional procedure in which melodic lines ('voices') enter consecutively and imitatively, according to strict laws. When all the voices have entered and combined to form a complex contrapuntal texture the 'exposition' is said to be over; the music may then continue in a freer style, though often material from the exposition is worked into the ensuing music.

R Bullivant, *Fugue* (London, 1971)

MLD

Fullerton-Cattell law *Psychology* Formulated by the American psychologist G S Fullerton (1859–1925) and the Anglo-American psychologist Raymond Bernard Cattell (1905–).

Proposed as an updated version of WEBER'S LAW, this states that error of observation increases with the square root of the stimulus intensity. The law places emphasis on errors of observation being more psychological than introspective (in contrast to Weber's findings) in the case of determining just noticeable differences in stimulus change.

R B Cattell, *Abilities: Their Structures, Growth and Action* (Boston, 1971)

NS

function principle *Psychology See* FACULTY THEORY.

functional analysis *Mathematics* The branch of mathematics, pioneered in the early 20th century by D Hilbert, S Banach, J von Neumann and others, which studies functions as elements of large vector spaces (in particular *Banach spaces*) by analogy with calculus and geometry in *n*-dimensional Euclidean space.

Typical directions include the relationships between the space X and the *dual space* X^* of continuous linear functionals on X, and operators between spaces. Standard tools include geometric-analytic results like the SUPPORT THEOREM, the BISHOP-PHELPS THEOREM, the EBERLEIN-SMULYAN THEOREM and the JAMES THEOREM. The functional analytic framework is crucial in many areas of abstract OPTIMIZATION THEORY like the CALCULUS OF VARIATIONS and OPTIMAL CONTROL.

AL

functional analysis (1950s) *Music* Proposed by the Austrian-born writer on music Hans Keller (1919–1985), this theory postulates that a piece of music embodies a melodic motif or 'basic idea' which unifies its structure by functioning on two levels: the foreground presents variations of the idea while at a background level the idea unfolds over time as a latent entity.

Keller's views are controversial but important as they are argued on a purely musical level: his analyses comprise a performance of excerpts from the piece in question, interspersed with specially composed 'analytic' passages revealing the thematic links.

H Keller, 'Functional Analysis of Mozart's G Minor Quintet', *Music Analysis*, vol. IV (1985), 73–94

MLD

functional grammar (1960s–) *Linguistics* A linguistic theory developed by the British linguist Michael Alexander Kirkwood Halliday (1925–); the designation is also used by other schools. The basis of Halliday's model is the FUNCTIONALIST LINGUISTICS of the PRAGUE SCHOOL.

The structure of a language, variety or text is shaped by the social and personal needs it is called on to serve. Halliday proposes that each utterance simultaneously performs three 'metafunctions': the 'ideational' (representation of ideas and experience, for example by choices in the TRANSITIVITY system); 'interpersonal' (speaker/hearer participation); 'textual' (*see also* COHESION). Functional factors determine choices from systems networks (*see* SYSTEMIC GRAMMAR). *See also* REGISTER.

M A K Halliday, *An Introduction to Functional Grammar* (London, 1985)

RF

functional sentence perspective (1929) *Linguistics* Initiated by the PRAGUE SCHOOL founder V Mathesius (1882–1945) and maintained after World War II by F Danes and J Firbas.

The ordering of information in a sentence in relation to the communicative context and purpose of the discourse. *See also* COMMUNICATIVE DYNAMISM, and THEME AND RHEME.

J Firbas, 'On the Dynamics of Written Communication in the Light of the Theory of Functional Sentence Perspective', *Studying Writing: Linguistic Approaches*, C R Cooper and S Greenbaum, eds (Beverly Hills, 1986), 40–71

RF

functionalism *Philosophy* Any theory analyzing something in terms of its function; that is, any theory claiming that the best, or only, way of defining something is in terms of what it does or the role it plays in the ongoing course of events.

Functionalism tends to define things in terms of their causes and effects, and, in particular, a functionalist in philosophy of mind defines mental states and properties in terms of their causes and their effects as seen in behaviour. Two such states or properties will be the same if they have the same causes and effects.

An objection alleged against such functionalism is that it cannot account for the 'inner' nature of conscious experience, for it seems possible in principle that two people might be subject to the same stimuli and exhibit the same behaviour while having different inner experiences; or one of them might have none at all (that is, be a zombie).

N Malcolm, '"Functionalism" in Philosophy of Psychology', *Proceedings of the Aristotelian Society* (1979–80)

ARL

functionalism (20th century) *Politics/Sociology* Theory of relation of parts to social whole.

Society is a system of interrelated institutions and processes, which are to be understood in terms of the function they perform for the system as a whole. These functions are not necessarily intended, and may even be contrary to the expressed intentions of those concerned. *See also* STRUCTURAL FUNCTIONALISM and SYSTEMS THEORY.

Geoffrey Roberts and Alistair Edwards, *A New Dictionary of Political Analysis* (London, 1991)

RB

functionalist linguistics (1930s, 1970s) *Linguistics* Main concepts come from the PRAGUE SCHOOL, the founder of which was V Mathesius (1882–1945).

Grammatical properties of texts are shaped by pragmatic, communicative and social purposes of communication. *See* FUNCTIONAL GRAMMAR, FUNCTIONAL SENTENCE PERSPECTIVE.

R Dirven and V Fried, eds, *Functionalism in Linguistics* (Amsterdam, 1987)

RF

functionalist theory (1920s) *Anthropology* Developed by the Polish-born anthropologist Bronislaw Malinowski (1884–1942) and by the British anthropologist A R Radcliffe-Brown (1881–1955).

The approach which emphasizes relations between elements of social structure within a given society, rather than the historical development of those elements. It was the established perspective of British anthropology from the 1920s to the 1950s. Functionalist theory is not currently fashionable, but it is inherent in the fieldwork methods of all social anthropologists. *See also* EVOLUTIONISM.

A Kuper, *Anthropology and Anthropologists: The Modern British School* (London, 1983)

AB

fundamental art (painting) (*c.*1960) *Art* E de Wilde coined this phrase to describe abstract paintings by a group of American and European artists. Characterized by some critics as post-CONCEPTUAL ART, its concentration on procedure and the physical nature of materials brings it close to MINIMALISM; emphasizing the means rather than the result.

Fundamental Painting, exh. cat. (Amsterdam, 1975)

AB

fundamental attribution error *Psychology* Term coined by the American psychologist Harold H Kelley (1921–).

A judgmental error whereby the behaviour of others is often explained in terms of dispositional qualities (traits, motives, needs and so on) rather than situational or circumstantial variables. The error lies in the under-emphasis of situational factors and has been observed in numerous experimental and naturalistic contexts. *See* ACTOR-OBSERVER BIAS and SELF-SERVING BIAS.

H H Kelley and J L Michela, 'Attribution Theory and Research', *Annual Review of Psychology*, vol. XXXI (1980), 457–501

NS

fundamental disequilibrium (1945–) *Economics* A term first used by the International Monetary Fund to describe a situation in which a persistent discrepancy exists between the official exchange rate of a currency and its actual purchasing power.

When national inflation rates vary, official rates of exchange will no longer mirror the value of a currency and an adverse balance of payments will develop until the exchange rates are altered. If the situation persists, it is considered fundamental rather than temporary. *See* INTERNAL AND EXTERNAL BALANCE.

J K Horsefield, *The International Monetary Fund 1945–65: Twenty Years of International Cooperation* (Washington, DC, 1969)

PH

fundamental theorem of algebra (1799) *Mathematics* First proved in the doctoral thesis of German mathematician and astronomer Carl Friedrich Gauss (1777–1855), who is widely regarded as one of the most prolific and influential mathematicians of all time.

It is the result that every polynomial whose coefficients are complex numbers has at least one complex root. Consequently, any such polynomial can be expressed as a product of linear polynomials over the field of complex numbers. *See also* FACTOR THEOREM and LIOUVILLE'S THEOREM.

T W Hungerford, *Algebra* (Springer, 1974)

MB

fundamental theorem of arithmetic *Mathematics* The theorem that every integer greater than 1 has a factorization as a product of prime numbers which is unique apart from the order of the factors.

A Baker, *A Concise Introduction to the Theory of Numbers* (Cambridge, 1984)

MB

fundamental theorem of calculus (18th century) *Mathematics* Proved by the German mathematician and philosopher Gottfried Wilhelm von Leibniz (1646–1716) while reading a paper of Pascal. It was also known to Newton in geometric form in 1676.

The theorem which shows the inverse relationship between an integral and a derivative. If f is continuous on an interval $[a,b]$ and if F is an indefinite integral of f in $[a,b]$ then

$$\int_a^b f(x)\,dx = F(b) - F(a).$$

Conversely, if f is continuous on an open interval I and F is defined by

$$F(x) = \int_a^x f(t)\,dt$$

then at each point x in the interval I,

$$I,\ F'(x) = f(x)\ .$$

H Anton, *Calculus with Analytic Geometry* (New York, 1980)

ML

fundamental theorem of natural selection *Biology See* NATURAL SELECTION.

fundamental theorem of projective geometry (19th century) *Mathematics* Essential for the methods of proof in projective geometry as formulated by Jean-Victor Poncelet (1788–1867).

A projectivity between the points of two lines is a function which uniquely associates each point on one line with one on the other. The theorem asserts that a projectivity is uniquely determined by three pairs of corresponding points.

H S M Coxeter, *Introduction to Geometry* (New York, 1961)

JB

fundamental theorem of space curves (1851) *Mathematics* Also known as the Serret-Frenet formula because it was proved by Joseph Alfred Serret (1819–85) and Jean-Frederic Frenet (1816–1900).

Let **u**, **v**, **w** be unit vectors in the directions of the tangent, principal normal and binormal at a point P on a curve in three-dimensional space, let k be the curvature at P, let t be the torsion at P and let **Dx** represent the vector **x** differentiated with respect to the arc length on the curve. Then the following equations hold:

$$\begin{aligned} \mathbf{Du} &= k\mathbf{v} \\ \mathbf{Dv} &= t\mathbf{w} - k\mathbf{u} \\ \mathbf{Dw} &= -t\mathbf{v}\ . \end{aligned}$$

H S M Coxeter, 'Differential Geometry of Curves', *Introduction to Geometry* (New York, 1961)

JB

Futurism (1909) *Art/Literary Theory* Avant-garde movement founded by the Italian poet and novelist Emilio Filippo Tommaso Marinetti (1876–1944). Futurism was inspired by modernity, speed, the machine, the sights and sounds of the 20th century; it was against sentimentalism and history.

In art, the first manifesto was followed by the *Manifesto of Futurist Painting* (1910) and a *Technical Manifesto* (1910), where those who signed (Boccioni, Balla, Carra and Severini) declared their wish to depict the figure and machines in motion. VORTICISM and RAYONISM bear some similarities to this movement.

In literature, the strongest influence was on Russian poetry of the 1910s, and the best practitioner was V Klebnikov (1885–1922). Futurism produced texts of fragmentary and infinitive syntax, and favoured sound effects and onomatopoeia. *See* CONSTRUCTIVISM.

F T Marinetti, 'Futurist Manifesto', *Le Figaro* (20 February 1909, Paris); M W Martin, *Futurist Art and Theory 1909–1915* (1968); J J White, *Literary Futurism: Aspects of the First Avant Garde* (Oxford, 1990)

AB, RF

fuzzy grammar (early 1970s) *Linguistics* Proposal by the American linguist J R Ross in the context of GENERATIVE SEMANTICS.

GRAMMATICALITY is regarded – as against Chomsky's 'either/or' criterion – as existing, and recognized by speakers, to various degrees. Fuzzy grammar offers rules that reflect this variability.

J R Ross, 'Nouniness', *Three Dimensions of Linguistic Theory*, O Fujimara, ed. (Tokyo, 1973), 137–258

RF

fuzzy set theory (1965) *Mathematics* A form of informal SET THEORY in which belonging is replaced by an index of membership between 0 and 1. Intended for practical decisions, the theory is little used in mathematics but has some currency in operations research.

L Zadeh, 'Fuzzy Sets', *Information and Control*, 8 (1965), 338–53

JB

G

Gaia hypothesis (1970s) *Biology/Geology* Proposed by British scientists James Ephraim Lovelock (1919–), an organic chemist, and Lynn Margulis (1938–), a molecular biologist. Named after Gaea (*gaa* in Greek), the goddess Mother Earth. The theory that the Earth is a living organism with a self-regulating mechanism that is yet undefined.

All animals, plants and human activities are believed to contribute to the system, which has checks and balances to ensure the continuance of life. The stability of atmospheric components over many eons is cited as evidence for this controversial theory.

J Lovelock, *The Ages of Gaia: A Biography of Our Living Earth* (Oxford, 1988)

KH

gain-loss theory of attraction *Psychology* Proposed by the American psychologist Elliot Aronson (1932–).

People are attracted to those who will provide them with the greater gain and not to those who will provide them with the greater loss. The theory is founded on an analysis of inter-relationships in terms of economists' concepts. While the theory can successfully predict the outcome of some simple relations, it is regarded as inadequate for more complex relationships. *See* LAW OF ATTRACTION.

E Aronson and D Linder, 'Gain and Loss of Self Esteem as Determinants of Interpersonal Attractiveness', *Journal of Experimental and Social Psychology*, vol. I (1965), 156–71

NS

galactic rotation theory (1913) *Astronomy* The galactic system has the appearance of a wheel with curved spokes. French mathematician Henri Poincaré (1854–1912) suggested that such a system could only be stable if it were in a state of rotation.

A rough calculation suggested that it would be necessary for the galaxy to rotate in its plane so that it completed one revolution every 500 million years. In fact, if the rotational motion of the galaxy is to counteract the gravitational attraction of the innermost stars, the innermost parts of the galaxy must rotate faster than the outer parts. A careful analysis of stellar motions has disclosed such a motion. The rotational speed at the distance of the Sun from the centre of rotation is between 200 and 300 km s^{-1}.

S Mitton, ed., *The Cambridge Encyclopaedia of Astronomy* (London, 1973)

MS

Galois theory *Mathematics* Named after French mathematician Evariste Galois (1811–32), this is the branch of ALGEBRA whose principal concern is the study of automorphism groups of fields.

The theory arose from Galois's studies on the solubility of polynomial equations by radicals, in which he exhibited a correspondence between particular fields and particular groups. Using Galois theory, it is possible to show that there is no general formula in terms of radicals for the solution of a polynomial equation of degree greater than 4. *See also* CARDANO'S FORMULA, GROUP THEORY and FIELD THEORY.

I N Herstein, *Topics in Algebra* (Wiley, 1975)

MB

gambler's fallacy *Psychology* Closely linked with the work on judgmental biases by the American psychologist Harold H Kelley (1921–).

The term refers to the mistaken belief that one may predict correctly in a completely chance or random situation. The bias is usually explained in terms of an overemphasis on the role of contextual factors in understanding events in the world. *See* ACTOR-OBSERVER BIAS, FUNDAMENTAL ATTRIBUTION ERROR and SELF-SERVING BIAS.

B P Bruns, *Compulsive Gambler* (New York, 1973)

NS

game theory (1947) *Mathematics* Created by American mathematician John von Neumann (1903–57) from analyses of games of poker, this is the finding of optimal strategies in competitions or conflicts, particularly those with only a finite number of outcomes.

The theory has been applied to such diverse disciplines as business, animal behaviour and war. It was also used in the development of the H-bomb. It is still in use today in all aspects of decision-making behaviour. *See also* MINIMAX THEOREM.

J L von Neumann and O Morgenstern, *Theory of Games and Economic Behaviour* (Princeton, 1947)

KD, NS

game theory (2) (20th century) *Politics/Economics/Psychology* Account of politics using the analogy of competitive games.

Individuals and institutions pursue their rationally predicted maximum self-INTERESTS. They do so in a manner analogous to games players trying to calculate not only their own advantages, but the likely moves of their opponents. The result, paradoxically, is frequently neither the maximum individual nor the maximum collective self-interest. *See also* CHICKEN, CO-OPERATIVE GAMES THEORY, PRISONER'S DILEMMA, TRAGEDY OF THE COMMONS and ZERO-SUM.

Robert Abrams, *Foundations of Political Analysis* (New York, 1979)

RB

gamma-permanence rule *Chemistry See* LANDÉ PERMANENCE RULE.

Gamow-Gurney-Condon theory of alpha particle decay (1928) *Chemistry/Physics* Named after its originators, the Soviet cosmologist George Gamow (1904–68) and, independently, Ronald Wilfrid Gurney (1898–1953) and Edward Uhler Condon (1902–74).

Alpha particle decay arises as a consequence of quantum-mechanical tunnelling through a potential barrier; that is, the alpha particle penetrates the potential barrier instead of going over the top as demanded by classical physics. The calculation assumes that the alpha particle is pre-formed in the nucleus before being emitted in the decay process. Once it has been emitted, the alpha particle gains its final kinetic energy by electrostatic repulsion as it moves away from the residual nucleus.

J Thewlis, ed., *Encyclopaedic Dictionary of Physics* (New York, Oxford and London, 1962)

MS

gap *Literary Theory See* INDETERMINACY.

Garcia effect/toxicosis *Psychology* Named after the American psychologist J Garcia (1917–86), this is an acquired syndrome in which an organism learns to avoid a particular food because of a conditioned aversion to its smell or taste. The original association must be with an internal, digestively-linked stimulus (either the smell or taste of the food substance), and the adverse outcome must be associated with alimentary function (for example, nausea). Even where the atoxic reaction is not experieced for a number of hours after eating, the toxicosis reaction may be formed in a single exposure if consumption of a novel food is followed by nausea and sickness.

J Garcia, K W Rusinak and L P Brett, 'Conditioned Food Aversion in Wild Animals. *Caveant Caonici*', *Operant-Pavlovian Interactions*, H Davis and H Hunitz, eds, (Hillsdale, N.J., 1977)

NS

garden path (1970) *Linguistics* Process identified by the American psychologist Thomas G Bever.

Processing a sentence from left to right, we make guesses of structure on the basis of early syntactic cues, and may be led up the garden path. For example, hearing the sentence 'The horse raced past the barn fell', we may misconstrue *the horse* as subject of 'raced'.

T G Bever, 'The Cognitive Basis for Linguistic Structures', *Cognition and the Development of Language*, J R Hayes, ed. (New York, 1970), 279–352

RF

gas laws *Chemistry See* AVOGADRO'S LAW, BOYLE'S LAW, CHARLES'S LAW, DALTON'S LAW OF PARTIAL PRESSURES, GENERAL GAS LAW and GRAHAM'S LAW OF DIFFUSION.

gastraea theory (1872) *Biology* First suggested by Ernst Heinrich Haeckel (1834–1919), a German zoologist known for his early support of Darwinism. This is the idea that all organisms pass through a stage in which the gastraea (embryo) consists of two layers, an inner one and an outer one. *Compare with* GERM LAYER THEORY, and RECAPITULATION.

KH

gatekeeping theory (20th century) *Politics/Feminism* Concept in SYSTEMS THEORY.

In a political system there are 'gatekeepers', individuals or institutions which control access to positions of power and regulate the flow of information and political influence. Feminists have adapted this theory to explain male control of language and knowledge.

Geoffrey Roberts and Alistair Edwards, *A New Dictionary of Political Analysis*, (London, 1991)

RB

gauge theories *Physics* Gauge theories are quantum field theories based on the use of a field that possesses one or more gauge symmetries. They have the property that under a gauge transformation, certain quantities may be re-gauged (re-scaled) without affecting the values of the observable field quantities. It is believed that gauge theories can provide the basis for a description of all elementary particle interactions. *See also* ELECTROWEAK THEORY, GRAND UNIFIED THEORIES, QUANTUM CHROMODYNAMICS and QUANTUM

Gauss-Markov least squares theorem (1900) *Mathematics* The minimum variance approach to estimation is due to Markov and the least squares method is due to the German mathematician Carl Friedrich Gauss (1777–1855).

The theorem of STATISTICS that if the random variables $X_1,\ldots,X_n$ are given the least squares estimate of β for the model

$$Y_i = \beta X_i + \epsilon_i$$

where

$$E[\epsilon_i] = 0, \quad \text{var}[\epsilon_i] = \sigma^2 ,$$

and

$$\text{cov}\,(\epsilon_i,\epsilon_j) = 0 ,$$

has uniform minimum variance of all unbiased linear estimates of β.

H T Nguyen and G S Rogers, *Fundamentals of Mathematical Statistics* (New York, 1989)

ML

Gauss's lemma *Mathematics* Named after German mathematician and astronomer Carl Friedrich Gauss (1777–1855), who is generally regarded as one of the most prolific and influential mathematicians of all time.

It is the result that if a polynomial with integer coefficients has a factorization over the rationals, then it has a factorization over the integers as well. *See also* EISENSTEIN'S CRITERION.

I N Herstein, *Topics in Algebra* (Wiley, 1975)

MB

Gauss's theorem *Mathematics See* DIVERGENCE THEOREM.

Gauss's theorem in electrostatics *Physics* This is an alternative form of Coulomb's inverse square law of force between point electric charges. It was formulated by the German mathematician Carl Friedrich Gauss (1777–1855).

The outward flux of electric field strength E over any closed surface S is equal to the algebraic sum of the enclosed charges Q_i, divided by the permittivity of free space ϵ_0, that is,

$$\oint E.\text{d}S = \sum Q_i/\epsilon_0 .$$

See COULOMB'S LAW OF ELECTROSTATIC FORCE.

J Thewlis, ed., *Encyclopaedic Dictionary of Physics* (New York, Oxford and London, 1962)

MS

Gay-Lussac's law *Chemistry See* CHARLES'S LAW.

Gay-Lussac's law of combining volumes (1808) *Chemistry* Proposed by French scientist Joseph Gay-Lussac (1778–1850), this law states that when gases combine with one another they do so in volumes that are in small whole number ratios.

For example, one volume of nitrogen (N_2) combines with three volumes of hydrogen (H_2) to form two volumes of ammonia (NH_3)

$$N_2 + 3H_2 \rightarrow 2NH_3$$

The law is strictly valid only for ideal gases.

R H Petrucci, *General Chemistry: Principles and Modern Applications* (New York, 1985)

MF

gaze (20th century) *Politics/Feminism* Feminist theory of presentation of women in culture.

Women are presented in works of art and literature, and particularly in film, through three principal male perspectives or 'gazes'. Each divides by gender, and treats women in a demeaning way as objects of male voyeurism.

Maggie Humm, *The Dictionary of Feminist Theory* (London, 1989)

HR

Gebrauchsmusik (1920s) *Music* German for 'functional music', this term was identified in about 1925 by, among others, the German musicologist Heinrich Besseler (1900–69).

It refers to music intended for practical use; for example in plays or films, but especially for amateur performance in the home or at school.

Simple and well-defined, it represents a reaction against expressionism and 'music for its own sake'. The term is favoured more in English-speaking countries where it is applied to music of the interwar period in Germany. Its best-known exponent, the composer Paul Hindemith (1895–1963), later preferred to call his music *Sing-und Spielmusik* ('music to sing and play').

T W Adorno, 'Ad vocem Hindemith', *Impromtus* (Frankfurt, 1968), 51–87

MLD

Geiger and Nuthall rule (1911) *Physics* Named after its discoverers, the German physicist Hans Wilhelm Geiger (1882–1945) and John Mitchell Nuthall (1890–1958).

There is a systematic dependence of the energy of alpha particles from different nuclides in a given radioactive series on the half-life of the nuclide emitting them. This relationship may be written

$$\ln (\lambda + c) = \ln R$$

where λ is the decay constant for a particular nuclide, c is a constant for a particular radioactive series and R is the range of the alpha particles in a given medium.

J Thewlis, ed., *Encyclopaedic Dictionary of Physics* (New York, Oxford and London, 1962)

MS

Geiger's law (20th century) *Physics* Named after the German physicist Hans Wilhelm Geiger (1882–1945), who discovered that in an experimental study of alpha particle behaviour, the speed v of an alpha particle and its range R in a given medium are connected by the empirical equation

$$R = \text{constant} \times v^3.$$

J Thewlis, ed., *Encyclopaedic Dictionary of Physics* (New York, Oxford and London, 1962)

MS

Gelber-Jensen controversy (1950s) *Biology* Named after Gelber and Jensen. A heated debate concerning whether *Paramecium aurelia*, a single-celled animal, can be trained to approach a wire that has been coated with a favourite food (bacteria) on previous occasions.

Gelber reported in 1952 that the *Paramecium* had learned to approach and cling to the bare wire. Jensen contended that the animals were merely responding to environmental stimuli which caused an automatic (unlearned) response to cling to objects. The controversy remains.

KH

Gelfond-Schneider theorem (1934) *Mathematics* Named after Russian mathematician

Alexander Osipovich Gelfond (1906–68) and the mathematician Theodor Schneider (1911–) who independently gave a demonstration.

It is the result in NUMBER THEORY whereby for an algebraic number α, not 0 or 1, and an algebraic irrational β, the number α^{β} is transcendental. For example,

$$e^{\pi} = (-1)^{-i} \text{ and } 2^{\sqrt{2}}$$

are transcendental. This theorem provides the answer to the seventh of HILBERT'S PROBLEMS and has been generalized by Baker. *See also* BAKER'S THEOREM.

A Baker, *A Concise Introduction to the Theory of Numbers* (Cambridge, 1984)

MB

gender and language (1960s) *Feminism/Linguistics* There had been earlier anthropological and anecdotal accounts of 'women's language', but improved methodology of SOCIOLINGUISTICS, and the impetus of FEMINISM, have made this a substantial topic since the 1960s.

(1) Study of differences in speech (syntax, vocabulary, pronunciation), conversational behaviour, and attitudes to language as between men and women. (2) Discussion of the proposition that gender roles are coded within language, with discriminatory or sexist consequences. (*See also* FEMINIST LINGUISTICS.)

(1) J Coates, *Women, Men and Language* (London, 1986); (2) D Cameron, ed., The Feminist Critique of Language: A Reader (London, 1990)

RF

gender theory (20th century) *Politics/Feminism/Biology* Theory of relative weight of NATURE AND NURTURE.

The biological differences between males and females – sexual differences – account for a relatively small part of the actual differences between men and women. Most of these differences are matters not of sex but of gender which, unlike sex, is socially formed and cultivated. Differences of gender, however, are used to justify inequalities between the sexes and the appropriation by males of the major part of power, leisure, time and property.

Ann Oakley, *Sex, Gender and Society* (London, 1972)

HR

gene (or genetic) map (1987) *Biology* The first complete map of a living organism, *Escherichia coli*, was completed in February 1987 by a research team at Columbia University (USA).

The order of nucleotides and genes on a chromosome or chromosome fragment, deduced from experiments in genetic recombination. A complete map provides a biochemical blueprint for reproducing an organism.

KH

gene selection *Biology See* KIN-SELECTION THEORY.

general equilibrium theory (19th century–) *Economics* First developed by French-born economist Léon Walras (1834–1910), this theory studies simultaneous equilibria in a group of related markets.

Attributed to Walras, who studied a theoretical economic system in which all consumers were utility maximizers and firms were perfectly competitive, the model shows that a unique stable equilibrium can exist under such conditions. Economists have since questioned whether such an equilibrium is stable, unique, and (if a general equilibrium does exist) whether there are many sets of prices at which markets will clear. *See* EQUILIBRIUM THEORY, PARTIAL EQUILIBRIUM THEORY, WALRAS'S STABILITY, SAY'S LAW, CLASSICAL MACROECONOMIC MODEL, ARROW-DEBREU MODEL, and THEORY OF THE CORE.

E R Weintraub, *General Equilibrium Theory* (London, 1974)

PH

general gas law *Chemistry* A combination of BOYLE'S LAW and CHARLES'S LAW, this is also known as the combined gas law or simply the gas law.

The pressure (p_1) and volume (V_1) of a gas at absolute temperature T_1 is related mathematically to its pressure (p_2) and volume (V_2) at temperature T_2 by

$$\frac{p_1 V_1}{T_1} = \frac{p_2 V_2}{T_2}$$

The law enables the volume of a gas to be calculated at a specific temperature and pressure so long as the volume of the gas is known at another temperature and pressure. The law can also be expressed as

$$\frac{pV}{T} = \text{constant}$$

where the exact value of the constant depends on the amount of gas. The law is valid only for ideal gases.

M Freemantle, *Chemistry in Action* (Basingstoke, 1987)

MF

general strike (20th century) *Politics/ Economics* Theory of political action of French writer on SYNDICALISM Georges Sorel (1847–1922).

Capitalism would be overthrown by the economic rather than the political actions of the working class. This revolutionary economic activity would be carried out in pursuit of the ultimate ideal of a universal, general strike. But the purpose of advocating such a strike was not to achieve it, but rather to inspire industrial workers to move in the direction of it. Thus the 'MYTH' of the general strike was to act for political activity in the same way that heaven acted for moral activity.

George Sorel, *Reflections on Violence* (trans. London, 1950)

RB

general systems theory *Psychology* Proposed by the Austrian-Canadian biologist Ludwig von Bertalanffy (1901–72).

The theory was proposed to formulate principles of functioning that would be characteristic of all biological systems. A distinction is made between systems affected by external forces and systems which are not. A further distinction is made between non-living and living systems. The theory emphasizes the importance of understanding the interaction of activity within the organism and that between the organism and its environment. Concepts from general systems theory have been taken up within different branches of psychology such as clinical and educational psychology.

L von Bertalanffy, *General Systems Theory* (New York, 1981)

NS

general theory of relativity *Physics See* RELATIVITY, THEORY OF GENERAL.

general will (1762) *Politics* Theory propounded initially by the French philosopher Jean-Jacques Rousseau (1712–78).

There is a general will of society as a whole which is distinct from the particular wills of individuals. The general will is both the truest interest of individuals, and the justification for government; although, paradoxically, individuals may consent to a government that thereafter limits their choices.

David Miller *et al.*, eds, *The Blackwell Encyclopaedia of Political Thought* (Oxford, 1987)

RB

general theory of employment, interest and money (1936) *Economics* The title of a book by English economist John Maynard Keynes (1883–1946), this theory represents a major contribution to modern economic thought. It attacked classical economics and put forward important theories on the consumption function, aggregate demand, the multiplier, marginal efficiency of capital, liquidity preference and expectations.

See ABSOLUTE INCOME HYPOTHESIS, AGGREGATE DEMAND THEORY, DEMAND FOR MONEY THEORY, DISEQUILIBRIUM THEORY, THEORY OF INCOME DETERMINATION, MARGINAL EFFICIENCY OF CAPITAL, NATURAL AND WARRANTED RATES OF GROWTH, OWN RATE OF INTEREST and PARADOX OF THRIFT.

J M Keynes, *The General Theory of Employment, Interest and Money* (New York, 1936)

PH

generalized phrase-structure grammar (GPSG) (late 1970s) *Linguistics* Developed by British linguist Gerald Gazdar and others.

Alternative to GOVERNMENT AND BINDING. GPSG has no TRANSFORMATIONS but a greatly refined PHRASE STRUCTURE GRAMMAR in which classes of CONSTITUENT are represented as clusters of features.

G Gazdar and E Klein and G K Pullum and I Sag, *Generalized Phrase Structure Grammar* (Oxford, 1985)

RF

generation time hypothesis *Biology* The notion that the evolution of proteins and deoxyribonucleic acid (DNA) is faster in species having unusually short generation times than it is in species with unusually long generation times. Generation time (Tg) is defined as the time required for a cell to complete one growth cycle

KH

generative grammar (1957) *Linguistics* Proposed and formulated by the American linguist Avram Noam Chomsky (1928–). The most influential theory in linguistics in the second half of the 20th century.

A generative grammar is an account of a natural language which identifies all and only the grammatical sentences of a language. It consists of a finite set of rules (which in the case of a classic Chomskyan grammar are phrase structure and transformational rules) which will define members of the infinite number of sentences possible in a natural language. Chomsky speaks of a sentence as a 'pairing of sounds and meanings'; thus the rules must cover PHONOLOGICAL, SEMANTICS and (centrally) syntactic structure. A generative grammar is both a formal account of sentences and also a description of a speaker's LINGUISTIC COMPETENCE.

N Chomsky, *Syntactic Structures* (The Hague, 1957)

RF

generative phonology (late 1950s–) *Linguistics* Developed by the American linguists Avram Noam Chomsky (1928–) and Morris Halle.

Non-segmental model of the sound patterns of language (*see also* PHONEMICS) developed within TRANSFORMATIONAL-GENERATIVE GRAMMAR. Derives the pronunciation of sentences by applying formalized phonological rules to syntactic surface structures.

N Chomsky and M Halle, *The Sound Pattern of English* (Cambridge, Mass., 1968)

RF

generative semantics (late 1960s–early 1970s) *Linguistics* Proposal by the American linguist George Lakoff and others.

Outgrowth of TRANSFORMATIONAL-GENERATIVE GRAMMAR which attempted to replace the base of the STANDARD THEORY with a very abstract semantic/logical level from which meanings were mapped into surface structures by syntactic rules. The theory of generative semantics was overtaken by other developments in generative grammar.

G Lakoff, 'On Generative Semantics', *Semantics*, D D Steinberg and L A Jakobovits, eds (Cambridge, 1971), 232–96

RF

generic cycles, theory of (1831) *Biology* A theory of the British naturalist philosopher Robert Knox (1793–1862).

The life cycle of a species can be compared to that of an individual. A species goes through four stages; vigorous spread; maximum phyletic activity (giving rise to daughter species); decline due to competition with these progeny; extinction. The broad analogy can be useful, but senility leading to extinction in a species has never been observed.

Rehbok, *The Philosophical Naturalists* (Wisconsin, 1983)

RA

genetic algorithms (20th century) *Mathematics* Algorithms to determine properties of entities defined by recursion. These are often based on stochastic ideas or on biological analogy.

JB

genetic assimilation *Biology See* BALDWIN EFFECT.

genetic drift (mid-20th century) *Biology* Idea derived from Sewall Wright (1889–1988), the American population theorist who first described the FOUNDERS EFFECT. This term refers to changes in genes which occur by chance rather than by natural selection, mutation or immigration.

The effect is most noticable in small, isolated populations, but in theory all populations are subject to such random fluctuations of gene frequencies.

F Magill, *Magill's Survey of Science: Life Science Series* (Englewood Cliffs, NJ, 1991)

KH

genetic engineering (20th century) *Biology/Medicine* The process that changes the genome of a living cell without involving sexual or asexual transmission of genetic information, especially when this manipulation is designed to alter the functions or products of that cell. Genetic engineering has many current and potential applications. In medicine, the technique is used to create normal bodily substances, such as hormones and growth factors, in the laboratory for medicinal use.

J Cherfas, *Man-Made Life: An Overview of the Science, Technology and Commerce of Genetic Engineering* (New York, 1982)

KH

genetic fitness *Biology See* INCLUSIVE FITNESS.

genetic revolution (1954) *Biology* Proposed by German-American evolutionary biologist Ernst Mayr (1904–), this is a model for the evolution of new species holding that the FOUNDERS EFFECT is characterized by a great increase in the homozygosity of individuals, followed by the loss of additive genetic variance because the population remains so small.

The forces of NATURAL SELECTION brought on by this homozygosity would then trigger a genetic revolution – a new species. Many biologists find problems with this model, although PUNCTUATED EQUILIBRIUM is considered to be a generalization of it. *Compare with* FOUNDER-FLUSH HYPOTHESIS. *See also* SPECIATION, THEORY OF and SPECIES, THEORY OF.

KH

genetical theory of social behaviour *Biology See* ALTRUISM.

geneticism *Psychology* Particularly associated with the work of the English scientist Sir Francis Galton (1822–1911).

This term denotes any form of thinking or philosophy which emphasizes the view that behaviour is genetically determined. The emphasis can be placed on individual genotype or on species-specific genes make-up. In debate, the term has often been used to describe the position of those who place excessive importance on the role of genetic factors in psychological processes.

D S Falconer, *An Introduction to Quantitative Genetics* (New York, 1960)

NS

genotype theory (20th century) *Biology* The theory that the first living cells arose from naked genes or from deoxyribonucleic acid (DNA).

The genotype theory has not achieved widespread acceptance among biologists because, for life to occur, proteins would have to have arisen contemporaneously with genes; yet the mechanism for this simultanous evolution has yet to be described. *Compare with* MINERAL THEORY, PHENOTYPE THEORY, PRECELLULAR EVOLUTION, RIBOTYPE THEORY and VIRAL THEORY. *See also* DYNAMIC STATE THEORY and ORIGINS OF LIFE.

M Barbieri, *The Semantic Theory of Evolution* (Chur, Switzerland, 1985)

KH

genre (traditional idea; term from *c.*1900) *Literary Theory* Originated with the Greek philosopher Aristotle (384–322 BC); in English, earlier 'species'; also 'kind'.

Genre (for example poetry, prose, drama) and sub-genre (for example tragedy, comedy) are terms in the classification of types of literary text; also by extrapolation non-literary, for example conversation, advertisement. Members of a genre have common characteristics of style and organization (for example stanza structure) and are found in similar cultural settings. No adequate taxonomy exists; the most powerful attempt being *Anatomy of Criticism* (Princeton, 1957) by Northrop Frye (1912–).

A Fowler, *Kinds of Literature* (Oxford, 1982)

RF

geocentric theory of planetary motion *Physics* A theory of the solar system or universe, such as the Ptolemaic model which has the Earth as its central point of reference.

GD

geographic speciation *Biology See* SPECIATION, THEORY OF.

geometry (*c.*400 BC) *Mathematics* Logical proof rather than measurement has been applied to geometrical problems at least since the time of Pythagoras.

Geometry is the study of space, which can be studied empirically as a branch of physics or conducted as an informal deductive system or formulated as a specific abstract geometry by axioms or constructions with numbers. The last method allows the study of geometries of dimensions greater than three and geometries having no resemblance to physical space, such as geometries with a finite number of points.

H S M Coxeter, Introduction to Geometry (New York, 1961)

JB

geometry of numbers (20th century) *Mathematics* The branch of NUMBER THEORY which relies on Euclidean geometry. The name is due to the Russian-born Swiss-German mathematician Hermann Minkowski (1864–1909) who first systematically exploited the observation that intuitive deductions related to the geometry of plane figures can sometimes yield results of great importance in number theory. The most important result in this context is MINKOWSKI'S THEOREM ON CONVEX BODIES. *See also* DIOPHANTINE ANALYSIS.

A Baker, *A Concise Introduction to the Theory of Numbers* (Cambridge, 1984)

MB

geomorphology (early 20th century) *Geology* The science of land form morphology and the processes causing it, this was largely established by the American William Morris Davis (1850–1934).

Occasionally, the term has been used to describe studies of the overall shape of the Earth, but these are now usually encompassed within the discipline of geodesy.

DT

geopolitics *Politics* Theory linked with Swedish political scientist Rudolf Kjellén (1864–1922) and German political geographer Karl Haushofer (1869–1946). Nations are understood as analogous to natural organisms, competing for geographical space. An archaic theory.

Roger Scruton, *A Dictionary of Political Thought* (London, 1982)

RB

germ layer theory (19th century) *Biology* Proposed by the Russian scientist Christian Heinrich Pander (1794–1865) and the Estonian naturalist Karl Ernst von Baer (1792–1876). The idea that an embryo has distinct structural layers, called germ layers, which invariably develop into corresponding differentiated tissues.

Modern biologists still accept that the embryo has three germ layers (endoderm, mesoderm and ectoderm), but they now believe that the development of these layers and the resulting tissues is highly complex. *Compare with* RECAPITULATION.

P B Medawar and J S Medawar, *Aristotle to Zoos: A Philosophical Dictionary of Biology* (Oxford, 1983)

KH

germ line theory *Biology See* GERM PLASM THEORY.

germ line theory of antibody diversity (1900) *Biology* Also called the multigene hypothesis, the first detailed model representative of this theory was proposed by the German bacteriologist Paul Ehrlich (1854–1915), winner of the 1908 Nobel prize for Physiology or Medicine. The term refers to any of several theories of immune response holding that antigens choose a pre-existing, complementary antibody and then amplify that antibody to produce a specific immune response.

The SIDE-CHAIN THEORY was the first such model. This theory implies that approximately 10 million antibodies are encoded in the germline, which raised the question of whether sufficient deoxyribonucleic acid (DNA) exists for such a task. Germline theories represent one of the two major perspectives for explaining the diversity of antibodies. *See also* ANTIBODY, THEORIES OF. *Compare with* SOMATIC MUTATION THEORY OF ANTIBODY DIVERSITY.

T Kindt and J Capra, *The Antibody Enigma* (New York, 1984)

KH

germ plasm theory (1892) *Biology* Also called Weismannism, and germ line theory, this was experimentally demonstrated by the German biologist and physician August Weismann (1834–1914). The theory holds that acquired characteristics are not inherited by offspring, because the germ cells of each generation descend directly from the germ cells of the previous generation.

That is, the cells responsible for reproduction are not affected by body cells because they are set aside early in the development of the individual. Also, germ cells are potentially immortal because they are transmitted from generation to generation. *Compare with* GERM LAYER THEORY and LAMARCKISM.

A Weismann, *The Germ Plasm: A Theory of Heredity* (London, 1893)

KH

germ theory of disease (19th century) *Biology* The theory (proven in the 19th century) that many diseases are caused by self-propagating, transmissible agents (germs), such as bacteria and viruses. The Italian physician and scholar Girolamo Fracastoro (1483–1553) early hypothesized the existence of tiny, infective, reproducing seeds.

G Fracastoro, *On Contagion, Contagious Diseases and Their Treatment* (1546); W Spink, *Infectious Diseases: Prevention and Treatment in the Nineteenth and Twentieth Centuries* (Minneapolis, Minn., 1979)

KH

Gerschgoren circle theorem *Mathematics* The theorem of perturbation theory that for matrices X and A if

$$X^{-1}AX = \text{diag}\,(d_1,\ldots,d_n) + F$$

and F has zero diagonal entries, then the spectrum of A is contained in the union of D_i where

$$D_i = \{z \in C : |z - d_i| \leq \sum_{j=1}^{n} |f_{ij}|\}.$$

Each D_i is called a Gerschgoren disc. It can be shown that if a Gerschgoren disc is isolated from the other discs, then it contains exactly one of the eigenvalues of A.

G H Golub and C F VanLoan, *Matrix Computations* (Baltimore, 1983)

ML

gestalt theory *Psychology* The forerunners of Gestalt Psychology were the German Max Wertheimer (1880–1943), Wolfgang Köhler (1887–1967) and Kurt Koffka (1886–1941).

'Gestalt' is a German term which has no direct translation, but it is taken to mean any of the following; form, configuration, shape and essence. Gestalt psychology takes the view that phenomena should be studied and treated as a whole rather than on an elemental or componential basis. A maxim developed from this theory is that the whole is different from the sum of its parts.

A D Ellis, ed., *A Sourcebook of Gestalt Psychology* (London, 1938)

NS

Gibbs's adsorption theorem (1878) *Chemistry* Theorem named after American mathematical physicist Josiah Willard Gibbs (1839–1903), one of the founders of statistical mechanics.

Solutes concentrated at the surface of a solvent tend to lower the surface tension of the solvent. The converse tendency also applies. The exact relationship between adsorption and surface tension was derived by Gibbs in 1878 and independently by British mathematical physicist Joseph John Thomson (1856–1940) in 1888. It is known as the Gibbs adsorption equation.

S Glasstone, *Textbook of Physical Chemistry* (London, 1948)

MF

Gibbs-Konowalow rule *Chemistry* Applying to the phase equilibrium of binary solutions, this states that when the compositions of two phases are identical, the equilibrium temperature is a maximum or minimum at constant pressure. The statement also applies to pressure at constant temperature.

MF

Gibbs's paradox *Chemistry* Named after American mathematical physicist Josiah

Willard Gibbs (1839–1903). When two gases with identical thermodynamic properties (such as CO and N_2) are mixed, work results. However, when two portions of the same gas are mixed, no work results.

MF

Gibbs's phenomenon (19th century) *Mathematics* Named after the American theoretical physicist and chemist Josiah Willard Gibbs (1839–1903), this is the behaviour of the partial sums of a FOURIER SERIES near a jump discontinuity of a function of bounded variation in which the partial sums converge to the function in the neighbourhood of zero, but not uniformly. The partial sums tend to approximate vertical segments longer that the jump by a proportion of

$$\frac{2}{\pi}\int_0^{\pi} \frac{\sin x}{x}\, dx \approx 1.17898\ldots.$$

M Hazewinkel, ed., *Encyclopaedia of Mathematics* (Dordrecht, 1988)

ML

Gibbs's theorem *Physics* Named after its discoverer, the American mathematical physicist Josiah Willard Gibbs (1839–1903), this hypothesizes that the entropy of a mixture of ideal gases is equal to the sum of the partial entropies of the individual gases.

J Thewlis, ed., *Encyclopaedic Dictionary of Physics* (New York, Oxford and London, 1962)

MS

Gibrat's rule of proportionate growth (1931) *Economics* Named after French economist Robert Gibrat (1904–80) (and sometimes called Gibrat's law), this states that the proportional change in the size of a company in an industry is the same for all such companies irrespective of their original size.

If a company with sales of £10m doubles in size over a period of time, it is likely the same will happen for a company beginning with sales of only £1m.

R Gibrat, *Les inégalités économiques* (Paris, 1931)

PH

Gibson effect *Psychology* Named after the American, J J Gibson (1904–80).

A group of negative perceptual aftereffects that set in after prolonged observation of spacial patterns (for example a scene of uniformly curved lines) as a result of visual adaptation to the viewed scene.

L Kaufman, *Sight and Mind: An Introduction to Visual Perception* (New York, 1974)

NS

Giffen paradox (*c*.1895) *Economics* Proposed by Scottish economist Sir Robert Giffen (1837–1910) from his observations of the purchasing habits of the Victorian poor, this paradox states that demand for a commodity increases as its price rises.

The paradox is explained by the fact that if the poor rely heavily on basic commodities like bread or potatoes, when prices are low they might still have some disposable income for purchases of other items. As bread or corn prices rise, these other purchases are no longer possible, thereby forcing the poor to concentrate all their purchasing power on the bread or corn. It should not be confused with products bought as status symbols or for CONSPICUOUS CONSUMPTION.

R Giffen, *Economic Inquiries and Studies* (London, 1904)

PH

given and new information (1960s–) *Linguistics* Various models in FUNCTIONALIST LINGUISTICS.

Some parts of a sentence express information which has already been given by preceding text, or by context. New information is added by the sentence. To be distinguished from THEME AND RHEME, although theme very often coincides with given and rheme with new.

W J Vande Koppe, 'Given and New Information and Some Aspects of the Structures, Semantics, and Pragmatics of Written Texts', *Studying Writing: Linguistic Approaches*, C R Cooper and S Greenbaum, eds (Beverly Hills, 1986), 72–111

RF

glacial theory (1779) *Geology* In contrast to strict DILUVIANISM, Horace Benédict de Saussure (1740–99) hypothesized that some unconsolidated sediments of northern Europe, particularly erratic blocks, were deposited as a result of erosion and deposition from advancing and retreating ice sheets.

The Scottish geologist James Hutton (1726–97) rejected this theory in favour of an ice flow mechanism, but the German Leopold von Buch (1774–1853) confirmed and amplified de Saussure's observations. The main theory was developed by a Swiss hunter, Jean-Pierre Perraudin, in 1815; extended by the Swiss civil engineer Ignace Venetz (1788–1859) in 1821 and 1829; and established globally by the Swiss naturalist Jean Louis Rodolph Agassiz (1807–73). Final acceptance of the theory came in about 1870. *See also* DRIFT THEORY.

A Hallam, *Great Geological Controversies* (Oxford, 1983)

DT

Gladstone and Dale law (1858) *Physics* Named after its discoverers, John Hall Gladstone (1827–1902) and T P Dale, this is a relationship between the refractive index n of a gas and its density ρ, when the temperature remains constant:

$$\frac{n+1}{\rho} = \text{constant} .$$

J Thewlis, ed., *Encyclopaedic Dictionary of Physics* (New York, Oxford and London, 1962)

MS

Glashow-Weinberg-Salam theory *Physics* *See* ELECTROWEAK THEORY.

global analysis, or analysis on manifolds (19th century) *Mathematics* Probably started by Bernhard Riemann (1826–66), this is the study of functions of which the variables range over the points of a geometrical figure, in particular over a manifold in a differential geometry.

JB

Gloger's law or rule *Biology* Named after C L Gloger, this law asserts that animals adapted to cool, dry environments tend to be lighter in colour than those living in warm, humid regions. Specifically, pigments are often black in warm, humid places, red and yellow in dry areas, and muted in cool climates. The reason for this tendency is unclear.

KH

glossematics (1930s) *Linguistics* Theory developed by the Danish linguists Louis Hjelmslev (1899–1965) and H J Uldall (1907–57).

Global system for the analysis of the expression and content levels of natural languages, basically within the Saussurean tradition of STRUCTURAL LINGUISTICS and SEMIOTICS, but with an extensive idiosyncratic terminology which has inhibited general use of the model. Some influence on STRATIFICATIONAL GRAMMAR and SEMANTICS.

L Hjelmslev, *Prolegomena to a Theory of Language*, F J Whitfield, trans. (Madison, 1953); F J Whitfield, 'Glossematics', *Linguistics Today*, A Martinet and U Weinreich, eds (New York, 1954), 250–58

RF

glottochronology (1950) *Linguistics* A 'decay dating' technique devised by the American linguist Morris Swadesh (1909–67). It is part of LEXICOSTATISTICS and is sometimes referred to by this name.

The technique is based on the assumption that around 20 per cent of the basic vocabulary of a language is replaced every 1000 years. This is used to calculate the time-lapse between different stages of a language, or the date of divergence of two related languages. The technique is controversial, but has been used widely in ANTHROPOLOGICAL LINGUISTICS and COMPARATIVE AND HISTORICAL LINGUISTICS.

R Anttila, *An Introduction to Historical and Comparative Linguistics* (New York, 1972), 395–98

RF

Gödel-Rosser theorem (1936) *Mathematics* An extension of Gödel's incompleteness theorem found by John Barkley Rosser (1907–).

In a formal system which determines the positive integers, there is a 'correct' proposition which can neither be proved nor disproved in the system. *See also* GÖDEL'S INCOMPLETENESS THEOREM and HILBERT'S PROGRAMME.

JB

Gödel's completeness theorem (1930) *Mathematics* Named after the Czech-born American mathematician Kurt Gödel (1906–78) who gave its first demonstration, this is the result in logic whereby a propositional theory is consistent if and only if it has a model.

When expressed in terms of logical arguments, it is the theorem that an argument is valid (that is, the conclusion is true whenever all the premises are true) if and only if the argument has a formal proof (that is, the conclusion follows from the premises by applying certain rules of inference). *See also* MODEL THEORY, GÖDEL'S INCOMPLETENESS THEOREM and GÖDEL-ROSSER THEOREM.

J E Rubin, *Mathematical Logic: Applications and Theory* (Saunders, 1990)

MB

Gödel's incompleteness theorem (1931) *Mathematics* Named after the Czech-born American mathematician Kurt Gödel (1906–78) who gave its first demonstration, this is the result in logic whereby if a system which formalizes the theory of the natural numbers is consistent, then this system contains a logical formula such that neither the formula nor its negation can be proved within the system. Thus one can never hope to capture truth entirely within the formal system. *See also* GÖDEL'S COMPLETENESS THEOREM, GÖDEL-ROSSER THEOREM and PEANO'S AXIOMS.

Encyclopedic Dictionary of Mathematics (MIT Press, 1987)

MB

going concern principle *Accountancy* A fundamental principle of accountancy which assumes that a firm will continue to trade in the future as a viable entity. Financial reports are prepared on this basis rather than stating the break-up value of the company and its assets. The principle is part of the EC's Fourth Directive and the *UK Companies Act, 1981*.

PH

Goldbach conjecture (1742) *Mathematics* Named after the Prussian-born mathematician Christian Goldbach (1690–1764) who proposed it in a letter to Leonhard Euler (1707–83).

This is the notorious unsolved problem in NUMBER THEORY that every even integer greater than 2 is the sum of two primes. In 1974, Chen showed that the conjecture is valid for sufficiently large even integers if one of the primes is replaced by a number with at most two prime factors.

A Baker, *A Concise Introduction to the Theory of Numbers* (Cambridge, 1984)

MB

golden mean *Mathematics* The proportion which results when a line segment is divided in such a way that the smaller is to the larger as the larger is to the whole. Numerically, it is

$$G = \frac{\sqrt{5} - 1}{2} = 0.618033988\cdots.$$

The inverse of this number

$$\frac{\sqrt{5} + 1}{2} = 1 + G = 1.618033988\cdots$$

is sometimes also referred to as the golden ratio. A golden rectangle has the property that the ratio of the difference of the sides to the smaller equals the ratio of the smaller to the larger; in classical aesthetic theory, this was considered to be uniquely pleasing to the eye. *See also* CONTINUED FRACTION ALGORITHM.

MB

golden rule *Philosophy* 'Do as you would be done by', or 'Treat others as you would have them treat you'. Apart from the New Testament (Matthew 7.12) the rule occurs as far back as Confucius (551–479 BC) in his *Analects* (15.23; compare 5.11).

ARL

golden rule of capital accumulation (1961) *Economics* A term used by English economist Ernest Phelps (1906–), this principle analyzes the best plan of economic growth which will give the optimal sustained level of consumption *per capita* in an economy.

It is assumed that each generation will save for future generations a proportion of the income which was in turn saved for it by its previous generation.

E S Phelps, 'The Golden Rule of Accumulation: A Fable for Growthmen', *American Economic Review*, vol. LII (September, 1961), 638–43

PH

Goldschmidt's law/rule (1930s) *Geology* Formulated by the Norwegian geochemist

Victor M Goldschmidt, this law asserted that if a block of rock's temperature and pressure were maintained, then the minerals would equilibriate to a specific assemblage.

This established the basis for investigations of phase relationships between different mineral compositions within rocks. If the different phases are preserved by rapid cooling or the sudden release of pressure, the mineral phase relationships can be used to assess the initial pressure-temperature conditions.

V M Goldschmidt, *Geochemistry* (Oxford, 1958)

DT

Goldstone's theorem (1961) *Physics* Named after its discoverer, Jeffrey Goldstone (1933–). In quantum field theory, the spontaneous breaking of any continuous symmetry is accompanied by the appearance of one or more species of massless particle, known as Goldstone bosons.

J Thewlis, ed., *Encyclopaedic Dictionary of Physics* (New York, Oxford and London, 1962)

MS

Gondwanaland hypothesis (1872) *Geology* The British geologists H B Medlicott, H F Blandford and, later, the Austrian, Edouard Suess (1831–1914) theorized that similar rock types and fossil flora found in India, Africa and Australia showed these areas once belonged to a single continent. Medlicott named this continent after the Kingdom of the Gonds in Madhya Pradesh, India, where he had first made his observations.

Gondwanaland is now thought to have comprised South America, Africa, Arabia, Malagasy, India, Australia, New Zealand, Antarctica, and most of SE Asia (including Thailand and parts of southern China). These continents were directly adjacent to each other some 300–200 million years ago, but separated at different times as the intervening Indian, South Atlantic and Southern Oceans developed by SEA-FLOOR SPREADING processes.

M G Audley-Charles and A Hallam, 'Gondwana and Tethys', *Geological Society of London Special Publication*, vol. XXXVII (London, 1983)

DT

good continuation, law of *Psychology* Formulated by the German-American psychologist Max Wertheimer (1880–1943).

One of the gestalt laws of perceptual organization, this states that lines which are broken or overlapped are seen as belonging together as long as they result in either a gentle curve or a straight line. That is, we generally assume a line to continue in the direction from which it started. The law is easily demonstrated. *See* GESTALT THEORY.

J E Hochberg, *Perception* (Englewood Cliffs, N.J., 1978)

NS

good shape, law of *Psychology* Formulated by the German-American psychologist Max Wertheimer (1880–1943).

One of the gestalt laws of perceptual organization, this states that we generally perceive stimulus units and patterns in a methodical and uniform fashion. The law is easily demonstrated. *See* GESTALT THEORY.

J E Hochberg, *Perception* (Englewood Cliffs, N.J., 1978)

NS

Goodman's paradox (20th century) *Philosophy* Linguistic theory (also known as the 'new riddle of induction') concerning the concept of confirmation or prediction, developed by the American philosopher Nelson Goodman (1906–).

According to Goodman, it is possible to define a vocabulary in such a way that, given a choice between two possibilities, it is as likely that the possibility which runs counter to previous experience will be chosen as will the more predictable one. To illustrate this apparent breakdown in reasoning, Goodman devised a new predicate (which he called 'grue'); given a choice between two separate provable statements that all emeralds were green and all emeralds were blue, he argued instead that all emeralds were in fact 'grue'. *See* INDUCTIVE PRINCIPLE and INDUCTION.

N Goodman, *Fact, Fiction and Forecast* (1954, revised 1965)

DP

gothic (1765) *Literary Theory* Initiated by the English writer Horace Walpole (1717–97) with his novel *The Castle of Otranto*. 'Gothic', in the sense of 'medieval', was used in contemporaneous reviews.

This GENRE of narrative fiction is characterized by violence, mystery, extreme negative emotion, and an atmosphere of gloom and foreboding. It was prolific in the late 18th and early 19th centuries, and is still practised in 20th century novel and film.

D Punter, *The Literature of Terror* (London, 1983)

RF

government and binding (1980) *Linguistics* Consolidated by the American linguist Avram Noam Chomsky (1928–).

Model of GENERATIVE GRAMMAR developed from, and replacing, EXTENDED STANDARD THEORY. Concerned with universal formal conditions on human language, and constraints on, or parameters of, variation. Incorporates features of EST including X-BAR THEORY. Government relates items to one another; binding determines which NPs in a sentence are co-referential.

N Chomsky, *Lectures on Government and Binding* (Dordrecht, 1981); V Cook, *Chomsky's Universal Grammar: An Introduction* (Oxford, 1988)

RF

gradualism (19th century) *Biology* Also called phyletic gradualism, this idea originated with the British evolutionary theorist Charles Darwin (1809–82). It proposes that species evolution occurs slowly through barely perceptible changes, rather than suddenly, as a result of cataclysmic events.

Modern gradualism holds that those individuals best adapted to their environment are the ones most likely to transmit beneficial genes to offspring, so that over time the frequencies of such beneficial genes will increase in the population. When the gene pool of the evolving population becomes sufficiently different from that of the original population, a new species has developed. Scientists continue to debate the universal application of gradualism. *See also* DARWINISM and HOPEFUL MONSTERS. *Compare with* PUNCTUATED EQUILIBRIUM.

N Eldredge, *Life Pulse: Episodes from the Story of the Fossil Record* (New York, 1987)

KH

graduation model of evolution *Biology See* GRADUALISM.

Graham's law of diffusion (1833) *Chemistry* Named after Scottish chemist Thomas Graham (1805–69), this law states that the rate of diffusion (r) of a gas is inversely proportional to the square root of its density (d)

$$r \propto \frac{1}{d}$$

For gases a and b with molar masses M_a and M_b under identical conditions of temperature and pressure

$$\frac{r_a}{r_b} = \sqrt{\frac{d_b}{d_a}} = \sqrt{\frac{M_b}{M_a}}$$

M Freemantle, *Chemistry in Action* (Basingstoke, 1987)

MF

Graham's law of effusion (1829) *Chemistry* An empirical law formulated by Scottish chemist Thomas Graham (1805–69), this states that the relative rates of effusion of different gases under the same conditions of temperature and pressure are inversely proportional to the square roots of their densities.

M Freemantle, *Chemistry in Action* (Basingstoke, 1987)

MF

grammar (traditional; 1960s) *Linguistics* (1) In traditional Graeco-Roman educational system, that part of the syllabus concerned with language structure. (*See also* RHETORIC and TRADITIONAL GRAMMAR.) (2) In GENERATIVE GRAMMAR from 1957 onwards, the linguistic knowledge which has been 'internalized' by a speaker of a language, and the linguist's description of that knowledge.

F R Palmer, *Grammar* (Harmondsworth, 1971); N Chomsky, *Aspects of the Theory of Syntax* (Cambridge, Mass., 1965)

RF

grammaticality (1957) *Linguistics* Otherwise called 'syntactic well-formedness'. Formulated by the American linguist Avram Noam Chomsky (1928–).

Whereas American STRUCTURAL LINGUISTICS had been willing to treat as data any sentence which occurred, Chomsky stipulated that grammar should account for only sentences judged 'grammatical' according to native speakers' INTUITIONS. He later distinguished grammaticality from acceptability: a sentence might be unacceptable because too difficult to process due to psychological limitations, but nevertheless grammatical.

N Chomsky, *Aspects of the Theory of Syntax* (Cambridge, Mass., 1965)

RF

grammatology (1967) *Linguistics* Term coined by I J Gelb in 1952, and elevated to major importance by the French philosopher Jacques Derrida (1930–).

The science of writing. In DECONSTRUCTION, this is founded on an analysis of the paradoxes in Saussurean linguistics (*see* PHONOCENTRISM, LOGOCENTRISM).

J Derrida, *Of Grammatology*, G C Spivak, trans. (Baltimore, 1977 [1967])

RF

grammetrics (1964) *Stylistics* Term coined by the British linguist P J Wexler; though ancestry of the idea lies in 18th century French metrical rules.

Analysis of verse which refers simultaneously to the metrical structure of the verse lines, and the syntactic structure of the sentences, clauses and phases which inhabit them. Grammetrics acknowledges the importance of syntax in determining rhythm.

P J Wexler, 'Distich and Sentence in Corneille and Racine', *Essays on Style and Language*, R Fowler, ed. (London, 1966), 100–17

RF

grand theory *Politics/History/Sociology* A class of theories rather than a particular one.

Grand theory is any theory which attempts an overall explanation of social life, history, or human experience. It is normally contrasted with EMPIRICISM, POSITIVISM, or the view that understanding is only possible by studying particular instances, societies, or phenomena.

Quentin Skinner, ed., *The Return of Grand Theory in the Human Sciences* (Cambridge, 1985)

RB

grand unified theories (GUTs) *Physics* Grand unified theories are quantum field theories of the electro-magnetic, weak and strong interactions, in which the known interactions are considered to be low-energy manifestations of a single unified interaction. The unification takes place at energies too great to be achieved in current particle accelerators. GUTs question the assumptions of thermodynamic equilibrium and particle-antiparticle symmetry in order to understand both the present predominance of baryons over antibaryons and the baryon/photon ratio of approximately 10^{-9}. GUTs predict that the process of proton decay into a positron and a neutral pion would be expected with predicted lifetimes of 10^{35} years, although so far searches for proton decay have been unsuccessful. As yet, there is no single grand unified theory that is universally accepted.

Paul Davies, ed., *The New Physics* (Cambridge, 1989)

GD

graph theory *Mathematics* A *graph* is a collection of points called *vertices*, some pairs of which are connected by *edges*. The study of graphs, an area of COMBINATORICS, dates back at least to the famous *Königsberg bridge problem*, solved by Euler, which asks for a *circuit* through the vertices of a given graph.

Other typical questions include finding *matchings* (collections of vertex-disjoint edges); *coverings* (collections of vertices including at least one of each edge; *see also* KÖNIG'S THEOREM); *colourings*, as in the famous *four colour problem* which asks whether an arbitrary map can be coloured with four colours so countries sharing borders have distinct colours; and *embedding questions* which ask when a given graph can be drawn on a surface without edges intersecting. Numerous applications include, for example, the *travelling salesman problem* (which asks for the shortest circuit around the vertices of a given graph in the plane) and network flow models in combinatorial OPTIMIZATION THEORY.

AL

Grassmann's laws of colour vision (1853) *Physics* Named after their discoverer, the German mathematician Hermann Günther

Grassmann (1809–77), these laws are concerned with the results of mixing coloured lights.

(1) Three parameters are necessary and sufficient to define the colour of a light. (2) All mixtures of coloured lights yield continuous scales. (3) The result of an additive mixture of coloured lights depends only on their visual appearance and is independent of the physical origin of their coloured aspect.

J Thewlis, ed., *Encyclopaedic Dictionary of Physics* (New York, Oxford and London, 1962)

MS

gravitation, law of universal (1687) *Physics* Discovered by the English physicist Sir Isaac Newton (1642–1727), this law states that two particles of masses m_1 and m_2, respectively, a distance r apart, attract each other with a force which acts along the line joining them and which has a magnitude F proportional to m_1m_2/r^2; that is

$$F = G\frac{m_1m_2}{r^2}$$

where G is a universal constant known as the gravitational constant and equal to $6.673 \times 10^{-11}\ m^3\ kg^{-1}\ s^{-2}$.

J Thewlis, ed., *Encyclopaedic Dictionary of Physics* (New York, Oxford and London, 1962)

MS

gravity model (1947) *Economics* Developed by American astronomer James Stewart, this is an examination of the interaction between two places relating to size, distance and importance. The relationship between the two places is seen as corresponding directly to product, and inversely to distance.

The gravity model is used in regional economics and transport studies. However, it is criticized for having little grounding in economic theory.

G A P Carrothers, 'A Historical Review of the Gravity and Potential Concepts of Human Interaction', *Journal of the American Institute of Planners*, vol. XXII (1956), 94–102

PH

great chain of being *Biology See* CHAIN OF BEING.

great chain of being (16th–18th centuries) *Literary Theory* View of the world as comprising a finely graded hierarchy of entities with 'God' at the top, 'man' in the middle (problematically between 'angels' and 'animals') and the lowest forms at the bottom. This world view has implications for literary structures and themes.

A O Lovejoy, *The Great Chain of Being* (Cambridge, Mass., 1936)

RF

great man theory *Politics/History* Theory associated with the Scottish writer Thomas Carlyle (1795–1881).

The most significant contribution to any society is made by outstanding individuals. It is such great men, rather than circumstances or broad social or historical movement, who are responsible for progress.

B E Lippincot, *Victorian Critics of Democracy* (Minneapolis, 1938)

RB

great man theory *Psychology* Proposed by the American, Robert Freed Bales (1916–).

This states that all great historical events and achievements can be attributed to the effort of one individual 'great man'. The theory opposes the naturalistic approach which places greater emphasis on the environmental context of the said achievement. *See* GREAT WOMAN THEORY.

R A Baron and D Byrne, *Social Psychology: Understanding Human Interaction* (Boston, 1984)

NS

great woman theory *Psychology* Proposed by the American psychologist Florence Denmark (1932–).

An explanation of leadership which emphasizes the significance of personality traits and qualities, and attempts to account for the readily observable sex difference in the number of men and women who are recognized leaders. It seems unlikely that sex, *per se*, is an important determinant of leadership; rather, that cultural and social factors are predominant. *See* GREAT MAN THEORY.

F Denmark, 'Styles of Leadership', *Psychology of Women Quarterly*, vol. II (1977), 99–113.

NS

Great Dying *Biology See* MASS EXTINCTION.

greatest happiness principle *Philosophy/ Politics See* UTILITARIANISM.

Green's theorem (19th century) *Mathematics* Named after British mathematician and physicist George Green (1793–1841).

(1) The theorem that if f and g are smooth functions, then

$$\iint (f\nabla g - g\nabla f)\cdot \mathbf{n}\mathbf{ds} = \iiint (\mathbf{f}\nabla^2\mathbf{g} - \mathbf{g}\nabla^2\mathbf{f})\mathbf{dV}$$

where **n** is a unit normal, S is the surface and V the volume of a closed surface.

(2) The theorem that the line integral of two continuous differentiable functions f and g around a closed curve is the same as the double integral of the partials taken over the planar region enclosed by the curve. That is,

$$\int_{\partial\omega} (f\mathrm{d}x + g\mathrm{d}y) = \iint_{\omega} \frac{\partial f}{\partial x} - \frac{\partial g}{\partial y}\,\mathrm{d}A$$

This is a special case of STOKES'S THEOREM.

H Anton, *Calculus with Analytic Geometry* (New York, 1980)

ML

Greenspoon effect *Psychology* Named after the American psychologist J Greenspoon (1921–).

An experimental effect found in some studies of verbal conditioning in which the speaker's use of certain classes of words (for example, plural nouns) may increase in frequency when reinforced by the listener making appropriate diffident gestures of assent. Behaviourists in particular have taken this finding as evidence that language can be brought under operant control; and further, that language learning takes place through a process of social reinforcement. *See* BEHAVIOURISM.

NS

Greenstein hypothesis *Biology* The theory that tumours possess a general convergence of enzyme patterns leading to biochemical uniformity in tumour tissues.

KH

Gresham's law (19th century) *Economics* Usually attributed to English businessman Sir Thomas Gresham (1519–79), this law is often summarized as 'Bad money drives out good.'

Gresham's observation concerned the likelihood that coins with bullion content equal or higher than their face value would be removed from circulation and melted down, leaving in circulation only coins with a metal value lower than their face value. The law has little useful application today.

PH

Griffith's theory of brittle strength (1921) *Physics* Named after its originator, Alan Arnold Griffith (1893–1963). Attempts to calculate the tensile (brittle) strength of simple crystals give results much higher than those observed. Griffith explained this by postulating the presence of minute cracks in the material (now known as Griffith cracks).

The effect of such a crack is to produce a very high concentration of stress at its tip, while the bulk stress can remain low. When a crack extends under load it releases strain energy which is used to provide the surface free energy of the new surfaces created. The crack lengthening becomes catastrophic, leading to brittle fracture, if the crack length exceeds a critical length L_c. The observed brittle strength σ_o ισ γιωεν βυ

$$\sigma_o = \frac{\gamma E}{L_c}$$

where γ is the specific surface free energy and E is Young's modulus for the material.

J Thewlis, ed., *Encyclopaedic Dictionary of Physics* (New York, Oxford and London, 1962)

MS

Grimm's law (1822) *Linguistics* Formulated by the German linguist Jacob Grimm (1785–1863). Otherwise called the 'First Germanic consonant shift'.

Exemplary early statement of regular sound changes between Indo-European and Germanic: p becomes f, b becomes p, and so on. *See* VERNER'S LAW.

R Anttila, *Introduction to Historical and Comparative Linguistics* (New York, 1972)

RF

Grotian theory (20th century) *Politics/Law* Named after the Dutch legal theorist Hugo Grotius (1583–1645).

Concepts of law and justice can be applied to international society in the manner implied in the work of Grotius, and can find expression in the practice of COLLECTIVE SECURITY.

Graham Evans and Jeffrey Newnham, *The Dictionary of World Politics* (Hemel Hempstead, 1990)

RB

Grotthuss' chain theory (1805) *Chemistry* Early theory explaining electrolysis, proposed by German chemist Theodor von Grotthuss (1785–1822).

An electric field orients electrolyte molecules in chains so that the negative parts of the molecules point to the positive electrode and the positive parts to the negative electrode. The electrodes attract the ends of the chains, thus releasing the positive part of one end molecule at the negative electrode and the negative part of the other end molecule at the positive electrode. The remaining parts of the end molecules then exchange partners with neighbouring molecules. The process continues through the whole length of the chain until a complete set of new molecules is formed. The theory was superseded by the ARRHENIUS DISSOCIATION THEORY in 1883.

MF

Grotthuss-Draper law (1817 and 1841) *Chemistry* Generalization first made on theoretical grounds by the German chemist Theodor von Grotthuss (1785–1822) in 1817, and later rediscovered by American scientist John William Draper (1811–82) in 1841 as the result of experiments on reaction of hydrogen and chlorine. The law is one of the two basic laws of photochemistry.

Only the radiation actually absorbed by a reacting system can initiate reaction. However, not all radiation absorbed brings about reaction. Some radiation may be re-emitted as fluorescence or heat. Light simply passing through the system does not initiate reaction. The other basic law of photochemistry is the EINSTEIN-STARK LAW.

P W Atkins, *Physical Chemistry* (Oxford, 1978)

MF

grounded theory (20th century) *Politics/Sociology* Attempt to relate empirical and theoretical social science.

Theories which seek to explain political or social phenomena must be 'grounded' in empirical observation, otherwise they are simple inventions. *See also* EMPIRICISM.

Barney G Glaser, *The Discovery of Grounded Theory* (New York, 1967)

RB

group selection (1962) *Biology* Most often associated with V G Wynne-Edwards, but dating back at least to Carr-Saunders' work in 1922, this is the mostly discredited idea that individual animal or human behaviour that is detrimental to the best interests of the individual, but that is in the best interests of the group, will be favoured through the mechanism of NATURAL SELECTION.

Wynne-Edwards applied this idea to reproduction, observing that some pairs of colonizing birds will refrain from reproducing if the colony's territory becomes overpopulated, thereby freeing more resources for the pairs that do reproduce. Contemporary biologists allow for the possibility of group selection in the natural world, but only in exceedingly rare cases. *See also* ALTRUISM, KIN-SELECTION THEORY, INCLUSIVE FITNESS and RECIPROCAL ALTRUISM.

V Wynne-Edwards, *Animal Dispersion in Relation to Social Behaviour* (Edinburgh, 1962)

KH

group theory (19th century) *Economics* Developed by Norwegian mathematician Marius Sophus Lie (1842–99) and later adopted by modern economists, this is a method of analyzing invariant relationships among economic variables where often the relationships are represented by differential equation systems.

R Sato, *Theory of Technical Change and Economic Invariance: Application of Lie Groups* (New York, 1981)

PH

group theory *Mathematics* The branch of ALGEBRA which is concerned with groups and their homomorphisms.

A group is a formal system consisting of a set, a distinguished element of the set (called the identity element) and an associative binary operation on the set, such that the set is closed with respect to the binary operation and every element of the set has an inverse in the set with respect to the binary operation and the identity element. For example, the integers form a group under addition with identity element zero. The concept of a group was first introduced in the early 19th century (albeit in a different form), but its rudiments can be found in antiquity. *See also* ISOMORPHISM THEOREM.

I N Herstein, *Topics in Algebra* (Wiley, 1975)

MB

group theory *Physics* Group theory is a mathematical technique for dealing with sets of elements or operations for which a law of combination may be defined. The effects of such a law satisfy certain conditions such as associativity, commutativity, possession of an identity element and so forth. In physics and chemistry, group theory is an important tool for analyzing symmetries such as the rotations and reflections of molecules. Gauge theories also make use of group theory to describe the fundamental interactions of subatomic particles.

GD

group theory (20th century) *Politics* Theory developed in the USA, of politics as the action of groups.

Neither individuals nor whole societies are significant political actors. The actions of groups in pursuit of their various INTERESTS are the sources of policy and the substance of politics. Group theory is thus a species of PLURALISM.

Bernard Crick, *The American Science of Politics* (Berkeley, 1959)

RB

groupthink theory *Psychology* Proposed by the American psychologist Irving Lester Janis (1918–).

A term referring to the manner in which pressure to conform to the group position may diminish an individual's capacity for rational thought and moral judgment. A group may unconsciously develop its own norms and illusions which eventually can hinder critical and rational decision making. Janis uses as one of his prime examples the Kennedy Administration's Bay of Pigs decision.

I L Janis, *Victims of Groupthink* (Boston, 1972)

NS

growth-pole theory *Economics* Rooted in the work of English economist and academic Sir William Petty (1623–87), and associated with French economist François Perroux (1903–87), this theory refers to the grouping of industries around a central core of other industries whose actions act as a catalyst to growth in the area.

Although the theory does not include geographical concentration of industries as part of its criteria, it has been applied to regional policy in Britain since the 1960s.

PH

growth principle *Psychology* Proposed by American psychologist Carl Rogers (1902–).

Rogers argues that physical and mental growth are best achieved in the absence of pressures, coercion and the fear of punishment. His client-centred therapies involve the therapist acting as a facilitator of growth and healing, allowing the client to re-possess the power of self-actualization. *See* ROGERIAN THEORY.

C R Rogers, *Client-Centered Therapy* (Cambridge, Mass., 1965)

NS

growth of the firm, theory of the (1959) *Economics* Posited by British economist Edith Penrose (1914–) and forming part of MANAGERIAL THEORIES OF THE FIRM, this analysis relates to economic expansion due to processes taking place within the firm.

Managers are presumed to reach their optimal rates of power and prestige by following a path towards product excellence and maximum growth. *See* THEORY OF THE FIRM, SATISFICING, X-EFFICIENCY, AGENCY THEORY, SCALAR PRINCIPLE, PARKINSON'S LAW and PETER PRINCIPLE.

E T Penrose, *The Theory of the Growth of the Firm* (Oxford, 1959); R Marris, *The Economic Theory of 'Managerial' Capitalism* (London, 1964)

PH

Gruneisen's law (1908) *Physics* Named after its discoverer, E A Gruneisen, this law states that the ratio of the expansivity of a metal to its specific heat capacity at constant pressure is a constant at all temperatures.

J Thewlis, ed., *Encyclopaedic Dictionary of Physics* (New York, Oxford and London, 1962)

MS

guild socialism (20th century) *Politics* Theory of PLURALISM and WORKERS' CONTROL developed in Great Britain.

Workers should control their own crafts and industries, but should be responsible to the INTERESTS of society as a whole through representative institutions based on occupation rather than geographical constituencies. This was a compromise between SYNDICALISM and SOCIAL DEMOCRACY.

Rodney Barker, *Political Ideas in Modern Britain* (London, 1989)

RB

Guldberg-Waage law *Chemistry See* MASS ACTION LAW.

Gurney-Mott process *Chemistry* Theory of the photographic process.

The photographic process has a two-stage mechanism. The primary electronic stage is a photoconductance process. A point within the silver halide gelatin absorbs a quantum of light. This releases a mobile electron, leaving a positive hole. These mobile defects move to trapping sites. The second stage is an ionic process. The negative charges of the trapped electrons are neutralized by positively charged silver ions forming silver atoms. The silver speck grows as further electrons are trapped.

MF

gynocritics *Literary Theory See* FEMINIST CRITICISM.

H

habituation theory of learning *Psychology* Not attributable to any one person, this term refers to a body of empirical evidence to do with the spontaneous cessation of a behavioural response previously linked with the presentation of an unconditional stimulus.

There are six main characteristics to the process: (1) after a lengthy absence of the unconditioned stimulus, the response will spontaneously recur; (2) presentation of the unconditioned stimulus after spontaneous behaviour has ceased will elongate its cessation; (3) presentation of another stimulus usually leads to spontaneous restoration of the behaviour; (4) the more often the unconditioned stimulus is presented the faster the process of habituation; (5) habituation may generalize to other stimuli which are like the original stimuli; (6) the stronger the unconditioned stimuli the slower the habituation process.

S H Hulse, H Egeth and J Deese, *The Psychology of Learning* (New York, 1980)

NS

Hadley cell model (1735) *Climatology* The English meteorologist George Hadley (1685–1768) attempted to explain the phenomenon of trade winds in terms of a thermally-driven convective cell in the atmosphere in equatorial and polar latitude. Air heated near the Equator would rise and then sink down near the poles.

This model failed, mainly because it did not take into account the effect of the Earth's rotation. Nonetheless, thermally-driven circulation is still the basic model for understanding the general pattern of the Earth's atmospheric circulation pattern, and is still used in the description of individual circulations driven by surface temperature distributions, as in equatorial and polar latitudes. *See also* ROSSBY MODEL and THREE-CELL MODEL.

G Hadley, 'Concerning the Cause of the General Trade Wind', Philosophical Transactions of the Royal Society of London, vol. XXXIX (1735), 58–73

DT

haecceitism *Philosophy* Literally: 'thisness-ism'. Theory deriving from Johannes Duns Scotus (*c.*1266–1308), with roots in Aristotle, that as well as ordinary general properties there are special properties (*haecceities* or thisnesses) necessarily associated each with just one individual.

Socrates has the property of Socrateity and Plato that of Platonity. Traditional Aristotelianism individuated objects by their matter, which as such (abstracted from all form) was unknowable. Properties, however, count as form rather than matter, and so Socrates and Plato could now – at least in principle – be distinguished by reason, not just by the senses. Recently, 'anti-haecceitism' has been used for the rejection either of primitive (that is unanalyzable) thisnesses or of primitive transworld identity (*see also* COUNTERPART THEORY).

A B Wolter, *The Philosophical Theology of John Duns Scotus*, M McC Adams, ed. (1990), ch. 4; see also item 152 in the bibliography

ARL

Haeckel's law of recapitulation *Biology See* RECAPITULATION.

Hahn-Banach theorem (1927, 1929) *Mathematics* Named after Hans Hahn (1879–1934) and Stefan Banach (1892–1945), this is the fundamental result in FUNCTIONAL ANALYSIS whereby a linear functional which is defined on a subspace of a vector space and dominated by a sublinear function defined on the entire space has a linear extension to the entire space which is still dominated by the sublinear function.

Béla Bollobás, *Linear Analysis* (Cambridge, 1990)

MB

hailstone sequence *Mathematics See* COLLATZ SEQUENCE.

Haldane-Oparin hypothesis *Biology See* OPARIN-HALDANE HYPOTHESIS.

Haldane's law or rule (20th century) *Biology* Formulated by British geneticist John Burdon Sanderson Haldane (1892–1964), this law states that when one sex is absent, rare or sterile, in the offspring of two different animal species, that sex is the one which produces gametes containing unlike sex chromosomes (for example, the X and Y chromosome together in the sperm cells of male mammals). Haldane's rule is known to apply to numerous mammals, birds and insects.

KH

Hall effect (1879) *Physics* Named after its discoverer Edwin H Hall (1855–1938).

When a current-carrying conductor or semiconductor is placed in a transverse magnetic field of flux density **B**, a transverse potential difference $\mathbf{E}_{\mathrm{H}}$ is set up across the conductor in the direction mutually perpendicular to the directions of the magnetic field and the current flow. This is given by

$$\mathbf{E}_{\mathrm{H}} = R\mathbf{Bj}t$$

where **j** is the current density and t is the thickness of the material in the direction of the Hall voltage. R is a constant, characteristic of the material, known as the Hall coefficient. The Hall voltage is developed because the moving charges that constitute the current move to the surface of the specimen until the electric field associated with this accumulated charge cancels the force produced by the magnetic field.

J Thewlis, ed., *Encyclopaedic Dictionary of Physics* (New York, Oxford and London, 1962)

MS

Hallwachs's effect (1888) *Physics* Named after its discoverer, German physicist Wilhelm Ludwig Franz Hallwachs (1859–1922), who showed that the irradiation of an initially uncharged and electrically isolated metallic body with ultra-violet light causes the body to acquire a positive charge.

In 1899, the British mathematical physicist Sir Joseph John Thomson (1856–1940) showed that the photoeffect observed by Hallwachs is a result of the emission of electrons induced by the ultra-violet light. *See* PHOTOELECTRIC EFFECT.

J Thewlis, ed., *Encyclopaedic Dictionary of Physics* (New York, Oxford and London, 1962)

MS

halo effect *Psychology* First empirically supported by the American psychologist Edward Lee Thorndike (1874–1949).

The extension of an overall impression of a person (or one particular outstanding trait) to influence the total judgment of that person. The effect is to evaluate an individual high on many traits because of a belief that the individual is high on one trait. Similar to this is the 'devil effect', whereby a person evaluates another as low on many traits because of a belief that the individual is low on one trait which is assumed to be critical.

E L Thorndike, 'A Constant Error on Psychological Rating', *Journal of Applied Psychology*, vol. IV (1920), 25–29

NS

halting problem *Mathematics* A TURING MACHINE is an idealized computer used conceptually in COMPUTABILITY THEORY, named after A Turing (1912–54). It consists typically of an infinitely long tape which is divided up into cells, in each of which may possibly be written one of a finite number of possible symbols. The tape is read by a tape head which may be in one of a finite number of states. Depending on its state and what it reads, the tape head may overwrite a cell and move to another cell. The input consists of a finite number of cells containing symbols, and the output consists of the written cells if and when the machine halts. The halting problem is then to determine for a given input if the machine will halt. This problem is unsolvable – no finite procedure can answer it for an arbitrary input.

AL

ham sandwich theorem *Mathematics* The theorem that, given three volumes in Euclidean three-dimensional space, there is at least one plane which bisects all three volumes simultaneously. So a ham sandwich can be sliced into two pieces so that each piece has equal amounts of ham and bread.

ML

hamartia *Literary Theory See* TRAGIC FLAW.

Hamilton's genetical theory of social behaviour *Biology See* ALTRUISM.

Hamilton's principle (1835) *Physics* Named after the Irish mathematician Sir William Rowan Hamilton (1805–65).

The evolution of a dynamical system from a time t_1 to a time t_2 is such that the action $S(t_1,t_2)$ is a minimum with respect to arbitrary small changes in the trajectory. The action is defined as the time integral of a function called the Lagrangian:

$$S = \int_{t_2}^{t_1} L \, dt$$

where, for conservative systems, $L = T - V$. T is the kinetic energy and V is the potential energy of the system.

Robert M Besançon, ed., *The Encyclopedia of Physics* (New York, 1985)

GD

Hammick and Illingworth's rules *Chemistry* Empirical rules concerning the direction of substitution in aromatic molecules.

Consider the following structure

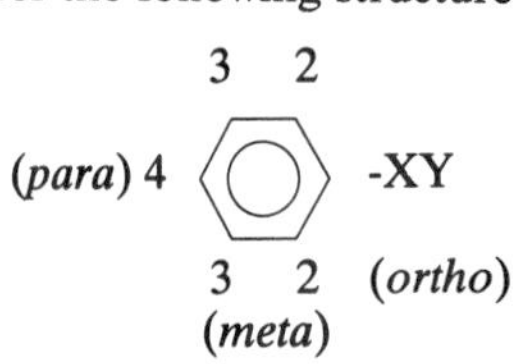

The group —XY directs substituents to the 3- (or *meta*-) position if either Y is in a higher group of the periodic table than X, or X and Y are in the same group and Y has the lower atomic mass. Examples are:

$$-XY = -CN,\ -NO_2,\ -SO_3H\ .$$

The group $-XY$ is 2,4-(*ortho*, *para*-) directing if either Y is in a lower group of the periodic table than A, or the group $-XY$ is a single atom. Examples are: $-NH_2$, $-OH$, $-Cl$. These two rules do not apply universally.

MF

Hammond principle (1955) *Chemistry* Also known as the Hammond postulate, this is the hypothesis that when a transition state leading to an unstable reaction intermediate (or product) has nearly the same energy as that intermediate, the two are interconverted with only a small reorganization of molecular structure.

Essentially the same idea is sometimes referred to as Leffler's assumption, which was put forward in 1953. This states that the transition state bears the greater resemblance to the less stable species (reactant or reaction intermediate/product).

Compendium of Chemical Terminology: IUPAC Recommendations, (Oxford, 1987)

MF

handicap principle (1975) *Biology* Proposed by A Zahavi, this is the theory that a male animal's possession of an apparent physical handicap (such as a large, encumbering tail) evolved because only especially strong, fit individuals could survive with such an affliction. In choosing such a male, a female would pass this high survival potential on to her offspring. Several weaknesses have been identified in this model, however, including the fact that not all sons can be expected to inherit both the handicap and the propensity to survive, and that the handicap would not be passed on to daughters in any case.

A Cockburn, *An Introduction to Evolutionary Ecology* (Oxford, 1991)

KH

haplodiploid hypothesis (1964) *Biology* Developed by W D Hamilton as a specific extension of KIN-SELECTION THEORY, this is the idea that an individual can increase his or her fitness for survival and that of close relatives (the individual's INCLUSIVE FITNESS) by taking action to enhance the fitness of close relatives.

Among bees, ants and wasps (*Hymenoptera*), females have a diploid number of chromosomes while males are haploid. Consequently, a female can better enhance her genetic contribution by helping her mother care for her sisters, with whom she shares three-quarters of her genes, than in producing offspring, with whom she would share only half her genes. *See also* ALTRUISM, MATERNAL-CONTROL HYPOTHESIS and MUTUALISTIC HYPOTHESIS.

D Dewsbury, *Comparative Animal Behavior* (New York, 1978)

KH

happening (1959) *Art* Ultimately related to performances in DADA and SURREALISM, this term was coined by American artist Allan Kaprow (1927–) in 1959. It refers to an assemblage of events which can occur in any environment, according to a plan but without rehearsal, and depending on audience participation for its development (for example, the Fluxus group and BODY ART). *See also* ENVIRONMENT ART.

A Kaprow, *Assemblage, Environments and Happenings* (1966)

AB

hard-sphere collision theory *Chemistry* This theory enables the rates of molecular gas-phase reactions to be calculated by assuming that the molecules collide as hard spheres. *See also* COLLISION THEORY.

MF

Hardy-Littlewood method (1920) *Mathematics* Named after the British mathematicians Godfrey Harold Hardy (1877–1947) and John Edensor Littlewood (1885–1977) who introduced it.

An analytic method arising from work on the GOLDBACH CONJECTURE, this forms the basis of much numerical work related to WARING'S PROBLEM. It is a powerful method which has been adopted to attack many problems in additive NUMBER THEORY.

R C Vaughan, *The Hardy–Littlewood Method* (Cambridge, 1981)

MB

Hardy-Schultze rule *Chemistry* This is an empirical rule concerning the influence of electrolytes on the precipitation of hydrophobic sols. It is also known as the Schultze-Hardy rule.

The effective entity in precipitation is the ion of opposite charge to that on the surface of the colloidal particle. Destabilizing power increases with the charge on the effective ion. The nature and valency of the ionic species with the same charge as that of the sol has relatively little effect on the critical coagulation concentration (also known as the precipitation or flocculation value).

W J Popiel, *Introduction to Colloid Science* (New York, 1978)

MF

Hardy-Weinberg equilibrium, law or principle (*c*.1908) *Biology* Also called the Castle-Hardy-Weinberg equilibrium after William Ernest Castle (1867–1962), American biologist. Independently described by Wilhelm Weinberg (1862–1937), a German physician, and Godfrey H Hardy (1877–1947), a British mathematician.

This law states that the frequencies of alleles in a randomly mating population will achieve equilibrium frequency, provided the population is large enough to be governed by Mendel's laws. This law demonstrates that meiosis and recombination of genes do not alter gene frequencies.

F Magill, *Magill's Survey of Science: Life Science Series* (Englewood Cliffs, NJ, 1991)

KH

harmonic analysis (19th century) *Mathematics* A technique for representing functions as sums of simpler functions as suggested by the work of Jean Baptiste Fourier (1768–1830).

The representation of a function $f(x)$ with the property that there exists a period q such that $f(x + q) = f(x)$ for all x by means of sums and integrals of trigonometrical functions.

JB

harmonization of impulses (1924) *Literary Theory* Idea proposed by English theorist Ivor Armstrong Richards (1893–1979); based on a yoking of Samuel Taylor Coleridge's IMAGINATION and early 20th century neurology.

Humans try to satisfy the maximum number of 'appetencies', and are led to a state of conflicting impulses; the poet's experiences are more integrated, and a poem has value to the extent that it can harmonize the impulses of readers. This is a problematic theory, but one which strongly influenced NEW CRITICISM in its preoccupation with tension and PARADOX.

I A Richards, *Principles of Literary Criticism* (London, 1924)

RF

Harrod-Domar growth model (20th century) *Economics* Named after English economist Roy Harrod (1900–78) and Polish-born American economist Evsey Domar (1914–), this model postulates three kinds of growth: (1) warranted growth (the rate of output at which firms feel they have the right level of capital and do not wish to expand or decrease investment); (2) natural rate of growth (corresponding to growth in the labour force); (3) actual growth (resulting from a change in aggregate output).

However, there are problems between actual and natural growth and warranted and actual growth. The factors that determine actual growth (propensity to save, investment) are autonomous from those factors determining natural growth (birth control, tastes of population, and so on). Disequilibrium arises in a situation in which warranted growth is different to the natural rate of growth; when equal steady growth is accompanied by full employment or a constant rate of unemployment occurs. *See* NATURAL AND WARRANTED RATES OF GROWTH, and SOLOW ECONOMIC GROWTH MODEL.

R F Harrod, *Towards a Dynamic Economics* (London, 1948); E Domar, *Essays in the Theory of Economic Growth* (New York, 1957)

PH

Hartree-Fock theory *Chemistry See* SELF-CONSISTENT FIELD THEORY.

Hausdorff maximality theorem (1914) *Mathematics* Named after its discoverer Felix Hausdorff (1868–1942).

Let A be a set with a partial order, that is, a relation like $<$ but such that $a < b$ and $b < a$ may both be untrue for some pairs of elements. Then A has a subset B such that for c and d in B either $c < d$ or $d < c$. This axiom is equivalent to ZORN'S LEMMA.

JB

Hawaiian radiation *Biology See* ADAPTIVE RADIATION.

Hawthorne effect (1924–36) *Psychology* Named after the industrial plant where it was first observed by the American psychologist G C Homuns.

This is a generalization stating that anything new will bring about a short-term improvement; for example, new programmes, working conditions and so forth. It suggests that any workplace change (such as a research study) makes people feel important and thereby improves their performance. In the original study, even control conditions designed to lower workers' productivity resulted in increased output. The existence of this effect makes an evaluation of any new programme a difficult process. By implication there can also be a negative Hawthorne effect.

D Bramel and R Friend, 'Hawthorne, The Myth of the Docile Worker and Class Bias in Psychology', *American Psychologist*, vol. XXXVI (1981), 867–78

NS

Hayflick limit (1960s) *Biology* Also invoked as the cell theory of ageing. Proposed by Leonard Hayflick of Stanford University, California, this is the concept that the cells of most animals grown in the laboratory will divide a limited number of times and then cease dividing.

More specifically, it is the observation that normal human diploid cells cultured *in vitro* possess a specific clonal lifespan, the Hayflick limit, that is similar to that seen in some protozoans. Ageing and senescence are hypothesized to be related to this limit: when cells have divided all they can, they die, and hence the animal ages and dies. Based on this assumption, the maximum theoretical lifespan of humans is about 110 years. Evidence supporting the connection between the Hayflick limit and ageing remains sparse, however. *See also* AGEING, THEORIES OF.

G Stine, *The New Human Genetics* (Dubuque, Ia, 1989)

KH

healing power of nature (5th–4th century BC) *Medicine* Described by followers of the Greek physicians Hippocrates (*c*.460–*c*.377 BC) and Galen (129–*c*. AD 200). The doctrine that nature has endowed the body with inherent powers to restore itself to health.

In its extreme form, this theory holds that fever, inflammation, diarrhoea and other symptoms of disease are ultimately beneficial and not to be interfered with by physicians.

P Mattson, *Holistic Health in Perspective* (Palo Alto, Calif., 1981)

KH

heart-weight rule *Biology* Also called Hesse's rule, this holds that the ratio of body weight to heart weight increases in animal species found in cold regions compared with those from warmer climates. This tendency is attributed to the need to maintain a greater temperature differential between the body and the environment in cold regions.

KH

heartland theory (20th century) *Politics* Theory of GEOPOLITICS described by the British geographer Sir Halford Mackinder (1861–1947).

The group or nation which dominates the 'heartland' can then extend its domination over a far wider area. This heartland has at various times been Central Asia, the high seas, and Eurasia.

Graham Evans and Jeffrey Newnham, *The Dictionary of World Politics* (Hemel Hempstead, 1990)

RB

heat death of the universe (19th century) *Physics/Astronomy* One conclusion from the second law of thermodynamics is that the entropy of a thermally isolated closed system can only increase when natural (irreversible) processes occur. Associated with this increase in the entropy is a redistribution of the energy of the system that will reduce temperature gradients in the system.

It is these temperature gradients that can be used to drive heat engines and provide useful work. It has been suggested that the universe may be treated as a closed, thermally isolated system and that time will come to an end when there are no more temperature gradients to operate heat engines. The universe will then be at a uniform temperature; a condition that has been termed the heat death of the universe. It is by no means certain that the universe can be treated as a thermally isolated closed system.

J Thewlis, ed., *Encyclopaedic Dictionary of Physics* (New York, Oxford and London, 1962)

MS

heat conduction in solids, theory of (20th century) *Physics* In metals and alloys there is a strong correlation between electrical and thermal conductivity which indicates that in these materials heat conduction is almost entirely by free electrons. In contrast, there is no such correlation for electric insulators and semiconductors. Heat transfer in these materials is mainly by high-frequency elastic waves, known as phonons as the energy is quantized. Conduction by phonons must also occur in metals, but their effect is greatly reduced as they are scattered by the free electrons. *See* WIEDEMANN-FRANZ-LORENTZ LAW and ELECTRON THEORY OF METALS.

J Thewlis, ed., *Encyclopaedic Dictionary of Physics* (New York, Oxford and London, 1962)

MS

Hebb's theory of perception learning *Psychology* Proposed by the Canadian psychologist Donald Olding Hebb (1904–).

Learning is postulated to be dependent on 'cell assemblies' in the brain which form the neural basis of complex, enduring patterns required for concept formation and so forth. These cell assemblies are related through a process of 'phase sequencing'. Practice has the effect of building up highly coordinated groups of cells in the brain cortex. The theory has been superseded by advances in the neuropsychology of perception and learning.

D O Hebb, *The Organization of Behavior: A Neuropsychological Theory* (New York, 1949)

NS

Heckscher-Ohlin trade theory (1919) *Economics* First developed by Eli Heckscher (1879–1952) and later developed by fellow Swedish economist Bertil Ohlin (1899–1979) in 1933, this is a theory to explain the existence and pattern of international trade based on a comparative cost advantage between countries producing different goods.

Heckscher and Ohlin state that this advantage exists because of the relative resource endowments of the countries trading. However, the Russian-born American economist Wassily Leontief (1906–) in 1954 examined US foreign trade and found that US exports were more labour intensive and imports were more capital intensive (the LEONTIEF PARADOX). *See* EQUALIZATION THEOREM.

E Heckscher, 'The Effect of Foreign Trade on Distribution of Income', *Readings in the Theory of International Trade* (1949); Bertil Ohlin, *Interregional and International Trade* (1933)

PH

hedonism *Philosophy* Set of doctrines shared between philosophical psychology and ethics.

Ethical hedonism says either that pleasure alone (or 'happiness', which is usually not distinguished from pleasure by hedonists) is ultimately good, or that every action should aim to maximize pleasure; in neither case need the pleasure be the agent's (a point that is often forgotten, as is the distinction between psychological and ethical hedonism).

Qualitative hedonism – associated especially with John Stuart Mill (1806–73) in his *Utilitarianism* (1861) – says that pleasures differ in quality as well as quantity, and that 'higher' ones should be preferred. This doctrine thus gravely complicates the task of aiming to produce the 'greatest' pleasure.

J C B Gosling, *Pleasure and Desire* (1969); J S Mill, *Utilitarianism* (1861), ch. 2 (reprinted in J Plamenatz, *The English Utilitarians* (1949), 137

ARL

hedonism *Psychology* A concept finding early expression in the work of English philosopher Jeremy Bentham (1748–1832).

The belief that behaviour is primarily motivated by an organism's avoidance of pain and its constant pursuit of pleasure. In motivation theories, it is believed that an organism's behaviour is constantly concerned with the reduction of tension and that the pleasure this reduction causes is observable.

N E Miller and J Dollard, *Social Learning and Imitation* (New Haven, Conn., 1941)

NS

hedonistic utilitarianism *Philosophy* Version of UTILITARIANISM specifying the good to be sought in terms of pleasure.

Most utilitarians until recently have adopted this version (which is the same as one version of ethical HEDONISM), but specifying that the pleasure concerned is that of people (or sentient beings) in general, neither limited to nor excluding that of the agent. Objections include doubts about whether pleasure is the only value, and indeed about what pleasure *is*, and about difficulties in measuring it. These difficulties lead one towards PREFERENCE UTILITARIANISM.

H Sidgwick, *The Methods of Ethics* (1874, 7th and final edition 1901); leading exponent

ARL

hegemony (20th century) *Politics* Theory of class domination associated with Italian Marxist Antonio Gramsci (1891–1937).

Previous Marxist theory had presented classes, especially the capitalist class, ruling via its domination of the state. Gramsci argued that domination of ideas and culture (hegemony) was equally effective, and that for the working class to challenge the existing order it would have – amongst other things – to challenge this cultural dominance with an alternative version of its own.

David Miller *et al.*, eds, *The Blackwell Encyclopaedia of Political Thought* (Oxford, 1987)

RB

hegemonic stability theory (20th century) *Politics* Theory of international politics. Nations achieve dominance in international systems, which they then must maintain by 'rewards' to less powerful nations. Such a system is, paradoxically, unstable.

Graham Evans and Jeffrey Newnham, *The Dictionary of World Politics* (Hemel Hempstead, 1990)

RB

Heine-Borel covering theorem *Mathematics* Named after German mathematician Heinrich Heine (1821–81) and French mathematician Félix Edouard Justin Emile Borel (1871–1956).

This is the fundamental result in TOPOLOGY whereby a subset of a Euclidean space is closed and bounded if and only if it is compact (every one-cover has a finite subcover). *See also* BOLZANO-WEIERSTRASS THEOREM.

W Rudin, *Principles of Mathematical Analysis* (McGraw-Hill, 1976)

MB

Heine's theorem *Mathematics* Named after German analyst Heinrich Heine (1821–81), this states that a continuous function on a compact subset A of a metric space to another metric space is uniformly continuous on A.

W F Trench, *Advanced Calculus* (New York, 1978)

ML

Heisenberg's theory of ferromagnetism *Physics* Named after the German theoretical physicist Werner Karl Heisenberg (1901–76), winner of the 1932 Nobel prize for Physics.

A theory in which ferromagnetism is explained in terms of the exchange forces between electrons in neighbouring atoms. The electrons are localized on individual atoms rather than being shared by the crystal as a whole. The exchange forces depend on relative orientations of electron spins: parallel spins are favoured so that all the spins in a lattice have a tendency to point in the same direction. *See also* COLLECTIVE ELECTRON THEORY and MAGNETISM, THEORIES OF.

Robert M Besançon, ed., *The Encyclopedia of Physics* (New York, 1985)

GD

Heisenberg's uncertainty principle (1927) *Physics* Named after the German theoretical physicist Werner Karl Heisenberg (1901–76), this is also known as the indeterminacy principle.

An inherent principle of quantum mechanics which states that at the microscopic level, it is impossible to know both the momentum p and position x of a particle with absolute precision. The product of the uncertainties in both measured values is of the order of magnitude of the Planck constant, $\Delta p_x \times \Delta x \geqslant h/4\pi$, where Δ is the root-mean-square value of the uncertainty. The principle is a consequence of the fact that any attempt at measurement must disturb the system under investigation, with a resulting lack of precision. The determination of both energy and time is also subject to the same uncertainty, $\Delta E \times \Delta t \geqslant h/4\pi$. One consequence of the uncertainty principle is that the macroscopic principle of causality cannot apply at the atomic level.

Paul Davies, ed., *The New Physics* (Cambridge, 1989)

GD

Heitler-London covalence theory (1927) *Chemistry* This is a theory for applying wave mechanics to chemical bonding, developed by German chemists W Heitler and Fritz London (1900–54).

The theory enables the binding energy and distance between the atoms of the hydrogen molecule (H_2) to be calculated. It assumes that the electrons are in atomic orbitals about each of the hydrogen nuclei, and that in the formation of a molecule the orbital part of the wave function is a linear combination of the products of the separate atomic orbitals.

S Glasstone, *Textbook of Physical Chemistry* (London, 1948)

MF

heliocentric theory of the solar system (3 BC) *Astronomy* A theory first proposed by Aristarchus of Samos (310–230 BC), this postulated that the Sun is at the centre of the solar system's motion; that is, the planets move in orbits around the Sun. The theory was revived in about 1515 (published in 1543) by the Polish-German astronomer Nicolaus Copernicus (1473–1543), who was looking for a simpler way of describing the motion of the planets relative to the Earth than that provided by the PTOLEMAIC THEORY OF THE SOLAR SYSTEM. Copernicus's proposal helped the German astronomer Johann Kepler (1571–1630) in his discoveries. *See* KEPLER'S LAWS OF PLANETARY MOTION.

S Mitton, ed., *The Cambridge Encyclopaedia of Astronomy* (London, 1973)

MS

Hellmann-Feynman theorem (20th century) *Physics* Named after Hellmann and the American physicist Richard Feynman (1918–88), joint winner of the 1965 Nobel prize for Physics.

For molecules and solids described by the Born-Oppenheimer approximation, the theorem states that the forces on the nuclei are those which would arise electrostatically if the electron probability density were treated as a static distribution of negative electricity.

GD

Helly's theorem (1923) *Mathematics* This result, in some sense dual to the CARATHÉODORY THEOREM, states that if a finite family of convex sets in n-dimensional Euclidean space has the property that any $n + 1$ sets have a point in common, then the whole family has a common point.

AL

hemispheric dominance and language *Linguistics See* CEREBRAL LOCALIZATION.

Hempel's paradox (20th century) *Philosophy* Also known as the confirmation paradox, it was discovered by Carl Gustav Hempel (1905–).

The statement 'All prime ministers live at 10 Downing Street' tends to be confirmed by finding a kennel containing a dog, because this is an example of a dwelling that is not 10 Downing Street which is the home of a non-prime-minister; which is a logically equivalent statement. However, the same process could be used to prove that all the roses in a large garden have green leaves by observing that all the plants with other-coloured leaves are not roses, a much easier process. The resolution of the paradox is to restrict the universe in which the search is made. However, the paradox is not a total surprise because non-mathematical induction is not a logical proof. For example, the assertion that water is a compound of hydrogen and oxygen is based on a study of a tiny proportion of the extant water, and it is conceivable that most of the rest is a compound of carbon and manganese.

JB

Henry's law (1803) *Chemistry* A law proposed by English chemist William Henry (1775–1836).

The mass (m) of a gas dissolved in a given volume of liquid with which it is in equilibrium is proportional to the pressure (p) of the gas at a given temperature,

$$m \propto p$$

and thus

$$m = Kp$$

where K is a constant of proportionality. The law is not valid when a chemical reaction occurs between the gas and the solvent.

M Freemantle, *Chemistry in Action* (Basingstoke, 1987)

MF

Herbartianism *Psychology* Advanced and named after the German philosopher Johann Friedrich Herbart (1776–1841).

The thesis proposed that ideas and concepts compete and struggle for acceptance. Herbart believed that new ideas had to be related to a previously acquired set of ideas. Herbart is regarded as having influenced psychoanalytic thought and to have contributed mathematical descriptions of human behaviour. His ideas have largely lost their influence in contemporary educational theories.

J F Herbart, *A Textbook of Psychology: An Attempt to Found the Science of Psychology on Experience, Metaphysics and Mathematics* (New York, 1891)

NS

hereditarianism *Psychology See* GENETICISM.

hereditary symbiosis *Biology See* SYMBIOTIC THEORY OF CELLULAR EVOLUTION.

heredity predisposition theory *Psychology* Proposed by the French naturalist Jean Baptiste Pierre Antoine de Monet, Chevalier de Lamarck (1744–1829).

Used in relation to pathological conditions to explain the conduct of an individual who appears to have inherited a predisposition towards a particular trait. The pathology is supposed to develop only in the appropriate environmental context. The most common pathology termed as arising from a heredity predisposition is schizophrenia.

L E Tyler, *Individuality* (San Francisco, 1981)

NS

heresy of paraphrase (1947) *Literary Theory* Formulated by the American NEW CRITIC Cleanth Brooks (1906–).

It is false to equate 'the real core of meaning' of a poem with its paraphraseable propositional content. Poems do not make statements true or false, but are complex unities of diverse meanings enriched by PARADOX, AMBIGUITY and IRONY. *See also* EMOTIVE AND SCIENTIFIC LANGUAGE.

C Brooks, *The Well Wrought Urn* (London, 1968 [1947])

RF

Hering theory of colour vision *Physics/ Psychology* Named after its originator, Karl Ewald Konstantin Hering (1834–1918), this is a three-substance but six-colour theory of colour vision.

It supposes a red-green substance, a yellow-blue substance and a white-black substance, each of which can be excited to respond in either catabolism or metabolism, corresponding to the sensations of white, yellow and red or black, blue and green, respectively. It was also postulated that there was an independent mechanism for the seeing of black and white. Since the experimental evidence is better discussed on a tri-receptor theory, Hering's theory is now obsolete. *See also* MUELLER'S THEORY OF COLOUR VISION.

J Thewlis, ed., *Encyclopaedic Dictionary of Physics* (New York, Oxford and London, 1962)

MS

hermeneutics *Philosophy* Literally, the study of interpretation. The term was originally associated with biblical studies, but a philosophical tendency has been developed especially by Friedrich Schleiermacher (1768–1834), Wilhelm Dilthey (1833–1911), and H G Gadamer (1900–).

Dilthey emphasized the need in human studies (*Geisteswissenschaften*) for an empathetic understanding (usually called by the German term, *Verstehen*) which went beyond mere external description. (Compare also the philosophy of Giovanni Vico (1668–1744). Gadamer has emphasized the way interpretation develops gradually by an interplay between the interpreter and the subject-matter, denying both that there is a single objectively correct interpretation and that we can never get beyond our own initial interpretation.

H G Gadamer, *Philosophical Hermeneutics* (1976); translated essays with introduction

ARL

hermeneutics (1654–) *Psychology* When applied in psychology, hermeneutics must accomplish by conscious effort and technique what ordinary conversationalists achieve effortlessly: an understanding of the contents of each other's 'minds'.

R E Palmer, *Hermeneutics: Interpretation Theory in Schleiermacher, Dilthey, Heidegger and Gadamer* (Evanston, Ill., 1969)

NS

Hertzfeld and Mayer theory of melting (1934) *Physics* Named after its originators, Karl Ferdinand Hertzfeld and Maria Goppert Mayer, this theory is a variant of LINDEMANN'S THEORY OF MELTING.

The amplitude of atoms' vibration in a solid increases as the temperature rises. The Hertzfeld-Mayer model assumes that each atom vibrates independently of its neighbours and that, when the amplitude of vibration reaches a certain critical fraction of the mean interatomic spacing, the vibrations will interfere to such an extent that the structure becomes mechanically unstable and melting occurs. Hertzfeld and Mayer assumed an interatomic potential of the Mie type and took as the critical amplitude of vibration of the atoms that corresponding to the point of inflexion on the interatomic potential curve.

J Thewlis, ed., *Encyclopaedic Dictionary of Physics* (New York, Oxford and London, 1962)

MS

Hess's law of constant heat summation (1840) *Chemistry* Named after German-born Russian chemist Germain Henri Hess (1802–50), the founder of thermo-chemistry.

The enthalpy change in a chemical reaction is the same whether it takes place in one or several stages. The enthalphy change of the overall chemical reaction is the algebraic sum of the enthalpy changes of the reaction steps.

M Freemantle, *Chemistry in Action* (Basingstoke, 1987)

MF

Hesse's rule *Biology* *See* HEART-WEIGHT RULE.

heteroglossia (1930s) *Linguistics/Literary Theory* Rendering of the Russian *raznorecie* in the work of Mikhail M Bakhtin (1895–1975).

A language is not a single entity; whatever one says might have been said in a different accent, variety or REGISTER, and is therefore always experienced in relation to alternative voices. *See also* DIALOGISM.

M M Bakhtin, trans. C Emerson and M Holquist, *The Dialogic Imagination* (Austin, 1981 [1934–41])

RF

hexachord (11th century) *Music* The hexachord system is attributed to the Italian music theorist Guido of Arezzo (10th–11th centuries). It is not discussed in his extant writings but he is cited as the originator by later theorists.

In medieval music, a succession of six pitches, separated by the intervallic distance tone, tone, semitone, tone, tone, that forms the basis of the Guidonean system of SOLMIZATION. Guido proposed that if seven hexachords are overlapped at particular intervals, the resulting range ('gamut') and organization of pitches may be used as a device for indicating the intervals required to sing plainchant. If a melody extends beyond the range of a given hexachord, the hexachord in use is said to 'mutate' (change to an overlapping hexachord) in order to accommodate the new note(s).

C G Allaire, *The Theory of Hexachords, Solmization and the Modal System* (Rome, 1972)

MLD

Hick-Hyman law *Psychology* Formulated by the American psychologists W E Hick (*fl.*1950s) and R Hyman (1928–).

As the number of options available to an organism increases, so will the reaction time increase. Reaction time is seen as a function of the amount of information available to the organism at the time a response is to be made. It is also noted that reaction time decreases when errors increase. The Hick-Hyman law is written as follows;

$$RT = a + bH ,$$

where RT is reaction time, a and b are constants and H is the amount of information measured in bits (binary digits).

M A Clee and R A Wicklund, 'Consumer Behavior and Psychological Reactance', *Journal of Consumer Research*, vol. VI (1980), 389–405

NS

Hicks-Hansen model *Economics* *See* IS-LM MODEL.

hidden variable theory *Physics* A theory arising out of the EPR paradox of 1935, identified by the German-Swiss-American mathematical physicist Albert Einstein (1879–1955), Podolsky and Rosen.

Einstein, Podolsky and Rosen proposed a 'thought experiment' involving two spinning particles that interact and then separate to a vast distance. This was an attempt to show the incompleteness of quantum theory in its attempts to give an objective description of reality. The hidden variable theory attempts to account for the non-physical and non-local correlations of the quantum description in a deterministic manner. The hidden variables are those components of the hypothetical complete state which are not contained in the quantum state. An experimental test of the theory was developed by Alain Aspect in 1982 which confirmed that there are indeed superluminal connections between distant regions of space-time, thus upholding the quantum view at the expense of Einstein's deterministic view of reality. *See also* SCHRÖDINGER'S CAT.

Paul Davies, ed., *The New Physics* (Cambridge, 1989)

GD

high politics (20th century) *History/Politics* An account of political history given by British historian Maurice Cowling (1926–).

Politics is carried on within a relatively restricted world of politicians, officials, and public figures. It is guided not so much by general principles as by the desire of the participants to survive politically in response to changing circumstances.

Maurice Cowling, *The Impact of Labour* (Cambridge, 1971)

RB

Hilbert's basis theorem (1888) *Mathematics* Named after the German mathematician David Hilbert (1862–1943).

This result essentially states that given a collection of infinitely many forms (that is, rational integral homogeneous functions in n variables with coefficients in a field) of any degree in the n variables, there is a finite number (a basis) $F_1,\ldots,F_m$ such that any form F of the collection can be written as

$$F = A_1F_1 + \ldots + A_mF_m$$

where $A_1,\ldots,A_m$ are suitable forms in the n variables with coefficients in the same field as the coefficients of the infinite system. Hilbert's theorem solved the principal problem in 19th-century invariant theory of showing that any form of given degree and given number of variables has a finite complete system of independent rational integral invariants and covariants. *See also* ALGEBRAIC GEOMETRY and BASIS THEOREM.

T W Hungerford, *Algebra* (Springer, 1974)

MB

Hilbert's *Nullstellensatz* (1893) *Mathematics* Named after the German mathematician David Hilbert (1862–1943), this result essentially states that every algebraic structure in a space of arbitrarily many homogeneous variables $x_1,\ldots,x_n$ can be represented by a finite number of homogeneous equations

$$F_1 = 0, \ldots, F_m = 0$$

so that the equation of any other structure containing the original one can be represented by

$$A_1F_1 + \cdots + A_mF_m = 0$$

where the As are homogeneous integral forms whose degree is such that the left-hand side of the equation is itself homogeneous. *See also* ALGEBRAIC GEOMETRY and HILBERT'S BASIS THEOREM.

M Reid, *Undergraduate Algebraic Geometry* (Cambridge, 1990)

MB

Hilbert's paradox (20th century) *Mathematics* This paradox concerning infinite numbers was found by David Hilbert (1862–1943) and is also called the infinite hotel paradox.

Suppose that a hotel with a (countably) infinite number of rooms is full but another guest arrives. Then if every guest moves from room n to room $n + 1$, room 1 becomes available for the newcomer. This can be modified for any finite number of new guests, but if an infinite party arrives, everyone can move from room n to room $2n$ and leave all the odd-numbered rooms vacant. This serves to illustrate the surprising, but not contradictory, properties of infinite numbers.

JB

Hilbert's problems (1900) *Mathematics* A list of 23 problems put forth by the German mathematician David Hilbert (1862–1943) at the Paris conference of the International Congress of Mathematicians. Hilbert, who was already famous for his work on the theory of invariants, considered them to be significant unsolved problems with which mathematicians of the 20th century should be concerned. Several of them remain unsolved today.

(1) To prove the CONTINUUM HYPOTHESIS.

(2) To investigate the consistency of the axioms of arithmetic. *See* GÖDEL'S THEOREM.

(3) To show that it is impossible to prove (using only the congruence axioms) that two tetrahedra having the same altitude and base area have the same volume. This was solved by Max Dehn in 1900.

(4) To investigate geometries in which straight lines are geodesics.

(5) To obtain the conditions under which a topological group is a Lie group. This was solved by A M Gleason in 1952, H Yamabe in 1953, and D Montgomery and L Zippin in 1955.

(6) To axiomatize mathematical physics.

(7) To establish the transcendence of certain numbers. *See* GELFOND-SCHNEIDER THEOREM.

(8) To investigate problems concerning the distribution of prime numbers; in particular, to prove the RIEMANN HYPOTHESIS.

(9) To establish a general law of reciprocity. Artin obtained one for abelian extensions of Q in 1927; the non-abelian case is still open.

(10) To find a method to determine the solubility of DIOPHANTINE EQUATIONS. Matijasevich showed in 1970 that no such method exists.

(11) To investigate the theory of quadratic forms over an arbitrary algebraic number field of finite degree.

(12) To construct class fields of algebraic number fields.

(13) To show that the general algebraic equation of the seventh degree cannot be solved using compositions of continuous functions of two variables.

(14) To determine whether the ring

$$K \cap k[x_1,\ldots,x_n]$$

is finitely generated over K, where K is a field, $k[x_1,\ldots,x_n]$ is a polynomial ring and

$$k \subset K \subset k(x_1,\ldots,x_n)$$

This was proved false by Nagata in 1959.

(15) To establish the foundations of ALGEBRAIC GEOMETRY.

(16) To investigate the topology of algebraic surfaces.

(17) To determine whether a rational function with real coefficients which is positive can be written as a sum of squares of rational functions. This was proved by Artin in 1927.

(18) To determine whether there exist nonregular space-filling polyhedra.

(19) To determine whether the solutions of regular problems in the CALCULUS OF VARIATIONS are necessarily analytic.

(20) To investigate the general boundary value problem. *See* DIRICHLET PROBLEM and NEUMANN PROBLEM.

(21) To show that there always exists a linear differential equation of the Fuchsian class with given singular points and monodromic group.

(22) To uniformize complex analytic functions by means of automorphic functions.

(23) To develop the CALCULUS OF VARIATIONS.

MB

Hilbert's programme (1920) Formulated by David Hilbert (1862–1943) as a process to give mathematics a sound logical foundation, it aimed to relate all mathematics to systems of axioms and then show by finite logical arguments that the axiom systems are consistent, in that they can never lead to contradictions.

Unfortunately, GÖDEL'S THEOREM showed that this programme is impossible, though the programme also created topics of COMPUTABILITY THEORY and PROOF THEORY which continue to be of theoretical and practical interest.

JB

Hildebrandt's rule (1915) *Physics* Named after its discoverer, the German chemist Georg Friedrich Hildebrandt (1764–1816) this empirical rule states that the molar entropy of vaporization of a liquid is a constant when the density of the vapour is maintained constant.

J Thewlis, ed., *Encyclopaedic Dictionary of Physics* (New York, Oxford and London, 1962)

MS

Hilt's law *Geology* This states that the deeper the coal, the deeper its rank (grade).

The law holds true if the thermal gradient is entirely vertical, but metamorphism may cause lateral changes of rank, irrespective of depth.

DT

histogen theory (1920–25) *Biology* The obsolete theory in botany that the embryonic tissue of plants (the meristem) possesses three main sections – the dermatogen, periblem and plerome – each of which gives rise to particular structures. The dermatogen was thought to produce epidermis; the periblem, the cortex (or outer surface, such as bark); and the plerome all other tissues inside the cortex. The TUNICA-CORPUS theory largely replaced this concept

KH

histoire* and *discours (1966) *Stylistics* Distinction drawn by the French linguist Emile Benveniste, but used in a slightly different sense (below) in studies of narrative.

In Benveniste the terms refer to two kinds of utterance (*see also* 'ÉNONCIATION'). But in narratology, for example Genette, they refer to two levels, roughly like 'FABULA' AND

'SJUZHET'; that is, the stuff of the story, and the way it is told.

E Benveniste, *Problems in General Linguistics* (Coral Gables, Fla., 1971 [1966])

RF

historical materialism (19th century) *Politics/History/Economics/Sociology* Historical theory initiated by Karl Marx (1818–83).

The material circumstances of society, the manner in which its members get a living and the control of their resources for doing so, are the chief influences upon its history. *See also* DIALECTIC and DIALECTICAL MATERIALISM.

Jon Elster, *An Introduction to Karl Marx* (Cambridge, 1986)

RB

historicism *Philosophy* Term used for different and indeed incompatible theories. It has two main senses. First, that historical events must be seen in their uniqueness and can only be understood against the background of their context. In this sense it is akin to the emphasis on *Verstehen* in Wilhelm Dilthey's HERMENEUTICS. The second sense is that of Karl Raimund Popper (1902–) who uses 'historism' for the above sense. Historicism for Popper is the view that history is governed by inexorable laws, which the historian tries to predict, and is thus assimilated to science in a way quite incompatible with any appeal to *Verstehen*. Popper's historicism (which he was concerned vigorously to oppose) also gives corporate wholes a life of their own which cannot be explained in terms of the individuals composing them. *See also* HOLISM.

K R Popper, *The Poverty of Historicism* (1957)

ARL

Hofmann's rule *Chemistry* Named after the German organic chemist August Wilhelm von Hofmann (1818–92), founder of the coal tar industry, this rule relates to the reaction of quaternary ammonium hydroxides (known as Hofmann degradation, Hofmann elimination, or Hofmann exhaustive methylation).

In this type of reaction, hydroxides decompose on strong heating to yield water, a tertiary amine and an alkene. According to Hofmann's rule, the least substituted alkene is formed preferentially.

D J Cram and G S Hammond, *Organic Chemistry* (New York, 1964)

MF

Hogarth's line (1753) *Art* A term introduced by the English painter William Hogarth (1697–1764). The so-called line of beauty is a graceful curve, proposed as the foundation of all good artistic design.

W Hogarth, *The Analysis of Beauty* (London, 1753)

AB

hole theory of electrons (1928) *Physics* Introduced by the English physicist Paul Adrien Maurice Dirac (1902–84), who had derived an equation describing the motion of an electron in an electromagnetic field that satisfied the requirements of both quantum mechanics and of special relativity. The equation gives rise to both positive and negative energy states of the electron.

Dirac proposed that all the negative energy states are normally filled so that the sea of electrons with negative energies is not observed. However, if the absorption of energy makes an electron leave the sea of negative energy states it is observed as a normal electron. The residual 'hole' has the appearance of a physical particle with a positive charge and a positive energy. Originally, Dirac suggested that this particle should be the proton, but it was soon realized that it is the positive electron or positron.

J Thewlis, ed., *Encyclopaedic Dictionary of Physics* (New York, Oxford and London, 1962)

MS

hole theory of liquids (1926) *Physics* A theory advanced by a number of authors including J Frenkel (1894–1954) and, independently, Henry Eyring (1901–). This theory treats a liquid close to its freezing-point as essentially a broken-up solid and takes as the most significant difference between the liquid and solid phases the increase of about 10 per cent that occurs in the specific volume on melting.

The change in molecular separation corresponding to this change in volume cannot be distributed uniformly through the liquid as it would lead to fragmentation of the liquid. Rather, the free volume excess is distributed partly in a discontinuous way as separate 'holes' in the liquid, and partly in a continuous way as a general increase of the average distance between the molecules in those small regions or cells which preserve their homogeneity.

J Thewlis, ed., *Encyclopaedic Dictionary of Physics* (New York, Oxford and London, 1962)

MS

holism *Philosophy* Any view which emphasizes the whole of something as distinct from its parts.

In particular, this doctrine says that the whole in question cannot be predicted from or explained in terms of its parts (*see also* EMERGENCE THEORIES); or else that the whole is more important than its parts (as in, for example, collectivist political theories which say that the interests of the individual must be subordinated to those of the state). In philosophy of science, holism says that empirical statements cannot be conclusively verified individually, but 'our statements about the external world face the tribunal of sense experience not individually but only as a corporate body' (W V O Quine). Holism contrasts with INDIVIDUALISM and REDUCTIONISM. *See also* METHODOLOGICAL THEORIES, VITALISM, ORGANICISM.

W V O Quine, 'Two Dogmas of Empiricism', *Philosophical Review* (1951), 41; reprinted with revisions in W V O Quine's *From a Logical Point of View* (1953).

ARL

holism *Psychology* A term coined by the American philosopher John Dewey (1859–1952).

Referring to any philosophical theory which has as its main interest the whole living organism, holism's basic principle (as developed within psychology) is that more can be understood or learned through an analysis of the organism as a whole entity rather than an analysis of its constituent parts. Examples of this philosophy are to be found in the works of Freud and Gestalt Psychology. Holism is in direct contrast with atomism and elementalism. *See* GESTALT THEORY.

S Coren, C Porac and L M Ward, *Sensation and Perception* (New York, 1984)

NS

holistic explanation *Philosophy* Explanation of a kind claimed to be especially required in the spheres of perceptual experience and the actions of a rational agent, where explanations cannot be given in terms of single factors (beliefs, desires, and so on) but only in terms of whole systems of such factors interrelated in complex ways. However, the elaboration of the features that identify holistic explanations as such is itself a difficult and disputed topic.

C Peacocke, *Holistic Explanation* (1979)

ARL

Holliday model (1964) *Biology* Proposed by Robin Holliday, this is a complex model for the recombination (breakage and reunion during crossing over) of two homologous chromosomes in cells with nuclei (eukaryotes).

The model posits the formation of the Holliday intermediate structure, a molecule that helps account for the stepwise process of reciprocal deoxyribonucleic acid (DNA) exchange between non-sister chromatids.

G Stine, *The New Human Genetics* (Dubuque, Ia, 1989)

KH

holophrastic speech (1960s) *Linguistics* Noticed by several researchers in child language much earlier this century, but popularized under this name by the American linguist David McNeill.

Infants go through a stage of speaking in single words, but each one-word utterance expresses a complex of ideas, equivalent to an adult's full sentence: the one-word utterances are *holo-phrases*. They are understood in relation to their context of utterance.

D McNeill, *The Acquisition of Language* (New York, 1970)

RF

homeopathy (1810) *Medicine* Proposed in Germany by Samuel Christian Friedrich Hahnemann (1755–1843), this is a holistic approach to medicine contending that drugs which produce symptoms similar to those of a disease can be expected to cure that disease (*similia similibus curantur*, 'like cures like'); and that minute doses of drugs provide the most powerful effect. Mainstream scientists dispute homeopathy and consider it irregular medicine.

SCF Hahnemann, *Organon der rationellen Heilkunde* (Principles of Rational Medicine) (1810); M Kaufman, *Homeopathy in America: The Rise and Fall of a Medical Heresy* (Baltimore, Md, 1971)

KH

homeostasis (1926) *Biology* Developed and named by the American physiologist Walter Bradford Cannon (1871–1945) this is the classic concept that the physiological system of higher animals is designed to maintain internal stability, owing to the co-ordinated response of the body's systems to any force tending to disturb its normal condition or functions. In other words, the theory that the body tends to maintain the 'MILIEU INTÉRIEUR' described by Bernard. *See* CONTROL THEORY.

M J Apter, *Cybernetics and Development*, (Oxford, 1966); W B Cannon, 'A Charles Richet: ses amis, ses collègues, ses élèves', *Les Editions Médicales*, Auguste Pettit, ed. (22 Mai, 1926); J Fulton, *Selected Readings in the History of Physiology* (Springfield, Ill., 1966)

KH

homology (1974) *Biology* Associated with Austrian zoologist and ethologist Konrad Lorenz (1903–89), although the idea predated his elaboration, this is the theory that a particular resemblance between two species, either in a structure or a behaviour, can be explained by their having descended from a common ancestor who possessed that particular trait.

For example, the wings of birds and bats are homologous as forelimbs, because they evolved from a common ancestor who possessed forelimbs. Judgments of homology can be controversial, especially when applied to behaviour. *Compare with* ANALOGY and CONVERGENT EVOLUTION.

R Owen, *On the Archetype and Homologies of the Vertebrate Skeleton* (London, 1848)

KH

homology theory (1927) *Mathematics* Developed from ideas by Leopold Vietoris (1891–), Paul S Alexandroff (1896–) and Eduard Čech (1893–1960).

In ALGEBRAIC TOPOLOGY, geometric figures can be distinguished by their infinite families of homology groups Hn which are defined in terms of n-dimensional simplices, the generalizations of triangles in the plane.

P J Hilton and S Wylie, *Homology Theory* (Cambridge, 1962)

MB

homonym *Linguistics See* POLYSEMY.

homophone *Linguistics See* POLYSEMY.

Hooke's law (1678) *Physics* Named after its discoverer, the English scientist Robert Hooke (1635–1703) this law established the relation between the magnitude of forces applied to bodies and the deformations that they produced: 'In every springing body . . . the force or power thereof to restore itself to its natural position is always proportionate to the distance or space it is removed therefrom. . . .'

In modern terms the law states that, for small strains (and at constant temperature), the strain is proportional to the stress producing it; that is,

$$\text{stress} = \text{constant} \times \text{strain}$$

where the constant is known as a modulus of elasticity.

J Thewlis, ed., *Encyclopaedic Dictionary of Physics* (New York, Oxford and London, 1962)

MS

hopeful monsters (1940) *Biology* Coined by Richard Goldschmidt (1878–1958), a German-born geneticist who emigrated to the US, this is a reference to Goldschmidt's theory that sudden jumps in evolution are necessary to explain speciation.

Goldschmidt believed that chromosomal mutations accumulate in populations until some threshold is breached, propelling the species across 'an unbridgeable gap' to a new species. While he expected that most mutants (called 'monsters') would fail to survive, under certain conditions mutants could be more successful than competing individuals. Such successful mutants leading to a new species were called 'hopeful monsters'. Modern geneticists reject this theory. *Compare with* GRADUALISM. *See also* SALTATION SPECIATION.

J Levington, *Genetics, Paleontology, and Macroevolution* (Cambridge, 1988)

KH

Hopf bifurcation theorem (1942) *Mathematics* Named after the Polish-born German topologist Heinz Hopf (1894–1971), this theorem of dynamical systems theory gives conditions for a map or flow depending on a parameter to undergo a Hopf bifurcation in which, as the parameter increases beyond a critical value λ_0, an invariant circle is born and the attracting fixed point becomes repelling. This theorem was developed in order to understand how turbulence arises.

R L Devaney, *An Introduction to Chaotic Dynamical Systems* (New York, 1989)

ML

horizon of expectations (1967) *Literary Theory* A key idea in the literary theory of the German critic Hans Robert Jauss (1921–), who used the term *Erwartungshorizont*. The phrase refers to the 'mental set' or predisposition that readers bring to a work of art, formed through their previous experiences of genre and style, and their beliefs and assumptions about meanings likely to be encoded in a particular species of work.

The horizon of expectations differs in different periods and cultures. The idea is not clearly defined in Jauss. (The term has an earlier history in philosophy and aesthetics. The most relevant and helpful usage being by E H Gombrich.)

H R Jauss, 'Literary History As A Challenge to Literary Theory' [1967], reprinted in *Toward an Aesthetics of Reception*, T. Bahti, trans. (Minneapolis, 1982); E H Gombrich, *Art and Illusion* [1960] (Princeton, 1972)

RF

horizontal evolution *Biology* The evolution that can occur when a population splits into two or more subpopulations which then go on to evolve into distinct species. Also, the simultaneous but parallel evolution of many sequences in the gene complement of a single species.

KH

hormic theory *Psychology* Proposed by Anglo-American psychologist William McDougall (1871–1938).

Often described as goal-orientated, 'purposeful' psychology. 'Hormic' comes from the Greek for animal impulses. McDougall criticized behaviourists such as John Broadus Watson (1878–1958) for defining psychology as the 'science of conduct'. He felt many behaviours were instinct-based and by 1932 had postulated 17 instincts. His work is mainly of historical interest today, but there are some contemporary psychologists who still advocate his theory.

W McDougall, *Body and Mind: A History and Defence of Animism* (London, 1911)

NS

Horner's law *Biology/Psychology* Discovered by the Swiss ophthalmologist Johann Friedrich Horner (1831–86).

The well established genetic principle that the most common form of colour blindness (red-green) is transmitted from male to male through unaffected females.

L M Hurvich, *Color Vision* (Sunderland, Mass., 1981)

NS

horseshoe map, Smale's (1961) *Mathematics* Named after American mathematician Stephen Smale (1930–), this is a map of the unit square to itself formed iteratively by stretching and bending it to form a horseshoe and then stretching and bending that horseshoe to form another inside the first and so on. The horseshoe map was the first example of a diffeomorphism which is structurally stable, but has infinitely many periodic points.

R L Devaney, *An Introduction to Chaotic Dynamical Systems* (New York, 1989)

ML

hot-spot hypothesis (1960s) *Geology* Primarily developed by the American W Jason Morgan, this theorizes that certain major isolated volcanoes (for example, in Hawaii) are associated with plumes of hot material rising from great depths, possibly as deep as the Earth's core.

Such hot-spots are thought to be more fixed in space than the Earth's continents, and hence provide a more consistent frame of reference for studying motions of the Earth's surface. (*See* CONTINENTAL DRIFT.) Whilst this model is still widely used, there are some doubts about the actual depths of the plumes involved.

W J Morgan, 'Convective Plumes in the Lower Mantle' *Nature*, vol ccxxx (1971) 42–3

DT

Hotelling's law (1929) *Economics* Formulated by American economist Harold Hotelling (1895–1973) from his observations on the stability of competition, this states that competitors differentiate their goods and services as little as possible in order to maximize demand from the public.

The law explains why retailers (department stores, newsagents and restaurants) tend to cluster together, and why airlines adopt similar flight schedules. (Hotelling also proposed the principle of MARGINAL COST PRICING.)

H Hotelling, 'Stability in Competition', *Economic Journal*, vol. XXXIX (1929), 41–57

PH

hour-glass model (1959) *Biology* First proposed by the British biochemist Norman Pirie (1907–).

Pirie viewed the origin of life in a diagram which comprised two cones, one on top of the other with their apexes meeting to make an hour-glass shape. The wide base of the lower cone represented the wide range of biological molecules in the primordial soup. These became less diverse as time moved on, represented by the narrowing of the cone. At the apex, life originates (calculated by Pirie to be one to two billion years ago), with one set of biological molecules and one morphological form. The upper cone represents the degree of diversity of morphology of organisms, which becomes greater as time moves on until one reaches the inverted base which represents the present day. The model is not accepted and, in the light of geological evidence, neither is the estimated time for the origin of life.

D M Moore, *Green Planet: The Story of Plant Life on Earth* (Cambridge 1982)

RA

Hubble's law (1929) *Astronomy* Named after its discoverer, the American astronomer Edwin Powell Hubble (1889–1953), this gives the rate of expansion of the universe (that is, the rate at which the galaxies are receding from each other) and was derived by interpreting the frequency shift of the light from distant galaxies as a Doppler shift.

The rate of separation of any two galaxies in the universe is proportional to their separation. If two galaxies are a distance R apart, their present relative speed of separation v is given by

$$v = HR$$

where H is a constant (known as Hubble's constant/parameter), equal to $2.32 \times 10199^{18}\ s^{-1}$. Hubble's constant is a constant in the sense that the proportionality between speed and separation is the same for all galaxies at a given time; but Hubble's constant changes with time as the universe evolves.

S Mitton, ed., *The Cambridge Encyclopaedia of Astronomy* (London, 1973)

MS

hubris *Literary Theory See* TRAGIC FLAW.

Hückel (4*n* + 2) rule *Chemistry* Named after E Hückel. Monocyclic systems of trigonally hybridized atoms that contain $(4n + 2)$ π-electrons (where n is a non-negative integer) exhibit aromatic character. This rule is derived from the Hückel molecular orbital calculations on planar monocyclic conjugated hydrocarbons $(CH)_m$ where m is an integer equal to or greater than three; according to which $(4n + 2)$ π-electrons are contained in a closed-shell system. Systems containing $4n$ π-electrons (such as cyclobutadiene and the cyclopentadienyl cation) are anti-aromatic.

Compendium of Chemical Terminology: IUPAC Recommendations (Oxford, 1987)

MF

Hückel molecular orbital theory (1931) *Chemistry* Named after its originator, E Hückel, the theory concerns the π-electron systems responsible for the special properties of conjugated and aromatic hydrogens. The secular equation for the π-electrons in ethylene is

$$\begin{vmatrix} \alpha - E & \beta \\ \beta & \alpha - E \end{vmatrix} = 0$$

This simplified equation is based on the following assumptions: first, overlap integrals S_{ij} are set to zero unless i = j, when $S_{ij} = 1$; secondly, all the diagonal elements in the secular equation are assumed to be the same; thus, the Coulomb integrals H_{ij} are all set equal to α; and finally, the resonance integrals H_{ij} are set equal to zero, except for those on neighbouring atoms which are set equal to β. *See also* HÜCKEL ($4n + 2$) rule.

R A Alberty and R J Silbey, *Physical Chemistry* (New York, 1992)

MF

Hullian theory *Psychology* Proposed by the American psychologist Clark Leonard Hull (1884–1952).

This theory proposed the deduction of hypotheses from postulates and then the empirical testing of such hypotheses. In other words it was a rigorous, experimental approach; the central concept of which was 'habit strength'. The theory postulated literally hundreds of variables and became unworkable. The approach had a major impact on the use of emperimental method in investigations of psychological processes.

C L Hull, *Principles of Behaviour* (New York, 1943)

NS

human capital theory (1960s) *Economics* With its roots in the work of British economists Sir William Petty (1623–87) and Adam Smith (1723–90), the theory was extensively developed by American economists Gary Becker (1930–) and Theodore Schultz (1902–). It postulates that expenditure on training and education is costly, and should be considered an investment since it is undertaken with a view to increasing personal incomes.

The human capital approach is often used to explain occupational wage differentials. English philosophers John Locke (1632–1704) and John Stuart Mill (1806–73), Scottish economist Adam Smith (1723–90) and German social theorist Karl Marx (1818–83) all argued that training, not natural ability, was important in understanding differentials. Human capital can be viewed in general terms, such as the ability to read and write, or in specific terms, such as the acquisition of a particular skill with a limited industrial application. Critics of the theory argue that it is difficult to separate human capital investment from personal consumption. *See* SEARCH THEORY.

G S Becker, *Human Capital: A Theoretical and Empirical Analysis with Special Reference to Education* (New York, 1964)

PH

human nature *Politics/Sociology/Philosophy* Theory of human individual and social character.

There is a 'natural' human character as there is a natural shape to a particular plant or a natural form to a particular animal. This human nature is prior to the particularities of any time or place.

David Miller *et al.*, eds, *The Blackwell Encyclopaedia of Political Thought* (Oxford, 1987)

RB

humanism (15th–16th century) *Art/Literary Theory* A formal term applied in the 19th century to the re-emergence and synthesis of classical thought during the Renaissance.

Although evident in Petrarch's studies of classical texts in the 14th century, it was particularly during the 15th century (in Florence, Venice and Naples) that scholars' rediscovery of classical civilization and literature led to the development of rational thought and approaches which were not dependent on Christian customs. In Florence, this intellectual shift was confirmed by the establishment of Marsilio Ficino's neo-Platonic academy in 1492.

AB

humanity, principle of *Philosophy* Principle named by R E Grandy in 1973 as a supplement to the principle of CHARITY.

It says that when interpreting another speaker we must assume not simply that he is intelligent and so on, but that his beliefs and desires are connected to each other and to reality in a way that makes him as similar to ourselves as possible. As with the principle of charity, this principle is not – in the view of thinkers like Willard Van Orman Quine (1908–) – only intended for interpreting remote civilizations (*see* INDETERMINACY OF REFERENCE AND TRANSLATION) since we apply it automatically in our daily intercourse.

R E Grandy, *The Journal of Philosophy*, (1973), 443; I Hacking, *Why Does Language Matter to Philosophy?* (1975), 146–50

ARL

Humboldt's law (1817) *Biology* Described by German naturalist Alexander von Humboldt, this describes the tendency for the upper tree limit (also called the tree line) to occur at lower and lower elevations as one moves away from the Equator, until the tree line reaches sea level above the Arctic Circle.

KH

Hume-Rothery's rule (1926) *Chemistry* Crystal structure rule proposed by English chemist William Hume-Rothery (1899–1968), and also known as the eight-*N* rule. This states that atoms of elements in groups numbered 4–7 of the periodic table have $8 - N$ nearest neighbours in the solid state, where N is the number of the group to which the element belongs. For example, carbon (in group 4) has four nearest neighbours in diamond.

S Glasstone, *Textbook of Physical Chemistry* (London, 1948)

MF

Hume's law *Philosophy* Derived from the Scottish philosopher David Hume (1711–76), an informal name for a distinction (rather like the FACT/VALUE DISTINCTION) between statements of fact and utterances with an 'ought' in them.

In his *Treatise of Human Nature* (1739–40) Hume claimed (as usually interpreted) that the latter could never be logically derived from the former, and this has been the subject of considerable debate in the last 30 years or so. As with the fact/value distinction, the present distinction has proved hard to maintain in its pristine clarity, though the issue is far from settled.

W D Hudson, ed., *The Is/Ought Question* (1969)

ARL

humoral action, concept of (1742) *Biology/Medicine* Proposed by Théophile de Bordeu (1722–76), French anatomist and endocrinologist. The theory that every gland and tissue in the body releases products into the blood which can affect the function of other glands and tissues.

Although de Bordeu borrowed the term 'humours' for these products, he also specifically rejected Galen's mystical notion of vital spirits. De Bordeu's theory predated the isolation of such hormones as testosterone. *See also* HUMORALISM, TESTICULAR EXTRACT THEORY and VITALISM.

J Fulton, *Selected Readings in the History of Physiology* (Springfield, Ill., 1966)

KH

humoralism (6th century) *Medicine* Systematically described by the Greek physician Hippocrates (*c.*460-377 BC) and promoted by the Greek physician Galen (*c.*129–*c.* AD 200). The doctrine that four basic body substances (blood, phlegm, yellow bile and black bile (melancholy)), called humours, are responsible for health, disease and temperament.

Galen went on to relate four temperaments to Hippocrates's four humours: an excess of blood led to a cheerful disposition; excess of bile gave rise to a melancholic individual; excess of yellow bile led to a fiery individual; and an excess of phlegm led to the phlegmatic temperament. Bloodletting represented the most important humoral treatment. This highly influential theory was discarded in the 19th century after the establishment of cell theory and the discovery of bacterial causes for disease. *See also* HUMORAL ACTION, CONCEPT OF, CONSTITUTIONAL THEORY.

Galen, *On Anatomical Procedures* (London, 1956); O Temkin, *Galenism: Rise and Decline of a Medical Philosophy* (Ithaca, New York, 1973)

KH, NS

Hund's rule (1925) *Chemistry* Rule for the electronic structure of an atom proposed by F Hund.

The orbitals of a sub-shell of an atom must be occupied singly by electrons with parallel spins before they can be occupied in pairs. For example, nitrogen has the electronic configuration $1s^2 2s^2 2p^3$. The three electrons occupying the 2p sub-shell must occupy the three separate 2p orbitals singly. They must also have parallel spins.

M Freemantle, *Chemistry in Action* (Basingstoke, 1987)

MF

Hurwitz's theorem (19th century) *Mathematics* Named after the German mathematician Adolf Hurwitz (1859–1919), this is the result whereby for any irrational number α, there are infinitely many distinct rationals p/q such that

$$\left|\alpha - \frac{p}{q}\right| < \frac{1}{\sqrt{5}q^2}$$

The constant $\sqrt{5}$ is the best possible, as can be verified by taking

$$\alpha = \frac{1 + \sqrt{5}}{2}.$$

See also CONTINUED FRACTION ALGORITHM and THUE-SIEGEL-ROTH THEOREM.

A Baker, *A Concise Introduction to the Theory of Numbers* (Cambridge, 1984)

MB

Huygens's principle (1678) *Physics* Named after the Dutch physicist Christiaan Huygens (1629–93), who introduced it, this stated that each point at the front of a wave may be regarded as a small source of wave motion. The waves produced by these small sources are called secondary waves. At a later time, the position of the main, or primary, wave is the envelope of the secondary waves.

J Thewlis, ed., *Encyclopaedic Dictionary of Physics* (New York, Oxford and London, 1962)

MS

hydraulic model *Biology See* LORENZ'S HYDRAULIC MODEL.

hydraulic theory *Psychology* Often associated with the work of the Austrian founder of psychoanalysis Sigmund Freud (1856–1939).

Theories subsumed under this title are any that view a psychological phenomenon in hydraulic terms. Hence, the assumption is made that the phenomenon behaves like fluids under pressure. Freud's psychoanalytic theory is often considered to some degree hydraulic because of the way in which energy is said to be controlled by the ego. Hydraulic theories in general are unable to account for a psychological phenomenon, although they enjoy some popularity as analogical explanations useful in the teaching of psychology.

A Meyer, *Psychobiology: A Science of Man* (Springfield, Ill., 1957)

NS

hydropathy (19th century) *Medicine* Promoted by German farmer Vincent Priessnitz (1799–1851), this is the doctrine that water can cleanse the body of the foreign matter that causes disease.

Priessnitz treated patients with wet bandages, special baths, high water intake, simple foods, clear mountain air and exercise. Although attacked as quackery at first, by the mid-19th century hydropathy was incorporated into conventional medicine, and elements of Priessnitz's approach persist today.

P Mattson, *Holistic Health in Perspective* (Palo Alto, Calif., 1981)

KH

hylomorphism *Philosophy* View that reality, or certain parts of it, is to be analyzed in terms of form and matter.

The form referred to usually means the immanent forms of Aristotle, not the transcendent Forms of Plato. The two main problems the doctrine is introduced to deal with are: the explanation of change (in terms of the persistence of a substrate while one property is replaced by another); and the relating of body and soul (by claiming that the soul is not a separate entity but is the form of the living body). This general approach was developed later in the Middle Ages and Renaissance, especially by St Thomas Aquinas (*c.*1224–74) who undertook to reconcile Aristotle with Christianity.

Aristotle, *Physics* (especially books 1, 2, 3, 5); *De Generatione et Corruptione* (On Coming to Be and Passing Away); *De Anima* (On the Soul), especially books 2, 3

ARL

hylozoism *Philosophy* Treatment of matter, or parts of the material world, as intrinsically alive. Where ANIMISM tends to view the life as taking the form of discrete spirits, and PANPSYCHISM tends to refer to strictly philosophical views like that of Gottfried Wilhelm Leibniz (1646–1716), hylozoism refers largely to views such as those of the earliest Greek philosophers (6th and 5th centuries BC).

Certain of these treated the magnet as alive because of its attractive powers (Thales), or air as 'divine' (Anaximenes), perhaps because of its apparently spontaneous power of movement, or because of its role as essential for life in animals. However, some have since claimed that 'hylozoism' should properly be used only where body and soul are explicitly distinguished, the distinction then being rejected as invalid.

J Glucker, 'Who Invented "Hylozoism"?', *Ionian Philosophy*, K J Boudoouris, ed. (1989)

ARL

Hyman's law (1959) *Psychology* Proposed by the American psychologist H Hyman (1928–).

This states that adults, especially parents, are more likely to influence a child's decisions on such things as future aspirations, academic choices and political views than are the child's. The validity of the law is largely determined by the relative impact of parental child-rearing practices, and as such it is highly malleable.

H M Hyman, *Political Socialization* (New York, 1959)

NS

hypercorrection (1960s) *Linguistics* Proposed by the American linguist William Labov for a finding in CORRELATIONAL SOCIOLINGUISTICS.

Some social groups display a much higher incidence of some sociolinguistic feature than would be predicted, indeed higher than the class above: they HYPERCORRECT, emulating the prestige form ('overt prestige') through insecurity or desired upward-mobility. Working-class men often emulate the covert prestige of stigmatized forms (for example 'dropped *h*'), for male solidarity.

W Labov, *Sociolinguistic Patterns* (Philadephia, 1972)

RF

hypercycle theory (1977) *Biology* Described by M Eigen and P Schuster. The theory that life evolved as a cyclically closed hierarchy of chemical reaction cycles, including those involving enzymes and transfer-ribonucleic acid (tRNA), surrounded by a membrane.

These chemical reactions, Eigen and Schuster say, were the functional equivalent of life, and these 'little bags' were the precursors to single cells. *See also* 'LITTLE BAGS' THEORY OF EVOLUTION. *Compare with* PROGENOTE THEORY.

M Eigen and P Schuster, 'The Hypercycle. A Principle of Natural Self-Organization', *Naturwissenschaften* vol. LXIV (1977) 541–65

KH

hype (20th century) *Politics* Political version of GRESHAM'S LAW.

People and institutions whose surface value ('hype') is less than their substance will be driven out by those of whom the reverse is true. 'An ounce of presentation is worth a ton of production.'

RB

hypothesis testing *Mathematics* The theory and methods used in STATISTICS to test one hypothesis about the probability distributions of a sample population (null hypothesis) against another (alternative hypothesis). Once a suitable test statistic and significance level have been chosen, the null hypothesis is rejected if the test statistic lies in a critical region for that significance level. *See also* LIKELIHOOD RATIO TEST and NEYMAN-PEARSON LEMMA.

R V Hogg and A T Craig, *Introduction to Mathematical Statistics* (New York, 1967)

ML

hypothetico-deductive method *Philosophy* Scientific method whereby science should set up testable hypotheses and then try to falsify them, rather than trying to confirm them directly by accumulating favourable evidence. Introduced by the English scholar William Whewell (1794–1866) and developed especially by the Austrian philosopher Karl Raimund Popper (1902–).

Those hypotheses which – despite severe tests – survive unfalsified are thereby confirmed for Whewell; Popper goes further and says they are merely 'corroborated', a notion which is supposed to avoid the logical invalidity associated with induction. *See also* FALSIFICATIONISM, INDUCTIVISM, DEDUCTIVISM and IMPROBABILISM.

K R Popper, *The Logic of Scientific Discovery* (1959); unrevised German original 1934

ARL

I

iatrochemistry (16th century) *Biology* Systemized by the Flemish chemist Johannes (Jean) Baptiste van Helmont (1579–1644), this is the theory that all body processes are chemical and caused by special ferments (gases) that are capable of transforming food into flesh.

Body heat was believed to be a by-product of fermentation. Van Helmont's theory also included the belief that the ferments are spirits, and that the soul (located in the pit of the stomach) controls vital processes.

R McGrew, *Encyclopedia of Medical History* (New York, 1985)

KH

iatrophysics (17th century) *Biology* Proposed by the Italian mathematician and physician Giovanni Alfonso Borelli (1608–1679), and the Italian anatomist Giorgio Baglivi (1668–1707). An empirical, mechanistic approach to explaining muscle function, glandular secretions, respiration, cardiac motion and neural responses. For example, Borelli believed that muscles contain a 'contractile element' triggered by a fermentation-like process.

KH

ideal gas law *Chemistry* Also known as the ideal gas equation. Ideal gases obey the following equation

$$pV = nRT$$

where V is the volume of n moles of an ideal gas with pressure p and temperature T, and R is the gas constant. *See also* GENERAL GAS LAW.

M Freemantle, *Chemistry in Action* (Basingstoke, 1987)

MF

ideal utilitarianism *Philosophy* Version of UTILITARIANISM which (in contrast to HEDONISTIC UTILITARIANISM) does not take pleasure to be the only, or even necessarily the main, value.

The version of English empiricist George Edward Moore (1873–1958) emphasized aesthetic values and certain personal relationships, and was especially influential on the Bloomsbury Set during the early 20th century.

G E Moore, *Principia Ethica* (1903)

ARL

idealism *Philosophy* Any view saying that reality is in some way mental, or depends intrinsically – and not just causally – on mind (not necessarily the human mind).

The term may also apply to features of some philosophy, but is connected for philosophers with 'idea' rather than, as in popular usage, with 'ideal' in the sense of goal of behaviour; nor does it apply now to PLATO'S THEORY OF FORMS (OR IDEAS) since though these are not material neither are they mind-dependent. Idealism may be opposed to MATERIALISM or to REALISM. Sometimes the term 'idealism' refers to the opinion that reality can only be described from some point of view, not in a way that transcends all points of view (compare PERSPECTIVISM). *See also* OBJECTIVE IDEALISM, SUBJECTIVE IDEALISM, TRANSCENDENTAL IDEALISM.

A C Ewing, ed., *The Idealist Tradition* (1957); selected idealist writings

ARL

idealization (Antiquity–) *Art* The theory that art not only reproduces nature, but perfects and improves upon it.

Since Aristotle (384–322 BC) and Plato (*c.*427–*c.*347 BC) there have been accounts of artists who reveal beauty in nature through their own work. The neo-Platonists revived Plato's *Theory of Ideas*, in which objects are imperfect copies which relate to a doctrine of Ideas and Forms. The 17th-century Italian theorist Bellori characterized the artist as the key to the revelation of beauty to the spectator, citing as an example the French painter Nicolas Poussin (1594–1665). In NEO-CLASSICISM, idealization can be understood as following a canon of perfection. *See* NEO-PLATONISM.

AB

ideational function *See* FUNCTIONAL GRAMMAR.

ideational theories of meaning *Philosophy* Theories which say that words have meaning by standing for ideas, thoughts or concepts, and so on.

Such theories are found in Aristotle's (4th century BC) early work *De Interpretatione* (*On Interpretation*), especially chapters 1–4; and in the writing of English philosopher John Locke (1632–1704). They have the advantage over NAMING THEORIES OF MEANING in that they provide a single kind of thing for diverse kinds of word to stand for, but share with such theories problems over what 'standing for' amounts to.

J Locke, *An Essay Concerning Human Understanding* (1690), book 3

ARL

identical direction, law of *Psychology* Formulated by the German physiologist Ewald Hering (1834–1918).

When an object is fixated visually, the visual axes of the two eyes are aligned so that they intersect at the fixation point. However, they do not terminate there but continue beyond the intersection so that, if the fixation point is appropriately defined within an experimental setting, two objects will not appear alongside one another but fixed into a single image on an imaginary median line between the two eyes.

L Kaufman, *Perception* (New York, 1979)

NS

identical elements theory/law *Psychology* Proposed by the American psychologist Edward Lee Thorndike (1874–1949).

This is one of five laws of learning proposed by Thorndike. It states that the degree of transfer of learning between two tasks is a function of the number of elements that the two tasks have in common. *See* THORNDIKE'S LAW OF ASSOCIATIVE SHIFTING, and BELONGINGNESS.

E L Thorndike, *The Fundamentals of Learning* (New York, 1932)

NS

identity equation *Mathematics* *See* PARSEVAL'S THEOREM.

identity, law of *Philosophy* One of the traditional THREE LAWS OF THOUGHT, the other two being the laws of CONTRADICTION and EXCLUDED MIDDLE.

'Everything is what it is and not another thing', or (where 'P' is any proposition) 'If P then P'. The English empiricist George Edward Moore (1873–1958) took the first quotation above as the motto for his book *Principia Ethica* (1903), attributing it to Bishop Joseph Butler (1692–1752). In comparison to the other two laws of thought, little attention has been paid to this one.

ARL

identity of indiscernibles *Philosophy* One part of LEIBNIZ'S LAW, saying that if what appear to be two or more things have all their properties in common they are identical and so only one thing.

In its widest and weakest form, the properties concerned include relational properties such as spatiotemporal ones and self-identity. A stronger version limits the properties to non-relational properties (that is, qualities), and would therefore imply that there could not be, for example, two exactly similar ball-bearings. Even the weaker version faces objections if we envisage two ball-bearings alone in an otherwise empty universe, or in corresponding positions in the two halves of a symmetrical universe:

what property would one of them have and the other lack? (To try to distinguish them by their relations to each other would presuppose that we could already distinguish them, and the same holds of the halves of the symmetrical universe.) *See also* principle of SUFFICIENT REASON.

M J Loux, ed., *Universals and Particulars* (1970)

ARL

identity theorem *Mathematics* Also known as the uniqueness theorem or the principle of analytic continuation, this is the result in COMPLEX FUNCTION THEORY which essentially says that if two analytic functions are equal on a small portion of a region, then they are equal on the whole region on which they are both analytic. This theorem is a good example of the startling differences between real analysis and complex analysis. *See also* REFLECTION PRINCIPLE OF SCHWARZ.

S G Krantz, 'Functions of One Complex Variable', *Encyclopedia of Physical Science and Technology* (Academic Press, 1987)

MB

identity theory *Psychology* Proposed by R Carnan (1891–1970).

The strong form of identity theory (type-identity) argues that every mental event or mental state is identical with a particular brain state, and in principle specifiable as a physiological (that is, physical) state. It supports the extension that when two persons share something mental (for example, an idea), they also have in common equivalent physical (brain) states. Without the extension that types of mental events are assumed to correspond to types of physical events, one merely has a token identity theory which is a considerably weaker position in which only specific individual (token) mental states are assumed to reflect these equivalences.

M Billig, *Social Psychology and Intergroup Relations* (London, 1976)

NS

identity theory of mind *Philosophy* Theory, coming primarily from Australia in the 1950s, that various mental phenomena are identical with certain cerebral or neurophysiological phenomena. (The names 'brain process theory' and 'central state materialism' are sometimes used for these two alternatives, respectively; more generally, the theory is called simply MATERIALISM OR PHYSICALISM, though both these terms have other uses too.)

Sometimes the mental phenomena in question are limited to pains or other sensations, and sometimes they cover thoughts, beliefs, desires, emotions, and so on, though this raises problems about whether reference to external objects is involved (*see also* EXTERNALISM). The identity is normally taken to be contingent rather than necessary, which itself raises problems (see Kripke). An important distinction is between type and token identity theories. The stronger, type, version correlates types of pain, thought, emotion, and so on with types of physical phenomena. The weaker, token, version claims merely that any given occurrence of a pain, thought, and so on is identical with some physical phenomenon (not necessarily of the same kind in each case). *See also* DOUBLE ASPECT THEORY, NEUTRAL MONISM, ANOMALOUS MONISM, INTERNALISM and EXTERNALISM.

C V Borst, ed., *The Mind/Brain Identity Theory* (1970); S Kripke, *Naming and Necessity* (1980) especially, pp. 144–55

ARL

identity theory of predication *Philosophy* Theory that subject/predicate statements are really identity statements, so that 'X is red' means 'X is identical with some red thing'. This is in effect the same as the theory Geach traces back to Aristotle (384–322 BC), but which is criticized even earlier by Plato (*c*.427–*c*.347 BC) in the *Sophist*, and treats predation in terms of a two-term theory, two-name theory, and two-class theory.

P T Geach, 'A History of the Corruptions of Logic', *Logic Matters* (1972)

ARL

identity theory of truth *Philosophy* Named by S Candlish (in *Mind*, 1989) and recently attributed to the idealist philosopher Francis Herbert Bradley (1846–1924). It is also seen as having strong affinities with the views of George Edward Moore (1873–1958) and Bertrand Russell (1872–1970) at one period in the development of their respective philosophies, and possibly with that of Gottlob Frege (1848–1925).

The theory says that the truth of a judgment consists in the identity of its content with a fact.

J Dodd and J Hornsby, 'The Identity Theory of Truth: Reply to Baldwin', *Mind* (1992); see also Baldwin's article in previous volume

ARL

ideology *Politics/Philosophy* Theory of knowledge and values. Term first used by the French philosopher Destutt de Tracy (1754–1836) in 1795 to denote the general science of ideas, but now principally associated with MARXISM.

The major version of the theory of ideology sees both perceptions and values as shaped by the social situation of the person, and especially of social classes. Ideology is partial truth presented as if it were universal truth. It justifies and sustains the privileges of dominant classes. A minor version of the theory, promulgated in the 1950s and 1960s, spoke of an END OF IDEOLOGY, and described as 'ideology' any view which claimed either general explanations or general solutions.

David Miller *et al.*, eds, *The Blackwell Encyclopaedia of Political Thought* (Oxford, 1987)

RB

ideoplastic (1917) *Art* Coined by Max Vorworm, this term refers to a type of representation which derives from the artist's knowledge of a subject and not from direct observation or memory of an object.

Applied specifically to the study of childrens' and primitive peoples' schematic drawings, the theory has been contested by GESTALT psychologists who argue that these works represent general features of primary perception.

M Vorworm, *Zur psychologie der primitiven Kunst* (1917)

AB

idiolect (1950) *Linguistics* A term coined by British phonetician Daniel Jones (1881–1967). The consistent and distinctive vocal speech style of an individual.

D Jones, *The Phoneme: Its Nature and Use*, 3rd edn (Cambridge, 1967 [1950])

RF

illocutionary act *Linguistics See* SPEECH ACT THEORY.

illuminance, law of *Physics* Illuminance is defined as incident luminous flux per unit area. If a point source of luminous intensity I is placed at a distance d from a plane surface, the normal to which passes through the source, the illuminance E varies according to the inverse square law:

$$E = \frac{I}{d^2}\,.$$

J Thewlis, ed., *Encyclopaedic Dictionary of Physics* (New York, Oxford and London, 1962)

MS

illusionism *Art* The use of pictorial techniques such as perspective and foreshortening to deceive the eye into believing that what is painted is real.

Popular in the Hellenistic period, especially in the painted fictive architecture at Pompeii, the technique was revived by Italian painter Andrea Mantegna (1431–1506) in his ceiling for the *Camera degli Sposi* (1474) in Mantua. Illusionist effects reached their height in 16th and 17th century Italian architecture, and in the peepshow cabinets of the 17th-century Dutch painters.

AB

image (18th century) *Literary Theory* A commonly used term with several meanings.

In the 18th century, an imaginary visual perception suggested by words, as in descriptive poetry; this is the basic sense in early 20th century IMAGISM; generalized in early 19th and early 20th century Romantic (*see* ROMANTICISM) poetry to non-visual sensory modes; also applied, unnecessarily and unhelpfully, to FIGURES OF SPEECH, particularly METAPHOR.

P N Furbank, *Reflections on the Word 'Image'* (London, 1970)

RF

imagination (1800s) *Literary Theory* Acute and influential discussion by English poet Samuel Taylor Coleridge (1772–1834).

The aesthetic concept of 'imagination' is quite distinct from the everyday sense of the word. Up to the 18th century, 'imagination' in literary discussion had to do with the production of images. Coleridge, drawing on German philosophers Immanuel Kant (1724–1804) and Friedrich von Schelling (1775–1854), redefined it as a property of perception, a synthesizing human power which could project order onto the heterogeneity of experience. Imagination is therefore the source of poetic CREATIVITY and metaphor-making. *See also* FANCY, ORGANIC FORM.

S T Coleridge, *Biographia Literaria* [1815], ch. 13, repr. in H Adams, ed. *Literary Theory since Plato* (New York, 1971), 470–71; W K Wimsatt (jnr) and C Brooks, *Literary Criticism: A Short History* (New York, 1967), ch. 18

RF

Imagism (1961) *Art* A term coined by H H Arnason in reference to the exhibition held at the Solomon R Guggenheim Museum in 1961 entitled 'American Abstract Expressionists and Imagists'. It referred to the work produced by artists contemporary with the Abstract Expressionists whose style was not expressionist, and is sometimes described as 'hard edge'. *See* ABSTRACT EXPRESSIONISM.

AB

Imagism (1912–17) *Literary Theory* Term coined by the American poet Ezra Pound (1885–1972); theoretical input also from English critic and poet Thomas Ernest Hulme (1883–1917).

School of poets around Pound whose programme opposed the imprecision and self-indulgence of ROMANTICISM. Emphasis on brevity, precision, and hard, clear, concrete images.

W C Pratt, *The Imagist Poem* (New York, 1963)

RF

imbalance theory (20th century) *Biology/Medicine* The theory that cancer is caused by a breakdown in an organism's mechanism for controlling cell growth. When the imbalance in favour of growth exceeds a certain threshold, a tumour results. Imbalance theory represents one of several explanations for cancer. *See* CANCER, THEORIES OF.

KH

imitation (*c*.373 BC) *Literary Theory* Discussed by the Greek philosopher Plato (*c*.427–*c*.347 BC).

A preoccupation of literary theory since Plato, used in several senses. MIMESIS, the fiction of a world, is not strictly imitation since no real world pre-existed to imitate. The word also refers to the representation of speech; and (a quite different meaning) to the imitation of models of good writing – a notion disfavoured with the post-Romantic stress on CREATIVITY.

M H Abrams, *The Mirror and the Lamp* (New York, 1953)

RF

immaterialism *Philosophy* Name coined by George Berkeley (1685–1753) for his own philosophy, now more usually called SUBJECTIVE IDEALISM.

Berkeley's choice of the term was to emphasize his own view that matter does not exist, but in calling his opponents (René Descartes (1596–1650), John Locke (1632–1704), and so on) materialists he was using 'materialist' in an unusually weak sense. Descartes and others did indeed accept that matter exists, but they did not deny the existence of other things too, notably souls or spirits.

G Berkeley, *A Treatise Concerning the Principles of Human Knowledge* (1710)

ARL

immiseration (19th century–) *Politics* Theory within MARXISM of steady decline of condition of working class.

The theory is now widely abandoned, save by those who argue that the working class has become relatively worse off. It is now more common to read of the opposite process of EMBOURGEOISEMENT.

RB

immunologic theory of ageing (1969) *Biology/Medicine* First stated by Walford, this is the theory that the integrity of the immune system is largely responsible for the lifespan of humans and animals.

Walford hypothesizes that the immune system not only maintains health, but also causes many of the diseases associated with ageing and that it may induce senescence itself. *See also* AGEING, THEORIES OF.

T Makinodan and E Yunis, *Immunology and Aging* (New York, 1977)

KH

impacted pluralism (20th century) *Politics* Pessimistic theory of PLURALISM.

A politics characterized by the pursuit by groups of their various INTERESTS which used to give something to everyone and achieve negotiated compromises, but which has changed to a situation where competing interests are deadlocked, and neither the public interest nor that of groups is served.

S H Beer, *Britain Against Itself* (London, 1982)

RB

imperfect competition *Economics* Developed by English economist Joan Robinson (1903–83), this term describes a market characterized by a large number of buyers and sellers, dealing with differentiated products, and in which there are no barriers to entry or exit.

Imperfect competition differs from PERFECT COMPETITION principally in that its products are highly differentiated. *See* MONOPOLISTIC COMPETITION.

J Robinson, *The Economics of Imperfect Competition* (London, 1933); P A Samuelson, 'The Monopolistic Competition Revolution', *Monopolistic Competition Theory: Studies in Impacts*, R E Kuenne, ed. (New York, 1967)

PH

imperfection, human *Politics* Element in CONSERVATISM.

Human nature is inherently flawed, the secular equivalent of the doctrine of original sin. It is both useless and dangerous to pursue ambitious schemes of social or political progress. Government should, rather, seek to preserve order and check humanity's natural viciousness.

David Miller *et al.*, eds, *The Blackwell Encyclopaedia of Political Thought* (Oxford, 1987)

RB

imperialism (20th century) *Politics/Economics* Economic explanation of overseas expansion of European nations. Originally developed by the maverick British economist J A Hobson (1858–1940) and the Russian Marxist V I Lenin (1870–1924).

Imperial expansion resulted from the exhaustion of domestic markets, industrial nations propping up their economies at the expense of those parts of the world which, through military domination, they subjected to economic exploitation. *See also* COLONIALISM.

David Miller *et al.*, eds, *The Blackwell Encyclopaedia of Political Thought* (Oxford, 1987)

RB

impersonality (1917) *Literary Theory* Theory of art propounded by the American poet and critic Thomas Stearns Eliot (1888–1965).

Powerful objection to the Romantic doctrine of poetry as the expression of the poet's individual self: 'The progress of an artist is a continual self-sacrifice, a continual extinction of personality.' The mind of the artist is not a subjectivity but a 'catalyst' that synthesizes poetic materials. This theory influenced NEW CRITICISM. *See also* INTENTIONAL FALLACY.

T S Eliot, 'Tradition and the Individual Talent', *Selected Essays* (London, 1951)

RF

implication-realization (1970s) *Music* Adapted and developed from GESTALT psychology by the American musicologists Leonard B Meyer (1918–) and Eugene Narmour (1939–), this is the theory that the first few notes of a given piece imply possible continuations which, if realized, form some sort of satisfactory 'closure'.

For example, a stepwise motion may suggest continuation in the same direction towards the tonic; while a leap may suggest a stepwise return home to the initial note. Interesting complications arise as realizations are postponed, left unfulfilled, or they set up new implications, both long and short-term. Meyer and Narmour's theories provide an important alternative to the Schenkerian notion that all pieces share the same overall closure pattern.

E Narmour, *Beyond Schenkerism* (Chicago, 1977)

MLD

implicature (1975) *Linguistics* A term coined by the American philosopher Herbert Paul Grice (1913–).

(1) *Conventional implicature* is what is usually implied over and above the meanings of propositions; (2) *Conversational implicature* is an unexpected meaning created when one of the conversational maxims (*see* CO-OPERATIVE PRINCIPLE OF CONVERSATION) is deliberately flouted.

H P Grice, 'Logic and Conversation', P Cole and J L Morgan, eds *Syntax and Semantics*, 3: *Speech Acts* (New York, 1975), 41–58

RF

implicit function theorem *Mathematics* The theorem in multivariable calculus or COMPLEX FUNCTION THEORY that gives conditions under which a particular relationship among a set of variables expresses one of these variables as a differentiable function (at least locally) of the others.

T M Apostol, *Mathematical Analysis* (Addison-Wesley, 1974)

MB

implied author (1961) *Literary Theory* Proposed by the American critic Wayne C Booth (1921–).

In narrative, the impression of the author's beliefs and values which readers gain from the text of the novel; the same biographical author may project different implied authors in different novels.

W C Booth, *The Rhetoric of Fiction* (Chicago, 1961)

RF

implied reader (early 1970s) *Literary Theory* Part of RECEPTION THEORY of the German critic Wolfgang Iser (1926–); parallels IMPLIED AUTHOR.

Skeleton of values implicit in a text in relation to which the real reader works to orient himself.

W Iser, *The Implied Reader* (Baltimore, 1974 [1972])

RF

impossibility of a gambling system, principle of the *Philosophy* Principle that a properly defined collective (*see* FREQUENCY THEORY OF PROBABILITY) will be random in a sense that makes it impossible to construct a system for predicting results with any greater probability than would be possible without the system. The principle was named by the German mathematician and philosopher Richard von Mises (1883–1953).

The key condition is that the limiting frequency of the characteristic concerned should be the same for all partial sequences we could select from the collective, provided only that whether a given term in the collective is taken into a partial sequence is independent of whether that term manifests the characteristic concerned.

R von Mises, *Probability, Statistics and Truth* 2nd edn (1939), 30–35

ARL

impossibility theorem (1966) *Economics* Formulated by American economist Kenneth Arrow (1921–), this theory asserts that it is impossible to devise a constitution or voting system which offers more than two reasonable choices to the individual, or that will guarantee to produce a constant set of preferences for a group which correspond to the preferences of the individuals making up that group.

Arrow proposed that no system could be both rational and egalitarian, and that even in a simple voting system the PARADOX OF VOTING will arise.

K J Arrow, *Social Choice and Individual Values* (New York, 1966); Robert Abrams, *Foundations of Political Analysis* (New York, 1979)

PH RB

impossibility theorem (1826) *Mathematics* First proved by Niels Henrik Abel (1802–29) who showed that there is no solution by radicals to the general quintic equation.

The term is also applied to any theorem which asserts the impossibility of attaining some objective, such as the trisection of angles by ruler and compass or finding a rational value for π. *See also* ARROW'S IMPOSSIBILITY THEOREM, GÖDEL'S THEOREM, COMPUTABILITY THEORY and GALOIS THEORY.

JB

impoverished theory of mind (20th century) *Psychology* Proposed by the British psychologist S Baron-Cohen.

A theory offered to account for the condition of autism in children, it suggests that such sufferers have an impoverished theory of mind; they have difficulty imagining others to hold beliefs, expectations, ideas and so on. On the other hand, these children can identify emotional states, perhaps because these can often be recognized from facial and other observable facets of a person's behaviour. The theory is regarded as offering a major advance in our understanding of autism.

K Ellis, ed., *Autism: Professional Perspectives and Practices* (London, 1990)

NS

impression management theory *Psychology* Proposed by the American sociologist Erving Goffman (1922–).

The theory analyzes social behaviour as being 'drama' within a 'dramatic' or dramaturgical model. The elements of the theory involve concepts of role, actor, audience, and script; and the interaction between them. The theory offers important insights into the interaction between self-esteem and social conduct.

E Goffman, *The Presentation of Self in Everyday Life* (New York, 1959)

NS

impressionism (1874) *Art* Originally a derogatory name given to a broad movement which emerged in Paris on the occasion of an independent exhibition in 1874. The name was suggested by one of the exhibits: *Impression: Sunrise* (1872) by French painter Claude Monet (1840–1926).

Although no manifesto for their theories was formulated, the Impressionists' aim was to achieve a greater NATURALISM by: the rendition of flickering light and shade on objects; the analysis of colour, based on the theories of Michel Eugène Chevreul (1786–1889); and '*plein air*' painting in a loose and free style.

R Shattuck, *The Banquet Years: The Arts in France 1885–1918* (1959)

AB

impressionism (1880s) *Music* A term borrowed from the visual arts, this refers to music that is deliberately ambiguous and undefined; especially compositions that seek to convey an overall mood or feeling, rather than to present contrasted musical events.

The effect is achieved largely through ambiguous tonal progressions, CHROMATICISM, minimal delineation of structural units and a strong emphasis on colour and texture. Impressionism is most often associated with the music of French composer Claude Debussy (1862–1918), though he felt more kinship with the aesthetics of SYMBOLISM.

C Palmer, *Impressionism in Music* (London, 1973)

MLD

improbabilism *Philosophy* Somewhat bizarre name occasionally given to the view that the scientist should look for the most improbable hypothesis, because it will be the easiest to refute, if false, but the most significant to accept if it survives testing.

Improbabilism is associated especially with Karl Raimund Popper (1902–); *see* HYPOTHETICO-DEDUCTIVE METHOD. For him, the only way a hypothesis can be 'probable' is by saying comparatively little, and so hedging itself against refutation.

ARL

***in-situ* (swamp) theory** *Geology* Coal is considered to have formed from plants which previously grew in the same location. Some fossil plants can still be seen in their original position and most, but not all, material in a coal sea is usually of very local derivation.

DT

inchworm theory (20th century) *Biology* A theory in genetics whereby the movement of messenger ribonucleic acid (mRNA) is compared to that of an inchworm creeping along a stem.

The theory holds that during enzymatic binding of aminoacyl-transfer RNA (tRNA) to the A-site of the ribosome, a kink is created in the mRNA. This kink may be formed by rotation of the tRNA, which is hydrogen-bonded to the mRNA, along its long axis. The kink causes the amino acid to be brought near the peptide in the P-site so that a peptide bond can be formed. When the kink in the mRNA straightens out (owing to GTP and translocase activity), the peptidyl-tRNA is translocated from the A-site to the P-site of the ribosome.

KH

included fragments, principle of *Geology* Any rock containing fragments of another must be younger than the rock from which the fragments were derived.

DT

inclusion-exclusion principle (20th century) *Mathematics* This elementary result has a modern formulation but it is the basis of older methods of counting.

To find the total number of elements in three sets *A*, *B* and *C* we first include them all by adding the numbers of elements in each set. However, this has counted twice each element which occurred in pairs of *A*, *B* and *C*, so we exclude these by subtracting the numbers of elements in both *B* and *C* and in both *C* and *A* and in both *A* and *B*. However, this has subtracted twice the number of elements that occurred in all three sets, so we include this number of elements by adding the number of elements that occur in all three sets. This result generalizes to any finite number of sets.

JB

inclusive fitness (1964) *Biology* Also called genetic fitness, this theory was proposed by W D Hamilton. This is the idea that an individual's fitness for survival consists of his or her own fitness plus that individual's influence on the fitness of his or her relatives.

More formally, the theory asserts that the impact an allele has on an individual's fitness (perhaps by making the individual behave in a certain way) depends not only on that individual's reproductive potential, but also on the potential of relatives sharing that allele. Under this theory, an individual can optimize inclusive fitness by promoting the survival of relatives, especially through altruistic behaviour towards them. Social insects appear to conform to this model. *See also* ALTRUISM, GROUP SELECTION and RECIPROCAL ALTRUISM.

D Dewsbury, *Comparative Animal Behavior* (New York, 1978)

KH

income determination, theory of (20th century) *Economics* Based on the income-determination model developed by English economist John Maynard Keynes (1883–1946), and later modified by American economist Paul Samuelson (1915–), this theory postulates that the level of NATIONAL INCOME is determined where aggregate demand equals aggregate supply.

The crucial component of aggregate demand is the consumption function. Keynes asserted that his theory was different from those of the classical economists in that the economy could be in equilibrium at any level of employment. *See* EQUILIBRIUM THEORY, GENERAL EQUILIBRIUM THEORY, PARTIAL EQUILIBRIUM THEORY, DISEQUILIBRIUM THEORY, CLASSICAL MACROECONOMIC MODEL and LAW OF MARKETS.

J M Keynes, *General Theory of Employment, Interest and Money* (New York, 1936)

PH

income distribution, theory of (18th century–) *Economics* Developed over a 250-year period by a wide variety of economists including Adam Smith (1723–90) and Karl Marx (1818–83), this theory analyzes the pattern of payment to the factors of production; namely rent, wages, profit and interest.

C J Bliss, *Capital Theory and the Distribution of Income* (New York, 1975)

PH

inconsistency principle *Psychology See* ADVANTAGE, PRINCIPLE OF.

incrementalism (20th century) *Politics* A theory of the manner in which governments make decisions.

Policies are changed bit by bit in response to changing circumstances, rather than as the result of grand plans or wide-ranging rational calculations. Those who fear change see incrementalism as a slippery slope to disaster; those who want REVOLUTION view it as an impediment to progress. *See also* CONSERVATISM.

C E Lindblom, *Politics and Markets* (New York, 1977)

RB

independent assortment, law of *Biology See* MENDEL'S LAWS.

independent migration of ions law *Chemistry See* KOHLRAUSCH'S LAW OF INDEPENDENT IONIC MOBILITIES.

indeterminacy (1970) *Literary Theory* Proposed by the German critic Wolfgang Iser (1926–).

In RECEPTION THEORY, texts have built into them sites of indeterminacy or 'gaps' which the IMPLIED READER is called on to resolve from his cultural knowledge.

W Iser, 'Indeterminacy and the Reader's Response to Prose Fiction' [1970], *Aspects of Narrative*, J Hillis Miller, ed. (New York, 1971)

RF

indeterminacy *Music* Associated with 20th-century music, and especially the work of American composer John Cage (1912–92), this is a precept demanding that some elements of the composition should be left undetermined by the composer. The result is called aleatory music.

It may be that the music is composed using chance procedures (for example, choosing notes according to the tossing of a coin), or that the performer determines exactly what is heard (for example, a composer may provide a set of pitches to be sounded in an unspecified order). An extreme example is Cage's *4′33″*, a composition in which only the length of the piece is specified: its content is provided by the performer(s) and/or audience.

M Nyman, *Experimental Music: Cage and Beyond* (London, 1974)

MLD

indeterminacy of reference and translation *Philosophy* Willard Van Orman Quine (1908–), the American mathematical logician, has claimed that when translating an alien language we construct hypotheses as to what is being said, or what items the words refer to; however, except in a few basic cases, it is impossible in principle to decide conclusively between different hypotheses which differ in ways affecting not only the meaning but even the truth of what is being said.

Quine's example is that a 'native' might utter the word 'gavagai', which we take to mean 'rabbit'; but it might be interpreted (with suitable adjustments to our interpretation of the rest of the sentence) as 'undetached rabbit part', or 'momentary temporal part of an enduring rabbit', and so on. Strictly, this indeterminacy applies to all translation or interpretation, even within one language. This makes Quine doubt whether there really is such a thing as synonymy, and is connected with his HOLISM. *See also* principles of CHARITY, HUMANITY and RADICAL INTERPRETATION.

W V O Quine, *Word and Object* (1960), ch. 2

ARL

indeterminacy principle *Physics See* HEISENBERG'S UNCERTAINTY PRINCIPLE.

indeterminism *Philosophy* The contradictory of DETERMINISM; that is, the theory that at least some events have no cause. An alternative formulation is that some event could, or might, have been different even if everything in the universe up to the time of its occurrence had been the same. (Here there are problems about the interpretation of 'could' and 'might'.)

Indeterminism has been claimed mainly for quantum events in physics and for human acts of freewill, where incompatibilists (*see* COMPATIBILISM) think it indispensable. Attempts have been made to use the indeterminism of quantum phenomena to ground that of human actions, but objections have been raised that more is needed if actions are to be more than random occurrences (*see also* LIBERTARIANISM).

G E M Anscombe, 'Causality and Determination', reprinted in E Sosa, ed., *Causation and Conditionals* (1975))

ARL

index *Stylistics See* KERNEL, CATALYSER, INDEX.

index laws (17th century) *Mathematics* For any multiplication for which the ASSOCIATIVE LAW holds, such as for matrices, the following laws hold whenever the expressions are defined:

$$\begin{aligned} x^a x^b &= x^{a+b}, \\ (x^a)^b &= x^{ab}, \\ x^{-a} &= (x^a)^{-1}, \\ x^0 &= 1 \\ (xy)^{-1} &= y^{-1}x^{-1} \end{aligned}$$

If the COMMUTATIVE LAW also holds, then

$$x^a y^a = (xy)^a$$

JB

index of refraction *Mathematics See* REFRACTIVE INDEX.

indifference curve (20th century) *Biology* A model for how animals 'decide' whether to perform a particular behaviour based on changes in two variables which can increase in intensity, one along the x-axis and the other along the y-axis, in one quadrant of a co-ordinate plane.

For example, the x-axis may measure the quantity of food available while the y-axis measures the risk involved in obtaining it. The indifference curve, which is concave and hyperbolic in shape, is drawn to predict the animal's behaviour at various levels of risk and food availability. Experiments with rats have tended to support this model.

C Barnard, *Animal Behaviour: Ecology and Evolution* (New York, 1983)

KH

indifference, principle of (16th century) *Mathematics/Philosophy* The fundamental principle of statistical theory that unless there is a reason for believing otherwise, each possible event should be regarded as equally likely.

In this crude form, the principle leads to paradoxes because we can group the alternatives in different ways: the next flower I meet might be blue or red, so its being blue has a probability of one-half; but it also might be blue or crimson or scarlet, so the probability of blue is only one-third). Evidently we require not mere absence of knowledge of reasons favouring one alternative over another, but knowledge of the absence of such reasons. But this may be hard to achieve, even in apparently symmetrical cases like the outcomes of throwing a die; for example, what do we do about the possibility of its standing on edge, or the fact that the paint on the 'six' side will be heavier than on the 'one' side? *See also* PROPENSITY THEORY OF PROBABILITY.

W C Kneale, *Probability and Induction* (1949), §§31, 34

ARL, JB

indigenous theory of fossil origins (18th century) *Biology* Anonymous.

Fossils of plants and animals are typical of the region in which they are found. This theory was antagonistic to the EXOTIC THEORY OF FOSSIL ORIGINS. In fact either can be true; some fossils are transported, some are not.

D M Moore, *Green Planet: The Story of Plant Life on Earth* (Cambridge 1982)

RA

indiscernibility of identicals *Philosophy* One part of LEIBNIZ'S LAW, named by Willard Van Orman Quine (1908–), the American mathematical logician. It says that if what appear to be two or more objects are in fact identical, there can be no property held by one and not by the others.

This must be distinguished from the substitutivity of identicals, which says that two names or true descriptions for the same object can always be intersubstituted; this is false for if 'Cicero' and 'Tully' name the same orator we cannot infer from 'Smith believes that Cicero was an orator' to 'Smith believes that Tully was an orator'. If we make this distinction (which Leibniz didn't) then the Cicero/Tully example will not be an objection to Leibniz's law properly stated.

R Cartwright, 'Identity and Substitutivity' in M K Munitz, ed., *Identity and Individuation* (1971); reprinted in R Cartwright, *Philosophical Essays* (1987)

ARL

individual psychology *Psychology See* ADLERIAN THEORY.

individualism *Politics* Theory opposed to COLLECTIVISM.

Human social life is to be understood in terms of the actions of individuals, who are the basic units of society. Complementarily, the basis of moral reasoning consists in the RIGHTS of individuals, rather than of groups, societies, or nations.

David Miller *et al.*, eds, *The Blackwell Encyclopaedia of Political Thought* (Oxford, 1987)

RB

individualistic model (1917) *Biology* First proposed by the American botanist Henry Allan Gleason.

No two vegetation communities are the same; the coexistence of component species occurs simply because of similar needs. Boundaries between community types as proposed in the ORGANISMAL MODEL and species associations are not as predictable in this model. Current opinion is somewhere between these two models.

RA

individuality theory *Psychology* Not attributable to any one originator, this is a term referring to a class of theory which attempts to account for individual differences.

It is often suggested that one can identify higher order psychological components as the determiners of lower order factors. Thus, individuals differ on each of the multiple dimensions of such psychological components (intelligence, personality and so forth), and the individuality of each person is identified by the person's multidimensional profile. Research has indicated three classes of factor (general, group, and specific) and the main factor theories consist of different combinations of these.

J P Guilford, *The Nature of Human Intelligence* (New York, 1967)

NS

individuation principle *Philosophy* The means by which separate items or individuals are distinguished. Debate over the years has centred upon the question of whether such individuation is achieved through some inherent characteristic or through some formal acceptance of a necessary 'uniqueness' belonging to every being and object.

DP

Indo-European (1786) *Linguistics* The resemblance between Sanskrit and European languages was noted by the English colonial judge Sir W Jones (1746–94). The German linguist Franz Bopp (1791–1867) first used the term in 1816; his compatriot August Schleicher (1821–68) consolidated the idea and introduced the family-tree presentation in 1861–62.

A 'family' of languages as apparently diverse as English, Hindi, Welsh, French, Dutch, Russian, and so on, derived from a hypothetical protolanguage: Proto-Indo-European. The latter is thought to have been spoken in south-eastern Europe *c*.3000 BC, disseminated by migration and developing into separate languages through lack of contact. A large effort in 19th century COMPARATIVE AND HISTORICAL LINGUISTICS went into the reconstruction of Indo-European.

W B Lockwood, *A Panorama of Indo-European* (London, 1972)

RF

indoleamine hypothesis of depression (1972) *Psychology* Proposed by A Copper, A J Prange (jnr), Peter C Whybrow and R Noguera.

There are two main biological theories of depression. The first is the genetic theory and the second is the biochemical theory. Within the biochemical theories there are three main areas: (1) pituitary adrenal axis theories; (2) biogenic amine theories; (3) electrolyte metabolism theories. The indoleamine hypothesis is part of the biogenic amine theories. There is evidence to suggest that unusual levels of indoleamines such as serotonium can cause depression.

G Winocur, *Depression: The Facts* (Oxford, 1981)

NS

induced-fit hypothesis (1973) *Biology* Developed by Daniel Edward Koshland (jnr) (1920–), American biochemist and longtime editor of the journal *Science*.

The theory that the active site of enzymes can be induced, when in close proximity to the appropriate substrate or substrates, to change shape to better fit the substrate while forming an enzyme-substrate complex. This theory represents a refinement of the outdated LOCK-AND-KEY MODEL.

KH

induction *Mathematics/Philosophy* Also known as mathematical induction or finite induction, this affirms that to prove that a certain property P holds for all natural numbers, it suffices to show that $P(1)$ is true and that $P(k + 1)$ is true whenever $P(k)$ is true.

Intuitively, one can think of this in terms of climbing a ladder, where proving $P(1)$ corresponds to getting on the ladder, and proving $P(k + 1)$ from $P(k)$ corresponds to taking the $(k + 1)$th step; once one knows how to move from one step of the ladder to the next, it is possible, in principle, to reach any step of the ladder.

There is an equivalent form of this principle, sometimes called complete induction, in which the inductive hypothesis above is replaced by the stronger assumption that $P(j)$ is true for all integers j less than or equal to k; this latter form of induction is often simpler to apply in practice. *See also* UNIFORMITY OF NATURE, INDUCTIVISM and GOODMAN'S PARADOX, WELL-ORDERING PRINCIPLE.

MB

inductivism *Philosophy* Claim that inference in accordance with some version of the inductive principle is, if not logically valid, at least rationally legitimate. Objections to it include those mentioned under UNIFORMITY OF NATURE and GOODMAN'S PARADOX.

R Swinburne, ed., *The Justification of Induction* (1974)

ARL

industrial democracy (20th century) *Politics* Theory of WORKERS' CONTROL.

Political democracy should be paralleled or complemented by industrial democracy. Individual factories or workshops should be managed by those answerable to, and elected by, the body of workers; and within whole firms or industries, processes similar to those in representative democracy should operate.

Carole Pateman, *Participation and Democratic Theory* (Cambridge, 1970)

RB

inert pair effect *Chemistry* Heavy elements in certain groups of the periodic table form compounds in which they exist with oxidation states two less than the common oxidation state for that group. For example, although the common oxidation state for elements in group 4 is +4, most elements in the group can also exist in oxidation state +2. This is because of the inert pair effect. In large atoms, such as those of tin and lead, some outer-shell electrons are not as well shielded as those in the inner core. They are therefore 'sucked' into the inner core of electrons and thus become inert.

M Freemantle, *Chemistry in Action* (Basingstoke, 1987)

MF

inevitability of gradualness (19th century) *Politics* Theory of SOCIALISM associated with FABIANISM.

Socialism was not the result of revolutionary conflict, or of confrontation with the existing order. It arose naturally out of the processes of industrial society as the most efficient way of managing them. Socialism was thus both inevitable and gradual.

Rodney Barker, *Political Ideas in Modern Britain* (London, 1978)

RB

infantile birth theories *Psychology* Proposed by the Austrian founder of psychoanalysis Sigmund Freud (1856–1939).

These are theories which children themselves generate about birth; they do not include cultural myths. Parents generally avoid giving children detailed descriptions of the birth process, therefore children fill in the gaps with their own theories. The most common of these theories are emergence through the navel or anus. *See* CLOACA THEORY.

O G Brim (jnr), 'Socialization Through the Life Cycle', *Socialization After Childhood: Two Essays*, O G Brim (jnr) and S Wheeler, eds (New York, 1966)

NS

inference, pragmatic (1970s) Various accounts. In conversation, A infers what B means by relating the propositions B utters to relevant MUTUAL KNOWLEDGE.

S C Levinson, *Pragmatics* (Cambridge, 1983)

RF

infinite divisibility (5th century BC) *Philosophy* Concept first discussed by the Greek philosopher Zeno of Elea, who argued that infinite divisibility was a logical impossibility.

According to the theory, nothing is infinitely divisible as all actual entities must be finite if they are to exist at all. To be infinitely divisible, an entity would have to be of infinite size.

DP

'infinite worlds' theory (1584) *Astronomy/Biology* Proposed by Giordano Bruno (1548–1600), an Italian philosopher who for this and other ideas was burned at the stake by the Inquisition in 1600.

He postulated that the universe is filled with planets (he called them 'infinite worlds') that are inhabited by intelligent life like that on Earth. Bruno's theory assumes that the laws of nature operate similarly throughout the universe, and if so, then those same laws could lead to the evolution of life on other, favourable planets. Modern scientists estimate that with 100,000 million stars in our galaxy and at least 10,000 million galaxies in the universe, even if there is only one Sun-like star per million stars and one Earth-like planet per million planets, then the universe contains at least a billion Earth-like planets. *See also* DIRECTED PANSPERMIA and PANSPERMIA.

G Bruno, *On the Infinite Universe and Worlds* (1584); M Barbieri, *The Semantic Theory of Evolution* (Chur, Switzerland, 1985)

KH

inflationary universe theory (20th century) *Astronomy* In some grand unified theories of the universe, the expansion of the universe is accelerated at about 10^{-35} seconds after the Big Bang, an effect referred to as inflation.

When the expansion of the universe lowers the temperature of the Big Bang to that at which the strong nuclear force becomes distinct from the weak nuclear force and the electromagnetic force, a phase change is possible in the mixture of quarks and leptons. This phase change releases a lot of energy in the form of electromagnetic radiation and the pressure of this radiation can cause a sudden and dramatic inflation of regions of space.

S Mitton, ed., *The Cambridge Encyclopaedia of Astronomy* (London, 1973)

MS

information processing theory *Psychology* Pioneered by the English mathematician Alan Mathison Turing (1912–54).

This concerns the way in which people attend to, select and internalize information, and how they later use it to make decisions and guide their behaviour. 'Information processing theory' denotes a scientific community with common presuppositions, problem priorities, a specialized scientific language, and a set of research methods. Information processing theory is a leading orientation in experimental psychology and has stimulated research in various areas of psychology, particularly memory, language, perception and problem solving.

B H Kantowitz, ed., *Human Information Processing* (Hillsdale, N.J., 1974)

NS

information theory *Economics/Mathematics/Physics* The mathematical study of information, its storage by codes (*see also* COMBINATORICS), and its transmission through channels of limited capacity (*see also* COMMUNICATION THEORY).

A fundamental idea is the *entropy* of a set of events,

$$H = -\sum_1^n p_k \log p_k,$$

where the p_ks are the probability of each event. This notion was introduced by C E Shannon in 1948 to measure the uncertainty of the set. *Channels* are considered as mechanisms which take a letter of an input *alphabet* and transmit letters of an output alphabet with various probabilities, depending on how prone they are to error or *noise*. Shannon was then able to measure the effectiveness of the channel, which he called *capacity*, using ergodic theory (*see also* BIRKHOFF ERGODIC THEOREM) and entropy.

AL

information theory (1940s) *Music* Adapted from science by various music theorists, notably the American Leonard B Meyer (1918–), this term refers to the notion that music can be analyzed statistically as a chain of events.

Each event (note, chord) suggests a possible continuation. If the predicted event occurs then no more information is collected; if the predicted event does not occur then the new event is added to the index of possible continuations, thus modifying the criteria for future predictions. The completed survey yields an objective record of repetition (unity) and diversity in the musical syntax. Though accurate answers are guaranteed, the process relies on the integrity of its user in deciding what information is meaningful.

R C Pinkerton, 'Information Theory and Melody', *Scientific American*, vol. CXCIV (1956)

MLD

inheritance of acquired characteristics *Biology See* ACQUIRED CHARACTERISTICS.

innate ideas *Philosophy* Ideas or concepts that we allegedly acquire or possess prior to experience can be called *a priori* (literally, 'from beforehand'). Sometimes, however, it is claimed that they are innate; that is, we have them from birth.

This is a stronger claim, since we might well (on a weaker view) acquire an idea independently of experiencing that *idea*, but not independently of having built up a stock of other ideas from experience. Two main issues therefore concern the sense, if any, in which we can already possess such ideas at birth, and what is to count as an idea in this context. John Locke's (1632–1704) vigorous attack on innate ideas focuses on the first idea, while in recent decades the American linguist Avram Noam Chomsky (1928–) has treated the second issue in terms of our innate knowledge of certain grammatical structures. Belief in innate ideas is one form of NATIVISM.

S Stich, ed., *Innate Ideas* (1975)

ARL

innateness hypothesis (1960s) *Linguistics* Proposed by the American linguist Avram Noam Chomsky (1928–).

The faculty of language is an 'innate idea', inbuilt knowledge of a set of LINGUISTIC UNIVERSALS, genetically available to a child; innate knowledge of language is cited as an explanation for the ease of first language acquisition.

N Chomsky, *Cartesian Linguistics* (New York, 1966)

RF

inoculative hypothesis (1962) *Psychology* Promoted by W McGuine, this hypothesis is part of the theory of attitudinal change. It postulates that an individual's attitude towards something is reinforced and strengthened when faced with either a weak or refutable counter-argument to his opinion.

D T Campbell, 'Social Attitudes and Other Acquired Behavioral Dispositions', *Psychology: A Study of a Science*, S Koch, ed., vol. VI (New York, 1963)

NS

input-output analysis (1941) *Economics* The Russian-born American economist Wassily Leontief (1906–) used matrix algebra and subsequently computer technology to trace the relationship between industries of the US economy.

Every output in an economy can be analyzed in terms of final consumption and all the inputs necessary for its production. By developing a matrix of various production functions, it is possible to track the knock-on effect of a change in demand for a given good or service on other industries or sectors. Lower automobile sales might mean a decline in steel demand, a fall in electricity or a drop in coal imports.

W W Leontief, *The Structure of the American Economy, 1919–29* (New York, 1941)

PH

inscriptionism *Philosophy* Also called inscriptivism. An inscription, in the relevant sense, is a word or phrase or sentence considered as written (or uttered) on some particular occasion. Inscriptionism is any view making significant use of such inscriptions, considered as contrasted with abstract entities such as meanings which they might be thought to represent.

One might, for example, claim that to have a belief is to be suitably related to some particular sentence one has uttered (perhaps just to oneself) on some particular occasion. Inscriptionism is a form of REDUCTIONISM, and claims the economy associated with that.

W V O Quine, *Word and Object* (1960), 214–15

ARL

insider-outsider wage determination (1980s) *Economics* Wage determination can be achieved inside a firm when increased productivity permits higher pay for existing workers. It can also be determined externally by the forces operating in the broader labour market.

A Lindbeck and D J Snower, *The Insider-Outsider Theory of Employment and Unemployment* (Cambridge, Mass., and London, 1989)

PH

instinct (20th century) *Biology* Developed by Konrad Lorenz (1903–89), an Austrian zoologist and ethologist who shared the Nobel prize for Physiology or Medicine in 1973. The concept that animal behaviours are dictated by stereotypical inborn tendencies ('fixed action patterns') which may be modified by experience, but none the less exert powerful influence on behaviour.

Despite some unresolved weaknesses in the concept, instinct remains prominent in the thinking of most animal behaviourists. *See also* ETHOLOGY.

D Dewsbury, *Comparative Animal Behavior* (New York, 1978)

KH

instruction theory *Psychology* Particularly associated with the work of the American psychologist Burrhus Frederic Skinner (1904–90).

The process of imparting knowledge systematically. Much of instructional technology was pioneered by behavioural psychologists; for example Edward Lee Thorndike (1874–1949) and B F Skinner, who both showed particular concern about educational science and the role of psychology in that process. *See* BEHAVIOURISM.

B F Skinner, 'The Science of Learning and the Art of Teaching', *Harvard Educational Review*, vol. XXIV, (1958), 86–97

NS

instructive theory (1930) *Biology/Medicine* Also called antigen template theory, and template theory. First detailed model proposed by Linus Carl Pauling (1901–), American chemist and winner of two Nobel prizes, the Chemistry prize and the Peace prize.

The general theory of antibody formation holding that the antigen instructs the lymphocytes to synthesize specific antibodies which, in the absence of the antigen, would never be formed or would be formed only very rarely, by chance. The information for antibody synthesis comes from the antigen (a template) rather than from the body's genetic material. This idea was toppled by the clonal selection theory. *See also* ANTIBODY, THEORIES OF. *Compare with* SELECTIVE THEORY OF ANTIBODY DIVERSITY.

T Kindt and J Capra, *The Antibody Enigma* (New York, 1984)

KH

instrumental theory of learning *Psychology* Proposed by the American psychologist Edward Lee Thorndike (1874–1949).

This refers to the theory that organisms learn to behave in certain ways from experiencing the consequences which follow a chosen behaviour. The theory is concerned with explaining the relationship between stimulus, response and reinforcement, but has been superseded by more sophisticated accounts. *See* OPERANT THEORY.

B F Skinner, *Beyond Freedom and Dignity* (New York, 1971)

NS

instrumentalism *Philosophy* Mainly now the theory that scientific laws and theories are not to be interpreted as stating truths, or as claiming objective correctness, but as instruments for the prediction of statements which can be tested by observation. It is in terms of usefulness rather than correctness that the laws are judged.

Instrumentalism has its greatest plausibility with theories about unobservable theoretical entities like quarks, and is contrasted with (one form of) REALISM. 'Instrumentalism' is also used for a development of PRAGMATISM by John Dewey (1859–1952), and for a view that values (either in general or in some sphere, like aesthetics) should be regarded as instrumental (for example, in promoting satisfaction). *See also* CONVENTIONALISM.

S E Toulmin, *The Philosophy of Science* (1953); develops an instrumentalist view

ARL

integral test *Mathematics* Attributed to the principle founders of calculus: the French mathematician Pierre de Fermat (1601–65), the British mathematician Sir Isaac Newton (1642–1727) and the German mathematician Gottfried Wilhelm Leibniz (1646–1716).

The test establishes convergence of an infinite series $\sum f(n)$ where f is continuous, decreasing and non-negative, by verifying that the corresponding integral

$$\int_1^\infty f(x)\,dx$$

is convergent. For example

$$\sum_{k=1}^{\infty} \frac{1}{k}$$

diverges since

$$\int_1^{\infty} 1/k = \lim_{b\to\infty} \ln(b)$$

which tends to infinity.

H Anton, *Calculus with Analytic Geometry* (New York, 1980)

ML

integration by parts *Mathematics* Attributed to the principal founders of calculus: the French mathematician Pierre de Fermat (1601–65), the British mathematician Sir Isaac Newton (1642–1727) and the German mathematician Gottfried Wilhelm Leibniz (1646–1716).

The technique of integration of two differentiable functions by the rule

$$\int f(x)g'(x)\,dx = [f(x)g(x)]_a^b - \int_a^b g(x)f'(x)\,dx$$

where

$$[f(x)\,g(x)]_a^b = f(b)\,g(b) - f(a)\,g(a)\,.$$

H Anton, *Calculus with Analytic Geometry* (New York, 1980)

ML

intention *Literary Theory See* FOUR KINDS OF MEANING.

intentional fallacy (1946) *Literary Theory* Proposed by the American critic William Kurtz Wimsatt (jnr) (1907–75) and the aesthetician Munro C Beardsley (1915–).

'A confusion between the poem and its origins'; Wimsatt and Beardsley condemn the practice of speculating on an author's intentions and raising them as a standard for the success of a poem. Intentions are not genuinely available, and they are superceded by the completed work itself. *See also* AFFECTIVE FALLACY.

W K Wimsatt, *The Verbal Icon* (Kentucky, 1954)

RF

intentional forgetting *Psychology See* MOTIVATED FORGETTING.

interaction principle *Psychology* A theory of the German-American psychologist Kurt Lewin (1890–1947).

Lewin claimed that behaviour should be thought of as a function of the individual and the individual's environmental context. The principle is sometimes summarized as $B = F(PE)$, where B is behaviour, F is a function, P denotes person and E denotes the environment. The interaction principle states that behaviour and environment have a reciprocal effect on each other and one cannot be investigated or explained without reference to the other.

T Blass, ed., *On Personality Variables, Situations, and Social Behavior* (Hillsdale, N.J., 1977)

NS

interactionism *Philosophy* Theory of how body and mind are related.

It can mean simply that there are two types of events, physical and mental, either of which can cause the other; for example, a pin-prick causes pain, and pain causes screaming. More usually, though, it is a philosophical theory grounding interactions of this kind in the existence of two substances, body and mind (or soul), where physical events take place in the one and mental events in the other, while the two series of events can interact causally, as above. In this form its main representative is René Descartes (1596–1650), but the theory goes back, in less developed forms, at least as far as Plato (*c*.427–*c*.347 BC).

R Descartes, *Meditations on the First Philosophy* (1641); especially Meditation 6

ARL

interests (20th century) *Politics* Account of politics derived from but not limited to MARXISM.

As members of groups, classes, and so on, people have interests which can be identified by the outside observer. These interests are real whether individuals or whole groups are aware of them or not, and enable commentators to detect success, EXPLOITATION, and so on, in the relations between one group and another. An obvious difficulty arises with attributing to a group an interest of which none of its members appear to be aware.

Geoffrey Roberts and Alistair Edwards, *A New Dictionary of Political Analysis* (London, 1991)

RB

interference theory of forgetting (1935) *Psychology* A term given to the substantial body of evidence reviewed by American psychologist G O McGeoch in 1935.

This evidence indicated that forgetting is caused by interference; interference being the process during which one set of information conflicts or interferes with another. For instance, the effective recall of information can be hindered by attending to other information. The theory has been superseded by others indicating important roles for metabolic processes and activities occurring during learning.

A D Baddeley, *The Psychology of Human Memory* (New York, 1976)

NS

interior monologue (1887/1922) *Stylistics* Technique used by the late 19th century French novelist Edouard Dujardin in 1887; term applied to James Joyce's *Ulysses* (1922).

Variety of STREAM OF CONSCIOUSNESS: psychological writing in which a character's thoughts are represented as direct thought, using first person, present tense sentences, subjective point of view, fragmented syntax; framed by third person narration: for example Leopold Bloom's fragmentary musings in *Ulysses*.

D Cohn, *Transparent Minds* (Princeton, 1973)

RF

interlanguage (1972) A term coined by linguist L Seliker.

Systematic errors by second-language learners, not identifiable as derived from either the first or the second language, interpreted as a sign of a developing system of rules.

L Seliker, 'Interlanguage', *International Review of Applied Linguistics* 10 (1972), 209–31

RF

intermediate compound theory of catalysis (1889) *Chemistry* Theory of catalysis by Swedish physical chemist Svante August Arrhenius (1859–1927).

A catalyst works by forming an intermediate compound. The catalyst (C) reacts with the reactant, known as the substrate (S), to form the intermediate compound CS. The intermediate compound then decomposes forming product P and unchanged catalyst C:

Step 1: $C + S \rightarrow CS$ (intermediate compound)

Step 2: $CS \rightarrow C + P$

The catalyst is thus consumed in the first step and regenerated in the next. The theory can also be applied to reactions involving two reactants.

M Freemantle, *Chemistry in Action* (Basingstoke, 1987)

MF

intermediate value theorem (19th century) *Mathematics* Also known as Bolzano's theorem, after the Czech analyst Bernhard Bolzano (1781–1848).

The result whereby a real function f which is continuous on $[a,b]$ takes every value between $f(a)$ and $f(b)$ for at least one argument between a and b.

H Anton, *Calculus with Analytic Geometry* (New York, 1980)

ML

internal and external balance *Economics* Economic analysis that attempts to balance full employment and price stability (the internal balance of an economy) with the balance of payments equilibrium (the external balance).

J E Meade, *The Theory of International Economic Policy*, vol. I (Oxford, 1951), ch. 10

PH

internal colonialism (20th century) *Politics/Sociology* Theory of the distribution of power and advantage within states.

Within apparently unified states there is a form of 'colonialism' practised by the centre against the periphery, as for example by London and the south-east of England against Scotland and Wales. Economic resources and power are concentrated at the centre, to the advantage of which the periphery is subordinated. Culture, values, and ways of life characteristic of the centre will be favoured above those of the periphery, which will be dismissed as less developed.

Michael Hechter, *Internal Colonialism* (London, 1975)

RB

internal relations, Doctrine of *Philosophy* Doctrine that all relations are internal to their bearers, in the sense that they are essential to them and the bearers would not be what they are without them.

Some relations are clearly internal in this sense (four would not be four unless it were related to two by being its square), and some relations are internal to their bearers under one description but not under another (a wife would not be a wife unless suitably related to a husband, but Mary would still be Mary had she not married). The doctrine that all relations are internal implies that everything depends on everything else for being what it is, and is therefore associated with monistic or OBJECTIVE IDEALISM.

G E Moore, 'External and Internal Relations', *Proceedings of the Aristotelian Society* (1919–20); reprinted in G E Moore, *Philosophical Studies* (1922)

ARL

internalism *Philosophy* Any view claiming that a certain phenomenon can, or must, be analyzed in terms belonging within a certain sphere. In particular, internalism applies to certain analyses of mental notions such as belief and knowledge.

An internalist analysis of believing, thinking of something, and so on, limits itself entirely to what is going on inside the believer's head; only on those terms could a theory like the IDENTITY THEORY OF MIND be true (though of course internalists need not hold it). An internalist view of knowledge says that for a belief to count as knowledge one must at least be aware of and/or able to present an adequate justification for it; it is not enough that one merely stand in certain causal relations to the fact in question. Various distinctions can be made between strong and weak (and so on) versions of internalism. *See also* EXTERNALISM.

J Dancy, *An Introduction to Contemporary Epistemology* (1985)

ARL

international morality (20th century) *Politics* Theory applying moral considerations to international politics.

International relations can and should be conducted with reference to shared ethical values. There is disagreement as to the usefulness of the theory, or the existence or power of such shared values.

Graham Evans and Jeffrey Newnham, *The Dictionary of World Politics* (Hemel Hempstead, 1990)

RB

International Phonetic Alphabet (1888) *Linguistics* Product of the Phonetic Teachers' Association – founded in 1886 by French linguist Paul Passy (1859–1940) – which became the International Phonetic Association in 1889.

IPA is the most widely used system of notation for the distinctive sounds of languages (that is, PHONEMES, not every phonetic distinction), consisting of alphabetic letters and diacritics.

The Principles of the International Phonetic Association (London, 1961 [1949])

RF

international system (20th century) *Politics* Theory of international politics as a system.

International relations are to be understood not as the interaction of individual states, but rather as the relations of parts of a system which has a reality distinct from the actions of particular nations. There is thus an analogy with SYSTEMS THEORY in the analysis of domestic politics.

Graham Evans and Jeffrey Newnham, *The Dictionary of World Politics* (Hemel Hempstead, 1990)

RB

internationalism (20th century) *Politics* An attempted application of ethical considerations to international politics.

A view opposed to REALISM in international relations, which seeks to understand international affairs by reference to INTERESTS wider than those of the particular national participants, and in terms of interests common to all.

Graham Evans and Jeffrey Newnham, *The Dictionary of World Politics* (Hemel Hempstead, 1990)

RB

interpersonal function *Linguistics See* FUNCTIONAL GRAMMAR.

interpersonal theory (1930s–) *Psychology* Developed during the 1930s by the American psychiatrist Harry Stack Sullivan (1892–1949).

Sullivan theorized that personality dynamics and disorders are primarily due to social forces and interpersonal situations. He felt that the performances (that is, interactions) in the inter-personal field of an individual were in fact the processes during which disorders are formed, identified, and finally treated. The theory enjoys a prominent position in the investigation of personality. *See* EMPATHY.

H S Sullivan, *The Interpersonal Theory of Psychiatry* (New York, 1953)

NS

interpretive semantics (1963) *Linguistics* Formalized by the American linguists Jerrold Jacob Katz (1932–) and Jerry Alan Fodor, in the context of TRANSFORMATIONAL-GENERATIVE GRAMMAR.

In STANDARD THEORY syntactic rules generate basic structures which are then semantically interpreted by replacing syntactic categories with lexical items and amalgamating their meanings by 'projection rules' which refer also to syntax; semantics is a secondary component of grammar. Challenged by GENERATIVE SEMANTICS. *See also* COMPONENTIAL ANALYSIS.

J J Katz and J A Fodor, 'The Structure of a Semantic Theory', *Language*, 39 (1963), 170–210

RF

intertextuality (1969) *Stylistics* Concept developed by the French semiotician Julia Kristeva (1941–).

A text is not an autonomous object with sharp boundaries, as in NEW CRITICISM, but a web of references to other texts by allusion, parody, quotation, stylistic nuance, and so on. These other texts are not 'sources', but ideological reference-points with which the text being read is in dialogue. *See also* DIALOGISM.

J Kristeva, *Desire in Language* (Oxford, 1980)

RF

interview problem *Mathematics* The sampling problem of determining how many applicants to screen successively and randomly for a job if the interviewer must hire a chosen candidate immediately following the interview.

The asymptotic solution is to use the strategy of interviewing $1/e \approx 36.78$ per cent of the candidates and then choosing the next candidate who is better than all the previous ones.

W Gellert, S Gottwald, M Hellwich, H Kästner and H Küstner, eds, *The VNR Concise Encyclopedia of Mathematics* (New York, 1975)

ML

intimacy, principle of *Psychology* Proposed within Gestalt Psychology by the German-American psychologist Wolfgang Köhler (1887–1967).

This principle states that the individual elements of a 'whole' are dependent on each to give the true picture of the 'whole'. These elements may not be taken away or replaced as this would change the perception of the whole. The principle has been chiefly investigated in visual perception, although it can usefully be applied to any aspect of behaviour. *See* GESTALT THEORY.

L Kaufman, *Sight and Mind: An Introduction to Visual Perception* (New York, 1974)

NS

introduction rule (20th century) *Mathematics* Any rule in a system of logic which specifies when a statement containing the introduced symbol may be valid. For example, the introduction rule may specify when A implies B can be inferred from A and B.

JB

intuition (1957) *Linguistics* Basic principle articulated by the American linguist Avram Noam Chomsky (1928–).

Data of linguistics should include introspective judgments of native speakers about certain properties of sentences, chiefly GRAMMATICALITY, AMBIGUITY, sentence-relatedness, and paraphrase.

N Chomsky, *Syntactic Structures* (The Hague, 1957)

RF

intuitionism *Philosophy* Any view holding that some of our knowledge is got by a direct process not depending on the senses and not open to rational assessment.

The objects of such knowledge may include: moral principles (whether as the basis of duty or as ultimate values); particular moral duties on a particular occasion (sometimes called perceptual intuitionism); space and time and their contents, so far as these are presented to us independently of anything contributed by the understanding (Immanuel Kant (1724–1804)), reality as it is itself, as opposed to reality processed by us for practical purposes (investigated by Henri Bergson (1859–1941)); things known by accumulated but forgotten experience or unconscious inference ('woman's intuition'; but this figures less prominently in philosophy); basic truths of logic and the principles of valid inference.

An important special case of the last example cited above is the mathematical intuitionism of L E J Brouwer (1881–1966) and A Heyting (1898–1980), a form of CONSTRUCTIVISM which insists that we should assert only what can be proved (by intuitively acceptable steps) and deny only what can be disproved. It therefore rejects the law of EXCLUDED MIDDLE and half of the DOUBLE NEGATION PRINCIPLE. *See also* ANTIREALISM.

D Pole, *Conditions of Rational Inquiry* (1961), ch. 1

ARL

invariance of domain theorem (1910) *Mathematics* Proved by L E J Brouwer (1881–1966).

A set U of points in an n-dimensional Euclidean geometry is open if for every point P in U there is an n-dimensional sphere S with centre P such that every point inside S belongs to U. The theorem asserts that if there is a set V in an n-dimensional Euclidean geometry such that there is a function mapping U on to V which is bicontinuous (has values which change by a small amount when all the variables change by small amounts), then V is also open. A consequence is that the dimension n of a Euclidean geometry is not altered by topological transformations. *See also* TOPOLOGY.

JB

invention (classical Greece and Rome) *Literary Theory* Greek *heuresis*, Latin *inventio*.

Not the modern sense of novel creation. Invention was the first stage of composition in RHETORIC: the finding of a subject from a stock of existing materials, for example traditional themes and stories. In neo-classical Europe, the term extended to poetic composition. *See also* CREATIVITY.

W Nash, *Rhetoric*, (Oxford, 1989)

RF

inverse function theorem *Mathematics* The theorem in multivariable calculus or COMPLEX FUNCTION THEORY that gives conditions under which a differentiable function has a differentiable inverse.

T M Apostol, *Mathematical Analysis* (Addison-Wesley, 1974)

MB

invisible hand (1776) *Economics* This phrase was used by Scottish economist Adam Smith (1723–90) in explaining his belief that the actions of individuals, although taken for their own economic benefit, are guided in such a manner as to benefit society as a whole.

'He intends only his own security, only his own gain. And he is in this led by an invisible hand to promote an end which was no part of his intention' (*An Inquiry into the Nature and Causes of the Wealth of Nations*, London, 1776).

PH

iron law of oligarchy (1912) *Politics* Theory of social organization set out by German social scientist Robert Michels (1876–1936) in his study of the internal politics of the German Social Democratic Party.

'He who says organization, says oligarchy.' As soon as people form organizations, power in those organizations gravitates upwards towards the permanent officials or officers. A second, subordinate law suggests that whatever purpose an organization was originally established to serve, the preservation of the organization itself, and of its oligarchy, will come to take precedence. *See also* ÉLITISM.

Geraint Parry, *Political Elites* (London, 1969)

RB

iron law of wages (19th century) *Economics* This theory has its roots in the work of classical economists, although the term was first used by German political economist Ferdinand Lassalle (1825–64). It postulates that wages will always revert to subsistence levels.

A rise in wages triggers an increase in the population, prompting a fall in wages back to subsistence levels. Also known as the subsistence theory of wages, this principle has no current relevance. *See* EQUILIBRIUM THEORY.

F J G Lassalle, *Open Letter to the National Labor Association of Germany* (Berlin, 1863)

PH

irony (traditional) *Literary Theory* From Greek *eiron*, a dissimulating comic character.

A mode of verbal expression in which one thing is said and the opposite is meant, unknown to the addressee. Some of the FIGURES OF SPEECH associated with irony were identified in traditional RHETORIC; it became a major preoccupation in the 20th century, along with other linguistic indirections such as PARADOX.

In 'dramatic irony', the playwright and audience know something of which the characters are ignorant.

D C Muecke, *Irony and the Ironic* (London, 1982)

RF

irradiation theory of learning *Psychology See* TRIAL AND ERROR THEORY OF LEARNING.

IS-LM model *Economics* Developed by English economist John Hicks (1904–) and American economist Alvin Hansen (1887–1975) to provide a framework for analyzing the factors determining the level of aggregate demand.

The model (also known as the Hicks-Hansen model) was adopted as a universal framework in studying macroeconomics because it seemed to incorporate different views of the working of the economy. It combines equilibria in the financial market with equilibria in the goods and services market, establishing an equilibrium level for demand. The framework represents investment-savings (IS) and the liquidity-money supply (LM), and can be used to illustrate how fiscal and monetary policies are employed to alter NATIONAL INCOME.

J Hicks, 'Mr. Keynes and the Classics: A Suggested Interpretation', *Econometrica*, vol. V (April, 1937), 147–59; J Dupuit, *On the Measurement of the Utility of Public Works* (Paris, 1844); J R Hicks, 'The Four Consumers' Surpluses', *Review of Economic Studies*, vol. XI (1943), 31–41; J F Muth, 'Rational Expectations and the Theory of Price Movements', *Econometrica*, vol. XXIX, 3 (1961), 315–35

PH

isolation species concept *Biology See* SPECIES CONCEPT.

isolationism *Philosophy/Politics* In aesthetics, the doctrine that a work of art can be appreciated independently of its cultural background, the circumstances of its production, the artist's intentions, and so on (*see also* CONTEXTUALISM). In politics, the doctrine that a nation's (or some particular nation's) interests are best served by minimizing its interference in affairs outside its own borders, or outside some specified sphere of interest.

ARL

isomorphism *Psychology* This is an hypothesis used in Gestalt Psychology, originally proposed by the German-American psychologist Wolfgang Köhler (1887–1967).

The theory argues that there is a formal point-by-point relationship between excitatory fields in the cortex and conscious experience. The correspondence is not presumed to be between the physical stimulus and the brain but between the perception of the stimulus and the brain. *See* GESTALT THEORY.

A D Ellis, ed., *A Sourcebook of Gestalt Psychology* (London, 1938)

NS

isomorphism theorem *Mathematics* Any one of a number of theorems in mathematics which asserts that two mathematical objects are isomorphic (that is, their structure is essentially the same).

For example, in GROUP THEORY, the *first isomorphism theorem* states that for any group G and any homomorphism θ with domain G, the group $G \ker\theta$ is isomorphic to the image of G under θ. The *second isomorphism theorem* states that $(G/N)/(K/N)$ is isomorphic to G/K where K is assumed to be a normal subgroup of G and N is assumed to be a normal subgroup of K. The *third isomorphism theorem* states that $(AB)/B$ is isomorphic to $A/(A \cap B)$.

I N Herstein, *Topics in Algebra* (Wiley, 1975)

MB

isoperimetric inequality *Mathematics* This is the inequality which relates the perimeter L of a plane figure R and its area A.

To be specific,

$$L^2 - 4\pi A \geqslant 0$$

with equality holding if and only if R is a circle. The inequality expresses the fact that among all plane figures with the same perimeter, the circle is the one with the largest area. *See also* CALCULUS OF VARIATIONS.

T W Körner, *Fourier Analysis* (Cambridge, 1988)

MB

isoperimetric problem (late 17th century) *Mathematics* The solution of this problem sparked a battle between the Swiss brothers James Bernouilli (1654–1705) and John Bernoulli (1667–1748) in 1691.

James withheld his solution, while John publicized an incorrect solution and claimed that his brother had none. In 1701 James presented his solution to the Paris Academy, but it remained in a sealed envelope until after his death. Even when James's solution was made public in 1706, John refused to admit his own error. The classical name for the problem is Dido's problem, named after the legendary first queen of Carthage.

A fundamental problem in the CALCULUS OF VARIATIONS, which goes back to the problem of finding a curve of given length which encloses a maximum area. According to Virgil, Dido was offered whatever area of land she could enclose with an ox hide on which to found her city. She solved the problem by turning the ox hide into cord and enclosing a large circle, which is the shape that maximizes the area.

M Hazewinkel, ed., *Encyclopaedia of Mathematics* (Dordrecht, 1988)

isoprene rule *Chemistry*. Isoprene has the structure

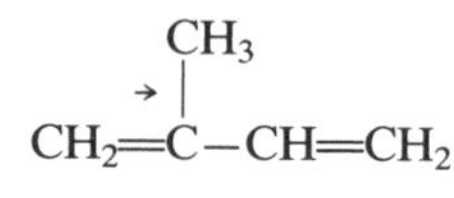

The isoprene unit occurs naturally in rubber and most terpenes, which are compounds found in the essential oils of many plants. Terpenes have carbon skeletons made up of isoprene units joined in a regular head-to-tail way. This is the isoprene rule.

R T Morrison and R N Boyd, *Organic Chemistry* (Boston, 1987)

MF

isostatic theory (*c*.1735–) *Geology* Derived from the Greek for equal balance, the general term 'isostasy' was coined in 1889 by the American geologist Clarence E Dutton (1841–1912), following much work by others in this field during the 18th and early 19th centuries. Isostatic theory explains the gravitational attraction of continental and oceanic areas, and accounts for apparent discrepancies between their actual and calculated deflective power.

Neither the AIRY HYPOTHESIS nor the PRATT HYPOTHESIS are sufficient, although both are partially true and of different importance in different areas. Generally, the rocks of the ocean basins are much denser than those of the continents, while 'roots' often exist beneath mountain areas. The crust is also sufficiently rigid, in some locations, to support small regional differences and thus give rise to isostatic anomalies.

M H P Bott, *The Interior of the Earth* (London, 1982)

DT

J

Jackson-Bernstein theorem *Mathematics* The Jackson theorems (1911) provide upper bounds for the error of a best uniformly approximating polynomial (algebraic or trigonometric) to a continuous function f (*see also* ALTERNATION THEOREM) in terms of the smoothness of F.

If f is continuously differentiable, for example, the bound is inversely proportional to the degree of the approximating polynomial. The prize-winning Bernstein theorems (1912) are converse results, allowing one to deduce smoothness properties of f from the rate of convergence of best uniform approximants as their degree increases.

AL

Jackson's law *Psychology* Formulated by the English neurologist John Hughlings Jackson (1835–1911).

This law states that functions or processes will deteriorate in relation to the length of time they are present in a system. In relation to learning, for instance, the law states that recently acquired information will be lost before deeply stored, older information. The law holds for a variety of learning experiences but does not explain the underlying processes which cause the phenomenon.

S H Hulse, H Egeth and J Deese, *The Psychology of Learning* (New York, 1980)

NS

Jacob-Monod model (1960) *Biology* Also called the operon model, this was proposed by French molecular biologist François Jacob (1920–) and biochemist Jacques Monod (1910–). They received the Nobel Prize for their work in 1965. The theory explains how gene activity is regulated and applies to bacteria and blue-green algae (prokaryotes), which are simple cells lacking nuclear membranes.

The model holds that protein synthesis is regulated not by the gene that codes for the protein (the structural gene), but by two other genes: an 'operator' gene next to the structural gene, and a 'regulator' gene elsewhere on the chromosome. Jacob and Monod called the structural, regulator and operator gene unit an 'operon'. A parallel model has been proposed for complex cells (eukaryotes) such as those found in animals. *Compare with* BRITTEN-DAVIDSON MODEL and REPEATED SEQUENCE.

E Keller, *A Feeling for the Organism: The Life and Work of Barbara McClintock* (San Francisco, 1983)

KH

Jacobi's theorem (19th century) *Mathematics* Named after German mathematician Karl Gustav Jacobi (1804–51), this states that rotations of the unit circle formed by adding λ multiples of 2π are dense in the unit circle if λ is irrational.

R L Devaney, *An Introduction to Chaotic Dynamical Systems* (New York, 1989)

ML

Jahn-Teller theorem (1937) *Chemistry* General theorem recognized and proved by H A Jahn and Hungarian-born American nuclear physicist Edward Teller (1908–). Any orbitally degenerate electronic structure of a non-linear molecule is intrinsically unstable. Another structure with lower molecular energy that eliminates the orbital degeneracy must exist.

For example, octahedral transition metal complexes such as those of copper (II) exhibit distortions from molecular geometry known as the Jahn-Teller effect.

K F Purcell and J C Kotz, *Inorganic Chemistry* (Philadelphia, 1977)

MF

James-Lange theory (1885) *Psychology* One of many THEORIES OF EMOTION, this is a combination of the views of American philosopher and psychologist William James (1842–1910), and the Danish psychologist Carl Georg Lange (1834–1900).

The theory proposes that physiological processes are fundamental in the experience of emotion and precede the conscious perception of the emotion. For instance, we cry therefore we are sad rather than we are sad therefore we cry. The theory has provoked considerable debate and continues to attract some critical support. *See also* ATTRIBUTION THEORY OF EMOTION.

G Mandler, *Mind and Body, Psychology of Emotion and Stress* (New York, 1984)

NS

James's theorem (1964) *Mathematics* A deep result in FUNCTIONAL ANALYSIS characterizing weak compactness in a real Banach space (*compare with* EBERLEIN-SMULYAN THEOREM).

A subset C is weakly compact exactly when it is closed (weakly) and every continuous linear functional attains its maximum on C (*compare with* SUPPORT THEOREM and the BISHOP-PHELPS THEOREM). A substantial reason for the importance of *reflexive spaces* (where $X = X^{**}$) is the useful characterization that every non-zero continuous linear functional is a support functional for the unit ball.

AL

Jones's polynomials (*c.*1980) *Mathematics* Named after their discoverer, Vaughan Jones, these are polynomials which can be associated with knots and which distinguish them better than *Alexander polynomials*. *See also* KNOT THEORY.

JB

Jordan curve theorem (19th century) *Mathematics* Named after French mathematician Marie-Ennemond Camille Jordan (1838–1922), this is the fundamental result, important in COMPLEX FUNCTION THEORY, whereby any simple closed curve has an inside and an outside; in particular, any simple closed curve divides the plane into two distinct regions with each having the curve as its boundary. While superficially obvious, this is hard to prove.

J R Munkres, *Topology: A First Course* (Prentice-Hall, 1975)

MB

Jordan-Holder theorem (1889) *Mathematics* Named after French mathematician Marie-Ennemond Camille Jordan (1838–1922), who proved a preliminary version, and Otto Holder (1859–1937), who completed the proof.

Each finite group G can be decomposed in a finite set of simple groups, each of which are uniquely defined by G, including their numbers of repetitions. This result relates the classification of all finite groups to the classification of FINITE SIMPLE GROUPS. *See also* GROUP THEORY.

JB

Josephson effect (1962) *Physics* Named after the Welsh physicist Brian David Josephson (1940–) who shared the 1973 Nobel prize for Physics.

Josephson showed theoretically that at low temperatures electric currents can flow across a very thin insulating layer separating two superconductors. In the absence of an applied voltage, the current leaks across the junction by means of electron tunnelling. When a voltage is applied, an AC current of frequency proportional to the voltage is produced. The effect was observed experimentally by Anderson, and such Josephson junctions are now used extensively in research and in fast switches for computers.

Paul Davies, ed., *The New Physics* (Cambridge, 1989)

GD

Jost's law *Psychology* Proposed by the German Adolph Jost.

Two phenomena are described by this law. First, if two associations are of equal strength but different ages, the older will lose strength or diminish more slowly with the further passage of time. Second, if two associations are of different strength and equal ages, further learning has greater value for the older one. The data supporting the law are meagre and it has fallen into disuse within psychological theory.

R L Klatzky, *Human Memory: Structure and Processes* (San Francisco, 1980)

NS

Joule's law (Mayer's hypothesis) (19th century) *Physics* Named after the English physicist James Prescott Joule (1818–89), who examined the law experimentally (1845), and the German physicist Julius Robert von Mayer (1814–78), who assumed it in a theoretical discussion (1842).

For a fixed mass of a dilute real gas (or an ideal gas at any pressure), the internal energy is a function of the temperature only. In particular, if U_m is the molar internal energy, for real gases

$$\lim_{p \to 0} U_m = U_m(T)$$

and for ideal gases

$$U_m = U_m(T) .$$

J Thewlis, ed., *Encyclopaedic Dictionary of Physics* (New York, Oxford and London, 1962)

MS

Joule's law of electric heating (1842) *Physics* Named after its discoverer, the English physicist James Prescott Joule (1818–89).

When electrons flow through a conductor, electrical energy is converted to internal energy (thermal energy) of the conductor in overcoming its resistance. If the resistance R is a constant, the power W dissipated in the conductor is proportional to the square of the current I flowing. In fact,

$$W = I^2R .$$

The thermal loss in a conductor is often called Joule heating.

J Thewlis, ed., *Encyclopaedic Dictionary of Physics* (New York, Oxford and London, 1962)

MS

Joule-Thomson effect (1852) *Physics* Named after its British discoverers, James Prescott Joule (1818–89) and William Thomson, 1st Baron Kelvin (1824–1907).

When a compressed gas passes through a small opening so that it expands adiabatically and with negligible change in flow speed, there is usually a difference in the temperature of the gas on the two sides of the orifice. Such a phenomenon is called a throttling process. For all gases there is a temperature, known as the inversion temperature and characteristic of the gas, above which a throttling process always produces a rise in gas temperature.

J Thewlis, ed., *Encyclopaedic Dictionary of Physics* (New York, Oxford and London, 1962)

MS

Jourdain's paradox (1913) *Philosophy* Named after its discoverer, the mathematician Philip Edward Bertrand Jourdain (1879–1919).

The paradox is equivalent to the sentence 'The second part of this sentence is true and the first part of this sentence is false', which contradicts itself whether the first half is true or not. However, the halves of the sentence refer indirectly to themselves, so any rule prohibiting statements that refer to themselves resolves the paradox.

JB

Julia set (early 20th century) *Mathematics* Named after the French mathematician Gaston Julia who, along with the French mathematician Pierre Fatou (1878–1929), first studied the set.

The closure of the set of repelling periodic points of a complex function f. The complement of the Julia set is called the Fatou set or stable set of f. The Julia set is usually a FRACTAL and iterates of f behave chaotically on the set. Recently, many complicated and beautiful pictures have been computer-generated for functions as simple as

$$f(z) = z^2 + c$$

for various values of c.

R L Devaney, *An Introduction to Chaotic Dynamical Systems* (New York, 1989)

ML

jumping genes (1931) *Biology* Proposed by American geneticist Barbara McClintock (1902–), winner of the Nobel Prize for

Physiology or Medicine in 1983. The name is derived from a reference to the theory that genes can move from one chromosome to another, and that such transpositions change the genetic programming of the cell.

McClintock first proposed her dynamic theory of the genome in 1931, but it was rejected by most mainstream scientists until molecular biological techniques verified it in the 1960s.

E Keller, *A Feeling for the Organism: The Life and Work of Barbara McClintock* (San Francisco, 1983)

KH

Junggrammatiker *See* NEOGRAMMARIANS.

Jungian aesthetics *Art* The theories of art proposed by Swiss psychiatrist Carl Gustav Jung (1875–1961) are concerned with the artist's use of archetypes, either consciously or unconsciously.

His assumption was that all artists are driven to creativity (which he deemed feminine) by their involvement with the Mother archetype. The experience of Jungian analysis prompted American painter Jackson Pollock (1912–56) to create a series of drawings which were subsequently published. *See* COLLECTIVE MEMORY.

M Philipson, *Outline of a Jungian Aesthetic* (Evanston, Ill., 1963)

AB

Jungian theory *Psychology* Proposed by the Swiss psychiatrist Carl Gustav Jung (1875–1961).

An analytical psychology associated with an approach to psychoanalysis that placed, relative to his contemporary and onetime colleague Sigmund Freud (1856–1939), little emphasis on the role of sex and sexual impulses. It focused instead on the hypothesized deep, inherited COLLECTIVE UNCONSCIOUS. Jungian analysis concerns itself with the rich interpretation of symbols and makes extensive use of dreams.

C G Jung, *Analytic Psychology: Its Theory and Practice (New York, 1968)*

NS

junk DNA *Biology See* SELFISH DNA.

just price (13th–15th century) *Economics* Associated with medieval schoolmen, particularly St Thomas Aquinas (1225–74), this concept has no allocative role but is rather a moral concept based on natural justice: the just price depends on moral judgment of a good or service.

V A Demant, ed., *The Just Price, An Outline of the Medieval Doctrine and Examination of its Possible Equivalent Today* (London, 1930)

PH

just war *Politics* Ethical assessment of war, often discussed by medieval Christian theorists.

Whilst war causes injury, some wars are justifiable (*ius ad bello*) if their object is morally justifiable, and if there is a proportionate relation between ends and means (*ius in bello*).

David Miller *et al.*, eds, *The Blackwell Encyclopaedia of Political Thought* (Oxford, 1987)

RB

justice *Politics/Law/Philosophy* Theory of the morally appropriate way of resolving social differences.

There is no one theory of justice. One view is that justice involves avoiding or preventing harm to people; another that it involves treating people according to their deserts; another that people should be treated according to their needs; another that they should be treated according to fair and impartial procedures. *See also* RAWL'S THEORY OF JUSTICE.

David Miller *et al.*, eds, *The Blackwell Encyclopaedia of Political Thought* (Oxford, 1987)

RB

K

K-theory (1950) *Mathematics* Initiated by John Henry Constantine Whitehead (1904–60).

The groups $K_0(A)$, $K_1(A)$ and $K_2(A)$ are defined for suitable categories A, and this provides a very general method for use in ALGEBRAIC TOPOLOGY and ALGEBRA. *See also* GROUP THEORY and CATEGORY THEORY.

M Atiyah, *K-Theory* (New York, 1967)

JB

Kakutani's fixed point theorem (1941) *Mathematics* A generalization of SCHAUDER'S FIXED POINT THEOREM which states that an upper semicontinuous set valued mapping T of a compact convex space K into itself, where T sends points of K into convex subsets of K, has a fixed point. This is an extension of BROUWER'S THEOREM. This result was further generalized to the case of locally convex topological linear spaces by Ky Fan in 1952. Kakutani's theorem has numerous applications in mathematical economics.

M Hazewinkel, ed., *Encyclopaedia of Mathematics* (Dordecht, 1988)

ML

Kalte Kunst *Art See* COLD ART.

Kaluza-Klein theory (1921) *Physics* Named after the German physicists Theodore Franz Eduard Kaluza (1885–1954) and Oskar Klein (1894–1977).

Kaluza was the first to propose a unified field theory which partly succeeded in unifying gravity with electromagnetism. The theory (developed by Klein) showed, on the basis of a five-dimensional space-time, that the appropriate curvature component in the fifth dimension corresponds to electromagnetism. That is, if Einstein's gravitational field equations are generalized to a five-dimensional space-time, they reproduce both ordinary gravitational physics (in four-dimensional space-time) and Maxwell's theory of electromagnetism.

Paul Davies, ed., *The New Physics* (Cambridge, 1989)

GD

KAM theory (1950s–60s) *Mathematics* Various versions of this result were proved by Kolmogorov in 1954, Moser in 1962 and Arnold in 1963.

The celebrated result on the stability of Hamiltonian systems which essentially states that an invariant torus is stable under sufficiently small perturbations provided that the frequency ratio is 'badly' approximable by rationals. Further, almost all (in the sense of Legesgue measure) frequency ratios have tori which are stable under sufficiently small perturbations. The theorem has been used to describe the stability of the solar system.

J Guckenheimer and P Holmes, *Nonlinear Oscillations, Dynamical Systems and Bifurcations of Vector Fields* (New York, 1983)

Karush-Kuhn-Tucker theorem *Mathematics* A fundamental characterization result in OPTIMIZATION THEORY, developed by W Karush in 1939 and H W Kuhn and A W Tucker in 1951.

It asserts, under a suitable regularity condition, the existence of LAGRANGE MULTIPLIERS for the problem of minimizing an objective function $f(x)$ subject to constraints

$$g_i(x) \leqslant 0 \text{ for } i = 1,..,r$$

and

$$h_j(x) = 0 \text{ for } j = 1,..,s\ ,$$

at an optimal solution $\bar{x}$; there exist non-negative multipliers $\lambda_1,..,\lambda_r$ (with $\lambda_i = 0$ if $g_i(\bar{x}) < 0$) and multipliers $\mu_1,\ldots,\mu_s$ so that $\bar{x}$ is a stationary point of the *Lagrangian*

$$f + \sum_1^r \lambda_i g_i + \sum_1^s \mu_j h_j\ .$$

These *necessary optimality conditions*, stronger than the FRITZ-JOHN CONDITIONS THEOREM, form the basis for computational techniques for these widely applicable problems.

AL

Kekulé structure (1865) *Chemistry* Theory of the structure of benzene proposed by German chemist Friedrick August Kekulé von Stradonitz (1829–96). This theory laid the basis for the development of aromatic chemistry.

A molecule of benzene has a cyclic structure with alternating double bonds:

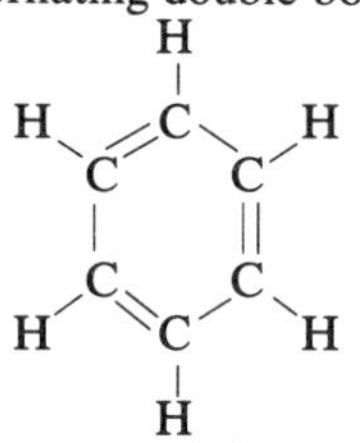

Kekulé later suggested that a benzene molecule existed as two rapidly alternating structures (hydrogen atoms not shown).

The benzene molecule is now known to exist as a resonance hybrid of these two limited forms.

M Freemantle, *Chemistry in Action* (Basingstoke, 1987)

MF

Keldysh theory *Chemistry* A theory of multiphoton ionization, this states that an atom is ionized by the rapid absorption of a sufficient number of photons. The theory predicts that the rate at which the atom ionizes depends principally upon the ratio of the mean binding electric field to the peak strength of the incident electromagnetic field. The rate also depends on the ratio of the binding energy to the energy of the photons in the field.

MF

Kelvin temperature scale (1854) *Physics* Named after the Scottish mathematician and physicist William Thomson, 1st Baron Kelvin (1824–1907), who introduced it.

Thomson realized that CARNOT'S THEOREM and its corollary allowed the setting up of temperature scales that did not depend on the properties of substances (absolute scales). In 1848 he suggested that an absolute scale be adopted that defined a temperature t through the equation

$$-\frac{Q_1}{Q_2} = \frac{\exp \tau_1}{\exp \tau_2}$$

where Q_1 is the heat absorbed per cycle by the Carnot engine from a reservoir at a temperature τ_1 and Q_2 that absorbed per cycle from the reservoir at a temperature τ_2. This scale was so different from existing scales that it was not acceptable. In 1854 he proposed the adoption of an alternative absolute scale on which a temperature T is defined through the equation

$$-\frac{Q_1}{Q_2} = \frac{T_1}{T_2}$$

where T_1 and T_2 are the temperatures of the two reservoirs. With the triple point of water being given the value 273.15 K, where K is the symbol of the unit of temperature, the kelvin, the Kelvin scale of temperature is completely defined. It has now been incorporated into the SI system of units and the temperature T is termed the thermodynamic temperature.

J Thewlis, ed., *Encyclopaedic Dictionary of Physics* (New York, Oxford and London, 1962)

MS

Kelvin's circulation theorem (19th century) *Physics* Discovered by William Thomson, 1st Baron Kelvin (1824–1907), this states that the circulation of a frictionless fluid acted on by forces which can be represented in terms of potential energies is not dependent on the time. This is also valid for fluids of uniform density which cannot be compressed, such as water.

J Thewlis, ed., *Encyclopaedic Dictionary of Physics* (New York, Oxford and London, 1962)

MS, JB

Kepler's laws of planetary motion (1601–19) *Astronomy* Named after the German astronomer and mathematician Johannes Kepler (1571–1630) who formulated them on the basis of observations made by the Danish astronomer Tycho Brahe (1545–1601).

(1) The planets move around the Sun in elliptical orbits, with the Sun at one focus. (2) The line joining the planet to the Sun sweeps out equal areas in equal times. (3) The square of the period of the planet is proportional to the cube of its mean distance from the Sun.

S Mitton, ed., *The Cambridge Encyclopaedia of Astronomy (London, 1973)*

MS

Kepler's rule (16th century) *Astronomy/Mathematics* Based on the harmony of the universe, by Johannes Kepler (1571–1630).

The spheres on which the planets (starting with Mercury) move around the Sun can be constructed by successively inscribing and circumscribing spheres about the regular polyhedra in the order octahedron, icosahedron, dodecahedron, tetrahedron, cube. Later observations by Kepler showed this was accurate to about 5 per cent. However, no further regular polyhedra exist to determine the spheres for Uranus, Neptune, Pluto and the minor planets. Also, according to Kepler's first law of planetary motion, which he discovered later, the planets do not move in circles about the Sun but on ellipses with the Sun at one focus. The basis for the 'proof' of the result is that it is easy to find numbers approximately equal to a few given numbers.

JB

kernel, catalyzer, index (1966) *Stylistics* Term coined by the French structuralist Roland Barthes (1915–80). His French terms are *noyau* or *fonction cardinal*, *catalyse*, *indice*.

Development of Vladimir Propp's NARRATIVE GRAMMAR of basic story structure. Kernels are actions and events which significantly project the story forward; catalyzers are subsidiary or filler events; indices give information about characters, atmosphere, and so on.

R Barthes, 'Introduction to the Structural Analysis of Narratives', *Image-Music-Text*, S Heath, trans. and ed. (London, 1977)

RF

kernel sentence (1957) *Linguistics* Proposed by the American linguist Avram Noam Chomsky (1928–) as part of his first version of TRANSFORMATIONAL-GENERATIVE GRAMMAR. Now merely of historical importance.

A basic simple active, affirmative, declarative sentence, to which obligatory TRANSFORMATIONS have been applied for number, tense, and so on, but no optional transformations (which would generate passive, negative, and so on).

N Chomsky, *Syntactic Structures* (The Hague, 1957)

RF

Kerr effects (1875) *Physics* Named after their discoverer, Scottish physicist John Kerr (1824–1907).

(1) The electro-optical effect. When a normally isotropic transparent material is placed in an electric field it becomes birefringent. In this condition, if n_1 and n_2 are the refractive indices for transmitted light of wavelength 1 polarized parallel and perpendicular to the electric field, respectively,

$$n_1 - n_2 = \lambda B E^2$$

where E is the applied electric field and B is a constant, characteristic of the material, known as the Kerr coefficient.

(2) The magneto-optical effect. When plane-polarized light is reflected from one pole face of a magnet it becomes elliptically polarized.

J Thewlis, ed., *Encyclopaedic Dictionary of Physics* (New York, Oxford and London, 1962)

MS

Kersten's theory (20th century) *Physics* Named after its originator, M Kersten, this theory (now superseded), explained the existence of non-zero coercive forces in ideal bulk ferromagnetic materials as arising from the effects of non-magnetic inclusions located in the material at the corners of a simple cubic lattice.

J Thewlis, ed., *Encyclopaedic Dictionary of Physics* (New York, Oxford and London, 1962)

MS

Keynsian economics *Economics* Named after the English economist John Maynard Keynes (1883–1946), Keynsian economics has influenced post-1945 economic management, particularly in its advocacy of government management of the economy.

Adherents believe that the macroeconomy tends towards extended business cycles, with high levels of unemployed factors. They assert that government management or stimulation of the economy to influence demand (through monetary or fiscal policy) can alleviate this problem of high unemployment. Monetary and fiscal policies thus stimulate the economy in times of slump by generating employment, and slow the economy down in times of inflation. *See* BUSINESS CYCLE, EQUILIBRIUM THEORY, GENERAL EQUILIBRIUM THEORY, and CLASSICAL MACROECONOMIC MODEL.

J M Keynes, *The General Theory of Employment, Interest and Money* (New York, 1936); J Hicks, *The Crisis in Keynsian Economics* (Oxford, 1974)

PH

Khatchian algorithm (1979) *Mathematics* The first polynomial-time method (*see also* COMPLEXITY THEORY) for solving linear programs (*compare with* DUALITY THEORY OF LINEAR PROGRAMMING). Also called the *ellipsoid method*, owing to its geometric basis in circumscribing the region of interest with ellipsoids. It has proved impractical compared with the traditional simplex method (which is not polynomial-time), although a somewhat analogous polynomial-time technique due to N Karmarkar in 1984 does appear competitive.

AL

kin-selection theory *Biology* The theory that animals engage in behaviours that promote the survival and reproduction of their close relatives, even when their own interests seem sacrificed, because doing so increases the transmission of their own genetic material (alleles) through NATURAL SELECTION. Sometimes used (mistakenly, according to some authorities) to mean GROUP SELECTION. *See also* ALTRUISM, HAPLODIPLOID HYPOTHESIS and RECIPROCAL ALTRUISM.

KH

kinematical theory of X-ray and electron diffraction (20th century) *Physics* This is a treatment of the process of X-ray or electron diffraction by crystals appropriate to very thin specimens, when the interactions between the incident beam and the diffracted beams may be neglected.

J Thewlis, ed., *Encyclopaedic Dictionary of Physics* (New York, Oxford and London, 1962)

MS

kinesics (1950s) *Linguistics* Developed by the American linguist R L Birtwhistell.

Study of patterns of body language and gesture which accompany speech; often using methods of linguistics.

R L Birtwhistell, *Kinesics and Context: Essays on Body Motion Communication* (Philadelphia, 1970)

RF

kinetic art (1920) *Art* Term first used in connection with the *Realistic Manifesto* (1920) of Russian brothers Naum Gabo (1890–1977) and Antoine Pevsner (1886–1962), but applied more frequently to a range of art styles current in the 1950s.

These include: styles in which optical illusions and visual ambiguity are emphasized, promoting instability in the spectator; objects whose appearance changes as the spectator moves position; and constructed works such as the mobiles by American sculptor Alexander Calder (1898–1976).

F Popper, *Origins and Development of Kinetic Art* (1968)

AB

kinetic theory of gases (1738–1892) *Chemistry* Mathematical model of an ideal gas used to account for the GAS LAWS (BOYLE'S LAW, CHARLES'S LAW, AVOGADRO'S LAW, DALTON'S LAW OF PARTIAL PRESSURE and GRAHAM'S LAW OF DIFFUSION). The theory developed from the ideas of: Swiss mathematician Daniel Bernoulli (1700–82) in 1738; Scottish chemist John James Waterston

(1811–83) in 1845, published 1892); German physicist Augustus Karl Krönig (1822–79) in 1856; and German physicist Rudolph Clausius (1822–88) in 1857. It was given precise mathematical form by Scottish physicist James Clerk-Maxwell (1831–79) in 1860, and Austrian physicist Ludwig Boltzmann (1844–1906) in 1868.

The fundamental equation of the kinetic theory of gases is

$$pV = \frac{1}{3} Nm\overline{c^2}$$

where p is pressure, V volume, N is the number of gas particles, m is the mass of a gas particle and $\overline{c^2}$ is the mean square speed of a gas particle; that is, the average value of $\overline{c^2}$.

The equation is based on five assumptions: (1) the gas consists of particles of negligible volume; (2) the particles are in continuous random motion; (3) the particles exert no attractive forces on each other; (4) the particles are perfectly elastic and thus, no kinetic energy is lost on collision; (5) the average kinetic energy of the particles is proportional to the absolute temperature.

M Freemantle, *Chemistry in Action* (Basingstoke, 1987)

MF

king's touch (13th century) *Medicine* Originating in France after Louis IX returned from the Crusades in 1254, and spreading to England between 1259 and 1272. The belief that the touch of royalty could cure tubercular adenitis, or scrofula. (This disease may have been the object of the King's touch because it produces open sores that are almost never fatal, but that typically disappear in time without treatment.)

England's Edward the Confessor (*c.*1002–66) and France's Robert the Pious (996–1031) were also credited with healing power. The belief was largely abandoned by the 18th century, with Queen Anne (1702–14) being the last practitioner in England and Charles X (1757–1836) performing the last ritual in France on 31 May 1825.

M Bloch, *The Royal Touch* (London and Montreal, 1973)

KH

king's two bodies (Middle Ages) *Politics/History* Medieval theory of divine authority of kings.

The king had two 'bodies', one of which was the ordinary mortal one, the other infused with the DIVINE RIGHT to govern which marked off the monarch from other men.

E H Kantorowicz, *The King's Two Bodies* (Princeton, 1957)

RB

kinship (19th century) *Anthropology* Several writers in the 19th century wrote on kinship, but the field came to be associated especially with American lawyer and anthropologist Lewis Henry Morgan (1818–81).

The branch of anthropology which includes the study of linguistic systems of classification for relatives (usage of words like 'uncle', 'cousin', and so on), the formation of kin groups (such as families and clans), and the relations between kin groups by marriage. It is widely held to be both the most difficult and the most distinctive branch of anthropology. *See also* ALLIANCE THEORY, EXTENSION OF KINSHIP TERMS, LINEAGE THEORY, and PRIMITIVE PROMISCUITY.

R Fox, *Kinship and Marriage* (Harmondsworth, 1967)

AB

Kirchhoff's laws for electrical circuits *Physics* Named after their discoverer, the German physicist Gustav Robert Kirchhoff (1824–87).

(1) The algebraic sum of all currents at a junction in a network is zero. (2) The algebraic sum of all potential drops around any closed path or network is zero. This must, of course, include sources of e.m.f.

J Thewlis, ed., *Encyclopaedic Dictionary of Physics* (New York, Oxford and London, 1962)

MS

Kirchhoff's law of thermal radiation (1859) *Physics* Named after its discoverer, the German physicist Gustav Robert Kirchhoff (1824–87), this law states that the ratio of the spectral emissive power to the spectral absorptive power of a body is a universal function of the wavelength of the radiation and of the temperature of the body.

J Thewlis, ed., *Encyclopaedic Dictionary of Physics* (New York, Oxford and London, 1962)

MS

Klein bottle (1874) *Mathematics* Named after German mathematician Christian Felix Klein (1849–1925) who, as the founder of modern geometry, unified Euclidean and non-Euclidean geometry.

A closed surface which has only one side and no interior. When it is cut lengthwise into two pieces, the result is two MÖBIUS STRIPS. Although it cannot be realized in three-dimensional space, it can be visualized as a three-dimensional tube in which one end is inserted through the wider end with the two ends sealed together.

I Peterson, *The Mathematical Tourist* (New York, 1988)

ML

Klein's Erlangen programme *Mathematics* *See* ERLANGEN PROGRAMME.

knapsack cryptosystem *Mathematics* First suggested by Martin Hellman and Ralph Merkle, this is a method of encoding and decoding secret messages based on a puzzle known as the knapsack problem: if the total weight of a packed knapsack and the weight of individual items which might be packed are known, find which items are likely to be inside.

For example, if the total weight is 37 and the possible items weigh 1, 2, 4, 8, 16, 32, then the answer must be $1 + 4 + 32 = 37$. Analogously, an encripted message could be 37 which could be decoded by someone who knows the possible numbers (the key). Simple knapsack cryptosystems were broken in the early 1980s and it is unknown whether a secure knapsack cryptosystem can be found.

I Peterson, *The Mathematical Tourist* (New York, 1988)

ML

kneading theory *Mathematics* The theory of chaotic dynamical systems which uses an elaborate version of *symbolic dynamics* to keep track of the orbit of a critical point of a map in order to understand better the dynamics of the map which is being iterated. *See also* CHAOS THEORY.

R L Devaney, *An Introduction to Chaotic Dynamical Systems* (New York, 1989)

ML

knot theory (1923) *Mathematics* Initiated by James Waddell Alexander (1888–1971).

A knot is a closed space curve, as if made by looping a rope in a complicated way and splicing the ends together; so the knots used to tether goats are not knots in this sense. The theory attempts to classify all knots which cannot be deformed in another without cutting the rope.

G Burde and H Zeischang, *Knots* (New York, 1985)

JB

Koch curve *Mathematics* A FRACTAL object formed by iteratively deforming a line segment of unit length. The middle third of the segment is removed and two sides of an equilateral triangle based on the removed segment are added. This process is then repeated for each of the four segments and again for the 16 segments and so on infinitely many times. The resulting limit is called the Koch curve or von Koch curve.

I Peterson, *The Mathematical Tourist* (New York, 1988)

JB

Kohlrausch's law of independent ionic mobilities (1875) *Chemistry* Named after German physicist Friedrich Kohlrausch (1840–1910).

The molar conductivity of an electrolyte at infinite dilution is the sum of the ionic mobilities of the ions forming the electrolyte. The ionic mobility of an ion is independent of the other ions.

M Freemantle, *Chemistry in Action* (Basingstoke, 1987)

MF

Kolmogorov-Smirnoff test (1944) *Mathematics* A non-parametric statistical test used for testing a hypothesis in which independent random variables $X_1,\ldots,X_n$ have a given continuous distribution function F. The test statistic is

$$\max_{i=1\ldots n} \left| \frac{i}{n} - F(x_i) \right| .$$

M Hazewinkel, ed., *Encyclopaedia of Mathematics* (Dordecht, 1988)

ML

Kolmogorov three series theorem *Mathematics See* THREE SERIES THEOREM.

Kondakov rule *Chemistry* Alkenes which add mineral acids readily react with chlorine or bromine to form unsaturated monohalides. Alkenes which do not add mineral acids readily form dihalides.

MF

Kondratieff cycles *Economics* Named after Russian-born economist Nikolai Kondratieff (1892–*c.*1931), this term refers to TRADE CYCLES of long duration.

Kondratieff studied American, British and French wholesale prices and interest rates from the 18th century, and found that the peaks and troughs in economic activity fell at regular intervals. J A Schumpeter applied the term 'Kondriatieff cycles' to cycles of 50–60 years in duration. Harvard economists conducted similar work into British wheat prices from the 13th century and found cycles lasting 54 years. *See* BUSINESS CYCLE, SUNSPOT THEORY, PRODUCT LIFE-CYCLE THEORY, ACCELERATION PRINCIPLE, FINE-TUNING, MULTIPLIER-ACCELERATOR MODEL, and POLITICAL BUSINESS CYCLE.

J J van Duijin, *The Long Wave in Economic Life* (London, 1983)

PH

Konigsberg bridge problem (18th century) *Mathematics* Named after the city of Konigsberg (now Kalingrad) in Prussia, the aim of this puzzle was to find a route starting from an arbitrary spot in the city and then crossing each of the seven bridges over the River Pregel (now the Pregolva) exactly once between the two banks and two islands and arriving back at the starting place.

The problem was solved by Leonhard Euler (1707–83), who proved that such a route was impossible unless each bank or island has an even number of bridges, which was not the case. This solution was the first result in the mathematical theory of graphs, but not necessarily the first bridge problem.

JB

Konowaloff rule *Chemistry* An empirical rule concerning vapour pressures. The vapour of a liquid mixture contains a higher proportion of the component which, on addition to the liquid, raises its vapour pressure.

MF

Koopmann's theorem *Chemistry* When an electron is removed from an orbital, the exact Hartree-Fock energy of the molecular orbital is the negative of the ionization potential. This theorem, although valid for most stable molecules, is strictly true only for closed shell molecules. *See also* SELF-CONSISTENT FIELD THEORY.

MF

Kopp's law (1864) *Physics* Following the suggestion in 1848, by A C Woestyn, that the heat capacity of a molecule of any compound is equal to the sum of the heat capacities of its constituents; the German physical chemist Hermann Franz Moritz Kopp (1817–92) showed that the specific heat capacity of water in combination in various hydrates is the same as that of ice. This general result is only approximately true.

J Thewlis, ed., *Encyclopaedic Dictionary of Physics* (New York, Oxford and London, 1962)

MS

Korovkin theorems *Mathematics* A class of theorems on uniform approximation. The most basic result is the theorem that if a sequence $\{L_n\}$ of positive linear operators on $C[a,b]$ is such that $L_n(x^k)$ converges uniformly to x^k for $k = 0,1,2,\ldots$, then for all continuous functions f the sequence $\{L_n(f)\}$ converges uniformly to f. An example of such a sequence is the Bernstein polynomials. Another basic result is that a sequence of non-negative linear operators on the periodic function space $C[-\pi,\pi]$ is such that $L_n(f)$ converges uniformly to f for 1, sin, cos, then $L_n(f)$ converges uniformly to f for all continuous functions f.

E W Cheney *Introduction to Approximation Theory* (New York, 1982)

ML

Korte's law *Psychology* Formulated by the German, A Korte (*fl.*1910s).

The law is proposed in relation to the phi phenomenon, that is apparent motion. Korte's law states that under the following conditions, apparent motion is not visible; when the light is low, when spatial distance is wide, and when the interstimulus interval is too short. The law acknowledges that compensation can take place. *See* PHI PHENOMENON.

M Wertheimer, 'Laws of Organization in Perceptual Forms', *A Sourcebook of Gestalt Psychology*, A D Ellis, ed. (London, 1938)

NS

Kossel-Sommerfeld law *Chemistry* The arc spectra of the atom and ions belonging to an isoelectronic sequence resemble each other, particularly in their multiplet structure.

MF

König's theorem (20th century) *Mathematics* Refers to one of two results in GRAPH THEORY, due to D König. Both can be phrased as network flow problems and hence proved via linear programming (*compare with* DUALITY THEORY OF LINEAR PROGRAMMING).

The first, sometimes called the marriage theorem (1916), is the result whereby given any set of n men, if each subset of k of them knows at least k women, then it is possible for each of the n men to marry a woman he knows.

The second result, called the König-Egerváry theorem (1931), states that in a bipartite graph (where the vertices split into two disjoint sets and each edge links the two), the size of a maximal matching equals the size of a minimal cover.

J E Rubin, *Mathematical Logic: Applications and Theory* (Saunders, 1990)

MB,AL

Kramers' degeneracy theorem (1930) *Physics* Named after its discoverer, the Dutch theoretical physicist Hendrik Anton Kramers (1894–1952), this theory proposes that the energy eigen-states of an odd number of spin ½ particles are at least doubly degenerate in the absence of an external magnetic field.

J Thewlis, ed., *Encyclopaedic Dictionary of Physics* (New York, Oxford and London, 1962)

MS

Krasnoselskii theorem *Mathematics* One of the first results of computational geometry. Suppose C is a closed bounded set in n-dimensional Euclidean space. We say a point *sees* another point in C if the line segment between the points lies in C. Krasnoselskii's result says that if any $n + 1$ points in C have a point which sees them all then there is a point which sees every point in C; or in other words C is *star-shaped* (*compare with* HELLY THEOREM).

AL

Krein–Milman theorem (20th century) *Mathematics* Named after Russian mathematicians Mark Grigor'evich Krein (1907–) and David Pinkhusovich Mil'man (1913–), this is the result in FUNCTIONAL ANALYSIS whereby a compact convex subset of a locally convex space is the closed convex hull of its extreme points which look like corners of the set.

Béla Bollobás, *Linear Analysis* (Cambridge, 1990)

MB

Kronecker's lemma (19th century) *Mathematics* Named after German algebraist, number theorist and philosopher of mathematics Leopold Kronecker (1823–91).

The result that if

$$\sum_{n=1}^{\infty} \frac{a_n}{n}$$

converges, then

$$\lim_{N \to \infty} \frac{1}{N} \sum_{n=1}^{N} a_n = 0.$$

K Knopp, *Theory and Application of Infinite Series* (New York, 1990)

ML

Kundt's law of abnormal dispersion (19th century) *Physics* Named after its discoverer, the German physicist August Adolph Kundt (1839–94), this law states that when the refractive index of a solution increases (for example, by increase of concentration), its optical absorption bands are displaced towards longer wavelengths. The law does not always hold in practice.

J Thewlis, ed., *Encyclopaedic Dictionary of Physics* (New York, Oxford and London, 1962)

MS

Kupka-Smale theorem (1963) *Mathematics* Proved independently by Kupka and the American mathematician Stephen Smale (1930–), this states that given an orientation preserving diffeomorphism *f* of the unit circle, there is a continuously differentiable Morse-Smale diffeomorphism *g* (that is, it has rational rotation number and all its periodic points are hyperbolic) which is arbitrarily close to *f*. A more general version exists for diffeomorphisms of *n*-dimensional manifolds.

R L Devaney, *An Introduction to Chaotic Dynamical Systems* (New York, 1989)

ML

Kutta and Joukowski hypothesis *Physics* Named after W M Kutta and N Joukowski who introduced it independently, this hypothesis provides a criterion for specifying the circulation around a closed contour at a large distance from an aerofoil.

The criterion must be satisfied if a satisfactory analytical study of the uniform flow of a fluid past a two-dimensional aerofoil is to be made. The hypothesis states that, in the irrotational flow of a uniform stream of inviscid fluid past an aerofoil with a cusped or wedge-shaped trailing edge, the circulation is chosen so that the fluid speed at the trailing edge is finite.

J Thewlis, ed., *Encyclopaedic Dictionary of Physics* (New York, Oxford and London, 1962)

MS

L

labour force participation *Economics* The ratio of the population in the labour force (employed, self-employed or unemployed and normally above 16 years of age) to total population.

This is an important area of study as it may be possible to explain why certain groups are employed in specific sectors. One of the most significant changes has been the growth of women's participation in the workforce (determined as over 15 years of age), which rose from 40 per cent of the total workforce in the mid-1960s, to 58 per cent in 1990 (Samuelson and Nordhaus). Wage-rates, changing attitudes to working women, education, later marriages, and lower birth-rates have all affected the level of female labour force participation. *See* DUAL LABOUR MARKET THEORY.

P Samuelson and W D Nordhaus, *Economics* (Mcgraw-Hill, 1992), 234; W G Brown and T A Finegan, *The Economics of Labor Force Participation* (Princeton, N.J., 1969)

PH

labour market discrimination *Economics* Discrimination in the labour market may take the form of different wage rates for equally productive workers with different personal characteristics (such as race, sex, age, religion, nationality, or education). It may also take the form of exclusion from jobs on the grounds of social class, union membership, or political beliefs.

American economist Gary Becker (1930–) has developed a theory of discrimination which examines the behaviour of employers who have displayed a 'taste' for discriminatory practices. *See* DUAL LABOUR MARKET THEORY, CROWDING HYPOTHESIS, SEGMENTED LABOUR MARKET THEORY, NON-COMPETING GROUPS, SEARCH THEORY and INSIDER-OUTSIDER WAGE DETERMINATION.

G Becker, *The Economics of Discrimination* (Chicago, 1957)

PH

labour theory of value (4th century BC–) *Economics* With roots in the work of the Greek philosopher Aristotle (384–322 BC), this theory became a central feature in analyses by such classical economists as the Scot Adam Smith (1723–90) and the Englishman David Ricardo (1772–1823). They stated that the value of a commodity was determined by the quantity of labour needed to produce it, the effort of the labour, or the amount of labour of others obtained in exchange.

The German theorist Karl Marx (1818–83) argued that labour might dictate the value of a good but the existence of capitalists extracting profits meant that labour did not get to keep all the value. The theory was superseded by the MARGINAL PRODUCTIVITY THEORY OF DISTRIBUTION at the end of the 19th century, which emphasized that many factors determined the value of a good. *See* MARGINAL UTILITY THEORY.

R Meek, *Studies in the Labour Theory of Value* (London, 1973)

PH

Ladd-Franklin theory of colour vision *Psychology* Proposed by the American C Ladd-Franklin (1847–1930).

The theory represents an attempt to reconcile the work of Franciscus Cornelis Donders (1818–89) and Hermann Ludwig Ferdinand von Helmholtz (1821-94) on colour vision. The theory assumes a system processing a complex photosensitive molecule which responds differently to different colours by releasing substances that stimulate different nerve endings. Dichromatic and achromatic vision are explained as having a less developed molecule. The theory provides a better account of colour blindness than does the YOUNG-HELMHOLTZ THEORY OF COLOUR VISION, but has been superseded. *See* DONDERS METHOD.

L M Hurvich, *Color vision* (Sunderland, Mass., 1981)

NS

Laffer curve (1980s) *Economics* Named after American economist Arthur B Laffer, who maintained that economic expansion could be achieved without government budget deficits.

The curve shows the tax-rate at which government tax revenue is maximized, after which it declines. It illustrates the relationship between average tax-rates and total tax revenue, and shows that above a certain average rate of tax, total tax revenue will fall. The curve implies that there is a maximum amount of tax that a government can raise, therefore there is a ceiling to the level of public goods which can be provided. This principle was also stated by French economist Jules Dupuit (1804–66) in 1844.

PH

Lagrange multipliers *Mathematics* If a point $\bar{x}$ minimizes an objective function $f(x)$ subject to constraints $h_j(x) = 0$ for $j = 1,\ldots,s$ then, as observed by J L Lagrange (1736–1813), under reasonable conditions there are multipliers $\lambda_1,\ldots,\lambda_s$ so that $\bar{x}$ is a stationary point of the *Lagrangian*

$$f + \sum_1^s \lambda_j h_j$$

(*see also* the KARUSH-KUHN-TUCKER THEOREM).

This fundamental idea of OPTIMIZATION THEORY provides the basis for most computational techniques for the problem. The multipliers may also be regarded as solutions of a dual problem (*see also* DUALITY THEORY OF LINEAR PROGRAMMING and FENCHEL DUALITY THEOREM).

AL

Lagrange's theorem (1770) *Mathematics* Named after French mathematician Joseph Louis Lagrange (1736–1813), this is the assertion that every positive integer is expressible as the sum of at most four square numbers. For example, $7 = 4 + 1 + 1 + 1$.

This theorem was explicitly stated by French scientist Claude Gaspard de Bachet (1591–1639) in 1621 and was probably known to Greek mathematician Diophantus of Alexandria (*fl.* AD 250) in a different form. The French mathematician Pierre de Fermat (1601–65) claimed to have a proof, but died before disclosing it. Lagrange's proof built on earlier work of Swiss-born mathematician Leonhard Euler (1707–83). *See also* WARING'S PROBLEM.

G H Hardy and E M Wright, *An Introduction to the Theory of Numbers*, 5th edn (Oxford, 1979)

MB

laissez-faire (18th century) *Economics* A doctrine proposed by the Physiocrats in France, whose principle of '*Laissez-faire, laissez-passer*' (literally, allow to act, allow to pass) was later adopted by such classical economists as the Scot, Adam Smith (1723–90). This approach advocates non-intervention or minimum intervention by government in the economic affairs of a country. *See* PHYSIOCRACY, MERCANTILISM, ECONOMIC LIBERALISM and NEW CLASSICAL MACROECONOMICS.

J Viner, 'The Intellectual History of Laissez-faire', *Journal of Law and Economics*, vol. III (1960), 45–69

PH

Lamarckian evolution *Biology See* LAMARCKISM.

Lamarckism (1801) *Biology* Also called autogenesis. Named after Jean Baptiste Pierre Antoine de Monet Lamarck (1744–1829), a French naturalist noted for his study and classification of invertebrates, this is a theory of evolution based on the inheritance of ACQUIRED CHARACTERISTICS and, especially, acquired habits.

Lamarck believed that if part of an animal's body is subjected to repeated environmental stress, that part will be modified and the change will be handed down to the individual's offspring. Although superseded by Darwin's theory of EVOLUTION, Lamarckism was an important predecessor to it. *See also* BALDWIN EFFECT, DARWINISM, GERM PLASM THEORY and LYSENKOISM.

J B Lamarck, *Zoological Philosophy* (1809; rep. Chicago, 1984)

KH

Lambert's law of absorption (1760) *Chemistry* A law named after the German mathematician Johann Heinrich Lambert (1728–77). It is also known as the Bouguer law and the Lambert-Bouguer law after the French physicist Pierre Bouguer (1698–1758) who also discovered it.

The law states that the intensity I of radiant energy passing through an homogeneous absorbing medium decreases exponentially according to the equation

$$I = I_0 \exp(-ax)$$

where I_0 is the incident intensity, a the absorption coefficient, and x the thickness of the medium. If the concentration of the absorbing molecules is taken into account, the resulting expression is known as the BEER-LAMBERT LAW.

GD

Lambert's law of emission (1760) *Physics* Named after the German mathematician Johann Heinrich Lambert (1728–77). (It is also known as the Lambert cosine law and the cosine law of emission.) The law states that the intensity (flux per unit solid angle) emitted in any direction from an element of a perfectly diffuse radiating surface is proportional to the cosine of the angle between the direction of radiation and the normal to the surface. Such a surface will appear equally bright from all directions.

GD

Lamy's theorem (17th century) *Physics* Named after French mathematician Bernard Lamy (1640–1715), this theory proposes that if a particle is in equilibrium under the action of three forces P, Q and R, then

$$\frac{P}{\sin a} = \frac{Q}{\sin b} = \frac{R}{\sin c}$$

where a is the angle between Q and R, b is the angle between R and P, and c is the angle between P and Q.

J Thewlis, ed., *Encyclopaedic Dictionary of Physics* (New York, Oxford and London, 1962)

MS

land art (1968) *Art* First conceived at the exhibition at the Dwan Gallery (1968), this movement is related to ENVIRONMENT ART and EARTH WORKS. It rejects modern commercialization, embracing instead ecological issues through the creation of art in and with nature.

Typical of this conceptual movement are the geometric arrangements in the landscape by British artist Richard Long (1945–) (for instance, *A Line made by Walking*, 1967), recorded in the museum space through photographic documentation, maps and site material.

AB

Land effect *Psychology* Discovered by the American inventor and physicist Edwin Herbert Land (1909–).

The effect produces the perception of colour from black and white photographs. In the simplest form, two black and white photographs are taken of a scene, one through a red filter and one through a blue-green filter. The two are projected together onto a screen, the former through a red filter and the latter through a green one. The resulting scene is perceived as being composed of a wide variety of colours, including blues that are not normally produced by a mixture of red and green.

NS

Landé's interval rule (1911) *Physics* Named after the physicist Alfred Landé (1888–1975).

The rule states that for a weak spin-orbit interaction in Russell-Saunders coupling, an energy level with definite spin, angular momentum and orbital angular momentum is split so that the levels have different total angular momentum; and the interval between successive levels is proportional to the larger of their total angular momentum values.

GD

Landé permanence rule *Chemistry* Named after the German physicist Alfred Landé (1888–1975), this rule is also known as the gamma-permanence rule. It states that for a series of states having the same spin and orbital angular momentum quantum numbers (or the same total angular momentum quantum numbers for individual electrons) but different total angular momenta, and having the same total magnetic quantum number, the sum of the energy level shifts produced by the spin-orbit interaction is independent of the strength of an applied magnetic field.

GD

Lane's law (19th century) *Astronomy* Named after its originator, the American physicist Jonathan Homer Lane (1819–80), this states that when a star contracts its internal temperature rises.

MS

Langevin's theory of diamagnetism (1905) *Physics* Named after the French physicist Paul Langevin (1872–1946) who proposed it, this theory examines the behaviour of an electron moving in a circular orbit when a magnetic field is established perpendicular to the plane of the orbit.

Unless the force law is the inverse cube, the radius of the orbit remains unchanged but the speed increases, giving a small change in the magnetic moment. The change in the angular velocity of the electron that occurs is always such as to produce a negative susceptibility; the atom with its circulating electron acts like an inductance in which an opposing e.m.f. is produced.

J Thewlis, ed., *Encyclopaedic Dictionary of Physics* (New York, Oxford and London, 1962)

MS

Langmuir isotherm (1916) *Chemistry* Named after the American chemist Irving Langmuir (1881–1957), for many years associate director of the General Electric Company research laboratory and winner of the Nobel prize for Chemistry in 1932.

For a solid surface, the fraction of active sites (θ) occupied by a monolayer of adsorbed gas molecules is given by the Langmuir adsorption isotherm

$$\theta = \frac{kP}{1 + kP}$$

where P is the pressure and k is a constant with different values as a function of temperature. The isotherm is of limited validity, usually as a result of surface inhomogeneity and interactions between adsorbed species.

P W Atkins, *Physical Chemistry* (Oxford, 1990)

GD

Langmuir's law *Physics See* CHILD'S LAW.

Langmuir's theory (1919) *Chemistry* A theory named after the American chemist Irving Langmuir (1881–1957) which supposed that electrons occupied imaginary shells surrounding an atom in accordance with the periods of the periodic system, and that the effect of primary valencies was supplanted by secondary valencies in the solid state. An extension of this theory is known as the Lewis-Langmuir theory or OCTET RULE.

GD

language and ideology *Linguistics See* SOCIAL SEMIOTIC, LANGUAGE AS.

language games (1795) *Stylistics* Idea of art as play proposed by Johann Christoph Friedrich von Schiller (1759–1805), *Letters on the Aesthetic Education of Man*: expression of *Spieltrieb* ('play-drive').

Literary MODERNISM and STRUCTURALISM have drawn attention to the extent and prevalence of language games (AMBIGUITY, PARADOX, puns, and so on) in the stylistic texture of literary works. Joyce's *Finnegans Wake* (1939) is the most striking and sustained example.

P Hutchinson, *Games Authors Play* (London, 1983)

RF

language of men (1800) *Literary Theory* Stylistic revolution proclaimed by the English poet William Wordsworth (1770–1850).

Wordsworth attacked 18th century POETIC DICTION as 'gaudy and inane phraseology'; he preferred 'language really used by men' as signifying more genuine and universal emotions. An important signpost toward modern poetry.

W Wordsworth, 'Preface to the Second Edition of *Lyrical Ballads*', *Critical Theory since Plato*, H Adams, ed., (New York, 1971), 433–43

RF

language of thought *Philosophy* Theory developed by J A Fodor, though going back to the English philosopher William of Ockham (*c*.1285–1349). It seeks to explain thinking by postulating a hypothetical language of thought (or mentalese) such that to have a belief or desire and so on is to be related in certain ways to one or more sentences of this language.

There are difficulties in spelling out these relations, and in saying how the items in mentalese – whatever form they may take – relate to the outer world which is being thought about. We must not rely on analogies with ordinary languages, since these presuppose thinking while the language of thought is supposed to explain it. Its reliance on discrete items ('words') to correspond to the various bits of our thinking contrasts it with CONNECTIONISM.

J A Fodor, *The Language of Thought* (1975)

ARL

language pathology (19th century–) *Medicine/Linguistics* General term for a broad field which contains several sub-areas and approaches; for example, aphasiology, CLINICAL LINGUISTICS, NEUROLONGUISTICS and speech therapy.

The study of the causes and characteristics of all kinds of speech and language disorders, and their remediation or mitigation by medical or therapeutic treatment of individuals. *See also* APHASIA and references.

D Crystal, *Introduction to Language Pathology*, 2nd edn (London, 1988)

RF

language typology (1928) *Linguistics* Principles articulated by the PRAGUE SCHOOL of linguists.

COMPARATIVE LINGUISTICS relates languages which are historically linked; language typology categorizes languages according to broad structural similarities and differences, without reference to genetic factors: 'tone languages', 'inflectional languages', and so on. *See also* LINGUISTIC UNIVERSALS.

J H Greenberg, ed., *Language Typology* (The Hague, 1974)

RF

langue and *parole* (1916) *Linguistics* One of the fundamental distinctions made by the French linguist Ferdinand de Saussure (1857–1913) which have had a lasting effect on modern linguistics.

Langue is the abstract system of language shared by people in a group, the 'language system' which is the linguist's basic object of description. *Parole* is the individual, concrete, act of speech. *See also* LINGUISTIC COMPETENCE.

F de Saussure, *Course in General Linguistics*, W Baskin, trans. (Glasgow, 1974 [1916])

RF

large deviation theory (20th century) *Mathematics* Statistical methods based on the speed with which the probability that a statistic T_n is greater than c_n where T_n is a function of n random variables and c_n tends to infinity.

JB

large numbers, law of (1713) *Mathematics/Economics* This theorem was proved by Swiss mathematician Jakob Bernoulli (1654–1705). This is the fundamental principle of STATISTICS that the sequence x_n/n tends to p where the random variables x_n have common mean p. This implies that the relative frequency of an event of probability p tends to p as the number of trials tends to infinity.

The weak law of large numbers asserts that the limit holds in measure by use of the weak convergence defined by Ernst Fischer (1875–1959); and the strong law of large numbers asserts that the limit holds pointwise by use of the strong convergence defined by Friedrich Riesz (1880–1956). These are correct versions to replace the law of averages; the erroneous idea that after repetitions of one outcome the others become more likely.

T W Anderson, *An Introduction to Multivariate Statistic Analysis* (New York, 1972)

JB

Larmor's theorem *Physics* Named after its discoverer, the Irish physicist Sir Joseph Larmor (1857–1942).

An accelerated charge radiates electromagnetic energy. Larmor showed that the rate of energy radiation $\mathrm{d}E/\mathrm{d}t$ from an accelerated charge momentarily at rest, or moving slowly relative to an observer, is given by

$$\frac{\mathrm{d}E}{\mathrm{d}t}=\frac{q^2a^2}{6\pi\epsilon_0c^3}$$

where q is the magnitude of the charge and a its acceleration, c is the speed of light in a vacuum and ϵ_0 is the permittivity of free space.

J Thewlis, ed., *Encyclopaedic Dictionary of Physics* (New York, Oxford and London, 1962)

MS

latent learning *Psychology* Proposed by the American psychologist Edward Chase Tolman (1886–1959).

One of Tolman's theories of learning which states that learning can occur in the absence of a foreseen goal or reward. It is also referred to as incidental learning. It is not directly observable but with the later introduction of a goal the evidence of latent learning becomes visible. The theory is criticized because it is always possible to identify goals retrospectively, but not prospectively. *See* EXPECTANCY THEORY OF LEARNING.

B Schwartz, *Psychology of Learning and Behaviour* (New York, 1989)

NS

law of constant extinction *Biology See* RED QUEEN HYPOTHESIS.

law of corresponding stages *Biology See* VON BAER'S LAW.

law of the minimum *Biology See* LEIBIG'S LAW.

law of mutual aid *Biology See* ALTRUISM.

law of recapitulation *Biology See* RECAPITULATION.

lawyer paradox (5th century BC) *Philosophy* Ascribed to the sophist philosopher Protagoras (*c.*490–420 BC).

A lawyer teaches law to a student without fee on condition that the student will pay him when he qualifies and wins his first case. However, when the student qualifies he takes up another profession. The lawyer sues him for his fees, on the grounds that if he wins, he is paid and if he loses, the student has won and so must pay by the agreement. The student is unperturbed because if he wins he need not pay the fees, and if he loses he does not owe them. There is some confusion concerning the agreement here, but logical rules preventing the application of a condition to itself certainly resolve the paradox.

JB

Lazarus-Schachter theory of emotion *Psychology See* ATTRIBUTION THEORY OF EMOTION.

L-cubed algorithm (1982) *Mathematics* Also called the 3-L algorithm after the mathematicians A K Lenstra, H W Lenstra (jnr) and L Lovász, this is an algorithm which has many applications in NUMBER THEORY. In particular, it can be used to factor a polynomial having rational coefficients into irreducible factors within a realistic time frame.

E Hlawka *et al.*, *Geometric and Analytic Number Theory* (Springer, 1991)

MB

Le Châtelier's principle (1888) *Chemistry* Named after the French chemist Henry Le Châtelier (1850–1936), but sometimes known as the Le Châtelier-Braun principle. In general terms, the principle states that if a stress or force is applied to a system at equilibrium, the system responds by displacing the equilibrium in a direction which tends to diminish the effect of the stress. The principle suffers from a number of important exceptions and is best replaced by the VAN'T HOFF PRINCIPLE.

P W Atkins, *Physical Chemistry* (Oxford, 1990)

GD

Le Châtelier principle (1947) *Economics* Named after French chemist Henry le Châtelier (1850–1936) by American economist Paul Samuelson (1915–), this principle deals with constraints on maximizing behaviour, explaining that short-run demands have lower elasticity than those in the long run since a longer time frame allows new factors and prices to change. (Le Châtelier had earlier formulated a reaction law governing the effects on equilibrium of pressure and temperature.)

P Samuelson, *Foundations of Economic Analysis* (Cambridge, Mass., 1947)

PH

leadership theories (1532) *Psychology* The earliest study of leadership regarded as distinctively psychological has been attributed to Niccolò di Bernardo dei Machiavelli (1469–1527) in his book *De Principatibus* or *Il Principe* ('The Prince').

Leadership refers to the management and direction of a group of people or an organization. Theories on leadership vary in their perspectives and emphasis. Some emphasize the notion of specific leadership traits while others suggest that different leaders are required for different situations. *See also* CONTINGENCY THEORY OF LEADERSHIP, FIEDLER'S CONTINGENCY MODEL OF LEADERSHIP, MCGREGOR'S X/Y THEORY OF LEADERSHIP, NORMATIVE DECISION THEORY OF LEADERSHIP and PATH-GOAL THEORY OF LEADERSHIP.

B M Bass, *Stogdill's Handbook of Leadership* (New York, 1981)

NS

learned helplessness (1970s) *Psychology* A term coined by the American M E P Seligman (1942–).

It refers to the way in which an organism acts when exposed to unavoidable situations which are harmful, distainful or painful. The effect is to prevent or retard learning in subsequent situations in which escape or avoidance is possible. It has been observed in humans and animals alike.

M E P Seligman, *Helplessness: On Depression, Development and Death* (San Francisco, 1975)

NS

learning-by-doing (1962) *Economics* This term refers to the hypothesis that labour learns through experience in the production process, thereby allowing economies of scale in future output. The increase in productivity diminishes over time. The first theoretical model of this kind was constructed by American economist Kenneth Arrow (1921–), but many empirical studies had been carried out in the early part of the 20th century.

K J Arrow, 'The Economic Implications of Learning by Doing', *Review of Economic Studies*, vol. XXIX (1962), 155–73

PH

learning theory (1940s–1950s) *Biology/ Psychology* Also referred to as the universal law of learning, this was proposed and refined by various animal behaviourists, especially Burrhus Frederic Skinner (1904–90), but also Hul, Tolman and Guthrie, with early evidence being reported by American psychologist Edward Lee Thorndike. A reference to the group of similar theories that seek to explain how animals learn.

In general, learning theory holds that learning is an evolved trait and that all animals learn in essentially the same way, so that principles discovered in one species in a particular situation can be generalized to all types of learning. This universal (or equipotential) concept of learning has yet to be proved. *See also* BEHAVIOURISM, EFFECT, LAW OF and HABITATION THEORY OF LEARNING.

D Dewsbury, *Comparative Animal Behavior* (New York, 1978)

KH

least action, principle of (1860) *Physics* A principle which states that for a system whose total dynamical energy is conserved, the trajectory of the system in configuration space is that path which has a stationary value of the action as compared to other paths between the same points for which the energy has the same constant value. *See also* HAMILTON'S PRINCIPLE OF LEAST ACTION.

GD

least cost location theory (1826) *Economics* Part of general location theory, pioneered by Prussian landlord Johann von Thünen (1783–1850) and later by German economist and sociologist Alfred Weber (1868–1958); this posits that agriculture and industry locate their activities as close to their markets as possible, thereby achieving the least cost of transport for the goods they produce. Von Thünen's hierarchy of activities took the shape of concentric rings around urban centres. *See* WEBER'S THEORY OF THE LOCATION OF THE FIRM, CENTRAL PLACE THEORY and GRAVITY MODEL.

P Hall, ed., *Von Thünen's Isolated State* (Oxford, 1966)

PH

least effort principle *Psychology* Proposed by the American psychologist G K Zipf (1902–50).

In theories of psychology, this principle states that given certain possibilities for action an organism will select the one requiring the least effort. Used in explanations of how rats learn mazes and children develop articulation skills. *See* ZIPF'S LAW.

G K Zipf, *Human Behavior and the Principle of the Least Effort* (Cambridge, Mass., 1949)

NS

least time, Fermat's principle of *Physics See* FERMAT'S PRINCIPLE OF LEAST TIME.

least work principle *Geology* This principle states that geomorphological processes operate in ways by which the least work is involved. A river, for example, will flow in a way whereby it expends least energy. This is directly analogous to the physical LAW OF MAXIMUM ENTROPY.

DT

least work, principle of *Physics* The deflections of individual parts of an elastically deformed structure subjected to applied loads are such that the load will be carried with the minimum storage of energy in the elastic members.

MS

Lebesgue's theorem *See* DOMINATED CONVERGENCE THEOREM.

Leffler's assumption *Chemistry See* HAMMOND PRINCIPLE.

left-hand rule *Physics See* FLEMING'S RULES.

legal positivism *Philosophy* Doctrine (or set of doctrines) stemming primarily from English jurist John Austin (1790–1859) in his *The Province of Jurisprudence Determined* (1832). It emphasizes what the law actually is rather than what it should be: it cannot, like natural law, be defined by reference to its content, but is what is commanded by the sovereign.

H L A Hart, *The Concept of Law* (1961); see especially p. 253

ARL

legal positivism *Politics/Law* Application of theory of POSITIVISM to law.

Law consists simply of the enforceable commands of government. It does not depend for its validity on any other criteria, and there are no religious or normative values by which it can be deemed invalid.

David Miller *et al.*, eds, *The Blackwell Encyclopaedia of Political Thought* (Oxford, 1987)

RB

legitimacy (20th century) *Politics/Philosophy/History* The view that systems of government either are or ought to be justified, and not simply based on coercion.

There are two versions of the theory of legitimacy, one deriving from political philosophy, the other from history and political science. The first seeks for principles which would oblige people to obey government, and then uses those principles to assess existing regimes as worthy or otherwise of being obeyed. The second treats a belief in the legitimacy of regimes as a common feature of government, however distasteful any particular regime may be to the observer. It then examines legitimacy as an historical phenomenon rather than engaging in moral appraisal. The two approaches are often thought to be incompatible, but are in fact complementary. *See also* POLITICAL OBLIGATION.

Rodney Barker, *Political Legitimacy and the State* (Oxford, 1990)

RB

legitimacy crisis (20th century) *Politics* Mixture of theories of LEGITIMACY and of CRISIS THEORY, associated in particular with the German social scientist Jurgen Habermas (1929–).

Modern capitalist societies are undergoing a legitimacy crisis whereby support for both government and economy is systematically eroded. Habermas's account of the impending legitimation or legitimacy crisis includes all forms of social relationships, both political and economics, and argues that the failure of existing institutions to meet the ethical criteria which would justify their acceptance is leading to a general crisis.

Rodney Barker, *Political Legitimacy and the State* (Oxford, 1990)

RB

Lehmann-Scheffé theorem (20th century) *Mathematics* Named after the statistician Lehmann and Scheffé, this states that an unbiased estimator that is a function of a complete sufficient statistic is a unique uniformly minimum variance unbiased estimator.

R V Hogg and A T Craig, *Introduction to Mathematical Statistics* (New York, 1967)

ML

Leibniz's alternating series test *Mathematics* *See* ALTERNATING SERIES TEST.

Leibniz's law *Philosophy* Name often given to either or both of the IDENTITY OF INDISCERNIBLES and the INDISCERNIBILITY OF IDENTICALS; called after German philosopher and mathematician Gottfried Wilhelm Leibniz (1646–1716). Leibniz himself seems to have held explicitly only the first, and to have treated it sometimes as necessary and sometimes as contingent.

H G Alexander, ed., *The Leibniz-Clarke Correspondence* (1956); see Introduction, pp. xxiii–iv for references

ARL

Leibniz's theorem (17th century) *Mathematics* Named after German philosopher, logician and mathematician Gottfried Wilhelm Leibniz (1646–1717), this states that the nth derivative of the product of two functions f and g, $[f \times g]^{(n)}$ is the binomial series

$$\sum_{k=0}^{n} \binom{n}{k} f^{(k)} g^{(n-k)}.$$

H Anton, *Calculus with Analytic Geometry* (New York, 1980)

ML

leniency effect *Psychology* Proposed by the Austrian-American psychologist F Heider (1896–).

As implied by its name, a judgmental error particularly likely to occur in personality assessment where known or sympathetic individuals are assessed more favourably then less familiar or less sympathetic individuals.

L J Cronbach, *Essentials of Psychological Testing* (New York, 1960)

NS

Leninism (20th century) *Politics* Theory of politics, SOCIALISM, and REVOLUTION of the Russian revolutionary V I Lenin (1870–1924).

States are tailor-made for the societies they govern. Socialists cannot therefore adapt a capitalist STATE to socialist or communist purposes, but must replace it with a new structure, the 'DICTATORSHIP OF THE PROLETARIAT'. In working for this end, the working class might not be sufficiently aware of their own true or best INTERESTS, and would therefore be led by a 'vanguard party'. Leninism has become a derogatory term to indicate élitist or undemocratic forms of SOCIALISM.

David Miller *et al.*, eds, *The Blackwell Encyclopaedia of Political Thought* (Oxford, 1987)

RB

Lenz's law (1834) *Physics* Named after its discoverer, the German physicist Heinrich Friedrich Emil Lenz (1804–65), this law states that any current set up by electromagnetic induction flows in such a direction as to oppose the change in magnetic flux responsible for the induction.

J Thewlis, ed., *Encyclopaedic Dictionary of Physics* (New York, Oxford and London, 1962)

MS

Leontief paradox (1953) *Economics* Russian-born economist Wassily Leontief (1906–) devised this contradiction of the HECKSHER-OHLIN TRADE THEORY.

Trade is determined by the relative abundance of factors of production in each economy. Leontief discovered that despite the USA being endowed with an abundance of capital, its exports were labour intensive and imports capital intensive. *See* FACTOR-PRICE EQUALIZATION THEOREM and RYBCZYNSKI THEOREM.

W W Leontief, 'Domestic Production and Foreign Trade: The American Capital Position Re-examined', *Proceedings of the American Philosophical Society*, vol. XCVII (September, 1953), 332–49

PH

lettrism (1944) *Art* A Paris-based movement founded by Romanian-French poet Isidore Isou (1925–), who proposed incorporating letters, numerals and non-Western calligraphy into painting; fusing art with poetry to create a music of letters. Other members included G Pomerand, M Lemaitre and R Sabatier. *See* TACHISM.

LL

levels of information processing theory *Psychology* Proposed by the Scottish psychologist Fergus I M Craik (1935–).

A theory of memory which, in contrast to the DUPLEX THEORY OF MEMORY, suggests that there may be a single system of memory with variations in levels of processing. Deep processing of some information requires greater interpretation, analysis and evaluation, leading to better and longer memory for that information. The theory is important in the way in which it emphasizes depth of processing. It cannot account for a variety of phenomena which can be successfully explained by duplex theory.

F I M Craik and R S Lockhart, 'Levels of Processing: A Framework for Memory Research', *Journal of Verbal Learning and Verbal Behavior*, vol. XI (1972) 671–84

NS

Levi's theorem *Mathematics* (1) Another name for the DOMINATED CONVERGENCE THEOREM. (2) The theorem that if f is a Lebesgue integrable function, then

$$\lim_{h\to\infty} \int_0^h |f(x+t) - f(x)| \, dt$$

is zero almost everywhere. *See* MONOTONE CONVERGENCE THEOREM.

G B Folland, *Real Analysis* (New York, 1984)

Lewis-Langmuir theory *Chemistry See* OCTET RULE

Lewis's colour theory (1916) *Chemistry* Named after the American chemist Gilbert Newton Lewis (1875–1946) who was one of the 20th century's greatest contributors to chemical bonding theory. This theory proposed that colour is due to the presence of odd electrons which can absorb those light rays which vibrate with the same frequency.

The theory has been superseded by the more rigorous explanations of the MOLECULAR ORBITAL THEORY and VALENCE BOND THEORY. *See also* OCTET RULE

GD

Lewis's theory of acids and bases (1923 and 1938) *Chemistry* Named after the American chemist Gilbert Newton Lewis (1875–1946). The theory defines an acid as a species which can accept an electron pair from another atom and a base as a species which can donate an electron pair to complete the valence shell of another atom.

Lewis acids include molecules such as BF_3 and $AlCl_3$ which can react with ammonia, for example, to form an addition compound or Lewis salt. Some species are bases in both the Lewis and BRØNSTED-LOWRY senses while all Brønsted-Lowry acids are also Lewis acids.

M Freemantle, *Chemistry in Action* (Basingstoke, 1987)

GD

lexical-functional grammar (LFG) (late 1970s) *Linguistics* Alternative to GOVERNMENT AND BINDING developed by the American linguist J W Bresnan and colleagues.

A grammatical model which abandons transformations and incorporates in the LEXICON information about functional relationships. The model preserves Chomsky's early claim (on which GENERALIZED PHRASE STRUCTURE GRAMMAR is agnostic) of the psychological reality of syntactic representations.

J W Bresnan, *The Mental Representation of Grammatical Relations* (Cambridge, Mass., 1982)

RF

lexicon (1960s) *Linguistics* Component of a GRAMMAR which specifies the semantic, syntactic and phonological properties of individual words. *See also* MENTAL LEXICON.

A Radford, *Transformational Grammar* (Cambridge, 1988), ch. 7

RF

lexicostatistics (1950s) *Linguistics* Techniques developed particularly in the 1950s, but with antecedents in anthropology of the 1930s. Statistical studies of vocabulary directed to various purposes, including GLOTTOCHRONOLOGY and other historical and comparative studies of the grouping and relationships of languages, studies of literary authorship, and so on.

I Dyen, *Linguistic Subgrouping and Lexicostatistics* (1975)

RF

lexie (1970) *Stylistics* Term coined by the French structuralist Roland Barthes (1915–80).

Lexies are fragments of a text, usually quite short, on which attention dwells and commentary focuses one by one in a critical reading. Not tied to specific linguistic units such as phrase, sentence, paragraph.

R Barthes, *S/Z*, trans. R Miller (London, 1975)

RF

lexis (1960s) *Linguistics* Vocabulary, particularly as studied in SYSTEMIC GRAMMAR; *see also* COLLOCATION.

C E Bazell and J C Catford and M A K Halliday and R H Robins, eds *In Memory of J R Firth* (London, 1966)

RF

l'Hôpital's rule (or l'Hospital's rule) (18th century) *Mathematics* Discovered by Swiss mathematician Jakob Bernoulli (1667–1748), but more generally associated with his student the French analyst and geometer Guillaume François Antoine de l'Hôpital, Marquis de St Mesmé (1661–1704) who wrote one of the first texts of calculus.

The rule states that the limit of an indeterminate quotient (that is, the quotient of two functions whose limits are both zero or both infinity) is the limit of the quotient of the derivatives of the two functions. For example,

$$\lim_{x \to 0} \frac{\sin x}{x}$$

is an indeterminate of the form 0/0, but it is equal to

$$\lim_{x \to 0} \frac{\cos x}{1} = 1.$$

H Anton, *Calculus with Analytic Geometry* (New York, 1980)

ML

Liapunov convexity theorem (1940) *Mathematics* The range of an n-dimensional vector measure is a closed, bounded subset of n-dimensional Euclidean space, and is convex if the measure is non-atomic.

For example, for any integrable functions $a_1,\ldots,a_n$ and measurable set D in $\mathbb{R}$ there is a measurable subset E of D with

$$\int_E a_i(s)\,ds = (1/2)\int_D a_i(s)\,ds$$

This result allows the substantial reduction of many problems in CONTROL THEORY. *See also* BANG-BANG PRINCIPLE.

AL

liberal democracy (19th century–) *Politics* Theory of limited government and individual rights under democracy.

In order for democracy to be effective or meaningful as 'rule by the people', there must be constitutional limits on government and constitutional guarantees of the civil and political rights of citizens. This will ensure, or at least encourage, freedom of expression, opinion, and publication, and the free, frequent and informed elections which are necessary for democracy to be other than a formal title.

David Held, *Models of Democracy* (Oxford, 1987)

RB

liberal feminism *Politics/Feminism* Individualist rights-based feminist theory.

Women suffer principally because of the denial of their rights as individuals. The solution to their unequal treatment is the creation of true equality of rights, which is likely to be achieved by rational persuasion.

Z R Eisenstein, ed., *The Radical Future of Liberal Feminism* (New York, 1981)

HR

liberalism *Politics* A broad body of political theory based on the significance of the individual.

Individuals are to enjoy liberty, including the liberty to own and produce wealth, to conduct themselves in private as they please, and to associate with others and publish and discuss opinions in public as freely as is consistent with the avoidance of harm to others. Government's principal responsibility is to safeguard this. Government is therefore a secondary and artificial activity, but a necessary one.

David Miller *et al.*, eds, *The Blackwell Encyclopaedia of Political Thought* (Oxford, 1987)

RB

libertarianism *Politics/Philosophy* In political theory, an extreme form of LIBERALISM. Individuals are free to pursue their own INTERESTS unqualified by any conception of public interest or public duty. The individual is the best and only judge of his or her own interests, and government and law should do no more than provide a minimal framework of order in which these interests can be pursued.

In philosophy (where it is often a view in the philosophy of mind or action) it is a claim that DETERMINISM is false for human actions, and that something more than mere INDETERMINISM is needed. This something may take the form of claiming that there is a special entity, the 'self', which is itself immune to causal influence, or at least to compulsion, and can intervene from the outside, as it were, in the causal chain of events. Chisholm distinguishes in this context between *immanent causation* (by agents) and *transeunt causation* (by events).

David Miller *et al.*, eds, *The Blackwell Encyclopaedia of Political Thought* (Oxford, 1987); R Chisholm, 'Freedom and Action', *Freedom and Determinism*, K Lehrer, ed. (1966)

RB, ARL

Liebig's law (1841) *Biology* Also called law of the minimum, this was proposed by the German theorist Justus von Liebig (1803–73) who greatly advanced the techniques of organic chemistry and is regarded as the founder of agricultural chemistry. The theory that the growth of a plant is limited by the nutrient present in the least quantity, assuming that all other essential nutrients are available in sufficient amounts.

For example, if crop yield is limited by insufficient nitrogen, then adding more nitrogen will enhance plant growth until growth is limited by the insufficiency of some other substance, such as phosphorus. In the soil, Liebig recognized nitrogen (N), phosphorus (P) and potassium (K) as minimum whereas phosphorus, nitrogen and silicon (Si) are minimum in the sea.

Although Liebig's law holds under some highly controlled, experimental conditions, it rarely if ever functions as predicted in the natural world. *See also* BLACKMAN CURVE. *Compare with* TOLERANCE, LAW OF.

J von Liebig, *Organic Chemistry in Its Applications to Agriculture and Physiology* (1841); F Salisbury, *Plant Physiology* (Belmont, Calif., 1992)

KH, GD

life-cycle hypothesis (1957) *Economics* Comprising the analysis of individual consumption patterns, this theory was developed by American economist Irving Fisher (1867-1947) and English economist Roy Harrod (1900-78), before later being extended by Japanese economist Albert Ando (1929–) and Italian-born economist Franco Modigliani (1918–).

The model assumes that individuals consume a constant percentage of the present value of their life income. This is dictated by preferences and tastes and income. Ando and Modigliani argued that the average propensity to consume is higher in young and old households, whose members are either borrowing against future income or running down life-savings. Middle-aged people tend to have higher incomes with lower propensities to consume and higher propensities to save. *See* FORCED SAVING.

A Ando and F Modigliani, 'Tests of the Life Cycle Hypothesis of Saving: Comments and Suggestions', *Oxford Institute of Statistics Bulletin*, vol. XIX (May, 1957), 99–124

PH

life style theory *Psychology* Not attributable to any one originator, this is a general approach which contends that a person's past decisions and experiences can be continually reviewed, modified, and/or changed. By exploring and assessing an individual's life style, it is possible to develop an understanding of that person as a self-consistent and self-directed entity. A central theme is that personal actions are forward-oriented, purposive, and determined by individual values; rather than simple physiological responses to the environment.

L Banith and D Eckstein, *Life Style: Theory, Practice and Research* (Dublique, Iowa, 1981)

NS

ligand field theory *Chemistry* Incorporating elements from the MOLECULAR ORBITAL THEORY and VALENCE BOND THEORY, Ligand field theory is concerned with the effects of co-ordinating groups (ligands) on the inner orbitals of a central metal atom.

As the ligands are brought up to a charged transition-metal ion from a distance, their mutual electrostatic repulsions raise the energy of all five metal d-orbitals (*crystal field theory*). Then, because of their directional character, the d-orbitals are split in energy as the ligands approach to within bonding distance. The nature of the particular ligands determines the splitting for a given central metal ion. Ligand field theory has been used very successfully to interpret luminescence, magnetism and paramagnetic resonance as well as visible absorption spectra.

P W Atkins, *Physical Chemistry* (Oxford, 1990)

GD

light quantum theory (1905) *Physics* A model of light proposed by German-born mathematical physicist Albert Einstein (1879–1955).

Einstein developed the quantum ideas put forward by German theoretical physicist Max Karl Ernst Planck (1858–1947), and suggested that not only is electromagnetic radiation emitted in packets or quanta (called photons) but is transmitted and absorbed in such packets. Monochromatic radiation of frequency ν behaves as if it comprises mutually independent energy quanta of magnitude $h\nu$, where h is Planck's constant. Einstein used this idea in discussion of the photoelectric effect, in which he assumed that a photon transfers all its energy to a single electron and the energy transfer by one light quantum is independent of the presence of other light quanta. *See* PHOTOELECTRIC EFFECT.

J Thewlis, ed., *Encyclopaedic Dictionary of Physics* (New York, Oxford and London, 1962)

MS

likelihood ratio test *Mathematics* The statistical test in which the null hypothesis is rejected for small values of

$$\lambda = \frac{p_0}{p_1}$$

where p_i is the maximum of the probabilities $P[\mathbf{X}|\ \theta]$, where θ ranges over the possibilities permitted by the respective hypotheses, and $\mathbf{X}$ is a vector of observations. This is a useful technique in HYPOTHESIS TESTING. *See also* NEYMAN-PEARSON LEMMA.

ML

Likert scale *Psychology* Originated by the American psychologist R A Likert (1903–81).

A technique for measuring attitudes. Likerts's scale consists of a series of statements to which a respondant must answer. The Likert five-point scale is normally as follows; strongly agree, agree, neither agree nor disagree, disagree, and strongly disagree. The data gained through Likerts' scale is easily amenable to factor analysis and has proved this scaling preferable to Thurstone's. There are two modified versions of the scale, one for the illiterate and one with no neutral category. *See* THURSTONE'S LAW OF COMPARATIVE JUDGMENT.

R A Likert, 'Techniques for the Measurement of Attitudes', *Archives of Psychology*, vol. CXL (1932), 1–55

NS

limited independent variety, principle of *Philosophy* Principle adopted by English economist John Maynard Keynes (1883–1946) to underpin his Bayesian approach to induction by finding a justification for assigning the relevant probabilities.

The principle says that, for at least that sphere we are investigating, the number of objects and qualities it contains may be infinite, but the number of independent groups into which they fall is finite – or at least that there is a non-zero probability of this. One of the objections to the principle is that there is no adequate reason to think it true, and that even if true it will not help us to assign actual figures to the probabilities unless we know how many independent groups there are; otherwise we cannot know that any progress we make by the Bayesian procedure is more than infinitessimal.

J M Keynes, *A Treatise on Probability* (1921), ch. 22, 9

ARL

limiting factor *Biology See* BLACKMAN CURVE.

Lindahl equilibrium (1919) *Economics* Named after Swedish economist Erik Lindahl (1891–1960), this theorizes that the provision of public goods reaches an equilibrium when everyone agrees on the level of goods to be provided, and their prices.

Thus, a set of Lindahl prices comprises individual shares of the collective tax burden of an economy. The sum of Lindahl prices is equal to the cost of supplying public goods. *See* SCITOVSKY PARADOX, MARGINAL COST PRICING, SOCIAL WELFARE FUNCTION and COMPENSATION PRINCIPLE.

PH

Lindemann's theorem (1882) *Mathematics* Named after German mathematician Carl Louis Ferdinand von Lindemann (1852–1939), but is sometimes called the Lindemann-Weierstrass theorem after German mathematician Karl Theodor Wilhelm Weierstrass (1815–97) who made more rigorous the original ideas of Lindemann.

It is the result in NUMBER THEORY whereby for any distinct algebraic numbers $\alpha_1,..,\alpha_n$ and any non-zero algebraic numbers $\beta_1,..,\beta_n$ it is the case that

$$\beta_1 e^{\alpha_1} + \cdots + \beta_n e^{\alpha_n} \neq 0 .$$

It is an immediate consequence of this result and the identity $e^{i\pi} = -1$ that the number π is transcendental. This solves the ancient Greek problem of constructing with ruler and compasses only a square with area equal to that of a given circle; in particular, the length $\sqrt{\pi}$ being transcendental cannot be classically constructed and so the quadrature of the circle is impossible. Lindemann's theorem also includes the transcendence of e (proved earlier by Hermite) and of $\log\alpha$ for algebraic α not zero or one. *See also* BAKER'S THEOREM and GELFOND-SCHNEIDER THEOREM.

A Baker, *A Concise Introduction to the Theory of Numbers* (Cambridge, 1984)

MB

Lindemann's theory (1921) *Chemistry* Also known as the Lindemann-Hinshelwood theory after the German-born British physicist Frederick Alexander Lindemann, 1st Viscount Cherwell (1886–1957) and chemist Sir Cyril Norman Hinshelwood (1897–1967) who shared the Nobel prize for Chemistry in 1956. The theory was proposed to account for the observed first-order kinetics of unimolecular gas-phase reactions.

Lindemann suggested that a molecule gained sufficient energy to be transformed to product by colliding with another molecule. This energized molecule could either lose its energy in a further collision or be transformed by unimolecular decay. The theory has limited validity and has been largely superseded by RRKM THEORY.

P W Atkins, *Physical Chemistry* (Oxford, 1990)

GD

Lindemann's theory of melting (1910) *Physics* Named after its originator, the British scientist Frederick Alexander Lindemann, Viscount Cherwell (1886–1957), this model is similar to that of Hertzfeld and Mayer but uses a different criterion for the onset of melting.

Melting is assumed to occur when the amplitude of vibration of the atoms reaches a certain fraction of the equilibrium interatomic spacing. Lindemann's model assumes that the vibrations of the atoms are harmonic, with a common frequency, and reasonable agreement is obtained with experiment if the amplitude of vibration is taken to be about 1/7 of the equilibrium interatomic spacing. *See* HERTZFELD AND MAYER THEORY OF MELTING.

J Thewlis, ed., *Encyclopaedic Dictionary of Physics* (New York, Oxford and London, 1962)

MS

lineage theory (1920s) *Anthropology* Associated especially with British anthropologist Sir Edward Evans-Pritchard (1902–73).

The abstract study of the segmentation of clans and lineages in societies where such kin groups are the basis of social structure and the primary political units. The classic example is the Nuer of southern Sudan. However, recent critics have noted that hardly any societies, even in that part of

Africa, actually work in the abstract manner claimed by lineage theorists. *See also* ALLIANCE THEORY and KINSHIP.

A Kuper, 'Lineage theory: a critical retrospect', *Annual Review of Anthropology*, vol. XI (1982), 71–95

AB

linear algebra *Mathematics* The branch of ALGEBRA which is concerned with the study of vector spaces or more generally with modules over a *ring*. The subject arose from the study of matrices and systems of linear equations.

I N Herstein, *Topics in Algebra* (Wiley, 1975)

MB

linear programming (1947) *Economics* First used by American mathematician George Bernard Dantzig (1914–), and widely adopted in logistical planning and the optimization of economic development planning; this is a mathematical technique for determining a range of maximum values at minimum cost while dealing with known constraints.

A car-maker would try to programme the optimum production of a range of models, where each has different raw material, labour, warehousing and sales requirements, all of which are limited in supply.

G B Dantzig, 'Programming in a Linear Structure', *Comptroller, USAF* (Washington, DC, February, 1948)

PH

linguistic competence (*c.*1960) *Linguistics* The distinction between linguistic competence and LINGUISTIC PERFORMANCE was formulated by the American linguist Avram Noam Chomsky (1928–) following the distinction of LANGUE and PAROLE made by the French linguist Ferdinand de Saussure (1857–1913), but with an important difference.

Knowledge of language versus use of language. The object of linguistics is linguistic competence, knowledge of a language possessed by 'an ideal speaker-listener'. Linguistic competence is designed as a scientific idealization, filtering out 'grammatically irrelevant conditions', errors produced in 'actual linguistic performance'. The theory puts Chomsky's linguistics into the realm of psychology, whereas Saussure's 'langue' did not make that move. The distinction relegates the study of linguistic performance (compare with 'parole'), actual language use, to such disciplines as SOCIOLINGUISTICS.

N Chomsky, *Aspects of the Theory of Syntax* (Cambridge, Mass., 1965)

RF

linguistic criticism (1970s–) *Stylistics* Widely practised approach; terminology stabilized by the English linguist Richard Fowler (1938–).

Linguistic analysis of literary texts whose aim is not merely to characterize style, but to arrive at a statement of the text's significances within its cultural and historical settings. Literary texts are seen not as objects but as interactions. The linguistic methodology is drawn mainly from M A K Halliday's FUNCTIONAL GRAMMAR. *See also* CRITICAL LINGUISTICS.

R Fowler, *Linguistic Criticism* (Oxford, 1986)

RF

linguistic determinism *Linguistics See* SAPIR-WHORF HYPOTHESIS.

linguistic phenomenology *Philosophy* Name sometimes used for the detailed and careful analysis of ordinary language undertaken by linguistic philosophy.

Though not unconnected with ordinary PHENOMENOLOGY – especially in the work of English philosopher Gilbert Ryle (1900–76) – it was an empirical rather than an *a priori* study, and did not involve 'bracketing' the world.

G Ryle, *Collected Papers*, 1 (1971); see especially ch. 10

ARL

linguistic philosophy *Philosophy* Also called ordinary language philosophy. A philosophical movement arising after World War II and lasting until the early 1960s (not to be confused with the philosophical subject called philosophy of language). A leading exponent was John Langshaw Austin (1911–60).

Partly as a reaction against the constraints of LOGICAL POSITIVISM, and influenced by Ludwig Wittgenstein's (1889–1951) slogan 'meaning is use', it insisted that philosophy should confine itself to analyzing concepts, words, and ways of speaking (conceptual analysis) and should not, like logical positivism, dictate the limits of meaningfulness. It should study actual linguistic practice in subjects like metaphysics, ethics, aesthetics, religion, and so on, non-censoriously, taking them at their own valuation and not pronouncing on substantive issues in these subjects. It should, for example, ask what 'right' and 'wrong' mean, but not what things are right and wrong. Though permissive towards non-philosophers where logical positivism had been constrictive, it was accused of being constrictive in its own activities, refusing to enter the arena and encouraging an 'anything goes' attitude, especially in ethics.

J L Austin, *Collected Papers* (1961)

ARL

linguistic relativity *Linguistics See* SAPIR-WHORF HYPOTHESIS.

linguistic universals (17th century; 1960s) *Linguistics* Preoccupation of early rationalist philosophers and linguists (*see* RATIONALISM IN LINGUISTICS); diminished by the empiricism of HISTORICAL LINGUISTICS and American STRUCTURAL LINGUISTICS.

(1) Chomsky revived the rationalist tradition in the 1960s, arguing that language must reflect the universal properties of mind. He distinguished 'formal universals' which are abstract specifications of the forms human languages must take (for example STRUCTURE-DEPENDENCY), and SUBSTANTIVE UNIVERSALS, sets of permissible features and parameters of variation. (2) A more empirical approach takes a large representative sample of historically unrelated languages and seeks common properties by observation; also the foundation of LANGUAGE TYPOLOGY.

N Chomsky, *Aspects of the Theory of Syntax* (Cambridge, Mass., 1965); J H Greenberg, ed., *Universals of Language*, 2nd edn (Cambridge, Mass., 1966)

RF

Liouville's theorem (19th century) *Mathematics* Named after French mathematician Joseph Liouville (1809–82), this is the result in elementary COMPLEX FUNCTION THEORY whereby a bounded entire function (that is, a function which is bounded and analytic on the entire complex plane) must be constant. An immediate and powerful consequence of this theorem is the FUNDAMENTAL THEOREM OF ALGEBRA.

S G Krantz, 'Functions of One Complex Variable', *Encyclopedia of Physical Science and Technology* (Academic Press, 1987)

MB

lipid-membrane hypothesis of chilling injury (1970s). Also known as the theory of chilling injury, this theory postulates that tropical and subtropical plants are damaged by cold temperatures because lipids in their cellular membranes solidify (crystallize) at some critical temperature that varies with the ratio of saturated fats to unsaturated fats in the membrane.

Plants induced towards better tolerance of cooler temperatures (as when gardeners 'harden off' tomato plants) are believed to develop a proportional increase in unsaturated fatty acids or sterols. While there is substantial support for this hypothesis, not all studies show results consistent with the theory.

KH

liquid drop model of the atomic nucleus (20th century) *Physics* A model of the nucleus first suggested by the German physicist Baron Carl Friedrich von Weizsäcker (1912–), and developed by the Danish physicist Niels Henrik David Bohr (1885–1962) and F Kalckar in 1937.

Since in all but the lightest nuclei the density of nucleons and the binding energy per nucleon are approximately constant, it is possible to model a nucleus by a drop of an incompressible liquid. The model allows an expression for the binding energy of a nucleus to be developed, leading to an expression for the mass, known as the semi-empirical mass formula. Volume and surface terms follow directly from the analogy with the liquid drop but other terms are added (hence the description 'semi-empirical') to allow for other effects that govern the stability of atomic nuclei. For a nucleus of mass number A and atomic number Z the complete expression for the binding energy B is

$$B = a_1A - a_2A^{2/3} - a_3Z^2A^{-1/3} - a_4(A - 2Z)^2/A \pm \delta$$

where a_1, a_2, a_3, and a_4 are constants that are evaluated by fitting to the experimental data and δ is positive for nuclei with both A and Z even; negative for nuclei with Z odd and A even; zero for nuclei with A odd.

J Thewlis, ed., *Encyclopaedic Dictionary of Physics* (New York, Oxford and London, 1962)

MS

lisible and *scriptible* (1970) The authorized translations are 'readerly' and 'writerly'. Distinction proposed by the French structuralist Roland Barthes (1915-80); relates to his WORK VS TEXT.

A 'readerly' piece of literature or 'work' is a closed, single-meaning product which is simply consumed by the reader; the 'writerly' text invites rewriting by an active reader who finds a plurality of new meanings in the SEMIOTIC codes which structure the text. The distinction seems to be about attitudes to reading rather than types of text.

R Barthes, *S/Z*, trans. R Miller (London, 1975)

RF

literariness (1921) *Literary Theory* Fundamental contribution to POETICS by the Russian linguist and literary theorist Roman Jakobson (1896–1982).

Poetics, or the science of literature, is not concerned with the history, criticism, or interpretation of specific works of literature, but with 'literariness'. This is a general property which 'makes a given work a *literary* work', a universal which underlies all individual literary texts. *See also* FORMALISM, RUSSIAN and POETIC PRINCIPLE.

R F R Jakobson, 'Recent Russian Poetry' [1921], *Selected Writings*, vol. v (The Hague, 1969)

RF

literary competence (1975) *Stylistics* Term coined by the literary theorist Jonathan Culler (1944–) on the analogy of LINGUISTIC COMPETENCE.

Literary interpretation is a specialized kind of reading which depends on more than one's knowledge of the language; it depends on literary competence, a familiarity with the conventions of literary reading, and ability to deploy them, which only those who have experienced a literary education possess. Dell H Hymes's COMMUNICATIVE COMPETENCE would be a better analogy.

J Culler, *Structuralist Poetics* (London, 1975)

RF

literary language (*c*.1915–) *Stylistics* The modern debate originates in the proposals of the Russian formalists (*see* FORMALISM, RUSSIAN) and of the PRAGUE SCHOOL (*see also* FOREGROUNDING).

In the 19th century the concept 'literature' acquired its modern meaning of high, creative, imaginative writing. Early stylistics argued that there is an identifiable 'literary language' distinct from 'ordinary language'. In a paper of 1958, Roman Jakobson (1896–1982) described a POETIC PRINCIPLE which claims to specify the linguistic properties of a 'poetic function', which is however to be found outside literature as well. Most contemporary stylisticians argue that the strict separation of two kinds of language is harmful; that literariness is a property of reading, not of language.

R A Carter and W Nash, 'Language and Literariness', *Prose Studies*, 6 (1983), 123–41

RF

literary mode of production (1970s) *Literary Theory* Concept in MARXIST CRITICISM developed by English critic Terry Eagleton following Pierre Macherey (1938–) (*see* ABSENCE).

Writing is a product of economic and social relations which can be of various kinds (modes): from individual patronage to mass publication with its economic, educational and ideological implications. The *idea* of literature is produced by the same institutional forces (*see also* LITERATURE).

T Eagleton, *Criticism and Ideology* (London, 1976)

RF

literary pragmatics (1980s) *Stylistics* Popularized by the literary linguist Roger D Sell (1942–) and his colleagues at Åbo Akademi, Finland.

Linguistic studies of literature which (in opposition to formalist studies) emphasize the relative and social rather than essential and material nature of literature, its relation with contexts social and historical, its production and reception, the activities of and effects on readers. The methods of linguistic PRAGMATICS may be used, but not necessarily. 'Literary pragmatics' is better regarded as an umbrella term than as a name for a single approach.

R D Sell, *Literary Pragmatics* (London, 1991)

RF

literary theory (1960s–) *Literary Theory*. Also 'theory'.

Literary criticism and history, and traditional stylistics, are attempts to describe and understand particular texts, authors, and periods. 'Theory' is more self-conscious about its assumptions, methods and terminology, relating a text (say) to an external intellectual context such as FEMINISM, MARXISM, sociology, linguistics, STRUCTURALISM, psychoanalysis, and so on. Theory is highly diverse, but often experienced by students as a single discourse: abstract, technical, sceptical. *See also* POETICS.

T Eagleton, *Literary Theory* (Oxford, 1983)

RF

literature (traditional) *Literary Theory* The idea of a single entity goes back to classical times as 'poetry'; 'literature' in this sense only from *c*.1800.

Creative or imaginative literature has traditionally been regarded as an aesthetically or morally distinct form, or use, of language. Many criteria have been proposed: characteristic structural or linguistic properties, type of effect on reader, fictionality, cultural role or content, and so on. There is in fact no satisfactory way of distinguishing literature from non-literature; the concept shifts according to the needs of a culture. *See also* POETICS.

R Fowler, 'Literature', *Encyclopedia of Literature and Criticism*, M Coyle *et al.*, eds (London, 1990), 3–26

RF

literature as discourse (1970s) *Stylistics* General reorientation in the 1970s, but the work of the British linguist Richard Fowler (1938–) and colleagues is often cited for basic theory and practice.

Theorizing literature as discourse, or social discourse, rather than material textual structure, allows patterns of language to be linked with social and historical context, with relationship between writer and readership. There is also a stress on the belief systems or ideologies which are embedded in a text's language through its social positioning.

R Fowler, *Literature as Social Discourse* (London, 1981)

RF

literature as play *Stylistics See* LANGUAGE GAMES.

'little bags' theory of evolution *Biology* Also called evolution in little bags, this is a reference to the set of theories proposing that life evolved from mixtures of organic chemicals that somehow became trapped in tiny vesicles. *See also* HYPERCYCLE THEORY and ORIGINS OF LIFE.

KH

Liusternik theorem (1934) *Mathematics* A result in non-linear FUNCTIONAL ANALYSIS extending the classical INVERSE FUNCTION THEOREM to infinite dimensions.

The inverse function theorem is a key result of calculus which allows one to consider the solution of the equation $f(x) = y$ near a current solution $f(\bar{x}) = \bar{y}$ as a well-behaved function $x(y)$ of the right-hand side, given certain smoothness conditions on f. Liusternik's result proves this in the case where f maps between Banach spaces and has a surjective Fréchet derivative at $\bar{x}$.

AL

loanable funds theory of the rate of interest (19th century–) *Economics* Developed by Swedish economist Knut Wicksell (1851–1926), this theory posits that interest rates are determined by the supply and demand of loanable funds in the capital markets.

The theory suggests that investments and savings determine the long-term level of interest rates, whereas short-term rates are determined by financial and monetary conditions in the economy. It was widely accepted before the work of English economist John Maynard Keynes (1883–1946). *See* TERM STRUCTURE OF INTEREST RATES.

D H Robertson, *Essays in Monetary Theory* (London, 1940)

PH

local sign theory *Philosophy* Theory, originated by German philosopher Rudolph Hermann Lotze (1817–81), that we assign a bodily location to the cause of a bodily sensation (for example, we come to treat a pain as 'in' our right hand) because of a special quality which the sensation has. This special quality we come to associate with the location of its cause through experience and inference.

G N A Vesey, *The Embodied Mind* (1965), ch. 4

ARL

localization theory *Psychology* Proposed by the French physiologist Pierre Jean Marie Flourens (1794–1867).

This theory assumes that each area within the brain is responsible for specific psychological skills. Thus, the location of brain injury is the most important factor; extent of injury is important only in so far as a larger lesion involves more area of brain and thus disrupts more skills. Localization theory is unable to account for why a specific deficit in psychological functioning can be caused by damage to different parts of the brain. *See* equipotential theory and MULTIPLE CONTROL THEORY.

B Kolb and I Q Whishaw, *Fundamentals of Human Neuropsychology*, 3rd edn. (San Francisco, 1990)

NS

lock and key theory (1960s) *Biology* A theory of the mechanism of enzyme action which evolved over the whole of the 20th century (emerging during the 1960s) and cannot be attributed to any one individual or small group.

An enzyme is a protein which has a particular shape at its site of action (the active site) into which substrate(s) must fit exactly. At this site the substrate(s) is/are held in such a way as to induce catalysis and make products that have an altered shape and so no longer fit the active site and therefore will be released. The mechanism has been likened to the enzyme being a lock into which the specific substrate, the key, fits. Enzymes are now envisioned as somewhat looser structures which move slightly to clamp substrates in place. This was first suggested in 1959 by the American molecular biologist D E Koshland in the induced fit hypothesis, which is simply a modification of the lock and key theory.

A G Loewy and P Siekevitz, *Cell Structure and Function* (London, 1970)

RA

lock and key model (*c.*1884) *Biology* Hypothesized by German organic chemist Emil Hermann Fischer (1852–1919), this is a model for the relationship between an enzyme and its substrate that assumes a lock-and-key biochemical fit between the two.

That is, the enzyme's active site is considered to be a rigid arrangement of charged groups that precisely matches complementary groups in the substrate, forming a highly stable enzyme-substrate complex. This model was modified and usurped by the INDUCED-FIT HYPOTHESIS. *See also* SELECTIVE THEORY OF ANTIBODY DIVERSITY and SIDE-CHAIN THEORY.

F Salisbury, *Plant Physiology* (Belmont, Calif., 1992)

KH

locutionary act *Linguistics See* SPEECH ACT THEORY.

logic and foundations (19th century) *Mathematics* Although the axioms of geometry have been studied since *c.*300 BC, the foundations of mathematics itself attracted interest only after the work on logic by George Boole (1815–64) and on SET THEORY by Georg Cantor (1845–1918).

The study of definitions of mathematics and associated logical systems in which to construct proofs and the consequent philosophical schools. *See also* SET THEORY and REALISM/INTUITIONISM.

JB

logical atomism *Philosophy* Theory, held briefly by Bertrand Russell (1872–1970) and Ludwig Wittgenstein (1889–1951) soon after World War I, that a proper description of reality would be in terms of atomic propositions, each containing a word standing for a quality or relation and one or more words standing for objects which had the quality or relation.

The objects must be basic and unanalyzable, whether they are objects of immediate experience (sense-data), as for Russell, or unspecified, as for Wittgenstein. Atomic propositions are true or false according to whether they do or don't correspond to and picture atomic facts. Molecular propositions are formed from them by negation and connectives (primarily 'and', 'or', and 'if . . . then') which are truth-functional; that is, given the truth or falsity of its atomic propositions, the truth or falsity of a molecular proposition could be inferred. The theory then claims that this apparatus is adequate for describing reality completely. However, problems arose in dealing with negation, general propositions, propositions about belief and other notions that appear to violate the EXTENSIONALITY THESIS, and there was some vacillation on whether any non-atomic facts could be admitted.

D F Pears, *Bertrand Russell and the British Tradition in Philosophy* (1967)

ARL

logical empiricism *Philosophy* Version of EMPIRICISM applying to the meanings of words or sentences, whereby they have meaning only if there are rules involving sense-experience for applying or verifying them; the rules may also constitute the meaning. (Analytic sentences – that is, roughly, those made true or false by logical considerations – are excepted.) Akin to, though some say slightly less rigorous than, LOGICAL POSITIVISM.

ARL

logical positivism *Philosophy* A 20th century development of POSITIVISM which emphasizes questions of language and meaning and the role of logical relations like entailment.

It originated in the VIENNA CIRCLE and continued mainly in English-speaking countries (with Holland and Scandinavia) until World War II, after which it was replaced by LINGUISTIC PHILOSOPHY in Britain and various movements in the USA and elsewhere. Its central tenet is the VERIFIABILITY PRINCIPLE, which in turn has its roots in David Hume's (1711–76) distinction – in the last paragraph of his *An Enquiry Concerning Human Understanding* (1748) – between 'abstract reasoning concerning quantity or number' and 'experimental reasoning concerning matter of fact and existence', all else being 'sophistry and illusion'.

A J Ayer, ed., *Logical Positivism* (1959)

ARL

logical relation theory of probability *Philosophy* Theory due especially to English economist John Maynard Keynes (1883–1946) in his *Treatise on Probability* (1921), Chapter 1. It says that the probability of a hypothesis is a logical relation (rather like logical entailment, only weaker) between a hypothesis and a body of evidence for it. Probability is thus made relative to evidence.

This could be avoided by considering all the evidence (requirement of total evidence), but there are difficulties in specifying this. The relation in question is hard to specify, and would not give an analysis of 'probably' anyway, for if we have some evidence which entails a conclusion we can assert the conclusion; but if it only makes it probable ('probabilifies' it) we can only say 'the conclusion is probable', without saying what this means (unless we are saying merely that the evidence exists, without saying what it does, that is without *using* it). The theory is also subject to various paradoxes.

H E Kyburg, *Probability and Inductive Logic* (1970), ch. 5

ARL

logicism *Philosophy* Theory, due to Gottlob Frege (1848–1925) and Bertrand Russell (1872–1970), that the concepts and theories of mathematics (in particular of arithmetic) can be derived from those of logic. This, if feasible, would support LOGICAL POSITIVISM and REDUCTIONISM in general.

Arithmetic was in fact reduced to set theory – developed by Georg Cantor (1845–1918) – as a first step, but set theory itself has never successfully been derived from pure logic, and the enterprise was frustrated by K Gödel's (1906–78) proof in 1931 that for any system rich enough to formalize arithmetic there will always be truths that can be stated in the system (and so form part of it) but cannot be proved within it.

S Körner, *Philosophy of Mathematics* (1960), chs 2, 3

ARL

logocentrism (1960s) *Literary Theory* Formulated by the Algerian-French philosopher Jacques Derrida (1930–).

Mistaken faith, in modern Western metaphysics, in certain transcendental ideas such as truth, meaning, and justice, which have been illusorily stabilized by being named, and encoded in linguistic signs. DECONSTRUCTION challenges this by techniques of DECENTRING.

J Derrida, *Of Grammatology*, G C Spivak, trans. (Baltimore, 1977 [1967])

RF

logogen (1969) *Linguistics* Theoretical construct proposed by the British psycholinguist John Morton (1933–).

In psycholinguistics, a theory of lexical access; that is, how readers/hearers recognize words in writing or speech and relate them to entries in their 'mental dictionary'. A logogen is a perceptual device (one for each word a person knows) which 'fires' when context confirms the word guessed. *See also* SEARCH MODEL.

M Garman, *Psycholinguistics* (Cambridge, 1990)

RF

Lombrosian theory *Psychology* Proposed by the Italian criminologist Cesare Lombroso (1836–1909).

Lombroso hypothesized that criminality is biological and that criminals can be identified by hereditary 'stigmata of degeneracy' such as low foreheads, close-set eyes, and small pointed ears. This theory is without medical or scientific foundation.

H J Eysenck, *Fact and Fiction in Psychology* (Harmondsworth, 1974)

NS

long waves *Economics See* KONDRATIEFF CYCLES.

Lorentz contraction hypothesis *Physics See* FITZGERALD-LORENTZ CONTRACTION.

Lorentz force law (1895) *Physics* Named after the Dutch physicist Hendrik Antoon Lorentz (1853–1928), who introduced the law.

A particle of charge e and velocity $\mathbf{v}$, passing through a region in which there is a magnetic field of induction $\mathbf{B}$ and an electric field of strength E, experiences a force $\mathbf{F}$ given by

$$\mathbf{F} = e(E + \mathbf{v} \times \mathbf{B}) .$$

J Thewlis, ed., *Encyclopaedic Dictionary of Physics* (New York, Oxford and London, 1962)

MS

Lorentz-Lorenz law (19th century) *Physics* Named after the Dutch physicist Hendrik Antoon Lorentz (1853–1928) and the Danish physicist Ludwig Valentin Lorenz (1829–91) who deduced the law; this states that at constant temperature, the refractive index n of all states of a dielectric is related to the density ρ by the equation

$$\frac{n^2 - 1}{n^2 + 2} = \text{constant} \times \rho .$$

For gases n is close to unity and this law leads to the GLADSTONE AND DALE LAW.

J Thewlis, ed., *Encyclopaedic Dictionary of Physics* (New York, Oxford and London, 1962)

MS

Lorenz attractor *Mathematics See* STRANGE ATTRACTOR.

Lorenz's hydraulic model (1950) *Biology* Also called the 'flush toilet' model, this was proposed by Konrad Zacharias Lorenz (1903–89), Austrian zoologist and ethologist.

An early model for animal behaviour comparable to a hydraulic flow system, such as a flush toilet. The fluid in the tank is analogous to action-specific energy (ASE) which accumulates spontaneously over time, increasing the animal's drive to perform a particular behaviour. When an appropriate stimulus occurs, the intensity of the animal's response is proportional to the power of the stimulus (which is analogous to the amount of weight involved in throwing open a valve) and the build-up of ASE. This model has been largely replaced by more complex concepts, such as TINBERGEN'S HIERARCHICAL MODEL. *See also* DEUTSCH'S FEEDBACK MODEL and ETHOLOGY.

D Dewsbury, *Comparative Animal Behavior* (New York, 1978)

KH

Lowry-Brønsted theory *Chemistry* *See* BRØNSTED-LOWRY THEORY.

Löwenheim-Skolem theorem (1915) *Mathematics* Also known as the Skolem-Löwenheim theorem after mathematicians Leopold Löwenheim (1878–1940) and Albert Thoralf Skolem (1887–1963).

This is the important result in MODEL THEORY whereby if a countable set of formulae has a model, then it has a countable model. *See also* SKOLEM PARADOX.

Encyclopedic Dictionary of Mathematics (MIT Press, 1987)

MB

Lucas's theory *Chemistry* A theory named after the American chemist Howard Johnson Lucas (1885–) which proposes that when a radical with a high electron affinity substitutes into an organic compound, the electron pair of a carbon atom is pulled towards the substituting radical and *vice versa*.

GD

Ludwig's theory (1842) *Biology/Medicine* Also called filtration theory of kidney function, this was originated by Carl Friedrich Wilhelm Ludwig (1816–95), an eminent German physiologist and physician known for his vigorous opposition to VITALISM. It is the theory of kidney function holding that a filtrate containing all the water-soluble constituents of the blood is separated at the glomerulus, and that a process of selective reabsorption occurs in the urinary tubule.

Significantly, Ludwig sought to explain kidney function solely in terms of chemical and physical processes, without the need to invoke a mystical 'life force' to account for the wonder of urine production. *See also* REDUCTIONISM. *Compare with* BOWMAN-HEIDENHAIN HYPOTHESIS and CUSHNY'S THEORY.

KH

lullaby effect *Psychology* First investigated experimentally by the Russian physiologist Ivan Petrovich Pavlov (1839–1946).

The term refers to the process whereby an organism becomes adapted to a novel stimulus when repeated frequently. For instance, the sudden onset of a rapping sound may elicit a startle response, but this diminishes if repeated and the stimulus rapidly loses its effectiveness. There are numerous theories to account for this phenomenon, which reflects a fundamental feature of human attention and learning.

R S Woodworth and H Schlosberg, *Experimental Psychology* (London, 1966)

NS

lump-of-labour theory of wages *Economics* In the short term, the demand for labour is fixed and employment can be created only by job-sharing and reducing existing working hours of workers. This theory ignores the role of macroeconomic policy in stimulating the economy.

M Allais and O Hagen, eds, *Expected Utility Hypotheses and the Allais Paradox* (Dordrecht, 1974)

PH

Lusin's theorem (early 20th century) *Mathematics* Named after Russian analyst, topologist and logician Nikolai Nikolaevitch Lusin (1883–1950), this states that a function which is measurable and is almost everywhere finite on a measurable subset of Euclidean space (which has finite measure) is continuous except on a set of arbitrarily small measure; that is, for any such f and $\epsilon > 0$ there is a continuous function g such that the measure of the set of points where f is not equal to g is of measure less than ϵ.

G B Folland, *Real Analysis* (New York, 1984)

ML

Lydekker's line (1891) *Biology* Named after the British naturalist R Lydekker (1849–1915).

The accepted line which defines the easternmost geographical limit of animals from the Oriental faunal region into the neighbouring Australian faunal region. A faunal region is a division of the Earth's surface defined by fauna more or less peculiar to it, such as marsupials in the Australian faunal region. A minimum of six faunal regions are recognized.

RA

Lypapanov's theorem (1940) *Economics* Named after Russian mathematician A A Lypapanov, this theory asserts that the range of a non-atomic totally finite vector-valued measure is both convex and compact.

A A Lyapunov, 'On Completely Additive Vector-Functions', *Izvestia Akademii Nauk SSSR*, vol. IV (1940), 465–78

PH

lyric (classical Greece) *Literary Theory* Originally a song performed to the accompaniment of a lyre.

A short expressive poem usually on a single topic (non-narrative), strongly personal and often emotional in tone (*see* PERSONA), and mellifluous or musical in style – hence the attribute 'lyrical' recalling the etymology of the word. In this extended sense, prose may be lyrical. Modern usage of the term is extremely inclusive, but it might be generally agreed that the height of achievement in lyric poetry was during the Elizabethan period.

D Lindley, *Lyric* (London, 1985)

RF

Lysenkoism (20th century) *Biology* Named after Trofim Denisovich Lysenko (1898–1976), a Russian geneticist, this was a revival of the Lamarckism school of genetics based on Lysenko's belief that characteristics acquired during life could be passed on to offspring.

Eventually, Lysenko discarded most aspects of the gene concept. Lysenkoism flourished in the Soviet Union because it was consistent with the Marxist doctrine of individual improvement through training and environment.

D Joravsky, *The Lysenko Affair* (Cambridge, Mass, 1970)

KH

Lysippan proportions (4th century BC) *Art* Named after Alexander the Great's official portraitist, the Greek sculptor Lysippus of Sicyon (*fl. c.*325 BC). According to ancient tradition, he introduced a new system of proportions for the body, superseding those of Polyclitus (*fl c.* 232 BC) in which the head of a statue was made smaller in order to give the impression of increased height.

AB

M

Machiavellianism *Politics* Pejorative term to describe beliefs or conduct allegedly derived from the views of Italian political theorist Niccolo Machiavelli (1429–1527).

The end justifies the means, and moral considerations should be subordinated to the achievement of material or political goals. Used in this way, the term is a parody of the views actually expressed by Machiavelli.

David Miller, ed., *The Blackwell Encyclopaedia of Political Thought* (Oxford, 1987)

RB

machine intelligence *Psychology See* ARTIFICIAL INTELLIGENCE.

machine translation (1950s) *Linguistics* Automatic translation of natural languages, using procedures of computational linguistics. Initially unsuccessful, it is now flourishing with advances in the theory and techniques of natural language processing.

H L Somers and W J Hutchins, *An Introduction to Machine Translation* (London, 1992)

RF

machinery question (1695) *Economics* English merchant and writer John Cary (d.*c*.1720) first dealt with the impact of the introduction of machinery on employment, an issue later examined by classical economists such as the Englishman David Ricardo (1772-1823).

Ricardo saw increased mechanization as reducing the wages fund, although the English philosopher John Stuart Mill (1806-73) argued that any fall in wages would be temporary. The issue is still important in understanding Third World economics. *See* TECHNOLOGICAL GAP THEORY.

J Cary, *A Discourse on Trade* (London, 1695); K Wicksell, *Lectures on Political Economy* (London, 1911)

PH

Mach's principle (1883) *Astronomy/Physics* Named after the Austrian philosopher and physicist Ernst Mach (1838–1916). Mach raised a number of conceptual objections to the laws of motion as laid down by Sir Isaac Newton (1642–1727).

He argued that unless there is a material background against which motion is to be measured, the concepts of rest and motion are meaningless. In general terms the principle means that the local physical laws observed on Earth are influenced by the large-scale distribution of matter in the universe. For example, it follows from this that the magnitude of the inertia of any body is determined by the masses in the universe and by their distribution. The German-born American mathematical physicist Albert Einstein (1879–1955) was greatly influenced by Mach's discussion, and proposed a universe filled with matter providing a background against which a local observer can measure motion and formulate laws of mechanics. *See also* BRANS-DICKE THEORY.

S Mitton, ed., *The Cambridge Encyclopaedia of Astronomy* (London, 1973); J V Narlikar, *Introduction to Cosmology* (Cambridge, 1993)

GD, MS

Maclaurin's theorem (18th century) *Mathematics* Named after Scottish mathematician and physicist Colin Maclaurin

(1698–1746), this states that a real-valued function $f(x)$ which has n derivatives in a neighbourhood of zero can be represented locally as a Maclaurin series of the form

$$f(x) = \sum_{k=0}^{n} \frac{f^{(k)}(0)}{k!}x^k + r_n(x)$$

where $f^{(k)}(x)$ is the kth derivative of $f(x)$ and $r_n(x)$ is the remainder which tends to zero as n tends to ∞. For example

$$\sin x = x - \frac{x^3}{3!} + \frac{x^5}{5!} - \frac{x^7}{7!} + \ldots$$

is the expansion to order 8. A Maclaurin series is a special case of a Taylor series. A complex function which is holomorphic inside the disc $|z| < \mathrm{r}$ can be expressed as

$$f(z) = \sum_{n=0}^{\infty} \frac{f^{(n)}(0)}{n!}z^n$$

for all z in the disc.

H Anton, *Calculus with Analytic Geometry* (New York, 1980)

ML

MacLean's theory of emotion *Psychology* Proposed by the American psychologist P D MacLean (1913–).

Based on PAPEZ'S THEORY OF EMOTION, this hypothesis further asserted that other areas of the limbic system (particularly the hippocampus and amygdaloid complex) are involved in emotion as well as the hypothalamus; and that the more primitive layers of the cortex play a major role in integrating information. This theory of emotion is part of MacLean's more general characterization of the triune brain.

J W Papez, 'A Proposed Mechanism of Emotion', *Archives of Neurology and Psychiatry*, vol. XXXV (1937), 725–44

NS

macroevolution (1940) *Biology* Coined by Richard Goldschmidt (1878–1958), German-born geneticist best known for his theory of HOPEFUL MONSTERS, the term refers to major evolutionary events or trends viewed from the perspective of a long timespan (geological time). Also, the evolutionary development of species and, especially, categories above the species level. *See also* EFFECT HYPOTHESIS. *Compare with* EVOLUTION and MICROEVOLUTION.

M Barbieri, *The Semantic Theory of Evolution* (Chur, Switzerland, 1985)

KH

macrolinguistics (1949) *Linguistics* A term coined by the American linguist George L Trager. In a terminology which was widely accepted in American STRUCTURAL LINGUISTICS of the 1950s, Trager divided up the field of linguistics thus:

macolinguistics

prelinguistics microlinguistics metalinguistics

Macrolinguistics is the most inclusive term, covering the whole field. Prelinguistics studies the material aspects of speech; that is, PHONETICS. Microlinguistics is 'linguistics proper' in the American tradition: strict attention to linguistic form – PHONOLOGY, MORPHOLOGY, SYNTAX – without regard to extralinguistic and metalinguistic features. Metalinguistics studies the relationship of language to setting and cultural context (*see* CONTEXT OF SITUATION).

G L Trager, *The Field of Linguistics* (Norman, Okla., 1949). There is an accessible synopsis in: D Crystal, *Prosodic Systems and Intonation in English* (Cambridge, 1969)

RF

macrostructure (1970s) *Stylistics* Modern term applied to a variety of kinds of proposed large textual structures.

The broad principle of a text's organization, as opposed to the local linguistic structure of its sentences, and the relationships between sentences (COHESION). Macrostructures may be defined, for example, in terms of NARRATIVE GRAMMAR, or SCHEMATA, or on the analogy of generative grammar (*see* STANDARD GRAMMAR) as deep structures.

T A van Dijk, *Some Aspects of Text Grammars* (The Hague, 1972)

RF

magic square (*c.*2000 BC) *Mathematics* Studied since ancient times. The magic square of order 3 was known in China in *c.*2000 BC.

A square array of integers arranged so that the sum of the integers in each vertical column, horizontal row and diagonal is the same. Dürer's famous engraving 'Melancholy' shows the magic square of order 4 below.

16	3	2	13
5	10	11	8
9	6	7	12
4	15	14	1

La Loubère (*c.*1670) discovered an algorithm for constructing magic squares of odd order n. There is still no general theory for the construction of magic squares. In fact, the number of magic squares of order $n \geqslant 5$ is unknown. There is only one square (up to symmetries) of order 3 and there are 880 for order 4.

I Peterson, *The Mathematical Tourist* (New York, 1988)

ML

magical number seven (1956) *Linguistics* Theory of American psychologist G A Miller (1920–).

The maximum number of units of information that humans can hold in short-term memory for processing at one time is about seven; economy is achieved by 'chunking' smaller into larger units (phonemes into words, words into phrases, and so on).

G A Miller, 'The Magical Number Seven, Plus or Minus Two: Some Limits on our Capacity for Processing Information', *Psychological Review*, 63 (1956), 81–96

RF

magic(al) realism (1925) *Art/Literary Theory* A term first used by German art historian Franz Roh (1890–1965) in reference to Post-Impressionism. Roh sought to distinguish this style of German painting from other contemporary objective styles (for example, *Neue Sachlichkeit*) current in the 1920s. Precise realism is contrasted with a mood of fantasy and the juxtaposition of incongruous details, as also seen in Metaphysical or Surrealist art (*see* SURREALISM).

The term was extended to fiction in 1952 by the Austrian novelist George Saiko (1892–1962). It describes a quasi-surrealist technique of writing in which clearly delineated realism is juxtaposed with fantasy, dreams and myths; and in which complicated narratives and shifts of time-sphere are common. The effect is often bizarre, puzzling or shocking. The leading exponent is held to be the Argentinian fiction-writer Jorge Luis Borges (1899–1988).

W Schmied, *Neue Sachlichkeit und Magischer Realismus in Deutschland 1918–1933* (Hanover, 1969); G Saiko, *Die Wirklichkeit hat doppenlten Boden. Gedanken zum magischen Realismus* (1952)

AB, RF

magmatic differentiation *Geology* A molten rock (magma) will usually change composition with time, giving rise to rocks of different compositions. This differentiation can occur at an atomic level (for example, ionic migration) and by the separating-out (for example, by gravitation or convective settling, filter pressing, and so on) of already-crystallized material from the remaining liquid fraction.

D Walker, 'New Developments in Magmatic Processes', *Reviews in Geophysics and Space Physics*, VOL. XXI (1983), 1372–84

DT

magnetism, theories of *Physics See* BAND THEORY OF FERROMAGNETISM, EWING'S THEORY OF FERROMAGNETISM, HEISENBERG'S THEORY OF FERROMAGNETISM, WEISS'S THEORY OF FERROMAGNETISM and LENZ'S LAW.

male-male competition (20th century) *Biology* A theory of reproductive strategy in which the male maximizes his fitness by physically dominating other males; by occupying choice territories; and/or by successful sperm competition – that is, neutralizing the sperm of other males, preventing other males from mating, or protecting his own sperm from neutralization by other males. *Compare with* FEMALE CHOICE. *See also* SEXUAL SELECTION.

D Dewsbury, *Comparative Animal Behavior* (New York, 1978)

KH

Malthusian population theory (1798) *Biology/Economics* Named after English economist the Reverend Thomas Robert

Malthus (1766–1834), who believed that population would increase at a geometric rate and the food supply at an arithmetic rate. This disharmony would lead to widespread poverty and starvation which would only be checked by natural occurrences such as disease, high infant mortality, famine, war or moral restraint.

The theory was eventually dismissed for its pessimism and failure to take into account technological advances in agriculture and food production. *See* DEMOGRAPHIC TRANSITION and SECULAR STAGNATION.

In biology, the theory asserts that the reproductive potential of virtually any organism or SPECIES greatly exceeds the earth's capacity to support all its possible offspring. Consequently, species diversity is preserved through mechanisms that keep population sizes in check, such as predation.

T R Malthus, *An Essay on the Principle of Population* (London, 1798); A Chase, *The Legacy of Malthus* (New York, 1977)

KH, PH

Malus's law (18th century) *Physics* Named after its discoverer, the French physicist Etienne Louis Malus (1775–1812).

When a beam of plane-polarized light of intensity I_0, produced by a polarizer, falls on an analyzer, the intensity I of the transmitted beam varies as the square of the cosine of the angle between the two planes of transmission.

$$I = I_0 \cos^2 \theta \, .$$

J Thewlis, ed., *Encyclopaedic Dictionary of Physics* (New York, Oxford and London, 1962)

MS

Malus's theorem (early 19th century) *Physics* Named after its discoverer, the French physicist Etienne Louis Malus (1775–1812), this proposes that the optical path between any two wavefronts is the same for any ray.

When a pencil of light rays crosses two or more media, so that it has been refracted one or more times, a new wavefront may always be found in one of the subsequent media by measuring off equal optical paths along all rays, starting from a wavefront in the first medium.

J Thewlis, ed., *Encyclopaedic Dictionary of Physics* (New York, Oxford and London, 1962)

MS

man made language (20th century) *Feminism/Linguistics* Concept of Australian feminist Dale Spender (1943–) to describe male domination of language.

Language is not a neutral medium in which any meanings one wishes can be expressed. It makes possible certain kinds of meanings, favourable to male power or PATRIARCHY, and makes difficult or impossible the expression of critical or dissenting views.

Dale Spender, *Man Made Language* (London, 1980)

RB

managerial revolution (20th century) *Politics* Theory associated with James Burnham (1905–).

The characteristic feature of modern industrial societies is the rise of managers as the effective wielders of power, particularly in the economy. In capitalist societies this means that owners or capitalists are losing power; in state socialist or communist societies, that politicians or the working class are losing it.

James Burnham, *The Managerial Revolution* (London, 1941)

RB

managerial theories of the firm (1960s) *Economics* A range of theories suggesting that managements in large oligopolistic organizations have supplanted the traditional goal of profit maximization. (New goals may, for example, focus on sales or asset growth maximization.) Managerial theories also recognize that power within the organization has shifted away from shareholders to management. *See* THEORY OF THE FIRM, AGENCY THEORY, THEORY OF THE GROWTH OF THE FIRM, ORGANIZATION THEORY and THEORY OF BUREAUCRACY.

W J Baumol, *Business Behavior, Value and Growth* (New York, 1959); R Marris, *The Economic Theory of 'Managerial' Capitalism* (London, 1964)

PH

managerialism (20th century) *Politics* Ideology of organizational power.

Organizations, both public and private, are best run when power is exercised hierarchically by managers who are distinct from the producers of goods or the providers of services, but have general power to dispose of the organization's resources. Managerialism is thus a late 20th-century flexible mutation from the theory of BUREAUCRACY, and is in opposition to theories of COLLEGIALISM.

James Burnham, *The Managerial Revolution* (London, 1941)

RB

Mandelbrot set (1980) *Mathematics* Named after Polish-born American mathematician Benoit Mandelbrot (1924–), this is the set of complex parameters c for which the JULIA SET of

$$f_c(z) = z^2 + c$$

is connected.

When plotted in the complex plane the set reveals under magnification a complicated structure of fine hairs and bulbs. Recently, there has been an explosion of popular interest in the detailed structures of the Mandelbrot set and the related Julia sets.

R L Devaney, *An Introduction to Chaotic Dynamical Systems* (New York, 1989)

ML

Manicheism *Philosophy* Religious system founded by Mani of Persia (*c.* AD 215–76) and emphasizing fundamental dualism of good and evil as independent principles, represented by spirit and body and symbolized by light and dark.

Sometimes treated as a Christian heresy, Manicheism is rather a separate religion with its roots in Zoroastrianism (founded by Zoroaster (or Zarathustra) of Persia probably in the 6th century BC). St Augustine (AD 354–430) briefly adhered to it, and the Albigensian creed of medieval France was strongly influenced by it. Its adherents sought to ally themselves to Good by a thorough asceticism, including in Albigensian times an explicit readiness to go to the stake.

ARL

Mann-Whitney-Wilcoxen test (1947) *Mathematics* Originally proposed by Wilcoxen in 1945 for samples of equal size, but later extended to samples of unequal size by Mann and Whitney, this is a distribution free statistical test based on comparing the distributions of the ranks of the difference of the scores between two distributions of data collected in two experimental conditions. A generalization of the Mann-Whitney test is the Kruskal-Wallis test for k samples.

H T Nguyen and G S Rogers, *Fundamentals of Mathematical Statistics* (New York, 1989)

ML

mannerism (16th century) *Art* The Italian art historian Giorgio Vasari (1511–74) first coined the term *maniera* in reference to an effete and highly developed style now known as Mannerism. This emerged in Italy from, and in reaction to, High Renaissance art, and was later disseminated in northern Europe.

Seen by many as representing the degeneracy of Renaissance art, Mannerism produced a bizarre and elegant style in which tension, juxtapositions of brilliant colour and virtuoso displays of the artists' powers of invention were encouraged. With the emergence of Counter-Reformation doctrine, its stylistic importance waned.

AB

Manneville-Pomeau phenomenon *Mathematics* The theory of the onset of chaos in which a dynamical system undergoes an *inverted saddle node bifurcation* from periodicity to aperiodic motion on a STRANGE ATTRACTOR. This is also termed *intermittent transition to turbulence*. *See also* CHAOS THEORY, RUELLE-TAKENS SCENARIO and FEIGENBAUM PERIOD-DOUBLING CASCADE.

D Ruelle, *Elements of Differentiable Dynamics and Bifurcation Theory* (San Diego, Calif., 1989)

ML

many worlds hypothesis (1957) *Physics* A hypothesis proposed by the American physicists Hugh Everett III, John Archibald Wheeler (1911–) and Neil Graham.

This theory is an attempt to solve the problem of measurement in quantum mechanics, typified in the paradox known as

SCHRÖDINGER'S CAT. The hypothesis assumes that the universe is described by a wave-function which contains all possible outcomes, each of which is actualized but in a different, and equally real, 'parallel universe'. With every measurement made by an observer, the universe splits into further new universes (the 'many worlds'), in a way which avoids the 'collapse' of the wave-function. There is no scientific proof for the hypothesis, which also fails to explain the special nature of measurement processes that lead to splitting universes. *See also* HIDDEN VARIABLE THEORY.

Paul Davies, ed., *The New Physics* (Cambridge, 1989)

GD

Maoism (20th century) *Politics* Theory of politics of Mao Zedong (Mao Tse-Tung) (1893–1976).

Adaption of MARXISM and STALINISM to the conditions of China, in particular to guerrilla war in largely peasant societies. It attempts to combine traditional Marxism with respect for the people and their ideas, as well as to abolish the profit motive in favour of moral incentives.

David Miller *et al.*, eds, *The Blackwell Encyclopaedia of Political Thought* (Oxford, 1987)

RB

Marbe's law (19th century)*Psychology* Formulated by the German, C Marbe.

The law states that the more common a particular response in word association tasks, the quicker that response is likely to be in comparison to responses which are less common.

J M Mandler and G Mandler, *Thinking: From Association to Gestalt* (New York, 1964)

NS

marginal cost pricing *Economics* This is the pricing of a product so that it covers the cost of producing one extra unit of the product. This pricing method is frequently used in public services and utilities where the aim is to maximize the economic welfare of the state.

In reality, deficits can arise for a firm with declining average total costs (and consequently falling marginal costs) as prices, if set to equal marginal cost, fail to cover fixed costs. *See* AVERAGE COST PRICING and MARK-UP PRICING.

R Rees, *Public Enterprise Economics* (London, 1984); P L Joskow, 'Contributions of the Theory of Marginal Cost Pricing', *Bell Journal of Economics*, 7, 1 (Spring, 1976), 197–206

PH

marginal efficiency of capital (1936) *Economics* Developed by English economist John Maynard Keynes (1883–1946), this term describes the rate of discount which would make the present value of expected income from fixed capital assets equal to the present supply price of the asset.

As investment increases, the rate of returns decreases because early investment was directed at the most lucrative possibilities; subsequent investment is channelled into less promising areas and the returns diminish. *See* CAPITAL ASSET PRICING MODEL, PORTFOLIO SELECTION THEORY, CAPITAL THEORY, ARBITRAGE PRICING THEORY and UNCERTAINTY.

J M Keynes, *The General Theory of Employment, Interest and Money* (New York, 1936)

PH

marginal productivity theory of distribution (1899) *Economics* First formulated by American economist John Bates Clark (1847–1938), this shows how capital or labour will be sought until the marginal revenue from employing either is equal to its marginal cost. The theory deals principally with demand for factors of production and disregards the supply side. *See* EULER'S THEORY and RETURNS TO SCALE.

J B Clark, *The Distribution of Wealth: A Theory of Wages, Interest and Profits* (New York, 1899)

PH

marginal utility theory (1870s) *Economics* Proposed in the late 19th century by the Marginalist group of economists, who used differential calculus to study the impact of small changes in economic quantities, marginal utility refers to the additional satisfaction a consumer derives from the consumption of one extra unit of a product.

Thus, an individual's demand for a product is determined not by the total utility of it but by its marginal utility. Therefore, the greater the supply of a product, the smaller its marginal utility. The Marginalists rejected the LABOUR THEORY OF VALUE which had previously been central to classical economics.

R D Black, A W Coats and C D W Goodwin, *The Marginal Revolution in Economics* (Durham, NC, 1973)

PH

marginal value theorem *Biology/Ecology* Proposed by Charnov, this asserts that a predator will continue to feed in a particular area (a 'patch') until the food supply is so depleted that the predator's food intake drops to a level equal to the average intake rate for the environment as a whole (the environment's 'marginal value'). At this point, the model predicts that the predator will move on to a new, undepleted patch.

KH

Mariotte's law (1676) *Physics* The name sometimes used in continental Europe for BOYLE'S LAW, after the French physicist Edmé Mariotte (1620–84).

GD

mark-up pricing (20th century) *Economics* Developed by Polish economist Michal Kalecki (1899–1970), this is an aspect of AVERAGE COST PRICING in which firms calculate the average cost of a product and add on a mark-up, or profit.

Research conducted in 1939 showed that the mark-up often remains constant irrespective of supply and demand conditions. Mark-up pricing is considered an alternative to MARGINAL COST PRICING, but has been cited as a contributory factor in COST-PUSH INFLATION.

M C Sawyer, *The Economics of Michal Kalecki* (Basingstoke, 1985); R E Hall and C Hitch, 'Price Theory and Business Behaviour', *Oxford Economic Papers*, vol. II (May, 1939), 12–45

PH

markedness (1931) *Linguistics* Originally formulated by a Russian member of the PRAGUE SCHOOL Nikolay S Trubetzkoy (1890–1938), but in more general use from the 1960s onwards.

An unmarked form is neutral, basic, unexpected; its marked version is derived, secondary, deviant. The marked form usually has a more complex linguistic expression. First developed for phonology (for example, in many languages voiced consonants are said to be marked in relation to their voiceless counterparts). Now in more general usage, and illuminating in many areas; for example FEMINIST LINGUISTICS, where study of paired terms such as 'mayor/ess', 'doctor/woman doctor' shows that the female term is regularly marked and therefore coded as deviant.

N S Trubetzkoy, *Principles of Phonology* (Berkeley, 1969)

RF

market socialism (20th century) *Politics/Economics* Advocacy of markets as an element of SOCIALISM.

Markets are effective ways of distributing goods and services, and responding to actual wants. But they are not neutral, and the results they give depend upon the structure of laws and the distribution of wealth within which they operate. Appropriate legal and economic structures can make markets an effective means of achieving the aims of socialists.

David Miller, *Market, State and Community* (Oxford, 1990)

RB

markets, law of *Economics* A recognition of the fact that any market (that is, a medium of exchange for buyers and sellers) operates with basic principles such as supply and demand leading to an equilibrium. Imperfections within the market would in time create monopolies and oligopolies.

(1) Market forces are those pressures generated by buyers and sellers which prompt changes in the price or volume of goods exchanged. (2) Market adjustment is a change in prices or quantities which occurs when supply and demand within the market change. (3) Market clearing is adjustment of supply and demand until an equilibrium is reached. (4) Market equilibrium is a state of rest for a market when the quantity of a good is constant and prices do not rise or fall, with the consequence that there is no incentive for buyers or sellers to change their behaviour. (5) Market failure is a malfunctioning of a market because of imperfection

within it. It can take the form of unemployment, skills shortages, balance of payments disequilibria and unexpected inflation. (6) Market orientation is the tendency for manufacturers to locate their factories close to the market where their goods will be sold. (7) Market share is the proportion of the total sales of a market that a single firm controls. (8) Market power is the ability of a buyer or seller to influence a price. (9) Market structure is the number of firms, buyers and producers in a market. They may be structured in a competitive, oligopolistic or monopolistic manner.

See AGGREGATE DEMAND THEORY, BERTRAND DUOPOLY MODEL, BILATERAL MONOPOLY, COLLUSION THEORY, COURNOT DUOPOLY MODEL, DEMAND THEORY, DUOPOLY THEORY, MONOPOLISTIC COMPETITION, EQUILIBRIUM THEORY, THEORY OF GROWTH OF THE FIRM, ORGANIZATION THEORY and SAY'S LAW.

PH

Markov process *Linguistics See* FINITE STATE GRAMMAR.

Markownikoff's rule (*c.*1869) *Chemistry* Named after the Russian chemist Vladimir Vasilevich Markownikoff (Markovnikov) (1837–1904), this is an empirical rule in organic chemistry relating to the addition reactions of unsymmetrical alkenes.

The rule states that in the addition of the species represented as RH, the negative part of the addendum (R) adds to the carbon atom bonded to the lesser number of hydrogen atoms. For example with $CH_3CH = CH_2$ (propene), of the two products possible, CH_3CHRCH_3 is formed in preference to $CH_3CH_2CH_2R$. It has been shown, however, that peroxy compounds can often reverse the direction of addition. *See also* ZAITSEV'S RULE.

I L Finar, *Organic Chemistry* (London, 1973)

GD

marriage theorem *Mathematics See* KÖNIG'S THEOREM.

Marshall-Lerner principle (1944) *Economics* Named after English political economist Alfred Marshall (1842–1924) and Romanian-born economist Abba Lerner (1905–82), this principle states the conditions under which a change in a country's exchange rate will improve its balance of payments.

In its simplest form, the theory states that the price elasticity of demand for imports and exports must be greater than unity for improvements to be effected in the balance of payments. *See* FUNDAMENTAL DISEQUILIBRIUM, INTERNAL AND EXTERNAL BALANCE.

A P Lerner, *Economics of Control: Principle of Welfare Economics* (New York, 1944)

PH

Marxism *Politics/Economics/Sociology* Theories derived from the work of Karl Marx (1818–83).

The influence of Marx and of Marxism may be judged from the fact that Marxism has been compared, in its enormous variety, to Christianity. Starting points, though not conclusions, for Marxism are an understanding of history as moved by CLASS STRUGGLE; of economic classes as the principal components of society; of politics as derived from clashes of economic INTERESTS; of capitalism as a system which denies fundamental human aspirations.

David Miller *et al.*, eds, *The Blackwell Encyclopaedia of Political Thought* (Oxford, 1987)

RB

Marxist criticism (1930s–) *Art/Literary Theory* Prominent exponents include: Christopher Caudwell (1907–37), Walter Benjamin (1892–1940), Georg Lukács (1885–1971), Théodor W Adorno (1903–69), Raymond Williams (1921–88), Pierre Macherey (1938–), Frederic Jameson, Terry Eagleton.

A number of critical approaches based on the DIALECTICAL MATERIALISM of Karl Marx (1818–83) and Friedrich Engels (1820–95) and other Marxist thinkers, notably Louis Althusser (1918–90). All proceed by relating literature to the political, economic and social circumstances of its production. Soviet and early Western 'vulgar' or 'mechanical' Marxist criticism viewed literature as an expression and documentation of class interest. Another tradition sees literature as uniquely free of the ideology which pervades discourse. Modern Marxist criticism operates with a more sophisticated view. *See also* CULTURAL MATERIALISM, LITERARY MODE OF

PRODUCTION, LITERATURE AS DISCOURSE, MECHANICAL REPRODUCTION.

T Eagleton, *Marxism and Literary Criticism* (London, 1976); M A Rose, *Karl Marx and the Visual Arts* (Cambridge, 1984)

RF

Marxist feminism (20th century) *Politics/Feminism* Development of MARXISM to add gender division to class conflict.

The class analysis of Marxism is inadequate rather than incorrect. It needs to be complemented by an understanding of the divisions, particularly in the household, of work and the control over work along lines of gender.

N O Keohane *et al.*, eds, *Feminist Theory* (Brighton, 1982)

RB

Marxist psychological theory *Psychology* Derived from the work of German social theorist Karl Heinrich Marx (1818–83).

Marx argued that society is in a constant state of change. He felt that individuals are products of their society and of the social forces imposed upon them. In all stratified societies, he argued, there is inherent potential for social conflict, with economic conditions affecting power relationships. Many contemporary psychological theories of social change and conflict have been influenced by Marx's analysis, although direct reference to his work and ideas is uncommon in psychology. *See* FROMM'S THEORY OF PERSONALITY.

R P Appelbaum, *Theories of Social Change* (Chicago, 1970)

NS

mass action law (1864) *Chemistry* Also known as the *Guldberg-Waage law* after the Norwegian chemist and mathematician Cato Maximilian Guldberg (1836–1902) and his brother-in-law Peter Waage (1833–1900). The law states that in a reversible system at constant temperature, the rate of the forward or reverse reaction is proportional to the product of the active masses (concentration) of the reactants or products, respectively.

The product of the active masses on one side of a chemical equation divided by the product of the active masses on the other side is a constant, independent of the amounts of each substance present at the beginning of the reaction.

M Freemantle, *Chemistry in Action* (Basingstoke, 1987)

GD

mass-action, law of *Psychology See* EQUIPOTENTIAL THEORY.

mass extinction *Biology* Also called the 'Great Dying' when referring to events of the late Cretaceous, this term refers to the apparent destruction of large numbers of species as evidenced by the fossil record, especially the mass extinction occurring during the Mesozoic era about 65 million years ago.

Dinosaurs died out at this time, along with 25 per cent of all families, 13 per cent of marine families, 50 per cent of marine genera, and up to 75 per cent of all marine species. Many theories exist to explain mass extinction, including those involving volcanoes and asteroids, but the theories remain controversial. *See also* ALVAREZ THEORY and EXTINCTION.

W Beck and K Liem and G Simpson, *Life: An Introduction to Biology*, 3d edn (New York, 1991)

KH

mass-luminosity law (20th century) *Astronomy* A theoretical relation between the mass and luminosity of a star, derived by the British astronomer Sir Arthur Stanley Eddington (1882–1944), this law is approximately true, at least for main sequence stars.

A simplified statement of the law is that the luminosity of a star is proportional to the cube of its mass. (Luminosity is the candlepower of the star on a scale that takes that of the Sun as unity.)

S Mitton, ed., *The Cambridge Encyclopaedia of Astronomy* (London, 1973)

MS

mass society (20th century) *Politics/Sociology* Theory of modern society. Old hierarchies have been replaced by a society in which everyone is an isolated individual. But because social order is unavoidable, it is created by herding people into organizations and movements led despotically from above.

David Miller *et al.*, eds, *The Blackwell Encyclopaedia of Political Thought* (Oxford, 1987)

RB

massed-spaced theory of learning *Psychology* Proposed by the American psychologist Robert Sessions Woodworth (1869–1962).

This is a general term referring to the body of evidence on the relative advantages of learning material in an intensive or 'massed' fashion, or with time lapses between each learning session. There are advantages and disadvantages to learning-scheduling according to the volume of material to be learned, the nature of the theory, the context in which the material is learned and performed, and individual differences in the learner.

R S Woodworth and H Schlosberg, *Experimental Psychology* (London, 1966)

NS

master race (20th century) *Politics* Version of RACISM associated with the German Nazi Party.

An extreme development of SOCIAL DARWINISM, the theory argued that northern Europeans were superior to other races, and constituted a 'natural' dominant race. The theory was employed to justify mass murder.

J Barzun, *Race: A Study in Superstition* (New York, 1965)

RB

master-slave hypothesis (1967) *Biology* Proposed by geneticist Harold Garnet Callan (1917–) working with his research assistant L Lloyd, this is the hypothesis that the chromosome consists of families of serially repeated genes, one of which is the 'master' while all the others are 'slaves'.

The master gene specifies the sequence of every slave gene by a process called rectification. For this reason, only mutations in the master gene are ever detected (according to the hypothesis). Difficulties in understanding rectification called the master-slave hypothesis into question and encouraged the development of competing theories.

R King, *Handbook of Genetics* (New York, 1976)

KH

material balances principle (20th century) *Economics* Developed in the early planning programmes of the Soviet Union, this principle attempts to balance supply and demand for a given commodity.

If an imbalance occurs, the planning process may require imports of additional raw materials or reductions in the amount of material used by other parts of the economy or organization.

J M Montias, 'Planning with Material Balances in Soviet-type Economies', *American Economic Review*, vol. XLIX (December, 1959), 963–85

PH

materialism *Philosophy* Any theory emphasizing the existence, priority, or value of matter or material objects; though the popular sense of emphasizing the value of material things is uncommon in philosophy.

Usually materialists say that matter alone exists, everything else (notably minds or spirits and their ideas and experiences) being analyzable in terms of matter (a form of REDUCTIONISM; *see also* IDENTITY THEORY OF MIND); or else, more weakly, that though minds and so on may be different from matter, they originated from matter and would not exist without it (a form of EMERGENCE THEORY). A slightly less weak materialism would add that such minds would vanish were matter to vanish (since they still depend on it causally). For a still weaker version *see* IMMATERIALISM. Materialists may also deny the substantive and irreducible existence of abstract objects like properties, numbers, propositions, and so on though this is usually less emphasized.

Modern physics has cast the notion of matter itself into some confusion, though in ways that have not so far greatly affected the above debates; problems concerning it, and in particular its relation to space, go back at least to René Descartes (1596–1650) and indeed to Plato (4th century BC). For dialectical and historical materialism *see* DIALECTIC.

A Quinton, *The Nature of Things* (1973)

ARL

materialist theory of history *Politics See* HISTORICAL MATERIALISM.

materiality *Accountancy* A principle of accounting which maintains that accountancy rules need not be strictly applied to unimportant, or immaterial, amounts of money. A sum is considered immaterial if its exclusion does not alter an outsider's assessment of the firm's accounts.

AV, PH

maternal-control hypothesis (1974) *Biology* Proposed by American zoologist Richard Dale Alexander (1929–), this is the hypothesis that the complex society (eusociality) of ants, bees and wasps (*Hymenoptera*) evolved because the mother insects adopted a reproductive strategy of producing some offspring for the express purpose of promoting the survival of other offspring. These helper offspring were sacrificed so that the favoured offspring might, in the long run, maximize the genetic contribution of their mother. *Compare with* HAPLODIPLOID HYPOTHESIS and MUTUALISTIC HYPOTHESIS.

D Dewsbury, *Comparative Animal Behavior* (New York, 1978)

KH

mathematical law(s) (1860) *Psychology* First used in psychology by the German experimental psychologist Gustav Theodor Fechner (1801–87).

The term refers to any mathematical formalism which describes an aspect of behaviour. The use of mathematics in this way can be traced back to Fechner's *Elements of Psychophysics* (1860), although more recent developments are often associated with the work of C E Shannon. *See* STATISTICAL LEARNING THEORY and STIMULUS SAMPLING THEORY.

R D Luce, R R Bush and F Galunter, eds, *Handbook of Mathematical Psychology* (Chichester, 1963)

NS

Matthiessen's rule (1864) *Physics* Named after its originator, the English physicist Augustus Matthiessen (1831–70), this states that the electrical resistivity of a metal which contains foreign (impurity) atoms in solid solution is nearly always greater than that of the pure metal.

It was first shown by Matthiessen that the increase in the resistivity of a metal due to a small concentration of another metal in solid solution is, in general, independent of temperature.

J Thewlis, ed., *Encyclopaedic Dictionary of Physics* (New York, Oxford and London, 1962)

MS

Maunder minima (1880s) *Astronomy/Climatology/Geology* Between 1645 and 1715 there is thought to have been a dramatic drop in the sun spot activity on the Sun, as recognized by the British solar astronomer Edward Walter Maunder (1851–1928).

The theory was confirmed by the American solar physicist Jack Eddy in 1976 from historical records of naked eye observations, auroral records, and reports of solar eclipses. This sun spot minimum coincides with the *Little Ice Age* and suggests a causative relationship.

M Waldmeier, *The Sunspot Activity in the Years 1610–1960* (Schulthess, 1961)

DT

Maupertuis' principle (1746) *Physics* Named after the French mathematician Pierre Louis Moreau de Maupertuis (1698–1759).

This states that the principle of least action is sufficient to determine the motion of a mechanical system. *See also* LEAST ACTION, PRINCIPLE OF.

maxims of conversation *Linguistics See* CO-OPERATIVE PRINCIPLE OF CONVERSATION.

maximum modulus theorem *Mathematics* Also known as the maximum principle and is a part of COMPLEX FUNCTION THEORY. It states that an analytic function which attains its maximum modulus in some open region must be constant. As a consequence, if f is analytic on a bounded region D (and continuous on the closure of D), then $|f(z)|$ achieves its maximum on the boundary of D. A powerful consequence of the maximum principle is SCHWARZ'S LEMMA.

S G Krantz, 'Functions of One Complex Variable', *Encyclopedia of Physical Science and Technology* (Academic Press, 1987)

MB

maximum value theorem (19th century) *Mathematics* Proved by the German analyst Karl Theodor Wilhelm Weierstrass (1815–

97), this states that a continuous real valued function on a closed bounded interval (or more generally a compact set) achieves its supremum (infimum).

For example, on the interval [0,1], the function x^2-1 achieves its maximum at $x = 1$, but on the open interval (0,1) it does not since for all $x < 1$ there is an $x < x' < 1$ for which $f(x') > f(x)$.

H Anton, *Calculus with Analytic Geometry* (New York, 1980)

ML

Maxwell-Boltzmann distribution law (1871) *Chemistry* Named after the Scottish physicist James Clerk Maxwell (1831–79) and the Austrian physicist Ludwig Boltzmann (1844–1906) who were both responsible for important work on the kinetic theory of gases. The law describes an exponential function giving the statistical distribution of velocities and energies of gas molecules at thermal equilibrium and obeying classical (non-quantum) mechanics.

In the case of indistinguishable particles, Maxwell-Boltzmann statistics reduce to Bose-Einstein statistics for bosons (photons, α-particles and all nuclei with an even mass number) and Fermi-Dirac statistics for fermions (electrons, protons and neutrons).

P W Atkins, *Physical Chemistry* (Oxford, 1990)

GD

Maxwell's demon (1868) *Physics* Named after the Scottish physicist James Clerk Maxwell (1831–79).

Maxwell's demon is an imaginary creature used to illustrate the statistical nature of the second law of thermodynamics. The demon operates a trapdoor between two compartments containing gas in thermal equilibrium, and is sufficiently dexterous to allow only fast molecules to move in one direction between compartments and slow molecules to move only in the opposite direction. In this way, the gas could be separated into hotter and cooler components without expending external work. The fallacy in the argument is shown by the fact that the demon can only determine the speeds of the molecules by interacting with them by means of radiation and would therefore cause other changes. *See also* HEISENBERG'S UNCERTAINTY PRINCIPLE.

JB

Maxwell's equations (1865) *Physics* Named after their formulator, the Scottish physicist James Clerk Maxwell (1831–79), these are a form of the basic equations of electromagnetism to represent correctly the relationship between the electric field E and magnetic induction B in the presence of electric charges and electric currents, with steady or varying amplitudes, in a vacuum or in matter.

The laws of Gauss and Faraday are unchanged, but Ampère's theorem has to be modified to allow for varying electric fields by the introduction of the displacement current. In modern vector form, the equations are

$$\mathrm{div}\, E = \rho/\epsilon_0 \qquad \text{(Gauss)}$$

$$\mathrm{div}\, B = 0 \qquad \text{(Gauss)}$$

$$\mathrm{curl}\, E = -\partial B/\partial t \qquad \text{(Faraday)}$$

$$\mathrm{curl}\, B = \mu_0\, [j + \epsilon_0(\partial E/\partial t)] \qquad \text{(Maxwell/Ampère)}$$

where ρ is the total electric charge density, j is the total current density, μ_0 is the permeability of free space and ϵ_0 is the permittivity of free space.

J Thewlis, ed., *Encyclopaedic Dictionary of Physics* (New York, Oxford and London, 1962)

MS

Maxwell's law of gas viscosity (1866) *Physics* Named after its discoverer, the Scottish physicist James Clerk Maxwell (1831–79), this states that at constant temperature, the coefficient of viscosity of a gas is independent of its density; that is, of its pressure.

The law fails at high pressures, where the mean free path of the molecules is comparable to the molecular diameter, and also at low pressures, where the mean free path is comparable to the dimensions of the container.

J Thewlis, ed., *Encyclopaedic Dictionary of Physics* (New York, Oxford and London, 1962)

MS

Maxwell's rule *Physics* *See* CORKSCREW RULE.

Mayer's hypothesis *Physics* *See* JOULE'S LAW.

McCullough effect *Psychology* Named after the American C McCullough (*fl.*1960s).

The term refers to any psychological phenomenon which follows removal of a stimulus, but in particular to the after image produced by saturating the eye with red and green patterns set at different angles. Usually, black horizontal lines are set against a bright red background and alternated every few seconds with vertical black lines set against a bright green background. Later, on presentation of black and white lines set at different angles, the horizontal lines appear tinged with green and the vertical lines with red.

T N Cornsweet, *Visual Perception* (New York 1970)

NS

McDougall's colour theory *Psychology* Proposed by the Anglo-American psychologist William McDougall (1871–1938).

This early theory of colour vision held that all colours were reducible to three basic or fundamental colours (reds, greens and blues), and that there are in the retina two distinct receptor mechanisms for light: rods for dim light and cones for normal and intense light. *See* DUPLEXITY THEORY OF VISION.

E B Goldstein, *Sensation and Perception* (Belmont, Calif., 1984)

NS

McGregor's x/y theory of leadership (1960) *Psychology* Proposed by the American academic Douglas Murray McGregor (1906–64).

The basic assumptions of theory x are that the typical person has a rational aversion to work and because of this will only work when ordered, threatened or coerced. The average person is therefore indolent, irresponsible, unambitious and inclined to value security above everything else. Theory y proposes that work is as rational as rest or play, and only the conditions make it aversive. Given the appropriate conditions, individuals demonstrate self-direction and control; therefore it should be the goal of an organization to offer the opportunity to achieve this. *See* LEADERSHIP THEORIES.

D McGregor, *The Human Side of Enterprise* (New York, 1960)

NS

McNaughton rule (19th century) *Psychology* Established by the case of the British defendant Daniel McNaughton, and also known as McNaghten, or M'Naghten rules.

This forms a principle for establishing criminal responsibility and state that to establish a defence on the grounds of insanity, it must be proved that the defendant at the time he/she committed the crime, either because of insanity of temporary insanity, did not know what he/she was doing. The issue of criminal insanity is highly controversial and subject to strong social and cultural influences.

J P Dwortzky, *Psychology*, 3rd edn (New York, 1985)

NS

mean, doctrine of the *Philosophy* The doctrine of Aristotle (384–322 BC) that moral virtue can be defined as a disposition concerned with choice and lying in a mean.

Any given virtue lies between two extremes, for example courage lies in a mean between rashness and cowardice. The mean, however, is not an arithmetical mean, but is 'relative to us'; that is, to our natural tendencies. Since we naturally tend more to cowardice than to rashness, the mean is nearer to rashness. (But sometimes Aristotle seems to say only that we should especially avoid our own pet vices.) The doctrine thus risks being vacuous, the mean being whatever point we ought to pursue; this being determined by the moral insight of the trained and practised 'man of practical wisdom'. But Aristotle could say that the doctrine reminds us that there are always two opposite errors which we must avoid.

Whether all virtues can be so classified, without artificiality or triviality, may be disputed, but there is no mean of the mean or extremes: we don't have to avoid excess of virtue, nor (Aristotle's example) can a man commit adultery with the right woman at the right time in the right way.

Aristotle, *Nicomachean Ethics*, book 2, chs 5–9

ARL

mean value theorem *Mathematics* Attributed to the Italian-born French mathematician and physicist Joseph Louis Lagrange (1736–1813).

(1) The theorem that the average value of a continuous function on an interval must be attained by that function; specifically, if a real function is continuous of a closed interval $[a,b]$ and differentiable on the open interval (a,b), then there is at least one point strictly between a and b where the first derivative equals

$$\frac{f(b)-f(a)}{b-a}.$$

Geometrically, this means that on a curve f there must be a place where the tangent line to the curve is parallel to the secant line joining a and b. The generalized mean value theorem states that if the functions f and g are differentiable on (a,b) and continuous on $[a,b]$ and the derivative of g is non-zero in the open interval (a,b), then there is at least one point c in (a,b) where

$$f'(c)[g(b)-g(a)] = g'(c)[f(b)-f(a)].$$

(2) The first mean value theorem for integrals states that if a function is continuous on a closed interval $[a,b]$, then there is at least one number c in $[a,b]$ such that

$$\int_a^b f(x)dx = f(c)(b-a).$$

The generalized mean value theorem extends this result to show that if, in addition, g is non-negative and integrable, then there is some c in (a,b) such that

$$\int_a^b f(t)g(t)dt = f(c)\int_a^b g(t)dt.$$

The second mean value theorem for integrals states that if f is monotone and g is integrable, then there is a value c in $[a,b]$ such that

$$\int_a^b f(t)g(t)dt = f(a)\int_a^c g(t)dt + f(b)\int_c^b g(t)dt.$$

H Anton, *Calculus with Analytic Geometry* (New York, 1980)

ML

meaning, theories of (traditional) *Linguistics/Philosophy* Discussed BC in both Greek philosophy and Indian linguistics. Much theoretical progress in latter half of the 20th century.

An elusive concept which has been theorized from many different perspectives: meaning as use, as behaviour, as intention, as concepts, as images, as truth-conditions, and so on. It is best to disperse the term into several concepts which can each be defined appropriately. *See* SEMANTICS and references; COMPONENTIAL ANALYSIS; and PROTOTYPE.

J D Fodor, *Semantics* (Hassocks, 1977)

RF

meaning-nn (1957) *Linguistics* Formulated by the American philosopher Herbert Paul Grice (1913–).

Distinct from natural meaning (smoke means fire), sentence meaning, word-meaning; meaning-nn (or 'non-natural meaning') is the meaning a speaker *intends* to convey and intends his or her interlocutor to recognize.

H P Grice, 'Meaning', *Philosophical Review*, 66 (1957), 377–88; reprinted in D D Steinberg and L A Jakobovitz, eds, *Semantics* (Cambridge, 1971), 39–48

RF

measure theory *Economics/Mathematics* This is a branch of mathematics dealing with the attribution of measure to subsets of a given set.

In economics, measure theory is useful in analyzing the influence that individuals or groups have on market operations. It also underpins probability theory and the measure of the frequency of phenomena. *See* INFORMATION THEORY and UNCERTAINTY.

P R Halmos, *Measure Theory* (Princeton, N.J., 1961)

PH

measure theory (1902) *Mathematics* Based on a definition by Henri Léon Lebesgue (1875–1941).

A measure is a definition of area and volume which generalizes those of Giuseppe Peano (1887) and Marie-Ennemond Camille Jordan (1892). This leads to a definition of the definite integral which generalizes that of Bernhard Riemann (1854) so that functions which are not integrable according to Riemann are integrable according to Lebesgue but the definitions give the same value if both are defined. Consequently stronger results about functions hold for measure theory.

W W Rogosinski, *Volume and Integral* (Edinburgh, 1952)

JB

mechanical reproduction (1936) *Literary Theory* Pioneering analysis of mass communication by the German Marxist critic W Benjamin (1892–1940).

Works of art which are mechanically reproduced, following the technologies developed from the mid-19th century onwards, become detached from tradition and ritual, lose their 'aura'; accessibility means a new, 'progressive', participation by 'the masses'.

W Benjamin, 'The Work of Art in the Age of Mechanical Reproduction', in *Illuminations*, H Zohn, trans., H Arendt, ed. (London, 1973), 219–53

RF

mechanism *Biology See* REDUCTIONISM.

mechanism *Philosophy* As a theory, rather than a device, the view that everything happens mechanically; that is, everything can ultimately be explained in terms of certain laws of nature which apply to the behaviour of matter in motion, as in the popular example of clockwork.

Ideally the laws should require as few terms as possible – perhaps just solidity, void, and motion (*see also* ATOMISM – and be intuitively clear, preferably by being reducible to logical laws. For traditional atomism it is a logical law that two atoms cannot occupy the same place, so it seems that a moving atom must push another one it hits. However, the pursuit of such an ideally simple mechanical model has proved illusory. Gravity and magnetism, at least, require pulling as well as pushing, and solidity itself has proved to be problematic. The properties and forces appealed to by modern physics have no intuitive appeal, and intuition therefore sees no reason why they should not be added to or replaced indefinitely. However, mechanism can still take the form of an appeal to simplicity and uniformity of explanation (*see also* OCKHAM'S RAZOR, and principle of PARSIMONY) to attack its traditional rivals such as FINALISM, ORGANICISM, VITALISM and EMERGENCE THEORIES with varying success.

ARL

mechanistic philosophy *Biology See* REDUCTIONISM.

mechanistic psychology *Psychology* Exemplified in the work of the German-American biologist Jacques Loeb (1859–1924).

Deriving from the philosophical doctrine of MECHANISM, mechanistic psychology refers to the idea that, although animals (including humans) are complex, they can ultimately be understood in mechanical terms. Hence mechanistic psychology is strongly deterministic and opposed to other philosophical ideas such as DUALISM, IDEALISM, and MENTALISM. More importantly, it reflects the possibility of understanding human entities according to the basic principles of physics.

G Murphy and J K Kovach, *Historical Introduction to Modern Psychology* (New York, 1972)

NS

mediation theory *Psychology* Proposed by the British philosopher and statistician Karl Pearson (1857–1936), this refers to any theoretical approach to learning that assumes there exist mediating processes between the stimulus (S) and the response (R).

The approach is closely associated with the use of statistical techniques to model the presence and effect of such mediating processes. A difficulty for mediation theory is to identify what the intervening process or processes might be. *See* DYNAMIC THEORY.

G Murphy and J K Kovach, *Historical Introduction to Modern Psychology* (New York, 1972)

NS

Meinong's jungle *Philosophy* The Austrian philosopher Alexius von Meinong (1853–1920) thought that since we can apparently refer to things that do not exist (the golden mountain, the prime number between eight and ten, and so on) such things must have some sort of being. This he called '*sosein*', or 'being so'. Meinong's jungle is a nickname for the repository of the multiplicity of things thus given a shadowy half-existence.

J N Findlay, *Meinong's Theory of Objects* (1933)

ARL

meliorism *Philosophy* Doctrine that the universe is becoming progressively and inevitably better.

This may be for religious reasons involving the working out of some grand design, or for reasons connected with late 18th-century optimism concerning inevitable progress and the perfectibility of man, inspired by scientific and technological progress and revolutionary political ideas. In theology it can also refer to the doctrine that God is benevolent but not omnipotent, and that we must therefore co-operate with Him. An anticipation of something like the first view can perhaps be seen in Plato's nephew Speusippus (*c.*409–337 BC) and certain Pythagoreans in the 4th century BC, according to Aristotle (*Metaphysics*, 1072.b.31).

ARL

membrane trigger hypothesis *Biology* The concept that some proteins can take on two conformations: one that is quite stable in water, and another that is triggered (induced) by contact with the hydrophobic cell membrane.

The soluble protein precursor is believed to change in conformation as it inserts itself into the bilayer membrane. This theory provides an alternative to the SIGNAL HYPOTHESIS.

KH

Mendeleyev's periodic law (1869) *Chemistry* Named after the Russian chemist Dmitri Ivanovich Mendeleyev (1834–1907) who was Professor of Chemistry at St Petersburg. Also known as the periodic law and independently formulated by the German chemist Julius Lothar von Meyer (1830–95). The law states that the physical and chemical properties of the elements are periodic functions of the atomic weights.

Mendeleyev's formulation predicted the existence of a number of then unknown elements which were subsequently discovered. However, his and other representations of periodicity fail to place elements 58–71 (lanthanides or rare earths) and 90–103 (actinides) correctly in the underlying geometric matrix. In tribute to his pioneering work, element 101 (mendelevium) was named after him.

D M Considine, *Van Nostrand's Scientific Encyclopedia* (New York, 1989)

GD

Mendelian genetics *Biology See* MENDELISM and MENDEL'S LAWS.

Mendelism (19th century) *Biology* Based on the ideas of Gregor Johann Mendel (1822–84), an Austrian monk considered the father of genetics. Modern Mendelism derives from work by American biologist Thomas Hunt Morgan (1866–1945) and others who identified chromosomes (and is also called PARTICULATE INHERITANCE). Mendelism is the study of heredity based on Mendel's writings; inheritance according to the chromosome theory of heredity. Mendel stated that the reproductive cells of living things contain 'factors' that transmit particular characteristics to offspring. *See also* TELEGONY.

R Olby, *The Origins of Mendelism* (London and Chicago, 1985)

KH

Mendel's laws (19th century) *Biology* Based on the ideas of Gregor Johann Mendel (1822–84), an Austrian monk considered to be the father of genetics. The two laws, which apply only to sexual reproduction, are the law of independent assortment, and the law of segregation.

(1) The law of independent assortment states that factors governing two different characteristics are inherited separately from one another. More formally, the random distribution of alleles to the gametes results from the random orientation of chromosomes during meiosis. Thus an animal may have brown fur, expressing the allele for brown colour at one particular gene locus, and also have short fur, expressing the allele for short fur at some other locus. The offspring of this animal will show an independent assortment of fur colour and length. (Modern genetics has shown, however, that the two loci involved must occur on separate chromosomes, or the law does not apply.)

(2) The law of segregation states that the two members of an allele pair or pair of homologous chromosomes separate during gamete formation, and each gamete receives only one member of the pair. *See also* MENDELISM.

R Olby, *The Origins of Mendelism* (London and Chicago, 1985)

KH

Menelaus's theorem (*c.*AD 100) *Mathematics* Named after Roman mathematician Menelaus of Alexandria, but apparently known to Euclid in the 3rd century BC.

The theorem that a line intersects the sides or their extensions of a triangle ABC at the points A', B', C' if and only if

$$\frac{AC'}{BC'} \cdot \frac{BA'}{CA'} \cdot \frac{CB'}{AB'} = 1 .$$

The theorem can be generalized to the case of a polygon.

M Hazewinkel, ed., *Encyclopaedia of Mathematics* (Dordrecht, 1988)

ML

Menger sponge *Mathematics* A FRACTAL object formed by starting with a cube, dividing it into 27 equal smaller cubes, and removing the centre cube as well as the centre cubes of each face and successively repeating this process for each of the remaining smaller cubes. The limiting object is the Menger sponge. Its *fractal dimension* is log20/log30 = 2.727... and is therefore closer to a three-dimensional solid body than a two-dimensional surface.

I Peterson, *The Mathematical Tourist* (New York, 1988)

ML

mental faculty theory (4th century BC–) *Psychology* This can be traced back to Aristotle (384–322 BC), with more recent examples including the work of the German anatomist Franz Joseph Gall (1758–1828).

Theories of mental faculties are subsumed under the term faculty psychology, and consider the mind to be divided into a number of separate faculties of abilities. Faculty psychology came under attack for reasoning that the mind performs different functions but that such a state does not imply the existence of distinct faculties for those functions. Today, faculty psychology is largely discredited. *See* FACULTY THEORY.

T H Leahey, *A History of Psychology: Main Currents in Psychological Thought* (Englewood Cliffs, N.J., 1980)

NS

mental lexicon (1970s) *Linguistics* Major concept in PSYCHOLINGUISTICS.

Storage in the mind of knowledge about words (*see also* LEXICON); how this knowledge is accessed in processing speech is an important research topic: *see* LOGOGEN, and SEARCH MODEL.

J Aitchison, *Words in the Mind: An Introduction to the Mental Lexicon* (Oxford, 1987)

RF

mental model (1983) *Linguistics* Proposed by the British cognitive scientist Philip Nicholas Johnson-Laird.

A mental representation of entities and their relationships being communicated about. Not all features will be mentioned in language; some are supplied by conventional knowledge.

P N Johnson-Laird, *Mental Models* (Cambridge, 1983)

RF

mentalism *Psychology* Proposed by the Irish empiricist philosopher Bishop George Berkeley (1685–1753).

This doctrine maintains that an adequate explanation of human behaviour is impossible without reference to mental phenomena. Mentalists believe the subject matter of psychology should be the mind, using introspection as the research tool. This use of introspection is now largely of historical significance only, but the arguments posed by mentalism are still popular.

T H Leahey, *A History of Psychology: Main Currents in Psychological Thought* (Englewood Cliffs, N.J., 1980)

NS

mentalism in linguistics (late 1950s) *Linguistics* Reorientation of the philosophical basis of linguistics accompanying the GENERATIVE GRAMMAR of the American linguist Avram Noam Chomsky (1928–).

Chomsky rejected the behaviourist view of language which had dominated American linguistics since the 1920s (*see* BEHAVIOURISM IN LINGUISTICS): language was not a set of learned habits under the control of stimulus-response mechanisms, but a mental faculty with an innate basis. *See also* EMPIRICISM IN LINGUISTICS.

N Chomsky, *Language and Mind*, 2nd edn (New York, 1972)

RF

Mercalli scale (20th century) *Geology* Initiated by the Italian seismologist Giuseppe Mercalli (1850–1914) at the beginning of this century, but refined and extended by the American seismologist Charles Francis Richter (1900–85) in 1956, this is a scale for measuring the intensity of an earthquake on the basis of observed phenomena (for example, chandeliers swinging and chimney pots falling).

This is the only method for evaluating the earthquakes of antiquity, but has been largely replaced today by studying seismic waves to estimate the energy released (magnitude), using the RICHTER SCALE

B Gutenberg and C F Richter, 'Magnitude and Energy of Earthquakes', *Annales Geofisica* vol. IX (1956), 1–15

DT

mercantalism (17th century–) *Politics/ Economics* Theory of the responsibility of the STATE to protect and promote national wealth by encouraging exports and limiting imports.

Since wealth is limited, trade between nations is a ZERO-SUM game, so one country can only benefit at the expense of another.

Mercantilism was advocated by a number of English, French and German writers, many of whom were merchants. Leading Mercantilists included Gerald Malynes (1586–1641), Thomas Mun (1571–1623) and John Locke (1632–1704). Mercantilists welcomed government involvement in economic matters as a means of stimulating the creation of wealth, and favoured such policies as high-import tariffs, prohibition of bullion exports, and exchange control. *See* 'LAISSEZ-FAIRE', PHYSIOCRACY, ECONOMIC LIBERALISM and NEW CLASSICAL MACROECONOMICS.

E F Heckscher, *Mercantilism* (London, 1935); David Miller *et al.*, eds, *The Blackwell Encyclopaedia of Political Thought* (Oxford, 1987)

RB, PH, AV

mere exposure effect *Psychology* First reported by the Polish psychologist Robert Boleslaw Zajonc (1923–).

The principle that repeated exposure to a neutral stimulus increases the attraction for that stimulus. This effect is not restricted to interpersonal attraction, but also affects the evaluation of various visual, verbal and auditory stimuli. Zajonc has also shown that such preferences develop without conscious recognition or discrimination. However, in many situations other factors may have a stronger effect on attitudes or attraction and overshadow the effect of mere exposure.

R B Zajonc, 'Feeling and Thinking: Preferences Lead to Inferences', *American Psychologist*, vol. xxxv (1980), 151–75

NS

mereology *Philosophy* Literally, 'theory of parts'. Term introduced by the Polish logician Stanislaw Leśniewski (1886–1939) to cover a theory which used the whole/part relation as a substitute for the class-membership relation to deal with the structure of classes in ways that would avoid various difficulties connected with the VICIOUS CIRCLE PRINCIPLE and the theory of TYPES. (The term also has a technical use within the theory itself.)

The point about the whole/part relation is that, unlike class-membership, it is transitive; that is if *a* is a part of *b*, and *b* is a part of *c*, then *a* is a part of *c*. The notion has also been used (by N Goodman, *The Structure of Appearance* (1951)) to deal with problems concerning stuffs (like water) or general qualities (like red): 'water' is taken to be a name for the total quantity of water in the universe (so that the Pacific Ocean counts as a *part* of water); similarly 'red', in naming the colour red, names the totality of red things, treated as a single large object split up over space.

P Simons, *Parts* (1987), ch. 1

ARL

meritocracy (1958) *Politics* Mixture of social prediction and description of English sociologist and writer Michael Young (1915–).

Social power, and particularly economic power, will in the future be held by those selected on the basis of measurable merit. In a society nominally egalitarian, these meritocrats will enjoy higher standards of living than their fellows by the manipulations of 'perks'. *See also* ARISTOCRACY.

Michael Young, *The Rise of the Meritocracy* (London, 1958)

RB

Merkel's law *Psychology* Formulated by the German anatomist Friedrich Siegismund Merkel (1845–1919).

The law states that equal sensations above threshold strength correspond to equal stimulus differences; that is, a linear relationship exists between increased stimulation above the threshold necessary to experience sensation. Merkel's law is now largely discredited, FECHNER'S LAW and WEBER'S LAW being much more widely accepted.

S S Stevens, *Psychophysics: Introduction to its Perceptual, Neural and Social Aspects* (New York, 1975)

NS

Mersenne primes (1644) *Mathematics* Named after the French monk, theologian, philosopher and mathematician Marin Mersenne (1588–1648) who studied them, these are prime numbers of the form $2^n - 1$, where n is necessarily prime. It is not known whether there are infinitely many of them.

Mersenne primes are often used to provide examples of large prime numbers; for instance, it is known that $2^{44497} - 1$ is the 27th Mersenne prime, a number with 13,395 digits. The largest known prime is the Mersenne prime $2^{756839} - 1$, a number with 227,832 digits. *See also* PERFECT NUMBER.

A Baker, *A Concise Introduction to the Theory of Numbers* (Cambridge, 1984)

MB

Mersenne's laws *Physics* Named after their discoverer, the French mathematician Marin Mersenne (1588–1648), these laws refer to the frequency of vibrations.

(1) For a given uniform string and given stretching force, the fundamental frequency of vibration varies inversely as the length of the string. (2) For a uniform string of given length and material, the fundamental frequency of vibration varies as the square of the stretching force. (3) The fundamental frequency of vibration of strings of the same length, subjected to the same stretching force, varies inversely as the square root of the mass per unit length of the string.

J Thewlis, ed., *Encyclopaedic Dictionary of Physics* (New York, Oxford and London, 1962)

MS

Merten's theorem *Mathematics* The theorem that the Cauchy product series

$$c_n = \sum\nolimits_{k=0}^{n} a_{n-k} b_k$$

converges if one factor is absolutely convergent while the other factor is just convergent.

E Hille, *Analytic Function Theory* (New York, 1982)

ML

meson theory of nuclear forces (1935) *Physics* A theory proposed by the Japanese physicist Hideki Yukawa (1907–81).

The theory explains the short-range (10^{-13} cm) nucleon-nucleon forces in terms of the exchange of a particle between them. Yukawa predicted the particle (meson) which was subsequently discovered. The π-meson (also called the nuclear force meson) is the one principally involved in nucleon-nucleon forces. They are the free-quanta counterparts in the meson field to photons in the electromagnetic field.

Paul Davies, ed., *The New Physics* (Cambridge, 1989)

GD

metafiction (1970s) *Literary Theory* Term originated with the American novelist W Gass (1924–).

Novels and stories which self-consciously proclaim their fictional nature by devices emphasizing artifice, interfering with the illusion of realism. Fiction about fiction. Early instances include Sterne's *Tristram Shandy* (1759–67) and Cervantes' *Don Quixote*; the mode became extremely popular in America in the 1960s. *See also* NOUVEAU ROMAN, POSTMODERNISM.

P Waugh, *Metafiction* (London, 1984)

RF

metagenesis *Biology See* ALTERNATION OF GENERATIONS.

metalanguage (1943) *Linguistics/Philosophy* Standard distinction, applied to linguistics by Danish linguist Louis Hjelmslev (1899–

1965); *see also* GLOSSEMATICS. Also discussed by Roman Jakobson (1896–1982) – *see* POETIC PRINCIPLE for reference.

Language about language: metalanguage is a system of notation, descriptive terms, and so on, for an 'object language'. Metalanguage may be related to natural language – terms like 'passive', 'auxiliary' – or an abstract notation as in symbolic logic.

J Lyons, *Semantics* (Cambridge, 1977)

RF

metalinguistics *Linguistics See* MACROLINGUISTICS.

metaphor (*c*.330 BC) *Literary Theory/ Linguistics* From the Greek words *meta pherein*, meaning 'carry over', 'transfer'. In the Western tradition, theory of metaphor originates with the Greek philosopher Aristotle (384–322 BC). Traditionally very important in discussions of literature, and a central problem in modern theories of language.

Reference to one object in terms of another, so that features of the second are transferred to the primary referent: 'The oat was merry in the wind' (Dylan Thomas). Metaphor is not merely a decorative touch in poetry; it has been claimed to be a central process by which humans construct the world through language. There are many unresolved theories: metaphor as comparison, as substitution, as interaction. *See* TENOR, VEHICLE AND GROUND for an example. *See also* METAPHOR AND METONYMY.

T Hawkes, *Metaphor* (London, 1972); A Ortony, ed., *Metaphor and Thought* (Cambridge, 1989)

RF

metaphor and metonymy (1956) *Linguistics* Theory of the expatriate Russian linguist Roman Jakobson (1896–1982).

Two terms are appropriated from RHETORIC to explain two types of APHASIA: 'contiguity disorder' (no syntax) produces metaphor-like utterances; 'similarity disorder' is characterized by relationships of continuity (metonymy) only. Jakobson generalizes, connecting metaphor and metonymy with PARADIGM AND SYNTAGM. This was the basis for his POETIC PRINCIPLE.

R Jakobson and M Halle, *Fundamentals of Language* (The Hague, 1956)

RF

metaphysical poets (1692) *Literary Theory* Term coined by John Dryden (1631–1700); established by Dr Samuel Johnson (1709–84) in his *Lives of the English Poets* (1779). The designation was pejorative until the poets concerned were rehabilitated in the early 20th century.

A group of 17th century poets – notably John Donne (*c*.1571–1631), also Andrew Marvell (1621–78) , George Herbert (1593–1633), Henry Vaughan (1622–95), and Richard Crashaw (*c*.1612–49) – whose verse was marked by far-fetched comparison (CONCEIT), abstruse scientific reference, irony, dramatic speech rhythms. 'Metaphysical' refers to the coupling of intellection and passion, much admired by Thomas Stearns Eliot (1880–1965). (*See* DISSOCIATION OF SENSIBILITY, MODERNISM and NEW CRITICISM).

H J C Grierson, ed., *Metaphysical Lyrics and Poems of the Seventeenth Century* (Oxford, 1921)

RF

Method, the (1940s) *Literary Theory* System of acting, and of training actors, derived from the views and practices of the Russian actor and director Konstantin Sergeivich Stanislavski (1863–1938).

The Method stresses genuineness and sincerity in performance, and condemns artificiality and mannerism. (In this respect it is the reverse of Brecht's ALIENATION EFFECT.) The actor is trained to live a role, drawing on the resources of his or her own experience. Method training is particularly associated with the Actors' Studio, founded in New York in 1947 by the American actor and director Elia Kazan (1909–).

K S Stanislavski, *An Actor Prepares*, E R Hapgood, trans. (London, 1937)

RF

methodological theories *Philosophy* The term 'methodological' is prefixed to terms – such as BEHAVIOURISM, HOLISM, INDIVIDUALISM, SCEPTICISM and SOLIPSISM – to indicate that the doctrine in question is being taken to prescribe a certain method rather than to make a substantive claim about reality. This is irrespective of whether or not the prescription is based on such a substantive claim (in the case of holism and individualism it usually is; but in that of behaviourism, not necessarily).

Methodological holism and methodological individualism form an important contrast pair in the philosophy of the social sciences, the former seeking explanations in terms of social wholes or structures and the latter seeking them ultimately in facts about individuals.

A Ryan, *The Philosophy of the Social Sciences* (1970), ch. 8

ARL

metric density theorem (20th century) *Mathematics* A fundamental theorem concerning integration (*see also* MEASURE THEORY), this theorem determines whether a set of points is measurable and therefore provides a necessary and sufficient condition for the existence of integrals in the sense of Lebesgue.

JB

metrical theory (1956) *Stylistics* Pioneer application of linguistics to literature; in 1956 a special edition of the literary journal *Kenyon Review* was devoted to 'English verse and what it sounds like'.

Metrics or prosody is the description of the patterns of extra linguistic regularity – various kinds of pattern – which make verse metrical. Early linguistic studies of metre employed the SUPRASEGMENTAL PHONOLOGY of George L Trager and H L Smith. GENERATIVE PHONOLOGY stimulated much work in the 1970s, principally by the American linguists Morris Halle and Samuel J Keyser (1935–), with rules mapping an abstract underlying metrical pattern to the words and syllables.

P Kiparsky and G Youmans, eds *Phonetics and Phonology* I: Rhythm and Meter (San Diego, 1989)

RF

Meyer's law of esterification (*c*.1870) *Chemistry* Named after the German chemist Viktor Meyer (1848–97) who was Professor of Chemistry at a number of universities including Heidelberg before he committed suicide. The law states that the presence of 1- and 2-*ortho* ring substituents on aromatic acids hinders and prevents esterification, respectively.

GD

miasmatic theory of contagion (5th–4th century BC) *Biology* Based on the Hippocratic notion of balance between humankind and the environment, this was the belief that epidemic outbreaks of infectious diseases are induced by a pathological state of the atmosphere.

This atmospheric corruption was believed to be caused by an unfavourable conjunction of weather conditions and local circumstances. By the Middle Ages, air was thought polluted by miasmas (from the Greek word for stain), which were disease-generating emissions from putrid organic matter and stagnant water. By the 1860s, the idea that diseases are communicable had taken hold, and theories such as this one were abandoned.

W Spink, *Infectious Diseases: Prevention and Treatment in the Nineteenth and Twentieth Centuries* (Minneapolis, Minn., 1979)

KH

Michael's continuous selection theorem (1956) *Mathematics* In non-linear FUNCTIONAL ANALYSIS one frequently considers set-valued functions F between topological spaces X and Y (so the values $F(x)$ are subsets of Y). We say F is *lower semicontinuous* if the *inverse image* $\{x|\ F(x) \cap B \neq \emptyset\}$ is open whenever B is open. In this case, if Y is a Banach space and X is, for example, either compact or a metric space while F has closed convex images then F has a *continuous selection* – a continuous single-valued function f from X to Y with the value $f(x)$ belonging to the image $F(x)$ for each x. This result has been applied widely throughout analysis, in particular in the study of fixed points.

AL

microevolution (1940) *Biology* Term coined by Richard Goldschmidt (1878–1958), a German-born geneticist best known for his theory of HOPEFUL MONSTERS. It refers to evolution characterized by potentially reversible gene frequency changes within a population. It refers also to observable changes that occur within the confines of a species (such as geographic races), or small changes observed over a relatively short timespan (a few generations). *See also* EFFECT HYPOTHESIS. *Compare with* EVOLUTION and MACROEVOLUTION.

M Barbieri, *The Semantic Theory of Evolution* (Chur, Switzerland, 1985)

KH

microlinguistics *Linguistics* *See* MACROLINGUISTICS.

micromolar theory *Psychology* Proposed by the American psychologist Frank Anderson Logan (1924–).

The theory states that quantitative dimensions of a response are defining properties and hence are learned. Therefore it is the person's reaction while learning which effects learning of a stimulus. Logan's theory is an assimilation of several others into a hybrid system. *See* CORRESPONDENCE THEORY/LAW.

F A Logan, 'A Micromolar Approach to Behaviour Theory', *Psychological Review*, vol. LXIII (1956), 63–73

NS

microscopic reversibility, principle of *Chemistry* Also known as the reversibility principle, the principle of detailed balancing and the principle of entire equilibrium; this states that in a system at equilibrium, all molecular processes and their reverse proceed at the same average rate.

Compendium of Chemical Terminology: IUPAC Recommendations (Oxford, 1987)

GD

Milankovitch hypothesis (1930) *Astronomy/Climatology/Geology* Proposed by the Yugoslavian geophysicist M Milankovitch (1879–1958), this theory evaluated the change in solar radiation which would be received by the Earth as a consequence of its own and the Sun's changing orbits.

These changes are primarily cycles of *c.*21,000 years (precession of the Spring point), *c.*41,000 years (tilt of the Earth's axis) and *c.*97,000 years (change in eccentricity of the Earth's path). Milankovitch's model is now widely accepted as accounting for climatic change during the last few hundred thousand years, and it may explain the Ice Ages of the last 2 million years. There is also increasing evidence for the effects of such cyclicity in the remote geological past.

N-A Mörner and W Karlen, *Climatic Changes on a Yearly to Millennial Basis* (Reidel, 1984)

DT

milieu intérieur (1854) *Biology/Medicine* Coined by Claude Bernard (1813–78), an eminent French physiologist regarded as the founder of experimental medicine, this term refers to the concept that higher animals live in a stable, internal environment of circulating organic liquid which bathes and nourishes the tissues; and that this environment provides independence and freedom of existence by regulating temperature, oxygen supply, chemical reserves and all other conditions necessary for life. Although a simple concept, the idea of *milieu intérieur* has had a profound influence on the development of physiological thinking.

J Fulton, *Selected Readings in the History of Physiology* (Springfield, Ill., 1966)

KH

Mill's canons (1834) *Psychology* Described in *A System of Logic*, written by the English philosopher John Stuart Mill (1806–73).

Mill states five canons prescribing methods of investigating and uncovering causal laws and connections. His logic anticipated the work of Wilhelm Wundt (1832–1920) and is fundamental to GESTALT THEORY. It is still regarded as a classic work on scientific method.

D N Robinson, *An Intellectual History of Psychology* (London, 1976)

NS

millenarianism *Politics* Theory of apocalyptic transformation based on prophecy contained in *Revelations*.

Millenarian theories share an expectation of some transforming change: either the coming of the Kingdom of God, or the just society, or harmony amongst peoples. They differ in almost every other respect.

Norman Cohn, *The Pursuit of the Millennium* (London, 1970)

RB

mimesis (*c.*373 BC) *Literary Theory* Originally applied to literature by the Greek philosopher Plato (*c*.427–*c*.347 BC), and developed by Aristotle (384–322 BC).

The Greek term *mimesis* is preferred to its English translation 'imitation' (which has several senses in literary theory) for conveying the essentially Aristotelian idea of art as creating an impression of a represented world. *See also* FICTION and IMITATION.

E Auerbach, W R Trask, trans., *Mimesis* (Princeton, 1953)

RF

mimicry (19th century) *Biology* Idea pioneered by the British naturalist Henry Walter Bates (1825–92), this term refers to a resemblance of two organisms which provides one or both of them protection from predators. *See also* BATESIAN MIMICRY and MÜLLERIAN MIMICRY.

P B Medawar and J S Medawar, *Aristotle to Zoos: A Philosophical Dictionary of Biology* (Oxford, 1983)

KH

mind-style (1977) *Stylistics* Term coined and defined by the British linguist Richard Fowler (1938–).

In literary fictions, a character may be experienced as having a particular kind of mental set towards the world. Fowler called this 'mind-style', and showed how the linguistic structures typically chosen for a character's thought and speech give the illusion of a distinctive style of thought and values.

R Fowler, *Linguistics and the Novel* (London, 1977)

RF

mind-twist hypothesis (1827) *Psychology* Proposed by W Cullen, this is an outdated hypothesis proposing mental disorders to be functional rather than organic.

For example, there is no known organic pathology responsible for such symptoms as hysteria, hypochondriasis, insanity or malingering.

C Bass, ed., *Somatization: Physical Symptoms and psychological illnesses* (Oxford, 1990)

NS

mineral theory (1966) *Biology* Also called the clay theory of evolution, this postulates that life evolved through NATURAL SELECTION from inorganic crystals, such as those found in clay. It is associated with John Bernal and especially with A G Cairns-Smith, who clearly stated the theory in 1966.

Cairns-Smith felt that the universal failure of experiments designed to support the evolution of nucleic acids as precursors to life provided support for his theory, which assumes that the earliest life forms had a simpler, solid-state biochemistry. Direct evidence for the mineral theory remains elusive, however. *Compare with* GENOTYPE THEORY, PHENOTYPE THEORY, PRECELLULAR EVOLUTION, RIBOTYPE THEORY and VIRAL THEORY. *See also* ORIGINS OF LIFE.

J Bernal, *The Physical Basis of Life* (1951); A G Cairns-Smith, 'The Origin of Life and the Nature of the Primitive Gene,' *Journal of Theoretical Biology* vol. x (1966), 53–88

KH

minimalism (1950–) *Art* An artistic trend, mainly in the USA, repudiating ACTION PAINTING and ABSTRACT EXPRESSIONISM. Minimalists use the minimum of means; favour neutral, blank canvases or mechanically produced unexpressive forms; and are opposed to all forms of ILLUSIONISM.

A variety of individuals and styles can be termed Minimalist, from certain works by Andy Warhol (1928–87) and Claes Oldenburg (1929–) to the Luminist and KINETIC ART of Julio Le Parc (1938–) and the Zero group; from the undifferentiated canvases of Yves Klein (1928–62) to the sculptures of Carl André (1935–), Robert Morris (1931–), Tony Smith and Donald Judd.

G Battcock, *Minimal Art. A Critical Anthology* (London, 1968)

LL

minimalism (1960s) *Music* A term derived from the use of minimal means to generate music. Often associated with the American composers Philip Glass (1937–) and Steve Reich (1936–), minimalist compositions are sometimes known as systems music.

The main element in work of this kind is repetition. A melodic and/or rhythmic motif is repeated with slight modifications (extension, contraction) over time; or the same motif may be repeated by two performers playing simultaneously at first, then at different speeds, phasing in and out of synchrony. Sometimes it is a unit of time that is repeated, in which a 'sound pattern' is gradually built up.

M Nyman, *Experimental Music: Cage and Beyond* (London, 1974)

MLD

minimax theorem *Mathematics* Any result giving conditions on a function F ensuring

$$\min_X \max_Y F(x,y) = \max_Y min_X F(x,y) .$$

For example, the von Neumann minimax theorem (1928) asserts this when X and Y are polyhedra and $F(x,y) = xAy$, where A is an $m \times n$ matrix. This may be derived either from the DUALITY THEORY OF LINEAR PROGRAMMING or the FENCHEL DUALITY THEOREM.

In GAME THEORY, X and Y represent two players' possible choices of strategy and F is the pay-off function. The common value is then the value of the game. The minimax principle asserts that either player should aim for a (possibly randomized) minimax strategy attaining this value.

AL

minimum, law of the *Biology See* LIEBIG'S LAW.

minimum structural change, principle of *Chemistry* According to this over-simplified hypothesis, either chemical species do not isomerize in the course of a transformation (that is, a substitution reaction), or the change of a functional group of a chemical species into a different functional group does not involve the making or breaking of more than the minimum number of bonds required to effect that transformation. For example, any new substituents should enter the precise positions previously occupied by the displaced group. The term molecular rearrangement is traditionally applied to any reaction that violates this principle.

Compendium of Chemical Terminology: IUPAC Recommendations (Oxford, 1987)

MF

Minkowski's inequality (late 19th century) *Mathematics* Named after the Russian-born, Swiss-German mathematician Hermann Minkowski (1864–1909), this term is applied to several generalizations of the triangle inequality to multiple sums and to other norms such as the linear normed space L^p where $1 \leqslant p < \infty$.

G B Folland, *Real Analysis*, (New York, 1984)

ML

Minkowski's theorem on convex bodies (1896) *Mathematics* Named after the Russian-born, Swiss–German mathematician Hermann Minkowski (1864–1909) who initiated the branch of NUMBER THEORY known as the GEOMETRY OF NUMBERS.

This theorem is the result whereby a convex body in Euclidean n-space which is symmetric about the origin and has volume exceeding 2^n contains an integer lattice point (a point all of whose co-ordinates are integers) other than the origin. (A convex body is a bounded open set of points in Euclidean space which contains $\lambda x + (1 - \lambda)y$ for all λ with $0 < \lambda < 1$ whenever it contains x and y.) *See also* DIOPHANTINE ANALYSIS.

A Baker, *A Concise Introduction to the Theory of Numbers* (Cambridge, 1984)

MB

mise en abyme (1948) *Literary Theory* The term originated with French novelist André Gide (1869–1951) and is exemplified in his novel *The Counterfeiters* (1926). Prominent in POSTSTRUCTURALISM during the 1970s.

'Placed in the abyss': the infinite regress of mirrors. In narrative, a story within a story, the internal story mirroring and therefore commenting on the framing story. Also in DECONSTRUCTION, the infinite deferral of meanings.

S Rimmon-Kenan, *Narrative Fiction: Contemporary Poetics* (London, 1983)

RF

Mitscherlich's law of isomorphism (*c*.1820) *Chemistry* Named after the German chemist Eilhard Mitscherlich (1794–1863) who was Professor of Chemistry at Berlin from 1822, this law states that substances which have similar chemical formulae will have similar chemical properties and crystalline forms (for example, sodium dihydrogenphosphate, $NaH_2PO_4.H_2O$, and sodium dihydrogenarsenate, $NaH_2AsO_4.H_2O$).

The law also implies that isomorphous substances have similar spectra. Mitscherlich was also the first to recognize the phenomena of dimorphism and polymorphism where, for example, the element sulphur can exist in two or more crystalline forms. The law is not rigidly correct.

GD

mixed cerebral dominance theory *Psychology* Proposed by the English neurologist John Hughlings Jackson (1835–1911).

The theory that speech disorders and some other maladjustments may be due wholly or in part to the fact that one cerebral hemisphere does not consistently lead the other in control of bodily movement. The theory has been superseded by more comprehensive and sophisticated accounts of the functional neuropsychology of the brain. *See* MULTIPLE CONTROL THEORY.

K N Walsh, *Neuropsychology* (London, 1978)

NS

Möbius strip *Mathematics* Named after German mathematician and theoretical astronomer August Ferdinand Möbius (1790–1868), the Möbius strip was presented in a paper discovered after his death. The 19th-century German mathematician Johann Benedict Listing discovered it independently.

It is an object having only one side and one edge formed by a strip connected at the ends with a half twist in the middle. It was developed to illustrate the properties of one-sided surfaces. *See also* KLEIN BOTTLE.

M Hazewinkel, ed., *Encyclopaedia of Mathematics* (Dordrecht, 1988)

ML

modal realism *Philosophy* Term used for the theory (going back to Gottfried Wilhelm Leibniz (1646–1716)) of 'possible worlds'; used to analyze necessity and possibility and similar notions, which are known as modal notions.

The actual world is regarded as merely one among an infinite set of logically possible worlds, some nearer to the actual world and some more remote. A statement is called necessary if it is true in all possible worlds, and possible if it is true in at least one. Possible worlds are usually regarded as real but abstract possibilities. However, for D K Lewis they are concrete worlds, like this one only without any spatial or temporal connections with it. *See also* COUNTERPART THEORY.

D K Lewis, *On the Plurality of Worlds* (1986)

ARL

modality (traditional; 1970s) *Linguistics* Ideas in this area have sources in TRADITIONAL GRAMMAR, modal logic and FUNCTIONAL GRAMMAR.

The speaker's attitude to his/her statement and to the addressee, most clearly expressed in the 'modal auxiliaries' *can*, *may*, *will*, *should*, *must*, and so on, and in adverbs and adjectives such as *possible/ly*, *necessary/ily*. More abstractly, systems of meanings in the areas of evaluation, possibility, permission and obligation.

M A K Halliday, *Introduction to Function Grammar* (London, 1985); J Allwood and L-S Andersson and Ö Dahl, *Logic in Linguistics* (Cambridge, 1977), ch. 7

RF

modality (4th century BC–) *Music* Derived from the Greek word for 'measure', this term describes the ordering of music which evolved in the Western world after Pythagorus (*fl.*6th century BC) had devised a pitched scale, based on scientific proportions.

This music has a fixed scale of pitches determined by its internal relationship of tones and semitones, usually with one note that functions as a tonic. The seven transposable modes favoured by the Greeks ascended from successive notes of Pythagorus's scale (roughly equivalent to the 'white notes' on a modern piano). By the 17th century only two modes remained in common use (on C and A); these being the forerunners of the major/minor key system. See TONALITY.

MLD

model theory (20th century) *Mathematics* The branch of logic which studies the semantic (rather than the syntactic) properties of formal theories. In particular, it is concerned with the concepts of truth, satisfaction, and validity, rather than the intrinsic property of formal deduction (*compare with* proof theory). *See also* GÖDEL'S COMPLETENESS THEOREM and LOWENHEIM-SKOLEM THEOREM.

Encyclopedic Dictionary of Mathematics (MIT Press, 1987)

MB

modern synthesis *Biology See* NEO-DARWINISM.

modern theory of urine secretion *Biology See* CUSHNY'S THEORY.

modernism (*c.*1900–30) *Literary Theory/Art* Artistic revolution affecting all the arts throughout Europe. Modernism is generally seen as an artistic response to a range of philosophical and social changes which undermined the securities on which 19th century literature was founded, including the impact of: writings by Charles Darwin (1809–92), Karl Marx (1818–83) and Sigmund Freud (1856–1939); urbanisation; cultural alienation; world war. (*See also* DARWINISM, MARXISM and FREUDIAN THEORY.)

In literature, key figures are the American poets Ezra Pound (1885–1972) (*Cantos*, from 1925) and Thomas Stearns Eliot (1888–1965) (*The Waste Land*, 1922) and the Irish writer James Joyce (1882–1941) (Ulysses, 1922). Pound called upon poets to 'make it new'. Modernism is marked by self-consciousness and by a plethora of technical innovations designed to disrupt the 'common sense' of traditional discourse, for example: ALIENATION EFFECT (which unsettles the relationship between audience and play); CONCRETE POETRY (which heightens the materiality of the sign, a modernist strategy with many expressions); IMAGISM (direct expression of ideas without the blurring of SYMBOLISM); STREAM OF CONSCIOUSNESS (new recognition of the free form of thought). Modernism was contemporaneous with the STRUCTURALISM of Ferdinand de Saussure (1857–1913) and with the formulation of DEAUTOMATIZATION by the Russian formalists (*see* FORMALISM, RUSSIAN).

Modernist architecture is usually associated with BAUHAUS, which was fundamental in the development of an international style (International Modernism) in which simplicity, functionalism and the use of new materials were paramount. In painting, a number of movements (such as CUBISM) can be grouped under this general term, but it more accurately refers to early 20th-century art as a whole.

M Bradbury and J McFarlane, eds *Modernism* (Harmondsworth, 1976); R H Wilenski, *The Modern Movement in Art* (1927)

AB, RF

modernization (20th century) *Politics* Theory of rational progress.

'Modern' societies are characterized by the rational use of scientific techniques, and by the application of reason to meet the common INTERESTS of all. Critics have argued that there are neither common problems nor common solutions, and that we have already entered a world better understood by POSTMODERNISM.

Allan Bullock, Oliver Stallybrass, and Stephen Trombley, eds, *The Fontana Dictionary of Modern Thought*, 2nd edn (London, 1988)

RB

mode *Linguistics See* REGISTER.

Modigliani-Miller theory of the cost of capital (1961) *Economics* Named after Italian-born American economist Franco Modigliani (1918–) and American economist Merton Miller (1923–), this theory states that the overall cost of capital remains constant as the financial gearing of a firm increases. Critics have suggested that the theory ignores the risk of bankruptcy as a firm's debt increases.

M H Miller and F Modigliani, 'Dividend Policy, Growth, and the Valuation of Shares', *Journal of Business*, vol. XXXIV (October, 1961), 235–64

PH

modularity (1970s) *Linguistics* Psychological framework for GENERATIVE GRAMMAR.

Avram Noam Chomsky (1928–) claims that language is a separate 'mental organ', not dependent on other mental abilities. Recent generativists see the mind as comprising a number of distinct but interacting 'modules', principally linguistic, cognitive, and perceptual. The term 'module' is sometimes used for the components within GOVERNMENT AND BINDING theory.

J A Fodor, *The Modularity of Mind* (Cambridge, Mass., 1983)

RF

modus ponens *Mathematics* Also known as *modus ponendo ponens*, or the rule of detachment, this is the rule of inference in logic which enables one to deduce the formula B from the formulas $A \rightarrow B$ and A. For example, given the statements

if it's raining this must be London

and

it is raining

one can validly detach the consequent of the conditional statement to infer

this must be London.

J E Rubin, *Mathematical Logic: Applications and Theory* (Saunders, 1990)

MB

modus tollens *Mathematics* Also known as *modus tollendo tollens*, this is the rule of inference in logic which enables one to deduce the formula $-A$ from the formulae $A \rightarrow B$ and $-B$. For example, given the statements

if it's raining this must be London

and

this is not London

we can infer

it can't be raining.

J E Rubin, *Mathematical Logic: Applications and Theory* (Saunders, 1990)

MB

Mohr-Coulomb theory of failure (1773) *Physics* Named after its originators, the French physicist Charles Augustin de Coulomb (1736–1806) and the German engineer Christian Otto Mohr (1835–1918), this theory explains the process of material rupture.

Coulomb, in 1773, suggested that failure of a material occurs when the maximum shear stress in the material is equal to some definite value, known as the shear strength of the material. A modification by Mohr assumes that when failure occurs, the normal and shear stresses across a plane in the material are connected by some functional relation. In three dimensions this leads to the result that failure always takes place on planes passing through the direction of the intermediate principal stress. This result is not always consistent with experiment.

J Thewlis, ed., *Encyclopaedic Dictionary of Physics* (New York, Oxford and London, 1962)

MS

Mohs's scale (1812) *Geology* Originated by the Austrian mineralogist Friedrich Mohs (1773–1839), this is a scale for evaluating the hardness of minerals.

This scale can be approximated by the scratch test of how well a mineral can be marked by a penknife, but can also be determined by precise impact methods. It remains the standard reference for hardness.

DT

molecular clock (1967) *Biology* Devised by Vincent Sarich and Allan Wilson, this is the notion that the rate of accumulation of genetic difference between two species is approximately proportional to the evolutionary time that has passed since the species diverged.

Using molecular clock calculations based on microcomplement fixation, Sarich and Wilson concluded that humans, chimpanzees and gorillas diverged from one another only 4–5 million years ago, instead of 14 million years as believed in the 1960s. Additional studies suggest that the divergence occurred 7 or 8 million years ago, lending support to this theory; although the Japanese geneticist Motoo Kimura (1924–) disagreed, saying that the notion was incompatible with NATURAL SELECTION.

E Staski, *Evolutionary Anthropology* (Fort Worth, Tex., 1992)

KH

molecular orbital theory *Chemistry* A mathematical method for describing the distribution of electrons in molecules as opposed to individual atoms.

In the simplest approach, the molecular orbital or wave function is treated as a linear combination of atomic orbitals. For example, the wave function for the hydrogen molecule-ion, H_2^+, which contains a single electron, is composed of equal contributions from the atomic orbitals of the two hydrogen atoms. Molecular orbitals are like atomic orbitals but are considered to be 'smeared out' over the whole region occupied by the component atoms of the molecule. There are three types of molecular orbitals: bonding, anti-bonding and non-bonding orbitals. *See also* VALENCE BOND THEORY.

P W Atkins, *Physical Chemistry* (Oxford, 1990)

GD

monetarism *Economics* A revival of the QUANTITY THEORY OF MONEY, monetarism asserts that increases in the money supply cause inflation ('too much money chasing too few goods').

Monetarism emerged as an important economic doctrine under the influence of American economist Milton Friedman (1912–92). Its adherents challenged the Keynsian approach to macroeconomics, emphasizing the importance of monetary policy in stabilizing the economy. 'LAISSEZ-FAIRE' is generally associated with monetarists, who avoid active manipulation of the economy by government. Believing that the money supply is the major determinant of nominal GNP, they tend to believe that fluctuations in the economy result from erratic growth in the money supply. *See* THATCHERISM, REAGANOMICS, and SUPPLY-SIDE ECONOMICS.

PH

monism *Philosophy* Any view claiming to find unity in a certain sphere where it might not have been expected.

The main forms of monism have been: a strong form, claiming that there is only one object (ELEATICISM, Baruch de Spinoza (1632–77), Georg Wilhelm Friedrich Hegel (1770–1831)); and a weaker form, claiming that there is only one kind of object, and in particular that matter and mind are not two independent kinds of thing (MATERIALISM, IDENTITY THEORY OF MIND, NEUTRAL MONISM, and SUBJECTIVE IDEALISM in which matter is rejected, though minds and ideas are considered different).

Other forms of monism may say, for example, that things are related together or unified in their being governed by a simple law or principle (Heraclitus (*fl.*500 BC); or that there is only one proper kind of explanation of COVERING LAW MODEL; or one basic ground for our duties (for example UTILITARIANISM). *See also* ANOMALOUS MONISM.

ARL

monism *Psychology* Advanced by the German experimental psychologist Gustav Theodor Fechner (1801–87) and the Scottish philosopher Alexander Bain (1818–1903).

Usually in reference to the mind-body problem, monists postulate that the human organism consists of a single unified identity and that empiricism is the methodological approach that should be used when studying the mind. Hence, they are reductionists in the sense of viewing or equating the mind to the activity of the brain and the nervous system. At this stage in the development of brain science, theories such as monism are interesting but speculative.

R J Hernstein and E G Boring, eds, *A Sourcebook in the History of Psychology* (London)

NS

monodromy theorem of Darboux (19th century) *Mathematics* Named after French mathematician Jean Gaston Darboux (1842–1917), this is the result in COMPLEX FUNCTION THEORY which essentially asserts that if a function is analytic on a region bounded by a simple closed curve, continuous on its closure and one-to-one on the boundary, then the function is one-to-one throughout the region. *See also* IDENTITY THEOREM.

S G Krantz, 'Functions of One Complex Variable', *Encyclopedia of Physical Science and Technology* (Academic Press, 1987)

MB

monologism (1920s) *Linguistics/Literary Theory* False ideology of language identified by the Russian critic Mikhail M Bakhtin (1895–1975).

Claim that there is only one point of view on a question, to be expressed one way alone, in words with single undisputed meanings; a position found in, for example, the classic 19th century realist novelist such as George Eliot, or in the modern news media. *See also* CLOSURE and MULTIACCENTUALITY.

V N Voloshinov (Bakhtin), trans. L Matejka and I R Titunik, *Marxism and the Philosophy of Language* (New York, 1973 [1929])

RF

monopolistic competition (1933) *Economics* Developed by American economist Edward

Chamberlin (1899–1967) and English economist Joan Robinson (1903–83), this term refers to competition between several firms producing an almost identical product in a market.

The demand for each good is not perfectly elastic. Monopolistic firms command brand loyalty and therefore are not price-takers. Under this form of competition, total product equals the sum of marginal cost and marginal revenue.

E H Chamberlin, *A Theory of Monopolistic Competition* (Cambridge, Mass., 1933)

PH

monopoly *Economics* A market in which one supplier dominates and sets price and quantity of the good.

The assumption in this model is that there are no substitutes and the firm is thus a price-maker. The firm may be motivated by profit maximization, and restrictive barriers to entry of the market prevent competition. Output is set at the point at which marginal revenue equals marginal cost. Critics of this system argue that prices are far higher and output much lower than under other forms of competition. As a result, legislation has been introduced to restrict the power of monopolies. *See* IMPERFECT COMPETITION, MONOPOLISTIC COMPETITION, BILATERAL MONOPOLIES and DUOPOLY THEORY.

PH

monopoly capitalism *Economics* An economy dominated by oligopolistic firms earning supernormal profits. The phrase can also mean a centrally planned economy with state-run monopolies organizing economic activity.

P A Baran and P M Sweezy, *Monopoly Capital: An Essay on the American Economic and Social Order* (New York, 1966)

PH

monotone convergence theorem (early 20th century) *Mathematics* Proved by French mathematician Henri-Léon Lebesgue (1875–1941) for the case where the limit function is integrable. When $\int f = \infty$, the theorem is due to B Levi.

The fundamental theorem of *Lebesgue integration theory* that if $\{f_n\}$ is an increasing sequence of non-integrable measurable functions which converges almost everywhere to a function f, then

$$\int f = \lim_{n\to\infty} \int f_n.$$

See also DOMINATED CONVERGENCE THEOREM and FATOU'S LEMMA.

G B Folland, *Real Analysis* (New York, 1984)

ML

Montague grammar (1960s) *Linguistics* Developed by the American logician R Montague (1930–70).

Formal semantic description of meaning in natural language, drawing on concepts of formal and modal logic, as a principled basis for syntax. It has influenced both GENERALIZED PHRASE STRUCTURE GRAMMAR and LEXICAL-FUNCTIONAL GRAMMAR.

R Montague, *Formal Philosophy*, R H Thomason, ed. and introd. (New Haven, 1974)

RF

Montel's theorem *Mathematics* Named after French mathematician Paul Montel (1876–1975), this element of COMPLEX FUNCTION THEORY states that non-normal families of functions take all, except possibly one, complex values near every point. JULIA SET theory depends on this finding.

J B Conway, *Functions of One Complex Variable* (New York, 1973)

ML

mood theory (20th century) *Politics/ Marketing* Concept in study of public opinion.

For the body of public opinion there are 'moods' which are not necessarily rationally derived, and are responses to other events or influences rather than spontaneous expressions of any coherent view.

Graham Evans and Jeffrey Newnham, *The Dictionary of World Politics* (Hemel Hempstead, 1990)

RB

Moore-Osgood theorem (1910) *Mathematics* Originally proved by Eliakim Hastings Moore (1862–1932) and generalized by William Fogg Osgood (1864–1943).

In its simplest form, the theorem asserts that if the limit as x tends to a of the sequence of functions $f_n(x)$ is $f_n(a)$ which is attained uniformly in n and the limit as n tends to infinity of $f_n(x)$ is $f(x)$ for all x then the limit as x tends to a and n tends to infinity is the same whichever order the limits are taken. This is a result with practical consequences and wide generalizations.

JB

moral hazard *Economics* This term refers to the idea that certain types of insurance systems might cause individuals to act in a more dangerous way than normal, causing a difference between the private marginal cost and the marginal social cost of the same action.

S Ross, 'The Economic Theory of Agency: the Principal's Problem', *American Economic Review*, vol. LXIII (1973), 134–39

PH

moral sense theories *Philosophy* Theories postulating a special moral sense which either enables us to perceive special moral qualities of virtue and vice in action (which thereupon affect us favourably or unfavourably), or else simply arouses feelings of approval or disapproval in us on contemplating the ordinary qualities of actions (it is not always clear which alternative is intended). These theories were popular in the 18th century, and are associated especially with Anthony Ashley Cooper, 3rd Earl of Shaftesbury (1671–1713) and Francis Hutcheson (1694–1746).

Though sometimes classed as a version of INTUITIONISM, the theories stood in contrast to contemporary intuitionism which claimed that we intuit moral facts about actions, rather than being sensitively affected by their qualities.

D D Raphael, *The Moral Sense* (1947)

ARL

Mordell's theorem (1922) *Mathematics* Named after British mathematician Louis Joel Mordell (1888–1972), this is the result in NUMBER THEORY and ALGEBRAIC GEOMETRY whereby the group of rational points on the ELLIPTIC CURVE $y^2 = x^3 + t$ in the real projective plane has a finite basis.

This is equivalent to the assertion that there is a finite set of rational points on the curve such that on starting from the set and taking all possible chords and tangents one obtains the totality of rational points on the curve. This result initiated a substantial amount of research into the nature of rational points on algebraic curves. The generalization of this theorem to curves of higher genus is known as the Mordell-Weil theorem.

A Baker, *A Concise Introduction to the Theory of Numbers* (Cambridge, 1984)

MB

Morera's theorem (19th century) *Mathematics* Named after Italian mathematician Giacinto Morera (1856–1909), this is the result that a continuous function with the property that its complex line integral over any closed curve is zero, must be analytic. *See also* CAUCHY'S INTEGRAL THEOREM, COMPLEX FUNCTION THEORY and GREEN'S THEOREM.

S G Krantz, 'Functions of One Complex Variable', *Encyclopedia of Physical Science and Technology* (Academic Press, 1987)

morpheme *Linguistics See* MORPHOLOGY.

morphology (19th century) *Linguistics* Developed in TRADITIONAL GRAMMAR, historical linguistics, and in STRUCTURAL LINGUISTICS.

Study of the internal structure of words in terms of segments called 'morphemes', realized by forms called 'morphs'. Morphology treats the division of words into root and inflection: *hat* + PLURAL = *hats*; the derivation of words by, for example, the addition of a derivational suffix: *drive* + *-er*; compounding, for example *night* + *club*. Morphs which can stand alone – for example *drive*, *club* – are 'free'. 'Bound' morphs – such as *-er*, *-s* – cannot stand alone.

P Matthews, *Morphology* (Cambridge, 1974)

RF

Moseley's law (1913) *Physics* Named after its discoverer, the English physicist Henry Gwyn Jeffreys Moseley (1887–1915), this law states that the frequency $f_{K\alpha}$ of the K_α lines in X-ray emission spectra from elements is given by

$$\sqrt{(f_{K\alpha})} = a(Z - 1)$$

where a is a constant that can be related to Bohr's model of the atom and Z is an integer that increases by unity on going through the periodic table and is now identified with the atomic number of the element.

J Thewlis, ed., *Encyclopaedic Dictionary of Physics* (New York, Oxford and London, 1962)

MS

Moser's twist theorem *Mathematics* Named after German-born American mathematician Jurgen Kurt Moser (1928–), this states that for an area preserving map subject to certain differentiability and eigenvalue conditions, there are infinitely many invariant circles around an elliptic point. On these circles, the map is an irrational rotation, but between these circles, the map may be chaotic.

R L Devaney, *An Introduction to Chaotic Dynamical Systems* (New York, 1989)

ML

Mössbauer effect (1958) *Chemistry* Named after the German-born American physicist Rudolph Mössbauer (1929–) who shared the Nobel prize for Physics in 1961 with Robert Hofstadter for their work into atomic structure. The Mössbauer effect is the resonant emission and re-absorption of gamma-ray photons by the excited states of the nuclei of atoms bound in a crystal without loss of energy through nuclear recoil.

The effect is an important tool in nuclear and solid state physics, as well as chemistry, as a result of the extreme sharpness of the gamma-ray transition.

D M Considine, *Van Nostrand's Scientific Encyclopedia* (New York, 1989)

GD

motherese (1970s) *Linguistics* Otherwise called 'caretaker language'. A special style of speech used by some people to address babies: short, simple sentences, high and rapidly shifting pitch, special vocabulary. Whether this facilitates language development is controversial.

L Gleitman and M Newport and H Gleitman, 'The Current Status of the Motherese Hypothesis', *Journal of Child Language*, 11 (1984), 43–79

RF

motion, Newton's laws of (1687) *Physics* Named after the English scientist and mathematician Sir Isaac Newton (1642–1727).

In his *Principia* Newton based his system of mechanics on the three fundamental laws of motion: (1) a body continues in a state of rest, or uniform motion in a straight line, unless it is acted on by an external force which compels it to change that state; (2) the rate of change of linear momentum is proportional to the applied force, and occurs in the direction in which that force acts. The acceleration a produced by a force F acting on a body of mass m is given by the equation $F = ma$; (3) the force exerted by a body A upon another body B and the force exerted by B upon A are simultaneous, equal in magnitude, opposite in direction, in the same straight line, and caused by the same mechanism. Einstein's theory of relativity is a more general system of mechanics, which reduces to Newton's laws for velocities small compared with that of light. *See* RELATIVITY, THEORY OF SPECIAL.

Robert M Besançon, ed., *The Encyclopedia of Physics* (New York, 1985)

GD

motivated forgetting *Psychology* Proposed by the Austrian founder of psychoanalysis Sigmund Freud (1856–1939).

This is a purposive process of forgetting whereby certain memories or motives are prevented from entering consciousness in order to satisfy unconscious needs to avoid such memories, motives and so on. The process through which we do this is called repression or, less frequently, intentional forgetting. *See* PARAPRAXIS.

D A Norman, *Memory and Attention: An Introduction to Human Information Processsing* (New York, 1977)

NS

motor theory of consciousness (1920s) *Psychology* Proposed by the American behaviourist John Broadus Watson (1878–1958).

This theory states that consciousness is an epiphenomenon, that what a person experiences is the product of their own actions and what a person perceives is determined by how they react to exterior stimuli. Hence, consciousness is not an entity in its own right. Motor theory of consciousness (or motor theory of thought, as it is sometimes called) is rejected by the majority of psychologists.

J B Watson, *Behaviorism* (New York, 1925)

NS

motor theory of meaning *Psychology* Proposed by the American behaviourist John Broadus Watson (1878–1958).

This structural psychology doctrine proposes that meaning consists of the images regularly associated with the sensory presentation or sensation. It is further proposed that certain sensory conditions will elicit certain conscious processes and motor responses that are related to the perceived object; for example, the meaning of the yellow dot in the glass bottle in the doctor's surgery is a tablet.

R J Herrnstein and E G Boring, eds, *A Sourcebook in the History of Psychology* (Cambridge, Mass.,)

NS

motor theory of speech perception *Psychology* A theory of speech perception proposed by A M Liberman (1917–).

This hypothesizes that speech is perceived by mapping the acoustic properties of an input onto some internal representation of speech. Little physiological evidence exists to support Liberman's theory.

G Beaumont, *Introduction to Neuropsychology* (Oxford, 1983)

NS

motor theory of thinking *Psychology See* MOTOR THEORY OF CONSCIOUSNESS.

Motzkin's theorem (1951) *Mathematics* A positive version of SYLVESTER'S THEOREM proved by T Motzkin. If $n > 1$ points in a Euclidean plane are not all on one line, then there exists a line containing exactly two of the points.

H S M Coxeter, 'Two-dimensional Crystallography', *Introduction to Geometry* (New York, 1961)

JB

mountain pass lemma *Mathematics* A theorem which gives conditions for a differentiable function (possibly unbounded) to have a critical point. If, for some value of a such that $\|a\| > 1$

$$f(0) < \inf \{f(x) : \|x\| = 1\} > f(a)$$

where f satisfies a 'growth condition' (such as f tends to infinity as x increases), then there is a critical point b such that

$$f(b) \geqslant \inf \{f(x) : \|x\| = 1\} \,.$$

This critical point is located in the 'mountain pass'.

ML

Movimento Spaziale *Art See* SPATIALISM.

Muller-Schumann law *Psychology* Formulated by the German, Johannes Peter Muller (1801–58).

This law holds that when two items have been associated (learned), it is more difficult to learn/form a new association between a third item than between either of the items individually before they had been paired.

J W Kling and L A Riggs, eds, *Experimental Psychology* (London, 1971)

NS

Mueller's colour theory *Psychology* Proposed by George Elias Mueller (1850–1934).

Similar to Hering's theory, Mueller proposed that all colours were reducible to two pairs of opposed or antagonistic colours with a chemical substance of reversible action in the retina for each pair. Mueller supplanted Hering's theory by proposing the involvement of certain cerebral processes.

E G Boring, *A History of Experimental Psychology* (Englewood Cliffs, N.J., 1957)

NS

Müllerian mimicry (1878) *Biology* Named after German zoologist J F T (Fritz) Müller (1822–97), this term refers to a physical resemblance (such as distinctive coloration) shown by two species sharing some characteristic adverse to predators, such as unpalatability.

Both species benefit from the mimicry because once a predator attacks a member of one species, it learns to avoid all individuals of similar appearance. *Compare with* BATESIAN MIMICRY.

L Brower, *Mimicry and the Evolutionary Process* (Chicago, 1988)

KH

multi-accentuality (1929) *Linguistics/Literary Theory* Linguistic principle articulated by the Russian critic Mikhail M Bakhtin (1895-1975).

All signs have an 'inner dialectic quality', potential to be accented or oriented differently according to the social relationships of speakers; *see also* MONOLOGISM, DIALOGISM and references.

V N Voloshinov (Bakhtin), trans. L Matejka and K Pomorska, *Marxism and the Philosophy of Language* (New York, 1973 [1929])

RF

multigene hypothesis *Biology See* GERMLINE THEORY OF ANTIBODY DIVERSITY.

multigroup neutron diffusion theory (20th century) *Physics* A theory of neutron transport in which it is assumed that the neutron population may be divided into a number of groups, each of which has a constant neutron energy.

J Thewlis, ed., *Encyclopaedic Dictionary of Physics* (New York, Oxford and London, 1962)

MS

multilateralism (20th century) *Politics* A theory of (nuclear) disarmament in opposition to UNILATERALISM, this argues that nuclear disarmament can only be achieved by negotiated mutual reductions in weapons by all states.

David Robertson, *The Penguin Dictionary of Politics* (London, 1986)

RB

multimodal theory of intelligence *Psychology* Proposed by A Binet (1857–1911).

The theory that intelligence is made up of several components or primary mental abilities. This theory was later expanded and postulated as many as 120 separate mental abilities. It was further proposed that these mental abilities could be described in terms of three dimensional character abilities; that is, contents (what the individual knows), operations (the process of what is known), and products (the end results).

R H Price, M Glickstein, D L Horton, and R H Bailey, *Principles of Psychology* (New York, 1982)

NS

multinomial theorem *Mathematics* The result which generalizes the BINOMIAL THEOREM to several variables. To be specific,

$$(x_1+x_2+\cdots+x_k)^n = \sum \frac{n!}{n_1!n_2!\cdots n_3!} x_1^{n_1} x_2^{n_2} \cdots x_k^{n_k}$$

where the sum is taken over all k- tuples of non-negative integers $(n_1,n_2,\ldots,n_k)$ such that

$$n_1 + n_2 + \cdots + n_k = n\,.$$

MB

multiple causation (4th century BC–) *Psychology* First proposed by the Greek philosopher Aristotle (384–322 BC).

A general term imported from philosophy and reflecting recognition of the fact that no one cause can account for the occurrence of a particular behavioural event. Rather, many factors lead to an event and these factors interact; that is, these different factors cannot be considered independently.

F Copleston, *A History of Philosophy: Greece and Rome*, vol. I (New York, 1962)

NS

multiple codon recognition *Biology See* WOBBLE HYPOTHESIS.

multiple control principle *Psychology* Advanced by the American psychologist Karl Spencer Lashley (1890–1958).

This principle maintains that any particular part of the brain is likely to be implicated in the performance of many different types of behaviour. Conversely, a single piece of behaviour involves a number of brain sites. The logical conclusion is that the brain functions as an integrated whole. The theory is challenged by evidence which suggests that there may be a considerable degree of specialization within the brain. *See* EQUIPOTENTIAL THEORY.

K Lashley, *Brain Mechanisms and Intelligence: A Quantitative Study of Injuries to the Brain* (Chicago, 1929)

NS

multiple factor theory *Psychology* Proposed by O Fenichle.

A theory and analytic method which postulates that more than one common factor can account for a given phenomenon; for example, intellligence is made up of different abilities or factors. Thus, it is probable that any psychological disorder is caused by multipe factors rather than a single factor operating in isolation.

R H Price, M Glickstein, D L Horton and R H Bailey, *Principles of Psychology* (New York, 1982)

NS

multiple intensity rules *Chemistry* Rules for the relative intensities of spectral lines in spin-orbit multiplet.

The sum of the intensities of all lines starting from a common initial level, or ending on a common final level, is proportional to $2J + 1$, where J is the total angular momentum of the initial level or final level, respectively.

MF

multiple proportions, law of (1804) *Chemistry* Originally proposed by the English chemist John Dalton (1766–1844), this law is also known as Dalton's law. It states that if two elements combine in more than one proportion, the weights of the first element which combine with a fixed weight of the second element are in the ratio of integers to each other.

For example, nitrogen combines with oxygen to form several different oxides: N_2O, NO, N_2O_3, NO_2 and N_2O_5. These contain weights of oxygen in the ratio 1:2:3:4:5, respectively. The discovery of non-stoichiometric compounds has rendered the law of limited validity.

M Freemantle, *Chemistry in Action* (Basingstoke, 1987)

GD

multiple response principle *Psychology* Proposed by the American psychologist Clark Leonard Hull (1884–1952).

The principle that an animal will react to a new or novel situation with a number of potential responses already within its behavioural repertoire.

G A Kimble, *Hilgard and Marquis Conditioning and Learning* (New York, 1961)

NS

multiplier (1931) *Economics* This is defined as the relationship between a change in the NATIONAL INCOME and the primary alteration in expenditure that brought it about.

In a simple model, the multiplier effect depends on the marginal propensity to consume. *See* MULTIPLIER-ACCELERATOR MODEL.

R F Kahn, 'The Relationship of Home Investment to Employment', *Economic Journal*, vol. XLI (June, 1931), 173–98

PH

multiplier-accelerator *Economics* Proposed by English economist Roy Harrod (1900–78) and American economist Paul Samuelson (1915–) as an extension of the work of English economists John Maynard Keynes (1883–1946) and R Kahn (1905–), this is a model analyzing economic fluctuations through the effects of the accelerator and multiplier models.

An increase in government expenditure may lead to an increase in consumer incomes which (through the multiplier effect) leads to an increase in output which in turn (through the accelerator process) raises investment. This process tends to work on a cyclical basis. *See* KONDRATIEFF CYCLES, SUNSPOT THEORY, PRODUCT LIFE-CYCLE THEORY, ACCELERATION PRINCIPLE, FINE-TUNING, POLITICAL BUSINESS CYCLE, and TRADE CYCLE.

R F Harrod, *The Trade Cycle* (Oxford, 1936); P Samuelson, 'Interactions Between the Multiplier Analysis and Principles of Acceleration', *Review of Economic Statistics* 21, 2 (May, 1939), 75–78

PH

Munsell's theory (1915) *Art* Named after its originator, Albert F Munsell, this is a system of colour notation based on the three responses of the human eye to colour: perception of hue, value and chroma. Faults in the theory's spacing of colour are now being corrected and the system improved.

J M Cleland *A Practical Description of the Munsell System* (1937)

AB

Muntz's theorem (20th century) *Mathematics* This gives more precise details of polynomials which approximately equal continuous functions. *See also* WEIERSTRASS APPROXIMATION THEOREM.

JB

music of the spheres (6th century BC) *Music* Also known as *musica mundana* from the Middle Ages onwards, this doctrine (attributed to Pythagoras's school) postulated that each planet produces a musical note; its pitch determined by the planet's speed of revolution and its distance from the earth; the combined notes of the planets forming a scale, or *harmonia*.

By the Middle Ages, theorists believed that although *musica mundana* was inaudible to humans, its numerical proportions governed the universe and the human soul.

K Meyer-Baer, *Music of the Spheres and the Dance of Death* (Princeton, N.J., 1970)

MLD

musical figures, doctrine of (Middle Ages) *Music* First codified by the German musicologist Joachim Burmeister (1564–1629) in his treatise *Musica Autoschediastike* (1601), this is the notion that music makes use of technical devices analogous to rhetorical figures of speech.

Some techniques operate on a purely musical level (for example, melodic repetition for emphasis) while others interact with text (thus the word 'fall' might be sung to a descending pattern of notes). Burmeister and others catalogued devices using labels borrowed from rhetoric, but where no oratorial equivalent existed they invented labels. Thus scholars disagree over how consciously composers copied rhetoric. *See also* AFFECTIONS, DOCTRINE OF THE.

F W Robinson and S G Nichols, eds, *The Meaning of Mannerism* (Hanover, N.H., 1972)

MLD

musique concrète (1948) *Music* Music composed and realized in an electronic studio using pre-existing ('concrete') sounds, rather than conceived abstractly, notated onto paper and then performed. The term was coined by the French composer Pierre Schaeffer (1910–) who first developed this type of music. The natural sounds are recorded onto tape and then manipulated as physical entities through tape splicing, looping and so on.

P Schaeffer, *A la Recherche d'une Musique Concrète* (Paris, 1952)

MLD

mutation theory (1903) *Biology* Proposed by Dutch botanist Hugo de Vries (1848–1935), this is the theory that new forms of living things can occur suddenly in a population. These randomly occurring genetic mutations may result in new species, or may provide new characteristics that may be acted upon by natural selection.

De Vries's observations were seen as a basic evolutionary mechanism for a time, but biologists now consider crossing-over of genes and recombination during meiosis to be the primary sources of the variation acted upon by natural selection.

H de Vries, *Mutationstheorie* (Mutation Theory) (1901, 1903); L Dunn, *A Short History of Genetics* (New York, 1965)

KH

mutation theory of cancer *Biology See* CHROMOSOME THEORY OF CANCER.

mutationism *Biology See* MUTATION THEORY.

mutual aid (19th century) *Politics* Theory of beneficient ANARCHISM described by Russian anarchist Peter Kropotkin (1842–1921).

Mutual assistance amongst members of the same species is as natural as competition. The species – and by implication the societies – that have prospered and will prosper are those characterized by such voluntary support amongst their members rather than by conflict or competition.

David Miller, *Anarchism* (London, 1984)

RB

mutual exclusion rule *Chemistry* A rule which states that for a molecule with a centre of symmetry, a transition is allowed in either its Raman scattering or its infra-red emission (and absorption), but not in both.

GD

mutual knowledge (1969/1970) *Linguistics* Several congruent proposals, including that of the American philosopher David Kellogg Lewis, and that of the sociolinguist William Labov ('AB-events').

Shared knowledge between speakers, which is assumed in communication: A has knowledge K that B knows, A knows that B knows K, B knows that A knows K and that A knows that B knows K, and so on.

N V Smith, ed., *Mutual Knowledge* (London, 1982)

RF

mutualistic hypothesis (1972) *Biology* Proposed by Michener, this is the idea that ants, bees and wasps (*Hymenoptera*) are social insects because group living (mutualism) conferred a defence against parasites and predators, and evolved prior to the development of the complex society maintained by these species.

Michener proposed this theory as an alternative to Hamilton's HAPLODIPLOID HYPOTHESIS, which he considered inadequate by itself to explain how eusociality evolved. *See also* MATERNAL-CONTROL HYPOTHESIS.

D Dewsbury, *Comparative Animal Behavior* (New York, 1978)

KH

mutually assured destruction (MAD) (20th century) *Politics* Theory of security through nuclear weaponry.

Since each STATE has enough nuclear weapons to destroy the other even if attacked first, any state beginning a nuclear war will cause the destruction of both itself and its opponent. This will deter it from so doing. *See also* BALANCE OF TERROR and DETERRENCE.

Graham Evans and Jeffrey Newnham, *The Dictionary of World Politics* (Hemel Hempstead, 1990)

RB

my man's different (20th century) *Feminism* Satirical account of individual response to FEMINISM.

Many women who are married to – or in other ways involved with or dependent upon – men accept the general feminist argument that women, as a whole, are exploited by men, as a whole. Their own case, however, they believe to be an exception because 'my man's different'.

HR

mysticism *Philosophy* A type of religious attitude (appearing in many guises and within many religions from antiquity onwards) emphasizing various practices – ascetic, contemplative, or other – for obtaining knowledge of and unification with God or spiritual reality by means not open to reason and not relying on dogma. Mystics claim to achieve this knowledge or unification by experiences which have a favourable affective quality and cannot be put into words (they are 'ineffable'); though it is claimed that mystics of widely differing traditions readily understand each other's writings.

H L Bergson, *The Two Sources of Morality and Religion* (1935); contains discussion of historical growth of mysticism, though in the service of Bergson's own philosophy

ARL

mysticism *Psychology* Exemplified in the teachings of Gautama the Buddha (563–483 BC).

A belief in spiritual sources of knowledge, which knowledge may be gained through contemplation and tuition as well as by sense experiences. Mysticism has been used in this way as a means to understand emotional states and cognition. Opponents of mysticism may use the term 'mystical' pejoratively to describe an hypothesis which cannot be substantiated empirically.

J Ferguson, *An Illustrated Encyclopaedia of Mysticism* (London, 1976)

NS

myth *Politics* Theory of the function of beliefs.

Stories about the destiny of nations, races, classes, or political groups have an important function which is quite independent of their truthfulness. The belief in some ultimate goal gives purpose; and the belief in some (mythical) identity as a race, a chosen political order, or a special nation gives social cohesion. One of the best known of such uses of myth is Georges Sorel's account of an ultimate and revolutionary GENERAL STRIKE.

Henry Tudor, *Political Myth* (London, 1972)

RB

myth criticism (1920s–) *Literary Theory* Sources in anthropology and JUNGIAN THEORY (psychology). Myths are symbolic narratives expressing the collective psychological concerns of a culture (Oedipus, The Green Man, and so on). Myth criticism interprets modern literature in terms of such (originally perhaps more 'primitive') structures. The high point of myth criticism was the work of the American critic Northrop Frye (1912–) with its totalizing system of only four very abstract myths underlying all particular manifestations. *See also* ARCHETYPE.

N Frye, *Anatomy of Criticism* (Princeton, 1957)

RF

N

NAIRU (non-accelerating inflation rate of unemployment) (1968) *Economics* Proposed by American economist Milton Friedman (1912–92), this term refers to the long-term rate of unemployment at which inflation neither rises nor falls, as upward and downward pressures on wage and price inflation are in equilibrium. The vertical PHILLIPS CURVE illustrates this situation.

M Friedman, 'The Role of Monetary Policy', *American Economic Review*, vol. LVIII (March, 1968), 1–17

PH

naive realism *Philosophy* Theory that we see the world as common sense supposes we do; that is, directly and without recourse to special intermediate 'sensations', 'sense-data', 'images' and so on which some other views involve (*see also* REPRESENTATIONALISM). We need not, however, always be free from error, any more than common sense thinks we are.

Properly speaking, naive realism is supposed to be the view of 'naive' (that is, not philosophically trained) common sense. But common sense has no systematically developed view, so that when naive realism is held as a philosophical theory and defended against objections it is often, and probably should be, called direct realism.

D M Armstrong, *Perception and the Physical World* (1961)

ARL

naive/sentimental (1795–96) *Literary Theory* Distinction proposed by the German poet and critic Friedrich von Schiller (1759–1805).

Naive poets are instinctual, spontaneous, representing the world as they experience it; sentimental or classical poets are rational and idealist. These tendencies are synthesized in the best poetry. This distinction is the foundation of others, for example CLASSICISM/ROMANTICISM, APOLLONIAN/DIONYSIAN.

F von Schiller, *On the Naive and Sentimental in Literature*, H Watanabe-O'Kelly, trans. (Manchester, 1981)

RF

naive set theory *Mathematics* The presentation of SET THEORY in the style of an informal mathematical theory regarding it as a given body of knowledge rather than the consequences of a collection of uninterpreted axioms.

The naive point of view, as originally put forward by the German mathematician Georg Ferdinand Ludwig Philip Cantor (1845–1918) seeks a formalization which is based on the intuitive conception of the nature of a set. While this approach cannot be rigorously sustained, it is quite sufficient for many purposes. *See also* AXIOMATIC SET THEORY and CANTOR'S PARADOX.

Encyclopedic Dictionary of Mathematics (MIT Press, 1987)

MB

naming theories of meaning *Philosophy* Also called denotative or referential theories. Theories which equate the meaning of a word with an object it stands for (like the

'FIDO'-FIDO THEORY), or else with the word's relation to such an object.

Proper names form the primary class, but general words can stand for abstract objects ('dog' for doghood, 'red' for the colour red, and so on). A word like 'dog', however, can also be taken to stand for different dogs on different occasions, which raises one of the standard objections to the theory: that it is unclear what 'standing for' amounts to, and how many different jobs it is supposed to do. *See also* USE THEORIES OF MEANING.

B Russell, *An Inquiry into Meaning and Truth* (1940), especially chs 1–7

ARL

Napier's rules *Mathematics* Named after Scottish mathematician John Napier (1550–1617) who invented logarithmic tables and invented the first mechanical calculator.

The rules result from the observation that in a right-angled spherical triangle, the cosine of any circular part (the remaining two angles, the hypothenuse and the complements of the sides containing the right angle) is equal to the product of the cotangents of the adjacent parts, and is also equal to the product of the sines of the opposite parts. The arguments of all trigonometric functions can be derived from these rules.

W Gellert, S Gottwald, M Hellwich, H Kästner and H Küstner, eds, *The VNR Concise Encyclopedia of Mathematics* (New York, 1975)

ML

narratee (1971) *Literary Theory* Term coined by the critic Gerald Prince.

The fictional audience or addressee within a narrative, to whom the story is told. May be overtly realized, as Marlow's shipmates in Joseph Conrad's 'Heart of Darkness' (1902), or implied. To be distinguished from reader and IMPLIED READER.

G Prince, *A Grammar of Stories* (The Hague, 1973)

RF

narrative (traditional) *Literary Theory* The theory of narrative has been much discussed in the 20th century, particularly from the 1960s onwards.

A story told in oral or printed mode by a NARRATOR to an audience or reader. For technical aspects, *see* DIEGESIS, 'FABULA' AND 'SJUZHET', FOCALIZATION, KERNEL, CATALYZER, INDEX, NARRATEE, NARRATIVE GRAMMAR, NARRATIVE MODALITY, NARRATOR, NATURAL NARRATIVE, TELLABILITY.

R Scholes and R Kellogg, *The Nature of Narrative* (London, 1966)

RF

narrative grammar (1928, 1960s) *Stylistics* Based on the work of the Russian linguist Vladimir Propp (1895–1970), developed in French STRUCTURALISM, especially by Tzvetan Todorov (1939–); also in STRUCTURAL ANTHROPOLOGY, especially by Claude Lévi-Strauss (1908–).

A story is 'like a sentence': it may be analyzed as having a structure of development analogous to the syntactic structures of sentences in language. Propp analyzed Russian fairy tales as structured sequences of 'functions of the *dramatis personae*'. Narrative grammarians have mapped a linguistic analogy onto Propp's sequences, deploying such concepts as noun, verb, transformation in their presentation of a story's structure.

V Propp, *Morphology of the Folktale* [1928], trans. L A Wagner and A Dundes (Austin, 1968); M J Toolan, *Narrative: A Critical Linguistic Introduction* (London, 1988)

RF

narrative modality (1976) *Stylistics* The idea originates with the Canadian linguist L Dolozel.

Refinement of NARRATIVE GRAMMAR; categories of modal logic such as knowledge, possibility, permission, prohibition, obligation (which are implicit in Propp's work) are applied to the categorization of types of narrative. *See also* MODALITY.

L Dolozel, 'Narrative Semantics', *Poetics and the Theory of Literature*, 1 (1976), 129–51

RF

narrator (traditional) *Literary Theory* The concept originates with the Greek philosopher Plato (*c*.427–*c*.347 BC). *See also* DIEGESIS.

The teller in the tale. The voice which seems to tell the story, which may be anonymous or embodied as a character. Narrators may be first-person (autodiegetic or homodiegetic) or third-person; that is, not participating in the story (heterodiegetic). In classic 19th century fiction the narrator is often presented as 'omniscient' (*see also* MONOLOGISM), thus 'reliable'. Booth discusses unreliable narrators.

W C Booth, *The Rhetoric of Fiction* (Chicago, 1961)

RF

NARU (natural rate of unemployment) (1968) *Economics* Proposed by the American economist Milton Friedman (1912–92), this term refers to the rate of unemployment which occurs when all markets are in an equilibrium position.

The rate cannot be reduced by an increase in aggregate demand. Unemployment cannot be reduced below the natural rate of unemployment as this would result in accelerating inflation; hence, the natural rate is also referred to as the non-accelerating inflation rate of unemployment (NAIRU). The vertical PHILLIPS CURVE illustrates this situation.

M Friedman, 'The Role of Monetary Policy', *American Economic Review* vol. LVIII (March, 1968), 1–17

AV, PH

Nash equilibrium (1950) *Economics* Named after mathematician John Nash, and central to GAME THEORY, this concept refers to a situation in which individuals participating in a game pursue the best possible strategy while possessing the knowledge of the strategies of other players. It works on the premise that the player cannot improve his/her position given the other players' strategy.

This theory is sometimes referred to as the *non-co-operative equilibrium* because each player chooses his/her own strategy believing it is the best one possible, without collusion, and without thinking about the interests of either his opponent or the society in which he/she lives. *See* CO-OPERATIVE GAMES THEORY, COLLUSION THEORY, OLIGOPOLY THEORY and ALLAIS PARADOX.

J F Nash, 'Equilibrium Points in n-Person Games', *Proceedings of the National Academy of Science, USA,* vol. XXXVI (1950), 48–49

PH

national income (17th–20th century) *Economics* National income is the total income, over a specified period of time, of all the inhabitants of an economy after allowing for capital consumption. The term can also be used to describe a monetary flow that shows net additions to wealth. National accounting was first conducted by English economist Sir William Petty (1623–87), and techniques were further developed during the 1930s.

Most measures of national income exclude non-market activities and certain social costs. *See* THEORY OF INCOME DETERMINATION, THEORY OF INCOME DISTRIBUTION, NATURAL AND WARRANTED RATES OF GROWTH, WAGNER'S LAW and BALANCED BUDGET MULTIPLIER.

W Beckerman, *An Introduction to National Income Analysis* (London, 1966)

PH

nationalism *Politics* Theory of the nation as the basis for government.

People are identified as members of historically and culturally distinct nations. The frontiers of states and the frontiers of nations should coincide; and where they do not, government should be divided or re-organized in order to achieve this.

David Miller *et al.*, eds, *The Blackwell Encyclopaedia of Political Thought* (Oxford, 1987)

RB

nationalization (20th century) *Economics* The process by which privately-owned industries or companies are taken over by the state.

Nationalization is often applied to monopolies such as water and gas utilities or transport services, in which social needs are deemed more important than profitability. Since the 1980s, there has been a trend to de-nationalize (privatize) industries in Europe, particularly in the UK. *See* PRIVATIZATION, 'LAISSEZ-FAIRE', PHYSIOCRACY and MERCANTILISM.

S Holland, ed., *The State as Entrepreneur* (London, 1972)

PH

nativism *Philosophy* Any view claiming that something is innate, such as ideas or perceptual faculties. *See also* INNATE IDEAS.

ARL

nativism *Psychology* Advanced in an early form by the Anglo-Irish philosopher William Molyneux (1656–1738).

This holds that the ability to perceive time and space is inborn. A distinction can be made between extreme nativism (which emphasizes the ability to perceive independently of experience), and contemporary nativism (which stresses the primary genetic influences on behaviour over learned influences). Extreme nativism is no longer considered defensible.

G E Weisfeld, 'The Nature-Nurture Issue and the Integrating Concept of Function', *Handbook of Developmental Psychology*, B B Wolman, ed. (Englewood Cliffs, N.J., 1982)

NS

natural and warranted rates of growth (1930s–40s) *Economics* Formulated by English economist Roy Harrod (1900–78) and Russian-born economist Evsey Domar (1914–) in response to the *General Theory* of English economist John Maynard Keynes (1883–1946). The natural rate is the maximum long-term rate of growth in the HARROD-DOMAR GROWTH MODEL where population growth increases the labour force and technical advances enable productivity gains. The warranted rate is the growth of NATIONAL INCOME which maintains the equality of planned saving and planned investment. *See* SOLOW ECONOMIC GROWTH MODEL.

PH

natural categories (1970s) *Linguistics* Research by Eleanor Rosch and colleagues following Berlin and Kay's work on BASIC COLOUR TERMS.

Language categorizes experience, simplifying our cognitive relationship with the world. The world is not unstructured, and many sets of terms name items and processes which are naturally salient for us: shapes, dimensions, types of animal, and so on. (It would also have to be granted – *see* SOCIAL SEMIOTIC – that many categories are social in origin.) *See also* BASIC LEVEL TERMS, PROTOTYPE.

H H Clark and E V Clark, *Psychology and Language* (New York, 1977)

RF

natural justice *Politics/Law* Theory of general or absolute legal principles.

There are principles of natural justice which are independent of historical circumstances or the details or conventions of particular legal systems. These principles provide that there shall be redress or protection against injury; and, at the level of legal procedure, adjudication shall be independent (hence that no one is judge in their own case) with both sides to a dispute being properly heard.

Roger Scruton, *A Dictionary of Political Thought* (London, 1982)

RB

natural law *Politics/Law* Theory of basis of human law.

There are natural laws, either in the sense of being embedded in the essence of human society or HUMAN NATURE, or as being expressed through its development over time. The justification for 'manufactured' law lies in its expression of natural law, and the task of the legislator is thus to express natural law in STATE law.

David Miller *et al.*, eds, *The Blackwell Encyclopaedia of Political Thought* (Oxford, 1987)

RB

natural monopoly economics *Economics* The existence of an industry (gas, electricity, water, for example) in which the average costs of production per unit fall as output increases.

A single firm operating in the industry (a monopoly) can produce output more efficiently than several competing firms under these circumstances. *See* MONOPOLY.

W W Sharkey, *The Theory of the Natural Monopoly* (New York and Cambridge, 1982)

PH

natural narrative (late 1960s) *Stylistics* Analysis developed by the American sociolinguist William Labov.

Stories told orally by people have a regular structure based on a sequence of elements: abstract, orientation, complicating action, evaluation, result or resolution, coda. Extended to literary narrative by Mary Louise Pratt (1948–). *See also* TELLABILITY.

W Labov, *Language in the Inner City* (Philadelphia, 1972)

RF

natural rate of unemployment *Economics See* NARU.

natural-response theory of language *Psychology* Often accredited to the German, Max Muller (1862–1919), this theory is also known as the Ding-Dong theory of language.

One of the hypotheses which seek to explain how humans first developed language. Natural-response theory postulates that language began with vocal expressions being assigned to objects found in the environment. The validity of the theory is highly suspect.

R H Brown, *A First Language* (Cambridge, Mass., 1973)

NS

natural rights *Politics* Theory of human RIGHTS.

Rights arise from the nature of human or social existence, in the same way as does NATURAL LAW. There is disagreement as to whether rights are surrendered, or transformed, or held in abeyance when people enter political, law-governed society.

David Miller *et al.*, eds, *The Blackwell Dictionary of Political Thought* (Oxford, 1987)

RB

natural selection (19th century) *Biology* Also called selection of the fittest. Its importance first recognized by the English evolutionary theorist Charles Darwin (1809–82), this is the concept that the genetic component of those individuals with superior characteristics for survival and reproduction will eventually overcome that of those individuals less well endowed.

Natural selection is a key mechanism for Darwinian evolution. Other forces affecting the characteristics of populations include the FOUNDERS EFFECT and migration. *See also* DARWINISM, EVOLUTION, NATURAL THEOLOGY, NEO-DARWINISM and SURVIVAL OF THE FITTEST.

C Darwin, *Origin of Species by Means of Natural Selection: Or, the Preservation of the Favoured Races in the Struggle for Life* (London, 1859); E Sober, *The Nature of Selection* (Cambridge, Mass., 1984)

KH

natural state model of species *Biology See* SPECIES ESSENTIALISM.

natural theology (13th century) *Natural Sciences/Theology* Discussed by Thomas Aquinas (*c.*1225–74) and advanced by John Ray (1627–1705), author of *Wisdom of God in the Creation* (1691), and the Reverend William Paley (1743–1805).

Derived from combining Platonic and Aristotelian philosophy with Christianity, this doctrine asserts that reason coupled with faith can prove the existence of the Christian deity. In natural theology, the orderliness of nature is seen as evidence of a divine architect. *Compare with* EVOLUTION and NATURAL SELECTION.

W Paley, *Natural Theology: or, Evidences of the Existence of the Deity, Collected from the Appearance of Nature* (London, 1802)

KH

naturalism *Philosophy* Any view holding that things in general, or things in some sphere under investigation, are all of one kind (as opposed to being of radically different kinds), and are amenable to study by scientific methods, without appeal to supernatural intervention or special kinds of intuition. In art or literature, any of a variety of views saying the artist should imitate the natural world or actual human behaviour.

In recent philosophy, naturalism has usually taken the form of an ethical theory, denying the FACT/VALUE DISTINCTION, and also denying that ethical facts are *sui generis* in nature; those ignoring either of these denials are said to commit the naturalistic fallacy. Ethical naturalism is akin to, but not identical with, DESCRIPTIVISM (George Edward Moore (1873–1958) was a descriptivist but not a naturalist).

P Foot, ed., *Theories of Ethics* (1967)

ARL

naturalism (19th century) *Art/Literary Theory* Naturalistic aesthetics arose out of 19th-century POSITIVISM, and were developed in literary theory above all by the French writer Émile Zola (1840–1902) who spoke of the 'experimental novel'.

Rejecting the emotional emphasis of ROMANTICISM and notions of idealism and stylization, it sought to represent natural objects as they appear, acting as the mirror for nature. Its theoretical basis is not far removed from that of REALISM, particularly in its claim for scientific objectivity of recording, but built in is a Darwinian ideology of the animal nature of the human, portraying violence and primitive passion. *See also* SOCIAL NOVEL. During the 17th century, the Italian art historian Bellori (1672) was the first to characterize work by the followers of Michel Angelo Merisi Caravaggio (1569–1609) as naturalistic.

G J Becker, ed., *Documents of Modern Literary Realism* (Princeton, 1963)

AB, RF

naturalized epistemology *Philosophy* A notion introduced explicitly by Willard Van Orman Quine (1908–) though with roots going back to David Hume (1711–76). The idea is that since it is impossible to achieve a satisfactory justification for our claims to knowledge we should cease to look for one, and construct a scientific account – in purely 'natural' terms and without reference to justification – of how in fact we come to hold the beliefs we do.

This issue of whether to pursue justification or scientific explanation must be distinguished, it has been claimed, from the question of whether justification itself should be sought by appeal to natural facts (such as causal relations), as opposed to, for example, rational considerations that would entail the conclusion to be justified. (*See also* CAUSAL THEORY OF KNOWLEDGE, EXTERNALISM and INTERNALISM.) A comparison can also be made between NATURALISM in epistemology and in ethics. *See also* PSYCHOLOGISM.

J Kim, 'What is "Naturalized Epistemology"?', *Philosophical Perspectives* (1988); W V O Quine, *Ontological Relativity and Other Essays* (1969)

ARL

nature and nurture (19th century) *Biology/Psychology* A phrase probably first coined in this sense by British scientist Sir Francis Galton (1822–1911) in 1874, but appearing as 'nature nurture' in William Shakespeare's *The Tempest* (1611), Act IV, Scene 1. Addressed in the work of the English philosopher John Locke (1632–1704), and evolutionist Charles Darwin (1809–80). It is also known as the heredity-environment problem and the nativism-empiricism debate.

This idea refers to the separate influences of heredity (nature) and environment (nurture) on a living thing. The term is often used in discussions of the relative impact of these two influences on an organism's behaviour, neither nature nor nurture alone is sufficient, both being necessary for the development of most human characteristics.

A R Jensen, *Genetics and Education* (New York, 1972); F Galton, *English Men of Science: Their Nature and Nurture* (1874); P B Medawar and J S Medawar, *Aristotle to Zoos: A Philosophical Dictionary of Biology* (Oxford, 1983)

KH, NS

'Nature red in tooth and claw' *Biology See* SURVIVAL OF THE FITTEST.

Nazarene Brotherhood (1809) *Art* Name given to an association of artists, initiated by the German painters Franz Pforr (1788–1812) and Friedrich Overbeck (1789–1869), who joined together in Rome as the *Lukasbruder* ('the Order of St Luke') in July 1809.

Basing their philosophy on the *Art-Loving Friar* by German author Wilhelm Heinrich Wackenroder (1773–98), they rejected academic art instruction and attempted to reinvigorate art by returning to the spirit of the Middle Ages. They also fostered the concept of art as a servant of religion. Their philosophy was important for the PRE-RAPHAELITE movement in England.

AB

Nazi art *Art See* ART OF THE THIRD REICH.

nebular hypothesis of planetary formation (1775) *Astronomy* Proposed by the German philosopher Immanuel Kant (1742–1804) and, independently, by the French mathematician and astronomer Pierre Simon, Marquis de Laplace (1749–1827).

This hypothesis pictured that the primaeval solar system as a chaotic nebulous cloud of dust specks and other particles. Gradually this matter began to cohere and formed a very large mass, that is, the Sun. Kant thought that the act of cohering would set the mass into rotation and that, as it contracted under gravitation, it would rotate faster and faster and throw off masses of material which would cool and condense to form the planets. Laplace postulated a nebulous mass that was in rotation from the beginning. Cooling, contraction and increased speed of rotation then followed. The rotating mass then flattened into a disc which left matter behind at its periphery as the main mass continued to shrink. The difficulty with this theory is that conservation of angular momentum indicates that the angular momentum of the present solar system must equal the angular momentum of the primaeval Sun, allowing for loss by radiation emission. Calculation indicates that the primaeval Sun would not have had enough angular momentum to break up. Laplace's problem is one of scale. His theory can explain the birth of suns out of nebulae but not the birth of planets out of suns. *See also* VON WEIZSÄCKER'S THEORY OF PLANETARY FORMATION.

S Mitton, ed., *The Cambridge Encyclopaedia of Astronomy* (London, 1973)

MS

necessitarianism *Philosophy* Term occasionally used for the view that everything that happens is necessitated. The view that every event has a cause is the same, unless causation is distinguished from necessitation. *See also* DETERMINISM.

R R K Sorabji, *Necessity, Cause and Blame* (1980), chapter 2; distinguishes causation from necessitation, though without using the term 'necessitarianism')

ARL

need-drive incentive pattern *Psychology* Identified by the German psychologist L W Stern (1871–1938).

If the gratification of a specific need is thwarted, the drive will seek out some other object to satisfy its need, and this object is usually the closest approximation to its previously chosen object.

W Stern, *Character and Personality* (London, 1938)

NS

need-hierarchy theory *Psychology* One of the most frequently cited theories of motivation, proposed by the American psychologist Abraham Harold Maslow (1908–70).

Motivation may be conceptualized as an invariant hierarchy of motives. The five main divisions in Maslow's theory are often represented in pyramid form, with physiological needs at the base followed by safety needs, love and belonging needs, esteem and self-esteem needs, and the need to self-actualize at the apex. Many would dispute this theory's invariant nature, claiming, for example, that it may be possible to possess self-esteem without feeling loved.

J P Dwortezky, *Psychology*, 3rd edn (New York, 1985)

NS

needs-press theory of personality *Psychology* Developed by American psychologist Henry Alexander Murray (1893–), based on a definition of behaviour (as a function of both personality and the environment) made by the German-American psychologist Kurt Lewin (1890–1947).

Press are external and represent features of objects that have implications for individuals in their efforts to fulfil need (need being conceptualized as internal). Murray postulated two types of press, alpha and beta. Alpha press are those environmental stimuli that can effect a behavioural response and can be perceived and reported by an objective, knowledgeable observer. Beta press are divided into two subtypes: private beta press (the unique and private views each individual has of the environment), and consensual beta press (the interpretations shared by a group of individuals about the same environment).

G G Stern, *People in Context* (New York, 1970)

NS

negation, performative (or speech act) theory of *Philosophy* Theory that analyzes negation in terms of a special kind of linguistic activity, negating or denying; so that to say, for example, 'It's not raining' may indeed be (as anyone would agree in straightforward cases) to deny that it is raining, but is also to utter a sentence which gets its meaning from that very fact.

The alternative view would say one can only deny something by using a sentence that has its meaning independently of any act of denial. The German logician Gottlob Frege (1848–1925) raised an objection, relevant to other speech act theories too, about what happens when 'not' occurs in a clause governed by a phrase like 'if' or 'I wonder whether'. He also asked whether utterances such as 'Christ is immortal' count as assertions or denials.

G Frege, 'Negation', *Translations from the Philosophical Writings of Gottlob Frege*, P T Geach and M Black, eds (1952)

ARL

negative capability (1817) *Literary Theory* Phrase coined by the English poet John Keats (1795–1821).

Property of the poet, who is 'capable of being in uncertainties, mysteries, doubts, without any irritable reaching after fact and reason'. Beauty and truth are non-rational, non-partisan.

J Keats, 'Letter to George and Thomas Keats', *Literary Theory since Plato*, H Adams, ed. (New York, 1971), 474

RF

negative income tax (1962) *Economics* This term was first used by American economist Milton Friedman (1912–92) to describe a form of income maintenance which aims to bring low-income households living below the subsistence level up to a minimum income level set by the government.

M Friedman, *Capitalism and Freedom* (Chicago, 1962)

PH

negative law of effect (–1931) *Psychology* Formulated by the American psychologist Edward Lee Thorndike (1874–1949).

Responses followed by a negative state of affairs are less likely to recur. Therefore one unlearns behaviour which gives rise to negative consequences. Thorndike 'repealed' the negative law of effect in 1931 because responses followed by negative states tend not to be unlearned but suppressed.

J P Dworetzky, *Psychology* 3rd edn (New York, 1985)

NS

negative utilitarianism *Philosophy* Version of UTILITARIANISM which replaces the maximization of good by the minimization of evil.

Supporters of the theory, who include Karl Raimund Popper (1902–), say that by aiming at removing evils rather than achieving positive goods we shall avoid the disadvantages of UTOPIANISM usually incurred by those who try to plan for a perfect world. To take an extreme example, it is alleged that one might well increase either the total happiness or the average happiness by simply killing off the unfit and others.

H B Acton and J W N Watkins, 'Negative Utilitarianism', *Proceedings of the Aristotelian Society*, supplementary volume (1963); K R Popper, *The Open Society and Its Enemies* (1945)

ARL

neo-classical growth theory (late 19th century) *Economics* Forming part of the broader NEO-CLASSICAL THEORY, this is an analytical framework in which emphasis is placed on the easy substitution of labour and capital in the production function to generate a steady-state of growth, and where all variables are growing at a constant, proportionate, rate.

This steady state eliminates the instability of the HARROD-DOMAR GROWTH MODEL. Neoclassical growth models identify the sources of growth as technical progress and population increases, with capital accumulation determining the capital-to-labour ratio in the steady state.

J E Meade, *A Neoclassical Theory of Economic Growth* (London, 1962)

PH

neo-classical theory (19th century) *Economics* Influenced by classical economic theory, this school of thought developed after World War II in opposition to the Cambridge School. It focuses on microeconomic theory and explores the conditions of static equilibrium.

Neoclassical theory is essentially concerned with the problems of an economy enjoying equilibrium at full employment. The theory is also concerned with savings-determined investment, marginal utility and marginal rates of substitution. Leading adherents and developers of the theory were

John Clark (1884–1963), Francis Edgeworth (1845–1926), Irving Fisher (1867–1947), Alfred Marshall (1842–1924), Vilfredo Pareto (1848–1923), Léon Walras (1834–1910), and Knut Wicksell (1851–1926). *See* NEOCLASSICAL GROWTH THEORY.

J F Henry, *The Making of Neoclassical Economics* (London, 1990)

PH

neo-classicism *Art* Term used later to describe a movement beginning in the 1750s as a revival of Antique and Renaissance forms and ideals (and 17th-century CLASSICISM) throughout European art.

In architecture, the classical orders and geometric forms were favoured by exponents such as Sir John Soane (1753–1837). In painting, classical subject matter (especially Roman and Greek history) was preferred. French painter Jacques Louis David (1748–1825) and Italian sculptor Antonio Canova (1757–1822) depicted high moral standards and virtue, turning for inspiration to Antique sculpture and the work of Raphael (1483–1520). In so doing they repudiated the frivolity of Rococo art. These processes were aided by archaeological discoveries of the time. Contemporary theorists (for example, Winkelmann) recommended 'imitating the Ancients', but not literally copying them (except for learning), whilst also advocating the artistic purification and ennoblement of Nature.

By the 19th century the ideal of a universal eternal art had been politicized by French emperor Napoleon Bonaparte (1769–1821), and literal imitation and revivalism had distorted the high ideals of the earlier generation. However, elements of neo-classicism continued to exert influence throughout the era of ROMANTICISM.

Arts Council of Great Britain, *The Age of Neo-classicism*, exh. cat. (London, 1972)

LL

neo-classicism (1920s) *Music* A compositional style, characterized by its revival of the controlled forms and techniques used by earlier composers, in reaction to the excesses of late-Romanticism.

The term often implies an element of parody, particularly through its distortion of established tonal techniques (for example in Stravinsky's *Pulcinella* (1919–20)), though it may also apply to atonal or serial music, such as Schoenberg's *Piano Suite op.25* (1921–23). The term is somewhat misleading as it most commonly refers to music based on Baroque models.

E Salzman, *Twentieth-Century Music: An Introduction* (Englewood Cliffs, N.J., 1967), 43–67

MLD

neo-conservatism (20th century) *Politics* A term referring to the views of North American conservatives in the second half of the 20th century.

Conservatism in the United States became radical and assertive from the 1960s onwards, combining liberal economics and a suspicion of the state with authoritarian moral attitudes and a bellicose foreign policy. Such views constituted part of the NEW RIGHT.

Kenneth Hoover and Raymond Plant, *Conservative Capitalism in Britain and the United States* (London, 1989)

RB

neo-corporatism (20th century) *Politics See* CORPORATISM.

neo-Darwinism (19th century) Also called modern synthesis and the synthetic theory of evolution, 'neo-Darwinism' was coined by George John Romanes (1848–94), a Canadian-born British evolutionist.

This term has had two related meanings. One focuses on the idea that NATURAL SELECTION alone can explain evolution, without the need for any other mechanisms. The other, perhaps more common, meaning combines natural selection with an understanding of genetics (MENDELISM) that was unavailable in Darwin's time. In this neo-Darwinism the thing selected for is the gene, and theories of gene frequency (such as the HARDY-WEINBERG EQUILIBRIUM) play an important role.

P Bowler, *Evolution: The History of an Idea* (Berkeley, Calif., 1984)

KH

neo-functionalism (20th century) *Politics* A version of FUNCTIONALISM in international relations theory.

An attempt to account for the development of functional relationships which transcend individual states, particularly in regionally limited systems such as Europe and the European Community.

Graham Evans and Jeffrey Newnham, *The Dictionary of World Politics* (Hemel Hempstead, 1990)

RB

neo-grammarians (later 19th century) *Linguistics* Self-styled '*Junggrammatiker*' (1878), they were a group at Leipzig University. The main members were Karl Brugmann (1849–1919) and H Osthoff (1847–1909).

Historical linguists who proclaimed that the laws of sound change operated mechanically and consistently, without exception; apparent exceptions were the product of speakers overregularizing by analogy.

R H Robins, *A Short History of Linguistics* (London, 1967)

RF

neo-Impressionism (1884) *Art* Name given to an artistic movement by the French critic Félix Fénéon (1861–1944) after the exhibition of *Le groupe des Indépendants* in 1884.

Based on a colour theory similar to DIVISIONISM and POINTILLISM (all three relying on optical mixtures), the first neo-Impressionist painting was *Baignade* (1884) by Georges Seurat (1859–91). By applying dots of pure pigment which fused at a distance, he achieved a colour-saturated and luminous surface.

AB

neo-liberalism (20th century) *Politics* Theories of revived classical liberalism in the 20th century. The market is the most effective (or least irrational) method of distributing goods and resources, and the role of the state should be limited to the maintenance of necessary order, legality, and stability.

Desmond King, *The New Right: Politics, Markets and Citizenship* (London, 1987)

RB

neo-Marxism (20th century) *Politics* Theories of European Marxists after the Russian Revolution. As with all uses of the prefix 'neo', the essential distinction is between contemporary or near contemporary views, and earlier ones (for example, NEO-CONSERVATISM, NEO-CORPORATISM and NEO-LIBERALISM). The theories designated as 'neo-' Marxist were concerned in particular with culture and IDEOLOGY, and with the role of capitalist states' welfare institutions in retarding rather than advancing SOCIALISM.

L Kolakowski, *Main Currents of Marxism*, 3 vols (Oxford, 1978)

RB

neo-Platonism *Philosophy* Movement initiated by Plotinus (AD 205–70) and carried forward by various philosophers of the next three centuries, having repercussions in the Renaissance especially among the Cambridge Platonists of the 17th century and, later, Georg Wilhelm Friedrich Hegel (1770–1831).

Neo-Platonism claimed to interpret Plato (*c*.427–*c*.347 BC), and to reconcile Aristotle (384–322 BC) with Plato, though modern scholars dispute its success in either claim. It borrowed from other schools, but opposed EPICUREANISM.

Reality is seen as hierarchically ordered, having at its apex a spiritual entity or 'hypostasis', usually called the One, which is unknowable and ineffable and can be described only indirectly (negatively by abstraction, or by analogy). Other hypostases are derived from this by a process called 'emanation' (hence emanationism as a name for neo-Platonism): first intellect (*nous*), then soul (*psyche*, including discursive reason), and finally matter. These stages involve manifestations of the One, but of decreasing perfection and actuality; the changing world around us, for instance, involves potentialities that are not always actualized, and evil can only arise as a privation and where there is change and unrealized potentiality.

A H Armstrong, ed., *Cambridge History of Later Greek and Early Medieval Philosophy*, 2nd edn (1970)

ARL

neo-Pythagoreanism *Philosophy* A revival in the 1st century BC and the next century or two of various features traditionally associated with the followers of Pythagoras (*fl.*6th century BC); *see* PYTHAGOREANISM.

Though of some minor importance as an influence on NEO-PLATONISM, the movement largely occupied itself with arithmetic and arithmology (attributing metaphysical and mystical properties to numbers), developing material from Plato (*c*.427–*c*.347 BC) and other philosophical movements along these lines.

D J O'Meara, *Pythagoras Revisited: Mathematics and Philosophy in Late Antiquity* (1989)

ARL

neo-Ricardian theory (1960s) *Economics* English economist David Ricardo (1772–1823) looked to labour theory in his unsuccessful search for an invariable measure of value. The Italian-born economist Piero Sraffa (1898–1983) attempted to solve the problems raised in Ricardo's and Marx's work by viewing prices in terms of wages, quantities of capital, and profits, rather than as labour time. He asserted that real prices (measured in terms of composite, or standard commodities) do not change unless technology does.

P Sraffa, *Production of Commodities by Means of Commodities* (Cambridge, 1960)

PH

neo-Romanticism (1970s) *Music* A compositional style employing the large forms, smooth orchestral writing and lush harmonies generally associated with the Romantics. Usually viewed as a reaction against the complexities of mid-20th-century music, it has gained little validity as a movement; more often it is associated with individual composers rejecting the complexity of their own former compositional style, as in the case of the Polish composer Krzysztof Penderecki (1933–). The return to a more direct, emotive musical language may be interpreted as a political statement, attempting to reach the masses after years of alienation, but is liable to be dismissed as mawkishness.

MLD

neptunism (18th–19th centuries) *Geology* Initiated by the German geologist Abraham Gottlib Werner (1749–1817), this theory proposed that the Earth's rocks had crystallized from an original 'sea' of aqueous solutions which included all Earth materials. This process was followed by erosion, resulting in sediments; and internal heating resulting in intrusions of volcanic rocks.

The theory was placed on a more scientific basis by the Scottish geologist James Hutton (1726–97), with no resort to supernatural causes. It is now recognized that the Earth resulted from the process described by the ACCRETION HYPOTHESIS, though many of Hutton's concepts still form the basis of physical and dynamic geology.

R Porter, *The Making of Geology: Earth Science in Britain, 1660–1815* (Cambridge, 1977)

DT

Nernst-Lindemann theory of heat capacities (1911) *Physics* Named after its originators, the German physical chemist Walther Hermann Nernst (1864–1941) and the British scientist Frederick Alexander Lindemann, Viscount Cherwell (1886–1957).

Nernst and Lindemann suggested as an empirical result that the single vibrational frequency ν of the atoms in the Einstein theory for the specific heat capacity of an insulating solid should be replaced by two frequencies of equal weight and having values ν and $\nu/2$. This gives good agreement with experiment over a wide range of temperatures, but is unsatisfactory as absolute zero is approached. *See* EINSTEIN'S THEORY OF HEAT CAPACITY.

J Thewlis, ed., *Encyclopaedic Dictionary of Physics* (New York, Oxford and London, 1962)

MS

Nernst's heat theorem (third law of thermodynamics) (1906) *Chemistry/Physics* Named after the German physical chemist Walther Hermann Nernst (1864–1941) who introduced it as his New Heat theorem.

The original statement was that, as the temperature T of a system in thermodynamic equilibrium approaches absolute zero, the change in the Helmholtz function F and in the internal energy U in an isothermal process tends to zero; that is,

$$\lim_{T\to 0} \frac{\mathrm{d}\Delta F}{\mathrm{d}T} = \lim_{T\to 0} \frac{\mathrm{d}\Delta U}{\mathrm{d}T} = 0\,.$$

This statement is rather restrictive in its application. A modern version of the law, formulated by German physicist Franz

Eugen Simon (1893–1956) in 1927, states that the contribution to the entropy of a system by each aspect that is in internal thermodynamic equilibrium tends to zero as the temperature tends to zero. One consequence of the third law is that it is impossible to reduce the temperature of any system, or part of a system, to absolute zero in a finite number of operations. This result is sometimes taken as the statement of the third law.

J Thewlis, ed., *Encyclopaedic Dictionary of Physics* (New York, Oxford and London, 1962)

MS

Nernst's law (1889) *Chemistry* Named after the German chemist Walther Hermann Nernst (1864–1941) who was Professor of Chemistry in Berlin from 1905, and was famous for his formulation of the THIRD LAW OF THERMODYNAMICS in his famous heat theorem of 1906. He won the Nobel prize for Chemistry in 1920. The law states that the zero-current electrode potential (E) of a reversible electrode immersed in a solution of an ion of valency z_+ is given by

$$E = E^\ominus + \left(\frac{RT}{z} + F\right) \ln a_+$$

where $E^\ominus$ is the standard electrode potential, R is the gas constant, F is the Faraday constant, and a_+ is the activity of the ion.

P W Atkins, *Physical Chemistry* (Oxford, 1990)

GD

nervism *Psychology* An hypothesis by the Russian physiologist Ivan Petrovich Pavlov (1849–1936).

This holds that all functions of the body are controlled by the nervous system. Greatly influenced by the belief of Ivan Mikhaylovich Sechenov (1829–1905) in the reflexological nature of all psychic activity, Pavlov regarded physiological rather than psychological activity to be the primary source of study and did not think behaviour could be explained in psychological terms.

I P Pavlov, *Lectures on Conditioned Reflexes: Twenty-five Years of Objective Study of Higher Nervous Activity (Behaviour of Animals)* (New York, 1928)

NS

network model (20th century) *Psychology* Proposed by the American psychologist M R Quillan, this is a generic group of hypothetical models of human semantic memory, based on the assumption that representations in memory are stored in a complex network of interactions and associations. What distinguishes this theory from basic associationistic models is that the 'link' between memory representations (the 'nodes' in the network) are labelled and note the nature of the relations.

M R Quillan, 'Semantic Memory', *Semantic Information Processing*, M Minsky, ed. (Cambridge, Mass., 1968)

NS

Neue Sachlichkeit (1920s) *Art* A German name (meaning 'new objectivity') for the group of artists led by Otto Dix (1891–1969) and George Grosz (1893–1959). They continued and emphasized the elements of social comment found in German EXPRESSIONISM to create works which condemned social hypocrisy, corruption and cruelty with unrelenting bitterness and realism. Their work was banned and the movement broke up following the rise of the Third Reich; however, Dix avoided persecution by painting seemingly innocuous landscapes.

H Osborne, *Oxford Companion to 20th-Century Art* (Oxford, 1981)

LL

Neumann-Kopp law (1831) *Chemistry* Named after the German chemist Franz Ernst Neumann (1798–1895) who extended the DULONG-PETIT LAW, and the German chemist Hermann Kopp (1817–92). The law states that the molecular heat (the molecular weight times the specific heat) of a compound is equal to the sum of the atomic heats of its constituent elements.

Kopp discovered while attempting to verify the Neumann law that the relationship was in fact more complex.

GD

Neumann's law (1831) *Physics* Named after its discoverer, the German physicist Franz Ernst Neumann (1798–1895).

The product of the relative molecular mass and specific heat capacity is constant for all compounds having the same general formula and being similarly co-ordinated (for example, Al_2O_3, Cr_2O_3), but the constant varies from one series to another. This is an extension of the law of the DULONG AND PETIT LAW.

J Thewlis, ed., *Encyclopaedic Dictionary of Physics* (New York, Oxford and London, 1962)

MS

Neumann's principle (1823 and 1830) *Chemistry* Named after the German chemist Franz Ernst Neumann (1798–1895), this principle requires that the symmetry elements of the point group of a crystal must be included in the symmetry elements of any physical property of the crystal.

GD

neural Darwinism (1980s) *Biology/Medicine* A term coined by American scientist Gerald Maurice Edelman (1929–), who won the Nobel prize for Physiology or Medicine in 1972, this is the theory that selection and adaptation shape the human mind throughout life in the same way that they shape the physiology and morphology of species.

According to this theory, neuronal groups that respond successfully to stimuli are selected for, while those that are counterproductive are gradually eliminated.

G M Edelman, *Neural Darwinism: The Theory of Neuronal Group Selection* (1987); M Shafto, *How We Know* (New York, 1986)

KH

neural networks (*c.*1985) *Computing* Techniques for organizing searches, for example, in problems of perception.

B T Khanna, *Foundations of Neural Networks* (London, 1989)

JB

neurolinguistics (1950s–) *Linguistics* A research field applicable to LANGUAGE PATHOLOGY and to PSYCHOLINGUISTICS.

The study of the neurological bases of language and speech, and the neurological processes involved in language use (including impaired language use). An interdisciplinary field, it involves linguistics (see CLINICAL LINGUISTICS), psycholinguistics and neurology. *See also* BROCA'S AREA and WERNICKE'S AREA for the classic 19th-century background and a textbook reference.

H and H A Whitaker, *Studies in Neurolinguistics*, vol. I (New York, 1976)

RF

neutral monism *Philosophy* Theory associated primarily with William James (1842–1910), who named it, and Bertrand Russell (1872–1970); though it has affinities to the views of Ernst Mach (1838–1916), Henri Bergson (1859–1941) and others.

Neutral monism says that mind and matter can both be reduced to a single type of thing, sometimes called 'neutral stuff'. This took the form of sensations or experiences, which constituted minds if thought of as arranged in one way, and matter if in another.

B Russell, *The Analysis of Mind* (1921)

ARL

neutral theory of molecular evolution (early 1970s) *Biology* Developed by Japanese geneticist Motoo Kimura (1924–), with support from Americans Jonathan Alan King (1941–), a molecular biologist and developmental geneticist, and Thomas Hughes Jukes (1906–), a biochemist. This is the controversial idea that most changes occurring in the amino acids of individuals will be neither good nor bad but neutral, and will persist in a population or vanish according to chance.

This theory assumes that changes in amino acids providing a survival advantage will be selected for through NATURAL SELECTION, while those that are harmful will be eliminated. If this idea is accepted, then mathematical proof shows that the rate of evolution for a given protein type will be roughly constant over millions of years, and that those proteins having the most complex form will evolve the slowest. This concept also provides the theoretical foundation for MOLECULAR CLOCK calculations.

M Kimura, *The Neutral Theory of Molecular Evolution* (Cambridge, 1983)

KH

new age (20th century) *Politics/Sociology* Semi-ethical, semi-religious theory of cultural renewal.

So called 'new-age theory' is not a coherent doctrine, but a mixture of cultural, ecological, pantheistic and communalist beliefs which from the 1980s, particularly in North

American and Western Europe, were presented as an alternative to 'consumerist' values. *See also* COMMUNALISM.

Monica Sjooo, *New Age and Armageddon* (London, 1992)

RB

new class (20th century) *Politics/Economics* Theory of élites in state socialist ('communist') societies of the Yugoslav writer Milovan Djilas (1911–).

Within state socialist societies a 'new class' of party officials – who exercise a command over resources similar to that exercised by capitalists – has arisen to frustrate the egalitarian intentions of the regimes' founders.

Milovan Djilas, *The New Class* (London, 1957)

RB

new classical macroeconomics (1970s) *Economics* Developed by American economists Robert Lucas (1937–) and Thomas Sargent (1943–), and British economists Patrick Minford (1943–) and Michael Beenstock (1946–). New classical macroeconomists argue that the economy will settle at a natural rate of unemployment and attempts to alter this equilibrium state will be counteracted by economic agents.

When the Keynsian dominance of macroeconomics ended in the early 1970s, several new schools of economic thought arose. There are three main facets to the new classical macroceconomics: (1) the real economic decisions of agents (for example, saving, consumption, or investment) are based on real not nominal or monetary factors; (2) agents are held to be continuously in equilibrium; (3) agents hold on to their rational expectations. The RATIONAL EXPECTATIONS HYPOTHESIS and the NAIRU (NON-ACCELERATING INFLATION RATE OF UNEMPLOYMENT) are important propositions. *See* KEYNSIAN ECONOMICS and 'LAISSEZ-FAIRE'.

K D Hoover, *The New Classical Macroeconomics* (Oxford, 1988)

PH

new complexity (1980s) *Music* Associated with the English composers Brian Ferneyhough (1943–) and James Dillon (1950–) and some of their contemporaries in France and Germany, this term denotes a type of music produced by combining extremely detailed compositional processes.

Micro- intervals, complex rhythms, a wide variety of timbres and 'effects', and an extreme range of subtly shaded dynamics are explored within a multi-layered framework. This produces an incredible density of sound, where details are lost and overall texture emerges as the music's most audible feature.

MLD

new criticism (*fl. c.*1930–60) *Literary Theory* Dominant American critical approach, founders of which were John Crowe Ransom (1888–1974) (*The New Criticism*, 1941), Allen Tate (1899–1979), Robert Penn Warren (1905–89); also Cleanth Brooks (1906–), William Kurtz Wimsatt (1907–75), René Wellek (1903–). It was indebted to ideas of the English critic Ivor Armstrong Richards (1893–1979) and the expatriate American poet Thomas Stearns Eliot (1888–1965).

Wanting to display the richness of poetry as against the rationality and scepticism of science, against modern faithlessness, and against the barrenness of urban life, these Southern American, conservative, Christian poets and critics established a practice of close reading of the language of 'the poem itself'. They rejected reference to historical context, the intentions of the author, and effects on the reader, as considerations in critical reading. Poems were autonomous and rich structures of meaning, full of tension, AMBIGUITY, PARADOX, IRONY; unparaphraseable. *See also* AFFECTIVE FALLACY, AUTONOMY OF LITERARY TEXT, HERESY OF PARAPHRASE, INTENTIONAL FALLACY and PRACTICAL CRITICISM. New Criticism's premises and aims were later widely attacked by STRUCTURALISM, POSTSTRUCTURALISM and LINGUISTIC CRITICISM.

C Brooks, *The Well Wrought Urn* (London, 1968 [1947]); M Krieger, *The New Apologists for Poetry* (Westport, 1977 [1957])

RF

new historicism (1980s) *Literary Theory* Largely American critical practice pioneered, and named in 1982, by Stephen J Greenblatt. (More recently termed 'poetics of culture', again by Greenblatt in 1989.)

A form of cultural criticism concerned with texts of all genres in their historical situations, particularly in the context of the relations of power and of exchange which constitute capitalism. New Historicism is not necessarily of the Left (as CULTURAL MATERIALISM tends to be).

H A Veeser, ed., *The New Historicism* (New York, 1989)

RF

new left (20th century) *Politics* Body of radical cultural and political theories of the 1960s and 1970s.

CONVERGENCE promoted oligarchy, orthodoxy, and the wilting of DEMOCRACY in both Eastern and Western Europe. New participatory and decentralized forms of political and social organization would be an antidote to this. FEMINISM and INDUSTRIAL DEMOCRACY were key elements.

Rodney Barker, *Political Ideas in Modern Britain* (London, 1978)

RB

new liberalism (20th century) *Politics* Adaptation of LIBERALISM to COLLECTIVISM.

The liberal demand for freedom was best met, in a DEMOCRACY, by the use of STATE power to enhance the material opportunities of citizens, thus promoting equal liberty.

Rodney Barker, *Political Ideas in Modern Britain* (London, 1987)

RB

new riddle of induction *Philosophy* *See* GOODMAN'S PARADOX.

new right (20th century) *Politics* Alliance of economic liberal and conservative theory.

In economic affairs, the STATE should encourage free markets and private economic enterprise. In cultural and moral affairs it should sustain traditional values in education and family life.

Desmond King, *The New Right: Politics, Markets and Citizenship* (London, 1987)

RB

new simplicity (1970s) *Music* Associated with a number of Dutch and Scandinavian composers, notably Louis Andriessen (1939–) and Poul Ruders (1949–), this term denotes a type of music that mixes minimalist techniques with other compositional styles. The resulting works clearly articulate their form and expose their methods of construction, evoking in the listener a feeling of comprehension.

MLD

Newlands's law of octaves (1863) *Chemistry* Named after the English chemist John Alexander Reina Newlands (1837–98) who worked in a sugar refinery at Victoria Docks, London. Newlands, pre-dating MENDELEYEV'S PERIODIC LAW, proposed a connection between every eighth (hence 'octaves') element arranged in increasing atomic weight – an idea for which he was ridiculed.

Eventually, the Royal Society awarded him the Davy medal in 1887 for his work. His rigid arrangement, however, did not allow for as yet undiscovered elements, and breaks down after the first two short periods of the periodic table.

GD

Newton-Kantorovich theorem *Mathematics* Named after British physicist, astronomer, and mathematician Sir Isaac Newton (1642–1727) and Russian mathematician Leonid Vital'evich Kantorovich (1912–), this contraction theorem gives precise conditions for NEWTON'S METHOD to converge to a root of the equation $f(x) = 0$.

Given a continuously differentiable function

$$f : C \subseteq \Re^n \to \Re^n$$

on the convex set $C_0 \subseteq C$ which satisfies the conditions

(a) $\|Df(x) - Df(y)\| \leq \gamma\|x - y\|$ for all $x, y \in C_0$

(b) $\|Df(x_0)^{-1}f(x_0)\| \leq \alpha$

(c) $\|Df(x_0)^{-1}\| \leq \beta$

for some x_0 in C_0, define the constants $h := \alpha\beta\gamma$ and

$$r_{1,2} := ((1 \mp \sqrt{1 - 2h}\ )/h$$

If $h \leqslant 1/2$ and $S_{r_2}(x_0) \subseteq C_0$ then the Newton iterates remain in $S_{r_1}(x_0)$ and converge to the unique zero of $f(x)$ in $C_0 \cap S_{r_2}(x_0)$.

J Stoer and R Bulirsh, *Introduction to Numerical Analysis* (New York, 1980)

ML

Newtonian mechanics (1687) *Physics* This is the name given to the set of laws and conditions systematized by English physicist Sir Isaac Newton (1642-1727) and published in 1687.

Newtonian mechanics gives the laws of motion, describes the motion of bodies in the everyday world, and effectively presents the concept of force. Embodied in this system is the Newtonian principle of relativity which states that the laws of mechanics are equally valid to all observers moving uniformly with respect to each other. *See* NEWTON'S LAWS OF MOTION.

J Thewlis, ed., *Encyclopaedic Dictionary of Physics* (New York, Oxford and London, 1962)

MS

Newton's law of cooling (1701) *Physics* Named after its discoverer, the English physicist Sir Isaac Newton (1642–1727), this states that when a body is losing heat under conditions of forced convection, the rate of loss of heat $\mathrm{d}Q/\mathrm{d}t$ is proportional to the surface area A of the body and to the difference between the temperature T of the body and the temperature T_s of its surroundings.

Then,

$$\frac{\mathrm{d}Q}{\mathrm{d}t} = hA\,(T - T_s)$$

where h is a constant for a particular arrangement, now known as the heat transfer coefficient.

J Thewlis, ed., *Encyclopaedic Dictionary of Physics* (New York, Oxford and London, 1962)

MS

Newton's law of universal gravitation *Physics See* GRAVITATION, LAW OF UNIVERSAL.

Newton's law of viscosity (1687) *Physics* Named after the English physicist Sir Isaac Newton (1642–1727) who first proposed that, under conditions of steady streamline flow, the shearing stress needed to maintain the flow of the fluid is proportional to the velocity gradient in a direction to the direction of flow.

Consider two parallel planes, each of area A, immersed in a fluid and a distance L apart. If the force needed to keep them moving with a relative velocity v is F, the shearing stress is F/A and the velocity gradient normal to the direction of flow is v/L. Then,

$$\frac{F}{A} = \eta\,\frac{v}{L}$$

where η is known as the coefficient of viscosity or dynamic viscosity of the fluid. Substances for which η is a constant at constant temperature, independent of (F/A), are known as Newtonian fluids.

J Thewlis, ed., *Encyclopaedic Dictionary of Physics* (New York, Oxford and London, 1962)

MS

Newton's method *Mathematics* Named after British physicist, astronomer and mathematician Sir Isaac Newton (1642–1727), this is an iterative method for approximating solutions to a non-linear equation $f(x) = 0$ by repeatedly computing

$$x_{\mathrm{NEW}} = x_{\mathrm{OLD}} - \frac{f(x_{\mathrm{OLD}})}{f'(x_{\mathrm{OLD}})}.$$

This is equivalent to approximating a function by its tangent and converges quadratically if the initial estimate is sufficiently close to the root. In more than one dimension Newton's method is

$$x_{\mathrm{NEW}} = x_{\mathrm{OLD}} - G^{-1}[f(x_{\mathrm{OLD}})]$$

where G is the matrix of partial derivatives of f evaluated at x_{OLD}. *See also* Newton-Kantorovich theorem.

J Stoer and R Bulirsh, *Introduction to Numerical Analysis* (New York, 1980)

newtsex model (20th century) *Biology* Also called Newtsex MK IV, this is the most useful of the four newtsex models developed by Houston. This is a complex, computer-generated model that accurately predicts the courtship behaviour of newts.

The model holds that negative feedback loops among the male's assessment of the female's readiness ('hope'), the male's movement toward the female ('creep'), the male's pulling away from the female ('retreat'), the male's ability to deposit a spermatophore ('deposit') and the oxygen debt of both animals determine when mating occurs.

KH

Neyman-Pearson lemma *Mathematics* Named after Jerzy Neymann and Karl Pearson, who is considered to be the founder of 20th century STATISTICS, this is the result whereby of all the tests of a given hypothesis with the same significance level the LIKELIHOOD RATIO TEST has the maximal power. Nearly all tests now in use for testing parametric hypotheses are likelihood ratio tests.

R V Hogg and A T Craig, *Introduction to Mathematical Statistics* (New York, 1967)

ML

Nicod's criterion *Philosophy* A criterion offered by French philosopher Jean Nicod (1893–1924) for when one proposition confirms another.

A hypothesis of the form 'All A are B' is confirmed by objects that are A and B, and disconfirmed by objects that are A and not B, objects that are not A being irrelevant. An advantage of this last clause is that it avoids certain paradoxes raised by Carl G Hempel (1905–); but a disadvantage (apart from its being limited in scope) is that a hypothesis will need different evidence to confirm or disconfirm it, according to the terms in which it is formulated.

C G Hempel, 'Studies in the Logic of Confirmation', *Mind* (1945); reprinted with additions in C G Hempel *Aspects of Scientific Explanation* (1965)

ARL

nihilism *Philosophy* Term invented or popularized by Russian novelist Ivan Sergeevich Turgenev (1818–83) in his novel *Fathers and Sons* (1861) for the rejection of all traditional values.

Literally meaning 'nothingism', the term can be applied to views saying that all knowledge is impossible, that all alleged metaphysical truths or values are illusory, or that ethical values cannot be given any foundation and so are arbitrary. 'Nihilism' has been applied particularly to a movement in Czarist Russia which held that any means were permissible in overthrowing the existing order (the value of overthrowing it being tacitly taken for granted), and to later offshoots and imitations of that movement elsewhere. The term is in fact seldom used in modern English-speaking philosophy.

ARL

NIMBY (Not In My Back Yard) (20th century) *Politics* Phrase attributed to Walton Rodger and widely linked to British Conservative politician and minister Nicholas Ridley (1929–93).

There are always those who attempt to exclude themselves from the consequences of policies which in general they support. A house owner might advocate new motorways, but not at the bottom of his or her own back-garden. NIMBY as an observation about the behaviour of both individual citizens and their parliamentary representatives was gleefully applied to many public figures.

RB

no-ownership theory of the mind *Philosophy* Theory that states of consciousness exist in their own right and are not owned by some substantive entity such as a mind, a person, or even a body (or brain). The theory fits with a BUNDLE THEORY of the self.

P F Strawson, *Individuals* (1959), ch. 3; critical

ARL

noblesse oblige *Politics* Theory of aristocratic duties. Nobility, distinction, ARISTOCRACY or good fortune are privileges which therefore involve corresponding obligations to public service.

RB

'no hair' theorem (1966) *Astronomy* A result discovered by Roy Kerr, the theorem is a mathematical argument proving that a black hole can possess, at most, three observable properties accessible to outside observers; namely, its mass, electric charge and angular momentum.

Two black holes with equal values of these three properties cannot be distinguished; that is, they have no superficial properties ('no hair') that makes them individually distinguishable.

S Mitton, ed., *The Cambridge Encyclopaedia of Astronomy* (London, 1973)

MS

nominalism *Philosophy* Any view which analyzes a given subject-matter in terms of words or language, derived from the Latin '*nomen*' meaning 'name', 'term' or 'word'.

A nominalist view of universals (*see* PLATONISM) says they are neither substantive realities (REALISM) nor mental concepts (CONCEPTUALISM). Rather, they are simply words which we apply to a group of objects; the members of the group owing their membership to resembling each other in some relevant respect. (But this leads to difficulties: is not resemblance itself a universal? And what about the respect in which the resemblance holds?)

A nominalist about definitions says there can only be *nominal definitions* (accounts of how a word is or should be used), not *real* definitions (analyses of a concept or thing which the word is supposed to stand for: '*res*' is Latin for 'thing').

A nominalist about modalities says there are only *de dicto*, not *de re*, modalities; that is, roughly, statements may be necessarily true, but things do not have necessary properties. (Necessarily, a husband has a wife; but no man has the property of necessarily having a wife.) *See also* CONCEPTUALISM, PARTICULARISM and REALISM.

M J Loux, ed., *Universals and Particulars* (1970)

ARL

non-accelerating inflation rate of unemployment *Economics See* NAIRU.

non-cognitivism *Philosophy* Theory that there is no such thing as knowledge of truths in a certain sphere because there are no such truths to be known. The sphere normally intended by the term is ethics, and non-cognitivists adopt a SPEECH ACT THEORY when analyzing what appear to be moral or value statements. EMOTIVISM and PRESCRIPTIVISM are forms of non-cognitivism, while DESCRIPTIVISM is a form of cognitivism.

ARL

non-common effects principle *Psychology* Proposed by the British psychiatrist A E Jones (1879–1958).

The fewer distinctive reasons an actor has for an action, and the less the reasons are stated in culture, the more informative is the action about the intentions and dispositions of the actor. The disposition of intention begun by an action is most readily seen by acknowledging the non-common consequences of alternative actions. The fewer non-common effects of the action and alternative actions, the more readily an attribution of intention of disposition can be made. *See* ATTRIBUTION THEORY.

F Heider, *The Psychology of Interpersonal Relations* (New York, 1958)

NS

non-competing groups *Economics* Identified by English economist John Stuart Mill (1806–73), and named by Irish political economist John Cairnes (1823–75), this term describes groups of individuals who are excluded from entering certain professions.

Originally viewed as the result of disproportionate education opportunities, this analysis has been extended to exclusion on the grounds of discrimination and trade union/craft barriers to entry to an industry. *See* DUAL MARKET LABOUR THEORY, CROWDING HYPOTHESIS, SEGMENTED LABOUR MARKET THEORY, LABOUR MARKET DISCRIMINATION, SEARCH THEORY and INSIDER-OUTSIDER WAGE DETERMINATION.

J E Cairns, *Some Leading Principles of Political Economy* (London, 1874)

PH

non-continuity in learning theory *Psychology* A theory of the American psychologist Burrhus Frederic Skinner (1904–90).

The theory argues that an animal in an experiment initially responds to a stimulus on the basis of irrelevant but discriminative features. It is the subsequent training (by reward or punishment) that determines the animal's response to the stimulus; hence, on the basis of being rewarded or punished for responding (or not) to the stimulus, it becomes a discriminating stimulus.

S H Hulse, H Eyeth and J Deese, *The Psychology of Learning* (New York, 1980)

NS

non-crossing rule *Chemistry* A rule which states that when plotted as a function of internuclear distance, the potential energy curves of two electronic states of a diatomic molecule do not cross unless the states have different symmetry.

GD

non-nested hypotheses (1960s) *Economics* Based on work by English statistician David Cox (1924–), this term refers to economic models or hypotheses which cannot be obtained from another model by the use of 'appropriate parametric restrictions or as a limit of a suitable approximation'. *See* ECONOMIC METHODOLOGY.

D R Cox, 'Tests of Separate Families of Hypotheses', *Proceedings of the Fourth Berkeley Symposium on Mathematical Statistics and Probability* (Berkeley, 1961)

PH

non-profit organization *Economics* This theory posits the existence of organizations pursuing different objectives (and working under different constraints) to the functional profit-making model.

Usually supported by private/public grants or donations, such organizations are implicity or explicitly non-profit making. Examples include health foundations, charities, and clubs.

R S Gassier, *The Economics of Non-profit Enterprise: A Study in Applied Economic Theory* (New York and London, 1986)

PH

normative decision theory of leadership *Psychology* Proposed by the Austrian-American sociologist Peter Blau (1918–).

This theory prescribes the conditions under which leaders should make decisions autocratically, or in consultation with the group members, or with group members fully participating. The theory assumes: (1) individual decisions are more time-effective than group decisions; (2) subordinates are more committed to a decision if they participate in its formulation; (3) complex and ambiguous tasks require more information and consultation for reaching high-quality decisions. *See* LEADERSHIP THEORIES and PATH-GOAL THEORY OF LEADERSHIP.

P M Blau, *Exchange and Power in Social Life* (New York, 1964)

NS

normative theory (20th century) *Mathematics* A branch of modal logic.

JB

nouveau réalisme (1960) *Art* A style named after the *nouveaux réalistes* ('new realists'), a group founded by French art critic Pierre Restany. Jean Tinguely (1925–), Yves Klein (1928–62) and Arman (1928–) were adherents, rejecting traditional REALISM and conventional forms of paintings for the use of real materials, new lighting techniques and the recycling of artefacts.

J Becker and W Vostell, *Happenings, Fluxus, Pop Art, Nouveau Réalisme* (Reinbek bei Hamburg, 1968); P Restany, *Nouveau Réalisme* (1960)

AB

nouveau roman (1950s–60s) *Literary Theory* Meaning the 'new novel', this was a French practice of novel-writing, the best-known exponent of which was Alain Robbe-Grillet (1922–).

Anti-realist movement in which linguistic FOREGROUNDING and structural deformations deliberately interfere with the transparency of the text, drawing attention to its artificiality and constructedness. *See also* METAFICTION, MODERNISM and REALISM.

S Heath, *The 'Nouveau Roman'* (London, 1972)

RF

nouvelle tendance (1961) *Art* An artistic movement, the name for which originated in exhibitions called 'Nove Tendencje' held at the Gal. Suvremene Umjetnosti in Zagreb from 1961 and subsequent similar exhibitions in Europe and South America. Common to all the strains of the movement was an affiliation with KINETIC ART and programmed art, for which pure visuality and the depersonalization were foremost principles.

AB

novel (18th century) *Literary Theory* The dominant literary form in modern times, and subject of intense theoretical discussion in the 20th century.

An extended fictional narrative in prose. The novel genre is extremely catholic as regards orientation, content, structure, and style; it has reflected a number of different social and intellectual trends through its history (*see* REALISM, NATURALISM, STREAM OF CONSCIOUSNESS) and has been a major medium of literary experimentation (*see* METAFICTION, MODERNISM). For technical details of relevant theories *see* FICTION, NARRATIVE and references.

J Halperin, ed., *The Theory of the Novel: New Essays* (New York, 1974); I Watt, *The Rise of the Novel* (Berkeley, 1957)

RF

novel of development *Literary Theory See* BILDUNGSROMAN.

novel of education *Literary Theory See* BILDUNGSROMAN.

NP-complete problem (*c.*1970) *Mathematics* A decision problem is one with a yes or no answer and it is in the class *P* if it has a polynomial time algorithm; that is, an algorithm for which the number of individual operations increases like a polynomial as a function of the input. A computation is in the class *NP* if the decision problem concerning a guessed solution is in *P*. A decision problem is *NP*-complete if it has a polynomial time algorithm if and only if $P = NP$.

JB

nuclear winter theory (20th century) *Physics* This is the view that, following a nuclear war, there would be an enormous reduction in the amount of sunlight reaching the Earth's surface as a result of the dust and smoke accumulated in the Earth's atmosphere. Consequently, there would be severe climatic changes ('nuclear winter'). This would have a catastrophic effect on plants and animals.

MS

nucleosynthesis *Physics* A theory which attempts to account for the creation of the lighter chemical elements through nuclear reactions.

Various elementary particles and antiparticles were created within a fraction of a second after the big bang; these produced photons of radiation through annihilation. Deuterium and helium nuclei were next to be synthesized in the following 100 seconds; most of the helium in the universe today was formed at this time. Some 10,000 years later, an ionized gas of free electrons, protons and helium nuclei was formed. Neutral hydrogen was formed about 300,000 years after the big bang when the temperature had dropped further. The heavier elements are synthesized in nuclear reactions occurring largely in stars and in supernova explosions.

See also BIG BANG THEORY, FRIEDMANN UNIVERSES and STEADY-STATE HYPOTHESIS.

Paul Davies, ed., *The New Physics* (Cambridge, 1989)

GD

nucleus, models of *Physics See* COLLECTIVE MODEL OF THE ATOMIC NUCLEUS, LIQUID DROP MODEL OF THE ATOMIC NUCLEUS, and SHELL MODEL OF THE ATOMIC NUCLEUS.

null hypothesis *Mathematics See* HYPOTHESIS TESTING.

ML

number theory *Mathematics* The branch of mathematics whose principal concern is the study of whole numbers and their extensions both algebraic and analytic. This includes the study of divisibility, primality, and factorization properties for integers, as well as the study of irrational and transcendental numbers. The subject is as old as mathematics itself and is intimately linked with nearly all its branches.

A Baker, *A Concise Introduction to the Theory of Numbers* (Cambridge, 1984)

MB

numerical analysis *Mathematics* The study of the computation of solutions to mathematical problems. Methods generally based on the discretization of equations have been developed for problems in areas including integration, differentiation, solution of linear and non-linear equations, ordinary and partial DIFFERENTIAL EQUATIONS, interpolation, COMBINATORICS and constrained optimization. With the rapid development of digital electronic computers since the 1940s, numerical algorithms are constantly being adjusted to take advantage of improved memory and speed in each new generation of computer.

M Hazewinkel, ed., *Encyclopaedia of Mathematics* (Dordrecht, 1988)

ML

Nyquist's noise theorem (1928) *Physics* Named after Harry Nyquist who deduced the result theoretically, the effect is sometimes called 'Johnson noise' after J B Johnson who verified the result experimentally.

Any electrical or electromechanical system in equilibrium with its surroundings will exhibit random fluctuations in the variables which describe the condition of the system; for example, the current flowing. If two terminals of the system offer an impedance $R + jX$ at a frequency ν, the mean square open-circuit thermal noise voltage at the terminals is given by

$$\langle \mathrm{d}e^2 \rangle = 4kTR\,\mathrm{d}\nu$$

where k is Boltzmann's constant and R is the molar gas constant.

J Thewlis, ed., *Encyclopaedic Dictionary of Physics* (New York, Oxford and London, 1962)

MS

Nyquist's sampling theorem (1950s) *Mathematics* A function which is 0 outside the ranges from a to b and from $-b$ to $-a$ can be recovered from a sample of values spaced $2b$ apart. This is crucial to INFORMATION THEORY.

JB

O

object language *Linguistics See* METALANGUAGE.

object relations theory *Psychology* Proposed by the Austrian psychologist M Klein (1882–1960), and developed from the ideas of the Austrian founder of psychoanalysis Sigmund Freud (1856-1939).

This theory deals with the interaction between an object and a subject. Freud viewed early development of humans as occurring primarily through the satisfaction and inhibition of instinctual drives. Hence, he saw interactions occurring simply in order to satisfy these drives. However, object relations theory sees interaction as the goal of maturation and a way in which the development of the self occurs in relation to the corresponding ability to differentiate between self and object.

M Klein, *The Psychoanalysis of Children* (London, 1932)

NS

objective correlative (1919) *Literary Theory* Term coined by the expatriate American poet and critic Thomas Stearns Eliot (1888–1965).

'A set of objects, a situation, a chain of events' which are used to symbolize a particular emotion in poetry, preferred by Eliot to the direct statement of emotions as in Romantic poetry.

T S Eliot, 'Hamlet and his Problems', *Selected Essays* (London, 1951)

RF

objective idealism *Philosophy* Associated with Georg Wilhelm Friedrich Hegel (1770–1831) and his followers, notably in England Francis Herbert Bradley (1846–1924). *See also* COHERENCE THEORY OF TRUTH.

This is a form of idealism whereby reality, though mental or spiritual, does not depend on the human mind in particular but comprises a single spiritual entity: the ABSOLUTE (hence the name 'absolute idealism' also given to this view). Reality is one, and individual minds and their contents are mere parts or aspects of this and have no separate existence.

A Quinton, *Absolute Idealism* (1972, reprint of Dawes Hicks lecture at the British Academy in 1971)

ARL

objectivism (*c*.1930) *Literary Theory* Term coined by the American poet William Carlos Williams (1883–1963) for an American school of poetry of the 1920s.

Related to IMAGISM, objectivism laid particular stress on precision and detail of image and construction. Its best practitioner was perhaps the American poet Louis Zukovsky (1904–).

R F M Heller, *Conviction's Net of Branches: Essays on the Objectivist Poets and Poetry* (Carbondale, 1985)

RF

objectivism *Philosophy* Any theory saying of a given subject-matter that it contains objects existing independently of human beliefs or attitudes, or that there are similarly independent truths in the area, or that there are methods of studying the area and arriving at truths within it which are not arbitrary and do not depend on the approach adopted or convenience of application and so on.

The contrast term is SUBJECTIVISM. A half-way house exists when *intersubjective* agreement is possible; that is, agreement which does not depend on the position or attitudes of those in dispute, but does presuppose the existence of conscious experience. There may be, for example, standard methods for establishing what colour something is, even if without sighted creatures things might have no colour (in the ordinary sense) at all.

ARL

oblique rotator theory of Ap stars *Astronomy* A theory to explain cyclic variations in the magnetic field strength of magnetic Ap stars.

The term Peculiar A star (Ap star) is used to describe a range of stars between spectral types B5 and F5. The chief common feature of these stars is their possession of a strong and often variable magnetic field. This behaviour is explained by assuming that the stellar magnetic poles are not close to the poles of the rotation axis of the star. Then, as the star rotates, the north and south poles of the star alternately come round to the observable side.

S Mitton, ed., *The Cambridge Encyclopaedia of Astronomy* (London, 1973)

MS

observer's paradox (1966) *Linguistics* Methodological problem in sociolinguistics named and circumvented by the American William Labov.

Difficulty of sampling *spontanteous* speech: the presence of the investigator disturbs the naturalness of the situation. The paradox is that 'our goal is to observe the way people use language when they are not being observed'.

W Labov, *Sociolinguistic Patterns* (Philadelphia, 1972)

RF

occasionalism *Philosophy* The idea is attributed to L de la Forge (1632–66) in his *Treatise on the Spirit of Man* (1665), but the chief occasionalist was Nicolas Malebranche (1638–1715). *See also* PSYCHO-PHYSICAL PARALLELISM and PRE-ESTABLISHED HARMONY.

Occasionalism says that there is only one true cause, God, who causes what seem to be effects to appear on the occasions when what seem to be causes appear. If I strike a match, God will cause the match to light; the striking being called an *occasional cause* of the lighting.

The theory arose largely because of difficulties experienced by René Descartes (1596–1650) in saying how mind and body – which on his dualist view he treated as totally different substances – could interact causally with each other. The theory was therefore primarily applied to mind/body interactions, but was extended more widely in so far as it was felt that matter was passive and could not itself contain powers or forces.

N Malebranche, *Dialogues on Metaphysics and on Religion* (1688; translated 1923), 7th Dialogue

ARL

occupation segregation *Economics* This term refers to an uneven distribution of male, female, ethnic, racial and religious groups in the labour force.

The most common example is the heavy concentration of women in nursing, retailing, and low-paid office employment. *See* CROWDING HYPOTHESIS, SEGMENTED LABOUR MARKET THEORY, DUAL LABOUR MARKET THEORY, LABOUR MARKET DISCRIMINATION, NON-COMPETING GROUPS, SEARCH THEORY and INSIDER-OUTSIDER WAGE DETERMINATION.

J A Jacobs, *The Sex Segregation of Occupations and the Career Patterns of Women* (Ann Arbor, 1983)

PH

Ockham's Razor *Philosophy See* PARSIMONY, PRINCIPLE OF.

octaves, law of *Chemistry See* NEWLANDS'S LAW OF OCTAVES.

octet rule (1919) *Chemistry* Also known as Lewis-Langmuir theory after the American chemists Gilbert Newton Lewis (1875–1946) and Irving Langmuir (1881–1957).

An octet is a chemically inert group of eight electrons in an atom or ion which form the most stable configuration of the outermost or valence electron shell. In forming a chemical bond, an atom or ion will attempt to complete its octet of electrons either by sharing electrons (covalence) or by gaining or losing electrons (electrovalence) from or to another atom or ion. Lewis believed that the electrons in a molecule such as Cl:Cl occupied the corners of two 'cubic atoms' bonded along a common edge by a pair of electrons. Many compounds, however, do not possess octets of electrons; for example, most transition-metal compounds, NO which is paramagnetic, BF_3 and SF_6. *See also* LANGMUIR'S THEORY and LEWIS'S COLOUR THEORY.

M Freemantle, *Chemistry in Action* (Basingstoke, 1987)

GD

officer problem *Mathematics* One of the first problems of the area of COMBINATORICS known as *design theory* (*compare with* BRUCK-RYSER-CHOWLA THEOREM), considered by Leonhard Euler (1707–83). It asks for ways of positioning 36 officers of 6 ranks and 6 regiments in a *Latin square*, requiring that each row and each column has a representative from each regiment and each rank.

AL

Ohm's acoustic law *Psychology* Formulated by the German physicist Georg Simon Ohm (1787–1854).

The law states that a complex tone is analyzed by the hearer into its frequency components. This analysis is normally unconscious, but a trained perceiver can learn to distinguish individual harmonics in a complex sound.

S Coren, C Porac, and L M Ward, *Sensation and Perception* (New York, 1984)

NS

Ohm's law of electricity (1827) *Physics* Named after its discoverer, the German physicist Georg Simon Ohm (1787–1854), this law states that the potential difference V across a metallic conductor is directly proportional to the current I through it, provided that the temperature remains constant. The constant of proportionality R is the resistance of the conductor:

$$V = IR .$$

J Thewlis, ed., *Encyclopaedic Dictionary of Physics* (New York, Oxford and London, 1962)

MS

Ohm's law of hearing (1843) *Physics/Psychology* Named after its originator, the German physicist Georg Simon Ohm (1787–1854), this law stated that the ear perceives only simple harmonic vibrations.

When a complex sound wave strikes the ear, the ear decomposes it into the same simple harmonic waves that would be obtained mathematically from Fourier analysis.

J Thewlis, ed., *Encyclopaedic Dictionary of Physics* (New York, Oxford and London, 1962)

MS

Okun's law (1970) *Economics* Named after American economist Arthur Okun (1926–80), this model states that the elasticity of the ratio of actual to potential output, with regard to a change in the employment rate, is a constant of roughly three.

Okun looked at the US GNP during the 1950s and 1960s and found that a one per cent rise in unemployment was associated with a three per cent decrement in the ratio of actual GNP to full capacity GNP. This became known as Okun's law.

A M Okun, *The Political Economy of Prosperity* (Washington, 1970)

PH

Olbers' paradox (1826) *Astronomy* Stated by the German physician and astronomer Heinrich Wilhelm Matthäus Olbers (1758–1840), the problem is: given a sky uniformly filled with stars and a light intensity falling off as the inverse square of the distance, why is the sky dark at night?

If the universe is uniformly populated with stars that have always been shining, the night should be as bright as the day. The paradox is resolved by two considerations: (1) the stars have not been shining for an infinite length of time; (2) because of the Doppler red shift of the light from distant galaxies caused by the speed of recession, the light from these distant galaxies is reddened. These two factors account for the observed brightness of the night sky.

S Mitton, ed., *The Cambridge Encyclopaedia of Astronomy* (London, 1973)

MS

oligarchy *Politics* Theory that power will normally be concentrated.

The term is used by Aristotle to describe the corrupt alternative to aristrocracy. It now commonly denotes the situation described by ÉLITISM and by the IRON LAW OF OLIGARCHY.

RB

oligopoly theory *Economics* First used by English humanist Sir Thomas More (1478–1535) in *Utopia* (1516), and later developed by the French economist Augustin Cournot (1801–77), this type of competitive model is characterized by a few suppliers producing a heavily differentiated good (differentiated through advertising, marketing and so on).

Cournot asserted that each firm set its price and output on the assumption that its rivals would not react at all. In this situation, each firm decreases its price and increases its output to control a larger market share. The result is a market in which prices are higher and output is lower than they would be in a more competitive market.

Other related theories suggest that: (1) firms recognize their interdependence and one firm sets the price with other firms following in its wake; (2) each firm acts as a leader; (3) firms assume that rivals will follow price changes and so are reluctant to alter them; (4) firms collude to set a market price and/or quantity. *See* COLLUSION THEORY, MONOPOLY, MONOPOLISTIC COMPETITION, ADMINISTERED PRICING, CONTESTABLE MARKETS THEORY and STRUCTURE-CONDUCT-PERFORMANCE THEORY.

J W Friedman, *Oligopoly Theory* (Cambridge, New York, 1983)

PH

omnipotency theory *Psychology* Proposed by the Austrian founder of psychoanalysis Sigmund Freud (1889–1939).

Freud postulated that thoughts alone can satisfy our wishes without use of an external object, but also uses the term 'omnipotency theory' to refer to the belief that our thoughts can influence external events. Both forms are assumed to be a natural stage in a child's development, and perhaps at the root of beliefs in magic and animism. Omnipotence of thought is seen in many forms of psychiatric disability but whether it is a stage in a child's development is difficult to establish.

O Fenichel, *The Psychoanalytic Theory of Neurosis* (London, 1946)

NS

oncogene hypothesis (1969) *Biology/Medicine* Also called the virogene-oncogene hypothesis, this was proposed by American scientists Robert Joseph Huebner (1923–), George Joseph Todaro (1937–) and colleagues. The theory postulates that at some point in the evolutionary past, viral genes were integrated into the deoxyribonucleic acid (DNA) of all species prone to cancer, and that these genes have been transmitted as part of these species' genome ever since.

Such genes can become activated by chemical carcinogens or radiation to create viral oncogenes, which then cause cancer by transforming normal cells into malignant ones. *See* CANCER, THEORIES OF. *Compare with* PROTOVIRUS THEORY and PROVIRUS HYPOTHESIS.

J Tooze, *Selected Papers in Tumour Virology* (Cold Spring Harbor Laboratory, 1974)

KH

one gene-one enzyme hypothesis (1940s) Proposed by George Wells Beadle (1903–) and Edward Lawrie Tatum (1909–75). According to this theory, each normal gene creates one enzyme which in turn causes a specific reaction. In the 1960s, this theory was replaced by the one gene-one polypeptide chain hypothesis after studies in molecular biology revealed that many enzymes are the products of more than one gene. *See* ONE GENE-ONE POLYPEPTIDE CHAIN HYPOTHESIS.

F Magill, *Magill's Survey of Science: Life Science Series* (Englewood Cliffs, NJ, 1991)

KH

one gene-one polypeptide chain hypothesis (1960s) *Biology* The modern restatement of the ONE GENE-ONE ENZYME HYPOTHESIS.

Enzymes are complex proteins consisting of many different polypeptide chains, each of which appears to be brought about by a single gene. Many genes do not code for any polypeptide chain, however.

F Magill, *Magill's Survey of Science: Life Science Series* (Englewood Cliffs, NJ, 1991)

KH

one-group theory (20th century) *Physics* A theory of neutron transport in which it is assumed that all the neutrons belong to the same energy group.

J Thewlis, ed., *Encyclopaedic Dictionary of Physics* (New York, Oxford and London, 1962)

MS

one over many principle *Philosophy* Principle which expresses the motivation underlying PLATO'S THEORY OF FORMS and similar doctrines. Where there are a number of objects of the same kind, or sharing a single property, it seems that there must be a single something which is this kind or property, and which therefore gets treated as an abstract non-material substance.

Strictly, the principle could be seen simply as giving a motive for postulating universals (*see* PLATONISM), whatever the status those universals then have; but the term is mainly used in a Platonic context. In the work of Plato (*c*.427–*c*.347 BC) himself, the principle is hardly explicit, the nearest reference to it being in his *Republic* (§596). It appears more explicitly in the lost but reconstructed work *De Ideis* (*On the Forms*) by Aristotle (384–322 BC).

ARL

one-trial learning theory *Psychology* Proposed by one of the most influential learning theorists, the American psychologist E R Guthrie (1886–1959).

Guthrie proposed that learning takes place on a single trial, and improvement with practice represents the acquisition of individual, simple components which make up more complex behaviours. Although praised for its simplicity, Guthrie's theory often succumbs to circular reasoning.

E R Guthrie, *The Psychology of Learning* (New York, 1935)

NS

Onsager reciprocal theorem (1930s) *Physics* Named after the Norwegian-born American physical chemist Lars Onsager (1903–76), winner of the 1968 Nobel prize for Chemistry.

Onsager's theorem relates to the thermodynamics of linear systems in which he showed that symmetric reciprocal relationships apply between forces and fluxes. A flow or flux of matter in thermodiffusion is caused by the force exerted by the thermal gradient. But conversely, a concentration gradient will cause a heat flow, an effect that has been experimentally verified.

GD

ontology *Philosophy* Generally, either the study of being, or a particular theory of what there is (as in 'Smith's ontology contains classes but not propositions', meaning that Smith believes there are such things as classes but not such things as propositions).

More specifically, part of the logical system underpinning the MEREOLOGY of Polish logician Stanislaw Leśniewski (1886–1939). The system uses three types of names – proper names ('Socrates'), common names ('dog'), and fictitious names ('Apollo', 'centaur') – and elaborates the relations between a complex set of connectives that can be applied to them; for example, 'is a' ('Socrates *is a* man'), 'overlaps' ('*Some* cats *are* pets'), and so on. The other part of Leśniewski's logical system is called *protothetic*, a sort of generalization of the standard logic of propositions.

C Lejewski, 'On Leśniewski's Ontology', *Ratio* (1958)

ARL

op art (1965) *Art* Name first used at a New York exhibition, 'The Responsive Eye', for works using geometrical abstraction to create optical illusions.

Lines (created with maximum precision) appear to move or ripple, the artists' intention being to give the viewer an intense visual experience. Prominent exponents of Op Art include French painter Victor Vasarely (1908–), Briton Bridget Riley (1931–) and German-born American Josef Albers (1888–1976).

LL

Oparin-Haldane hypothesis (20th century) *Biology* Also called the Haldane-Oparin hypothesis. Named after Russian biochemist Alexandr Ivanovich Oparin (1894–1980) and British geneticist John Burdon Sanderson Haldane (1892–1964). The idea that life began with simple organic compounds formed from atmospheric gases, followed by more complex compounds formed in the seas through a variety of reactions. The macromolecules that are the building blocks of life were then formed from these compounds and reactions.

KH

open and closed texts (1979) *Stylistics* Distinction proposed by the Italian semiotician Umberto Eco (1932–).

The distinction hinges on whether or not a text targets a specific kind of reader, imposes a single interpretation: the 'open' text does not. *See also* 'LISIBLE' AND 'SCRIPTIBLE' in Roland Barthes (1915–80), a similarly difficult distinction.

U Eco, *The Role of the Reader* (London, 1979)

RF

open class words *Linguistics See* PIVOT GRAMMAR.

open mapping theorem *Mathematics* The result in COMPLEX FUNCTION THEORY which asserts that a non- constant analytic function is an open mapping; that is, it is a function which sends open sets in the domain to open sets in the range. Similar results can also be found in the areas of FUNCTIONAL ANALYSIS and TOPOLOGY.

J E Marsden, *Basic Complex Analysis* (Freeman 1973)

MB

open society (1945) *Politics* Theory of government and social order of the Austrian philosopher Karl Popper (1902–).

An open society, as opposed to a planned or goal-directed society, is one in which governments have no ultimate aims, and hence one where people enjoy the maximum freedom to experiment and innovate.

Karl Popper, *The Open Society and its Enemies* (London, 1945)

RB

open theory (20th century) *Mathematics* A theory in logic is open if it considers only propositions with variables which are bound only by the universal quantifier 'for all'. Such a theory cannot discuss the man in the moon because this involves the quantifier 'there exists', but can consider all men x such that x is in the moon.

JB

operant theory *Psychology* A theory of the American psychologist Burrhus Frederic Skinner (1904–90).

Operant theory analyzes the interaction between the organism and its environment into a three-term sequence. A successful experimental analysis identifies the environmental cues (discriminative stimuli) which determine the occurrence of behaviour (operant) and of the environmental events (reinforcers) necessary for the establishment and maintenance of the behaviour. The theory has had a fundamental impact on psychology. It is chiefly criticized for its exclusion of unobservable, inferred psychological states.

B F Skinner, *The Behaviour of Organisms* (New York, 1938)

NS

operationalism (or operationism) *Philosophy* Theory due to American physicist Percy Williams Bridgman (1882–1961) and saying that scientific concepts must be defined in terms of the operations by which they are measured or applied.

The theory is akin to the VERIFIABILITY PRINCIPLE in its strongest form, identifying meaning with method of verification; but applies to concepts rather than sentences or propositions, and is in the spirit of LOGICAL POSITIVISM. It has untoward results, such as that length as measured by astronomers is a different concept from length as measured by ordinary people using a tape-measure.

P W Bridgman, *The Logic of Modern Physics* (1927)

ARL

operationalism *Psychology* Operationalism was espoused mainly by the radical behaviourists, because of their emphasis on the directly observable. However, many concepts in psychology are abstract and the idea that the measurement of a concept provides an adequate definition is now discredited. Operationalization of concepts in psychology still has a major role to play in theory construction.

T H Leahey, 'The Myth of Operationalism' *Journal of Mind and Behaviour*, vol. I (1980), 127–43

NS

operator theory (20th century) *Mathematics* A generalization of 19th-century ALGEBRA into mathematical analysis, started by David Hilbert (1862–1943) among others.

A linear operator (or, simply, operator) is a mapping of a vector space *V* onto another which preserves addition and multiplication by scalars. If *V* is finite-dimensional then a linear operator can be represented by a matrix. Because many operations in analysis, such as integration, can be regarded as linear operators acting on an infinite-dimensional vector space, operator theory has become an important part of modern analysis.

N Dunford and J T Schwartz, *Linear Operators* (New York, 1957)

JB

operon model *Biology See* JACOB-MONOD MODEL.

opponent-process theory of motivation (20th century) *Psychology* Proposed by the American psychologist R L Solomon.

This theory assumes that the functioning of an intact organism is predicated on the maintenance of a moderated position of 'motivational normality'. An opponent process is produced by a swing towards either pole on a motivational dimension, and operates to restore balance to the system; for example, fear will produce a tendency toward ecstasy and vice versa. The fundamental principle of this theory is that stimulation of one pole produces a simultaneous inhibition of the other.

R Solomon, 'The Opponent-Process Theory of Acquired Motivation', *American Psychologist*, vol. xxxv (1980), 691–712

NS

opponent theory of colour vision *Psychology* Formulated by the German physiologist Ewald Hering (1834–1918), but more accurately known as the opponent process theory of colour vision.

This proposes that there are three sets of colour receptor systems: red-green, blue-yellow, and black-white. Each of the receptors is assumed to be sensitive to light of all wavelengths but particularly sensitive to light of a certain wavelength. This theory also assumes that when one receptor is stimulated (for example the red receptor with red light) this has an inhibiting effect on the opposite receptor (in this example, the green receptor).

E Hering, *Outline of a Theory of the Light Sense*, L M Hurvich and D J Jameson, transs (Cambridge, Mass., 1964)

NS

opportunity cost *Economics* The sacrifice made when selecting one product or service over another. The popular political slogan 'Guns or butter?' suggests that national defence is the price to be paid for not fulfilling high consumer expectations, and vice versa. *See* COST BENEFIT ANALYSIS.

J M Buchanan, *Cost and Choice* (Chicago, 1969)

PH

optimal foraging theory (1980s) *Biology* Developed by John R Krebs of Oxford and David W Stephens of Amherst in *Foraging Theory* (1987). This is a mathematical model which proposes that NATURAL SELECTION encourages species to make foraging as efficient as possible, because better foragers are more likely to reproduce. The model uses the average rate of energy gain per unit of foraging time to quantify 'fitness'. Optimal foraging theory remains controversial.

KH

optimal stimulation, principle of *Psychology* Proposed by C Leuba.

This is the postulate that an organism tends to learn those responses or reactions that produce an optimal level of stimulation or excitation. Either drive reduction or drive arousal may lead to the optimal level of stimulation. *See also* DRIVE REDUCTION THEORY.

S Sahakian, *Learning: Systems, Models and Theories*, 2nd edn (Chicago, 1976)

NS

optimal tariff theory (1906) *Economics* Originating in the work of English economist Charles Bickerdike (1876–1961), this illustrates that it is possible for a country to improve its terms of trade by imposing a tariff or tax on certain imported goods.

The effect of this tariff is, *ceteris paribus*, a fall in demand for the taxed good; the country thus conducts a smaller level of trade on more favourable terms. The optimal is reached when the gain from the better terms of trade offsets the losses from the smaller volume of trade.

C F Bickerdike, 'The Theory of Incipient Taxes', *Economic Journal*, vol XVI (December 1906), 529–35; W M Corden, *Trade Policy and Economic Welfare* (Oxford, 1974)

PH

optimization theory *Mathematics* The body of mathematics concerned with maximizing or minimizing *objective functions* of many variables subject to constraints (*see also*, for example, FERMAT'S PROBLEM).

It developed partly through the classical CALCULUS OF VARIATIONS and optimal control and *approximation theory*, where the variables are functions; and partly through the more computational concerns of linear programming and its generalizations, where we consider real variables (or discrete, as in the case of the *travelling salesman problem*; *see also* GRAPH THEORY). Typical questions include the existence and uniqueness of optimal solutions and their characterization (*see also* the FRITZ JOHN CONDITIONS THEOREM, KARUSH-KUHN-TUCKER THEOREM, PONTRYAGIN MAXIMUM PRINCIPLE, BELLMAN PRINCIPLE OF OPTIMALITY and ALTERNATION THEOREM), relationships between primal and dual problems (*see also* DUALITY THEORY OF LINEAR PROGRAMMING and FENCHEL DUALITY THEOREM), and computational techniques. *Mathematical programming* and *operations research* are terms used for overlapping subjects.

AL

option pricing theory (20th century) *Economics* This is the analysis of the pricing of contracts to buy or sell (termed a call option, and a put option) a commodity or security within a stated period of time at a specific price. An early pioneer in this field was French economist Louis Bachelier (1870–1946) but the subject received more recent analysis during the 1970s. *See* ADAPTIVE EXPECTATIONS, RATIONAL EXPECTATIONS and RANDOM WALK HYPOTHESIS.

F Black and M J Scholes, 'The Pricing of Options and Corporate Liabilities', *Journal of Political Economy*, 81, 3 (May, 1973) 637–54

PH

oral-formulaic poetry (1930) *Literary Theory* Concept developed by the American scholars Milman Parry and Albert Bates Lord; in 1953 applied to Anglo-Saxon verse by Francis Peabody Magoun (jnr).

In pre-literate cultures, poetry constructed in oral performance on the basis of selections from a stock of conventional phrases and variants. These 'formulae' provided a skeleton of figurative, narrative and cultural references which could be filled out by the bard on the spot; if rhythmically repetitive, they gave the basis of the metrical structure too.

A B Lord, *The Singer of Tales* (Cambridge, Mass., 1960)

RF

orbit stabilizer theorem *Mathematics* The result in GROUP THEORY whereby if a group G acts on a non-empty set X, then the cardinality of the orbit of an element in X is the index of the stabilizer of that element in G.

J J Rotman, *An Introduction to the Theory of Groups* (Allyn-Bacon, 1984)

MB

organic form (*c*.330 BC; 1759; 1800s; 1930–60) *Literary Theory* Theory derived from: Greek philosopher Aristotle (384–322 BC); English poet Edward Young (1683–1765); English poet Samuel Taylor Coleridge (1772–1834); and NEW CRITICISM.

'Organic' views of literature are based on two metaphors, zoological and botanical. Aristotle's organicism views a poem as an integrated system of parts like the organs of a living body. The second image, natural growth, is clearly stated in Young: 'The mind of a man of genius is a fertile and pleasant field. . . . An original may be said to be of a vegetable nature; it rises spontaneously from the vital root of genius; it grows, it is not made' (*see also* ORIGINALITY). Organicism was taken on board wholeheartedly in ROMANTICISM; the clearest definition of organic form is by Coleridge. It persists in the 20th century as a structural principle among the CHICAGO CRITICS and as a source of metaphors in NEW CRITICISM.

S T Coleridge, 'Shakespeare's Judgement Equal to his Genius' [*c*.1808], *Critical Theory since Plato*, H Adams, ed. (New York, 1971), 460–62

RF

organic theory, of the state *Politics* Theory of the STATE as analogous to a natural organism.

The state is better understood as a 'natural' rather than a 'mechanical' phenomenon, with different institutions performing different functions, and the good health of the whole being attributable as much to the good working of the whole as to the contribution of any particular part. *See also* SOCIAL DARWINISM.

Greta Jones, *Social Darwinism and English Thought* (Brighton, 1980)

RB

organic unities, principle of *Philosophy* Principle that a whole may have a value which is different from, and not predictable on the basis of, the values of its parts. The attractiveness, for example, of a picture cannot normally be predicted from that of each colour-patch taken separately.

The principle was made much of by George Edward Moore (1873–1958), who distinguished his use from earlier, non-ethical, uses. *See also* ORGANICISM and HOLISM.

G E Moore, *Principia Ethica* (1903), especially §§18–22

ARL

organicism *Philosophy* A version of (or perhaps little more than an alternative name for) HOLISM, emphasizing the analogy with living organisms, whose parts only are what they are because of, and can only be understood in terms of, their contributions to the whole. *See also* ORGANIC UNITIES.

ARL

organismal model (1916) *Biology* First proposed by the American ecologist Frederic Clements (1874–1945).

Communities of organisms together have some properties of a single or super organism, the individuals becoming analogous to tissues or cells of a single organism. As such this provides a basis for classifying community types which is still used today, but ecologists tend not to view such ecosystems as superorganisms and accept the blurred boundaries that the INDIVIDUALISTIC MODEL proposes.

RA

organization, principle of *Psychology* Proposed by the Danish psychologist E J Rubin (1886–1951) and later adopted by the influential German-American gestalt theorist Wolfgang Köhler (1887–1967).

The principle states that in perception the relationship between the individual parts is more important than the parts alone; it is the integrated whole that is critical. Köhler emphasized that this integration occurs in pre-consciousness, leading to a conscious perception. The integration of parts has become an important concept in other areas of psychology, for example cognitive psychology. *See* GESTALT THEORY.

A D Ellis, ed., *A Sourcebook of Gestalt Psychology* (London, 1938)

NS

organization theory (1970s) *Economics* A modern theory of the firm which states that the goals and activities of a firm are the results of its organizational structure.

This challenges the traditional assumption of profit maximization by management, which is now seen as content to earn just satisfactory profits. *See* THEORY OF THE FIRM, SATISFICING, AGENCY THEORY, THEORY OF THE GROWTH OF THE FIRM, MANAGERIAL THEORIES OF THE FIRM and THEORY OF BUREAUCRACY.

O E Williamson, *Markets and Hierarchies, Analysis and Antitrust Implications: A Study in the Economics of Internal Organizations* (New York, 1975); J B Barney and W G Ouchi, eds, *Organizational Economics* (San Francisco and London, 1986)

PH

orgone theory *Psychology* Proposed by the Austrian-American psychoanalyst Wilhelm Reich (1897–1957).

Biological energy flows freely and fully in the healthy body, and psychopathology or physical pathology can result if this energy is blocked. This energy Reich referred to as orgone and the associated threrapeutic breakthrough was achieved by the client achieving the ultimate orgastic release comprising involuntary bodily convulsions. Reich led his theory in what many considered a farcical direction, and was disowned by the psychoanalytical community.

W Reich, *Character Analysis* (New York, 1961)

NS

original horizontality, law of (1669) *Geology* Implicit in the work of the Danish naturalist Nicolas Steno (1638–86), this law states that when subaqueous sediments are deposited they usually lie horizontally as their angles of rest are very low.

The law can be used to determine how far sedimentary layers have been tilted from their original depositional position. Whilst precise only for clays, it is generally valid within some 10° for coarser grained sediments. *See also* STENO'S LAW.

DT

originality (1759) *Literary Theory* Declaration by English poet Edward Young (1683–1765). *See also* INVENTION.

An 'original' is an organic growth stemming from the genius of the poet, not, as in the classical view, a skilled remaking of an existing subject. An important early statement of a reorientation in literary value towards the creative genius of the poet, it is a precursor to ROMANTICISM. *See also* ORGANIC FORM.

E Young, 'Conjectures on Original Composition' [1759], *Critical Theory since Plato*, H Adams, ed. (New York, 1971), 338–47

RF

origins of life (Antiquity–) *Biology/Philosophy* The theory or theories that seek to explain how biomolecules, subcellular structures, and ultimately living cells came into existence. Many myths, stories and hypotheses have been proposed. Some are still under investigation, while others remain contested or persist as statements of religious faith. *See also* DYNAMIC STATE THEORY, 'LITTLE BAGS' THEORY OF EVOLUTION, MINERAL THEORY, PHENOTYPE THEORY, PRIMORDIAL SOUP and PROGENOTE THEORY.

G Blandino, *Theories on the Nature of Life* (New York, 1969)

KH

Orphism (1912) *Art* Term used by the French poet Guillaume Apollinaire (1880–1918) to describe paintings exhibited by Robert Delaunay (1885–1941) at the 'Section d'Or' exhibition in 1912 and in Berlin in 1913. Apollinaire perceived the romantic aspect of this non-representational colour abstraction, which had some links with CUBISM and music. However, Delaunay was later to trace its links with NEO-IMPRESSIONISM and SYNCHROMISM painting, and to stress the importance of SIMULTANÉISME.

V Spate, *Orphism* (Oxford, 1980)

AB

Orphism *Philosophy* Name for a complex strand in Ancient Greek religious thought, contrasting with the more familiar strand of the Olympian deities (Zeus, Apollo, and so on). A body of religious writings from the 7th and succeeding centuries BC was attributed to the mythical singer Orpheus and his followers.

In Classical times, Orphic ideas were connected with current mystery religions. They concerned purification and initiation rites, and doctrines to do with reincarnation and post-mortal punishment and reward. Philosophically, their main importance lies in their influence on PYTHAGOREANISM and Plato (*c.*427–*c.*347 BC), notably in his *Meno* and *Gorgias*. However, modern scholars have doubted how much can be attributed to a specifically Orphic movement distinguishable from the wider background of mystery religions generally.

I M Linforth, *The Arts of Orpheus* (1941); takes a somewhat sceptical approach

ARL

orthogenesis (19th century) *Biology* A term coined by Wilhelm Haacke (1855–1912) and associated with T G H Eimer (1843–98), this describes a specific trend in evolutionary change observed in a related group of organisms (for example, the increase in body size apparent in the fossil record of the evolution of the horse). Also, the notion that mystical forces propel evolution toward a preordained outcome, especially to a 'higher' state of being. Modern biologists believe instead that evolutionary forces operate opportunistically, without any sense of ultimate goal. *See also* CHAIN OF BEING.

R Milner, *The Encyclopedia of Evolution: Humanity's Search for Its Origins* (New York, 1990)

KH

orthogenesis (1893) *Psychology* A term first applied by W Haacke to the theory that evolution is directed along definite lines and eliminates the necessity for evolutionary forces, that is autonomous evolutionary forces.

In more recent years, the term has been related to the view that personality has intrinsic resources for normal development unless subjected to distorting forces from without. This idea also underpins progressive education and social and cultural development.

W J Mager and J B Dusek, *Child Psychology – A Developmental Perspective* (Lexington, 1979)

NS

Ostwald's dilution law (1888) *Chemistry* Named after the German chemist Friedrich Wilhelm Ostwald (1853–1932) who was professor at Leipzig and won the Nobel prize for Physics in 1909. This law states that for a sufficiently dilute solution of univalent electrolyte of concentration c, the dissociation constant, K_a, is related to the degree of ionization (α) by

$$K_a = \frac{\alpha^2 c}{1 - \alpha}$$

The law is valid if ionic activities are used rather than concentrations.

M Freemantle, *Chemistry in Action* (Basingstoke, 1987)

GD

Oudeman's law *Chemistry* The law states that as the concentration of solutions of the various salts of an acid or base are reduced to zero, the molecular rotations of the solutions tend towards the same limiting value.

GD

outsider (1950s) *Literary Theory* Term coined by the British writer Colin Wilson (1931–).

It refers to a creative personality whose way of life and thought is outside the norms of conventional society; or a character in fiction who is alien to the culture in which s/he lives. Meursault in French novelist Albert Camus's *L'Etranger* (1942) is an often-cited example of the latter usage.

C Wilson, *The Outsider* (London, 1957)

RF

overcoding (1976) *Stylistics* Proposed by the Italian semiotician Umberto Eco (1932–).

A level of secondary significance in aesthetic works, over and above the primary meanings they express. Overcoding is achieved through manipulating the expression and content levels by STYLISTIC and RHETORICAL devices.

U Eco, *A Theory of Semiotics* (Bloomington, 1976)

RF

overextension (traditional; 1970s) *Linguistics* In a child's language development, the application of a word to a broader range of categories than in adult language, as when a child says 'doggie' for dogs and horses.

H H Clark and E V Clark, *Psychology and Language* (New York, 1977), ch. 13

RF

overgeneralization (1960s) *Linguistics* Universal process in language development whereby a child imposes regular forms on irregular lexical items: 'buyed', 'goed', 'foots', 'mouses'; or analogous overgeneralizations in phonology or syntax. In generative approaches to first-language development, overgeneralization is taken as evidence that the child is following a RULE.

J Aitchison, *The Articulate Mammal*, 3rd edn (London, 1989)

RF

overkill (20th century) *Politics* Theory in military strategy which asserts that if a state has so much weaponry that it can defend itself, destroy any aggressor (and perhaps the world) and still have weapons to spare, it is capable of overkill.

Not surprisingly, the concept has been met with a mixture of derision and dismay, and has passed into common usage as a synonym for an unnecessary and excessive response in any situation; not so much a sledgehammer to crack a nut, as a flame-thrower which defeats the object of eating the nut.

David Robertson, *The Penguin Dictionary of Politics* (London, 1986)

RB

overlexicalization (1976) *Linguistics* A term coined by English linguist Michael Alexander Kirkwood Halliday (1925–).

In an ANTILANGUAGE, a profusion of terms for some area of experience which is a preoccupation of, or problematic for, the anti-society; by extension, the same phenomenon in society generally; for example the over-lexicalization of women in English. The term is also used in LINGUISTIC CRITICISM. *See also* RELEXICALIZATION and UNDERLEXICALIZATION.

M A K Halliday, 'Antilanguages' [1976]; repr. in *Language as Social Semiotic* (London, 1978), 164–82

RF

overload theory (20th century) *Politics* Application of CRISIS THEORY to the modern state.

Modern states have more demands made upon them than they can meet. They are, by analogy with machines, overloaded and break down, or inefficiency results. The Left sees overload as a failure of capitalism; the Right of welfare socialism.

Geoffrey Roberts and Alistair Edwards, *A New Dictionary of Political Analysis* (London, 1991)

RB

overspecialization (1932) *Biology* An idea proposed by the British geneticist J B S Haldane (1892–1964).

An evolutionary trend may evolve to the point at which the lineage concerned is at an adaptive disadvantage. This is not an accepted theory, although the term is often used to describe species which cannot survive a change in the environment to which they are closely adapted. The idea was reiterated, however, in Sir Ronald A Fisher's runaway sexual selection hypothesis. In this theory, Fisher envisioned the preference by females of some species for males with extreme traits, such as the tail of a peacock; and that there should be a selective process that would eventually produce males disadvantaged by such traits that have become too extreme. In reality, other selection pressures appear to act against this selection before such an unhappy state of affairs arises.

J B S Haldane, *Causes of Evolution* (London, 1932)

RA

Owenism (19th century) *Politics* Views of – and following – the Welsh manufacturer and socialist Robert Owen (1771–1858).

Human character is shaped by circumstance. It can therefore be transformed by good working conditions, proper housing, and education. Owen himself attempted to promote this by variously exhorting industrialists, establishing model communities, and encouraging pressure from below through trade unions and co-operatives.

David Miller *et al.*, eds, *The Blackwell Encyclopaedia of Political Thought* (Oxford, 1987)

RB

own rate of interest (1932) *Economics* First outlined by the Italian-born economist Piero Sraffa (1898–1983), this term was later coined by English economist John Maynard Keynes (1883–1946) in his *General Theory*. It is defined as the percentage change in a current commodity price compared with its known future price in the market.

Every commodity has its own rate of interest, be it oil, coal or wheat. Each of these can be affected by other commodity interest rates. Thus, future wheat production will be influenced by future oil prices.

P Sraffa, 'Dr Hayek on Money and Capital', *Economic Journal*, 42 (March, 1932), 42–53; J M Keynes, *The General Theory of Employment, Interest and Money* (New York, 1936)

PH

P

P = NP problem *Mathematics See* NP-COMPLETE PROBLEM.

pacemaker theory of ageing *Biology/Medicine See* PROGRAMMED AGEING.

pacifism *Politics* Ethical objection to violence.

Violence and the taking of life are morally wrong. People should therefore refuse to engage in or support military activity, and no cause can justify the use of military force. There can therefore be no such thing as a JUST WAR.

David Miller *et al.*, eds, *The Blackwell Encyclopaedia of Political Thought* (Oxford, 1987)

RB

Padé approximates (late 19th century) *Mathematics* The first general problem concerning the interpolation of given values of a function at specified points by rational approximation was considered by the French mathematician and physicist Baron Augustin Louis Cauchy (1789–1857). Padé approximates are a field concerned with finding the best rational approximation to a power series. Numerous algorithms have been constructed for calculating them.

M Hazewinkel, ed., *Encyclopaedia of Mathematics* (Dordrecht, 1988)

ML

pain principle *Psychology* Proposed by the Austrian Sigmund Freud (1856–1939) as part of his psychoanalytic theory of personality.

The pain principle was Freud's early version of the thanathos concept. It states that we strive for death and hence return to the form from which we originated. The pain principle has proved extremely difficult to operationalize and test empirically. *See* PLEASURE PRINCIPLE.

NS

paint pot problem *Biology See* BLENDING INHERITANCE.

Paley's watchmaker (early 19th century) *Biology* Named after the Reverend William Paley (1743–1805) whose book *Natural Theology* (1802) employed the analogy, 'The Watch on the Heath'.

A reference to the concept that God must exist because the intricacies of design evident in nature could not exist without a designer. Paley's analogy compared finding a stone on a heath (field), which could have 'lain there forever', to finding a watch, which 'must have had a maker'.

R Dawkins, *The Blind Watchmaker: Why the Evidence of Evolution Reveals a Universe Without Design* (New York, 1986)

KH

Paneth's adsorption rule *Chemistry* Named after the Austrian chemist Friedrich Adolf Paneth (1887–1958) who became Professor of Chemistry at Durham University in 1939 and was later director of the Max Planck Institute in Germany. The law states that a radio-element is strongly adsorbed on a material of opposite surface charge if it can form a relatively insoluble compound with the adsorbing material.

GD

pangenesis (1868) *Biology* Idea attributed to the Greek physician Hippocrates (*c*.460–*c*.370 BC), with the term coined by English evolutionary theorist Charles Darwin (1809–82).

Obsolete theory of development which claimed that small particles from different parts of the body, called pangenes, were collected together in the body's reproductive cells and transmitted to offspring. A blending of inherited characteristics from the two parents would have resulted. The GERM PLASM THEORY replaced pangenesis.

F Magill, *Magill's Survey of Science: Life Science Series* (Englewood Cliffs, NJ, 1991)

KH

panpsychism *Philosophy* Literally, 'all-soul-ism'. The view that matter is intrinsically alive, or is made up from basic entities which are so. Various forms of such a view are found in the philosophies of Gottfried Wilhelm Leibniz (1646–1716), Alfred North Whitehead (1861–1947), and John McTaggart Ellis McTaggart (1866–1925) among others. *See also* HYLOZOISM.

ARL

panspermia (1908) *Astronomy/Biology* Also called cosmic insemination. A term coined by the Swedish chemist Svante August Arrhenius (1859–1927), this refers to the idea that life was introduced on Earth by microorganisms travelling through space and colonizing every habitable planet.

Some supporters of this view believe that life began only on planets, from which it was disseminated as seeds to other planets via space; while others believe that space itself provided the environment (womb) that produced life. Because no evidence of the theory has been found during space explorations, and because the Big Bang theory implies a specific starting point for the universe – weakening the notion of eternal organic life that is the basis for panspermia – this theory is not widely accepted. *See also* DIRECTED PANSPERMIA.

M Barbieri, *The Semantic Theory of Evolution* (Chur, Switzerland, 1985)

KH

pantheism *Philosophy* Literally, 'all-god-ism'. The view that God and the universe are identical; or that there is no transcendent God outside the universe who created it, but the universe itself is divine. Among philosophers, Baruch de Spinoza (1632–77) is a prominent exponent of such a view, and it appears also in STOICISM. The term itself was coined in 1705 by Irish writer John Toland (1670–1722).

ARL

pantonality *Music See* ATONALITY.

Papez's theory of emotion *Psychology* Proposed by the American psychologist J W Papez (1883–1958).

One of the first theoretical attempts to delineate the specific cortical mechanisms underlying emotion. Papez proposed three interlocking systems (sensory, hypothalamic and thalamic), all of which he hypothesized to be combined with the cortex where the 'psychological product' of emotion emerged. This represents a modification of Cannon's theory but has not withstood anatomical study. The theory has been influential in implicating the hypothalamus and focusing attention on the integrative role of the cortex. *See* CANNON-BARD THEORY OF EMOTION.

J W Papez, 'A Proposed Mechanism of Emotion', *Archives of Neurology and Psychiatry*, vol. xxxv (1937), 725–44

NS

Pappus's theorem (*c*.AD 300) *Mathematics* Named after its discoverer Pappus of Alexandria, this states that if the six vertices of a hexagon lie alternately on two lines, then the three points of intersection of the opposite sides are collinear. Pappus proved this by methods involving length and angle, but they do not appear explicitly in the statement. In fact, his theorem is an axiom of projective geometry in its axiomatic form, but is provable in the form of the geometry based on ALGEBRA provided that the elements in the co-ordinates satisfy the COMMUTATIVE LAW.

H S M Coxeter, *Introduction to Geometry* (New York, 1961)

JB

para bellum *Politics* Theory, or phrase, justifying military preparations.

The best way to preserve peace is to prepare for war since by so doing you make attack on yourself unlikely. *See also* DETERRENCE.

Graham Evans and Jeffrey Newnham, *The Dictionary of World Politics* (Hemel Hempstead, 1990)

RB

paraconsistency *Philosophy* View that there are important paraconsistent logical theories; that is theories that do not allow (as classical logic does: *see* RELEVANCE LOGICS) that a contradiction has every proposition among its logical consequences.

A system which contains contradictory proposition is inconsistent. But if it does not also contain every proposition (as it would for classical logic) it avoids being trivial. It is claimed that various scientific and mathematical theories are in fact of this nature, and so can be logically analyzed by paraconsistent logics, and also shown to be respectable as they stand (that is, we need not assume that the inconsistencies in them are merely aberrations that must be removed before they can be properly studied). *See also* DIALETHEISM.

G Priest and R Routley and J Norman, eds, *Paraconsistent Logic* (1985)

ARL

paradigm (20th century) *History* Theory of development of scientific knowledge of American historian and philosopher Thomas Kuhn (1922–).

Advances and changes in science and knowledge occur as the result of 'paradigmatic shifts' whereby one way of looking at the world is replaced relatively quickly by another, rather than by any slow process of rational reappraisal.

Geoffrey Roberts and Alistair Edwards, *A New Dictionary of Political Analysis* (London, 1991)

RB

paradigm and syntagm (1916) *Linguistics* Basic distinction drawn by the French linguist Ferdinand de Saussure (1857–1913). Paradigmatic relations are 'associative' relations for Saussure.

The syntagm is the horizontal dimension, the sequence of units chained together; paradigm is the set of units which might have been chosen at any point in the chain. In the work of English linguist Michael Alexander Kirkwood Halliday (1925–), paradigm is 'choice', syntagm 'chain'.

F de Saussure, *Course in General Linguistics*, W Baskin, trans. (Glasgow, 1974 [1916])

RF

paradigmatism *Philosophy* View that where one or more objects are of a certain kind or have a certain property, this is to be explained by postulating a non-material abstract entity to serve as a paradigm of which they are copies; in other words, universals (*see* PLATONISM) are to be regarded as (or replaced by) paradigms. The term usually refers to one way of looking at PLATO'S THEORY OF FORMS, a way exemplified particularly by his *Republic* §§596a–7e and *Timaeus* 27d–29d, 48d–52d. It is this feature of Plato's forms that seems to give rise to the THIRD MAN ARGUMENT.

ARL

paradox (traditional; 1947) *Literary Theory* Modern usage by New Critic Cleanth Brooks (1906–).

A contradictory statement from which a valid inference may be drawn; in RHETORIC, the FIGURE OF SPEECH 'oxymoron' (for example Milton's 'living death'). Brooks extended the term to other kinds of indirections and DEVIATIONS, declaring 'the language of poetry is the language of paradox'. *See* NEW CRITICISM.

C Brooks, *The Well Wrought Urn* (New York, 1947)

RF

paradox of thrift *Economics* The classical school of economists held the view that since what was saved was later invested, there could not be excessive saving.

The paradox was revised by English economist John Maynard Keynes (1883–1946) in the 1930s, who asserted that thrift is virtuous only up to a point. If an individual increases the proportion of income he saves, his reduced expenditure on goods will lower total demand in the economy. His thrift is laudable up to the point businessmen in the economy wish to borrow his savings for investment. *See* FORCED SAVING, LOANABLE FUNDS THEORY OF THE RATE OF INTEREST, LIFE-CYCLE HYPOTHESIS and RELATIVE INCOME HYPOTHESIS.

J M Keynes, *The General Theory of Employment, Interest and Money* (New York, 1936)

PH

paradox of value *Economics* A long-established principle with its roots in Greek philosophy, this paradox was popularized by Scottish economist Adam Smith (1723–90). It states that price is determined by scarcity rather than usefulness.

Water is an essential of life, but because of its abundance has a relatively low price. Diamonds, on the other hand, have little use in comparison with water, but because of their relative scarcity command high prices.

A Smith, *An Inquiry into the Nature and Causes of the Wealth of Nations* (London, 1776)

PH

paradox of voting (1950) *Economics* Developed by American economist Kenneth Arrow (1921–), this states that if there are more than two choices facing voters in a majority democratic selection process, a stalemate will result. (The paradox is also known as Arrow's theorem.) *See* IMPOSSIBILITY THEOREM.

K J Arrow, 'A Difficulty in the Concept of Social Welfare', *Journal of Political Economy*, vol. LVIII (1950), 328–46; *Social Choice and Individual Values* (New York, 1966)

PH

paradoxical cold *Psychology* Identified by the German physiologist Max von Frey (1852–1932).

The phenomenon whereby spots on the skin which respond to a cold stimulus are subsequently stimulated and give rise to a sensation of cold. The phenomenon, which is difficult to obtain, requires the application of a fairly high temperature in order to arouse the cold spots (about 45°C). *See* CONCENTRATION THEORY OF COLD and SPOT THEORY.

R S Woodworth and H Schlosberg, *Experimental Psychology* (London, 1966)

NS

paralanguage (1958) Formulated by the American linguist George L Trager.

Expressive features of speech style such as tempo, huskiness, pitch range, loudness, associated conventionally with mood, attitude and personality.

D Crystal and R Quirk, *Systems of Prosodic and Paralinguistic Features in English* (The Hague, 1964)

RF

parallel evolution (1960–65) *Biology* Also called parallelism, this term refers to the development of similar characteristics in related organisms owing to their being subject to similar selection pressures and their need to adapt to such pressures in a similar way. Thus, different species in the same genus living far apart may exhibit the same adaptations to their similar environments. Not all biologists accept that this mechanism exists, and examples are apparently relatively rare.

W Beck and K Liem and G Simpson, *Life: An Introduction to Biology* (New York, 1991)

KH

parallel law *Psychology* Discovered by the German experimental psychologist Gustav Theodor Fechner (1801–87), who developed the field of psychophysics.

The law states that if two stimuli of different intensities are presented to a receptor simultaneously, the absolute sensory intensity diminishes but the ratio of differences remains unchanged.

S S Stevens, *Psychophysics: Introduction to its Perceptual, Neural and Social Aspects* (New York, 1975)

NS

parallel postulate (*c*.300 BC) *Mathematics* The fifth of the axioms of Euclidean geometry proposed by Euclid of Alexandria. It states that if two lines are cut by a third, the two will meet on the side of the third on which the sum of the interior angles is less than two right angles.

As early as AD 150, Ptolemy expressed the view that this axiom was not obvious and tried to deduce it from the others. Later work included the enunciation of an alternative axiom which is equivalent to the parallel postulate when the first four axioms hold and is more intuitive than this. A good example of such an axiom was found by John Playfair (1748–1819) and is called Playfair's axiom: through a given point which is not on a line l there is exactly one line parallel to l. That it is impossible to deduce the parallel postulate from the others was shown by Nicholai Ivanovich Lobachevsky (1793–1856) in 1826 and independently by Janos Bolyai (1802–60) in 1825 when they proved that there is a 'non-Euclidean geometry' in which the other axioms of Euclidean geometry hold but the parallel postulate is replaced by 'through a point not on a line l there are at least two lines parallel to l'. This is also known as *hyperbolic geometry*. Carl Friedrich Gauss (1777–1855) claimed to have found this geometry earlier, but he certainly did not write an account of it. B Riemann constructed another non-Euclidean geometry, also known as *elliptic geometry*, in which the parallel postulate is replaced by 'any line through a point not on a straight line *l*, intersects the line *l*'.

M Kline, 'Non-Euclidean Geometry', *Mathematical Thought from Ancient to Modern Times* (New York, 1972)

JB

parallelism *Biology* *See* PARALLEL EVOLUTION.

parallelism *Psychology* Associated with the work of the German and Scottish philosophers Gustav Wilhelm Leibniz (1646–1716) and Alexander Bain (1818–1903), and with the empirically-oriented psychology of the late 19th and early 20th centuries.

Parallelism was an attempt to solve the mind-body problem by assuming that the two entities were merely separate 'tracks' and that for every mental or psychic event, there was a corresponding physical event and vice versa. The theory has since been superseded by more sophisticated accounts.

I Pavlov, *Conditioned Reflexes* (London, 1927)

NS

parallelism (1960s) *Stylistics* Repetition of some aspect of linguistic structure, such as a PHONEME, a clause or part of a clause. (Take for example Bacon's phrase 'Studies serve for delight, for ornament, and for ability.') Parallelism underpins many FIGURES OF SPEECH in traditional RHETORIC. The theoretical basis was expounded in Roman Jakobson's POETIC PRINCIPLE, and phenomena based on parallelism were much investigated in STYLISTICS during the 1960s.

G N Leech, *A Linguistic Guide to English Poetry* (London, 1969)

RF

parallelogram law *Mathematics* The norm identity for an inner product space:

$$\|x+y\|^2 + \|x-y\|^2 = 2\|x\|^2 + 2\|y\|^2$$

for all vectors x and y. This law extends the parallelogram rule to an inner product space and only holds when the norm is induced by an inner product

$$<x,x> = \|x\|^2.$$

W Gellert, S Gottwald, M Hellwich, H Kästner and H Küstner, eds, *The VNR Concise Encyclopedia of Mathematics* (New York, 1975)

parallelogram rule *Mathematics* The law for finding the resultant of two vectors by constructing a parallelogram in which each of the parallel sides represents the direction and magnitude of the given vectors and the diagonal represents the direction and magnitude of the resultant.

W Gellert, S Gottwald, M Hellwich, H Kästner and H Küstner, eds, *The VNR Concise Encyclopedia of Mathematics* (New York, 1975)

ML

parapatric speciation *Biology* Coined by Bush. The concept that new species can arise when genetically distinct organisms in a population obtain access to an ecological niche within the population's geographic range, causing individuals to become reproductively isolated from others without being geographically isolated. Several species, such as the Old World mole rat, are believed to have arisen in this way.

Compare with ALLOPARAPATRIC SPECIATION, ALLOPATRIC SPECIATION, STASIPATRIC SPECIATION and SYMPATRIC SPECIATION. *See also* SPECIATION, THEORY OF and SPECIES, THEORY OF.

C Barigozzi, *Mechanisms of Speciation* (New York, 1982)

KH

parapraxis *Psychology* Proposed by the Austrian founder of psychoanalysis Sigmund Freud (1856–1939).

In psychoanalytic theory, parapraxis is the disruption of a specific action or mental process by a determinant which has become unconscious as a result of regression. Disruption takes the form of omission (for example, forgetting) or substitution (for example, speech error) and is sometimes termed 'Freudian Slip'. The particular action process which is disrupted may be associated indirectly or symbolically with the repressed material. Freud distinguished between degrees of awareness of the source of the error, from being aware of the disturbed intentions to being quite unaware of the former intentions and a resistance to acknowledge it.

S Freud, 'Psychopathology of Everyday Life' (vol. XVI) and 'Introductory Lectures' (vol. XV), *Standard Edition of the Complete Works of Sigmund Freud* (London and New York, 1901 and 1916)

NS

parapsychology *Psychology* Pioneered by the American psychologist Joseph Banks Rhine (1895–1980).

The branch of psychology concerned with the investigation of psychic phenomena (psi). Two main areas are usually distinguished: (1) extrasensory perception (ESP); (2) psychokinesis (PK), also termed parakinesis. ESP refers to perception of information seemingly unavailable to the senses, and PK is the ability of an individual to influence external events without direct intervention. Among other things hindering its acceptance, para-psychology suffers by being unable to replicate its findings reliably.

J B Rhine, *New World of Mind* (New York, 1953)

NS

parental investment, theory of (1972) *Biology* Developed by Robert Trivers as an extension of ideas suggested by Bateman; this is the theory that when the effort or resources contributed by one sex towards the survival of offspring exceeds the contribution of the other sex, individuals of the sex making the lesser investment must compete among themselves for access to the sex making the greater investment.

According to this theory, FEMALE CHOICE and MALE-MALE COMPETITION exist because in most species, the female makes a greater parental investment in the survival of offspring than the male. *See also* SEXUAL SELECTION.

KH

Pareto efficiency (1906) *Economics* Named after Italian sociologist and economist Vilfredo Pareto (1848–1923), this term is defined as the efficiency of a market which is unable to produce more from the same level of inputs without reducing the output of another product.

V Pareto, *Manuale d'economia politica* (Milan, 1906)

PH

Pareto optimality (1906) *Economics* Named after Italian sociologist and economist Vilfredo Pareto (1848–1923), this is a situation which exists when economic resources and output have been allocated in such a way that no-one can be made better off without sacrificing the well-being of at least one person. *See* SOCIAL WELFARE FUNCTION, COMPENSATION PRINCIPLE, COST-BENEFIT ANALYSIS and SCITOVSKY PARADOX.

V Pareto, *Manuale d'economia politica* (Milan, 1906)

PH

parity, law of conservation of *Physics* The principle of parity invariance states that the laws of physics hold equally true for both right-handed and left-handed co-ordinate systems for all the phenomena described by classical physics. The parity of the total wave function describing a system of elementary particles is conserved in strong and electromagnetic interactions. However, weak interactions do not exhibit parity invariance, for example, beta decay.

Paul Davies, ed., *The New Physics* (Cambridge, 1989)

GD

Parkinson's law (1958) *Business* Formulated by English author and historian Cyril Northcote Parkinson (1909–93), after his study of staffing levels in the British Admiralty during the 1930s–50s, this states that work expands to fill the time available for its completion.

Even if the law is only partially valid, there are major implications for the efficiency of organizations, management efficiency, and motivation of the labour force. *See* X-EFFICIENCY.

C N Parkinson, *Parkinson's Law or The Pursuit of Progress* (London, 1958)

AV, PH

Parseval's theorem (identity equation) *Mathematics* Named after the French mathematician Marc Antoine Parseval de Chênes (1755–1836), this element of FOURIER SERIES states that for a square integrable function f with Fourier coefficients $\{c_n\}$,

$$\sum_{n=0}^{\infty} |c_n|^2 = \int_0^{2\pi} f(x)^2 dx.$$

More generally, if E is an orthonormal basis for a Hilbert space H, then if $h \in H$

$$\|h\|^2 = \sum \{| < h, e > |^2 : e \in E\}.$$

J B Conway, *A Course in Functional Analysis* (New York, 1985)

ML

parsimony, principle of *Philosophy* Also called Ockham's Razor. Principle that one should not multiply entities unnecessarily, or make further assumptions than are needed, and in general that one should pursue the simplest hypothesis.

Adoption of this principle, though seemingly obvious, leads to problems about the role of simplicity in science, especially when we are choosing between hypotheses that are not (or are not known to be) equivalent. There are often different and clashing criteria for what is the simplest hypothesis, and it is not clear whether a simpler hypothesis is *pro tanto* more likely to be true; and if not, what justification other than laziness there is for adopting it.

Philosophy of Science (1961); journal containing symposium on simplicity

ARL

parsimony, principle of *Psychology See* ECONOMY PRINCIPLE .

parsing (traditional) *Linguistics* Origin in Greek grammar, 4th–3rd century BC; *see* TRADITIONAL GRAMMAR.

Syntactic analysis breaking down sentences into their component parts such as noun, verb, and so on; CONSTITUENT STRUCTURE analysis. Term has renewed currency in PSYCHOLINGUISTICS and in research in ARTIFICIAL INTELLIGENCE AND LANGUAGE; *see*, for example, AUGMENTED TRANSITION NETWORK GRAMMAR.

M Garman, *Psycholinguistics* (Cambridge, 1990), ch. 6

RF

partial equilibrium theory (19th century–) *Economics* Developed by French economist Augustin Cournot (1801–77) and English political economist Alfred Marshall (1892–1924), this theory examines the conditions of equilibrium in an individual market or in part of a national economy.

Partial equilibrium theory usually looks at the relationship between two economic variables, assuming other variables are constant in value. *See* GENERAL EQUILIBRIUM THEORY, CLASSICAL MACROECONOMIC MODEL, and LAW OF MARKETS.

A Marshall, *Principles of Economics* (London, 1890)

PH

partial pressures, law of *Chemistry See* DALTON'S LAW OF PARTIAL PRESSURE.

particularism *Philosophy* The view that only particulars exist, and more specifically that the properties and relations of particulars are themselves particulars, not universals (*see* PLATONISM).

A particular has a certain unity in space and time. It cannot appear as a whole at separated places simultaneously (though its parts may be scattered, as when an object is dismantled for repair), and normally (there are exceptions) can appear as a whole at different times only if those times form a period, without gaps. Particulars need not be solid, and their parts may constantly change (shadows and flames are particulars, and so are events and actions). They are primarily contrasted with universals, which can be instantiated; no particular can be instantiated.

Particularism is similar to NOMINALISM, but a particularist might accept abstract particulars like numbers, seen – as a Platonist would see them – as abstract objects; they are particulars in so far as they vacuously satisfy (that is, do not positively break) the requirements about being unified in space and time: one can have instances of twoness or duality, which is a universal, but not instances of two. A nominalist, however, would normally reject numbers seen as abstract objects.

D M Armstrong, *Universals and Scientific Realism*, 1, *Nominalism and Realism* (1978)

ARL

particulate inheritance (19th century) *Biology* Also called beanbag genetics. Based on the ideas of Austrian pioneer geneticist Gregor Johann Mendel (1822–84), this is the theory that genetic factors from the father and mother do not blend, but instead retain their discreteness from generation to generation. *See also* MENDELISM.

R Olby, *The Origins of Mendelism* (London and Chicago, 1985)

KH

partition law *Chemistry See* DISTRIBUTION LAW.

Pascal's mystic hexagram theorem (1639) *Mathematics* Proved at the age of 16 by Blaise Pascal (1623–62) before he turned his attention to literature and theology, this states that the points of intersection of the three pairs of opposite sides of a hexagon which is inscribed in a conic are collinear.

From this result Pascal deduced the results about conics proved by Apollonius of Perga (*c.*255–190 BC). One needs to be a mystic to appreciate the mystical properties of the theorem.

H S M Coxeter, 'Projective Geometry', *Introduction to Geometry* (New York, 1961)

JB

Pascal's principle or law (17th century) *Physics* Named after its originator, the French mathematician Blaise Pascal (1623–62), this law applies to hydraulics.

The application of a force at a distant site is often achieved by means of a fluid in a tube connected to a piston and cylinder arrangement. Pascal's principle states that the change in pressure that results when the control piston is moved occurs almost simultaneously everywhere in the fluid. The pressure change is, in fact, transmitted at the speed of sound in the fluid.

J Thewlis, ed., *Encyclopaedic Dictionary of Physics* (New York, Oxford and London, 1962)

MS

Pascal's triangle *Mathematics* Named after French mathematician and physicist Blaise Pascal (1623–62).

A triangular array of integers with a 1 at the top and with each number below the sum of the two numbers directly above it. The kth element in the nth row is the coefficient of $x^k x^{n-k}$ in the expansion of $(x + y)^n$. *See also* BINOMIAL THEOREM.

W Gellert, S Gottwald, M Hellwich, H Kästner and H Küstner, eds, *The VNR Concise Encyclopedia of Mathematics* (New York, 1975)

ML

Pascal's wager *Philosophy* Argument for adopting a divinely favoured way of life – named after French philosopher, mathematician, physicist and pious gambler Blaise Pascal (1623–62) who stated it in his *Pensées* (§233) – but apparently stemming from Islam.

One statement of it (not Pascal's) is this. Let the utility of a policy be the gain it promises multiplied by the probability of getting it. Let the gain of a life of pleasure be some finite quantity X, certainly achievable. But if God exists he may punish our sins with infinite suffering in Hell. Let the probability that God both exists and will do this be as small as you like, but finite (above zero). The product of an infinite loss and a finite probability is still infinite. Therefore the utility of choosing pleasure over virtue is X minus infinity; that is, is infinity. But the utility of choosing virtue is at worst (that is, if God will not reward us, and death is the end) minus X (that is, the certain loss of X units of pleasure), which is finite. Therefore, it must be rational to choose virtue; that is one should act as if God exists because any positive probability of going to Hell and suffering infinite punishment cannot be balanced by the finite advantages gained by sin. Unfortunately, if the argument proves anything it proves too much, since it could be reiterated for infinitely many different hypotheses about God's preferences and intentions.

P T Landsberg, 'Gambling on God', *Mind* (1971)

ARL, JB

Paschen's law *Physics* Named after its discoverer, the German experimental physicist Louis Carl Heinrich Friedrich Paschen (1865–1947), this law states that for a given gas and electrode material, the breakdown potential difference between large plane and parallel electrodes depends only on the product of the gas pressure and the electrode separation.

J Thewlis, ed., *Encyclopaedic Dictionary of Physics* (New York, Oxford and London, 1962)

MS

Pasteur-Liebig controversy (late 19th century) *Biology* The intense controversy between French chemist and microbiologist Louis Pasteur (1822–95) and German chemist Justus Baron von Liebig (1803–73).

Pasteur believed that fermentation and related processes were caused by the metabolic activities of living cells; Liebig believed that they were caused by chemicals and self-perpetuating instabilities induced in solutions by exposing them to air. Modern microbiology proved Pasteur correct.

R Dubos, *Louis Pasteur, Free Lance of Science* (New York, 1976)

KH

pastoral (3rd century BC–) *Literary Theory* Genre originated with the *Idyls* of the Greek poet Theocritus. The term derives from the Latin word *pastor*, meaning 'shepherd'.

Poetry or prose depicting idealized, imaginary, rural situations; the activities and relationships of shepherds and shepherdesses. Pastoral nostalgically evokes an untroubled 'Golden Age'; it is sometimes an allegory of or satire on the present. The settings, topics and style are highly conventionalized; the conventions were much discussed in the 16th century.

W Empson, *Some Versions of Pastoral* (London, 1935)

RF

paternalism *Politics* Justification of central or professional power.

There will in any STATE be those who have a better insight into the needs of society than do ordinary people, and this insight should be employed to shape and implement government policy for the mass of the population.

David Miller *et al.*, eds, *The Blackwell Encyclopaedia of Political Thought* (Oxford, 1987)

RB

path-goal theory of leadership (20th century) *Psychology* Proposed by the American psychologist R J House.

This contends that the leader must motivate subordinates by: (1) emphasizing the relationship between the subordinates' own needs and the organizational goals; (2) clarifying and facilitating the path subordinates must take to fulfil their own needs as well as the organization's needs. The theory also attempts to predict the effect that structuring behaviour will have under different conditions. *See* FIEDLER'S CONTINGENCY MODEL OF LEADERSHIP, LEADERSHIP THEORIES and NORMATIVE DECISION THEORY OF LEADERSHIP.

R J House, 'A Path-Goal Theory of Leadership Effectiveness', *Administrative Science Quarterly*, vol. XVI (1971), 321–38

NS

pathetic fallacy (1856) *Literary Theory* Term coined by the English writer John Ruskin (1819–1900), and derived from the Greek word *pathos* meaning 'feeling'.

Attribution of human feelings and motives to inanimate objects such as landscapes and buildings; a process common in Gothic and Romantic writings, for example the anthropomorphism of William Wordsworth.

J Ruskin, 'Of the Pathetic Fallacy', *Modern Painters*, vol. III, part IV, ch. 12 [1856], repr. in *Literary Theory since Plato*, H Adams, ed. (New York, 1971), 616–23

RF

patriarchy (17th and 20th centuries) *Politics/Feminism* Rule or domination by kings, fathers, or men in general.

Either the argument that authority is derived by kings and aristocrats from God, whose fatherhood of all they represent on earth; or the view that power is divided along lines of gender and in favour of men. The original theory of patriarchy was employed in the 17th century to justify the rule of monarchs. Twentieth-century feminism has used the term to describe a division of power and advantage along lines of gender. The term thus means rule by men and – in the general absence of kings but the continuing presence of men – this is now its more common usage.

G J Schochet, *Patriarchalism in Political Thought* (Oxford, 1975); Maggie Humm, *The Dictionary of Feminist Theory* (London, 1989)

RB, HR

pattern theory *Psychology See* FILTER THEORY.

Pauli exclusion principle (1925) *Chemistry/Physics* Named after the Austrian-born American theoretical physicist Wolfgang Pauli (1900–58) who won the Nobel prize for Physics in 1945. The principle states that no pair of identical particles can simultaneously occupy the same quantum or energy state.

For example, electrons occupying the same orbital must have opposite spins. The exclusion principle is fundamental to the application of quantum theory to the periodic table of chemical elements and to the structure of complex atoms and molecules. It applies to fermions (including electrons, protons and neutrons) but not to bosons (such as photons and α-particles).

P W Atkins, *Physical Chemistry* (Oxford, 1990)

GD

Pauli's g-sum rule *Physics* Named after the Austrian-born American theoretical physicist Wolfgang Pauli (1900–58).

The rule states that, independent of the coupling scheme, for all states arising from a given electron configuration, the sum of the (Landé) g-factors for levels with the same J value is a constant.

GD

Pavlov's dog (1897) *Biology* Named after the experiments of Ivan Petrovitch Pavlov (1849–1936), Russian physiologist and winner of the 1904 Nobel prize for Physiology or Medicine.

Pavlov proved that the animals can be conditioned to salivate at the sound of a bell (conditioned reflex) after many instances of being presented with the bell's sound and food simultaneously. He extended his theory to human behaviour, deducing that psychosis is a conditioned response to painful stimuli. Following this theory, Russian psychiatrists treated patients by placing them in calm, quiet surroundings.

D Abbott, *Biologists* (New York, 1983)

KH

Pavlovianism *Psychology* Originated by the Russian physiologist Ivan Petrovich Pavlov (1849–1936).

Psychological states and processes are presumed to be identical with physiological states and processes. This view argues that investigation of the physiology and neurology of the brain are the only approaches likely to prove scientifically fruitful for the science of psychology. Pavlov's work proved particularly influential in the early development of BEHAVIOURISM.

I Pavlov, *Conditioned Reflexes* (London, 1927)

NS

PCism *Politics See* POLITICAL CORRECTNESS.

peaceful co-existence (20th century) *Politics* Optimistic theory of international relations.

Extension of liberal ideas on toleration to international society, so that ideological differences do not cause diplomatic or military conflict.

Graham Evans and Jeffrey Newnham, *The Dictionary of World Politics* (Hemel Hempstead, 1990)

RB

peak-load pricing *Economics* This is a policy of raising prices when the demand for a service is at its highest. The most recent analysis of this pricing policy stems from American research in the 1960s and 1970s.

Peak-load pricing is often used by electricity and telephone utilities as a means of reflecting the investment they have made to meet peak demand for their services. *See* AVERAGE COST PRICING, MARK-UP PRICING, MARGINAL COST PRICING and COST-PUSH INFLATION.

P L Joskow, 'Contributions to the Theory of Marginal Cost Pricing', *Bell Journal of Economics*, 7, 1 (Spring, 1976); O E Williamson, 'Peak

Load Pricing and Optimal Capacity under Indivisibility Constraints', *American Economic Review*, vol. LVI (1966), 810–27

PH

Peano's axioms (1879, 1889) *Mathematics* Named after Italian mathematician Giuseppe Peano (1858–1932) who published a formulation of them which was widely read. They were actually first stated by German mathematician Julius Wilhelm Richard Dedekind (1831–1916).

These are a set of axioms which formalize the theory of arithmetic; that is, the theory of the natural numbers. Among the axioms is the assumption that each number has a unique successor. Also included is the axiom of INDUCTION. Although it was not known to Dedekind at the time, these axioms have models which are very different from the ordinary natural numbers. These so-called non-standard models of arithmetic are non-Archimedean and have the peculiar property of possessing infinitely many 'infinite' integers. *See also* COMPACTNESS THEOREM.

Encyclopedic Dictionary of Mathematics (MIT Press, 1987)

MB

Pell's equation (17th century) *Mathematics* Mistakenly ascribed to British mathematician and astronomer John Pell (1610–85) who apparently made no contribution to the topic. It should properly be ascribed to French lawyer and amateur mathematician Pierre de Fermat (1601–65).

This is the DIOPHANTINE EQUATION of the form

$$x^2 - dy^2 = 1$$

where d is a positive integer which is not a perfect square. Fermat conjectured that there is at least one solution in integers x, y other than $x = \pm 1$, $y = 0$; the conjecture was proved by Lagrange in 1768. The complete set of positive solutions can be obtained by applying the CONTINUED FRACTION ALGORITHM to the number $\sqrt{d}$.

A Baker, *A Concise Introduction to the Theory of Numbers* (Cambridge, 1984)

MB

Peltier effect (1834) *Physics* Named after its French discoverer Jean Charles Athanase Peltier (1785–1845).

When an electric current flows across the junction between two different metals or semiconductors, a quantity of heat is produced that is proportional to the total quantity of electric charge that crosses the junction. The heat may be evolved or absorbed, depending on the direction of the current flow. This effect is additional to the Joule heating produced by the current. *See* JOULE'S LAW OF ELECTRIC HEATING.

J Thewlis, ed., *Encyclopaedic Dictionary of Physics* (New York, Oxford and London, 1962)

MS

penetration theory *Chemistry* A theory which proposes that mass transfer of a solute across an interface to a stirred liquid is brought about through diffusive penetration of the solute. The rate of mass transfer is proportional to the square root of the diffusion coefficient since the liquid surface is continually being renewed.

GD

penis envy *Psychology* Proposed by the Austrian founder of psychoanalysis Sigmund Freud (1856–1939).

Part of Freud's theory of psychoanalysis, the concept is primarily used in respect of women who, as young children, developed envy of the male external genitalia during development through the electra phase. Penis envy is now largely discredited within psychology, partly because of its implication that women are inherently inferior to men. Penis envy is also used to denote young males' envy of adult male external genitalia.

C S Hall, *A Primer of Freudian Psychology* (Cleveland, 1954)

NS

Penrose tiling (1964) *Mathematics* Named after British mathematician and physicist Roger Penrose, this is a tiling pattern which covers the whole plane in which the tiles lie in the shape of regular pentagons with five sides of equal length. Penrose used a pair of shapes known as a kite and dart which can be created by cutting a rhombus into two pieces so that the longer diagonal of the rhombus is divided in the golden mean. Penrose and John Conway showed that there are infinitely many ways to cover the whole plane with kite and dart tiles, none of which is periodic. Penrose tilings have been used to produce interesting results in crystallography.

I Peterson, *The Mathematical Tourist* (New York, 1988)

ML

perception *Psychology* Not attributable to any one originator, this term refers to the way we see the world around us.

Psychologists are interested in understanding how organisms, especially humans, detect and interpret stimuli about the world. The two main theories on perception are the classical approach and that contained within GESTALT THEORY.

P J Barber and D Legge, *Perception and Information* (London, 1976)

NS

perfect competition (20th century) *Economics* A competitive system in which a large number of firms produce a homogenous product for a large number of buyers.

All the firms share the same product/market knowledge and enjoy free entry/exit to and from the industry. They are price-takers and sell as much of the product as possible at the market price. Output is set where marginal cost equals marginal revenue. In the long run, average revenue equals marginal cost and firms enjoy only normal profits. *See* IMPERFECT COMPETITION, MONOPOLISTIC COMPETITION, MONOPOLY and DUOPOLY THEORY.

G Stigler, 'Perfect Competition, Historically Contemplated', *Journal of Political Economy*, vol. LXV (1957), 1–17

PH

perfect number *Mathematics* A natural number which is equal to the sum of its divisors other than itself.

For example, 6 is a perfect number since $6 = 1 + 2 + 3$. It can be shown that an even number is perfect if and only if it has the form

$$2^{p-1}(2^p - 1)$$

where both p and $2^p - 1$ are primes. Thus

$$28 = 2^2(2^3 - 1)$$

It is a notorious unsolved problem whether there are infinitely many perfect numbers or any odd perfect numbers. *See also* MERSENNE PRIME.

A Baker, *A Concise Introduction to the Theory of Numbers* (Cambridge, 1984)

MB

perfection, principle of *Philosophy* Also called the *principle of the best.* Principle of German philosopher and mathematician Gottfried Wilhelm Leibniz (1646–1716) that the actual world is the best of all possible worlds. Bertrand Russell (1872–1970) argued that Leibniz did not fully distinguish this principle from that of SUFFICIENT REASON.

B Russell, *A Critical Exposition of the Philosophy of Leibniz* (1900), §§14–15

ARL

performance, linguistic *See* LINGUISTIC COMPETENCE.

performative (or ditto) theory of truth *Philosophy* Theory developed by English philosopher Peter Frederick Strawson (1919–) in and after 1949 from Ramsey's REDUNDANCY THEORY OF TRUTH, and in opposition to the CORRESPONDENCE THEORY.

To call something true is to perform the act of agreeing with it, endorsing it, appraising it and so on. Like the EMOTIVE THEORY OF TRUTH this is a SPEECH ACT THEORY, and like similar theories in ethics and elsewhere is open – at least *prima facie* – to the objection that it cannot cover cases like 'If that is true, then . . .' where the relevant speech act is not being performed.

G Pitcher, ed., *Truth* (1964)

ARL

period-luminosity law for Cepheid variables (1912) *Astronomy* A Cepheid variable is a single star whose brightness varies periodically. The American astronomer Henrietta Swan Leavitt (1868–1921) found a relationship between the luminosity and the period of these stars: the brighter the star the greater is its period. This result enables the distance of distant Cepheids to be found from their apparent brightness once the distance of a nearby Cepheid has been determined by some other means.

S Mitton, ed., *The Cambridge Encyclopaedia of Astronomy* (London, 1973)

MS

periodic law *Chemistry See* MENDELEYEV'S PERIODIC LAW.

peripheral theory of drive *Psychology See* PERIPHERALISM and MOTOR THEORY OF MEANING.

peripheralism *Psychology* Proposed by the American psychologist John Broadus Watson (1878–1958).

A theoretical perspective which focuses on peripheral processes as explanatory devices. Supported by behaviouists, this view emphasizes events at the periphery of an organism, for example sense organs and skeletal muscles, rather than the functions of the central nervous system. Watson took this perspective to the extreme and argued that thinking was merely subvocal laryngeal movement and that all emotions were mechanical glandular responses. The perspective is no longer taken seriously within psychology. *See* BEHAVIOURISM.

J B Watson, *Behaviorism* (New York, 1925)

NS

Perky effect *Psychology* Identified by the psychologist C W Perky (*fl.* 1910s).

The effect refers to a perceptual error observable under experimental conditions. Typically, a person was asked to look at a blank screen and imagine an object, for example a banana. Unknown to the subject, a picture of the object which they were asked to imagine had been back-projected onto the screen and the brightness of the picture slowly increased until it could be clearly seen by anyone in the room. Generally, subjects failed to report the picture, attributing the projected image to their own vivid imagination.

P D Slade and R P Bentall, *Sensory Deception: A Scientific Analysis of Hallucination* (London, 1988)

NS

perlocutionary act *Linguistics See* SPEECH ACT THEORY.

permanent income hypothesis (1957) *Economics* Developed by American economist Milton Friedman (1912–92), in its simplest form this theory states that the choices consumers make regarding their consumption patterns are determined not by current income but by their longer-term income expectations.

Measured income and measured consumption contain a permanent (anticipated and planned) element and a transitory (windfall/unexpected) element. Friedman concluded that the individual will consume a constant proportion of his/her permanent income; and that low income earners have a higher propensity to consume; and high income earners have a higher transitory element to their income and a lower than average propensity to consume. *See* ABSOLUTE INCOME HYPOTHESIS, RELATIVE INCOME HYPOTHESIS and LIFE-CYCLE HYPOTHESIS.

M Friedman, 'A Theory of the Consumption Function', *National Bureau of Economic Research* (Princeton, N.J., 1957)

PH

permanent revolution (20th century) *Politics* Theory of REVOLUTION of the Russian revolutionary Leon Trotsky (1879–1940).

Socialist revolution and the overthrow of capitalism will only be complete when they are worldwide. There should therefore be no pause in revolutionary activity simply because power has been seized in a single country. This theory is thus opposed to that of SOCIALISM IN ONE COUNTRY.

David Miller *et al.*, eds, *The Blackwell Encyclopaedia of Political Thought* (Oxford, 1987)

RB

perpetual motion *Physics* Perpetual motion is motion which, once started, will continue indefinitely and, if it were possible, would be realized through a perpetual machine.

Two such kinds of machine are distinguished: (1) one that supplies an endless output of work without any input of fuel or other energy, therefore creating its own energy and so violating the FIRST LAW OF THERMODYNAMICS; (2) a device which can extract energy from a body at a constant temperature and convert it into work without the use of a body at a different temperature, therefore violating the SECOND LAW OF THERMODYNAMICS.

J Thewlis, ed., *Encyclopaedic Dictionary of Physics* (New York, Oxford and London, 1962)

MS

Perron-Frobenius theorem (*c.*1900) *Mathematics* Named after its discoverers, this theorem states that for a square matrix A with all elements strictly positive real numbers, there is a vector $\mathbf{x}$ with positive elements which is unique except for multiplication by positive numbers; such that $A\mathbf{x} = k\mathbf{x}$ and the absolute value of k is the greatest for any number l such that $A\mathbf{v} = l\mathbf{v}$ for some vector $\mathbf{v}$. There are generalizations of this result.

JB

perseveration effect *Psychology* Identified by the German psychologist G E Mueller (1850–1934).

This term is applied to the consolidation of memory. The idea is that preservation of a neural process is necessary for a permanent trace to be established. Therefore, a new experience creates neural activity in the brain and if this experience is to be remembered a process which changes the brain must occur. This rather basic idea has been elaborated and refined in subsequent neuropsychological investigations of memory.

D J Lewis, 'Psychology of Active and Inactive Memory', *Psychological Bulletin*, vol. LXXXVI (1979) 1054–83

NS

persona (classical; 20th century) *Literary Theory* From the Latin '*persona*' which was the mask worn by actors in drama, hence '*dramatis personae*'.

By extension, the image of a speaking subject, an implied 'I' relating to an implied addressee, created by the language of a literary work, particularly a LYRIC poem or a NARRATIVE. *See also* IMPLIED AUTHOR; and 'tone' in FOUR KINDS OF MEANING.

R Ellmann, *Yeats: The Man and the Masks* (London, 1948)

RF

personal construct theory *Psychology* Proposed by the American psychologist G A Kelly (1905–67).

A theory of personality based on the concept of 'personal construct', defined as thoughts and feelings about how we see some things as being alike and different from others. Constructs are organized into personal construct systems which can be regarded as intuitive theories about the world and ourselves. Personal construct theory has fostered an extensive range of investigations and the development of effective therapeutic techniques.

G A Kelly, *A Psychology of Personal Constructs* (New York, 1955)

NS

personal is political (20th century) *Politics/Feminism* Deliberately paradoxical phrase intended to subvert the limitations of the PUBLIC PRIVATE DIVIDE.

The distinction made by liberals between the public world and the private world is mistaken. It functions, moreover, to sustain the oppression of women since it presents the structured inequalities of the household as the result of free individual choice.

Carol Pateman, *The Disorder of Women* (Cambridge, 1989)

HR

personalism *Philosophy* The view that persons, divine or human, play the primary role in the structure of the universe. Personalism exists in a wide variety of forms, and is closely related to IDEALISM (the term *personal idealism* is often used) or to theism. What they have in common is that the notion of a person is treated as basic and not to be analyzed in terms of, or seen as emerging out of, other entities.

R T Flewelling, 'Personalism', *Twentieth Century Philosophy*, D D Runes, ed. (1943)

ARL

personalist theory of probability (20th century) *Mathematics* Developed by de Finetti and Ramsey (although early philosophers recognized a 'subjective' element of probability), this theory investigates probability methods which employ preferences when objective probabilities are not known.

T L Fine, 'Probability as Pragmatics Necessity', *Theories of Probability* (New York, 1973)

JB

personification *Literary Theory See* ALLEGORY.

personology *Psychology* A theory of personality originated by American psychologist Henry Alexander Murray (1893–).

Personology emphasizes the treatment of the person as a whole, and the need to conduct comprehensive studies of the person. The most distinctive feature of the theory is its emphasis on the organization of motivational concepts. Past and present factors both carry weight in determining behaviour, and unconscious motivation is a predominant concern.

C S Hall and G Lindzey, *Introduction to Theories of Personality* (Chichester, 1985)

NS

perspective *Art* The study and theory of the scientific representation of three-dimensional objects on a two-dimensional plane, now considered to belong to the science of geometry.

Although perspective painting existed in Roman art, it is only since the Renaissance that artists and theorists have attempted to give a theoretical basis (by the application of linear diagrams) to the problem of depicting solids and volume. Generally, this has involved the central-projection method, although more complex and sophisticated forms (such as parallel or axonometric perspective) also exist.

AB

perspective realism *Philosophy* Form of REALISM holding that the nature of an object depends on its relations to other objects. For example, a penny not only looks round from one perspective and elliptical from another but is round with respect to one and elliptical with respect to the other, no perspective having any special privilege. This enables us to say that the penny is seen as it is, because how it appears from a certain perspective is part of how it is.

E B McGilvary, *Toward a Perspective Realism* (1956)

ARL

perspectivism *Philosophy* Theory associated especially with Friedrich Wilhelm Nietzsche (1844–1900), José Ortega y Gasset (1883–1955), who named it, Edward Sapir (1884–1939), B L Wharf (1897–1941), Willard Van Orman Quine (1908–) and Thomas Samuel Kuhn (1922–).

Perspectivism says that there can be radically different and incommensurable conceptual schemes (ultimate ways of looking at the world) or perspectives, one of which we must (consciously or unconsciously) adopt, but none of which is more correct than its rivals.

For Sapir and Whorf our own scheme is dependent on the language we use. Like some other forms of RELATIVISM, perspectivism is open to the objection that it cannot cater for itself: is the view that there are different conceptual schemes itself something arising only within one, non-mandatory, conceptual scheme? *See also* INDETERMINACY OF REFERENCE AND TRANSLATION, though it has been claimed (see Hacking) that this is inconsistent with incommensurability.

M Krausz and J W Meiland, eds, *Relativism: Cognitive and Moral* (1982); I Hacking, *Rationality and Relativism*, M Hollis and S Lukes, eds, (1982)

ARL

perturbation molecular orbital (PMO) theory *Chemistry/Physics* A technique for solving complex quantum mechanical problems in terms of solutions for simpler problems.

Perturbation theory yields good approximate solutions if the system is assumed to differ only slightly from a simpler system for which the problem can be solved. The neglected difference can be dealt with as a perturbation of this simpler undisturbed system to give an approximate or analytic solution or to provide suitable algorithms for a numerical solution.

D M Considine, *Van Nostrand's Scientific Encyclopedia* (New York, 1989)

GD

Peter principle (1969) *Business* Named after Canadian sociologist Laurence Peter (1920–90), who conducted extensive research into business organizations, the principle states that in an organization, people are promoted to the level of their incompetence and remain there.

L Peter and R Hull, *The Peter Principle: Why Things Always Go Wrong* (London, 1969)

AV, PH

phase rule (1878) *Chemistry* Also known as the Gibbs phase rule after the American mathematical physicist Josiah Willard Gibbs (1839–1903) who was professor at Yale from 1871 and who contributed enormously to the study of thermodynamics.

The rule states that in an equilibrium system composed of C components and P phases, the number of degrees of freedom (F), or the *variance* of the system, is given by

$$F = C - P + 2$$

For example, the system ice and water in equilibrium has one component and two phases and thus one degree of freedom. This means that for the two phases to co-exist at any chosen temperature there is no freedom in the choice of pressure and *vice versa*.

P W Atkins, *Physical Chemistry* (Oxford, 1990)

GD

phase sequence hypothesis *Psychology* Proposed by the Canadian psychologist Donald Olding Hebb (1904–).

'Phase sequence' defines a series of nerve-cell systems that are in a functional relationship. Hebb developed this idea in his theory of cell assemblies, trying to explain how quite large brain damage does not necessarily result in a drop in IQ. His theory has illustrated the importance of early experience in the development of brain and intelligence.

D O Hebb, *The Organization of Behavior* (New York, 1949)

NS

phatic communion (1923) *Linguistics* A term coined by Polish anthropologist Bronislaw Malinowski (1884–1942) to refer to utterances such as 'How are you?' and 'What a lovely day' whose main function is to make and maintain links between speakers.

B Malinowski, 'The Problem of Meaning in Primitive Languages', supplement to C K Ogden and I A Richards, *The Meaning of Meaning* (London, 1923)

RF

phenomenalism *Philosophy* Literally, 'appearance-ism'. Any theory which explains a given subject-matter in terms of appearances, without needing to postulate anything else (*see also* REDUCTIONISM), much as facts about the average man are reduced to facts about ordinary men. The most notable 19th century phenomenalist was John Stuart Mill (1806–73).

Phenomenalists in the 20th century (for example, Alfred Jules Ayer (1910–89)) usually take a more linguistic approach, reducing sentences about one kind of thing (in particular, material objects) to sentences about sense-data (things allegedly given to us immediately in sensations and so on). This approach suits LOGICAL POSITIVISM because of the difficulties of directly verifying sentences about material objects.

The difference between phenomenalism and SUBJECTIVE IDEALISM can be seen by comparing two interpretations of how George Berkeley (1685–1753) accounts for ideas of unperceived material objects. On the idealist (and usual) interpretation such ideas exist in the mind of God. On the phenomenalist interpretation they do not exist at all, but God ensures we have the relevant ideas whenever (in common sense terms) objects come back into view. No successful phenomenalist analysis of material objects, however, has ever been completed.

J D Mabbott, 'The Place of God in Berkeley's Philosophy', *Locke and Berkeley*, C B Martin and D M Armstrong, eds (no date; original article 1931)

ARL

phenomenological theory *Physics* This term describes any theory which expresses mathematically the results of observations of phenomena without paying detailed attention to their fundamental significance.

J Thewlis, ed., *Encyclopaedic Dictionary of Physics* (New York, Oxford and London, 1962)

MS

phenomenology *Philosophy* Literally, 'the description or study of appearances'. Any detailed study of a phenomenon can be called a phenomenology, but the theory normally so called is associated with Franz Brentano (1838–1917) and (especially)

Edmund Husserl (1859–1938) and their followers, including several existentialists.

'Phenomena' for Husserl were the objects of experience or attitudes (in the sense in which even a non-existent fortune can be the object of my wish). These he treated as essences, aiming to give an analysis of them not unlike the 'conceptual analysis' of linguistic philosophy. The analysis, however, was *a priori*, and he aimed (at any rate in his later works) to avoid PSYCHOLOGISM, laying aside (*bracketing*) ideas derived from empirical science. *See also* LINGUISTIC PHENOMENOLOGY.

H Spiegelberg, *The Phenomenological Movement* (1960)

ARL

phenotype theory (1924) *Biology* Proposed by the Russian biochemist Aleksandr Ivanovich Oparin (1894–1980), this is the theory that primitive cells – the presumed precursors of life – evolved from coagulated particles (coacervates) in the emulsion of proteins.

This theory was contemporaneous with the idea that proteins are life-like (*see* DYNAMIC STATE THEORY) and was largely rejected when this idea proved to be false. *Compare with* GENOTYPE THEORY, MINERAL THEORY, PRECELLULAR EVOLUTION, RIBOTYPE THEORY and VIRAL THEORY. *See also* ORIGINS OF LIFE.

M Barbieri, *The Semantic Theory of Evolution* (Chur, Switzerland)

KH

phi phenomenon *Psychology* Proposed by the German psychologists Kurt Koffka (1886–1941), and Max Wertheimer (1886–1943).

A form of apparent movement produced when two stationary lights are flashed successively. When the interval between the two lights is optimal (approximately 150 milliseconds), then one perceives an apparent movement of light from one location to the other. This term was used by Wertheimer to refer to a 'pure' irreducible experience of motion, independent of other factors (for example, colour, size, spatial location); and he considered it a good example of second sense. *See* KORTE'S LAW.

A D Ellis, ed., *A Sourcebook of Gestalt Psychology* (London, 1938)

NS

Phillips curve (1958) *Economics* Named after British economist William Phillips (1914–75), this charts the significant relationship between the percentage change in money wages and the rate of unemployment. Its main implication is that low inflation and low unemployment are incompatible, and so governments have to choose the best combination of both. Phillips also showed that at an unemployment rate of 2.5 per cent, stable wages could be maintained.

A W H Phillips, 'The Relationship Between Unemployment and the Rate of Change of Money Wage Rates in the United Kingdom, 1861–1967', *Economica NS*, vol. xxv (November, 1958), 283–99

PH

Philo's theory of combustion *Chemistry* Named after the Byzantine scientist Philo (*c*.2nd century BC) who wrote a treatise on military engineering. He was one of the first to record the fact that a burning candle enclosed by a glass sphere inverted over a trough of water caused the air inside it to contract.

GD

philology (19th century) *Linguistics* Popular term with two senses, little used within linguistics today.

(1) 'Comparative philology', equivalent to COMPARATIVE AND HISTORICAL LINGUISTICS and so used by some of its pioneers. (2) Study of the relationships between language and (aesthetic) culture, chiefly classical.

R H Robins, *General Linguistics: An Introductory Survey* (London, 1964)

RF

phlogiston theory (1702) *Chemistry* A theory proposed by the German chemist and physician Georg Ernst Stahl (1660–1734) in his elaboration of an earlier theory of combustion by the chemist Johann Joachim Becher (1635–82). All combustible matter was supposed to contain phlogiston, a principle of fire akin to heat.

Metals were thought to consist of calx (now called oxide) as well as phlogiston. On heating, the phlogiston was released and the calx remained. To account for the observed weight changes, phlogiston would have had to have negative weight. The theory was finally displaced in the 1770s when the true role of oxygen in combustion was discovered by Antoine Lavoisier (1743–94). *See also* CALORIC THEORY.

M Freemantle, *Chemistry in Action* (Basingstoke, 1987)

GD

phoneme (1894) *Linguistics* Named by Polish linguist J Baudouin de Courtenay, the idea was developed somewhat earlier by English phonetician Henry Sweet (1845–1912) and subsequently by his compatriot Daniel Jones (1881–1967) and by American STRUCTURAL linguists.

Minimal unit in PHONEMICS; the smallest unit of the sound system. A segment of speech which can distinguish meanings, for example /b/ and /p/ in 'back' and 'pack' ('back' and 'pack' are called a 'minimal pair'). A phoneme has variants or 'allophones' which may be in 'complementary distribution' (for example /k/ is pronounced with the tongue in different positions in 'keep', 'calm' and 'cool', depending upon the nature of the following vowel); or in 'free variation', for example the last sound /p/ in 'tap' may be released or not.

An apparently commonsense notion, the phoneme is in fact a profoundly problematic theory, and has been abandoned in GENERATIVE PHONOLOGY.

D Jones, *The Phoneme: Its Nature and Use* 3rd edn, (Cambridge, 1967 [1950])

RF

phonemics (1930s–) *Linguistics* Study of PHONOLOGY (and alternatively so called) in STRUCTURAL LINGUISTICS; chiefly American.

Analysis of phonology is dependent centrally on the notion of the PHONEME. A language is seen as possessing a structured set of phonemes; 'segmental' phonemes arranged in linear patterns, and SUPRASEGMENTAL phonemes.

B Bloch and G L Trager, *Outline of Linguistic Analysis* (Baltimore, 1942)

RF

phonetics *Linguistics See* ACOUSTIC PHONETICS, ARTICULATORY PHONETICS, and AUDITORY PHONETICS.

phonocentrism (1960s) *Linguistics* Formulated by Algerian-French philosopher Jacques Derrida (1930–).

The privileging of speech over writing, leading to the latter being regarded as a derived form or a straightforward notation, rather than an independent mode of language. *See also* GRAMMATOLOGY and LOGOCENTRISM.

C Norris, *Deconstruction*, rev. edn (London, 1991)

RF

phonology (late 19th century–) *Linguistics* Various models throughout this century.

Phonology is the study of the communicatively significant sound patterns of language, as opposed to the 'raw' sounds of speech which are treated by the various branches of PHONETICS. *See also* GENERATIVE PHONOLOGY, PHONEME, and SUPRASEGMENTAL PHONOLOGY.

E C Fudge, ed., *Phonology: Selected Readings* (Harmondsworth, 1973)

RF

photochemical equivalence, law of *Chemistry See* EINSTEIN-STARK LAW.

photoconductive effect *Physics* This is the name given to the increase in the electrical conductivity of non-metallic solids which is produced by the motion of additional free electron carriers when electromagnetic radiation is absorbed.

Photoconductivity is associated with the simultaneous excitation of an electron into the conduction band and the creation of a 'hole' in the valence band. The spectral dependence of photoconductivity is characterized by the existence of a threshold frequency of the radiation, corresponding to a photon energy E_g. Only for photons with energies greater than E_g is photoconductivity detectable.

J Thewlis, ed., *Encyclopaedic Dictionary of Physics* (New York, Oxford and London, 1962)

MS

photoelectric effect (1888) *Physics* Use of this term is customarily reserved for the emission of electrons from substances (usually metals) when irradiated with light of a frequency greater than a certain minimum threshold value.

Electrons liberated in this way are called photoelectrons. While an electrical effect on irradiating metals was observed by German physicist Wilhelm Hallwachs (1859–1922) in 1888, the effect was first shown to be the result of the emission of electrons by the British physicist Sir Joseph John Thomson (1856–1940) in 1899. The theoretical explanation of the observed maximum kinetic energy of the photoelectrons ejected by light of a given frequency and of the existence of a threshold frequency was given by the German-born mathematical physicist Albert Einstein (1879–1955) in 1905. *See* HALLWACHS'S EFFECT.

J Thewlis, ed., *Encyclopaedic Dictionary of Physics* (New York, Oxford and London, 1962)

MS

photoperiodism (1920) *Biology* Developed by American agricultural scientists Wightman W Garner and Henry A Allard, this is the concept that plants are capable of measuring the length of daylight and nighttime, and that some physiological response (for example, flowering) is governed by these measurements.

Subsequent research demonstrated that insects, mammals, birds and other living things are capable of photoperiodism, although the mechanism remains unknown. *See also* BIOLOGICAL CLOCK.

F Salisbury, *Plant Physiology* (Belmont, Calif., 1992)

KH

photo-realism (1960s) *Art* Also known as Super Realism, this is a painting style which emulates the most sharply focused photographs; using precise, undifferentiated technique with no particular psychological meaning. It grew from POP ART and is best represented by the New York-based artist John Salt.

G Battcock, *Super Realism: A Critical Anthology* (London, 1975)

LL

photovoltaic effect *Physics* This term refers to the generation of an e.m.f. in a material as a result of the absorption of light. The effect can occur in solids, liquids and gases but has been mostly studied in semiconducting solids.

Three phenomena are involved in the effect: (1) the generation of equal numbers of positive and negative charge carriers by the absorption of the light, a process known as photoionization; (2) the migration of one or both types of photoliberated charge carriers to a region of the material where separation of the positive and negative charges can occur; (3) the presence of charge separation mechanisms that prevent recombination.

J Thewlis, ed., *Encyclopaedic Dictionary of Physics* (New York, Oxford and London, 1962)

MS

phrase structure grammar (1957) *Linguistics* Traditional type of grammatical analysis, formalized by the American linguist Avram Noam Chomsky (1928–).

Phrase structure rules identify the ways sentences can be built up from units such as Verb Phrase, Noun Phrase, Noun, Adverb by REWRITING RULES. This is a form of PARSING, or CONSTITUENT STRUCTURE analysis. Chomsky claimed that a phrase structure grammar was an essential part of a generative grammar, but not sufficient; TRANSFORMATIONS also were needed.

N Chomsky, *Syntactic Structures* (The Hague, 1957)

RF

phrenology (late 18th century) *Medicine* Originated by the Austrian anatomist Franz Josef Gall (1758–1828), this is a pseudoscience holding that the brain is the source of thought and will, and that its configurations dictate an individual's personalities and abilities.

Because the brain is not immediately accessible in a living person, observations and measurements of the exterior skull were used to deduce an individual's capabilities. Although always controversial, phrenology was discredited by experimental work in physiology and neurology in the mid-1800s.

D de Guistino, *Conquest of Mind: Phrenology and Victorian Social Thought* (London and Totowa, NJ, 1975)

KH

phyletic evolution *Biology* A theory defined variously, depending on the perspective adopted by the evolutionist. It is often considered synonymous with GRADUALISM, in which changes occur nearly continuously through the lineage (phylum). It is also used to describe evolutionary change (not necessarily speciation) through a phyletic lineage.

N Scott-Ram, *Transformed Cladistics, Taxonomy and Evolution* (Cambridge, 1990)

KH

phyletic gradualism *Biology See* GRADUALISM.

phylogenesis *Psychology* a term applied to psychology by the Swiss psychiatrist Carl Gustav Jung (1875–1961).

Phylogenesis refers to the origin and development of a species as a whole. It is generally limited to biological inheritance but within psychology was extended by Jung to include the development of psyche and archetypes.

S J Gould, *Ontogeny and Phylogeny* (Cambridge, Mass., 1977)

NS

phylogenetic systematics *Biology See* CLADISTICS.

physicalism *Philosophy* Term variously used. For Rudolf Carnap (1891–1970), a member of the VIENNA CIRCLE, it said that all scientific statements could be reduced to statements about ordinary physical objects (or else spatiotemporal points), such sentences having to be publically verifiable. For others it has meant that any meaningful statement can be translated into the language of physics. Currently, physicalism is most often used for the IDENTITY THEORY OF MIND, sometimes including BEHAVIOURISM. All these theories are forms of REDUCTIONISM.

R Carnap, *The Unity of Science* (1932); translated with Introduction by M Black (1934)

ARL

Physiocracy (18th century) *Economics* The French school of economics, led by physician and economist François Quesnay (1694–1774), whose theory maintained that a natural order existed and it was the duty of the state to preserve this order.

The Physiocrats believed that land was the only source of wealth and, therefore, agriculture was the only sector that should be taxed. Industry and commerce should be excluded from state intervention, giving rise to their policy of 'LAISSEZ-FAIRE' which was in direct opposition to MERCANTILISM. *See also* ECONOMIC LIBERALISM and NEW CLASSICAL MACROECONOMICS.

F Quesnay, *Tableau économique* (Paris, 1758); R L Meek, *The Economics of Physiocracy: Essays and Translations* (London, 1962)

PH

Picard's theorems *Mathematics* Named after French analyst, group theorist and mechanist Charles Emile Picard (1856–1941).

(1) Picard's first or 'little' theorem: a nonconstant entire function can miss at most one finite complex value in its range.

(2) Picard's second or 'big' theorem: an extension of the first which states that an analytic function takes all finite values except possibly one in every neighbourhood of an essential isolated singularity. *See also* CASORATI-WEIERSTRASS THEOREM.

J B Conway, *Functions of One Complex Variable* (New York, 1973)

ML

picaresque (16th century) *Literary Theory* Term derived from the Spanish word *picaro*, meaning 'rogue'.

A type of NARRATIVE which is essentially a sequence of escapades involving a single character, generally a likeable rogue living by her (more usually his) wits; realistic in detail, and usually comic or satirical. In English, see for example Daniel Defoe's *Moll Flanders* (1722), and Mark Twain's *Huckleberry Finn* (1884).

R Alter, *Rogue's Progress: Studies in the Picaresque Novel* (Cambridge, Mass., 1964)

RF

Pick's theorem (1899) *Mathematics* Named after its discoverer, Georg Pick. Choose lattice points in the plane with regular horizontal and vertical spacing. The area of a polygon with lattice points as vertices, and which does not cross itself, is

$$\frac{b}{2} + c - 1$$

where b is the number of lattice points on the boundary and c is the number of lattice points inside.

JB

picture plane *Art* Part of the theory of PERSPECTIVE, as it applies to painting. This is the plane of the surface (canvas, panel) upon which the artist depicts objects, and from which these objects appear to recede or project.

AB

picture theory of meaning *Philosophy* Theory which treats declarative sentences (as against commands and so on) as pictures of facts (if true) or possible facts (otherwise). A notable example of the theory is *Tractatus* (1921) by Ludwig Wittgenstein (1889–1951).

Each element in the sentence (bar certain connectives and so on) stands for something, be it an object or a quality or a relation, and so on; and the connectives and the way the words are put together correspond to the way the objects and qualities and so on are related in the envisaged situation. The theory is thus a form of CORRESPONDENCE OR RELATIONAL THEORY OF MEANING analogous to NAMING THEORIES OF MEANING for singular terms. One objection is that it does not seem to cater for the difference between merely picturing a scene and stating that that scene is part of reality.

E Daitz, 'The Picture Theory of Meaning', *Mind* (1953); reprinted in A Flew, ed., *Essays in Conceptual Analysis* (1956)

ARL

picturesque (18th century) *Art* Closely related to 18th- and 19th-century concepts of the SUBLIME and ROMANTICISM, the term is derived from the Italian word *pittoresco* meaning 'relating to a painter'.

It originally referred to landscape scenes which appeared to be copied from paintings by French artists like Claude Lorraine (1600–82) and Nicolas Poussin (1594–1665), but later developed into the reverse whereby painting directly mirrored the beauty of nature. The landscaped gardens of English designer 'Capability Brown' (1716–83) were largely inspired by the notion of the picturesque.

AB

piecemeal activity, law of *Psychology* Proposed by the American psychologist Edward Lee Thorndike (1874–1949).

This law refers to the fact that an aspect of a stimulus may become more important than other aspects, and hence that aspect may evoke a response even though the other aspects are not present. The existence of this law has been demonstrated empirically.

E L Thorndike, *Animal Intelligence* (New York, 1911)

NS

piecemeal social engineering (20th century) *Politics/Sociology* Theory of proper method of social reform employed by Austrian philosopher Karl Popper (1902–).

Rather than engage in grand schemes, government should deal with problems as they emerge, and respond to social or economic deficiencies in an *ad hoc* manner. Piecemeal social engineering was a concept favoured by those attracted to the idea of an END OF IDEOLOGY.

Roger Scruton, *A Dictionary of Political Thought* (London, 1987)

RB

pigeon-hole principle *Mathematics* Also known as the drawer principle, the letter-box principle, and Dirichlet's principle after the French-born German mathematician Peter Gustav Lejeune Dirichlet (1805–59).

This is the fundamental enumerative principle that if there are n holes containing $n + 1$ pigeons (letters), then there must be at least two pigeons (letters) in some hole. *See also* DIRICHLET'S THEOREM ON DIOPHANTINE APPROXIMATION.

MB

Pigou effect (1943) *Economics* Named after English economist Arthur Pigou (1877–1959), the analysis was firmly rooted in the classical school of economics and was subsequently overshadowed by the work of English economist John Maynard Keynes (1883–1946). The term may be defined as the impact of a change in the money supply on consumption.

Pigou maintained lower prices would encourage consumption, thereby boosting total income and employment. Implicit in the Pigou effect (also known as the real balance effect) was the belief that an improved employment situation was achievable through lower wages. The effect can be shown in diagrammatic form as a shift in the IS curve. *See* IS-LM MODEL.

A C Pigou, 'The Classical Stationary State', *Economic Journal*, vol. LIII (December, 1943), 343–51

PH

pinching theorem *Mathematics See* SANDWICH THEOREM.

Piper's law *Psychology* Proposed by the German psychologist H E Piper (1877–1915).

Piper proposed that for moderate-sized uniform areas of the retina outside the fovea (a small pit in the retina, densely packed with cones and the area of clearest vision), 'the absolute threshold is inversely proportionate to the square root of the area stimulated'.

N R Carlson, *The Physiology of Behavior*, 3rd edn (Boston, Mass., 1986)

NS

pitch-class set theory (1960s) *Music* Developed by the American music theorist Allen Forte (1926–) from existing mathematical theories, this is a way of expressing and organizing pitch formations and then analyzing them.

A pitch class comprises all pitches with the same name, irrespective of register; for example, all Cs belong to the pitch class C. Each pitch class is allotted a number so that C = 0, C sharp = 1, D = 2 and so on up to B = 11. Groups of notes may then be expressed in numerical terms; thus the chord D-E-A-F would be 2, 3, 9, 5. A set is placed in numerical order, or 'normal form' (2, 4, 5, 9), and transposed so as to start on zero, yielding its 'prime form' (0, 2, 3, 7). Sets may then be more easily compared and significant relationships uncovered. The theory is valuable for helping to make sense of non-serial atonal music but relies on the ability of the analyst to extract, at the outset, significant groups of notes for comparison.

A Forte, *The Structure of Atonal Music* (New Haven and London, 1973)

MLD

pivot grammar (1963) *Linguistics* Proposal by the American psycholinguist M D S Braine.

Children's two-word utterances are structured: one word comes from a small class of *pivot* words fixed in first position ('*More* juice') or second ('Pants *off*'); the other is drawn from an *open* class which can occur either first or second. A child's vocabulary expands in the open class. Later research, led by L Bloom, has cast doubt on pivot grammar.

M Braine, 'The Ontogeny of English Phrase Structure: the First Phase', *Language* 39 (1963), 1–13; repr. in C A Ferguson and D I Slobin, eds, *Studies of Child Language Development* (New York, 1973), 407–21

RF

Planck's quantum theory (1900) *Physics* Named after the German theoretical physicist Max Karl Ernst Planck (1858–1947), winner of the 1918 Nobel prize for Physics.

Planck's work on thermodynamics and black-body radiation led him to abandon classical dynamical principles and to formulate the quantum theory, which assumed that electromagnetic energy comes in discrete amounts or quanta, each of which has an energy $h\nu$, where ν is the frequency of the radiation and h is the Planck constant. This successfully accounted for the distribution of energy as a function of frequency in black-body radition, a result that was inexplicable in classical theory.

Robert M Besançon, ed., *The Encyclopedia of Physics* (New York, 1985)

GD

Planck's radiation law (1900) *Physics* Named after its discoverer, the German theoretical physicist Max Karl Ernst Ludwig Planck (1858–1947).

The energy density u_λ of the electromagnetic radiation in an evacuated cavity with opaque walls at a temperature T, in the wavelength range λ to $\lambda + d\lambda$, is given by

$$u_\lambda = (8\pi hc\lambda^{-5})/[\{\exp(hc/\lambda kT)\} - 1]$$

where h is Planck's constant, c is the speed of light in a vacuum and k is Boltzmann's constant.

J Thewlis, ed., *Encyclopaedic Dictionary of Physics* (New York, Oxford and London, 1962)

MS

planetesimal theory of planetary formation (1905) *Astronomy* Proposed by the American geologist Thomas Chrowder Chamberlin (1843–1928) and American astronomer Forest Ray Moulton (1872–1952), this is a modification of the tidal theory of planetary formation. (*See* TWO-BODY THEORIES OF PLANETARY FORMATION.)

Chamberlin and Moulton proposed that in the past the Sun had been liable to eruptions of considerable intensity. A wandering star then passed so close to the Sun that the eruptions were intensified and vast puffs of gas broke clear of the Sun's atmosphere and eventually condensed into small solid bodies called planetesimals. These bodies were much smaller than planet size, but in due course numbers of them came together and cohered to form the present planets. The problem with this theory is that masses of gas of planetesimal size would not condense: their internal pressure would cause them to expand and scatter the material into space.

S Mitton, ed., *The Cambridge Encyclopaedia of Astronomy* (London, 1973)

MS

plate tectonics (1967) *Geology* According to this theory, the upper shell of the Earth is formed of mobile plates comprising the colder and more brittle rocks at the Earth's surface (lithosphere). Although arguably originated by the British geologist Arthur Holmes in the 1930s, the term 'plate' was coined by the British and American earth scientists Dan P McKenzie and Robert L Parker.

Each of the six main plates has its own direction and rate of motion, and may comprise both oceanic and continental lithosphere. Various convective models have been posited to account for this motion. Areas where plates converge (destructive margins) and diverge (constructive margins) are marked by volcanic activity; and where they move in opposite directions, by earthquakes. In addition to explaining the Earth's present surface activity, plate tectonics provides a conceptual model within which its past geological history can be evaluated. *See also* CONTINENTAL DRIFT and SEAFLOOR SPREADING.

X Le Pichon, J Francheteau and J Bonnin, *Plate Tectonics* (Amsterdam, 1973)

DT

plate theory *Chemistry* A theory relating to the operation of distillation columns and gas chromatography columns.

A typical distillation column contains closely-spaced perforated ceramic plates through which vapour distillate rises through a series of bubble caps while liquid condensate returns to the distillation vessel. A theoretical plate is defined as one on which complete equilibrium is reached between the vapour rising from it and passing to the plate above, and the liquid leaving it and passing to the plate below. In practice, the efficiency of real plates is less than 100 per cent, with the result that the number required to achieve a given degree of separation or absorption is greater than the theoretical number.

D M Considine, *Van Nostrand's Scientific Encyclopedia* (New York, 1989)

GD

Plateau's problem (1873) *Mathematics* Named after Joseph Plateau (1801–83) who posed the problem and related it to one about fluids in equilibrium.

What is the surface of minimal area which passes through a given curve in three-dimensional Euclidean space? This can be approximated by modelling the curve in stiff wire, dipping it into a soap solution and viewing the resultant bubble in a variety of positions (in order to reduce the small effect of gravity).

JB

Plato's theory of Forms (or Ideas) *Philosophy* Theory developed by Plato (*c*.427–*c*.347 BC) in his middle-period dialogues (especially *Phaedo*, *Symposium*, *Republic*) and criticized by himself in his *Parmenides* (*see* THIRD MAN ARGUMENT). The language of the theory occurs in his earlier dialogues, but its interpretation is disputed, as is his reaction in later dialogues to the *Parmenides* criticisms: did he modify the theory, abandon it, or treat the criticisms as applying only to a distorted version of it?

Forms (usually given a capital F) were properties or essences of things, treated as non-material abstract, but substantial, entities. They were eternal, changeless, supremely real, and independent of ordinary objects which had their being and properties by 'participating' in them. But Plato puzzlingly treated them as both universals (*see* PLATONISM), suggesting they were immanent in things, and paradigms (*see* PARADIGMATISM), suggesting they were transcendent and themselves had the properties they represented: Beauty is beautiful (but Change changes – despite being, as a Form, unchanging). Aristotle (384–322 BC) too believed in forms (with a small F), but no longer as transcendent objects. 'Idea' is a misleading synonym for 'Form'; Forms were objects of knowledge but in no way themselves mental or in the mind. *See also* ONE OVER MANY PRINCIPLE.

W D Ross, *Plato's Theory of Ideas* (1951)

ARL

Platonic mathematics (4th century BC) *Mathematics* This view of mathematics was put forward by the philosopher Plato (427–347 BC).

Mathematics has an independent objective existence, so mathematical results are true whether or not we know or can prove them or, indeed, whether or not the earth exists. Consequently, mathematical results are discovered, not created. Despite the existence of mathematical philosophies that appear inconsistent with this view, it is hard to find any working mathematician who does not believe it, if perhaps only in some modified form.

JB

Platonism *Philosophy* Strictly, the philosophy of Plato (*c.*427–*c.*347 BC), but the word is often applied to any view which treats a given subject-matter as involving substantial, though abstract, entities (irrespective of Plato's own view on the topic in question). Such subject-matters have included numbers, propositions, universals (roughly, things named by words ending in '-hood', '-ness', '-ty'). Platonism is thus a form of REALISM.

ARL

play theory *Biology* A reference to the group of theories that seek to explain why young animals and children engage in play.

Researchers hypothesize that play exists because it allows practice of specific motor patterns and behaviours relevant to later life; that it generates new behaviours and introduces novel stimuli; and that it builds socialization skills. All these hypotheses, however, possess weaknesses that have precluded their widespread acceptance by animal behaviourists.

C Barnard, *Animal Behaviour: Ecology and Evolution* (New York, 1983)

KH

pleasure principle *Psychology* Proposed by the Austrian founder of psychoanalysis Sigmund Freud (1856–1939).

A psychoanalytic concept suggesting that an organism avoids pain and seeks immediate gratification. The id is said to function according to the pleasure principle. Early in the development of the id, before it has developed the ability to distinguish between internal and external reality, hallucinations and fantasies are often used to satisfy the pleasure principle. *See* PAIN PRINCIPLE.

C S Hall, *A Primer of Freudian Psychology* (Cleveland, 1954)

NS

Pleistocene overkill hypothesis (1960s) *Biology* Advanced by American ecologist Paul S Martin (1928–) in 1967, this is the hypothesis that excessive hunting by early humans caused the mass extinctions of large mammals at the end of the Pleistocene epoch (70,000 to 10,000 years ago). This unproven theory competes with others emphasizing climatic changes or other factors. The cause of the great Ice Age extinctions remains an active area of research. *Compare with* ALVAREZ THEORY and EXTINCTION.

P Martin, *Quaternary Extinctions: A Prehistoric Revolution* (Tucson, Ariz., 1984)

KH

plenitude, principle of *Philosophy* Principle that if the universe is to be as perfect as possible it must be as full as possible, in the sense that it contains as many kinds of things as it possibly could contain. The world of nature must be as rich as possible. This is connected with the idea, used by St Anselm (1033–1109) in his *ontological argument* for God's existence, that existence is a perfection.

Another version of the principle refers to events rather than to kinds of object. It says that there can be no possibilities that remain as possibilities (and are not foreclosed) but are unrealized throughout eternity; in this form, which goes back at least to Aristotle (384–322 BC), the principle is given some credibility by PROBABILITY THEORY: the probability that the proverbial monkey at a typewriter will type a Shakespeare sonnet straight off may be minute, but if he remains typing for long enough the probability increases indefinitely and it becomes increasingly surprising if he does *not* type one.

A O Lovejoy, *The Great Chain of Being* (1936)

ARL

pluralism (20th century) *Politics* Descriptive and prescriptive theory of power in modern societies.

Either: since people gain their identity from the groups of which they are members, political power should wherever possible be distributed to functional groups; or, in modern democracies, bargaining between groups contributes as much to policy as does formal DEMOCRACY. *See also* IMPACTED PLURALISM.

David Miller *et al.*, eds, *The Blackwell Encyclopaedia of Political Thought* (Oxford, 1987)

RB

plurality of causes *Philosophy* Principle saying that, though the same cause must have the same effect each time, the same effect need not have the same cause each time. (Of course the cause on one occasion may be complex and involve many contributory factors; but could these be replaced by different factors when the effect next occurs?)

The principle seems plausible, but is this because the effect is only vaguely described? Death can have many different causes, but could the precise death undergone by Smith at midnight last Thursday? Is it true that, given a complete description of the universe at one moment, we could in principle know what its future will be but not what its past was (ignoring for convenience problems about self-prediction and so on)?

ARL

plutocracy *Politics* Now little used, this term describes a situation where political power, in fact rather than in law or constitutional theory, lies with the wealthy.

RB

plutonism (1785) *Geology* The idea of the Earth's internal heating dates back to Antiquity, but the Scottish geologist James Hutton (1726–97) proposed that this process was primarily responsible for the cyclical creation of rocks and continents.

The Earth was originally an immense ocean with a few islands of primary rock. Heating consolidated the rock strata and melted the sediments (forming granites), and also uplifted the land (with volcanoes as escape valves). Once continents had formed, weathering, erosion and deposition eventually led to renewed heat accumulation and so the production of new continents from the debris of the old. This model contradicted NEPTUNISM in considering granites to be crystallized from molten rocks; and whilst similar to VULCANISM was less concerned with the origin of basaltic rocks. *See also* PLATE TECTONICS.

A Hallam, *Great Geological Controversies*, (Oxford, 1983)

DT

poet as legislator (1821) *Literary Theory* Formulation of English poet Percy Bysshe Shelley (1792–1822). 'Poets are the unacknowledged legislators of the world' was an extreme statement of the special creative powers of the poet claimed by ROMANTICISM. This polemic was based on a serious theory of language: poets' language is 'vitally metaphorical', lending a fresh shape to the world.

P B Shelley, 'A Defense of Poetry' [1821, publ. 1840], in *Literary Theory since Plato*, H Adams, ed. (New York, 1971), 499–513

RF

poetic diction (18th century) *Literary Theory* The basis is in the *Poetics* of Aristotle (*c*.330 BC), was elaborated in classical and medieval RHETORIC, and became an issue in English poetry and criticism of the 18th century.

Classically, poetry was to be marked off from speech by special linguistic devices, including archaisms, Latinate terms, special epithets, and so on. Extreme practice of poetic diction in the 18th century led to a dull, cliché-ridden style, attacked by William Wordsworth (1770–1850). *See* DECORUM and LANGUAGE OF MEN.

B Groom, *The Diction of Poetry from Spenser to Bridges* (Toronto, 1955)

RF

poetic justice (1678) *Literary Theory* Term coined by English antiquarian and critic Thomas Rymer (1641–1713).

Condition on TRAGEDY that the protagonists should be punished in proportion to their virtues or vices. In the work of Aristotle (*c*.330 BC), though, the hero is punished in excess of what is merited by his TRAGIC FLAW.

T Rymer, 'Tragedies of the Last Age' [1678], *Critical Essays of the Seventeenth Century*, J E Spingarn, ed., II (Oxford, 1909)

RF

poetic licence (traditional) *Literary Theory* Important in severe neo-classical periods of writing, for example 18th century England.

(1) Alterations to ordinary linguistic patterns to make language fit a strict metrical form, for example shortening words, changing word order. (2) Creative stylistic DEVIATIONS. (3) Use of fantastic or implausible content in poetry, allowed by Aristotle (*c*.330 BC).

G Leech, *A Linguistic Guide to English Poetry* (London, 1969)

RF

poetic principle (1958) *Stylistics* Theory formulated by the Russian linguist Roman Jakobson (1896–1982). Based on PARADIGM AND SYNTAGM.

Language is structured on two dimensions: the horizontal dimension along which PHONEMES and words are combined to form sentences (The cat sat on the mat); and the vertical, sets of equivalent terms or patterns from which selection is made ('cat' could have been 'kitty', 'mat' could have been 'rug'). The poetic function of language 'projects the principle of equivalence from the axis of selection into the axis of combination', thereby drawing attention to relationships of meaning, and giving extra meanings. Our mundane example is quasi-poetic because of the chiming equivalence of three words. PARALLELISM is the essence of the poetic principle, the basis of many figures of thought and sound. *See also* COUPLING.

R Jakobson, 'Linguistics and Poetics', *Style in Language*, T A Sebeok, ed. (Cambridge, Mass., 1960), 350–77

RF

poetics (traditional and 1960s) *Literary Theory* Traceable to Aristotle (384–322 BC), the term was given new impetus by STRUCTURALISM's use of the French word '*poétique*'.

Science or theory of literature which aims to define the essence of literature, and to establish it as an ontologically distinct entity. It should be distinguished from critical and other studies of particular texts, and from studies relating texts to contexts. *See also* LITERARY THEORY and LITERATURE.

Aristotle, *Aristotle's Theory of Poetry and Fine Art*, S H Butcher, trans. (New York, 1955); T Todorov, 'Poétique', *Qu'est-ce que le structuralisme?*, O Ducrot *et al.*, eds (Paris, 1968), 99–166

RF

poetics of culture *Literary Theory See* NEW HISTORICISM.

poetry (5th–4th centuries BC) *Literary Theory* Derived from the Greek word *poesis*, meaning 'making'; Aristotle's *Poetics* or *On the Art of Poetry* (330 BC) set the traditional meaning.

Traditionally, an artistic text which creates a fictional world by MIMESIS; in the Aristotelean approach, poetry could be subdivided into a number of kinds – EPIC, TRAGEDY, and so on – each kind with its own laws of composition. 'Poetry' remained the general term for verbal art until the 18th century; in the 19th, it was supplanted by 'LITERATURE', and 'poetry' became specialized to verse or metrical composition as opposed to prose. Because of its long centrality in literary aesthetics, 'poetry' remains uniquely charged with evaluation, particularly the values of CREATIVITY, ORIGINALITY and IMAGINATION in ROMANTICISM.

M H Abrams, *The Mirror and the Lamp* (New York, 1953)

RF

Poetzl effect *Psychology* identified by the Austrian psychologist O Poetzl.

The effect describes a phenomenon reported by subjects who had taken part in a tachistoscope experiment (that is, using an instrument which presents visual stimuli for very brief periods of time; in some cases lasting less than a second). The subjects reported perceptions of stimuli in their dreams which were presented during the experiment but not reported by them at the time.

G A Miller, *Psychology: The Science of Mental Life* (London, 1973)

NS

Poincaré's conjecture (1904) *Mathematics* Advanced by Jules Henri Poincaré (1854–1912) in his investigations of the topological properties of surfaces.

Every simply connected three-dimensional compact manifold is topologically equivalent to a three-dimensional sphere. The technical terms are essential to this conjecture, which is the subject of much work.

JB

Poincaré's lemma (*c.*1880) *Mathematics* Named after its discoverer Jules Henri Poincaré (1854–1912).

Every closed differential form on a simply connected region has a function from which it can be obtained by differentiation. The technical terms are essential for this result, but the result aids the solution of partial DIFFERENTIAL EQUATIONS.

JB

Poincaré's theorem (1890) *Physics* Named after the French mathematician Jules Henri Poincaré (1854–1912).

The theorem relates to the concept of cyclic recurrence or 'eternal return' and says that, given sufficient time, any isolated system of limited size will return to its initial state and will continue to do so endlessly given an infinite amount of time. For a complex system like the universe, the 'recurrence time' is many times its present estimated age (10–20 billion years).

Poincaré's theorem effectively counters the notion of an ever-forward moving 'arrow of time', and also the idea of evolution except in the most trivial sense.

GD

point of view (1900s) *Literary Theory* An aspect of FICTION given its modern importance by the American novelist Henry James (1843–1916), and especially his commentator, the English critic Percy Lubbock (1869–1965).

The angle or perspective (spatial, temporal, ideological, moral) from which a story is narrated; the kind of narrator (omniscient, unreliable, and so on) and his relationship with CHARACTERS. *See also* FOCALIZATION, NARRATOR, SHOWING AND TELLING and references.

P Lubbock, *The Craft of Fiction* (New York, 1972 [1921])

RF

pointillism (*c.*1886) *Art* A technique closely related to DIVISIONISM, first systematically applied by French painter Georges Seurat (1859–91) in his *La Grande Jatte* (1886).

Following the theories of Charles Blanc and Ogden N Rood, Seurat developed a technique of applying pigment of complementary colour in small dots which are additively mixed by the eye at a certain distance. This method thereby achieved more luminous colour and ensured a more controlled handling of tone.

C Blanc, *Grammaire des Arts du Dessin* (1867); O N Rood, *Modern Chromatics* (1879)

AB

Poiseuille's law (1840) *Physics* Named after the French physicist Jean Léonard Marie Poisseuille (1797–1869) who verified it following earlier experimental work by G H L Hagan (1797–1884) in 1839. The law is sometimes called the Poisseuille-Hagan law.

When a Newtonian liquid of viscosity η flows steadily under streamline conditions through a horizontal tube of constant radius

r and length L, the volume V of liquid crossing any section of the tube in unit time is given by

$$V = \frac{(p_1 - p_2)\pi r^4}{8\eta L}$$

where $(p_1 - p_2)$ is the pressure difference between the ends of the tube of length L.

J Thewlis, ed., *Encyclopaedic Dictionary of Physics* (New York, Oxford and London, 1962)

MS

Poisson's ratio (1830) *Physics* Named after its discoverer, French mathematician Siméon Denis Poisson (1781–1840).

When a cylindrical body with free (unloaded) prismatic surfaces undergoes a change in length there is an accompanying change in lateral dimensions. An extension produces a reduction in cross-sectional area while a compression produces an increase. When the deformation is elastic the ratio

$$\nu = -\frac{\text{lateral strain}}{\text{longitudinal strain}}$$

is found to be a constant, characteristic of the material, and is known as Poisson's ratio.

J Thewlis, ed., *Encyclopaedic Dictionary of Physics* (New York, Oxford and London, 1962)

MS

polar decomposition theorem (20th century) *Mathematics* If A is an arbitrary linear transformation on a finite-dimensional inner product space, then there is a (uniquely determined) positive transformation P, and there is an isometry U, such that $A = PU$. If A is invertible, then U also is uniquely determined by A. The representation $A = PU$ is called the polar decomposition of A because it is analogous to the complex factorization

$$z = e^{i \arg (z)}|z| .$$

See also SPECTRAL THEOREM.

P R Halmos, *Finite Dimensional Vector Spaces* (Van Nostrand, 1958)

JB

polarization identity *Mathematics* (1) The identity which holds for a complete inner product space

$$4 < x,y > = \|x+y\|^2-\|x-y\|^2 +\imath\|x+\imath y\|^2-\imath\|x-\imath y\|^2.$$

(2) The corresponding identity for a real inner product space is

$$4 < x,y > = \|x+y\|^2 - \|x-y\|^2.$$

See also PARALLELOGRAM LAW.

J B Conway, *A Course in Functional Analysis* (New York, 1985)

police state *Politics* Theory of purposeful government.

In popular speech a police state is a despotic, arbitrary and tyrannical one. The technical use of the term is different. A police state is one where the institutions and officials of the state act not in response to public pressures or demands but as the agents of some clear purpose of government.

Marc Raeff, 'The Well-Ordered Police State', *American Historical Review* (1975)

RB

polis *Politics* Theory, originating in ancient Greece, of political community.

Life as a citizen in the *polis* was the highest form of human activity, in which all (adult males) participated, and all were equal. The idea of the *polis* remains potent for those who see politics as potentially a fulfilling activity in its own right, and who support the concept of CITIZENSHIP.

David Robertson, *The Penguin Dictionary of Politics* (London, 1986)

RB

politeness (1978) *Linguistics* Area of PRAGMATICS theorized and described by Penelope Brown (1944–) and Steven C Levinson.

Verbal and KINESIC behaviour signal negative and positive politeness, threatening or saving one's addressee's 'face'. Development of Herbert Paul Grice's CO-OPERATIVE PRINCIPLE; 'face' derived from American sociologist Erving Goffman.

P Brown and S C Levinson, *Politeness* (Cambridge, 1987 [1978])

RF

political business cycle (1867) *Economics* Initially attributed to German political economist Karl Marx (1818–83) but later revised by, among others, Polish-born engineer and economist Michal Kalecki (1899–1970); this theory attributes economic fluctuations to politicians who manipulate fiscal and monetary policies (choosing between employment or inflation) in order to get elected/re-elected.

It is argued that in the period leading up to an election, policies are introduced to reduce unemployment regardless of the inflation which may result; however, after a successful election the party will introduce deflationary policies. This throws the economy into disequilibrium. Kalecki argued that big business is opposed to government experiments to create full employment through spending (as seen in the Depression of the 1930s in every country except Nazi Germany). Their dislike stems from a distrust of government interference in employment, a dislike of the areas in which spending is directed (public investment and subsidized consumption), and a dislike of the social and political changes arising from the creation of full employment.

See BUSINESS CYCLE, KONDRATIEFF CYCLES, SUNSPOT THEORY, PRODUCT LIFE-CYCLE, ACCELERATION PRINCIPLE, FINE-TUNING, MULTIPLIER ACCELERATOR MODEL and TRADE CYCLE.

M Kalecki, 'Political Aspects of Full Employment', *Political Quarterly*, vol. XIV (October–December 1943), 322–31; P Minford and D Peel, 'The Political Theory of the Business Cycle', *European Economic Review*, vol. X (1982), 252–70

PH

political correctness (1980s) *Politics* Pejorative term for an overly rigid or token adherence to a liberal canon of beliefs. This novel orthodoxy first surfaced in the USA as lobbyists for disadvantaged minorities began exerting significant political influence; and, consequently, as issues of gender, ethnicity and physical disability were widely promoted as legitimate subjects for sympathetic academic scrutiny. Opponents regard political correctness as an arbitrary pretext for the misuse of power, and as self-contradictory in its stifling of free thought and expression.

A related derogatory term, PCism, refers to loaded, proscriptive or euphemistic use of language in an effort to redefine, or not cause offence, or to appear politically correct. Examples include 'people of colour' meaning non-whites, 'differently abled' meaning handicapped, and 'vertically challenged' meaning short.

AJM

political culture (20th century) *Politics* Explanation of the character of politics.

A nation or a society is characterized by a political culture, into which its children are inducted, and by learning which they participate and preserve values and institutions.

Geoffrey Roberts and Alistair Edwards, *A New Dictionary of Political Analysis* (London, 1987)

RB

political obligation *Politics/Law* Theory of the moral requirement to obey law.

A moral or normative obligation exists, or can be created, to obey the law or lawfully established authorities. The basis of this obligation is a matter for wide dispute, and ranges from the free CONSENT of the governed to DIVINE LAW.

David Miller *et al.*, eds, *The Blackwell Encyclopaedia of Political Thought* (Oxford, 1987)

RB

Polya's theory of enumeration (1937) *Mathematics* Named after its discoverer George Polya (1887–1985), this powerful method calculates the number of ways in which boxes containing figures, each with numerical content, can produce a fixed total content. This is important for counting mathematical graphs or chemical compounds of prescribed kinds.

G Polya and R C Read, *Combinatorial Enumeration of Groups, Graphs and Chemical Compounds* (Berlin, 1987)

JB

polyarchy (20th century) *Politics* Theory of politics described by American political scientist Robert Dahl (1915–).

In modern democracies the people cannot literally rule in the classic sense of democracy, but if there are free and open democratic procedures, and a variety of groups and organizations, then the many may be said to rule. Polyarchy can be said to exist when DEMOCRACY mingles effectively with PLURALISM.

Robert A Dahl, *Polyarchy* (New Haven, 1971)

RB

Polyclitan school (*fl. c.*232 BC) *Art* Named after the Greek sculptor Polyclitus of Samos (*fl.* 5th century BC) whose works were characterized by equilibrium, rhythm and perfection. His greatest statues, the *Diadumenus* and the *Doryphorus* established the canon for correct proportions of the ideal male form.

AB

polyphony (1929) *Literary Theory* Theory of the novel proposed by the Russian critic Mikhail M Bakhtin (1895–1975).

In the polyphonic novel, the voices and consciousnesses of the characters are allowed free play and not 'objectivized', not fixed by the MONOLOGIC evaluation of the author. *See also* DIALOGISM and references.

M M Bakhtin, *Problems of Dostoevsky's Poetics*, C Emerson, trans., (Minneapolis, 1984 [1929])

RF

polysemy (1897) *Linguistics* This word was coined by the French linguist Michel Bréal (1832–1915). Polysemy exists when a word has more than one meaning; for example, *mean* as 'middle', 'inferior', and 'stingy'.

When two different words and meanings coincide, they are called homonyms; for example, *ear* and *ear (of corn)*. Homonyms which coincide in sound but not in spelling are called homophones; for example, *shoot* and *chute*. As Lyons points out, it is very difficult to draw these distinctions on close scrutiny.

J Lyons, *Semantics* (Cambridge, 1977)

RF

polytonality (20th century) *Music* The notion that many keys or tonalities may be traced as simultaneously present in a piece of music. Theoretically it is an extension of BITONALITY but in practice it is closer to ATONALITY since each key undermines the validity of the other keys.

MLD

Pomeranchuk's theorem (1958) *Physics* Named after its originator, Isaak Yakovlevich Pomeranchuk (1913–66), this posits that the cross-sections of particles and their anti-particles, incident on the same target, should approach the same, constant, value as the incident energy becomes very large.

J Thewlis, ed., *Encyclopaedic Dictionary of Physics* (New York, Oxford and London, 1962)

MS

Pontryagin maximum (minimum) principle *Mathematics* One of the fundamental results in optimal control. In outline, at an optimal solution of a control problem it asserts the existence of adjoint variables satisfying an adjoint equation, transversality conditions, and so on; such that the optimal solution maximizes the corresponding Hamiltonian (*compare with* the Lagrangian in the KARUSH-KUHN-TUCKER THEOREM). In principle, as with LAGRANGE MULTIPLIERS in OPTIMIZATION THEORY, this allows one to check optimality of a solution.

AL

Pop art (1959) *Art* This term was used for the first time in 1959 by the British-born critic Lawrence Alloway (1926–) in an article entitled 'The Arts and Mass Media'.

From the 1960s onwards, it was used to describe the figurative work of such British and American artists as Richard Hamilton (1922–), Eduardo Paolozzi (1924–) and Andy Warhol (1928–87). Their works, executed in a slick and humorous vein, typify the movement's interests in popular media, cars, film and urban culture; as well as in challenging the idea of 'high art' through their methods of production and reproduction.

M Livingstone, *Pop Art* (London, 1991)

AB

population thinking (1980s) *Biology* Coined by evolutionary biologist Ernst Mayr (1904–) of Harvard University, this is the perspective that a species can be more accurately understood as a reproducing population (replete with the natural differences found in individuals), than as individuals who vary from some 'ideal type'.

Thus, rather than collecting perfect museum specimens to represent a species, modern biologists observe entire populations and make generalizations based on those observations. According to Mayr, population thinking has largely replaced 'typological' thinking. *Compare with* ESSENTIALISM.

E Soper, *Conceptual Issues in Evolutionary Biology: An Anthology* (Cambridge, Mass., 1984)

KH

populism (19th century–) *Politics* Attribution of virtue and/or authority to the 'common people'.

'Ordinary people' are possessed of simple qualities and straightforward insights, which distinguish them from politicians and members of other élites, all of whom are characterized as devious and self-seeking. Politics should thus be carried out on behalf of 'the people'.

David Miller *et al.*, eds, *The Blackwell Encyclopaedia of Political Thought* (Oxford, 1987)

RB

portfolio selection theory (1952) *Economics* Analysis of asset selection which maximizes the return and minimizes risk for an investor. An early pioneer in this field was American economist Harry Markowitz (1927–).

One method of portfolio selection is asset diversification, which spreads risk over a mixture of equities, government bonds, commodities, precious metals, property and currencies. The final selection of a portfolio will be determined by available information, attitudes towards risk and the income objectives of the investor. *See* CAPITAL ASSET PRICING MODEL, MARGINAL EFFICIENCY OF CAPITAL, ARBITRAGE PRICING THEORY and UNCERTAINTY.

H Markowitz, 'Portfolio Selection', *Journal of Finance* 7, 1 (March, 1952), 77–97

PH

positive law *Politcs See* LEGAL POSITIVISM.

positivism *Philosophy/Psychology* A movement in the general tradition of EMPIRICISM and pioneered specifically by the French writer Auguste Comte (1798–1857), though under the influence of the social reformer Claude Henri, Compte de Saint-Simon (1760–1825), whom he served as secretary.

The main features of positivism were an insistence on a scientific approach to the human, as well as the natural, world; and a tendency to organize and classify, in particular the developmental stages of the sciences and of human thought in general. The sciences formed a hierarchy, which was also reflected in their historical development from mathematics through physics and biology to sociology (whose name, like that of positivism, Comte invented). Human thought itself developed through three stages: religious, metaphysical, and scientific. (Comte did not, however, entirely reject the value of the first two of these stages.) Positivism later developed into EMPIRIOCRITICISM, and then LOGICAL POSITIVISM (often loosely called just 'positivism'). LEGAL POSITIVISM is only partly connected to this development.

Positivism has had a profound influence on psychology, especially on BEHAVIOURISM, OPERATIONALISM and ideas about valid theory construction. Positivism, although judged useful by psychology, places many restrictions on its subject matter by disparaging METAPHYSICS and MENTALISM.

E Brunswick, 'The Conceptual Focus of Some Psychological Terms', *Journal of Unified Science* (1929), 36–49; L Kolakowski, *Positivist Philosophy* (1966, trans. 1968)

ARL, NS

possibilism (19th century) *Politics* Non-revolutionary revision of 'purist' SOCIALISM.

The argument that immediate gains for the working class by conventional means such as parliamentary politics, tactical voting and trade union negotiation were not a compromise with the existing order, but a preferable alternative to maintaining political purity and consequent political ineffectiveness. A term originally applied within the late 19th century French Socialist Party.

David Stafford, *From Anarchism to Reformism* (London, 1971)

RB

post-Impressionism (1910) *Art* A name first used by British critic and painter Roger Fry (1886–1934) for artists of the same or next generation as the Impressionists, who rejected the latters' preoccupation with surface effects of light and colour. (*See* IMPRESSIONISM.)

Three artists – Paul Cézanne (1839–1906), Paul Gauguin (1848–1903) and Vincent Van Gogh (1853–90) – while benefiting from the technical achievements of the Impressionists and sharing their disdain for academic art, sought in highly individualistic ways to achieve greater order and harmony within their paintings, and to explore the expressive significance of subjects. Their work had great significance for all early 20th-century art movements. The name loosely refers also to the work of Georges Seurat (1859–91) and NEO-IMPRESSIONISM.

LL

post-materialism (20th century) *Politics* A theory of the eclipse of material values.

In the second half of the 20th century, particularly amongst the young, the search for material utility which had underlain both capitalist and communist analyses was replaced by a priority to non-material values: quality of life, culture, the preservation of the environment. The theory has been used to explain the growth of environmental or green politics.

Geoffrey Roberts and Alistair Edwards, *A New Dictionary of Political Analysis* (London, 1991)

RB

post-modernism (1970s–) *Art* Term coined in its artistic sense by the American critic and architect Charles Jencks (1939–).

In its application to art and, above all, architecture, Jencks identified an eclectic revival of the classical tradition, where an often ironic or ambiguous use of its elements are applied to architecture and design. Important in this trend is the dissolution of the traditional divisions between the arts. It has been criticized for being superficial and insubstantial. *See* MODERNISM.

C Jencks, *Post Modernism. The New Classicism in Art and Architecture* (London, 1987)

AB

post-modernism (20th century) *Politics/Literary Theory* Theory of cultural and political variety.

The account of the world given by MODERNISM no longer works in the final decades of the 20th century. Instead of single sets of values or political loyalties, there is a wide variety of groups and classes, aims and ideologies.

David Harvey, *The Condition of Postmodernity* (Oxford, 1989)

RB

post-painterly abstraction (1964) *Art* This was the title given to a major exhibition of American painting organized by Clement Greenberg at the County Museum of Art, Los Angeles in 1964. The use of the phrase 'painterly' derived from Heinrich Wolfflin's '*malerisch*', and was used by Greenberg in relation to a group of artists who broke away from ABSTRACT EXPRESSIONISM.

Resisting the gestural brushwork of that style, these 'New Abstractionists'' compositions are rationally planned and executed in unmodulated areas of colour, eschewing ILLUSIONISM and relief surfaces.

C Greenberg, *Post-Painterly Abstraction*, Los Angeles County Museum of Art exh. cat. (Los Angeles, 1964)

AB

postremity principle *Psychology* Advanced by one of the most influential learning theorists, the American psychologist E R Guthrie (1886–1959).

The principle that the most probable response of an organism is the last response the organism made in the same situation. The principle has been found to work well in animals, especially under restricted laboratory conditions, but is not particularly applicable to human learning and behaviour.

E R Guthrie, *The Psychology of Learning* (New York, 1935)

NS

poststructuralism (1970s) *Literary Theory* A cluster of theoretical movements in the USA and Europe, though chiefly France where the main protagonists are: Roland Barthes (1915–60), Jacques Derrida (1930–), Michel Foucault (1926–84), and Julia Kristeva (1941–). In American usage, 'poststructuralism' is synonymous with DECONSTRUCTION.

A liberating reaction against the structuralist view of literature, poststructuralism developed out of elements already in STRUCTURALISM. The text is seen not as a closed system of meanings, but a weaving of CODES. A major determinant of meanings is INTERTEXTUALITY. The authority of the author is dismissed (*see* AUTHOR, DEATH OF THE), and the reader becomes an active producer of meanings.

J V Harari, ed., *Textual Strategies* (London, 1979)

RF

potential theory (17th century) *Mathematics* Initiated by Sir Isaac Newton (1642–1727) as a study of the gravitational effect of a solid body such as a sphere, and generalized by Pierre Simon Marquis de Laplace (1749–1827).

The study of harmonic functions; that is, twice differentiable solutions of Laplace's equation

$$D_1^2V + D_2^2V + D_3^2V = 0$$

or

$$D_1^2V + D_2^2V = 0$$

where D_i represents partial differentiation with respect to the variable x_i. Harmonic functions represent the potential of gravitational and potential fields and have applications to functions of a complex variable.

JB

Poussinism *Art* This term refers to a preference for drawing and design over colour. Exemplified by the work of Nicholas Poussin (1594–1665) and his followers amongst the artistic establishment in France in the 1670s, Poussinism was opposed to RUBENISM.

LL

power *Politics/Sociology* Theory of individuals' and groups' ability to achieve their ends.

In social relations, power is exercised by persons or institutions acting in such a way that their INTERESTS or wishes prevail over those of others. They are thus responsible for the consequences. There is dispute, however, over the extent to which they need to be aware of these consequences for it to be usefully said that power is being exercised.

David Miller *et al.*, eds, *The Blackwell Encyclopaedia of Political Thought* (Oxford, 1987)

RB

power and solidarity (1960) *Linguistics* Identified by the American linguists R Brown and A Gilman as social meanings conveyed by second-person pronouns such as French *tu* and *vous*, German *du* and *Sie*.

Several languages provide a choice of pronoun for addressing a second person: the options encode different social relationships between the two speakers. Using French *tu* when speaking to a child or a servant signifies the speaker's power; but reciprocal *tu* between equals signifies solidarity, familiarity, like-mindedness. Mutual use of *vous* means distance, impersonality. Power and solidarity are important theoretical concepts in SOCIOLINGUISTICS because they underlie many other expressions of interpersonal relationships, for example naming, DIGLOSSIA.

R Brown and A Gilman, 'The Pronouns of Power and Solidarity', *Style in Language*, T A Sebeok, ed. (Cambridge, Mass., 1960), 253–76; also repr. in several readers in sociolinguistics, for example P P Giglioli, ed., *Language and Social Context* (London, 1972), 252–82

RF

power corrupts (19th century) *Politics* Theory, or at least aphorism, of the English historian Lord Acton (1834–1902).

'All power corrupts, but absolute power corrupts absolutely'. The sentence expresses a fundamental suspicion of the ability of any person to be trusted with unaccountable discretion, however noble their expressed motives.

RB

power law *Psychology* Formulated by the American psychologist Stanley Smith Stevens (1906–73).

The law may be written thus: ¥ = ks^n, where ¥ is the sensory experience of the stimulus, s is the stimulus, k is a constant, and n is the exponent for that particular transmitting continuum. Stevens supported the idea that his equation represented the true psychophysical law for dealing with metathetic continuums, and that Fechner's logarithmic law only applied to prothitic continuums. Therefore the qualitative change in experience caused by a quantitative change in a stimulus in a metathetic continuum is best represented by the power law. *See* FECHNER'S LAW.

S S Stevens, *Psychophysics: Introduction to its Perceptual, Neural and Social Aspects* (New York, 1975)

NS

Poynting's theorem (1884) *Physics* This theory is named after its discoverer, the English physicist John Henry Poynting (1852–1914).

The energy flow in an electromagnetic wave is perpendicular to the direction of both the electric field (of strength **E**) and the magnetic field (of induction **B**) and has an instantaneous intensity (power per unit area) *S* given by

$$\mathbf{S} = \frac{\mathbf{E} \times \mathbf{B})}{\mu_0}$$

where μ_0 is the permeability of free space. For a sinusoidal time-dependence of **B** and **E**, the average intensity $\langle \mathbf{S} \rangle$ is given by

$$\langle \mathbf{S} \rangle = \frac{\mathbf{E}_0\mathbf{B}_0}{\mu_0}$$

where $\mathbf{E}_0$ and $\mathbf{B}_0$ are the amplitudes of **E** and **B**, respectively.

J Thewlis, ed., *Encyclopaedic Dictionary of Physics* (New York, Oxford and London, 1962)

MS

practical criticism (1930s) *Literary Theory* Pedagogic practice first formulated by English critic Ivor Armstrong Richards (1893–1979), and also associated with Frank Raymond Leavis (1895–1978). It strongly influenced NEW CRITICISM.

(1) Underpinned by the theory of FOUR KINDS OF MEANING, a close analysis of literary texts which aims to retrieve the author's meanings without reference to social and historical context. (2) A classroom and examination practice designed to improve literary reading. (3) More generally, any close reading of an individual literary text.

I A Richards, *Practical Criticism* (London, 1929)

RF

practice theory of play *Psychology* Proposed by the Swiss psychologist K Groos, this is the generalization that play is adaptive and that, especially in lower organisms, play in young life gives the organism the opportunity to practise and engage in behaviours which in later life will be of use to them.

In relation to humans, practice theory of play predicts that play prepares children for their adult roles. This theory is certainly applicable to lower organisms but is highly controversial when applied to humans; however, the hypothesis has been applied, with limited success, to children's play.

K Groos, 'The Play of Mammals: Play and Instinct', *Play*, J S Bruner, A Jolly and K Sylva, eds (Harmondsworth, 1976); first published 1896

NS

pragmatic (or pragmatist) theory of truth *Philosophy* American scientist and philosopher Charles Sanders Peirce (1839–1914) defined truth as 'the opinion which is fated to be ultimately agreed to by all who investigate' (see Buchler). His disciple William James (1842–1910) held that truth was indeed agreement with reality, but that what counted as 'agreeing with reality' was what worked, in the sense of ultimately satisfying us. He even allowed that our emotions could legitimately influence this (*see also* EMOTIVE THEORY OF TRUTH).

The theory is particularly tempting in the advanced sciences, where any kind of simple correspondence with reality seems to be precluded by the nature of the concepts and methods involved.

J Buchler, ed., *The Philosophy of Peirce* (1940), 38; W James, *The Meaning of Truth* (1909)

ARL

pragmatics (1938) *Linguistics* Defined by the American philosopher C W Morris in 1938, though current and important in linguistics from *c.*1970. The main sources of modern linguistic pragmatics are in ordinary language philosophy of the 1960s.

Study of the relationships between language and its users, and language and the non-linguistic world: what acts people perform using language, how they use language to link with the world, how they use their MUTUAL KNOWLEDGE to interpret discourse. To be distinguished from SEMANTICS, though the boundary is not sharp. Main theories subsumed in linguistic pragmatics are SPEECH ACTS, CO-OPERATIVE PRINCIPLE, IMPLICATURE, RELEVANCE, POLITENESS, 'DEIXIS', and INFERENCE.

S C Levinson, *Pragmatics* (Cambridge, 1983)

RF

pragmatism *Philosophy* Theory, originally developed by American scientist and philosopher Charles Sanders Peirce (1839–1914), that the meanings of concepts and propositions lay in their possible effects on our experiences and practices. He also originated the PRAGMATIC THEORY OF TRUTH. Peirce was thinking mainly of scientific or intellectual concepts, and called his own view pragmaticism when his follower William James (1842–1910) broadened the theory to cover the senses and emotions. Another notable pragmatist was the American educationalist John Dewey (1859–1952).

More generally, pragmatist features can be found in many philosophers otherwise not closely related, such as Henri Bergson (1859–1941), Frank Plumpton Ramsey (1903–30), and many writers associated with CONVENTIONALISM, INSTRUMENTALISM, OPERATIONALISM and POSITIVISM.

A Rorty, ed., *Pragmatic Philosophy* (1966)

ARL

pragnanz *Psychology* A principle of organization defined by the German psychologist Max Wertheimer (1886–1943).

Wertheimer stated that the perception of experienced forms tends to form a structure, and this structure is the simplest available. Pragnanz failed to account for everything originally accredited to it by Wertheimer, but has been of immense value to the study of sensory perception. Pragnanz is also termed the law of precision. *See* GESTALT THEORY.

A D Ellis, ed., *A Sourcebook of Gestalt Psychology* (London, 1938)

NS

Prague School (founded 1926) *Linguistics* The main early members were: V Mathesius (1882–1946), Roman Jakobson (1896–1982), Nikolay S Trubetzkoy (1890–1938), J Mukarovsky (1891–1975).

Highly productive circle of Czech and Russian scholars, generally functionalist in orientation, responsible for several foundational concepts in linguistics and in literary theory. Very creative in the 1930s, and again active after World War II. *See* COMMUNICATIVE DYNAMISM, FOREGROUNDING, FUNCTIONAL SENTENCE PERSPECTIVE, MARKEDNESS, POETIC PRINCIPLE, THEME AND RHEME.

J Vachek, *The Linguistic School of Prague* (Bloomington, 1966)

RF

Pratt hypothesis (1861) *Geology* John Henry Pratt, the British Archdeacon of Calcutta, calculated that the actual deflection of a surveyor's plumb bob by the mass of the Himalayas was not as much as it should have been if these mountains were simply a mass of rock superimposed on a rigid crust. This he initially attributed to a curvature of the meridian, but later proposed that it was due to density variations within the crust resulting from differential cooling rates.

Most modern definitions of this hypothesis err from that actually suggested by Pratt. They state that it applies to major differences between the density of rocks in areas of different elevation, with all areas of the Earth having the same weight at a depth of about 110 km. This (revised) model is close to the accepted explanation for continental and oceanic areas' gravitational attraction, but is inadequate when applied to those mountainous areas for which it was originally posited. *See also* AIRY HYPOTHESIS and ISOSTATIC THEORY.

DT

prebiotic evolution *Biology See* PRECELLULAR EVOLUTION.

Prebisch-Singer thesis (1960s) *Economics* Named after Argentine economist Raul Prebisch (1901–86) and German-born British economist Hans Singer (1910–); this asserts that, given the permanent tendency for the terms of trade to go against agricultural products, it is in the interest of developing countries to erect protective tariffs behind which they can industrialize.

R Prebisch, *The Economic Development of Latin America and its Principle Problem* (New York, 1960); H W Singer, 'The Distribution of Gains between Investing and Borrowing Countries', *American Economic Review*, vol. XL (May, 1950), 473–85

PH

precellular evolution (1860s) *Biology* Also called prebiotic evolution, this theory was fuelled by the work of French chemist and microbiologist Louis Pasteur (1822–95) on spontaneous generation and the publication of English evolutionary theorist Charles Darwin's *Origin of Species* (1859).

The theory asserted that life evolved from lower forms which were based on non-living chemical combinations that arose spontaneously over a long period of time. Specific kinds of precellular evolution have been proposed, including GENOTYPE THEORY,

PHENOTYPE THEORY, MINERAL THEORY, RIBOTYPE THEORY and the VIRAL THEORY. *Compare with* SUDDEN LIFE.

KH

precision, law of *Psychology See* PRAGNANZ.

prediction theory (20th century) *Mathematics* A modern theory to replace augury, astrology and the extension of graphs, this is a PROBABILITY THEORY designed to give a probability for an extrapolation using simple mechanisms treating equiprobable events with indifference.

JB

pre-established harmony, doctrine of *Philosophy* Doctrine primarily associated with Gottfried Wilhelm Leibniz (1646–1716) that there is no causation in the world but that each event arises when it does because it was pre-programmed to do so by God when the universe began. The doctrine is often illustrated by the image of the two clocks – attributed to Arnold Geulincx (1625–69) – which keep perfect time though neither influences the other, because their maker designed them to do so. The doctrine differs from OCCASIONALISM because God does all the causing in a single act at the start of the universe, not on each relevant occasion as it arises.

ARL

preference utilitarianism *Philosophy* Version of UTILITARIANISM which contrasts with both HEDONISTIC UTILITARIANISM and IDEAL UTILITARIANISM by specifying the end to be pursued in terms neither of pleasure nor of other specific values, but in terms of maximizing the satisfaction of desires or preferences, whatever their objects. This answers at least some of the objections to the rival versions mentioned above, and some form of preference utilitarianism is probably the commonest form of utilitarianism in recent years.

J J C Smart and B Williams, *Utilitarianism For and Against* (1973)

ARL

preformation (17th century) *Biology* Advocated by Italian physiologist Marcello Malpighi (1628–94), among others, this is the notion that an organism exists preformed in the germ, and that development is the unfolding of this pre-existent form.

That is, the fertilized egg was believed to contain a miniature replica of an adult (the homunculus) that needed only to grow in size to produce a foetus. By the 19th century, microscopic studies provided substantiation for epigenesis and preformation fell out of favour. *Compare with* VON BAER'S LAWS.

KH

Premack principle *Psychology* Defined by the American psychologist David Premack (1935–).

If two behaviours differ in their probability of occurrence, the engagement in the more probable will serve to reinforce engagement in the less probable. In other words, Premack showed that behaviours for any situation are arranged hierarchically according to which are more or less likely to occur. Hence, the Premack principle changed the emphasis of operant conditioning from 'How do reinforcements strengthen behaviour?' to 'What factors determine the response-choice of an organism in a given situation?'

D Premack, 'Toward Empirical Behaviour Laws', *Psychological Review*, vol. lxvi (1959), 219–33

NS

Pre-Raphaelite (1848–56) *Art* Term adopted by seven British artists who organized themselves into the 'Pre-Raphaelite Brotherhood' in 1848. They wished to return to the simplicity of painting which had existed before Italian artist Raphael (1483–1520).

Founding members included Dante Gabriel Rossetti (1828–82), John Everett Millais (1829–96) and William Holman Hunt (1827–1910). Their compositions were inspired by poetry, Arthurian legend and religious themes, and executed in minute detail with an intricately drawn and coloured style.

AB

pre-recognition hypothesis *Psychology* Proposed by the American psychologist Joseph Banks Rhine (1895–1980).

This theory suggests that the development of unverbalized expectations of an event occurs as a result of previous experiences in similar situations. Since it is impossible to know everything about a person's previous experiences, the hypothesis cannot be disproved. *See* PARAPSYCHOLOGY.

J R Anderson and G H Bower, *Human Associative Memory* (Washington D.C., 1974)

NS

prescriptivism *Philosophy* Like EMOTIVISM, which it grew naturally out of in the 1950s, a form of SPEECH ACT THEORY which analyzes value judgments and especially moral judgments, this time in terms of prescriptions.

When I tell you that lying is wrong I am telling you not to lie, though I am also committing myself not to lie, and to issue the same prescription (or at least issue none that conflicts with it) to anyone else; *see also* UNIVERSALIZABILITY. Prescriptivism contrasts with DESCRIPTIVISM, and shares many of the features of, and objections to, EMOTIVISM.

R M Hare, *The Language of Morals* (1952)

ARL

presence and absence hypothesis (1909) *Biology* Developed by English geneticist and embryologist William Bateson (1861–1926), this mistaken hypothesis held that a recessive trait is transmitted as the absence of a dominant gene, rather than as a genetic entity in its own right.

The nomenclature used today, in which a dominant allele is designated by an uppercase letter while a recessive allele for the same characteristic is designated by a lowercase letter, was introduced by Bateson on the basis of this hypothesis.

A Sturtevant, *A History of Genetics* (New York, 1965)

KH

prestige, overt/covert *Linguistics See* HYPERCORRECTION.

presupposition (1970s) *Linguistics* Logical concept much discussed in PRAGMATICS.

An implication or assumption of a sentence which follows whether the presupposing sentence is true or false; for example 'The King of France is bald' presupposes that there is a king of France. An elusive and controversial theory.

S C Levinson, *Pragmatics* (Cambridge, 1983)

RF

Prévost's theory of exchanges (1792) *Physics* Named after its originator, the Swiss physician Pierre Prévost (1751–1839), this theory posits that a body radiates the same amount of electromagnetic energy at a given temperature, whatever the temperature and nature of its surroundings.

J Thewlis, ed., *Encyclopaedic Dictionary of Physics* (New York, Oxford and London, 1962)

MS

price discrimination (20th century) *Economics* This term describes the sale of identical goods or services in different markets at different prices. Early analysis of this phenomenon was undertaken by English economist Arthur Pigou (1877–1959).

Pricing is usually linked to ability-to-pay; thus, students or pensioners may pay less than others for social services. On a larger scale, modern pharmaceuticals companies frequently sell the same compounds at radically different prices in different (especially European) countries because the local market demand will allow it. *See* ABILITY-TO-PAY PRINCIPLE and EQUAL SACRIFICE THEORY.

A C Pigou, *Economics of Welfare* (London, 1920)

PH

primacy, law of *Psychology* Formulated by the German psychologist Hermann Ebbinghaus (1850–1909).

In a free recall situation, the materials presented first in the series are recalled more easily than materials near the centre. For example, if an individual is presented with a series of 20 words at a rate of one per second, the person will recall more words from the beginning of the series than from the centre. The law of primacy is also termed the primacy effect, the principle of primacy, and the primacy component. *See also* RECENCY, LAW OF.

E G Boring, *A History of Experimental Psychology* (New York, 1957)

NS

prime number theorem (19th century) *Mathematics* The celebrated theorem that $\Pi(x)$, the number of primes less than x, is asymptotic to $x/(\log x)$, where $\log x$ denotes the natural logarithm of x; that is,

$$\lim_{x\to\infty} \frac{\Pi(x)\log x}{x} = 1.$$

Proposed independently by the German mathematician Carl Friedrich Gauss (1777–1855) in 1792 and the French mathematician Adrien-Marie Legendre (1752–1833) in 1798, it attracted the attention of several mathematicians including Chebyshev (1821–94) and Riemann (1826–66). Its proof in 1896, given independently and almost simultaneously by the French mathematicians Jacques Hadamard (1865–1963) and Charles Jean de la Vallée-Poussin (1866–1962), was one of the crowning achievements of 19th-century analytic number theory. In 1949, Atle Selberg (1917–) and Paul Erdös (1913–) gave a proof which is elementary in the sense that it makes no use of the ZETA FUNCTION or of COMPLEX FUNCTION THEORY and in principle is accessible to anyone familiar with elementary calculus.

T M Apostol, *Introduction to Analytic Number Theory* (Springer, 1976)

MB

primitive promiscuity (1860s) *Anthropology* Hypothesized, probably independently, by lawyer-anthropologists J J Bachofen (1815–87), a Swiss, Lewis Henry Morgan (1818–81), an American, and John F McLennan (1827–81), a Briton.

The supposed original phase of human society in which there was no marriage, but rather indiscriminate mating between males and females of a group. Due in part to recent comparative studies of primate mating patterns, the existence of primitive promiscuity is no longer thought likely. *See also* KINSHIP.

A Kuper, *The Invention of Primitive Society: Transformations of an Illusion* (London, 1988)

ABA

primitivism (19th–20th century) *Art* A term used to describe the work of untrained 'Sunday painters' whose naively produced works fall outside the canon of aesthetic principles.

This style is distinguished by extreme detail, overcrowded compositions, the use of brilliant and saturated colour and little attention to correct perspective. A popular movement in Eastern Europe, its style is also typified in the work of the American Edward Hicks (1780–1849) and the Frenchman Henri 'Douanier' Rousseau (1844–1910).

AB

primordial soup (pre-1890s) *Biology* The idea that the waters of the Earth where life first appeared were rich in the nutrients that sustained life, like a 'primordial soup' or broth. In 1899, the discovery that some bacteria can grow with only carbon dioxide and inorganic salts provided counter-evidence for this theory, as such bacteria could have originated under much harsher conditions than primordial soup. *See also* ORIGINS OF LIFE.

M Barbieri, *The Semantic Theory of Evolution* (Chur, Switzerland, 1985)

KH

principle of thermodynamic similarity *Physics* This principle states that any pure substance can be characterized thermodynamically by the values of the four independent quantities (thermodynamic variables): relative molecular mass; critical pressure; critical temperature; and critical density.

J Thewlis, ed., *Encyclopaedic Dictionary of Physics* (New York, Oxford and London, 1962)

MS

prior entry law *Psychology* Formulated by the English psychologist Edward Bradford Titchener (1867–1927).

For any two cotemporaneously presented stimuli, the one upon which attention was focused first will be perceived as occurring first. This is essentially a judgmental error by which what was seen first is assumed to have been presented first.

E G Boring, *A History of Experimental Psychology* (New York, 1957)

NS

prisoner's dilemma (20th century) *Politics/Mathematics* Instance within GAME THEORY.

Two prisoners given the chance of reduced sentences if they incriminate each other – even though with no confessions at all they might not be convicted – will hedge their bets whereas solidarity would have been in their best INTERESTS. Basing their behaviour on a calculation that others will not act as sensibly as themselves, people make decisions which are not as beneficial to either themselves or others as those that, in ideal circumstances, they might make. *See also* CHICKEN, TRAGEDY OF THE COMMONS, and ZERO-SUM.

Robert Abrams, *Foundations of Political Analysis* (New York, 1979)

RB

private language argument (1953) *Philosophy/Linguistics* Debate initiated by the Austrian philosopher Ludwig Wittgenstein (1889–1951) in his posthumous *Philosophical Investigations*.

This debate concerns the question of whether there could possibly exist a private language; that is, a language which is 'necessarily unteachable' because the meanings of words known by an individual are based on private and undemonstrable experiences of their referents.

Wittgenstein argued that the speaker of such a language could not, even in principle, check whether he was using such words correctly; from which it followed (he thought) that there would be no such thing as a correct, as against incorrect, use of such words, so that the speaker would not really be saying anything. The soundness of the argument is disputed, but its importance is that words referring to immediate experiences (like 'pain') have been thought to have meaning in this way, by empiricists in particular. If Wittgenstein is right, such notions as the EGOCENTRIC PREDICAMENT and SOLIPSISM and many forms of SCEPTICISM become incoherent.

L Wittgenstein, *Philosophical Investigations* (1953), §§243–315; O R Jones, ed., *The Private Language Argument* (London, 1971)

RF, ARL

privatization (20th century) *Politics/Economics* Explanation and/or justification of shift of functions from government to market.

Services are better provided and functions better carried out by market enterprises than by public bodies. At the same time the expectations of citizens are shifted from political solutions – which are collective – to economic ones, which are individual. *See* NATIONALIZATION, 'LAISSEZ-FAIRE', PHYSIOCRACY, and MERCANTILISM.

Andrew Gamble, *The Free Economy and the Strong State* (London, 1988)

RB

probabilism *Philosophy* Name for various theories, including: first, the view that certainty is unattainable and that we should therefore seek and be satisfied with mere probabilities (a mild form of SCEPTICISM); secondly, the view that science can give positive probabilities to hypotheses and need not content itself with FALSIFICATIONISM. These two views can be thought of as approaching the same position from different ends. In Roman Catholic doctrine, probabilism is the view that when in doubt about how we ought morally to act we may follow any reputable authority.

ARL

probabilistic functionalism *Psychology* A theory of perception which has been extended to become a grand theory of behaviour and is attributed to American psychologist E Brunswick (1903–55).

Brunswick's theory emphasizes that perception is a process of identifying the most useful or functional aspects of a stimulus in order to react correctly to the environment. The extension of this theory to behaviour in general suggests that behaviour is best understood by considering its probable success in attaining its goals.

E Brunswick, 'The Conceptual Framework of Psychology', *International Journal of Unified Sciences*, vol. x (1952), 1

NS

probability theory (17th century) *Mathematics* This originated with mathematical problems connected with dice throwing which were discussed in letters exchanged by French mathematician and physicist Blaise Pascal (1623–62) and French lawyer and magistrate Pierre de Fermat (1601–65) who worked on mathematics as a hobby.

It is the study of random or chance events. Originally, problems concerned with permutations, combinations and binomial coefficients were studied. Laplace (1812) systematized many of these methods. In the early 20th century, work primarily due to

Kolmogorov gave a measure-theoretic basis. *See also* STATISTICS, GAME THEORY, INFORMATION THEORY, and BAYES'S THEOREM.

W Feller, *An Introduction to Probability Theory and Its Applications* (New York, 1966)

ML

probability theory of learning *Psychology* Proposed by the American psychologist R Duncan Luce (1925–).

(1) The principle that when a choice of behaviours exists, the probability of a response tends towards the probability of reinforcement. (2) The recognition of the relative probabilities of the occurrence of several events with which an individual is faced; the individual must choose which event is most likely to occur. *See* MATHEMATICAL LAW(S).

G A Kimble, *Hilgard and Marquis Conditioning and Learning* (New York, 1961)

NS

process art (late 1960–early 1970s) *Art* In its use of poor materials, this movement developed against a background of increasing concern with environmental and ecological issues. As in the 'ARTE POVERA' movement, nature itself became art and the representation and symbolization of nature were rejected. The very insubstantiality and ephemeral nature of some materials was highlighted.

D Wheeler, *Art Since the Midcentury – 1945 to the Present* (London, 1991)

AB

process philosophy *Philosophy* Any of a variety of theories emphasizing that the basic reality in the universe is not objects or substances but processes. Objects are mere temporary bodies in the general flux, and are not sharply separated from one another; and real time is continuous and not an accretion of instantaneous moments. Process philosophy can be seen in Heraclitus of Ephesus (writing *c*.500 BC), and its leading modern exponents include William James (1842–1910), Henri Bergson (1859–1941) and Alfred North Whitehead (1861–1947).

W James, *A Pluralistic Universe* (1909), lecture 6; clear, readable, and sympathetic exposition of some of Bergson's philosophy, which had Bergson's full approval

ARL

product life-cycle theory (20th century) *Economics/Marketing* Long-term patterns of international trade are influenced by product innovation and subsequent diffusion. A country that produces technically superior goods will sell these first to its domestic market, then to other technically advanced countries. In time, developing countries will import and later manufacture these goods, by which stage the original innovator will have produced new products.

On a smaller scale, individual products pass through distinct phases: after a period of research and development, and trial manufacture, there is a period of introduction characterized by slow growth and high development costs. This is followed by a period of growth as sales and profits rise. A phase of maturity and saturation is then experienced as sales level off and the first signs of decline occur. The final phase is decline, characterized by lower sales and reduced profits, and perhaps final disappearance from the market. The duration of each stage of the cycle varies with the product and the type of management supporting it. *See* TECHNOLOGICAL GAP THEORY.

M V Posner, 'International Trade and Technical Change', *Oxford Economic Papers*, vol XIII (October, 1961), 323

PH

product rule *Mathematics* Attributed to the principle founders of calculus: the French mathematician Pierre de Fermat (1601–65), the British mathematician Sir Isaac Newton (1642–1727) and the German mathematician Gottfried Wilhelm Leibniz (1646–1716).

This rule states that the derivative of the product of two differentiable functions is

$$\frac{d}{dx}(f(x)g(x)) = f(x)\frac{dg}{dx}(x) + g(x)\frac{df}{dx}(x).$$

See also LEIBNIZ'S THEOREM.

H Anton, *Calculus with Analytic Geometry* (New York, 1980)

ML

production, theory of (18th century–) *Economics* Analysis concerned with transforming factor inputs into outputs according to a production function. Production is dependent on technology, the mix of factor inputs, factor prices and marginal productivity.

The Modern Cambridge School envisages an economic model which encompasses sociological, historical and psychological factors as well as pure economic ones. It attacks neo-classical economics for its use of the aggregate production function particularly in growth theory. *See* NEO-CLASSICAL GROWTH THEORY, CAMBRIDGE CAPITAL CONTROVERSIES, COBB-DOUGLAS PRODUCTION FUNCTION, and ROUNDABOUT METHOD OF PRODUCTION.

PH

profits, theories of (19th century–) *Economics* The early classical economists believed that profits would eventually decline. The Scottish economist Adam Smith (1723–90) viewed the growth rate of capital accumulation (which took place at a rate higher than total output) as the cause. The English economist David Ricardo (1772–1823) argued that a decline in the general rate of profit was caused by the diminishing marginal productivity of land. The German political economist Karl Marx (1818–83) believed that the rate of profit was influenced by the growth of competition between capitalists.

A Smith, *An Inquiry into the Nature and Causes of the Wealth of Nations* (London, 1776); D Ricardo, *On the Principles of Political Economy and Taxation* (London, 1817); K Marx, *Capital*, I–III (Harmondsworth, 1976–81)

PH

progenote theory (late 1970s) *Biology* Proposed by American evolutionist and biophysicist Carl R Woese (1928–), this theory asserts that the very first living cells (called progenotes) had low-molecular-weight ribosomes and consequently showed little biological specificity.

This means that a gene was translated into statistically probable proteins, rather than into identical proteins as is true of cells today. Woese's theory also holds that NATURAL SELECTION favoured an increase in ribosomal weights because larger ribosomes provide more precision in protein synthesis; and that progenotes evolved into higher life forms when their ribosomes reached molecular weights of 1 to 2 million. Like other theories on the ORIGINS OF LIFE, this theory contains weaknesses that have hampered its widespread acceptance thus far. *See also* HYPERCYCLE THEORY and WOESE'S DOGMA.

KH

programmed ageing (1881) *Biology* (Also referred to as the ageing clock, biological death clock, clock theory, and pacemaker theory of ageing.) Attributed to August Weismann (1834–1914), the German biologist, physician and founder of the GERM PLASM THEORY, this idea has been advocated in various forms by many other scientists. It asserts that ageing represents an unfolding of events as orchestrated by the inherited material of the organism, rather than being the result of random events occurring in the body during life. *See also* AGEING, THEORIES OF.

J Behnke and C Finch and G Moment, *The Biology of Aging* (New York, 1978)

KH

progress *Politics/History* Optimistic theory of history.

Human societies move towards higher and higher forms of culture or organization. There is much disagreement as to the character of this improvement, as there is also much scepticism about it.

David Miller *et al.*, eds, *The Blackwell Encyclopaedia of Political Thought* (Oxford, 1987)

RB

progression, law of *Psychology* Formulated by the Frenchman, J Delbouf (1831–96).

The generalization that successive increments in sensation are an arithmetical progression. Alternatively stated, for two stimuli of some general kind to be discriminated or told apart, the difference between them has to be a constant proportion of the smaller.

E C Carterette and M P Friedman, eds, 'Psychophysical Judgement and Measurement', *Handbook of Perception*, vol. II (New York, 1974)

NS

progressive endosymbiotic theory *Biology* *See* SERIAL ENDOSYMBIOTIC THEORY.

progressive evolution *Biology* The theory that organisms become increasingly complex over evolutionary time.

KH

prokaryotic theory of cellular evolution (1866) *Biology* Proposed by German naturalist Ernst Heinrich Haeckel (1834–1919), this is the concept that living cells lacking nuclei (bacteria) were the first to appear on earth. These cells then gave rise to protista (such as algae and protozoans), which gave origin to early plants and animals.

See also ORIGINS OF LIFE. *Compare with* AKARYOTIC THEORY, EUKARYOTIC THEORY, SERIAL SYMBIOSIS THEORY and SYMBIOTIC THEORY OF CELLULAR EVOLUTION.

M Barbieri, *The Semantic Theory of Evolution* (Chur, Switzerland, 1985)

KH

proletarianism *Politics* A socialist version of POPULISM.

The belief that the working class is both the repository of particular cultural and intellectual virtues, and the only feasible source of political progress under capitalism. More frequently used against those who are alleged to hold this view than as a term of self-description.

proof theory (20th century) *Mathematics* The branch of logic which studies the syntactic (rather than the semantic) properties of formal theories. In particular, it is concerned with formal deductions rather than the concepts of truth, satisfaction and validity. *Compare with* MODEL THEORY.

Encyclopedic Dictionary of Mathematics (MIT Press, 1987)

MB

propaganda by the deed *Politics* Theory of political tactics within ANARCHISM.

Selectively violent acts against government, by provoking a more violent repressive response, can reveal to people at large the 'true' nature of the STATE as a coercive institution. Such propaganda 'by the deed' achieves results which more conventional persuasion cannot.

David Miller, *Anarchism* (London, 1984)

RB

propensity theory of probability (16th century) *Mathematics* First used by Geralmo Cardano (1501–76), this is the traditional theory of probability based on *a priori* judgements of apparently equiprobable events, although equiprobability can be illusory. For example, as there are 7 days in the week the chance of 1 January 2000 being Sunday appears to be 1/7, but a century never starts on a Sunday, so the probability is 0.

JB

propensity theory of probability *Philosophy* Theory mainly associated with Karl Raimund Popper (1902–), though it goes back to Charles Sanders Peirce (1839–1914). Popper introduces it to replace the FREQUENCY THEORY OF PROBABILITY in view of an objection he brings to that.

Probabilities are propensities, not of objects under study but of the experimental arrangements which we keep constant during repeated experiments. Though not directly observable, Popper claims that propensities are no more mysterious than forces or fields, and that their existence can be falsified (*see* FALSIFICATIONISM). This, however, may be doubted, if the run of relevant events is potentially infinity; we may be forced to say that *probably* the propensity exists, or (if the propensity is simply identified with the evidence for it) that it is rational to act on it because probably it will continue. In neither case is 'probably' explained, and the theory seems open to most of the objections to the frequency theory, except the rather recherché one Popper brings it in to avoid.

C S Peirce, *Collected Papers*, 2 (1932), 404–14; K R Popper, 'The Propensity Interpretation of Probability', *British Journal for Philosophy of Science* (1959)

ARL

property *Politics/Economics* Account of RIGHTS over the material world.

Justifications of property rights vary, as do recommendations on the obligations and restraints which should accompany such rights. Disagreement exists as to whether such rights belong in the first place to individuals or to society.

David Miller *et al.*, eds, *The Blackwell Encyclopaedia of Political Thought* (Oxford, 1987)

RB

property is theft (19th century) *Politics* Argument of French anarchist Pierre-Joseph Proudhon (1809–65).

In translation the claim is self-contradictory, since without the concept of PROPERTY 'theft' is meaningless. What Proudhon meant was that purely legal ownership without any of the responsibilities which went with it was a theft from those who actually worked the land, or the raw materials, from which the owners profited.

George Woodcock, *Anarchism* (Harmondsworth, 1963)

RB

proportion *Art* The mathematical and geometrical rules conditioning the relationship of the parts to each other and to the whole.

Since the days of the Ancient Egyptian civilization, artists have employed some system of proportional canon. The Greek canon was established with the famous statue of *Doryphorus* by Polyclitus of Samos (*fl.*5th century BC), and was modified by Lysippus (*fl. c.*360–316 BC). Further modifications were introduced by: the Roman Vitruvius (*fl.*46–30 BC), the medieval mason Villard de Honnecourt (*fl.*1225–35 AD), Italian architect and theorist Leon Battista Alberti (1404–72), Leonardo da Vinci (1452–1519) and Albrecht Dürer (1471–1528), each reflecting particular aesthetic aims.

AB

propositional act *Linguistics See* SPEECH ACT THEORY.

prototype (1970s) *Linguistics* Cognitive-semantic theory developed by American psychologist Eleanor Rosch and colleagues.

In COMPONENTIAL ANALYSIS, the meaning of a word is a set of features; if an object possesses all the attributes implied by the features, it can be referred to by that word. Prototype theory recognizes that the boundaries between categories are not so clear-cut, identifying categories by their 'best instances'; for example robin for bird, oak for tree.

E Rosch and B B Lloyd, *Cognition and Categorization* (Hillsdale, NJ, 1978)

RF

protovirus hypothesis (1970s) *Biology/Medicine* Proposed by Howard Martin Temin (1934–), American molecular biologist and a winner of the Nobel prize for Physiology or Medicine in 1975. This is the idea that deoxyribonucleic acid (DNA) that 'goes wrong' is the source of new disease-causing viruses.

More formally, the theory asserts that retroviruses emerged when parts of RNA were reverse-transcribed through cellular enzymes into DNA products that became integrated into the host cell genome. In time these DNA products grew capable of producing infectious virus particles, and after millions of years they became scattered throughout the genetic material of many animals, including humans. *Compare with* CELLULAR ORIGIN THEORY, ONCOGENE HYPOTHESIS and PROVIRUS HYPOTHESIS.

H Fraenkel-Conrat, *Virology* (Englewood Cliffs, NJ, 1988)

KH

Proust's law *Chemistry See* CONSTANT COMPOSITION, LAW OF.

Prout's hypothesis (1815) *Chemistry* Named after the English chemist and physiologist William Prout (1785–1850). This hypothesis was in fact twofold, one part proposing the integral nature of atomic weights, and the other concerning the the unity of matter and proposing that all atoms are built up from hydrogen atoms.

The first part of the hypothesis was soon shown to be incorrect with the discovery that chlorine had an atomic weight of 35.5. Subsequent work by the English physicist Francis William Aston (1877–1945), however, revealed the existence of chemical isotopes and showed that there was indeed some basis for Prout's original ideas.

C C Gillespie, *Dictionary of Scientific Biography* (New York, 1981)

GD

provirus hypothesis (1964) *Biology* Proposed by Howard Martin Temin (1934–), American molecular biologist and a winner of the Nobel prize for Physiology or Medicine in 1975. This is the idea that some cancers are caused by the infection of a cell by a single-stranded ribonucleic acid (RNA) virus which, early on in the infection, is transcribed into a double-stranded deoxyribonucleic acid (DNA) intermediate (known as copy DNA). This becomes integrated into the host cell where it acts as a provirus.

The provirus, which retains all the information of the RNA virus but is now part of the host cell's chromosomes, goes on to produce more RNA virus or to trigger malignant cell growth. Because information appears to flow from RNA to DNA, instead of from DNA to RNA as the CENTRAL DOGMA holds, these RNA tumour viruses are called retroviruses. This represents one of several theories regarding carcinogenesis. *See also* ONCOGENE HYPOTHESIS. *Compare with* PROTOVIRUS HYPOTHESIS.

J Tooze, *The Molecular Biology of Tumour Viruses* (Cold Spring Harbor Laboratory, 1973)

KH

proximity, law of *Psychology* Formulated by the German-American psychologist Wolfgang Köhler (1887–1967).

One of the gestalt laws (or principles) of organization, this states that stimuli which are physically or temporally proximate will be perceived as belonging together in a group. *See* GESTALT THEORY

W R Uttal, *On Seeing Forms* (London, 1988)

NS

prudence principle *Accountancy* This asserts that accounts should be prepared on a conservative basis and that future profit or revenue should not be included in the accounts before they are realized. Adequate provision for losses should be made when these are foreseen.

AV, PH

pseudaposematic colouration *Biology See* BATESIAN MIMICRY.

pseudo-forgetting *Psychology* Defined by the American psychologist Elizabeth F Loftus (1944–).

The process whereby we think we know something (such as a telephone number), attempt to recall it and when we fail, conclude we have forgotten it. The fact is we may never have stored the material or it may have been incorrectly stored. Either way the information cannot be considered forgotten.

E F Loftus and G R Loftus, 'On the Permanence of Stored Information in the Human Brain', *American Psychologist*, vol. xxv (1980), 409–20

NS

psychic energy *Psychology* Proposed by the Austrian founder of psychoanalysis Sigmund Freud (1856–1939).

Also called mental energy, psychic energy is posited as the dynamic force behind all mental processes. The id is said to be the main source of this energy, though it is the ego that gains control of it in order to control and organize personality. The concept of psychic energy is also to be found in the analytic psychology of Carl Gustav Jung (1875–1961).

J P Dworetzky, *Psychology*, 3rd edn (New York, 1985)

NS

psychic resultants law *Psychology* Formulated by the German experimental psychologist Wilhelm Wundt (1832–1920).

The psychic resultants law (or principle of creative synthesis) has been referred to as Wundt's 'Mental Chemistry'. It asserts that many ideas can be knitted together by association into a more complex idea. However, the nature of the complex resultant is to be understood by the fact that all the components are still actually present.

E G Boring, *A History of Experimental Psychology* (Englewood Cliffs, N.J., 1957)

NS

psychical distance (1912) *Literary Theory* Term coined by the British philosopher E Bullough (1880–1934); otherwise called 'aesthetic distance'.

The aesthetic mode of response to literature involves a variable degree of distancing between the self and the work, a lack of direct subjective involvement with the work's concerns. *See also* DISINTERESTEDNESS.

E Bullough, 'Psychical Distance as a Factor in Art and an Aesthetic Principle', *British Journal of Psychology*, vol. v (1912), 87–98; excerpted in E Vivas and M Krieger, *The Problems of Aesthetics* (New York, 1953), 396–405

RF

psychoanalytic criticism (1919) *Literary Theory* Various models today, but generally underpinned by the theories of the founder of psychoanalysis Sigmund Freud (1856–1939). The French theorist Jacques Lacan (1901–81) has been a major influence on recent psychoanalytic and FEMINIST CRITICISM.

Classically, in psychoanalytic criticism literature is regarded as analogous to dream or fantasy and therefore an expression of drives or desire. Psychoanalytic critics have analyzed literary works as corroborating their hypothesized pathologies; or as evidence for the psyches of the works' authors. In the 1950s and 1960s, attention turned to the activities of readers, and their relationships with texts (*see* READER RESPONSE CRITICISM). Psychoanalytic concepts have also illuminated the structure of figurative processes in literature, and the ideological relationship between meanings in literature, and (more broadly) discourse within a society (for example patriarchy).

E Wright, *Psychoanalytic Criticism: Theory in Practice* (London, 1984)

RF

psychoanalytical theory *Psychology* Founded by the Austrian psychiatrist Sigmund Freud (1856–1939).

According to Freud, personality can be conceptualized as containing three parts: the id, ego, and superego. The id contains the reservoir of instinctual drives; the superego develops from and contains parental and societal standards of morality. The ego must regulate the relations between the id and ego and also present the visible personality. Freud's theory is therefore dynamic in the sense that a conflict exists which must be regulated by the ego. Freud's theory influenced a number of other theorists (for example, Carl Gustav Jung (1875–1961), Alfred Adler (1870–1937), and Karen Horney (1885–1952)), and he has unquestionably exercised the most influence on the history of psychology. *See* FREUDIAN THEORY.

J P Dworetzky, *Psychology*, 3rd edn (New York, 1985)

NS

psycholinguistics (1960s–) *Linguistics* Branch of linguistics developed principally in the USA.

Study of language and mind: the acquisition of LINGUISTIC COMPETENCE and its deployment in the production, perception and comprehension of language; also the relationship between language and thought. It was much facilitated by mentalist claims of GENERATIVE GRAMMAR: *see* MENTALISM IN LINGUISTICS. More recently, psycholinguistics has incorporated findings of modern cognitive psychology.

M Garman, *Psycholinguistics* (Cambridge, 1990)

RF

psychologism *Philosophy* The habit of treating philosophical or theoretical problems as though they were psychological ones, to be solved by methods such as introspection. Properly speaking it is only a theory when engaged in deliberately rather than, as more often, unconsciously or through confusion, though the distinction is not sharp.

Psychologism is common in the early empiricists like John Locke (1632–1704) and also (especially in arithmetic) John Stuart Mill (1806–73), and indeed in some British universities psychology and philosophy were only finally disentangled after World War II (paradoxically, perhaps, because philosophy has since moved towards closer contact with the sciences, and the line between them – and hence what counts as psychologism – has become more blurred).

Psychologism was, however, vigorously criticized by Francis Herbert Bradley (1846–1924), Gottlob Frege (1848–1925), and Edmund Husserl (1859–1938) in his later works. In any case psychologism should not be confused with NATURALIZED EPISTEMOLOGY, which explicitly *replaces* philosophical questions with psychological ones.

ARL

psychologist's fallacy *Psychology* Proposed by the Austrian psychologist F Heider (1896–).

The tendency of a psychologist to project his/her own point of view or interpretation onto another person. This can occur, for example, when an introspectionist reads unwarranted interperetations into an observer's report of a piece of behaviour, or when a psychologist makes an incorrect inference about a client in therapy.

J T Tedeschi, S Lindskold and P Rosenfeld, *Introduction to Social Psychology* (St Paul, 1985)

NS

psychophysical parallelism *Philosophy* Doctrine that mental and physical events are of entirely different kinds, so that while mental events can cause other mental events and physical events can cause other physical events they cannot cause each other but occur in parallel series.

If I touch a hot stove, feel a pain, withdraw my hand, and decide to be more careful in future, the touching causes the withdrawal but not the decision, while the pain (itself caused, if at all, by some preceding mental event) causes the decision but not the withdrawal. As this example suggests, it is hard to keep the two chains properly separated. The doctrine is similar in nature and in motivation to OCCASIONALISM, but occasionalism attributes all causation to God.

ARL

Ptolemaic theory of the solar system (2nd century) *Astronomy* The astronomer and geographer Claudius Ptolemy (*c.*AD 90–168) of Alexandria proposed that the planets describe closed orbits about the Earth.

To fit the observed motions, it was necessary to invoke complex orbits for the planets. In the simplest case, the planet was assumed to move uniformly on a circle (the epicycle) that itself moved uniformly on a larger circle, called the deferent.

S Mitton, ed., *The Cambridge Encyclopaedia of Astronomy* (London, 1973)

MS

public choice (20th century) *Politics/Economics* Application of economic models to politics.

Political and governmental decisions can be understood as rational choices of persons and institutions seeking to maximize their self-interest.

Patrick Dunleavy, *Democracy, Bureaucracy and Public Choice* (Brighton, 1991)

RB

public interest *Politics/Sociology* Critical revision of the theory of INTERESTS.

More important than an attempt to identify individual or group interests is the identification of the public interest or COMMON GOOD which is shared by members of society as a whole.

David Miller *et al.*, eds, *The Blackwell Encyclopaedia of Political Thought* (Oxford, 1987)

RB

public language *Linguistics See* RESTRICTED AND ELABORATED CODES.

public private divide *Politics* Theory underlying much liberal political thought.

There is a division between the public world of paid employment and public affairs, and the private world of the household or family. For mainstream liberal theorists, the private world is the arena of free individual choice; for feminist and Marxist critics, the division is a sham which in fact sustains either class or gender inequalities.

Carol Pateman*The Disorder of Women* (Cambridge, 1989)

HR

public utility pricing (1938) *Economics* First raised as an economic issue by American economist Harold Hotelling (1895–1973), this term refers to the setting of prices for goods and services in order to maximize the benefit to the community.

Such pricing for rail, telephone, water and electricity (which could not be carried out in normal market conditions) takes into account future demand, the state of future technology and likely costs of factors of production in the future. *See* SOCIAL WELFARE FUNCTION, PARETO OPTIMALITY, SCITOVSKY PARADOX, COST-BENEFIT ANALYSIS and COMPENSATION PRINCIPLE.

H Hotelling, 'The General Welfare in Relation to Problems of Taxation and of Railways and Utility Rates', *Econometrica*, 6, 3 (July, 1938), 242–69

PH

pulsating (oscillating) universe hypothesis (20th century) *Cosmology/Physics* The kinetic energy of the expanding universe is being transformed to gravitational potential energy and, if the mass density of the universe exceeds the so-called critical mass density, the expansion will slow down and eventually stop. After this stage, the universe will begin to contract with a reversal of the events that took place during the expansion. After some time (possibly about 10^{32} years), the universe will end in a 'big crunch', followed by a new 'big bang' (not necessarily identical to the previous one). The universe will then oscillate; that is, successively expand, stop and contract. *See* EXPANSION OF THE UNIVERSE.

S Mitton, ed., *The Cambridge Encyclopaedia of Astronomy (London, 1973)*

MS

puncept (1988) *Literary Theory* Neologism coined by critic Gregory L Ulmer (1944–).

In DECONSTRUCTION, the use of puns and word-play to DECENTRE the fixed meanings of established concepts. 'Puncept' exemplifies its own meaning.

G Ulmer, 'The Puncept in Grammatology', *On Puns: The Foundation of Letters*, J Culler, ed. (Oxford, 1988), 164–89

RF

punctuated equilibrium (1972) *Biology* Proposed by American paleontologists Niles Eldredge and Stephen Jay Gould (1942–), this is a model for evolution in which species are relatively stable and long-lived. New species appear during concentrated outbursts of speciation, which are followed by the successes and failures of the various species. This theory remains controversial. *Compare with* GRADUALISM.

N Eldredge, *Time Frames: The Rethinking of Darwinian Evolution and the Theory of Punctuated Equilibria* (New York, 1985)

KH

punctuated evolution *Biology See* PUNCTUATED EQUILIBRIUM.

purchasing power parity (1916) *Economics* With its roots in 17th century MERCANTILISM, this notion was developed by Swedish economist Gustav Cassel (1866–1945). It asserts that exchange rates are in equilibrium when the domestic purchasing power of currencies are the same.

A FFr10 = £1 rate would be in equilibrium if FFr10 bought the same quantity of goods and services in France as £1 bought in Britain. This is a useful concept when comparing international living standards. *See* FUNDAMENTAL DISEQUILIBRIUM and INTERNAL AND EXTERNAL BALANCE.

G Cassel, 'The Present Situation of the Foreign Exchanges – I', *Economic Journal*, vol. XXVI (March, 1916), 62–65

PH

purism (1) *Art Purismo* was an early 19th-century movement in Italian art and literature, advocating a return to the pure forms of the Renaissance.

Whilst allied to NEO-CLASSICISM, it sought greater spirituality and was closer to the NAZARENE BROTHERHOOD. The theorist was Antonio Bianchini; the leading painter, T Minardi (1787–1871), and the leading sculptor, Pietro Tenerani (1789–1869).

E Lavagnino, *L'Arte Moderna* (Milan, 1956)

LL

purism (2) (early 20th century) *Art* Movement based on the aesthetic of 'machine art', founded in Paris by Amédée Ozenfant (1886–1966) who, with Charles-Edouard Jeanneret (Le Corbusier) (1887–1965), published *Après le Cubisme* in 1918 and collaborated on a periodical called *L'Esprit Nouveau* (1920–25).

Purists admired machine-made objects and the avoidance of emotion, and had similar aims to the German BAUHAUS movement. After 1925, the most important work in this style was Le Corbusier's architecture.

C Green, 'Purism', *Concepts of Modern Art*, T Richardson and N Stangos, eds (London, 1974)

LL

Purkinje effect (1825) *Physics/Psychology* Named after its discoverer, the Czech physiologist Jan Evangelista Purkinje (1787–1869).

The sensitivity of the eye is different under conditions of daytime vision and night-time vision. This is because daytime vision is largely through the cones of the fovea (which are most sensitive to longer wavelength red-yellow light) while night vision is largely through the extra-foveal rods which are most sensitive to blue-green light (short wavelengths).

M D Vernon, *Experiments in Visual Perception*, Harmondsworth, 1970); J Thewlis, ed., *Encyclopaedic Dictionary of Physics* (New York, Oxford and London, 1962)

MS, NS

purposive psychology *Psychology See* EXPECTANCY THEORY.

Pygmalion effect *Psychology* Named after a 1914 play of the same name by Irish author George Bernard Shaw (1856–1950).

A term often used synonymously with SELF-FULFILLING PROPHECY, this refers to the frequently observed phenomenon whereby people come to behave in ways that correspond to others' expectations of them. The effect is of particular concern in psychological investigations and therapies where subjects may exhibit behaviour which they perceive the psychologist expects of them. *See* ROSENTHAL EFFECT.

R Rosenthal and L Jacobson, *Pygmalion in the Classroom: Teacher Expectations and Pupils' Intellectual Development* (New York, 1968)

NS

Pyrrhonism *Philosophy* An extreme form of SCEPTICISM, associated with Pyrrho of Elis (*c.*365–275 BC) and developed by his followers, notably Aenesidemus (1st century BC) and Sextus Empiricus (2nd century AD).

Pyrrhonism's distinguishing feature lay in its application of scepticism to itself: not only could we not know anything, but we could not even know that we could not know anything. Unlike the 'dogmatic' sceptics of the Academy, therefore (who did claim to know this last fact), Pyrrho advocated complete suspension of judgment and hoped to obtain a tranquil peace of mind thereby (an outlook with echoes in the philosophy of David Hume (1711–76)). Pyrrho wrote nothing (not surprisingly), but fortunately for our historical knowledge his followers did. 'Pyrrhonism' is occasionally used loosely for scepticism in general.

A A Long and D N Sedley, *The Hellenistic Philosophers* (1987); translations with commentary

ARL

Pythagoras's theorem (5th century BC) *Mathematics* Named after Greek mathematician Pythagoras (*c.*569–500 BC) who founded a school whose motto was 'all is number'. The word mathematics ('that which is learned') is said to be Pythagorean. The school had a strict code of secrecy, so it is unknown who actually discovered this theorem.

The theorem that in a right triangle, the square of the length of the hypothenuse is equal to the sum of the squares of the lengths of the other two sides.

W Gellert, S Gottwald, M Hellwich, H Kästner and H Küstner, eds, *The VNR Concise Encyclopedia of Mathematics* (New York, 1975)

ML

Pythagoreanism *Philosophy* Ideas held over the next two centuries by followers of Pythagoras of Samos (6th century BC).

Pythagoras is said to have founded a semi-religious brotherhood which developed doctrines about reincarnation and purification. He is also credited with noticing that simple harmonies (octave, fifth and so on) are associated with simple arithmetical ratios. He or his followers (who reverently attributed their own ideas to him) developed a metaphysics of limit (good) and unlimited (bad), and in some way saw number as lying at the basis of reality; but the mathematical ideas traditionally attributed to them (including 'Pythagoras' theorem') are now thought unlikely to belong to Pythagoreans earlier than Archytas of Tarentum (*fl.* 400 BC). Scientific ideas, such as that the Earth is not at the centre of the universe, may have a Pythagorean origin, but were probably developed for religious rather than scientific reasons. *See also* ORPHISM (I).

J A Philip, *Pythagoras and Early Pythagoreanism* (1966)

ARL

Q

quantification theory or calculus (19th century) *Mathematics* The rules for calculating with quantifiers (such as 'there exists') in predicate calculus.

JB

quantity theory of money (1885) *Economics* Developed by the Americans Simon Newcomb (1835–1909) and Irving Fisher (1867–1947), the latter of whom's original equation stated in simple terms that the amount of money in circulation equals money NATIONAL INCOME; that is,

$$MV = PT$$

where M is money stock, V is velocity of circulation, P is average price level and T the number of transactions. The equation assumes that the velocity of circulation of money is stable (at least in the short term) and that transactions are fixed by consumer tastes and the behaviour of firms. This theory was superseded by Keynsian analysis.

Members of the Cambridge School were concerned with the volume of money held given the number of transactions carried out. They argued that the greater the number of transactions, the greater the amount of money held. English economist Arthur Pigou (1877–1959), in particular, asserted that the nominal demand for money was a constant percentage of nominal income. In the Cambridge Equation, PT is replaced by Y (the income velocity of circulation). The equation is:

$$V = \frac{Y}{M}$$

where M is money stock in economy, Y income velocity of circulation and V average velocity of circulation.

Monetarists argue that an increase in prices would not lead to inflation unless the government increased the money supply. *See* COMMODITY THEORY OF MONEY and MONETARISM.

S Newcomb, *Principles of the Political Economy* (New York, 1885); I Fisher, *The Purchasing Power of Money* (New York, 1911); M Friedman, ed., *Studies in the Quantity of Money* (Chicago, 1956)

PH

quantum chromodynamics (QCD) *Physics* QCD is a relativistic quantum-mechanical GAUGE THEORY of the strong interactions based on the exchange of mass-less gluons between quarks and antiquarks. QCD is analogous to quantum electrodynamics (QED) in the field theory of electromagnetic interactions, but with the gluon replacing the photon and the 'colour' quantum number instead of electric charge. The theory has been tested successfully in high-energy experiments involving muon-nucleon scattering and proton-antiproton collisions. *See also* QUANTUM ELECTRODYNAMICS.

Paul Davies, ed., *The New Physics* (Cambridge, 1989)

GD

quantum electrodynamics (QED) *Physics* QED is a relativistic quantum-mechanical GAUGE THEORY of photon-mediated electromagnetic interactions.

Local symmetry operations (for example, change of phase of a wave function) are required not to affect the mathematical description of the behaviour of a charged particle, thereby yielding the electromagnetic force. The theory has been verified experimentally and has led to highly accurate predictions. *See also* QUANTUM CHROMODYNAMICS.

Paul Davies, ed., *The New Physics* (Cambridge, 1989)

GD

quantum evolution (1944) *Biology* Coined by American palaeontologist George Gaylord Simpson (1902–84), this is the theory that a relatively rapid change from an ancestral state to a new, distinctly different state can occur as an exceptional event in the evolutionary history of a small population.

This theory, which is believed to explain the origin of most major groups of organisms, is thought to account for the rarity of missing links in the fossil record. *See also* PUNCTUATED EQUILIBRIUM.

G Gaylord, *Tempo and Mode in Evolution* (1944); V Grant, *The Evolutionary Process: A Critical Study of Evolutionary Theory*, 2nd edn (New York, 1991)

KH

quantum field theory *Physics* A FIELD THEORY describing the behaviour of particles by representing them as fields whose normal modes of oscillation are quantized. All the physical observables of a system are represented by appropriate operators which obey certain commutation relations. The quantized field can be considered as an assembly of particles each of which is characterized by its own energy, momentum, charge and so forth, the total energy, etc., of the field being built up additively from the contributions of the individual particles. *See also* ELECTROWEAK THEORY, GAUGE THEORIES, QUANTUM CHROMODYNAMICS and QUANTUM ELECTRODYNAMICS.

Paul Davies, ed., *The New Physics* (Cambridge, 1989)

GD

quantum Hall effect (1980) *Physics* In 1980 it was discovered by K von Klitzing, G Dorda and M Pepper that the Hall resistance of silicon MOS-FET semiconductor devices, operated at liquid helium temperature and in strong magnetic fields with flux densities of about 1 T, varied in quantized steps. The most pronounced effects were observed with high-purity gallium arsenide devices. Strictly, it is the conductance that increases in quantized steps and these steps involve the ratio h/e^2, where h is Planck's constant and e is the magnitude of the charge on the electron. The relationship is so accurate that it is the basis of an atomic method of determining the unit of resistance: the ohm. *See* HALL EFFECT.

J Thewlis, ed., *Encyclopaedic Dictionary of Physics* (New York, Oxford and London, 1962)

MS

quantum speciation *Biology See* SALTATION SPECIATION.

quantum theory *Physics See* PLANCK'S QUANTUM THEORY.

quark theory (1964) *Physics* In 1964 it was pointed out independently by Murray Gell-Mann (1929–) and George Zweig (1937–) that all known hadrons (the particles that experience the strong nuclear force) could be constructed out of simple combinations of three particles, known as quarks, and their anti-particles.

These quarks have fractional electric charges, taking the charge on the electron as the unit. The theory supposes that a baryon is composed of three quarks (which need not be of the same type) while a meson is made up of a quark-anti-quark pair (which need not be of the same type). Quarks are bound together by a very strong force, mediated by particles called gluons. The strong nuclear force that binds protons and neutrons in the atomic nucleus is a vestige of this quark-gluon interaction.

Quarks are fermions and must, therefore, obey the PAULI EXCLUSION PRINCIPLE. To satisfy this condition it has been found necessary to endow the three original quarks (known as 'up', 'down' and 'strange' quarks, respectively) with another property known as 'colour'. It has also proved necessary to postulate three additional quarks (and their anti-quarks) known as the 'charm', 'truth' and 'beauty' quarks.

J Thewlis, ed., *Encyclopaedic Dictionary of Physics* (New York, Oxford and London, 1962)

MS

queuing theory (1970s) *Economics* Developed as an extension of probability theory, this deals with the analysis of congestion and delay in economic modelling. It features in stock control in the shape of the Lifo (Last in first out) and Fifo (First in first out) principles, and in financial markets. *See* INFORMATION THEORY.

D Gross and C M Harris, *Fundamentals of Queuing Theory* (New York, 1985)

PH

queuing theory (20th century) *Mathematics* The study of queues which assumes a distribution of time between customers and a distribution of time in serving, and then attempts to estimate waiting times and queue lengths.

The queuing theory of George Mikes (1912–90) is that queuing is the favourite activity of the British.

JB

quotient rule *Mathematics* Attributed to the principal founders of calculus: the French mathematician Pierre de Fermat (1601–65), the British mathematician Sir Isaac Newton (1642–1727) and the German mathematician Gottfried Wilhelm Leibniz (1646–1716).

The rule that the derivative of the quotient of two different functions is

$$\frac{d}{dx}\left(\frac{f(x)}{g(x)}\right) = \frac{g(x)\frac{df}{dx}(x) - f(x)\frac{dg}{dx}(x)}{[g(x)]^2}$$

where $g(x) \neq 0$. *See also* PRODUCT RULE.

H Anton, *Calculus with Analytic Geometry* (New York, 1980)

ML

R

Raabe's test *Mathematics* The test for convergence of a positive term series that if

$$\frac{a_{n+1}}{a_n} < 1 - \frac{A}{n}$$

for some constant $A > 1$ and for sufficiently large n, then Σa_n converges.

K Knopp, *Theory and Application of Infinite Series* (New York, 1990)

ML

racial memory *Psychology* Proposed by the Swiss psychiatrist Carl Gustav Jung (1875–1961).

Most often associated with Jung's analytic psychology, racial memory consists of thoughts, feelings and inferences which are believed to be passed on from generation to generation and which influence the behaviour of the individual. According to Jung racial memory is a part of the COLLECTIVE UNCONSCIOUS. Freud also believed in racial memory, but focused on a different aspect of it. The idea of racial memory is highly speculative.

C G Jung, *Analytical Psychology: Its Theory and Practice* (New York, 1968)

NS

racism *Politics/Sociology* Theory of the biologically determined basis of human social character. The term 'racism' is used critically of those employing such theory, rather than as a term of self description.

Humans are divided into biologically distinct groups whose characteristics are passed on by inheritance. Differences in ability, taste, aptitude and culture are thus explained by race. The theory of race, which developed in the 19th century, was widely discredited after its employment by the Nazi regime as justification for the mass murder of Jews, gypsies, and others deemed inferior. It has been re-employed in the second half of the 20th century by some conservative and right wing thinkers. *See also* EUGENICS.

David Miller *et al.*, eds, *The Blackwell Encyclopaedia of Political Thought* (Oxford, 1987)

RB

Rademacher's theorem (20th century) *Mathematics* Named after Hans Adolph Rademacher (1892–1969) who discovered it, this shows that if a function $f(x)$ of a variable x such that, for x and y in a suitable subset, the modulus of $f(x) - f(y)$ is at most k times the modulus of $x - y$ for a constant k, then $f(x)$ is differentiable almost everywhere. The theorem generalizes to normed vector spaces, when the modulus is replaced by the norm.

JB

radical empiricism *Philosophy* Name given by American William James (1842–1910) to his own pragmatist philosophy. *See also* NEUTRAL MONISM.

W James, *Essays in Radical Empiricism* (1912)

ARL

radical feminism (20th century) *Politics/Feminism* Theory of the primacy of sexual division and oppression in human society.

The most widely found division of advantage, power, and material well-being is between men and women. This system of oppression, termed PATRIARCHY, is not derived from other systems such as capitalism but is distinct from them. In consequence, women should organize on their own for the overthrow and transformation of the existing order.

Maggie Humm, *The Dictionary of Feminist Theory* (London, 1989)

HR

radical interpretation *Philosophy* A notion similar to that of Wilfred Van Orman Quine (1908–) (INDETERMINACY OF TRANSLATION), thought of primarily in connection with Donald Davidson (1930–) and his truth-conditional theory of meaning (*see* CONVENTION T).

To construct axioms suitable for deriving a theory of meaning for an alien language, we must interpret the utterances of its speakers. It is here that we have to engage in radical interpretation, aided now in choosing (but never conclusively) between the alternatives by the principle of CHARITY, to which was later added that of HUMANITY. Unlike Quine, however, Davidson does not envisage different speakers having radically different conceptual schemes (*see* PERSPECTIVISM). Rather, radical interpretation requires one to choose (upon occasion) between attributing false beliefs to one's interlocutor and taking his utterances to mean something different from what they seem to mean at first sight; neither of these involves attributing to him a different conceptual scheme.

D Davidson, 'Radical Interpretation', *Dialectica* (1973); reprinted in D Davidson, *Inquiries into Truth and Interpretation* (1984)

ARL

radicalism *Politics* Theory that political action must aim at fundamental change.

It is necessary to identify the root (*radix*) of current institutions and practices in order to begin afresh on superior foundations. Although radicalism is normally opposed to conservatism, the policies and theories of the NEW RIGHT appeared to unite both, to the frequent consternation of both enemies and friends.

RB

radioactive dating (20th century) *Physics* Since for any sample of a radioactive nuclide, half the undecayed atoms present decay in a fixed period of time, the ratio of undecayed atoms to decayed atoms may be used to measure the time elapsed; provided that no decayed atoms were present initially.

A useful nuclide for radioactive dating for times up to a few thousand years is ^{14}C, which has a half-life of 5730 years and occurs in the atmosphere with an abundance relative to the common isotope ^{12}C of 1.3×10^{-12}. A living organism exchanges CO_2 with the environment which gives a $^{14}C/^{12}C$ ratio in the organism equal to that in the atmosphere. When the organism dies this exchange ceases and the ^{14}C decays. The supply of ^{14}C in the atmosphere is replenished by the effects of cosmic radiation, and the concentration is thought to have been approximately constant for the past 40,000 years. *See* DECAY LAW FOR RADIOACTIVITY.

J Thewlis, ed., *Encyclopaedic Dictionary of Physics* (New York, Oxford and London, 1962)

MS

Radon-Nikodym theorem (20th century) *Mathematics* Named after the Czech-born Austrian mathematician Johann Radon (1887–1956) and Otton Nikodym.

The theorem that given a positive sigma finite measure μ and a signed sigma finite measure η which is absolutely continuous with respect to μ, there exists a function f such that

$$\int_E f\, d\mu = \eta(E)$$

where E is a measurable set. The function f is called the Radon-Nikodym derivative of η with respect to μ and is denoted by $d\eta/d\mu$.

G B Folland, *Real Analysis* (New York, 1984)

ML

Radon's theorem (*c*.1913) *Mathematics* Named after its discoverer Johann Radon (1887–1956), this states that any set of $n + 2$ points in n-dimensional Euclidean space can be partitioned into two sets so that the convex sets enclosed by level surfaces joining the points in each set do not overlap.

JB

raison d'état *Politics* Justification of overriding STATE power.

There are circumstances when the need to ensure the security or well-being of the state or the nation justifies governments ignoring the normal considerations of law or morality.

David Robertson, *The Penguin Dictionary of Politics* (London, 1986)

RB

Raman effect (1928) *Physics* This effect, which was predicted by Adolf Gustav Stephan Smekal (1895–1959) in 1923 and observed by the American physicist Chandrasekhara Venkata Raman (1888–1970) in 1928, is the phenomenon of light scattering in which the light undergoes a wavelength change in the scattering process, and the scattered light has no phase relationship with the incident light.

Raman observed that when solids and liquids were strongly illuminated with monochromatic light the spectrum of the light scattered contains frequencies not present in the exciting light and which are characteristic of the scattering medium.

J Thewlis, ed., *Encyclopaedic Dictionary of Physics* (New York, Oxford and London, 1962)

MS

Ramsay-Young law (1885) *Chemistry* Named after the Scottish chemist Sir William Ramsay (1852–1916) and Sidney Young (1857–1937) who was noted for his formulation of the boiling point laws. This empirical law states that for two chemically similar compounds having the same vapour pressure at different absolute temperatures, the ratio of those temperatures is independent of this vapour pressure.

GD

Ramsey pricing (1927) *Economics* Named after English economist Frank Ramsey (1903–60), this policy is concerned with prices that maximize the sum of industry consumer surplus and profits. *See* AVERAGE COST PRICING, MARGINAL COST PRICING, COST-PUSH INFLATION.

F Ramsey, 'A Contribution to the Theory of Taxation', *Economic Journal*, 37 (March, 1927), 47–61

PH

Ramsey theory (1928) *Mathematics* A method of counting devised by Frank Plumpton Ramsey (1903–30).

Consider the graphs consisting of edges (lines) joining vertices with at most one edge joining any two vertices and no loops joining a vertex directly to itself. For two positive integers k and m there is a number $R(k,m)$, a 'Ramsey number', such that any such graph with $R(k,m)$ vertices either has a set of k vertices all connected to each other by edges or has a set of m vertices which have no connection with each other.

JB

random hits theory of ageing
See SOMATIC MUTATION THEORY OF AGEING.

random walk hypothesis (1900) *Economics* First identified by French economist Louis Bachelier (1870–1946) from the study of the French commodity markets, this theory asserts that the random nature of commodity or stock prices cannot reveal trends and therefore current prices are no guide to future prices.

The short-term unpredictability of factors means that they appear to walk randomly on a chart, and the best guide to tomorrow's weather (or stock prices) is today's weather. *See* RATIONAL EXPECTATIONS, ADAPTIVE EXPECTATIONS and EFFICIENT MARKET HYPOTHESIS.

PH

range theories of probability (19th century) *Mathematics/Philosophy* Developed by French mathematician Laplace (1749–1827) Certain theories analyzing probability in terms of ranges of alternatives. W C Kneale (1906–) introduces such a theory to deal with paradoxes that face the CLASSICAL THEORY OF PROBABILITY when the relevant range of alternatives is infinite, and his theory consists basically of an extension to the classical theory to cover those cases.

W C Kneale, *Probability and Induction* (1949), §35

ARL

Ranschburg effect *Psychology* Identified by the German psychologist P Ranschburg, this is the generalization that under tachistoscopic viewing conditions (that is, short visual exposure using specialist instrumentation), more stimuli can be recognized if all are different than if some are similar.

R S Woodworth and H Schlosberg, *Experimental Psychology* (London, 1966)

NS

Rao-Blackwell theorem (1947) *Mathematics* The theorem of STATISTICS that given an unbiased estimator T and a sufficient statistic S, the estimator

$$T' = E(T \mid S = s)$$

(the conditional expectation) is unbiased and has smaller variance than T.

R V Hogg and A T Craig, *Introduction to Mathematical Statistics* (New York, 1967)

ML

Raoult's law (1882) *Chemistry* Named after the French chemist François Marie Raoult (1830–1901) who was Professor of Chemistry at Grenoble. The law states that for solutions of non-electrolytes, the elevation of the boiling point, the depression of the freezing point and the lowering of the vapour pressure are proportional to the mole fraction of the solute.

Solutions whose components are chemically very similar, such as benzene and toluene, obey Raoult's law very closely. Strong deviations from Raoult's law are shown by solutions of dissimilar species; for example, carbon disulphide and acetone. *See also* HENRY'S LAW.

P W Atkins, *Physical Chemistry* (Oxford, 1990)

GD

rate law *Chemistry* A mathematical technique for describing the time-dependence of a chemical reaction.

According to the law of MASS ACTION, the rate of a chemical reaction is proportional to the product of the active masses (molar concentration) of the reactants, and is by definition the amount of reactant consumed (or the amount of product formed) in unit time. The equation relating reaction rate and molecular concentration is called a rate law and involves a constant of proportionality known as a rate constant. If the reaction involves several steps, the overall rate is determined by the slowest (or rate-determining) step. A rate law may be derived either empirically or mathematically on the basis of a proposed mechanism.

P W Atkins, *Physical Chemistry* (Oxford, 1990)

GD

rate of living theory (1908) *Biology/Medicine* Proposed by German physiologist Max Rubner (1854–1932), this is the concept that the lifespan of a mammal, and perhaps also its senescence, is related to its rate of physiological living – that is, how 'fast' the mammal lives.

Rubner observed that the total number of calories burned per gram of bodyweight during a mammal's lifespan is roughly the same for all mammal species, despite wide variations in body size and life expectancy. It is also true, for example, that the hearts of a 3.5-year-old mouse and a 70-year-old elephant will have beaten roughly the same number of times – approximately 1.1 ± 10^9 beats for the mouse, and 1.0 ± 10^9 beats for the elephant – by the time of the animals' deaths. *See also* AGEING, THEORIES OF.

J Behnke and C Finch and G Moment, *The Biology of Aging* (New York, 1978)

KH

rate of return pricing Accountancy A variation of full cost pricing, this formulates a pricing policy on a projected rate of return on capital.

It includes the uses of a mark-up on costs to achieve a pre-set return on capital investment, and is particularly useful when dealing with a strong customer such as a state-run body. *See* MARK-UP PRICING.

AV, PH

ratio test (19th century) *Mathematics* Attributed to the French mathematician and physicist Baron Augustin Louis Cauchy (1789–1857), this is a test for convergence of an infinite series Σu_k with positive terms. The limit of the quotient of successive terms

$$\lim_{k \to \infty} \frac{u_{k+1}}{u_k} = \rho$$

is evaluated and if $\rho < 1$ the series converges, if $\rho > 1$ the series diverges and if $\rho = 1$ the test gives no conclusion. For example,

$$\sum_{k=1}^{\infty} \frac{1}{k!}$$

converges since

$$\rho = \lim_{k\to\infty} \frac{1/(k+1)!}{1/k!} = \lim_{k\to\infty} \frac{1}{k+1} = 0 < 1.$$

H Anton, *Calculus with Analytic Geometry* (New York, 1980)

ML

ratio theorem *Mathematics* The theorem of GEOMETRY stating that if a directed line segment

$$\overline{AB}$$

is divided by a point p in the ratio $m{:}n$, then the position vector $\mathbf{p}$, of p can be written as

$$\mathbf{p} = \frac{m\mathbf{a} + n\mathbf{b}}{m+n}.$$

W Gellert, S Gottwald, M Hellwich, H Kästner and H Küstner, eds, *The VNR Concise Encyclopedia of Mathematics* (New York, 1975)

ML

rational choice (20th century) *Politics* Theory underlying PUBLIC CHOICE analysis.

The behaviour of individuals, groups and institutions can be understood as a series of rational choices designed, in an analogy with economics, to maximize their utility.

Patrick Dunleavy, *Democracy, Bureaucracy and Public Choice* (London, 1991)

RB

rational expectations theory (1960) *Economics* Formulated by American economist John Muth (1930–), this states that individuals and companies, acting with complete access to the relevant information, forecast events in the future without bias.

Errors in their forecasts are assumed to result from random events. This theory has emerged as an important aspect of new classical economics. *See* ADAPTIVE EXPECTATIONS, RANDOM WALK HYPOTHESIS and NEW CLASSICAL MACROECONOMICS.

PH

rational indices, law of (1784) *Geology/Physics* Formulated by the French mineralogist René Just Haüy (1743–1822), this law states that all planes which can occur as the faces of a crystal have intercepts on the chosen crystal axes which (when expressed as multiples of certain unit lengths along those axes) have ratios that are rational numbers.

J Thewlis, ed., *Encyclopaedic Dictionary of Physics* (New York, Oxford and London, 1962)

MS

rational intercepts, law of *Chemistry* A law proposed by the German chemist Christian Samuel Weiss (1780–1856) and also known as the law of rational indices. The law states that the intercepts which the planes of a crystal make on the axes of the crystal lattice are in a simple ratio to one another.

These have been replaced by easier-to-handle indices devised by the Welsh-born chemist William Hallowes Miller (1801–80), and these are named after him.

P W Atkins, *Physical Chemistry* (Oxford, 1990)

GD

rational ratios of intercepts (indices), law of (1669) *Geology* Although first established by the Danish naturalist Nicholas Steno (1638–86), this theory is also known as Hauy's law, after the French mineralogist, R J Hauy (1743–1822). It states that the faces of a crystal cut its crystallographic axes in whole number values; 1, 2, 3, and so on.

DT

rational root theorem *Mathematics* The result whereby if the rational number p/q, with p and q coprime, is the root of a polynomial having integer coefficients, then p divides the coefficient of the constant term and q divides the coefficient of the leading term. Hence, by considering all the possible divisors of the leading term and the constant term, it is possible to determine all the rational roots of a polynomial with integer coefficients. *See also* FACTOR THEOREM.

MB

rationalism *Philosophy* Any theory emphasizing reason or intuition (usually in contrast to the senses, or, in ethics, to feelings and emotions), whether as the basis for acquiring knowledge, or as the basis for justifying moral judgments.

In these uses it contrasts with EMPIRICISM, and has similar varieties. The *a priori* is to rationalism what the *a posteriori* (or empirical) is to empiricism. In ethics, rationalism tends to be classed with INTUITIONISM (though different from it) in opposition to MORAL SENSE THEORIES. The late 19th and

20th centuries' use of 'rationalism' to contrast with belief in religious revelation is not common in philosophy. *See also* CONTINENTAL RATIONALISTS.

W von Leyden, *Seventeenth Century Metaphysics* (1968), especially ch. 3

ARL

rationalism in linguistics (17th century; 1960s) *Linguistics* Version of MENTALISM IN LINGUISTICS argued by American linguist Avram Noam Chomsky (1928–).

Drawing on the theory of innate knowledge propounded by the rationalist philosophers, especially René Descartes (1596–1650), and the linguistics of the Port-Royal grammar (1660), Chomsky opposed the BEHAVIOURISM and EMPIRICISM of modern linguistics. *See also* INNATENESS HYPOTHESIS and INTUITION.

N Chomsky, *Cartesian Linguistics* (New York, 1966)

RF

rationing (20th century) *Economics* Rationing is a deliberate attempt, frequently undertaken by governments, to allocate scarce supplies in the face of high demand. Severe rationing during the 1940s, 1950s and 1970s prompted American and European economists to study this subject and its consequences for product substitution.

Rationing does not exist in a free market because the excess demand would be countered by a rise in prices.

J Tobin, 'A Survey of the Theory of Rationing', *Econometrica*, vol. xx (1952) 512–53

PH

Rawls' theory of justice (1972) *Economics* Named after the American philosopher John Rawls (1921–), this theory sees justice as fairness, and its intuitive idea is that the well-being of society depends on co-operation.

It is based on the traditional theories of social contract as represented by English philosopher John Locke (1632–1704), Swiss philosopher Jean-Jacques Rousseau (1712–78) and the German philosopher Immanuel Kant (1724–1804). *See* ENTITLEMENT THEORY.

J Rawls, *A Theory of Justice* (Oxford, 1972)

PH

Rayleigh-Jeans radiation law (1900) *Physics* John William Strutt, 3rd Baron Rayleigh (1842–1919) applied classical physics to the problem of the energy radiated by a black body or the radiation within an evacuated cavity. He showed that, for a cavity with walls at a temperature T, the energy per unit volume in the wavelength range λ to $\lambda + d\lambda$ is

$$\frac{8\pi kT}{\lambda^4}\,d\lambda$$

where k is Boltzmann's constant. In his original derivation, Rayleigh missed a factor of 2 which was inserted by Sir James Hopwood Jeans (1877–1944) five years later. The equation holds for long wavelengths but not for short ones, and Rayleigh introduced an *ad hoc* cut-off factor

$$\exp\frac{-c'}{\lambda T}.$$

With this modification the expression is known as Rayleigh's law.

J Thewlis, ed., *Encyclopaedic Dictionary of Physics* (New York, Oxford and London, 1962)

MS

Rayleigh's scattering law (1871) *Physics* Named after its discoverer, the English physicist John William Strutt, 3rd Baron Rayleigh (1842–1919), this law states that when light is scattered from particles that have a refractive index different from that of the surrounding medium, and have linear dimensions considerably smaller than the wavelength of the incident light, the scattered intensity is proportional to the incident intensity, to the square of the volume of the scattering particle and inversely proportional to the fourth power of the wavelength of the light.

J Thewlis, ed., *Encyclopaedic Dictionary of Physics* (New York, Oxford and London, 1962)

MS

rayonism (1913) *Art* Theory devised by the Russian painter Mikhail Larionov (1881–1964) and his wife Natalia Goncharova (1881–1962). Larionov's manifesto (1913) stated that 'Rayonism is a synthesis of Cubism, Futurism amd Orphism', and much of his thesis was closely linked to Futurist ideas. Colours are dispersed as light rays emanating from objects; demonstrated by parallel lines or intersecting beams of light which are manipulated by the artist to create form. The movement was short-lived. *See* CUBISM, FUTURISM and ORPHISM (2).

H Osborne, *Abstraction and Artifice in Twentieth-Century Art* (1979)

AB

reactance theory (20th century) *Psychology* Proposed by the American psychologist Jack Williams Brehm.

When a person perceives that a specific behavioural freedom is threatened, a motivational state called psychological reactance is aroused. The theory proposes two important hypotheses: (1) the greater the attempt to limit a person's freedom to engage in a specific behaviour, the more attractive that behaviour becomes; (2) if a person perceives that attempts are being made to force a particular attitude upon him or her, that person will tend to adopt the opposite attitude. Considerable empirical support exists for reactance theory.

S S Brehm and J W Brehm, *Psychological Reactance: A Theory of Freedom and Control* (New York, 1966)

NS

reader response criticism (1970s) *Literary Theory* A common feature of criticism and theory since the 1970s has been the growth in importance of the reader or audience. Reader response criticism has been variously realized, but the work of N Holland – *The Dynamics of Literary Response* (1968) – is exemplary. *See also* POSTSTRUCTURALISM and RECEPTION THEORY.

S Suleiman and I Crosman, eds *The Reader in the Text* (Princeton, 1980)

RF

readerly (text) *Stylistics See* 'LISIBLE' AND 'SCRIPTIBLE'.

readiness, law of *Psychology* Proposed by the American psychologist Edward Lee Thorndike (1874–1949).

One of the necessary additions to Thorndike's LAW OF EFFECT, introduced to account for why learning occurs or fails to occur, the law of readiness was never clearly defined. Basically, it states that in order for an organism to learn it needs to attend to specific stimuli of consequence in a situation. This law is now considered to be of only historical interest.

G H Bower and E R Hilgard, *Theories of Learning* (Englewood Cliffs, N.J., 1981)

NS

Reaganomics *Economics* Named after ex-actor and former American president Ronald Reagan (1911–), who was an advocate of SUPPLY-SIDE ECONOMICS. Reagan stressed the need to reduce taxes, deregulate the economy and modernize US defence as part of his policy.

The monetarist economist Milton Friedman (1912–92) acted as his policy adviser (1981–90), and Reagan followed a domestic policy of tax reduction and deficit financing. This resulted in economic growth in the economy (1983–86), but by the end of Reagan's administration, the US was suffering from high budget and trade deficits. *See* MONETARISM and THATCHERISM.

M Boskin, *Reagan and the Economy* (San Francisco, 1989)

PH

real bills doctrine (18th century) *Economics* Developed by Scottish economist Adam Smith (1723–90), this asserts that there can never be an inflationary excess issue of commercial bills and other paper money because each bill represents a real transaction. The doctrine was later criticized for failing to recognize that the same sum of money can support many bills.

A Smith, *An Inquiry into the Nature and Causes of the Wealth of Nations* (London, 1776)

PH

real self *Politics/Philosophy* Theory of people's wants and wills found in IDEALISM.

People have a real self or 'real will' which is what they would want if they reflected in a fully rational way on their INTERESTS. It will frequently differ from their expressed will, or what they say and believe that they want. Some other person or institution, such as government, may therefore know what people's 'will' is better than they do themselves.

Andrew Vincent and Raymond Plant, *Philosophy, Politics and Citizenship* (Oxford, 1984)

RB

realism (*c.*1850) *Literary Theory/Art* Realism in the NOVEL and in painting was identified and defended in mid-19th century France and applies paradigmatically to the practices of the novelists Stendhal (1783–1842), Honoré de Balzac (1799–1850) and Gustave Flaubert (1821–80); by extension, to much 19th-century English fiction.

It is a mode of fictional representation which gives an illusion of a world experienced as a reader might experience life. Techniques vary, but they include a high level of specification of material details, concentration on mundane incidents and problems, focus on a single life, straightforwardness of temporal POINT OF VIEW, consistency of FOCALIZATION, and backgrounding of NARRATOR ('impersonality'). *See also* SOCIAL NOVEL, NATURALISM and 'VERISMO'.

Gustave Courbet (1819–77) epitomizes the Realist painter; his paintings, produced in naturalistic and verist style, depicting everyday life rather than grand historical subjects.

R Wellek, 'The Concept of Realism in Literary Scholarship' [1960], *Concepts of Criticism*, S G Nicholls, ed. (New Haven, 1962), 222–55; G J Becker, ed., *Documents of Modern Literary Realism* (Princeton, 1963)

AB, RF

realism *Philosophy/Psychology* Often associated with the work of Scottish philosopher Thomas Reid (1710–96), and German philosopher Immanuel Kant (1724–1804). Usually used in either of two ways: (1) the view that abstract concepts have a real existence and can be studied empirically; (2) the doctrine that the physical world has a reality separate from that of the mind.

Over the reality of universals (*see* PLATONISM) and other abstract objects, realism contrasts mainly with NOMINALISM and CONCEPTUALISM (*see also* RESEMBLANCE THEORIES OF UNIVERSALS). In dealing with the reality and status of things around us, it contrasts with IDEALISM and PHENOMENALISM. It contrasts with ANTIREALISM on the possibility of truths independent of our powers of verifying them or manifesting knowledge of them. All this suggests that realism (like 'real') is mainly defined by contrast. As with many philosophical terms, 'realist' can apply to some features of a view to other features of which some contrasting term applies. *Compare* CAUSAL REALISM, CRITICAL REALISM, MODAL REALISM, NAIVE REALISM and PERSPECTIVE REALISM.

D N Robinson, *An Intellectual History of Psychology* (London, 1976)

ARL, NS

realism *Politics* View of international relations as the pursuit of INTERESTS.

States pursue their own security and prosperity in international relations, whatever their apparent aims, alliances or moral claims.

Graham Evans and Jeffrey Newnham, *The Dictionary of World Politics* (Hemel Hempstead, 1990)

RB

realism/intuitionism (20th century) *Mathematics* Realism is the mathematical philosophy asserting that mathematical statements have a reality independent of their proposer which determines whether the statement is true or false; intuitionism asserts that a mathematical statement can be called true only if it has been proved in a finite number of steps.

JB

realistic grammar (1960s) Controversial assumption made in earlier versions of TRANSFORMATIONAL GENERATIVE GRAMMAR and LEXICAL-FUNCTIONAL GRAMMAR.

The form of a grammar should be 'psychologically' realistic; that is, the rules should reflect the mental organization of language and the psychological processes involved in its use. Most proposals of this kind (for example, the DERIVATIONAL THEORY OF COMPLEXITY) have been seriously challenged in PSYCHOLINGUISTICS.

M Halle, J Bresnan and G A Miller, eds, *Linguistic Theory and Psychological Reality* (Cambridge, Mass., 1978)

RF

reality principle *Psychology* Proposed by the Austrian founder of psychoanalysis Sigmund Freud (1856–1939).

In PSYCHOANALYTICAL THEORY the ego is said to function according to the reality principle because it mediates between the unconscious demands of the id and the realities of the social environment. The validity of the reality principle is difficult to establish because of its metaphysical nature.

C S Hall, *A Primer of Freudian Psychology* (Cleveland, 1954)

NS

rebound effect (20th century) *Psychology* Identified by the American psychologist W Dement.

A period of deprivation or inhibition will be followed by an increase in that physiological function. For instance, one or more nights of REM (rapid eye movement) sleep deprivation will be followed by an increase in REM on subsequent nights.

E Hartmann, *The Functions of Sleep* (New Haven and London, 1973)

NS

recapitulation (19th century) *Biology* Also known as the biogenetic law; and as Haeckel's law of recapitulation (after its proponent Ernst Heinrich Haeckel (1834–1919), a German biologist). This term refers to the re-enactment of evolutionary stages during the embryonic development of an organism; an idea often stated as 'ontogeny recapitulates phylogeny'.

Although the term is still used informally, recapitulation as originally proposed has been rejected by modern biologists. Any resemblance embryos have to their evolutionary ancestors is now viewed as adaptation, not as evidence of ancestry. *Compare with* GERM LAYER THEORY.

S Gould, *Ontogeny and Phylogeny* (Cambridge, Mass., 1977)

KH

recapitulation theory *Psychology* Proposed by the American psychologist Granville Stanley Hall (1844–1924).

The theory is applied in two ways: (1) to biological and physiological factors (see above); (2) to cognitve and perceptual development. Although if applied literally the theory is not of much value, the concept has sometimes been found to be a useful conceptualization tool for theory development.

G S Hall, *The Content of Children's Minds* (1883)

NS

recency, law of *Psychology* Formulated by the German psychologist Hermann Ebbinghaus (1850–1909).

A component of the serial position effect, the law of recency (also referred to as recency effect and the principle of recency) states that in a free recall situation, items presented at the end are more likely to be recalled than those presented near the centre. For example, if a list of words is read out at one-second intervals to an individual, he or she will recall words towards the end of the list with greater ease than words presented near the centre.

E G Boring, *A History of Experimental Psychology* (New York, 1957)

NS

reception theory (late 1960s) *Literary Theory* Developed by the German theoreticians Hans Robert Jauss (1921–) – who called it *Rezeptionsästhetik* – and Wolfgang Iser (1926–); it was strongly influenced by the phenomenononology of Roman Ingarden (1893–1970).

The fictional text does not have a determinate meaning; it presents the reader with INDETERMINACY and the need to construct meaning by the application of schemata. The reader's reception of the text in effect produces it. *See also* IMPLIED READER.

W Iser, *The Act of Reading* (Baltimore, 1978 [1976])

RF

reciprocal altruism (1971) *Biology* Proposed by Robert Trivers, this theory suggests that individuals engage in self-sacrificing acts because of an implied social contract. The contract is such that the individual enjoying a benefit from another now will likewise engage in altruism to benefit the other later.

Examples include the warning cries of birds and primates, symbiotic grooming behaviour, and jumping into the water to rescue a drowning person. *See also* ALTRUISM, INCLUSIVE FITNESS, GROUP SELECTION and KIN-SELECTION THEORY.

D Dewsbury, *Comparative Animal Behavior* (New York, 1978)

KH

reciprocal proportions, law of (1792) *Chemistry* Law attributed to German chemist Jeremias Benjamin Richter (1762–1807). Also known as the law of EQUIVALENT PROPORTIONS.

The law states that when two elements both form chemical compounds with a third, a compound of the first two contains them in the relative proportions they have in compounds with the third one or in simple multiples of them. For example, nitrogen and hydrogen combine to form ammonia, NH_3, in the ratio (of weights) 14:3. Oxygen and hydrogen combine to form water, H_2O, in the ratio 16:2 (24:3). Nitrogen and oxygen combine to form N_2O_3, in the ratio 28:48 (14:24) and other oxides which are multiples of this ratio.

M Freemantle, *Chemistry in Action* (Basingstoke, 1987)

GD

reciprocal theorem *Physics* A theorem found in several branches of physics. For example, in studies of elasticity, it appears as Betti's theorem; being named after Enrico Betti (1823–92) who proposed it in 1872. *See* BETTI'S RECIPROCAL THEOREM.

reciprocity, law of (1862) *Physics* Discovered by the German physicist Robert Wilhelm Eberhard Bunsen (1811–99) and the British chemist Henry Enfield Roscoe (1833–1915), this law states that when all other conditions are kept constant, the exposure time needed to give a certain photographic density is inversely proportional to the intensity of the radiation. The law does not hold at high and low light intensities, when reciprocity failure is said to occur.

J Thewlis, ed., *Encyclopaedic Dictionary of Physics* (New York, Oxford and London, 1962)

MS

reciprocity principle *Psychology* Proposed by the Austrian psychologist F Heider (1896–).

According to this principle, social attraction is mutual; that is, if one individual knows that another person likes him or her, then he/she will tend to like the other person. Heider set out his theory in *The Psychology of Interpersonal Relations*, which book has had a profound effect on our understanding of interpersonal perception. *See* BALANCE THEORY.

F Heider, *The Psychology of Interpersonal Perception* (New York, 1958)

NS

recognition concept of species (1980s) *Biology* Proposed by Australian entomologist Hugh Paterson, this is a model for SPECIES and SPECIATION which holds that recognizing appropriate mates is essential to the development of a new species, and that a species constitutes what Paterson calls 'the most inclusive population of individual biparental organisms which share a common fertilization system'.

This model is one of many proposed refinements to Mayr's theory of speciation, positing that an isolated population must develop mechanisms for recognizing mates that are similarly adapted, or the potential species will fail to develop into a new species because of dilution of the gene pool by other species.

M Ereshefsky, *The Units of Evolution: Essays on the Nature of Species* (Cambridge, Mass., 1992)

KH

rectangular distribution (1972) *Biology* Part of the punctuated equilibrium theory by the American biologists N Eldredge and S J Gould (1941–).

The absence of many evolutionary links between species in the fossil record can be explained if speciation occurs very rapidly after long periods of little evolutionary change. This view is antagonistic to phyletic gradualism which proposes that evolution occurs continuously in slow, minute steps. Evidence suggests that both processes in fact occur.

N Eldredge and S J Gould, 'Punctuated Equilibrium: An Alternative to Phyletic Gradualism', *Models in Palaeobiology*, T T M Schopf, ed. (San Francisco, 1972)

RA

recursion (1957) *Linguistics* A property of language identified by the American linguist Avram Noam Chomsky (1928–).

An infinite number of sentences are possible in a natural language. Chomsky relates this 'creativity' to the fact that there is 'no longest sentence': recursion permits this. Some syntactic rules can apply over and over again to their own output, permitting the generation of an infinite number of sentences. For example, the pair of rules

$$S \rightarrow NP\ VP$$
$$VP \rightarrow V\ (NP)\ (S)$$

allow the infinite EMBEDDING of sentences within one another and thus the infinite extension of any one sentence.

A Radford, *Transformational Grammar* (Cambridge, 1988)

RF

recursion principle (13th century) *Mathematics* First used by Leonardo Fibonacci (*c.*1170–1250), this is the application of a function to its own values to generate an infinity of values. *See also* RECURSIVE FUNCTION THEORY and FIBONACCI NUMBERS.

JB

recursive function theory (1937) *Mathematics* A theory initiated by Alan Mathison Turing (1912–54) concerning computability based on the use of functions defined by the recursion principle. *See also* COMPUTABILITY THEORY, RECURSION PRINCIPLE and TURING MACHINE.

JB

Red Queen hypothesis (1974) *Biology* (Also called the law of constant extinction.) Named after the Red Queen in Lewis Carroll's *Through the Looking Glass* (1872), who said: 'Now here, you see, it takes all the running you can do to keep in the same place'.

This is the idea that an evolutionary advance by one species represents a deterioration of the environment for all remaining species, placing selective pressure on those species to advance just to keep up. Van Valen's law – named after Leigh Van Valen – is a mathematical model based on his study of extinction rates in various lineages.

L Van Valen, 'A New Evolutionary Law', *Evolutionary Theory*, vol. I (1974), 1–30

KH

reduced cue *Psychology* A concept advanced in the work of American psychologists Burrhus Frederic Skinner (1904–90) and E R Guthrie (1886–1959).

Referring to any portion of a reduced stimulus (cue), the term is used in the learning principle which holds that after repeated stimulus-response trials the reponse can be elicited by a portion of the original stimulus or cue.

B F Skinner, *The Behaviour of Organisms* (New York, 1938)

NS

reducibility, axiom of *Philosophy* Axiom introduced by English philosopher and mathematician Bertrand Russell (1872–1970) in connection with the ramified theory of TYPES. It says that any higher-order property or proposition can be reduced to an equivalent first-order one.

The ramified theory caused difficulties for defining real numbers (using Dedekind sections) and for the process known as mathematical induction (roughly: if a property belongs to the first term in a series, and to the successor of any term to which it belongs, then it belongs to them all). Russell introduced the axiom to deal with these problems, but it was widely felt to be unfounded, and was later dispensed with by F P Ramsey (1903–30) in Chapter 1 of his *Foundations of Mathematics* (1931).

B Russell, 'Mathematical Logic as Based on the Theory of Types', *American Mathematical Monthly* (1908); reprinted in R C Marsh, ed., *Logic and Knowledge* (1956) and in J van Heijenoort, ed., *From Frege to Gödel* (1967)

ARL

reductionism (19th century) *Biology/Philosophy* Also called mechanism, or mechanistic philosophy. Associated with Carl Ludwig (1816–95), Hermann von Helmholtz (1821–94), Ernst von Brücke (1819–92) and Emil du Bois-Reymond (1818–96).

The theory that life can be understood entirely in terms of the laws of physics and chemistry. Modern bioscience approaches biology from this perspective. *Compare with* VITALISM.

KH

reductionism *Philosophy* Also called reductivism. The reducing of certain kinds of entities, or of theories, or even of whole sciences, to other, more basic, ones; entities that are reduced may be replaced ('Father Christmas is really Daddy') or simply explained ('Water is really H_2O').

PHENOMENALISM, for instance, reduces material objects, or sentences about them, to experiences, or sets of sentences about these. Similarly, one version of PHYSICALISM claims to reduce the other sciences to physics by showing that all their concepts and theorems can be expressed in terms of physics without loss of information. Contrasting approaches include HOLISM and EMERGENCE THEORIES, though intermediate positions can be held. Reductionism in general appeals to empiricists, nominalists, and others who use OCKHAM'S RAZOR to achieve a sparse ONTOLOGY (or list of what there is).

E Agazzi, ed., *The Problem of Reductionism in Science* (1991)

ARL

redundancy rule (1960s) *Linguistics* Procedure developed in GENERATIVE GRAMMAR.

In COMPONENTIAL ANALYSIS (a form of LEXICAL SEMANTICS) and in the DISTINCTIVE FEATURE approach to PHONOLOGY, it is redundant to specify some features. For example, all nasal sounds are 'voiced' (the vocal cords vibrate); all lexical items containing the feature HUMAN (for example *girl*, *teacher*) also contain the feature ANIMATE. A redundancy rule states these generalizations, thus simplifying the phonological or semantic description of the item concerned.

N Chomsky, *Aspects of the Theory of Syntax* (Cambridge, Mass., 1965)

RF

redundancy theory of truth *Philosophy* Also called the no-truth theory. Influenced by the difficulties in formulating a CORRESPONDENCE THEORY OF TRUTH, Frank Plumpton Ramsey (1903–30) proposed in 1927 that to call a proposition true is to do no more than assert the proposition. One objection is that this seems too thin a theory to cover all our uses of the notion of truth.

A J Ayer, *The Concept of a Person and Other Essays* (1963), ch. 6

ARL

reduplication hypothesis (early 1900s) *Biology* Proposed by English geneticist and embryologist William Bateson (1861–1926), author of *Mendel's Principles of Heredity* (1909).

This is the mistaken idea that SEGREGATION OF GENES does not occur at the time of meiosis – the special cell division occurring in gametes just before they mature into eggs or sperm, which reduces the number of chromosomes to the haploid number – but prior to meiosis, and not necessarily at the same time for each pair of genes. Apparently Bateson developed this theory because of his unwillingness to accept that segregation occurs during meiotic division.

A Sturtevant, *A History of Genetics* (New York, 1965)

KH

re-engagement theory of ageing *Psychology* *See* ACTIVITY THEORY OF AGEING.

reflection of light, laws of (2nd century BC–) *Physics* Understood at least as early as the 2nd century BC, these laws state that: (1) the incident ray, reflected ray and normal to the reflecting surface all lie in one plane; (2) the angle of incidence (the angle between the incident ray and the normal) is equal to the angle of reflection (the angle between the reflected ray and the normal).

J Thewlis, ed., *Encyclopaedic Dictionary of Physics* (New York, Oxford and London, 1962)

MS

reflection principle of Schwartz *Mathematics* Named after German mathematician Hermann Amandus Schwartz (1843–1921).

This is the result in COMPLEX FUNCTION THEORY whereby, in its simplest form, an analytic function f which is defined on a region A in the upper half plane containing an interval $[a,b]$ of the real axis and which is continuous on $[a,b]$, has an analytic continuation to the reflection of A in the real axis (that is, to the set of complex numbers z such that $\bar{z} \in A$) which can be given in the natural way, namely

$$f(\overline{z}) = \overline{f(z)}$$

See also IDENTITY THEOREM.

S G Krantz, 'Functions of One Complex Variable', *Encylopedia of Physical Science and Technology* (Academic Press, 1987)

MB

reflux theory *Geology* When dense, saline water passes through a sediment, it can result in the alteration of pre-existing minerals and the deposition of others. The mechanism behind this is not yet fully understood, but is thought to be mainly driven by evaporation causing the formation of denser, more saline waters which then percolate downwards into the underlying sediments.

DT

refraction of light, laws of *Physics* The second law was discovered in 1621 by the Dutch mathematician Willebrod van Roijen Snell (1580–1626) and is known as Snell's law. (In France it is known as Descartes's law.)

(1) The incident ray, refracted ray and normal to the surface at which refraction occurs all lie in one plane. (2) The angle of incidence i (the angle between the incident ray and the normal) and the angle of refraction r (the angle between the refracted ray and the normal) are related by the equation

$$\frac{\sin i}{\sin r} = \text{constant}$$

for the two media concerned.

J Thewlis, ed., *Encyclopaedic Dictionary of Physics* (New York, Oxford and London, 1962)

MS

regionalism (1930s) *Art* An artistic movement that emerged during the 1930s and 1940s from the American 'Ash Can School' of the early decades of the century. Its subject matter was drawn from American urban provincial life.

Foremost among its exponents were Thomas Hart Benton (1889–1975), John Sloan (1871–1951) and Edward Hopper (1882–67), whose occasionally xenophobic works also highlighted the loneliness and isolation of modern urban life.

AB

register (1964) *Linguistics* Originally formulated by the British linguist Michael Alexander Kirkwood Halliday (1925–).

A variety of language according to use. Different styles of language are appropriate in different settings, determined by FIELD (subject and activity), MODE (medium) and TENOR (relationship between participants: registers of advertising, law, classroom, and so on). A later version (1978) proposes that register determines 'semantic potential' – what one can say in a situation. *See also* FUNCTIONAL GRAMMAR.

M A K Halliday, *Language as Social Semiotic* (London, 1978)

RF

regularity theory of causation *Philosophy* Theory – primarily associated with, and originated by, David Hume (1711–76) – which analyzes causation in terms of nothing but regular sequence (together, in Hume's case, with priority in time and contiguity in time and, where relevant, space). The basic form of the theory says that one event causes another if it is followed by it and is such that events of the first kind are regularly followed by events of the second kind.

The point of the theory is to dispense with any mysterious causal necessity, and also to distinguish causal connections from logical connections: one event cannot logically necessitate another (Hume was the first to clarify this distinction). The theory is therefore reductionist in nature, and versions of it appeal to empiricists and positivists. Recently, however, the extent to which Hume himself adhered to the theory has come into dispute. *See also* CAUSAL REALISM.

J L Mackie, *The Cement of the Universe* (1974); sophisticated modern version reckoning to stay within the frame of the theory

ARL

regulation (19th century–) *Economics* This term describes government intervention in the price, sale and production decisions of a firm.

Regulation is often a response to chaotic growth, abuses of monopoly powers and price-fixing, and is seen as a method of consumer protection. *See* 'LAISSEZ-FAIRE', PHYSIOCRACY, MERCANTILISM, ECONOMIC LIBERALISM and NEW CLASSICAL MACROECONOMICS.

M A Utton, *The Economics of Regulating Industry* (Oxford, 1986)

PH

regulatory capture *Economics* An organization's evasion of control by a regulatory body, often an anti-trust or takeover body.

Organizations will attempt to dilute the effectiveness of regulatory bodies through political control, and by developing superior information and more effective staff.

PH

reinforcement *Biology See* EFFECT, LAW OF.

reinforcement *Psychology* A term used importantly in the work of the Russian physiologist Ivan Petrovich Pavlov (1849–1936) and the American psychologist Clark Leonard Hull (1884–1952).

The term 'reinforcement' is used in many different ways and consequently has a variety of meanings: (1) the procedure by which an event (which either occurs naturally in the environment or is experimentally arranged) serves to strengthen a performance or response; (2) a process whereby, when a response is followed by a reinforcement, something takes place in a person's central nervous system to make learning occur; (3) a term used in classical conditioning whereby the unconditioned stimulus (UCS) functions as a reinforcer of the conditioned stimulus (CS); (4) any circumstance found pleasurable; (5) any act or event which reduces a drive; (6) a behaviour which has a momentary higher probability of occurrence than other behaviour(s) or response(s) to the same stimulus; (7) feedback about the correctness of one's behaviour.

B F Skinner, *The Behaviour of Organisms* (New York, 1938)

NS

reinforcement retroactive paradox *Psychology* Identified by the American psychologist Burrhus Frederic Skinner (1904–90).

This is the problem of how reinforcement can strengthen a response tendency after the response has ceased. Unlike learning a new response or behaviour, the retention (or remembering) of material already learned can be affected by subsequent learning as well as prior learning. Such effects are called retroaction or proaction. *See* INTERFERENCE THEORY OF FORGETTING.

S H Hulse, H Egeth and J Deese, *The Psychology of Learning* (New York, 1980)

NS

relative autonomy *Politics* Account of modern STATE found within recent MARXISM employed by the Greek theorist Nicos Poulantzas (1936–79).

States work within the limits set by the socio-economic structures within which they operate and which they function to sustain. Their autonomy is thus relative to this overall constraint.

Tom Bottomore, *A Dictionary of Marxist Thought*, 2nd edn (Oxford, 1991)

RB

relative deprivation (20th century) *Politics/Sociology* A theory of the causes of social and political discontent.

People are roused to political action as a result not of absolute changes in their material conditions but of changes relative to the circumstances of those with whom they compare themselves.

W G Runciman, *Relative Deprivation and Social Justice* (London, 1966)

RB

relative income hypothesis (1949) *Economics* Proposed by James Semble Duesenberry (1918–) but subsequently overtaken by other studies on the behaviour of saving and consumption, this theory states that an individual's attitude to consumption and saving is guided more by his income in relation to others than by an abstract standard of living.

'Keeping up with the Joneses' may be a more powerful incentive than the pursuit of wealth for its own sake. *See* PERMANENT INCOME HYPOTHESIS, ABSOLUTE INCOME HYPOTHESIS, RELATIVE INCOME HYPOTHESIS and LIFE-CYCLE HYPOTHESIS.

J S Duesenberry, *Income, Saving and the Theory of Consumer Behavior* (Cambridge, Mass., 1949)

PH

relativism *Philosophy* Strictly, any doctrine that something exists, has a property, or obtains, relative to something else.

Two forms of relativism have been common, cognitive and moral; both of them are different from SUBJECTIVISM, though some versions are *also* subjectivist. *Cognitive relativism* may say that all beliefs are true, or true for their holders (the view Plato attributes to Protagoras in his dialogue *Theaetetus*); or it may take the form of

PERSPECTIVISM (or *cultural relativism*), perhaps limited to the advanced sciences, where straightforward reputation may be rare. As cognitive relativism relativizes truth, so *moral relativism* relativizes rightness or moral values, but must be distinguished from merely saying that what is (absolutely) right for one to do depends on one's role or the circumstances one is in. A thoroughgoing relativist must also avoid concluding that it is (absolutely) right to live and let live. (For linguistic relativism, *see* SAPIR-WHORF HYPOTHESIS.

M Krausz and J W Meiland, eds, *Relativism: Cognitive and Moral* (1982)

ARL

relativity, theory of general (1915) *Physics* A theory proposed by the German-Swiss-American mathematical physicist Albert Einstein (1879–1955).

The theory is a generalized form of the special theory of relativity dealing with relative motion between accelerated frames of reference, and leads to the field theory of gravity in particular. In non-inertial (accelerated) systems, certain fictitious forces make their appearance which also have a connection with the forces due to gravity, where the acceleration produced is independent of the mass. The principle of equivalence says that the inertial (non-accelerated) mass is the same as the gravitational (accelerated) mass. A further principle used in the general theory is that the laws of mechanics are the same in inertial and non-inertial frames of reference.

Paul Davies, ed., *The New Physics* (Cambridge, 1989)

GD

relativity, theory of special (1905) *Physics* A theory proposed by the German-Swiss-American mathematical physicist Albert Einstein (1879–1955).

The theory gives a unified account of the laws of mechanics and electromagnetism. Einstein rejected the Newtonian concepts of absolute space and time and the 19th century idea that an electromagnetic aether existed with respect to which motion could be determined absolutely. He made two fundamental postulates: (1) the laws of nature are the same in all inertial (non-accelerated) frames of reference, and (2) the velocity of light is the same in all such frames. These postulates predict that a body moving relative to a stationary observer will appear to increase in mass, and also to contract in length in the direction of motion by an amount (the Lorentz contraction) that becomes appreciable as the velocity approaches that of light. The theory also leads to the concept of the equivalence of mass and energy. *See also* FITZGERALD-LORENTZ CONTRACTION and RELATIVITY, THEORY OF GENERAL.

Paul Davies, ed, *The New Physics* (Cambridge, 1989)

GD

relaxation principle *Psychology* Proposed by the Hungarian Sándor Ferenczi (1873–1933). This is a modification of psychoanalysis based on the principle that a loving and tender approach to clients allows for more release of repressed feelings than does strict Freudian analysis. Ferenczi also believed in mutual analysis between the therapist and the client, a practice which caused him to be shunned by psychoanalytic therapists.

A Grunbaum, *The Foundations of Psychoanalysis: A Philosophical Critique* (Berkeley, California, 1984)

NS

relevance (1980s) *Linguistics* PRAGMATIC theory developed by Dan Sperber and Deirdre Wilson.

Humans communicate by INFERENCING procedures which draw on MUTUAL KNOWLEDGE, and achieve economy of effort by focusing on the most relevant information. *See also* CO-OPERATIVE PRINCIPLE OF CONVERSATION.

D Sperber and D Wilson, *Relevance* (Oxford, 1986)

RF

relevance logics *Philosophy* Logical systems based on the principle that logical consequence, or entailment, only holds between propositions which are relevant to each other.

They were developed, notably by A R Anderson and Nuel D Belnap (1920–), as a reaction to the claim of Clarence Irving Lewis and C H Langford (in *Symbolic Logic*

(1932), chapter 8) that logical entailment is the same as *strict implication*, where this is defined so that (where P and Q are propositions) P strictly implies Q if and only if it is logically impossible (that is to say a contradiction) for P to be true and Q false. A contradiction therefore strictly implies any proposition, and any proposition strictly implies a logical truth. These so-called *paradoxes of strict implication* seem counterintuitive, and relevance logics restrict entailment to apply more narrowly than strict implication. The need to do this, however, is disputed. *See also* CONNEXIVE IMPLICATION, PARACONSISTENCY.

J Bennett, 'Entailment', *Philosophical Review* (1968); general survey, questioning need for relevance logics

ARL

relevant alternatives (theory of) *Philosophy* Theory used in defending FALLIBILISM against the charge that it leads to SCEPTICISM.

Where P and Q are propositions, P counts for this purpose as an alternative to Q if it is inconsistent with Q, and counts as a relevant alternative if to know that Q we must also know that not-P. Variant formulations exist, but the point is that if P is a sceptical hypothesis (for example that you are now only dreaming you are reading this entry), it can be ignored if it can be shown to be not 'relevant' in the above sense. We shall then of course need some way of deciding when P *is* relevant to Q in this sense.

S Cohen, 'How to be a Fallibilist', *Philosophical Perspectives* (1988)

ARL

relexicalization (1976) *Linguistics* A term coined by English linguist Michael Alexander Kirkwood Halliday (1925–).

Substitution of invented, unofficial words in certain areas of vocabulary in an ANTI-LANGUAGE. *See also* OVER-LEXICALIZATION, UNDERLEXICALIZATION.

M A K Halliday, 'Antilanguages' [1976], repr. in *Language as Social Semiotic* (London, 1978), 164–82

RF

reliablilism *Philosophy* Theory that a belief can be called justified if it is formed by a process that is reliable, that is normally produces true beliefs.

This is an externalist account of justification if it is not insisted that the believer be aware of the method's reliability. This appeal to reliability may also contribute to an analysis of knowledge, though the questions of when a belief is justified and when it amounts to knowledge are different. This is because to have knowledge we may need more than justified belief (for example the belief must at least be true, and even a method that normally produces true beliefs might on some occasion produce a belief that was indeed true but only by accident: would that still amount to knowledge?); also some knowledge, for example of some of our inner states, may not need justification.

A I Goldman, 'What is Justified Belief?', *Justification and Knowledge*, G S Pappas, ed. (1979)

ARL

remainder theorem *Mathematics* The result whereby the remainder upon dividing a polynomial $p(x)$ by $(x - c)$ is equal to $p(c)$ whenever c and the coefficients of p belong to a field. *See also* FACTOR THEOREM and FIELD THEORY.

MB

Rensch's laws *Biology* These four rules state that: (1) species living in colder climates have relatively larger litters and egg clutches than similar species living in warm climates; (2) birds have relatively shorter wings and mammals shorter fur in cold climates than in warm ones (also called the wing rule); (3) land snails tend to have brown shells in cold climates, while those in warm climates will have white shells; (4) strong sunlight and arid conditions are associated with relatively thicker shells.

KH

rent seeking (1974) *Economics* The term was first used by American economist Ann Krueger (1934–) for a theory developed by American economist Gordon Tullock (1920–) in 1967. Tullock's theory addressed the active creation of monopolies, with the aim of achieving supernormal profits or market control, in competitive conditions. *See* MONOPOLY THEORY and MONOPOLISTIC COMPETITION.

A O Krueger, 'The Political Economy of the Rent-Seeking Society', *American Economic Review*, vol. LXIV (1974), 291–303

PH

reorganization theory *Psychology* Proposed by the German psychologist Max Wertheimer (1886–1943).

Derived from gestalt psychology and applying mainly to cognitive and perceptual processes, this theory states that learning involves the modification or altering of mental structures (the neural make-up). Given recent advances in our ability to observe functioning brains, it has been possible to show that learning causes changes in brain structure at the neural level. *See* GESTALT THEORY.

B Kolb and I Whishaw, *Fundamentals of Human Neuropsychology*, 3rd edn (San Francisco, 1990)

NS

repetition, law of Proposed by one of the most influential learning theorists, the American psychologist E R Guthrie (1886–1959).

This hypothesis posits that, all things being equal, a function or a goal or a behaviour is facilitated by being used or exercised and is weakened by disuse. A limitation of the law is that, where it is not observed, it is assumed that 'other things' were not equal but does not indicate how. *See* FREQUENCY THEORY.

J R Anderson and G H Bower, *Human Associative Memory* (Washington D.C., 1975)

NS

representation *Politics* Theory of political participation in complex, large, or modern societies.

The only way in which most people can participate in the government of their societies is by being represented by a relatively small number of people who will in some sense act on their behalf. There is disagreement about the most appropriate way in which this representation should take place, or the representatives be selected.

David Miller *et al.*, eds, *The Blackwell Encyclopaedia of Political Thought* (Oxford, 1987)

RB

representationalism *Philosophy* Also called *representativism* or *representative theories* of perception, memory, thinking, and so on. Any theory holding that these activities (perception is usually meant) involve the existence of mental objects (such as images or 'sense-data') which facilitate the activity by representing the external object.

We may be said to perceive the representative instead of perceiving the object (which is then *inferred* to exist – but on what grounds?); or to perceive the object indirectly by perceiving the representative directly (but what do 'directly' and 'indirectly' amount to?) A representative theory of memory may say we have an image which represents the past event (but how can we know it does?), as against saying we are somehow in direct contact with the past (despite its no longer existing). Representatives, therefore, which also are not always easy to find, may end up as barriers rather than bridges to what they are supposed to represent.

D W Hamlyn, *Sensation and Perception* (1961); mainly historical

ARL

reproduction theory of imagery *Psychology* Proposed by O Kulpe (1862–1915).

This rather basic theory of imagery states that an image is a copy or point-by-point reproduction of the original stimulus. In 1890 William James (1842–1910), the American philosopher, described this discredited theory as being 'as mythological an entity as the jack of spades'.

A Richardson, *Mental Imagery* (New York, 1969)

NS

reproductive phases theory *Psychology* Proposed by C Buhler (1893–1974).

A theory of development through the lifespan which states that the primary psychosocial phases of life parallel the primary biological phases. These phases are: progressive growth (birth to 15), emergence of sexual reproductive activity (ages 16–25), stability (ages 26–45), loss of reproductive capacity (46–65) and biological decline (66 to death). While criticized as excessively vague, the theory is valued for its emphasis on the significance of reproductive activities for development and evolution.

C Buhler and F Massarik, eds., *The Course of Human Life* (New York, 1968)

NS

republicanism *Politics* Theory of government by citizens.

In a republic – which is not necessarily presidential – government is in the hands of citizens, or of those accountable to them rather than in the control of a monarch or despot.

David Miller *et al.*, eds, *The Blackwell Encyclopaedia of Political Thought* (Oxford, 1987)

RB

repulsion theory *Biology* Also called field theory, this postulates that the pattern of origin of leaf bud (primordia) at the shoot tip (apex) is regulated by inhibitory substances synthesized by the apex and the older primordia.

A new primordium arises in a position where the concentration of these substances has fallen below a certain threshold. Although no inhibitory substances have been found, various experiments support this theory. *Compare with* AVAILABLE SPACE THEORY.

KH

re-registration (1983) *Stylistics* Term coined by the British stylisticians R A Carter and Walter Nash (1926–).

The incorporation within a literary text of expressions characteristic of a non-literary REGISTER. Out of its normal practical context, the style of the quoted register acquires new components of meaning. *See also* SOVEREIGNTY.

R A Carter and W Nash, 'Language and Literariness', *Prose Studies*, 6 (1983), 123–41

RF

resemblance, law of *Psychology* Formulated by the English philosopher Thomas Hobbes (1588–1679).

The principle that a thought, idea or feeling tends to bring to mind another that resembles it in some respect. The law is better considered as an empirical generalization than as an explanatory device and is linked with the associationist approach to human cognition. *See* SIMILARITY, LAW OF and ASSOCIATIONISM.

J R Anderson and G H Bower, *Human Associative Memory*, (Washington D.C., 1975)

NS

resemblance theories of universals *Philosophy* Some nominalists dispense with substantive universals (*see* PLATONISM) in treating the ONE OVER MANY PROBLEM by saying that what unites a group of objects of the same kind is that they resemble one of their number taken as a standard.

Objections to this are that resemblance itself seems to be an eliminable universal, and so does the respect in which objects resemble the standard one (that is, resemblance seems to presuppose rather than explain universals). Ludwig Wittgenstein (1889–1951) in his *Philosophical Investigations* (1953) dispensed with standard instances by taking a *family resemblance* view, where objects in the group have nothing in common to all of them, but any one of them has much in common with a large number of the others (the vagueness is deliberate), even though members on opposite sides of the central area of the 'family' may have nothing relevant in common. The cluster theory is somewhat similar.

M A Simon, 'When is a Resemblance a Family Resemblance?', *Mind* (1969); D Gasking, 'Clusters', *Australasian Journal of Philosophy* (1969)

ARL

residue theorem of Cauchy *Mathematics* Named after the French mathematician and physicist Baron Augustin Louis Cauchy (1789–1857).

This is the result in COMPLEX FUNCTION THEORY whereby if a function f is analytic on a simply connected region A except for a finite number of isolated singularities, then the complex line integral of f over any simple closed curve in A which does not pass through the singularities, is equal to $2\pi i$ times the sum of the residues (coefficient of the $(z - a)^{-1}$ term in the *Laurent expansion* of f where a is a singularity) of f at the singularities inside the contour. This theorem is a generalization of CAUCHY'S INTEGRAL THEOREM and is useful when evaluating definite integrals of real functions.

J E Marsden, *Basic Complex Analysis* (Freeman 1973)

MB

resonance, theory of (1931) *Chemistry* A theory due to the American chemist Linus Pauling (1901–), winner of the Nobel prizes for Chemistry (1954) and Peace (1962). More appropriately called mesomerism, resonance theory is a way of describing molecular structures which cannot be represented by any single Lewis structure.

According to the theory, the stability of a molecule results from a mixture (hybridization) of states with differing valence electron distribution but of equal (or nearly equal) energy. The stabilization of such a system over the non-resonating forms is the resonance energy. For example, the stability of the benzene molecule results from strong resonance contributions from Kekulé structures, weaker contributions from Dewar structures and very weak contributions from ionic structures.

P W Atkins, *Physical Chemistry* (Oxford, 1990)

GD

resonance theory of hearing *Psychology* Proposed by the German physiologist Hermann von Helmholtz (1821–94).

A theory of hearing to account for the perception of sound. Perceived pitch is hypothesized to be determined by the place within the organ or corti where hairs are stimulated. Loudness and tonal discrimination is determined by the number of neurons fired by the incoming stimulus. Helmholtz's theory cannot account for the hearing of certain frequencies of sound.

J P Dworetzky, *Psychology*, 3rd edn (New York, 1898)

NS

resonance theory of learning *Psychology* Proposed by the American psychologist Harry Frederick Harlow (1905–81).

This theory proposes that items belonging to a certain set are more likely to be recalled or responded to during the time that set is being dealt with or being responded to. The theory can account for only a small set of the learning processes that are known to be implicated in human learning and thinking.

H F Harlow, 'The Formation of Learning Sets', *Psychological Review*, vol. LVI (1949), 51–65

NS

response-unit hypothesis *Psychology* Proposed by the American psychologist Orval Hobart Mowrer (1907–82).

Proposed to account for conflicting evidence relating to the extinction or disappearance of behaviour when it is no longer reinforced. The hypothesis addressed the difficulty of defining a behavioural response: in animal studies a response would normally be defined in terms of an animal depressing a bar; Mowrer proposed a redefinition based on a larger behavioural sequence. The hypothesis could not account for the evidence in a comprehensive fashion.

G A Kimble, *Hilgard and Marquis Conditioning and Learning* (New York, 1961)

NS

restoration theory (20th century) *Psychology* Proposed by the American psychologist Ian Oswald. A theory of sleep which argues tham both REM (rapid eye movement) and non-REM sleep serve a restorative, replenishing function. It has also been used to account for high levels of REM sleep in babies.

A major objection to the theory comes from evidence indicating that sleep, especially REM sleep, is characterized by high levels of physiological arousal and therefore uses substantial amounts of energy.

I Oswald, *Sleep* (Harmondsworth, 1966)

NS

restricted and elaborated codes (*c.*1960) *Linguistics* Proposed by British sociologist Basil Bernstein (1924–) and much debated in education. Also called PUBLIC LANGUAGE and FORMAL LANGUAGE.

Restricted code is a variety of language allegedly characterized by simple syntax, limited evaluative vocabulary, idioms, frequent personal pronouns, and unexplicit reference. Elaborated code is said to have more complex and logical syntax, better indications of judgment and reference, more impersonality. Middle class children have both codes; some working class children have only the restricted code and fail at school because of the cognitive limitations of elaborated code (LINGUISTIC DEFICIT or LINGUISTIC DEPRIVATION hypothesis). The theory cited little empirical evidence, and is out of favour today.

B Bernstein, *Class, Codes and Control*, 1 (London, 1971)

RF

Retger's law *Chemistry* The law states that the physical properties of crystalline mixtures of isomorphous compounds vary in proportion to their percentage compositions.

GD

Réti's theory of the thematic process (1950s) *Music* American musicologist Rudoph Réti (1885–1957) postulated that musical composition represents an evolutionary process: the composer begins with a musical motif which he subsequently varies to form themes, chords and key relationships. The motif may even determine structure.

Réti's views convincingly challenge the notion that composers, such as Beethoven, started with a formal plan and then proceeded by filling it in.

R Réti, *The Thematic Process in Music* (New York, 1951)

MLD

reticular activating system (RAS) (1958) *Psychology* A term introduced by G Maruzzi and H W Magoun.

This refers to a body of brain tissue called the reticular formation, the arousal and alerting functions of which were discovered by Maruzzi and Magoun. The RAS is now considered to be only that part of the reticular formation of reticular cells which extends from the brain stem to all parts of the cerebral cortex.

N R Carlson, *Physiology of Behavior* (Boston, 1991)

NS

retinex theory *Psychology* Proposed by the American inventor and physicist Edwin Herbert Land (1909–).

This postulates three separate visual systems (retinexes), one primarily responsive to long-wavelength light, one to moderate, and the third to short-wavelength light. Each one is represented as an analogue to a black-and-white picture taken through a specific filter, each producing maximum activity in response to red, green and blue light for the long, moderate, and short-wavelength retinexes, respectively. The theory enjoys substantial empirical support. *See* LAND EFFECT.

E H Land, 'The Retinex Theory of Color Vision', *Scientific American*, vol. CCXXXVII (1977), 108–28

NS

retributivism *Philosophy* Theory of punishment whereby all or part of the purpose of punishment is the infliction of pain or disadvantage on an offender which is in some sense commensurate with his offence and which is inflicted independently of reform or deterrence.

For a weak theory the commensurate amount need not be inflicted but may be, and a limit is placed up to which reformative or deterrent punishment may go but beyond which it may not. A strong theory insists that the punishment must be inflicted, but again places a limit beyond which it may not go. Retributivism opposes excessive harshness as much as excessive leniency, and opposes the violation of the offender's rights in the interests of social expediency or personal spite and so on. Mitigating circumstances, diminished responsibility, and so on are taken into account before determining the commensurate amount, but there are still problems in determining this, and the strong retributivist, especially, must justify violating the presumed moral ban on inflicting unnecessary pain.

J Feinberg, 'The Expressive Function of Punishment', *Doing and Deserving* (1970)

ARL

retroaction theory of forgetting *Psychology* *See* INTERFERENCE THEORY OF FORGETTING.

retrograde evolution *Biology* The theory that viruses emerged from intracellular parasitic micro-organisms which, over time, lost the genetic programming for independent metabolism because of their habitat. *See also* CELLULAR ORIGIN THEORY and PROTOVIRUS HYPOTHESIS.

KH

returns to scale (18th century–) *Economics* The long-term relationship between outputs and the amount of inputs required to generate them.

If inputs are increased by half, economies of scale occur where a higher proportionate increase in production is achieved. Diseconomies of scale occur where output is increased by less than half. Classical economists were preoccupied with the diminishing returns to scale of land, whereas post-Marshallian studies examined increasing returns to scale. *See* EQUILIBRIUM THEORY.

A Marshall, *Principles of Economics* (London, 1890); P Sraffa, 'The Laws of Returns under Competitive Conditions', *Economic Journal*, vol. XXXVI (December, 1926), 535–50

PH

revealed preference theory (1938) *Economics* Pioneered by American economist Paul Samuelson (1915–), this is a method by which it is possible to discern consumer behaviour on the basis of variable prices and incomes.

A consumer with a given income will buy a mixture of products; as his income changes, the mixture of goods and services will also change. It is assumed that the consumer will never select a combination which is more expensive than that which was previously chosen. The theory deliberately ignores measures of utility and indifference. An empirical utility theory, it superseded cardinal utility in consumer theory. *See* SOCIAL WELFARE FUNCTION.

P A Samuelson, 'A Note on the Pure Theory of Consumers' Behaviour', *Econometrica NS*, 5 (1938), 353–54

PH

reversibility principle *Physics* This states that if a ray of light travels from one point to another through an optical system along a particular path, a ray can also proceed in the reverse direction along the same path.

J Thewlis, ed., *Encyclopaedic Dictionary of Physics* (New York, Oxford and London, 1962)

MS

revised extended standard theory *Linguistics* *See* EXTENDED STANDARD THEORY.

revisionism (19th century–) *Politics* An adaptation of MARXISM, originally associated with the German socialist Eduard Bernstein (1850-1932).

Capitalism was not in crisis, and its replacement by SOCIALISM was likely to be a matter of peaceful development and adaptation. A British version of this body of ideas is FABIANISM.

David Miller *et al.*, eds, *The Blackwell Encyclopaedia of Political Thought* (1987)

RB

revolution *Politics/History* Theory of historical change.

In its earlier version, the theory of revolution dealt not with violent or insurrectionary change, but with cycles whereby systems or institutions grew, matured, and declined or collapsed. Its 19th- and 20th-century version is either a view that significant changes in social, political or economic arrangements only occur as a result of disruptive – though not necessarily violent – upheaval, or the advocacy of such methods.

David Miller *et al.*, eds, *The Blackwell Encyclopaedia of Political Thought* (Oxford, 1987)

RB

rewriting rule (1957) *Linguistics* Formalization by the American linguist Avram Noam Chomsky (1928–).

Form of rule devised for PHRASE STRUCTURE GRAMMAR; for example S → NP VP, NP → Det N. The symbol on the left of the arrow was 'rewritten' as the sequence of symbols on the right, which are the constituents of the unit symbolized on the left. Of historical interest.

N Chomsky, *Syntactic Structures* (The Hague, 1957)

RF

rhetoric (5th century BC) *Literary Theory* Formalized in the *Rhetoric* (*c*.330 BC) of the Greek philosopher Aristotle (384–322 BC).

In classical education, rhetoric was the art of persuasion or of speaking well, and was one of the 'seven liberal arts' (others are GRAMMAR and dialectic). It was much concerned with the organization of an argument and the ornamentation of language (*see* FIGURES OF SPEECH). Since the 1970s there have been interesting suggestions for reformulating traditional rhetoric in modern linguistic terms.

W Nash, *Rhetoric* (Oxford, 1989); H F Plett, 'Rhetoric', *Discourse and Literature*, T A Van Dijk, ed. (Amsterdam, 1985), 59–84

RF

Ricardian equivalence theorem (1974) *Economics* Named by American economist Robert Barro (1944–) after English economist David Ricardo (1772–1823), this theory asserts that government deficits are anticipated by individuals who increase their saving because they realize that borrowing today has to be repaid later. One of the theory's central points is that the individual can unravel government policy. *See* CROWDING OUT.

D Ricardo, *On the Principles of Political Economy and Taxation* (London, 1817); R Barro, 'Are Government Bonds Net Wealth?', *Journal of Political Economy*, vol. LXXXII (1974), 1095–175

PH

Ricco's law *Psychology* Formulated by the Italian, Annibale Ricco (1844–1919). For very small areas of the retina (less than 10° of arc) the absolute threshold is inversely proportional to the area stimulated. This law has also been found to hold reasonably well for thermal thresholds on the skin.

S Coren, C Porac and L M Ward, *Sensation and Perception* (New York, 1984)

NS

Richardson's equation (1901) *Physics* Named after its discoverer, the British physicist Sir Owen Willans Richardson (1879–1959), the theory was extended in 1923 by Saul Dushman and the equation is sometimes called the Richardson–Dushman equation.

The relationship between the current I emitted per unit area of a heated metal surface and the temperature T of the surface is

$$I = AT^2 \exp \frac{-\phi}{kT}$$

where A and ϕ are constants for any given surface and k is Boltzmann's constant.

J Thewlis, ed., *Encyclopaedic Dictionary of Physics* (New York, Oxford and London, 1962)

MS

Richter scale (1935) *Geology* By analyzing the amplitudes of seismic waves detected at standardized seismometers, the American seismologst Charles Francis Richter (1900–85) could quantify the energy released by an earthquake and express it in a logarithmic scale.

Such amplitudes were difficult to quantify until nuclear weapon testing provided the opportunity for appropriate calibrations in terms of absolute energy released. The quantification of any individual earthquake is now routine, but in general they remain complex and poorly understood phenomena. *See also* MERCALLI SCALE.

C F Richter, 'An Instrumental Earthquake Scale', *Bulletin of the Seismological Society of America*, vol. XXV (1935), 1–32

DT

Richter's law (1791) *Chemistry* A law named after the German chemist Jeremias Benjamin Richter (1762–1807). The law states that an equivalent weight of an acid will exactly neutralize an equivalent weight of a base.

GD

Riemann-Lebesgue lemma *Mathematics* Named after German mathematician Georg Friedrich Bernhard Riemann (1826–66) and French mathematician Henri-Léon Lebesgue (1875–1941).

The classical result of analysis whereby the nth Fourier coefficient of a Lebesgue integrable function approaches zero as n approaches infinity; that is

$$\lim_{t \to \alpha} \int_I f(x) \exp(\imath tx) dx = 0$$

for any interval I of the real line and any real variable t.

J B Conway, *A Course in Functional Analysis* (New York, 1985)

ML

Riemann's hypothesis (1860) *Mathematics* Named after the German mathematician Georg Friedrich Bernhard Riemann (1826–66), this is the conjecture that all the non-trivial zeros of the ZETA FUNCTION must lie on the line

$$\Re(z) = \tfrac{1}{2}.$$

(The trivial zeros occur at the negative even integers.)

There is a fair amount of evidence in favour of this hypothesis; however, it remains unproven to date. If established, it would have many important consequences in prime NUMBER THEORY.

A Baker, *A Concise Introduction to the Theory of Numbers* (Cambridge, 1984)

MB

Riemann's mapping theorem (1851) *Mathematics* Named after German mathematician Georg Friedrich Bernhard Riemann (1826–66), this is the astonishing result in COMPLEX FUNCTION THEORY whereby every simply connected region A other than the complex plane itself is conformally equivalent to the open unit disc D; that is, there is a bijective analytic function (called a conformal map) which sends the region A to the set D of complex numbers z with $|z| < 1$.

As a consequence, any two simply connected regions which are not equal to the entire complex plane are conformally equivalent. While it is of little practical use, this theorem nevertheless has considerable theoretical importance.

S G Krantz, 'Functions of One Complex Variable', *Encyclopedia of Physical Science and Technology* (Academic Press, 1987)

MB

Riemann's theory of phrase structure (late 19th century) *Music* German musicologist Hugo Riemann (1849–1919) postulated that music is built up from regular 'weak-strong' stress patterns, representing the passing of energy through 'growth' to 'stress point and decay'.

These weak-strong units occur at all levels of the music, forming a conceptual 'grid' system of equal units. Music conformimg rigidly to the grid would be unbearably dull, so in practice composers upset the pattern using various techniques such as omitting/repeating stresses, expanding/reducing durations, and dovetailing phrases. Riemann's theory is controversial but provides a useful starting-point for analyzing musical metre, a neglected area of study.

H Riemann, *System der Musikalischen Rhythmik und Metrik* (Leipzig, 1903)

MLD

Riesz-Fischer theorem (20th century) *Mathematics* Named after Hungarian mathematician Frigyes Riesz (1880–1956) and British mathematician Ronald Aylmer Fisher (1890–1962). The theorem of analysis that every square summable sequence is the sequence of Fourier coefficients of a unique Hilbert function which is square integrable.

It follows that a necessary and sufficient condition for a given summable function to be an element of L_2 is that the series of the squares of the absolute values of its Fourier coefficients should be convergent.

L V Kantorovich and G P Akilov, *Functional Analysis* (Oxford, 1982)

ML

Riesz representation theorem (1909) *Mathematics* Named after Hungarian mathematician Frigyes Riesz (1880–1956), this is a result which shows that the continuous linear functionals on $C(X)$, the space of real valued continuous functions whose domain is the compact Hausdorff space X, may be identified isometrically with differences of regular Borel measures on X by the function

$$\Psi(f) = \int_X f d\mu.$$

In its simplest form, the linear functionals on $C([0,1])$ are identified with functions of bounded variation on $[0,1]$.

Béla Bollobás, *Linear Analysis* (Cambridge, 1990)

MB

right hand rule *Physics* *See* FLEMING'S RULES.

rights (20th century) *Politics/Law* Theory of entitlement to benefit or possession.

Either because of NATURAL LAW or divine or other moral principles, or as a result of actual practice often expressed in POSITIVE LAW, individuals and groups enjoy rights. These are either actual and enforceable entitlements to control over their own persons and over material goods and services (a form of PROPERTY), or claims to such entitlements.

David Miller *et al.*, eds, *The Blackwell Encyclopaedia of Political Thought* (Oxford, 1987)

RB

rites of passage (1908) *Anthropology* Coined, in the French form *rites de passage*, by the German-born, Dutch-French anthropologist Arnold van Gennep (1873–1957).

Rituals which differentiate one phase of life from another, such as baptism, marriage or initiation into a secret society. Some anthropologists have emphasized the 'liminal' nature of such rites; in other words, that a person undergoing them is characteristically in neither one phase nor the other but inbetween. The concept has continued to be widely accepted.

A van Gennep, *The Rites of Passage* (London, 1960 [French edn, 1908])

ABA

Ritz's combination principle (1908) *Chemistry/Physics* Named after the Swiss theoretical physicist Walter Ritz (1878–1909), it is also known as the combination principle. Originally an empirical formulation, the principle states that the wavenumber (v) of a spectral line of an atom is a combination of two terms as given by

$$v = \frac{R}{x^2} - \frac{R}{y^2}$$

where x remains constant for any given series, y assumes different integral values to give the lines in that series, and R is the Rydberg constant. Not all of the corresponding frequencies are found in the actual spectrum because of forbidden transitions or ones which occur infrequently.

P W Atkins, *Physical Chemistry* (Oxford, 1990)

GD

Roche's limit (1850) *Astronomy* This principle is named after its discoverer, the French mathematician Edouard Albert Roche (1820–83). Considering a small body moving around a much larger body in an orbit of gradually reducing radius, Roche showed that if the bodies have equal densities the small body will be broken up as soon as the radius of its orbit falls to 2.45 times the radius of the larger body. This distance is known as Roche's limit. It has been suggested that Saturn's rings are the broken-up fragments of a former satellite.

S Mitton, ed., *The Cambridge Encyclopaedia of Astronomy* (London, 1973)

MS

Rogerian theory *Psychology* Proposed by the American psychologist Carl Rogers (1902–).

This theory of personality is probably better known for its associated method of psychotherapy. The method is often called 'non-directive' or 'client-centred' and argues for an approach to the client based on unconditional positive regard for their thoughts and feelings. The theory and therapy enjoy considerable popularity, particularly among the counselling professions. *See* GROWTH PRINCIPLE.

C R Rogers, *On Becoming a Person* (Boston, 1961)

NS

role confusion *Psychology* A term used by the German-American psychoanalyst Erik Homburger Erikson (1902–).

Erikson identified eight developmental stages. He felt role confusion arose from an individual's failure to establish a sense of identity during the fifth psychosocial stage of development (adolescence). Erikson claimed that an individual would find it difficult to sustain a stable life if he had failed to form a sense of identity.

E H Erickson, *Identity, Youth, and Crisis* (New York, 1968)

NS

role enactment theory *Psychology* A theory of hypnosis proposed by the American psychologist Theodore R Sarbin (1911–).

This hypothesizes that a hypnotized person is not actually in an altered state of consciousness, but rather is so involved or engrossed in the suggestions of the hypnotist that he/she behaves or acts in accordance with the role set out by the hypnotizer. It is difficult to prove or disprove this theory, although many researchers believe social psychological variables as well as the interaction between the hypnotizer and subject play a part in the state we call hypnosis.

T R Sarbin, *Hypnosis: A Social Psychological Analysis of Influence Communication* (New York, 1972)

NS

role-role theory (1956) *Psychology* Developed by the American sociologist Erving Goffman (1922–).

Role-role theory has been used to analyze and explain social interaction behaviours in more formal situations, such as in institutions (for example, the role of doctors, nurses and patients within a hospital). Roles are used to explain and account for the regularities and patterns of social interactive behaviour. *See* IMPRESSION MANAGEMENT THEORY.

P Collette, ed., *Social Acts and Social Behaviour* (Oxford, 1977)

NS

Rolle's theorem *Mathematics* Named after French mathematician Michel Rolle (1652–1719), this states that a real function f which is continuous on the closed interval $[a,b]$ and differentiable of the open interval (a,b), and such that $f(a) = f(b)$, then there is at least one point between a and b where the derivative is zero. The mean value theorem follows from this result.

H Anton, *Calculus with Analytic Geometry* (New York, 1980)

ML

romance (12th century) *Literary Theory* Etymologically, 'romance' comes from *Romans*, the word used to refer to the Old French language, which was derived from Latin but distinct from it.

The term was originally applied to French and German chivalric romances of the 12th and 13th centuries; in English, *Sir Gawayne and the Green Knight* (*c.*1375) and Sir Thomas Malory's *Morte d'Arthur* (*c.* 1470) belong to the GENRE. When used more broadly as a genre term, romance is often contrasted with EPIC; courtly manners (*see also* COURTLY LOVE), fanciful setting, marvels, idealism rather than REALISM, and a narrative of quest are common elements. The term has been weakened by being generalized to fanciful and idealist fiction of other periods, including modern 'romantic fiction' of escapism and sexual relations.

G Beer, *The Romance* (London, 1970)

RF

Romanticism (*fl.* *c.*1790–1830) *Literary Theory/Art* Movement in the arts and in artistic theory, developed principally in Germany and England. In English literature, it is mainly associated with the poets William Blake (1757–1827), William Wordsworth (1770–1850), Samuel Taylor Coleridge (1772–1834), Lord Byron (1788–1824), Percy Bysshe Shelley (1792–1822) and John Keats (1795–1821). Romantic artists include William Blake (again), Joseph Mallord William Turner (1775–1851) and Caspar David Friedrich (1774–1840).

The artist becomes the central focus of his own work, which is expressive and self-reflexive in tone and subject. Individuality, creative freedom and ORIGINALITY are championed, and an authority which goes with heightened perceptivity and IMAGINATION (*see* POET AS LEGISLATOR). Poetry is organic and spontaneous. Rationalism is disfavoured, passion and themes of the subconscious advanced (*see also* GOTHIC).

Romanticism's criteria for poetic language have extended its scope into the modern period: fundamental is Wordsworth's search for a living speech (*see* LANGUAGE OF MEN) and distaste for POETIC DICTION: a programme which implies linguistic contact between art and life, and a language of poetry in which FIGURES OF SPEECH are functional rather than decorative.

M H Abrams, *The Mirror and the Lamp* (New York, 1953)

AB, RF

Romanticism (late 18th century) *Music* A term popularized, as a musical concept, by the German writer Ernst Theodor Wilhelm Hoffmann (1776–1822) in his 1813 essay 'Beethovens Instrumentalmusik'. It refers to the valuing of feeling over reason.

Romantic composers abandoned principles of form and structure in favour of freer writing, asserting instinctual needs. As musical content became more programmatic (exploring psychological, heroic, nationalistic and supernatural subjects), harmonic language grew increasingly chromatic in order to express emotion.

A Whittall, *Romantic Music* (London, 1987)

MLD

Romeo and Juliet effect *Psychology* Not attributable to any one originator, this is a loosely formulated term within social psychology referring to the increase in attractiveness between two people that may arise when parents or others attempt to keep them apart.

E Berscheid and E H Walster, *Interpersonal Attraction* (Reading, Mass., 1978)

NS

root test *Mathematics* Attributed to French mathematician and physicist Baron Augustin Louis Cauchy (1789–1857), this is a test for convergence of an infinite series $\Sigma\, a_k$ in which the limit

$$\rho = \lim_{k \to \infty} (|a_k|)^{1/k}$$

is evaluated and if $\rho < 1$ the series converges and if $\rho > 1$ the series diverges. If $\rho = 1$ the test is inconclusive. For example, the series

$$\sum_{k=1}^{\infty} 1/[\ln(x+1)]^k$$

converges since

$$\lim_{k \to \infty} 1/\ln(k+1) = 0 < 1.$$

This test is more powerful than the RATIO TEST.

H Anton, *Calculus with Analytic Geometry* (New York, 1980)

ML

Rosenthal effect *Psychology* Named after the American psychologist Robert Rosenthal (1933–).

Rosenthal conducted extensive research into the manner in which one's beliefs, biases and expectations can have an influence on a phenomenon under investigation. His work has been particularly important in alerting psychologists to the manner in which they and others may act in accordance with motives and expectations of which they have no conscious awareness. *See* PYGMALION EFFECT and SELF-FULFILLING PROPHECY.

R Rosenthal and D B Rubin, 'Interpersonal Expectancy Effects: The First 345 Studies', *The Behavioural and Brain Sciences*, vol. III (1978) 377–86

NS

Rossby model (1940s) *Climatology* Swedish-American meteorologist C G Rossby (1898–1957) modified the HADLEY CELL MODEL by proposing an additional surface temperature-driven circulation at the poles, separated by an intermediate circulation rising in moderately high latitudes and descending near the tropics where it meets the equatorial circulation system. This forms the basis of the THREE-CELL MODEL.

Rossby is now mainly recognized for his work on upper atmospheric circulation. His model claims upper atmosphere westerly winds tend to form a wavelike pattern but also contain jet streams. Such upper atmospheric regimes are thought to control low and high pressure zones' direction of movement.

C G Rossby, 'On the Nature of the General Circulation of the Lower Atmosphere', *The Atmospheres of the Earth and Planets*, G P Kuiper, ed. (Chicago, 1949)

DT

rotational sum rule *Chemistry* A rule which states that for a molecule which behaves as a symmetric top, the sum of the line strengths corresponding to transitions to or from a given rotational level is proportional to the statistical weight of that level; that is, to $2J + 1$, where J is the total angular momentum quantum number of the level.

GD

Rouché's theorem (19th century) *Mathematics* Named after French mathematician Eugène Rouché (1832–1910), this result in COMPLEX FUNCTION THEORY allows one to estimate the number of zeros of a given analytic function by comparing it to a suitably chosen simpler analytic function. To be specific, the theorem states that if f and g are analytic on a simply-connected domain containing a simple closed curve upon which

$$|f(z) - g(z)| < |f(z)|$$

then f and g have the same number of zeros (counting multiplicities) inside the contour.

S G Krantz, 'Functions of One Complex Variable', *Encyclopedia of Physical Science and Technology* (Academic Press, 1987)

MB

roundabout method of production (1889) *Economics* Advanced by Austrian economist Eugen von Böhm-Bawerk (1851–1914), this term describes the use of capital goods to increase future productivity of the factors of production.

Production efficiency often entails diverting labour and capital away from the immediate method of production in order to achieve a better method. A house painter may paint a door by hand, but a more efficient method would entail investing in spraying equipment, or developing an automated door-painting factory.

E von Böhm-Bawerk, *Positive Theorie des Kapitales* (Innsbruck, 1889)

PH

Routh's rule (1860) *Physics* Named after its originator, the British mathematician Edward John Routh (1831–1907), this rule is used in calculating the moments of inertia about an axis through the centre of mass for many regular shapes of body.

The moment of inertia I is given by M times the sum of the squares of the two semi-axes perpendicular to the axis of rotation, divided by 3, 4 or 5 for cuboids, cylinders and ellipsoids, respectively. M is the mass of the body.

J Thewlis, ed., *Encyclopaedic Dictionary of Physics* (New York, Oxford and London, 1962)

MS

RRKM theory (1928) *Chemistry* Named after the American chemists Rice, Ramsperger, Kassel and Marcus.

Rice, Ramsperger and Kassel in 1928 pinpointed the basic defect in the LINDEMANN THEORY which had been proposed to account for the observed first-order kinetics of unimolecular gas-phase reactions. Lindemann had assumed that the energized molecules would all have the same lifetime, irrespective of their degree of internal energy. The RRK model proposed that the probability of energized molecules being transformed to product increased with increasing excess energy and so had less chance of being deactivated. The theoretical equations obtained from this model are in good agreement with experiment. At high pressures, ABSOLUTE REACTION RATE THEORY can be applied directly, and the later refinement due to Marcus (RRKM) includes the rotational as well as the vibrational modes of the molecules.

P W Atkins, *Physical Chemistry* (Oxford, 1990)

GD

RSA system (1978) *Mathematics* Named after its inventors R L Rivest, A Shamir and L Adleman, this is the most widely used public key cryptosystem presently available. The system is based on modular arithmetic where the modulus is chosen to be a product of two distinct prime numbers, each of which typically has at least 100 digits in its decimal expansion. The security of the system requires that these prime factors be chosen in a particular form and depends on the fact that there is at present no fast algorithm to factor numbers with a large number of digits in their decimal expansion.

S Berkovits, 'Cryptography', *Encyclopedia of Physical Science and Technology* (Academic Press, 1987)

MB

Rubenism *Art* The view that colour is of equal importance to drawing and design; exemplified in the work of the Flemish artist Peter-Paul Rubens (1577–1640) and championed by those opposed to POUSSINISM in French academic circles.

LL

Ruelle-Takens scenario (1971) *Mathematics* Discovered by French mathematician David Ruelle and Dutch mathematician Floris Takens, this postulates that turbulence may not be quasi-periodic with many frequencies, but may be described by the presence of strange attractors This theory has been experimentally confirmed. *See* STRANGE ATTRACTOR, CHAOS THEORY, MANNEVILLE-POMMEAU PHENOMENON and FEIGENBAUM PERIOD-DOUBLING CASCADE.

D Ruelle, *Chance and Chaos* (Princeton, N.J., 1991)

ML

rule (traditional; 1960s) *Linguistics* The idea of a linguistic rule has become central and controversial in GENERATIVE GRAMMAR.

In TRADITIONAL GRAMMAR, a rule was a statement about regularity in a sentence whose force was *prescriptive*: correct speech or writing depended on following the rules. Prescriptivism was condemned in American STRUCTURAL LINGUISTICS which claimed to be *descriptive* – any observed structure was legitimate. In generative grammar, speakers are supposed to know (unconsciously) and follow 'rules' which, perhaps paradoxically, include speakers' INTUITIONS of GRAMMATICALITY. *See also* LINGUISTIC COMPETENCE.

N Smith and D Wilson, *Modern Linguistics: The Results of Chomsky's Revolution* (Harmondsworth, 1979)

RF

rule of law (20th century) *Politics/Law* A theory most familiarly associated with the English jurist Albert Venn Dicey (1835–1922).

People are governed by law rather than capriciously or arbitrarily, when all people including government and its officials are equally subject to law and when people are punishable only for an established breach of law.

David Miller *et al.*, eds, *The Blackwell Encyclopaedia of Political Thought* (1987)

RB

rule utilitarianism *Philosophy* Also called *restricted* or *indirect utilitarianism*. Version of UTILITARIANISM which says (in its main formulation) that our duty is not to aim for that act which will produce in fact the best overall consequences (because of the impossibility or impracticability of predicting these) but to follow that rule which would have the best consequences if generally followed. Objections include: the apparent pointlessness of mechanically following a rule which on *this* occasion will clearly not have the best results; the pointlessness or counterproductiveness of following a rule which would be the best if everyone followed it but which one knows not everyone will; and the difficulty in the end of even distinguishing rule utilitarianism from act utilitarianism.

B A Brody, 'The Equivalence of Act and Rule Utilitarianism', *Philosophical Studies* (1967)

ARL

Rutherford's theory of the atom (1911) *Physics* Named after its originator, the British physicist Ernest Rutherford, 1st Baron Rutherford of Nelson (1871–1937). It is also known as the Rutherford-Bohr atomic theory after the Danish physicist and 1922 Nobel prizewinner Niels Bohr (1885–1962).

Rutherford proposed a model of the atom (based on classical physics) after Geiger and Marsden had observed that some α-particles were scattered through large angles by thin gold and silver foils. The Thomson 'plum-pudding' model predicted only small-angle scattering. According to Rutherford's model almost all the atom's mass is contained in a very small volume called the nucleus, which also contains all the positive charge. The electrons necessary to maintain the electrical neutrality of the atom as a whole (and to give the atom its chemical characteristics) orbit the nucleus in circular paths, the centripetal force being provided by the Coulomb attraction between the positively charged nucleus and the negatively charged electrons.

The major defect of Rutherford's theory is that, classically, charges that are accelerated radiate electromagnetic energy. This loss would cause the electrons in Rutherford's model to spiral into the nucleus in a time of about 10^{-8} seconds, while emitting a continuous spectrum of radiation and not the line spectrum that is observed. Bohr modified the theory to restrict the electrons to certain allowed orbits. Both the Rutherford and Bohr theories have been replaced by quantum mechanics. *See* BOHR'S THEORY OF THE HYDROGEN ATOM.

J Thewlis, ed., *Encyclopaedic Dictionary of Physics* (New York, Oxford and London, 1962)

MS, GD

Rybczynski theorem (1955) *Economics* Named after Polish-born English economist Tadeusz Rybczynski (1923–), this theory posits that when one of two factors of production is increased there is a relative increase in the production of the good using more of that factor. This unfortunately leads to a corresponding decline in that good's relative price. *See* HECKSCHER-OHLIN TRADE THEORY

T Rybczynski, 'Factor Endowments and Relative Commodity Prices', *Econometrica NS*, vol. XXII (1955), 336–41

PH

S

Sabine's law (*c*.1910) *Physics* Named after its originator, the British scientist Wallace Clement Ware Sabine (1868–1919), this law states that the reverberation time *t* of a hall is related to the total absorption *A* of the hall and the volume *V* of the hall by the relation

$$t \propto \frac{V}{A}.$$

The reverberation time is the time in which the sound intensity falls to 10^{-6} of its initial value.

J Thewlis, ed., *Encyclopaedic Dictionary of Physics* (New York, Oxford and London, 1962)

MS

St Petersburg paradox (18th century) *Economics See* BERNOULLI'S HYPOTHESIS.

Saint-Venant's principle (1855) *Physics* Named after its originator, the French mathematician Adhemar Jean Claud Barre de Saint-Venant (1797–1886), this principle states that the strains produced in a body by the application to a small part of its surface of a system of forces (which is statically equivalent to zero force and zero couple) are of negligible magnitude at distances which are large compared with the linear dimensions of the part.

For example, the state of stress in a long bar bent by couples applied to its ends is practically independent of the distribution of the forces of which the couple is the resultant.

J Thewlis, ed., *Encyclopaedic Dictionary of Physics* (New York, Oxford and London, 1962)

MS

salt-gene hypothesis *Biology See* SODIUM HYPOTHESIS.

salt hypothesis *Biology See* SODIUM HYPOTHESIS.

saltation speciation (1940) *Biology* Also called quantum speciation, this theory was proposed by German-American geneticist Richard Goldschmidt (1878–1958). It suggests that a new species can arise in a relatively short period of time.

Polyploidy, the multiplication of entire chromosome complements in plants, is an example of saltation that is accepted by most biologists. In its extreme form, saltational speciation holds that a new species can arise in a single generation that is so different from other species that it must be placed in a new genera, family and class. This extreme interpretation has been rejected by modern biologists. *See also* HOPEFUL MONSTERS, SPECIATION, THEORY OF and SPECIES, THEORY OF.

C Barigozzi, *Mechanisms of Speciation* (New York, 1982)

KH

sampling theory (20th century) *Mathematics* The theory of STATISTICS concerned with choosing a sample from a population (especially a stratified sample which is controlled to give a correct representation of other factors) so that the distribution of statistics for the sample approximates that for the population. A biased sample is one in which the selected share a property which alters the distribution.

JB

sandwich theorem *Mathematics* Also known as the pinching theorem or squeeze rule, this is a technique for finding the limit of a function by 'squeezing' the function between two simpler functions whose limits are known. Let f, g and h be functions satisfying

$$g(x) \leqslant g(x) \leqslant h(x)$$

for all x in an open interval containing a point a, except possibly at a. If

$$\lim_{x \to a} g(x) = \lim_{x \to a} h(x) = L,$$

then

$$\lim_{x \to a} f(x) = L.$$

H Anton, *Calculus with Analytic Geometry* (New York, 1980)

ML

Sapir-Whorf hypothesis (*c*.1920) *Linguistics* Developed by the American anthropological linguists Edward Sapir (1884–1939) and Benjamin Lee Whorf (1897–1941).

Study of American Indian languages (structurally very different from Indo-European on which traditional grammars were based), led to the proposal that languages vary substantially and unpredictably (LINGUISTIC RELATIVITY). It was claimed that such structural differences encoded radically different world-views (Whorf contrasted English and Hopi). LINGUISTIC DETERMINISM hypothesizes the consequence that speakers of different languages must see the world in different ways.

B L Whorf, *Language, Thought and Reality*, J B Carroll, ed. (Cambridge, Mass., 1956)

RF

Sarkovski's theorem *Mathematics* A theorem of discrete dynamical systems theory about the behaviour of a function f and its iterates,

$$f(f(f \ldots (fx))) = f''(x)$$

which states that a continuous real-valued function which has a periodic point of period k also has a periodic point of period l where $k \rhd l$ in the ordering defined below.

$3 \rhd 5 \rhd 7 \rhd \ldots \rhd 2 \cdot 3 \rhd 2 \cdot 5 \rhd \ldots$
$\rhd 2^2 \cdot 3 \rhd 2^2 \cdot 5 \rhd \ldots \rhd 2^3 \cdot 3 \rhd 2^3 \cdot 5 \rhd \ldots$
$\rhd 2^3 \rhd 2^2 \rhd 2 \rhd 1.$

In particular, a function with a period 3 point also has periodic points of all other periods. This theorem is used to prove results about the chaotic behaviour of one-dimensional maps on the real line.

R L Devaney, *An Introduction to Chaotic Dynamical Systems* (New York, 1989)

ML

satisficing *Economics* The pursuit by a firm of satisfactory profits instead of maximum profit because the company has as its goal some other objective such as maximum market share, sales or management satisfaction.

The concept of 'satisfactory' profits is subjective and varies from firm to firm. *See* THEORY OF THE FIRM, THEORY OF THE GROWTH OF THE FIRM, ORGANIZATION THEORY, MANAGERIAL THEORIES OF THE FIRM and THEORY OF BUREAUCRACY.

S G Winter, 'Satisficing, Selection and the Innovating Remnant', *Quarterly Journal of Economics*, vol. LXXXV (1971), 237–61

AV, PH

saturation curve *Biology See* TOLERANCE, LAW OF.

satyagraha (20th century) *Politics* A theory of CIVIL DISOBEDIENCE or DIRECT ACTION expounded by the Indian political leader Mohandas (Mahatma) Gandhi (1869–1948).

Peaceful resistance to the policies of government is both politically effective and confronts a STATE which is inherently violent with an alternative manner of social conduct.

David Miller *et al.*, eds, *The Blackwell Encyclopaedia of Political Thought* (Oxford, 1987)

RB

sausage machine (1978) *Linguistics* Proposal by the American psycholinguists L Frasier and J D Fodor.

Syntactic PARSER which assumes that in analyzing utterances for comprehension, people divide sentences into parcels roughly six words long, and hypothesize the structure of that segment before moving on to the next 'sausage'. *See also* MAGICAL NUMBER SEVEN.

A Garnham, *Psycholinguistics* (London, 1985)

RF

Say's law (1803) *Economics* Named after Jean-Baptiste Say (1767–1832), this argued that an economy is self-regulating provided that all prices, including wages, are flexible enough to maintain it in equilibrium.

In a more simplistic, and somewhat inaccurate form, the law states that supply creates its own demand and over-production is impossible. This theory has major implications for how governments respond to periods of high unemployment or widespread underemployment. Say's law was accepted as a major plank in classical macroeconomic theory until English economist John Maynard Keynes (1883–1946) challenged its applicability in modern economies. Also known as Say's law of markets. *See* EQUILIBRIUM THEORY, GENERAL EQUILIBRIUM THEORY, PARTIAL EQUILIBRIUM THEORY.

J B Say, *Traité d'économie politique*, vol I, (Paris, 1803); T Sowell, *Say's Law: An Historical Analysis* (Princeton, N.J., 1972)

AV, PH

scala natura *Biology See* CHAIN OF BEING.

Scalar principle BUSINESS Managerial control and responsibility should flow uninterrupted from the top to the bottom of the organization.

The principle seeks to establish single personal links through a company, with each employee having just one superior and one subordinate. *See* THEORY OF THE FIRM and THEORY OF THE GROWTH OF THE FIRM.

AV, PH

scale-and-category grammar (late 1950s–early 1960s) *Linguistics* Developed by the British linguist Michael Alexander Kirkwood Halliday (1925–).

Theory of grammatical structure developing the linguistics of John Rupert Firth (1890–1960); precursor to SYSTEMIC GRAMMAR. The basic categories are unit, structure, class, and system; the scales comprise 'rank' (hierarchical relationships of constituents), 'exponence' (realization of the categories in actual language data) and 'delicacy' (levels of detail of description). *See also* FUNCTIONAL GRAMMAR.

M A K Halliday, 'Categories of the Theory of Grammar', *Word*, 17 (1961), 241–92

RF

scapigliatura (late 19th century) *Art* Movement in Italian art and literature, centred around Milan and typified by the novels of G Rovani. A reaction against the UTILITARIANISM of bourgeois culture, its adherents attempted to revive the spirit of ROMANTICISM. They rebelled against academic mannerisms, and artists such as L Conconi, L Bazarro, C Talone and L Previati painted in a loose Impressionistic manner close to DIVISIONISM.

P Nardi, *La Scapigliatura* (Bologna, 1924)

LL

scepticism *Philosophy* Literally, the habit of being given to enquiry.

The sceptic does not take things for granted. He may deny the existence of God, other minds than his own, a world of material objects behind what is immediately given to our senses, anything other than himself and his experiences (*see also* SOLIPSISM), even his own mind as anything but a set of experiences (David Hume (1711–76)), objective moral values, the possibility of getting any knowledge other than by the senses (*see also* EMPIRICISM), or by the senses (Plato sometimes), or of the past, or by the INDUCTIVE PRINCIPLE, or even by reason itself (Hume sometimes). Alternatively, the sceptic may simply doubt these things rather than deny them outright, and scepticism may be simply a methodological theory.

Among the Greeks, Plato's Academy came under the influence of Sceptics for two centuries starting with Arcesilius (*c.*316–*c.*242 BC) and renewed by Carneades (*c.*214–*c.*129 BC), and directed primarily against STOICISM. A more extreme form of scepticism was PYRRHONISM, one of whose adherents, Sextus Empiricus (2nd century AD), is our main source of knowledge for ancient Scepticism. *See also* PRIVATE LANGUAGE ARGUMENT.

P Unger, *Ignorance: A Case for Scepticism* (1975)

ARL

Schachter and Singer's theory of emotion *Psychology See* ATTRIBUTION THEORY OF EMOTION.

Schauder's basis problem (20th century) *Mathematics* Proposed by Jules P Schauder (1899–1940).

Is it true that every separable Banach space (such as certain vector spaces of continuous functions) has a basis such that every element can be written as a possibly infinite linear combination of basis elements? Although this is true for the well-known separable Banach spaces, it is now known not to hold in general.

JB

Schauder's fixed point theorem (1930) *Mathematics* The theorem that a continuous operator mapping a convex compact subset of a Banach space into itself has at least one fixed point. The question of how many fixed points remains open, and there is no algorithm for finding them. *See also* BROUWER'S THEOREM.

L V Kantorovich and G P Akilov, *Functional Analysis* (Oxford, 1982)

ML

schema (pl. schemata) (1932; 1970s) *Linguistics/Psychology* First defined in the relevant sense by the English psychologist Frederic Charles Bartlett (1886–1969); adapted for psycholinguistics in the 1970s.

Bartlett's definition: 'An active organization of past reactions, or past experiences'. In psycholinguistics: a structured area of knowledge held in memory which can be activated in comprehending a text, supplying inferences, helping construct a MENTAL MODEL or textual world (*see* COHERENCE). *See also* FRAME and SCRIPT for examples.

F C Bartlett, *Remembering* (Cambridge, 1932)

RF

scheme (traditional) *Literary Theory* Part of RHETORIC.

One of the two sub-divisions of FIGURES OF SPEECH (the other is TROPE). Schemes are based on patterning of linguistic form, largely syntactic and/or phonological PARALLELISM and repetition. For example, *chiasmus* is an ABBA mirror-pattern: 'Smooth flow (A) the Waves (B), the Zephyrs (B) gently play (A)' (Pope).

W Nash, *Rhetoric* (Oxford, 1989)

RF

Schleiden's theory of plant fertilization (1837) *Biology* Promoted by Matthias Jakob Schleiden (1804–81), a German botanist best known for his contribution to CELL THEORY. This is the mistaken idea that pollen serves as the egg of the plant, and that sexual reproduction does not occur in botany. While the theory was discredited during the 1840s, Schleiden's proposal stimulated much research during this period.

P Davies, *Historical and Current Aspects of Plant Physiology* (Ithaca, New York, 1975)

KH

Schnirelmann's density theorem *Mathematics* Named after the Russian mathematician Lev Genrikhovich Schnirel'man (1905–38), this is the result in NUMBER THEORY whereby for any two sets of non-negative integers, S and T, it is the case that

$$d(S + T) \geqslant \min\{1, d(S) + d(T)\}$$

where $d(S)$ is the Schnirelmann density of S (that is, the infimum of the ratio $S(n)/n$ where $S(n)$, for $n \geqslant 1$, is the number of elements in S which are no greater than n). This result is sometimes called the alpha-beta theorem when the notation

$$d(S) = \alpha, d(T) = \beta$$

is used. It allows one to deduce that if a set S has positive density, then every number may be represented as a sum of a fixed number of elements of S, for example, sinθ. The sum of two squares has density $6/\pi^2$ and LAGRANGE'S THEOREM follows.

Schreier-Nielsen theorem (1921, 1927) *Mathematics* Named after mathematicians Otto Schreier (1901–29) and Niels Nielsen (1865–1931), this is the result, of fundamental importance in GROUP THEORY, whereby every subgroup of a free group is itself free.

J J Rotman, *An Introduction to the Theory of Groups* (Allyn-Bacon, 1984)

MB

Schrödinger's cat (1930s) *Physics* A paradox proposed by the Austrian physicist Erwin Schrödinger (1887–1961), joint winner of the 1933 Nobel prize for Physics.

The paradox highlights the differences in interpretation between classical physics, the COPENHAGEN INTERPRETATION and the MANY WORLDS HYPOTHESIS. Schrödinger's 'thought experiment' involves a cat placed inside a box along with a vial of poisonous gas which is released when an atom of radioactive material – also in the box – decays. Radioactive decay is a quantum mechanical process for which only a probability of occurrence can be predicted. The common sense (classical physics) view says that the cat is either alive or dead, regardless of whether we look inside the box or not. Quantum physics, however, treats the system comprising the box and its contents as a wavefunction combining two possible, and mutually exclusive, outcomes – the cat is both alive and dead at the same time. Schrödinger's equation says that until the lid of the box is lifted the fate of the cat is not determined. The many worlds hypothesis takes the view that the process of radioactive decay splits the world into two universes each with a different cat – one alive and one dead. The act of lifting the lid splits these two worlds yet again, and so on. *See also* COPENHAGEN INTERPRETATION and MANY WORLDS HYPOTHESIS.

Peter Coveney & Roger Highfield, *The Arrow of Time* (London, 1991)

GD

Schroeder-Bernstein theorem (19th century) *Mathematics* Named after German mathematician Friedrich Wilhelm Karl Ernst Schröder (1841–1902) and Russian mathematician Sergei Natanovich Berstein (1880–1968). It is also known as the Cantor-Bernstein theorem after German mathematician Georg Ferdinand Ludwig Philip Cantor (1845–1918).

This is the result in SET THEORY whereby two sets have the same cardinality if there is an injective mapping from each set to the other.

P R Halmos, *Naive Set Theory* (Springer 1974)

MB

Schultze-Hardy rule *Chemistry See* HARDY-SCHULTZE RULE.

Schur's lemma (1911) *Mathematics* Named after its discoverer Issai Schur (1875–1941). For a square matrix A with complex numbers as elements, let A^* represent the transpose of the complex conjugate. There exists a square matrix U such that U^* is the inverse of U and $U^*AU = B$ where B has only zeros below the main diagonal and the diagonal elements are the eigenvalues of A. In particular, if $AA^* = A^*A$ then B has only diagonal elements.

JB

Schwartz's lemma *Mathematics* Named after German mathematician Hermann Amandus Schwartz (1843–1921), this is a tool for many elegant and useful geometric results in COMPLEX FUNCTION THEORY.

It is the consequence of the MAXIMUM MODULUS THEOREM which essentially says that an analytic function f which maps the unit disc (that is, the set of complex numbers z with $|z| < 1$) into itself and satisfies $f(0) = 0$ is either a rotation or satisfies $|f(z)| < |z|$ in the punctured disc and has

$$|f'(z)| < 1$$

See also RIEMANN MAPPING THEOREM.

S G Krantz, 'Functions of One Complex Variable', *Encyclopedia of Physical Science and Technology* (Academic Press, 1987)

MB

Schwartz's principle *Mathematics See* REFLECTION PRINCIPLE OF SCHWARTZ.

MB

scientific history (19th century–) *History* History viewed as an exact science often, though not exclusively, within MARXISM.

Historical knowledge can be gained by the study of objective fact. It is thus not dependent on the values or whims of the observer.

S H Rigby, *Marxism and History* (Manchester, 1990)

RB

Scitovsky paradox *Economics* Named after the Hungarian-born American economist Tibor Scitovsky (1910–), this states that in welfare economics there is no increase in social welfare by a return to the original part of the losers.

If Allocation X is changed to Allocation Y, those who suffer in the move could still gain enough by returning to X even after forgoing some of the difference between the two allocations. Thus, a medium-rate taxpayer might easily be persuaded to pay more taxes if the threat of paying much higher taxes is removed. *See* SOCIAL WELFARE

FUNCTION, PARETO OPTIMALITY and COST BENEFIT ANALYSIS.

T Scitovsky, *The Joyless Economy: An Inquiry into Human Satisfaction and Consumer Dissatisfaction* (Oxford, 1976)

AV, PH

screening hypothesis *Economics* This precept maintains that education is a filter, or screen, by which innate talent is identified.

The purpose of education is seen as confirmation of an individual's capability to be trained on the job rather than the conferring of skills to a worker. It is used as an alternative to human capital theory. *See* HUMAN CAPITAL THEORY and SEARCH THEORY.

AV, PH

script (1977) *Linguistics/Psychology* Type of SCHEMA proposed by American psychologists Roger Carl Schank (1946–) and Robert P Abelson.

A commonly assumed temporal ordering for some kind of event, for example, 'meal in a restaurant', 'trip to the beach'; activated in text construction and understanding.

R C Schank and R P Abelson, *Scripts, Plans, Goals and Understanding* (Hillsdale, NJ, 1977)

RF

sea-floor spreading (1963) *Geology* Established by the British geographers Drummond Matthews (1931–) and Frederick John Vine (1939–88) (following concepts propounded in the 1930s by the British geologist Arthur Holmes (1890–1965)), this theory describes the mechanism of modern ocean floors' continuous formation and relative motion.

Convective forces cause molten rocks to be extruded around oceanic ridges. In cooling, these are magnetized by the Earth's magnetic field and, over time, are pushed away before their eventual descent into the mantle. Sites where this subduction occurs are marked by oceanic trenches and seismic and volcanic activity. Evidence for this theory is contained in the magnetic anomalies of older rocks, reflecting past changes in the Earth's polarity.

C M R Fowler, *The Solid Earth* (Cambridge, 1990)

DT

search model (1976) *Linguistics* Developed by the psycholinguist K I Forster, this is a model of how words are recognized.

In contrast to the LOGOGEN model, which guesses a specific word, the search model suggests that hearers/writers receiving a perceptual signal entertain several candidate words and search through their mental dictionary files to narrow down to one appropriate word.

M Garman, *Psycholinguistics* (Cambridge, 1990)

RF

search theory (1962) *Economics* This is the analysis of how buyers and sellers acquire information about market conditions and how potential market participants are brought together. Its application to labour markets was pioneered by the American economist George Stigler (1911–).

The theory recognizes the principle that both employers and workers need to invest time and other resources to meet if mobility in the labour market is to continue. *See* DUAL LABOUR MARKET THEORY and CROWDING HYPOTHESIS. *See also* SEGMENTED LABOUR MARKET THEORY, LABOUR MARKET DISCRIMINATION and INSIDER-OUTSIDER WAGE DETERMINATION.

G J Stigler, 'Information in the Labor Market', *Journal of Political Economy*, vol. LXX (October, 1962), 94–105; D. Mortensen, 'Job Search and Labor Market Analysis', *Handbook of Labour Economics*, R Layard and O Ashenfelter, eds (Amsterdam, 1984)

AV, PH

second best, theory of (1956) *Economics* Proposed by Canadian economist Richard Lipsey (1928–) and Australian economist Kelvin Lancaster (1924–), this theory assumes that if one of the conditions necessary to achieve Pareto-optimality is missing then the 'second best' position can only be reached by departing from all the other Paretian conditions. *See* PARETO EFFICIENCY.

R G Lipsey and K Lancaster, 'The General Theory of Second Best', *Review of Economic Studies*, vol. XXIV (October, 1956) 11–32

AV, PH

second law of thermodynamics (19th century) *Physics* The second law of thermodynamics is a generalization of experience, and may be stated in a number of equivalent ways. An axiomatic statement of the second law is CARATHÉODORY'S PRINCIPLE. Three statements applying to the large-scale behaviour of matter follow.

(1) When thermal contact between two closed systems A and B causes system B to become hotter and system A to become colder, no matter where that contact is made, there is no process that can cause A to become hotter and B to become colder that does not involve a work interaction (F C Frank). (2) It is impossible to devise a machine which, working in a cycle, produces no effect other than the extraction of a certain quantity from its surroundings and the performance of an equal amount of work on its surroundings (Lord Kelvin). (3) When two systems are placed in thermal contact, the direction of energy transfer in the form of heat is always from the system at the higher temperature to the system at the lower temperature (R J E Clausius).

J Thewlis, ed., *Encyclopaedic Dictionary of Physics* (New York, Oxford and London, 1962)

MS

section d'or *Art* A French term meaning 'golden section', this refers to an irrational proportion known since the time of Euclid (*c*.3rd century BC) and once thought to possess a hidden harmonic proportion in tune with the universe. It may be defined as a line divided in such a way that the smaller part is to the larger as the larger is to the whole.

AB

secular stagnation theory (20th century) *Economics* Analysis of a protracted economic depression characterized by a falling population growth, low aggregate demand and a tendency to save rather than invest.

The bulk of modern economic endeavour has been to avoid this depression or to curb the impact of it on people. Some recent theorists have suggested that stagnation is a desirable state, in which environmental resources are not unduly exploited. *See* MALTHUSIAN POPULATION THEORY.

A Hansen, *Full Recovery or Stagnation?* (New York, 1938)

AV, PH

Seebeck effect (1821) *Physics* Named after its discoverer Thomas Johann Seebeck (1770–1831).

When two wires of different metals are joined at their ends to form a closed circuit, and the two junctions are maintained at different temperatures, an electromagnetic field is set up and, consequently, an electric current flows around the circuit.

J Thewlis, ed., *Encyclopaedic Dictionary of Physics* (New York, Oxford and London, 1962)

MS

segmented labour market theory (1970s) *Economics* The labour market consists of various sub-groups which have little cross-over capability.

Wage determination differs from group to group. A common example is DUAL LABOUR MARKET THEORY. *See* DUAL LABOUR MARKET THEORY AND CROWDING HYPOTHESIS.

F Wilkinson, ed., *The Dynamics of Labor Market Segmentation* (New York, 1981)

AV, PH

segregation of genes *Biology See* MENDEL'S LAWS.

selection of the fittest *Biology See* NATURAL SELECTION.

selective theory of antibody diversity (*c*.1900) *Biology* A casual reference to any theory of antibody formation in which the animal is believed to produce pre-existing antibodies selectively that bind to the antigen. Many specific models of this type have been proposed. *See also* ANTIBODY, THEORIES OF and LOCK-AND-KEY MODEL. *Compare with* INSTRUCTIVE THEORY.

KH

self-consistent field (SCF) theory (1932) *Chemistry* Also known as the Hartree-Fock method after the English mathematician and physicist Douglas Rayner Hartree (1897–1958) and the Russian theoretical physicist Vladimir Alexandrovich Fock (1898–).

The original Hartree method uses an iterative variational technique to solve the Schrödinger equation for a many-electron atom by finding a product of single-electron wave functions which are themselves solutions of the Schrödinger equation. The numerical solutions provided by the initial guessed orbitals are used to obtain further improved sets of orbitals until the process yields no significant difference between the results when the orbitals are said to be self-consistent. The Hartree-Fock method is a further refinement which uses determinants rather than products, thereby introducing exchange terms into the Hamiltonian.

GD

self-fulfilling prophecy *Psychology* Defined by the American sociologist William Isaac Thomas (1863–1947) in the dictum: 'If men define situations as real, they are real in their consequences.'

Events tend to turn out as one has hypothesized, not because of some great insight but because one behaves in a manner to achieve this outcome. A large body of evidence exists in various areas of psychology supporting the self-fulfilling prophecy. *See* PYGMALION EFFECT and ROSENTHAL EFFECT.

R Rosenthal, *Experimenter Effects in Behavioral Research*, (New York, 1976)

NS

self-perception theory *Psychology* Proposed by the American psychologist Daryl J Bem (1938–).

A person's attitudes, beliefs and self-characterizations are largely determined by self-observation. This theory has had a large influence upon our ideas concerning the process of attitude change. According to this theory, attitude change results by changing a person's behaviour.

D J Bem, 'Self Perception Theory', *Advances in Experimental Social Psychology*, L Berkowitz, ed. (New York, 1972)

NS

self-serving bias *Psychology* Identified from the work of American psychologist Harold H Kelley (1921–) on social perception.

This is a judgmental bias whereby people tend to deny responsibility for failure and take credit for success. Failure is often explained with reference to situational factors, and success with reference to dispositional qualities. For instance, success at the roulette wheel may be attributed to skill, and failure to bad luck. *See* ACTOR-OBSERVER BIAS, FUNDAMENTAL ATTRIBUTION ERROR and GAMBLER'S FALLACY.

H H Kelley and J L Michela, 'Attribution Theory and Research', *Annual Review of Psychology*, vol. XXXI (1988), 457–501

NS

self-suggestion *Psychology See* AUTO-SUGGESTION.

selfish DNA (20th century) *Biology* Also called junk DNA, this idea was popularized by British evolutionist Richard Dawkins (1941–) in the 1970s, although August Weismann (1834–1914) described the idea around 1900. This theory postulates that functionless deoxyribonucleic acid (DNA), called selfish or junk DNA, uses the cells of organisms to replicate itself.

Put another way, organisms exist as 'survival machines' for their genes. Although the existence of DNA with no known function is undisputed, this theory explaining its existence remains controversial.

R Dawkins, *The Selfish Gene* (Oxford, 1976)

KH

semantic atomism *Philosophy* Theory that the meaning of a phrase or sentence can be analyzed into, and can be constructed out of, the meanings of its constitute words. These meanings can be accounted for independently, and function as atoms of meaning. Similarly, the theory will analyze the meaning of complex sentences in terms of the meanings of their parts. The opposite view is called *semantic holism*.

ARL

semantic feature *Linguistics See* COMPONENTIAL ANALYSIS.

semantic field (1924; 1930s) *Linguistics* Term introduced by G Ipsen; approach developed by the German linguist J Trier and his students in the 1930s.

A conceptual area (for example kinship, cuisine) which is mapped by a structurally related set of vocabulary items, terms delimiting one another in the fashion of Ferdinand de Saussure's VALUE.

S Ullmann, *Semantics* (Oxford, 1962)

RF

semantics (early 20th century) *Linguistics/ Philosophy* Numerous contributors from both disciplines.

The study of meaning in natural language. 'Meaning' is an elusive concept which modern linguists tackle by dispersing into other fundamental ideas such as IMPLICATURE, MEANING-NN, SENSE AND REFERENCE, VALUE. *See* MEANING, THEORIES OF, SEMANTICS, LEXICAL, SEMANTICS, STRUCTURAL and SEMANTICS, TRUTH-CONDITIONAL.

J Lyons, *Semantics* (Cambridge, 1977)

RF

semantics, lexical (1930s, 1960s) *Linguistics* Various movements.

Studies of the meanings of words (as opposed to sentences and utterances): attempts to 'decompose' semantic structure of words (*see* COMPONENTIAL ANALYSIS); to study structural relationships between their senses (SENSE AND REFERENCE, SENSE-RELATIONS); to relate words to their habitual contexts (COLLOCATION); and to conceptual domains (SEMANTIC FIELD).

D A Cruse, *Lexical Semantics* (Cambridge, 1986)

RF

semantics, structural (1963) *Linguistics* Term deployed by the British linguist John Lyons (1932–).

Approaches to word-meaning which attempt to formalize (for example by positing logical relationships) the word-relationships which underpin F de Saussure's VALUE. Applied by Lyons to SEMANTIC FIELD theory and to his own analysis of SENSE RELATIONS.

J Lyons, *Structural Semantics* (Oxford, 1963)

RF

semantics, truth-conditional (1967) *Philosophy/Linguistics* A variant of the CORRESPONDENCE THEORY, and akin to the REDUNDANCY THEORY. It was developed by the Polish logician Alfred Tarski (1902–83), and applied to language by British philosopher D Davidson. (*See also* MONTAGUE GRAMMAR.)

Semantic theory for sentences rather than words (*see also* SEMANTICS, LEXICAL). We know the meaning of a sentence if we know the conditions under which it would be true.

The basic idea is that a sentence like 'Snow is white' is true in English (and '*La neige est blanche*' is true in French) if and only if snow is white. Truth is thus relative to a language, and Tarski applies the theory primarily to certain formal languages, developed so as to deal with the 'liar' paradox (is 'I am now lying' true or false?). For this and other reasons the full statement of the theory is very complicated, which is itself one of the objections to it.

M Black, 'The Semantic Definition of Truth', *Analysis* (1948); reprinted in M Black, *Language and Philosophy* (1949), and in M Macdonald, ed., *Philosophy and Analysis* (1954); R Kempson, *Semantic Thought* (Cambridge, 1977)

ARL, RF

semi-empirical molecular orbital theory *Chemistry* This theory is used to calculate molecular orbitals. The energies and forms of atomic orbitals are defined by a set of predetermined parameters. The minimum energy of an orbital is calculated using a self-consistent field. Various methods of calculation are used depending on the form and number of integrals used to define the atomic orbital. These methods include the complete neglect of differential overlap (CNDO), the intermediate neglect of differential overlap (INDO) and the neglect of diatomic differental overlap (NDDO).

MF

semiotic square (1966) *Stylistics* Developed by the French structuralist Algirdas Julien Greimas.

Analysis of signs by setting them in three kinds of logical opposition: contrariety, contradiction, and contrast. It is an extension of Ferdinand de Saussure's notion of VALUE, and is applied to myths, narratives, and so on, to analyze thematic and other relationships.

R Schleifer, *A J Greimas and the Nature of Meaning* (Beckenham, 1978)

RF

semiotics, semiology (the impact came from posthumous publications in 1916 and 1931) *Stylistics/Linguistics* Two traditions, one deriving from the French linguist Ferdinand de Saussure (1857–1913), the other from the American philosopher Charles Sanders Peirce (1839–1914). The definition below presents literary semiotics as it has been broadly understood from the 1960s onwards.

Semiology was proposed by Saussure as 'a science that studies the life of signs within society'. A society is made up of a multiplicity of language-like 'codes', in all media (speech, literature, architecture, clothes, vehicles, cooking, and so on), which establish objects such as texts, buildings, cars, and so forth as 'signs' having cultural meanings over and above their constructions and functions.

U Eco, *A Theory of Semiotics* (London, 1977)

RF

senescence *Biology See* AGEING, THEORIES OF.

senescence gene hypothesis (1882) *Biology* First proposed by the German evolutionist and cellular physiologist August Weismann (1834–1914), this is the theory that the genes for ageing become part of the genome during evolution, either through NATURAL SELECTION or through the random action of GENETIC DRIFT. Some scientists hypothesize that senescence genes are benign early in life, but develop destructive expression over time. *See also* PROGRAMMED AGEING.

E Schneider, *The Genetics of Aging* (New York, 1978)

KH

sensationalism *Philosophy* Also called *sensationism*, it is associated with Ernst Mach (1838–1916) and various other empiricists of the 18th, 19th, and early 20th centuries.

Either the theory that only sensations exist in what appears to be the material world, everything else being constructed by the methods of PHENOMENALISM; or the theory that all our knowledge must start with sensations, which are free from interpretation and judgment, and about which there is no room for error, the rest of our knowledge being derived from this by inference or by hypotheses which are confirmed by further sensations.

P Alexander, *Sensationalism and Scientific Explanation* (1963)

ARL

sense *Literary theory See* FOUR KINDS OF MEANING.

sense and reference (1892) *Philosophy/Linguistics* Distinction between *Sinn* and *Bedeutung* made by the German mathematician Gottlob Frege (1848–1925).

The meaning of an expression (sense) is a property of language, and is not to be equated with the object or concept the expression may be used to refer to: 'the morning star' and 'the evening star' have different meanings ('senses') but both refer to the planet Venus.

J Lyons, *Semantics* (Cambridge, 1977)

RF

sense relations (1960s) *Linguistics* Theory of structural semantics (*see* SEMANTICS, LEXICAL) principally associated with the British linguist John Lyons (1932–).

The SENSES of words are systematically related by such logical relationships as antonymy ('big/small'), complementarity ('male/female'), converseness ('above/below'), hyponymy ('tulip/flower').

J Lyons, *Semantics* (Cambridge, 1977), ch. 8

RF

sensibility (second half of 18th century) *Literary Theory* Literary reaction against 17th century RATIONALISM and STOICISM.

Awareness of and sympathy for the sufferings of others characterized the 'man of feeling', celebrated in such works as Laurence Sterne's *A Sentimental Journey* (1768) and Henry MacKenzie's *The Man of Feeling* (1771). Ridiculed by Jane Austen, sensibility is now out of favour as producing self-indulgent sentimentality.

L I Bredvold, *The Natural History of Sensibility* (Detroit, 1962)

RF

separation axioms *Mathematics* In TOPOLOGY, these are any of a number of additional assumptions which guarantee to a varying degree that points in a topological space can be separated into disjoint open sets. *See also* URYSOHN'S LEMMA and TYCHONOFF CONDITION.

J R Munkres, *Topology: A First Course* (Prentice-Hall, 1975)

MB

separation of powers *Politics* Constitutional theory principally associated with French political theorist Charles-Louis de Secondat Montesquieu (1689–1755).

Liberty and good government are secured by the powers of government being separated amongst distinct, autonomous, but co-ordinated institutions; usually executive, legislature, and judiciary.

David Miller *et al.*, eds, *The Blackwell Encyclopaedia of Political Thought* (1987)

RB

separation theorem of Mazur (20th century) *Mathematics* Also known as the geometrical form of the Hahn-Banach theorem (1927), this was found by Hans Hahn (1879–1934) and Stefan Banach (1892-1945) and was proved by Stefan Mazurkiewicz (1888–1945).

In *n*-dimensional Euclidean geometry, two sets of points which have the property that any line joining two points in the set lies entirely in the set are separated by a hyperplane, the *n*-dimensional analogue of a plane in three-dimensional space.

JB

separatism (20th century) *Politics/Feminism* Radical theory of FEMINISM.

Since existing institutions are both shaped in the INTERESTS of men and dominated by them, women can only achieve the abolition of gender based oppression by distinctively female and feminist forms of organization and methods of political action. This involves women forming and working within their own institutions for collective action.

Maggie Humm, *The Dictionary of Feminist Theory* (London, 1989)

HR

sequence rules (1880s) *Chemistry* In 1884, the Irish-born chemist Sir William Thomson, Baron Kelvin of Largs (1824–1907), introduced the concept of chirality for organic molecules which were not superimposable on their mirror images. Cahn developed a set of sequence rules for specifying the absolute configuration of such chiral molecules.

The first requires that the four groups bonded to the asymmetric centre be arranged in decreasing atomic number of the atoms by which they are bonded to the asymmetric carbon atom. Atoms of the same atomic number (isotopes) are arranged in decreasing mass number. A lone pair of electrons is considered a substituent and given a priority lower than H. Atoms of the same number and mass are ordered by applying the sequence rules to the atoms bonded to them and so on. Multiple bonds are treated as multiple single bonds; for example,

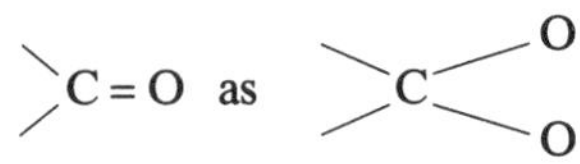

I L Finar, *Organic Chemistry* (London, 1973)

GD

serial endosymbiotic theory *Biology See* SERIAL SYMBIOSIS THEORY.

serial position effect *Psychology See* PRIMACY LAW and RECENCY, LAW OF.

serial symbiosis theory *Biology* Also called progressive endosymbiotic theory, and serial endosymbiotic theory, this postulates that certain structures (organelles) of cells (such as plastids, mitochondria, and possibly cilia and flagella) arose from bacteria and blue green algae (prokaryotic organisms) living in symbiosis within a eukaryotic host cell.

Plastids are thought to have originated from blue green algae, while mitochondria came from aerobic bacteria. This theory relies upon studies of the structural similarities between prokaryotes and some organelles of eukaryotes. *Compare with* AKARYOTIC THEORY, EUKARYOTIC THEORY, PROKARYOTIC THEORY and SYMBIOTIC THEORY OF CELLULAR EVOLUTION

KH

serialism (1920s) *Music* Name subsequently given to techniques first developed by the Austrian-born composer Arnold Schoenberg (1874–1951). The term is used to describe TWELVE-NOTE TECHNIQUE, though it may also denote similar techniques relating to other musical parameters.

The French and American composers Olivier Messiaen (1908–92) and Milton Babbitt (1916–) were the first to use a series of 12 different durations, each duration assigned to a particular pitch. Other parameters that may be similarly ‘serialized’ include instrumentation, timbre, dynamics, tempi and modes of attack. *See also* TOTAL SERIALISM.

G Perle, *Serial Composition and Atonality* (Berkeley, Calif., 1962)

MLD

set theory *Mathematics* The branch of mathematics, initiated by German mathematician Georg Ferdinand Ludwig Philip Cantor (1845–1918), which is concerned with the study of sets.

Cantor considered a set to be 'a collection of definite well-distinguished objects of our intuition or thought' (*Math Ann* 46 (1895)); however, this naive concept soon led to several logical paradoxes. Cantor's ideas were later refined to become part of ZERMELO-FRAENKEL SET THEORY. *See also* AXIOMATIC SET THEORY, NAIVE SET THEORY, and CANTOR'S PARADOX.

Encyclopedic Dictionary of Mathematics (MIT Press, 1987)

MB

set theory *Music See* PITCH-CLASS SET THEORY.

Sewall Wright effect *Biology See* FOUNDERS EFFECT.

sexism *Politics/Feminism* An assumption of fundamental male superiority.

The observable differences between men and women are neither accidental nor socially constructed but are the expression of fundamental and immutable differences which justify the different treatment of men and women and the privileges of the former. Sexism is normally an implication underlying many different theories rather than an articulated argument in itself.

Susan Moller Okin, *Women in Western Political Thought* (Princeton, 1979)

HR

sexual selection (1871) *Biology* A term coined by Charles Robert Darwin (1809–82), eminent English naturalist and evolutionary theorist, in *The Descent of Man and Selection in Relation to Sex*.

The presumed force of natural selection causing individuals to develop reproductive strategies which maximize the likelihood of their producing the fittest possible offspring. Darwin suggested competition between males and competition for females as the two most likely mechanisms of sexual selection. *See also* BATEMAN'S PRINCIPLE, FEMALE CHOICE and MALE-MALE COMPETITION.

P Bateson, *Mate Choice* (Cambridge, 1983)

KH

shadow pricing (1970s) *Economics* Use of linear programming techniques in situations in which a price cannot be charged or where the price does not reflect the effort made in producing the good.

In general this policy attempts to achieve an optimum allocation of resources in the absence of an effective price system. *See* COST-BENEFIT ANALYSIS and OPPORTUNITY COST.

I M D Little and M F D Scott, eds, *Using Shadow Prices* (London, 1976)

AV, PH

Shangri-La phenomenon (1978) *Biology* A term coined by the American biologist Gairdner B Moment, in *The Biology of Aging*, this is the popular idea that there exists somewhere in the world a small band of people who, for reasons of diet, culture or climate, are able to live fruitful lives through 140 or more years of age.

The habitats of such people which are cited most often are the Caucasus mountains in south Russia, Vilcabamba in the Equadorian Andes, and the Karakorum mountains of northern Kashmir. Scientific investigation into claims of extreme longevity among these peoples has so far failed to provide any evidence for longevity claims. *See also* AGEING, THEORIES OF.

J Behnke and C Finch and G Moment, *The Biology of Aging* (New York, 1978)

KH

Shannon's theorem (1948) *Physics* Named after the American applied mathematician and pioneer of information theory Claude Elwood Shannon (1916–).

Information theory is concerned with the encoding, transmitting and interpretation of messages. Shannon proposed a purely mathematical definition of information for any probability distribution within a system, and used this definition to calculate the probability of the information being found amongst the random noise. Shannon borrowed the concept of entropy from statistical mechanics as a way of expressing the degree of uncertainty present in the information.

GD

shared knowledge *Linguistics See* MUTUAL KNOWLEDGE.

Sheldon's constitutional theory *Psychology* Proposed by the American psychologist William Herbert Sheldon (1898–1970).

This theory postulates a strong relationship between personality development and physique. According to Sheldon there are three basic body types with three temperament dimensions. Although successful in uncovering a relationship between body type and temperament, Sheldon's theory is unable to account for the complex variability in personality. *See* CONSTITUTIONAL THEORY.

H J Eysenck, *Fact and Fiction in Psychology* (Harmondsworth, 1974)

NS

shell model of the atomic nucleus (1947) *Physics* Proposed by Maria Goppert Mayer and by O Haxel, J H D Jansen and Hans E Suess. In this model it is assumed that each nucleon (proton or neutron) moves independently in a potential well which represents the averaged effect of its interactions with all the other nucleons.

The wavefunctions and energies of the individual nucleons are obtained by solving the Schrödinger equation for the potential chosen. Discrete energy levels are obtained which are filled up by nucleons in order of increasing energy.

J Thewlis, ed., *Encyclopaedic Dictionary of Physics* (New York, Oxford and London, 1962)

MS

shift of level principle *Psychology* Identified by the German-American psychologist Kurt Koffka (1886–1941).

This refers to the phenomenon whereby when a change of circumstances alters the position of two stimuli on a continuum, the two tend to keep the same relation to each other. *See* GESTALT THEORY.

R H Price, M Glickstein, D L Horton and R H Bailey, *Principles of Psychology* (New York, 1982)

NS

shifters (1922; 1957) *Linguistics* Named by Danish linguist Otto Jespersen (1860–1943) in his book *Language*, but the theoretical importance of the idea was developed by Roman Jakobson (1896–1982). *See also* 'DEIXIS'.

Expressions such as the personal pronouns 'I', 'you', 'she', and so on, which depend for their meaning on the communicative context and message of the moment, 'shifting' according to context.

R Jakobson, 'Shifters, Verbal Categories, and the Russian Verb' [1957, drafted 1950], *Selected Writings*, 2 *Word and Language* (The Hague, 1971), 130–47

RF

shifting balance theory (1931) *Biology* Also referred to as adaptive landscape, this concept was proposed by American population theorist Seawall Wright (1889–1988) and remains controversial. It asserts that evolution often requires combining genes that, if expressed individually, would be harmful to the organism.

Wright believed that this kind of evolution is most likely in a population with many subpopulations, one of which may happen across a favourable combination of genes that would confer an advantage to that subpopulation, enabling it to increase and ultimately improve the entire population. The adaptive landscape provides a metaphor for this theory. *Compare with* FISCHER'S FUNDAMENTAL THEOREM.

G Kelsoe and D Schulze, *Evolution and Vertebrate Immunity* (Austin, Tex., 1987)

KH

shifting, law of *Psychology* Formulated by the American psychologist Edward Lee Thorndike (1874–1949).

This states that it is relatively easy to elicit a response which an animal is capable of performing in any situation and to which it is sensitive, and thereby form an association between the response and the characteristics of that situation. The law corresponds to Pavlov's classical conditioning. *See* PAVLOVIANISM.

B J Wolman and S Knapp, *Contemporary Theories and Systems in Psychology* (New York, 1981)

NS

short-circuiting law *Psychology* Formulated by the Canadian psychologist Donald Olding Hebb (1904–).

The law states that the neurophysiological mechanism underlying the process of a physical activity or a mental process tends to become automatic. By 'automatic', it is meant that no conscious effort of attention is used to perform a particular mental activity. Although it has been established that activities and mental events become automatic, it is has not been proved that short-circuiting is the underlying neurophysiological process responsible.

G Beaumont, *Introduction to Neuropsychology* (Oxford, 1983)

NS

showing and telling (1920s) *Literary Theory* Distinction particularly associated with the aesthetics of the American novelist Henry James (1843–1916).

In narrative fiction, an event or a character may be presented dramatically, apparently without narrator's comment ('shown'); or presented with an obvious narrative presence ('told'). Modern preference is for 'showing'.

W C Booth, *Rhetoric of Fiction* (Chicago, 1961)

RF

side-chain theory (1900) *Biology* Proposed by Paul Ehrlich (1854–1915), a German bacteriologist and a winner of the 1908 Nobel prize for Physiology or Medicine. This is a model for immunologic specificity, holding that immunocytes come equipped with diverse side-chain groups of chemical receptors on their surface which interact with antigens, either singly or in groups.

Once the side-chain group and the antigen interact, the complex is released from the cell, which triggers the synthesis of more side-chain groups. Ehrlich's model, the first specific SELECTIVE THEORY OF ANTIBODY DIVERSITY, has been abandoned, although the general concept is widely accepted. *See also* ANTIBODY, THEORIES OF and LOCK-AND-KEY MODEL.

G Bell and A Perelson and G Pimbley, *Theoretical Immunology* (New York, 1978)

KH

Sierpinski gasket *Mathematics* Named after the Polish mathematician Waclaw Sierpinski.

A FRACTAL object formed using a recursive procedure in which inverted equilateral triangles are repeatedly removed from an initial equilateral triangle. The resulting figure, after this has been repeated infinitely often, is the Sierpinski gasket.

I Peterson, *The Mathematical Tourist* (New York, 1988)

ML

sieve (of Eratosthenes) *Mathematics* Named after Greek mathematician and astronomer Eratosthenes of Cyrene (*c.*276–195 BC).

Eratosthenes's sieve for primes is a systematic procedure for isolating the prime numbers. Much more sophisticated sieves are used today in prime number theory. *See also* PRIME NUMBER THEOREM and DIRICHLET'S THEOREM ON ARITHMETICAL PROGRESSIONS.

MB

sign (1916) *Linguistics* Formulated by the French linguist Ferdinand de Saussure (1857–1913).

A sign is a union of a signified (concept) and signifier (vocal sound or writing). The relation between these components is 'arbitrary' in that signifiers vary in different languages: for example, dog, *chien*, *Hund*. There is no intrinsic connection (such as onomatopoeia) between signifier and signified. *See also* VALUE.

F de Saussure, *Course in General Linguistics*, W Baskin, trans. (Glasgow, 1974 [1916])

RF

sign language (18th century–) *Linguistics* Taught in various countries since the 18th century, it became an important research topic in linguistics from the 1970s onwards.

A system of gestures, visually received, for communication among the deaf. It has its own structure, not based on vocal language. Different countries have different sign languages.

M Deuchar, *British Sign Language* (London, 1984)

RF

signal detection theory *Psychology* Proposed by the American psychologist John Arthur Swets (1928–).

The detection of stimuli involves decision processes as well as sensory processes. According to the theory, an organism's capacity to detect a stimulus is determined by the intensity of the stimulus and the level of 'noise' (irrelevant stimulus activity) in the organism. The theory was imported into psychology from engineering through the work of Swets, W P Tanner and T G Birdsall, evolving from practical attempts to understand and improve the monitoring of complex displays such as radar.

J A Swets, W P Tanner and T G Birdsall, 'Decision Processes in Perception', *Psychological Review*, vol. LXVIII (1961), 301–40

NS

signal hypothesis (1970s) *Biology* Proposed by G Blobel and D D Sabatini, this is an explanation for how newly synthesized proteins are selected and transported to their proper locations in the cell.

Specifically, proteins that are to be secreted, sent to lysosomes (digestive organelles), or incorporated into the cell membrane are encoded by a messenger ribonucleic acid (mRNA) containing a nucleotide sequence that acts as a signal. This signal is picked up by the 'signal recognition particle', which causes a temporary pause in protein synthesis by the ribosome involved. The resulting complex then 'docks' itself to the proper site on the endoplasmic reticulum, where protein synthesis is resumed. This complex theory constitutes the prevailing explanation for this phenomenon, and is supported by a substantial body of evidence. *Compare with* MEMBRANE TRIGGER HYPOTHESIS.

P Sheeler, *Cell and Molecular Biology* (New York, 1987)

KH

signalling (1970s) *Economics* The provision of information for decision-making, especially regarding price. Price changes signal producers that supply and demand are no longer in equilibrium. *See* ADVERSE SELECTION.

A M Spence, *Market Signalling: Information Transfer in Hiring and Related Processes* (Cambridge, Mass., 1973)

AV, PH

signifier and signified *Linguistics See* SIGN.

similarity, law of *Psychology* Formulated by German-American psychologist Wolfgang Köhlererman (1887–1967).

One of the Gestalt Psychology laws of perceptual organization which state that physically similar objects tend to be grouped together. This is the generalization that stimuli are more likely to elicit similar stimuli than unfamilar stimuli. *See* GESTALT THEORY.

J P Dworetzky, *Psychology*, 3rd edn (New York, 1985)

NS

similarity paradox *Psychology* Identified by the American psychologists Ernest B Skaggs (*fl.*1920s–1940s) and Edward S Robinson (1893–1937).

This is the paradox in learning whereby when things are made as similar as possible from trial to trial there exist, on the one hand, conditions for maximal interference (like things interfering with each other more than unlike things); and, on the other hand, the best practical conditions for effective learning. The SKAGGS-ROBINSON HYPOTHESIS was proposed as a solution to this paradox.

J A McGeogh, *The Psychology of Human Learning* (New York, 1942)

NS

simple quark model of hadrons *Physics* Hadrons are elementary particles which can take part in strong interactions. They are of two kinds: mesons with zero or integer spin and consisting of a quark-antiquark pair, and baryons with half-integer spin and consisting of three quarks. The constituents are bound together within the hadron by the exchange of particles known as gluons, which are neutral massless gauge bosons. Quarks are elementary fermions with spin $\frac{1}{2}$, baryon number $\frac{1}{3}$, strangeness 0 or -1 and charm 0 or $+1$. They have six 'flavours' (each of three 'colours') with charges of either $\frac{2}{3}$ or $-\frac{1}{3}$ of the proton charge. Hadrons themselves do not show fractional charges, since the quarks form combinations with either integral or zero values.

Paul Davies, ed., *The New Physics* (Cambridge, 1989)

GD

simplest path law *Psychology* Formulated by the German psychologist Max Wertheimer (1886–1943).

This is a principle of Gestalt Psychology which holds that behaviour always follows the simplest path open to the organism at that particular time. It has been used to explain how children learn articulation skills and how economic systems operate. *See* GESTALT THEORY.

R L Solomon, 'The Influence of Work on Behaviour', *Psychological Bulletin*, vol. XLV (1948), 1–4

NS

simplicity, law of *Psychology* Formulated by the German-American psychologist Wolfgang Köhlererman (1887–1967).

The law of simplicity is derived from the term *pragnanz* in Gestalt Psychology, which, roughly translated, means 'good figure'. It is a law of perceptual organization stating that we perceive the result of stimulus patterns in the simplest form possible. *See* GESTALT THEORY.

E B Goldstein, *Sensation and Perception*, 3rd edn (Belmont, Calif., 1989)

NS

Simpson's paradox (1951) *Mathematics* Named after its discoverer E H Simpson.

In two trials of two brands of hair restorer, Vulpes reactivated 1 hair in 10 and 14 hairs in 40, with success rates 10 per cent and 35 per cent; whereas Lupus reactivated 5 in 40 and 4 in 10, with success rates 12.5 per cent and 40 per cent. The conclusion that Lupus is more successful is not supported by the totals: Vulpes reactivated 15 hairs in 50, with success ratio 30 per cent, while Lupus reactivated 9 hairs in 50, with success ratio 18 per cent. Therefore STATISTICS can support one conclusion when taken separately, and the contrary when taken together.

JB

Simpson's rule (18th century) *Mathematics* Named after British analyst, geometer, algebraist and probabilist Thomas Simpson (1710–61), although the result was well known in Simpson's lifetime.

The method for approximating a definite integral

$$\int_a^b f(x)dx \approx \frac{\delta}{3}[f(a)+4f(a+\delta)+2f(a+2\delta)+ 4f(a+3\delta)+2d(a+4\delta)+\ldots+f(b)]$$

where

$$\delta = (b-a)/2n.$$

The error of the approximation is

$$M(b-a)^{2n}/180n^4$$

where M is the maximum of the absolute value of the fourth derivative of f on the interval. This method is much more accurate than the TRAPEZOIDAL RULE and is exact for cubic functions.

H Anton, *Calculus with Analytic Geometry* (New York, 1980)

ML

simulated anealing (*c.*1985) *Computing* Named for its physical analogy, this is a random walk technique to avoid local extrema while searching for global extrema.

JB

simultanéisme (*c.*1912–13) *Art* Term adopted by the French painter Robert Delaunay (1885–1941) from an 1839 publication by colour theorist Michel Eugène Chevreul (1786–1889).

Delaunay sought to create an ABSTRACT ART that was dependent on colour alone to suggest form and movement. Simultaneity became crucial in the arts just before World War I. Multiple awareness, knowledge of things happening concurrently, and the concept of the continuous present were its key features.

M E Chevreul, *De la loi du contraste simultane des couleurs et de l'assortment des objets colorés* (1839); R Delaunay, *Du cubisme à l'art abstrait. Documents inédits*, P Francastel, ed. (Paris), 157

AB

simultanéisme (*c.*1900–20) *Literary Theory* French school of poets, led by H-M Bazun, who aimed for a coincidence or simultaneity of image and sounds, and mingled human speech and other noises in their verse.

Guillaume Apollinaire (1880–1918) also adopted the term in relation to his 'Calligrame' poetry.

H-M Bazun, *La Trilogie des forces* (1908–14)

RF

sincerity conditions *Linguistics See* SPEECH ACT THEORY.

single-gene hypotheses (20th century) *Biology* Developed by Collins and Fuller, who studied seizure susceptibility in mice, this is the group of hypotheses holding that one particular gene is responsible for a given characteristic or behaviour.

For example, the ability to taste phenylthiocarbamide (PTC) in humans is associated with the presence or absence of a single gene. Many single-gene hypotheses exist; some have been verified experimentally while others have been rejected after extensive study.

D Dewsbury, *Comparative Animal Behavior* (New York, 1978)

KH

single-line evolution *Biology See* ANAGENESIS.

situationism (1960–) *Art* This term is applied to works (particularly, large monochromatic paintings) where the spectator's aesthetic response is determined by the intensity and saturation of colour. In Great Britain, the Situation group exhibited together in 1960, its adherents including B Cohen, R Denny, J Hoyland and W Turnbull. Their work was strongly related to Colour Field painting in America. Situationism is also applied to all-encompassing sculpture, where a whole site is used; for example, the work of Carl André (1935–).

LL

situationism *Philosophy* Ethical doctrine that our moral duty cannot be rigorously subjected to general rules, but must take account of each situation as it arises.

Unlike ANTINOMIANISM it does not reject such rules altogether, but insists on flexibility in applying them. Unlike *casuistry* it does not insist on breaching rules only if some other rule can be found which takes precedence, but appeals rather to love as its supreme guiding principle. It may, however, be accused of similarly leading to uncertainty, or even moral anarchy, with inconvenient comparisons being rejected because of the alleged uniqueness of the present case.

J Fletcher, *Situation Ethics* (1966)

ARL

situationism (20th century) *Politics* Advocacy of comprehensive revolution.

Radical and not always precise proposal for total revolutionary transformation of every aspect of life, beginning with individual experience; much in evidence in France in May 1968.

Alan Bullock, Oliver Stallybrass, and Stephen Trombley, eds, *The Fontana Dictionary of Modern Thought*, 2nd edn (London, 1988)

RB

Skaggs-Robinson hypothesis *Psychology* Formulated by Ernest B. Skagg (*fl.*1920s–40s) using his own work and that of Edward S Robinson (1893–1937) dealing with the effect that similarity has on the recall of successively presented material. Maximum similarity between sets of material leads to maximal recall, and as similarity decreases so does recall until maximum dissimilarity which increases recall again. This hypothesis has been proposed as a solution to the SIMILARITY PARADOX.

E S Robinson, *Americal Journal of Psychology*, vol. XXXIX (1927), 297–312

NS

skaz (1919) *Stylistics* Concept analyzed and named by the Russian Formalist (*see* FORMALISM, RUSSIAN) Boris M Eichenbaum (1886–1959).

A first-person narrative style which foregrounds the oral, personal, aspects of delivery by puns, unusual or vernacular vocabulary, sound-gestures and intonation.

B Eichenbaum, 'The Structure of Gogol's "The Overcoat"', trans. B Paul and M Nesbitt, *Russian Review*, 22 (1963), 377–99

RF

skin effect *Physics* This is the phenomenon whereby very high frequency alternating currents tend to flow on the surface of a conductor.

J Thewlis, ed., *Encyclopaedic Dictionary of Physics* (New York, Oxford and London, 1962)
MS

Skolem paradox (1929) *Mathematics* Named after its discoverer Thoralf Skolem (1887–1963).

As a consequence of the theorem that Skolem proved with L Lowenheim in 1922, he showed that the construction of the real numbers can give an arithmetic in which the numbers can be associated one each with the positive integers, a property the set of real numbers does not have. This shows that the number of elements in a set varies with the theory in which it is defined, a valuable principle to apply to a bank overdraft.
JB

sleeper effect *Psychology* Identified by the American psychologist C I Hovland (1912–61).

This effect was devised to describe the 'hidden' impact that a mass communication or propaganda message can have on its audience. The attitude change produced by the message is frequently not detectable until a period of time has passed, hence the term 'sleeper effect'.

C I Hovland, A A Lumsdaine and F D Sheffield, *Experiments in Mass Communication* (Princeton, N.J., 1949)
NS

slender body theory of fluid flow *Physics* This is the simplest form of the general theory of the flow of a compressible, inviscid (non-viscous) fluid past a body.

The theory requires that the body have a pointed nose and base and a smooth surface; also, the ratio of the maximum thickness of the body to its length must be much less than unity and the angle between the tangent plane at any point on the body and the direction of motion must be small.

J Thewlis, ed., *Encyclopaedic Dictionary of Physics* (New York, Oxford and London, 1962)
MS

slot and filler grammar *Linguistics* *See* TAGMEMICS.

Slutsky's theorem (1915) *Economics* Named after its proposer, Soviet economist Evgeny Slutsky (1880–1948), this hypothesis was later developed by English economists John Hicks (1904–89) and Roy Allen (1906–). In its simplest form: Price effect = income effect + substitution efect.

Slutsky asserted in 1915 that demand theory is based on the concept of ordinal utility. This idea was developed by Hicks who separated the consumer's reaction to a price change into income and substitution effects.

E Slutsky, 'On the Theory of the Budget of the Consumer', *Readings in Price Theory*, K E Boulding and G J Stigler, eds (1953)
PH

Slutsky's theorem (20th century) *Mathematics* Named after Russian mathematician Evgeny Evgenievich Slutsky (1880–1948), this element of PROBABILITY THEORY asserts that if $X_1, X_2, \ldots X_n, X$ is a sequence of random variables such that

$$\lim_{n\to\infty} P[X \leq x] = P[X \leq x]$$

where $P[X \leqslant x]$ is continuous everywhere for some random variable x, then

$$\lim_{n\to\infty} P[g(X_n) \leq y] = P[g(X) \leq y]$$

for any continuous function g.

H T Nguyen and G S Rogers, *Fundamentals of Mathematical Statistics* (New York, 1989)
JB

Smale horseshoe map *Mathematics* *See* HORSESHOE MAP.

small number, law of the (20th century) *Politics/Sociology* Theory of the German social scientist Max Weber (1864–1920) regarding the influence of small groups in key positions.

'The ruling minority can quickly reach understanding among its members; it is thus able at any time quickly to initiate that rationally organized action which is necessary to preserve its position of power.' A concise statement of practical ÉLITISM.

Max Weber, *Economy and Society*, Guenther Roth and Claus Wittich, eds, 2 vols (London, 1978)
RB

small numbers, law of (20th century) *Mathematics* (1) Independent trials where the kth probability of success p_k behaves like λ/kth probability of success up to the nth trial behaves like a Poisson distribution with mean λ. (2) Informally, mathematicians use the law to say that it is dangerous to infer patterns from a few cases.

Encyclopedic Dictionary of Mathematics (MIT Press, 1987)

JB

small is beautiful principle (1973) *Economics* Developed by the German-born British economist and businessman Ernst Schumacher (1911–77), who advocated 'intermediate' and alternative technologies.

The principle challenged the tradition of large organizations, which Schumacher claimed were inefficient and a danger to the environment. He proposed small working units, communal ownership, and the use of local labour and resources. He placed the emphasis on people rather than the product.

E F Schumacher, *Small is Beautiful: A Study of Economics as if People Mattered* (London, 1973)

AV, PH

Snell's law *Physics See* REFRACTION OF LIGHT, LAWS OF.

snowblitz theory *Geology* A theory to explain the formation of ice sheets.

If snow persisted in lowland areas during a summer, it would reflect sunlight back into space and so maintain a lower temperature in the area than before. This would enable the thickness of snow to increase subsequently.

DT

social accounting theory (20th century) *Accountancy* This is an examination of the NATIONAL INCOME and expenditure accounts, showing all transactions during a given period of time, in different areas of the economy.

PH

social adjustment theory *Psychology* Not attributable to any one individual, this is an approach to the study of attitude formation and change.

Theories subsumed under this approach include ASSIMILATION-CONTRAST THEORY and ADAPTATION LEVEL THEORY, and all view change in terms of the manner in which an individual lives in and adapts to his/her social environment. These theories have adopted an experimental methodology and many use techniques derived from psychophysics.

K G Shaver, *Principles of Social Psychology* 3rd edn (London, 1987)

NS

social breakdown model of ageing *Psychology See* DISENGAGEMENT THEORY OF AGEING.

social choice (20th century) *Politics* Application of RATIONAL CHOICE THEORY to the provision of welfare services and public goods and the relation between citizens' preferences and public policy.

There are difficulties in rationally translating people's choices into public policy, as indicated in the IMPOSSIBILITY THEORY. Social choice is a theory not so much of the translation of individual into collective decisions as of the difficulty or impossibility of achieving such a connection.

Geoffrey Roberts and Alistair Edwards, *A New Dictionary of Political Analysis* (London, 1991)

RB

social comparison theory *Psychology* Proposed by the American psychologist L Festinger (1919–).

This posits that one uses other people as a basis for comparison in order to evaluate one's own judgments, ability, attitudes and so on. Social comparisons are particularly important when other objective standards are not available. The theory suggests a preference for associating with others who have similar or slightly superior abilities because such interactions yield the most informative social comparisons. Although the details of the theory are frequently criticized, few doubt the importance of its fundamental principles.

L Festinger, 'A Theory of Social Comparison Process', *Human Relations*, vol. VII, (1954), 117–40

NS

social contract *Politics/Law* Theory of POLITICAL OBLIGATION or authority.

The duty to obey government and law, and the right to govern and make law, arises from a contract or agreement either between ruler and ruled, or amongst the ruled. It is not normally suggested that there ever was an actual historical contract, but that it would be reasonable to behave as if there had been.

P Riley, *Will and Political Legitimacy* (Cambridge, Mass., 1982)

RB

social credit (20th century) *Politics* A theory of extended and distributed purchasing power.

Individuals would be benefited, and the economy stimulated, by the distribution of 'social credits.' The theory had little impact, and a short life.

John L Finlay, *Social Credit* (London, 1972)

RB

social Darwinism (19th century) *Politics/ Sociology* Attempted application of theory of the SURVIVAL OF THE FITTEST to public policy.

The provision of social services allowed the 'unfit' to survive, and reproduce children who inherited their social characteristics. Such services, therefore, however well-meaning, damaged society. *See also* EUGENICS.

David Miller *et al.*, eds, *The Blackwell Encyclopaedia of Political Thought* (Oxford, 1987)

RB

social democracy (19th century–) *Politics* Theory of democratic socialism.

Although social democracy was used to describe 19th century Marxist parties in continental Europe, the phrase established a distinctive, non-Marxist and non-revolutionary meaning after the 1917 Russian Revolution. SOCIALISM can and should be achieved by democratic means, thus extending the meaning of DEMOCRACY from simple political POWER to include social and economic power exercised by the electorate through the state.

David Miller *et al.*, eds, *The Blackwell Encyclopaedia of Political Thought* (Oxford, 1987)

RB

social exchange theory *Psychology* Proposed by the American psychologist Harold H Kelley (1921–) and John W Thibaut (1917–).

The basic principle states that much social interaction involves reciprocal exchange of material things, of ideas, of emotions and of behaviour. This theory expresses the reciprocal rule that one individual gives and another returns, in some measure, what is given. Status or resource differences between donors and recipients ensure the continuity of the reciprocal relationship by causing an imbalance and placing one or other in debt. *See* EQUITY THEORY.

J W Thibaut and H H Kelley, *The Social Psychology of Groups* (New York, 1959)

NS

social facilitation theory *Psychology* Proposed by the American psychologist N Triplett, this is a general theory indicating that the presence of another person has a motivational effect upon the performance of a subject by enhancing dominant responses.

This facilitation effect occurs with relatively well-learned/automatic behaviours; the presence of others can inhibit or interfere with behaviours that are not so well-learned or that are highly complex. In 1965, the Polish psychologist Robert Boleslaw Zajonc (1923–) argued that the presence of another person increases the drive of the subject, leading to improved performance when correct responses are dominant and worse performances when errors are dominant. However, Zajonc's explanation is far from conclusive.

R B Zajonc, 'Social Facilitation', *Science*, vol. CXLIX, (1965) 269–74

NS

social group theory *Psychology* Not attributable to any one individual, this is a general term used for psychological theories which address group processes.

A social group consists of a number of individuals who define themselves to be members of a group. They interact frequently, are defined as members by people outside the group, and share certain characteristics and behaviours. They expect certain behaviours of their fellow members which they do not expect of non-members. The investigation of group processes and effects is central to social psychology.

R K Merton, *Social Theory and Social Structure* (New York, 1968)

NS

social learning theory *Psychology* Proposed by the Canadian psychologist Albert Bandura (1925–).

This postulates that social behaviour develops mainly as a result of observing others ('models') and of reinforcing specific behaviours. Considerable research has been conducted within a paradigm described by Bandura. The theory is regarded as particularly successful in reconciling the importance of reinforcement with concepts that depict a thinking, knowing person with his expectations, beliefs and choices.

A Bandura, *Social Learning Theory* (Englewood Cliffs, N.J., 1977)

NS

social markers in speech (1970s) *Linguistics* A social psychological development of SOCIOLINGUISTICS in the 1960s.

Speech contains systems of phonetic, syntactic and lexical markers of social group (for example gender, ethnicity, age, personality, and so on) and social situation. These are expressed unconsciously, or may be manipulated towards a desired effect. *See also* ACCOMMODATION THEORY and ACT OF IDENTITY.

K R Scherer and H Giles, eds *Social Markers in Speech* (Cambridge and Paris, 1979)

RF

social networks and language (1975) *Linguistics* Methodological innovation in SOCIOLINGUISTICS, pioneered by Lesley Milroy.

A speech community is a network of relationships, of various kinds and strengths, between people. The fieldworker avoids the OBSERVER'S PARADOX by entering the network as a 'friend of a friend' and becomes an 'insider'; thus the spontaneity of speech she observes is not disturbed by her presence.

L Milroy, *Language and Social Networks* (Oxford, 1980)

RF

social novel (1903) *Literary Theory* Named by French literary historian Louis Cazamian.

Certain novels (published *c.*1830–60) which concern urban, industrial life, feature working-class characters, and explore moral, political and social issues arising from economic and technological changes within these domains. Prominent writers of social novels include Charles Dickens (1812–70), George William MacArthur Reynolds (1814–79), and Mrs (Elizabeth) Gaskell (1810–65).

K Tillotson, *Novels of the Eighteen-Forties* (London, 1954)

RF

social semiotic, language as (1970s) *Linguistics* Perspective on language developed by the British linguist Michael Alexander Kirkwood Halliday (1925–)

The structure of a language, and the structure of its varieties or REGISTERS, embodies the culture's categorization and evaluation of its experience; reciprocally, language shapes and maintains a society's ideologies.

M A K Halliday, *Language as Social Semiotic* (London, 1978)

RF

social welfare function (1938) *Economics* Introduced by American economist Abram Bergson (1914–) as a rejection of the cardinal utility approach to welfare economics, this term is understood as meaning the determination of a society's taste for different economic states.

There are two approaches to the welfare function: first, that it is an imposed structure; second, that it devises a single constitutional/voting system which changes the rankings of the individual into a single society ranking. *See* IMPOSSIBILITY THEOREM.

A Bergson, 'A Reformulation of Certain Aspects of Welfare Economics', *Quarterly Journal of Economics* 52, 2 (February 1938) 310–34

PH

socialism *Politics* Theory of the social nature or aspects of production and of its consequences.

Socialism is characterized by enormous variety. It generally involves the argument that economic production has an essential social as distinct from individual element, and that this requires public investment and justifies a public share in and distribution of rewards.

David Miller *et al.*, eds, *The Blackwell Encyclopaedia of Political Thought* (Oxford, 1987)

RB

socialism in one country (20th century) *Politics* Theory of REVOLUTION associated with the Russian revolutionary and politician Joseph Stalin (1879–1953) and with STALINISM.

Against the theory of PERMANENT REVOLUTION, it was argued that socialism could and should be built first in one country (the Soviet Union), thus accommodating MARXISM to Russian NATIONALISM.

David McLellan, *Marxism After Marx* (London, 1980)

RB

socialist realism (1934) *Art* The official Marxist artistic and literary movement established in the USSR in 1934.

Used as a propaganda device and painted in a naturalistic and idealized style, its art was faithfully and unflinchingly to depict history and phenomena relative to Marxism, and to promote the concept of the classless society.

AB

sociobiology (1975) *Biology/Sociology* Developed and popularized by American zoologist Edward O Wilson (1929–) in his text *Sociobiology: The New Synthesis* (1975), this is the study of animal and human social behaviour in the light of evolutionary biology.

Sociobiology encompasses the idea that evolution by natural selection applies to behaviour as it applies to physiology and morphology, and that genes affect behaviour and will be selected for if the behaviour enhances the 'fitness' of the organism for its environment. A relatively new field of study, sociobiology remains controversial, especially when applied to human behaviour.

David P Barash, *The Hare and the Tortoise* (London, 1987)

KH

sociolinguistics (1960s–) *Linguistics* Also called 'sociology of language'.

Sociolinguistics is the broad term for studies of relationships between linguistic structure and social structure, and of individual relationships mediating social structures. Linguistic analysis of data often focuses on pre-selected variables (for example phonemes, vocabulary, naming practices); the techniques of experimental social science (sample, interview, questionnaire, quantitative analysis, and so on) may be more or less prominent. Many different methodological models are practised, and in many different domains. The classic paradigm is CORRELATIONAL SOCIOLINGUISTICS. *See also* ACCOMMODATION, ACT OF IDENTITY, ANTI-LANGUAGE, REGISTER, SOCIAL MARKERS IN SPEECH.

W Downes, *Language and Society* (London, 1984; revised ed. forthcoming)

RF

Soddy and Fajans rule *Physics See* DISPLACEMENT LAW IN RADIOACTIVE DECAY.

sodium hypothesis (1904) *Medicine* Also called the salt-gene hypothesis, and the salt hypothesis, this theory was first proposed by French physicians Léon Ambard and E Beaujard, although clinical observations reported in the 18th century provided a basis for their work. The theory asserts that in genetically susceptible individuals, high blood pressure (hypertension) can be caused by a high salt (or high sodium) diet.

Although accepted as conventional medical wisdom throughout much of the 20th century, recent research has led many physicians to question the salt-hypertension link.

G Porter, 'Chronology of the Sodium Hypothesis and Hypertension', *Annals of Internal Medicine* vol. XCVIII (1983), 720–23

KH

Sohncke's law of brittle fracture (1869) *Physics* Named after its discoverer, the German physicist Leonhard Sohncke (1842–97), this law states that when a single crystal ruptures by brittle fracture (that is, with little or no plastic deformation), rupture occurs when the normal component of the stress acting across the plane of rupture (cleavage plane) reaches a critical value characteristic of the material. The law does not hold when significant plastic deformation occurs before fracture.

J Thewlis, ed., *Encyclopaedic Dictionary of Physics* (New York, Oxford and London, 1962)

MS

solipsism *Philosophy* Literally, 'only-oneself-ism'. An extreme form of *scepticism*, saying that nothing exists beyond oneself and one's immediate experiences. Seldom held deliberately, it is more likely to be fallen into by those who find themselves in the EGOCENTRIC PREDICAMENT, perhaps through holding solipsism as a METHODOLOGICAL THEORY; that in enquiring into a certain area it is best, or inevitable, to start from oneself and one's immediate knowledge. One objection is that one cannot even state it coherently, since to do so requires that one possess concepts one could never obtain were solipsism true: what could words like 'I' or 'my' mean except against a background of what is not me or mine? *See also* PRIVATE LANGUAGE ARGUMENT.

ARL

solmization (antiquity) *Music* A theoretical system in which single syllable labels (for example 'doh, ray, mi' . . .) are assigned to individual pitches, indicating the relative position of each pitch within a fixed pattern, such as a scale.

It differs from fixed-pitch scales ('A, B, C . . .') in that any pitch may be designated 'doh'; after that, the pitch of each other labelled note may be deduced by its distance from doh, an interval that remains constant. *See also* TONIC SOL-FA.

MLD

Solow economic growth (1970s) Named after American economist Robert Solow (1924–), this principle highlights the relationship of technological change to growth.

As the rate of return falls, firms turn to more capital intensive methods of production; therefore the rate of investment increases. However, it is possible to show a situation in which the rate of return would be such that firms would reduce their level of capital intensity, causing investment and the rate of return to decline; this is called capital reswitching. *See* ROUNDABOUT METHOD OF PRODUCTION.

R M Solow, *Growth Theory: an Exposition* (Oxford, 1970)

PH

somatic mutation hypothesis or theory (1914) *Biology* Proposed by German zoologist Theodor Boveri (1862–1915) in his book, *Zur Frage der Entstehung Maligner Tumoren* (*The Origin of Malignant Tumours*) (1914). The theory that cancer is caused by at least one mutation in the cells of the body.

The existence of widespread chromosomal abnormalities in cancer cells, plus the correlations observed between the mutagenicity and carcinogenicity of radiation and chemicals, provide substantial support for this hypothesis. Somatic mutation hypothesis also refers to the theory that antibodies are formed by the hypermutation of certain genes, or the expansion of clones in which advantageous mutations have occurred. *Compare with* CHROMOSOME THEORY OF CANCER.

KH

somatic mutation theory of ageing (1959) *Biology/Medicine* Also called RANDOM HITS THEORY OF AGEING. Proposed by American Leo Szilard, a nuclear physicist known for his work on the first nuclear reactor, this theory is based on observations of radiation-induced cell damage (made from the late 1930s onwards) and asserts that ageing is caused by an accumulation of mutations in the genomes of cells in the body (the somatic cells).

These changes in the genetic material are believed to inactivate the chromosome gradually until it loses the ability to function, causing cell death. *See also* AGEING, THEORIES OF. *Compare with* ERROR CATASTROPHE THEORY OF AGEING.

J Behnke and C Finch and G Moment, *The Biology of Aging* (New York, 1978); E Schneider, *The Genetics of Aging* (New York, 1978)

KH

somatic mutation theory of antibody diversity (1959) *Biology* Any of several theories holding that antigens interact with the immune system to provide information for the synthesis of antibodies. A modern version was proposed by Lederberg based on ideas from the 1930s.

The mutation of an antibody gene or genes, specifically those in the V-region of the germline, is the proposed mechanism for generating antibody diversity. The theory implies that a mutational frequency of at least 1 x 10199^5 per base pair per division is required. This theory counters the idea that animals carry millions of genes to form a great variety of antibodies in advance of encountering a specific antigen, and represents one of the two major schools of thought on antibody diversity. *See also* ANTIBODY, THEORIES OF. *Compare with* GERMLINE THEORY OF ANTIBODY DIVERSITY.

G Kelsoe and D Schulze, *Evolution and Vertebrate Immunity: The Antigen-Receptor and the MHC Gene Families* (Austin, Tex., 1987)

KH

somatic recombination model (1967) *Biology* Proposed by G M Edelman and J A Gally, this theory asserts that new antibodies are formed by recombinations of genes responsible for the synthesis of antibodies.

Edelman and Gally postulate the existence of about 50 antibody genes which evolved in tandem. Because tandem genes would be so similar in components and length, they suggest that the mispairing between two deoxyribonucleic (DNA) strands might occur often enough that non-identical genes would lie side-by-side. Crossing over at any point along these genes would create a new gene. This theory, originally considered daring but now enjoying experimental support, represents an alternative to both the GERMLINE THEORY OF ANTIBODY DIVERSITY and the SOMATIC MUTATION THEORY OF ANTIBODY DIVERSITY. *See also* SIDE-CHAIN THEORY.

G M Edelman and J A Gally, 'Somantic Recombination of Duplicated Genes: An Hypothesis on the Origin of Antibody Diversity', *Proceedings of the National Academy of Sciences (USA)*, vol. LVII (1967) 353 ff.

KH

Sommerfeld law for doublets *Chemistry* Named after the German physicist Arnold Johannes Wilhelm Sommerfeld (1868–1951) who was Professor of Physics at Munich and applied the quantum theory to spectral lines.

For relativistic as well as regular doublets, the law states that the splitting in frequency is

$$\frac{\alpha^2 R(Z-\sigma)^4}{n^3(l+1)}$$

where α is the fine-structure constant, R is the Rydberg constant of the atom, Z is the atomic number, σ is the screening constant, n is the principal quantum number and l is the orbital angular momentum quantum number.

GD

S-O-R theory *Psychology* Proposed by the American psychologist Robert Sessions Woodworth (1869–1962).

This is a change to the STIMULUS-RESPONSE (S-R) FORMULA proposed by Woodworth. Woodworth opposed BEHAVIOURISM to the degree that it neglected the organism in behaviour. Therefore he introduced O for the organism into the S-R formula to produce S-O-R. Woodworth contributed to the rediscovery of the organism's role in its own behaviour.

R S Woodworth, *Dynamic Psychology* (New York, 1918)

NS

sotto in su *Art* An Italian term meaning 'from below upwards', used from the 15th to the 18th century to describe ceiling painting in which the extreme foreshortening of figures and architecture promotes the impression of objects suspended in space.

AB

sound pattern theories of hearing *Psychology* Proposed by the German, J R Ewald (1855–1921).

This is one of Ewald's theories of hearing developed in an attempt to explain how physical sound vibrations give rise to the neural impulses of hearing. Sound-pattern theory proposes that different patterns of vibrations which impinge on the basilar memory (a delicate membrane in the cochlea of the inner ear) are the important factor here. The theory is outdated.

E G Boring, *A History of Experimental Psychology* (New York, 1957)

NS

sovereignty *Politics* Theory of the location of legal power.

In any territory there must be a sovereign, in the sense of an institution or person with the power to make ultimate and unchallengeable legal decisions. There is much disagreement as to the location and character of sovereignty. *See also* AUSTINIANISM.

David Miller *et al.*, eds, *The Blackwell Encyclopaedia of Political Thought* (Oxford, 1987)

RB

sovereignty (1983) *Stylistics* Term coined by the British stylisticians R A Carter and Walter Nash (1926–).

The dissociation of a literary text from any practical communicative context, allowing the new meanings produced by RE-REGISTRATION. *See also* AUTONOMY OF LITERARY TEXT.

R A Carter and W Nash, 'Language and Literariness', *Prose Studies*, 6 (1983), 123–41

RF

space *Physics* That in which material bodies exist, a three-dimensional continuum. The Greeks, and Aristotle in particular, conceived the notion of absolute space. Newton thought in terms of a Cartesian reference system filling the universe, in a state of absolute rest and to which the motion of all objects could be referred. Others later refined his ideas to invoke the so-called 'frame of the fixed stars', which took as its reference the most distant stars seen from Earth.

Newtonian mechanics, however, cannot logically support the concept of absolute space: Newton's laws describe the same physics in all such frames of reference. The idea was so deeply ingrained, however, that 19th-century physicists continued to search for the 'aether', assumed to pervade absolute space and to be the medium through which electromagnetic radiation was propagated. The advent of relativity theory superseded these ideas and necessitated the introduction of the concept of space-time. *See also* SPACE-TIME.

GD

space-time (1908) *Physics* A fundamental concept of relativity theory first described by the Russian-born German mathematician Hermann Minkowski (1864–1909).

Space-time is a four-dimensional coordinate system used in certain formulations of relativity theory, in which three dimensions represent the space coordinates and the fourth dimension is time. By analogy with ordinary space, 'events' take the place of point co-ordinates, with the interval between two events equivalent to the distance between two points in three-dimensional space. In relativity, geometrical properties are defined in terms of space and time as if these were really just aspects of a single space-time continuum. These properties are not invariant if a transformation is made between different reference frames which are moving uniformly relative to each other. *See also* SPACE and TIME.

Paul Davies, ed., *The New Physics* (Cambridge, 1989)

GD

spatialism (*c.*1946) *Art* Movement founded by the Italian artist Lucio Fontana (1899–1968) as the *Movimento Spaziale*, its tenets were repeated in manifestos between 1947 and 1954.

Combining elements of CONCRETE ART, DADA and TACHISM, the movement's adherents rejected easel painting and embraced new technological developments, seeking to incorporate time and movement in their works. Fontana's slashed and pierced paintings exemplify his theses.

L Fontana, *Manifesto Blanco* (Buenos Aires, 1946); G Giani, *Spazialismo* (Milan, 1956)

AB

special creation (Antiquity–) *Biology/Geology* Also called creationism, this is the doctrine that life was created directly by God, often (but not always) accompanied by the idea that species have remained essentially unchanged since creation.

L Godfrey, *Scientists Confront Creationism* (New York, 1983)

KH

special theory of relativity *Physics See* RELATIVITY, THEORY OF SPECIAL.

speciation, theory of (18th century) *Biology/Philosophy* Also called geographic speciation, this theory is most often associated with Ernst Mayr (1904–), an evolutionary biologist at Harvard University (although many other biologists have theorized on the subject). It asserts that new species arise among sexually reproducing organisms because geographic isolation enables a small subgroup to diverge genetically from the larger, established population.

This isolated subgroup may diverge owing to particular environmental pressures, mutations or GENETIC DRIFT. For this subgroup to become a new species, its unique characteristics must allow it to survive in its isolated location, and its gene pool must be protected from mixing with that of other species. *See also* ALLOPARAPATRIC SPECIATION, ALLOPATRIC SPECIATION, COHESION SPECIES CONCEPT, GENETIC REVOLUTION, PARAPATRIC SPECIATION, SALTATION SPECIATION, SPECIES ESSENTIALISM, STASIPATRIC SPECIATION and SYMPATRIC SPECIATION.

M Ereshefsky, *The Units of Evolution: Essays on the Nature of Species* (Cambridge, Mass., 1992)

KH

specie-flow mechanism *Economics* Outlined by Scottish philosopher David Hume (1711–76), among others, as an argument against the Mercantilist view that a nation should have a permanent balance of payments surplus. This is a corrective mechanism in the Gold Standard by which deficits and surpluses are eliminated by an induced flow of specie.

In its earliest form (as reported by Hume), the drain of gold would lead to a decrease in the quantity of money held by the deficit country; this in turn would lead to a fall in the general level of prices in relation to other countries. This would make the deficit country's exports more attractive and imports more expensive, resulting eventually in a favourable balance of trade. International MONETARISM is an analysis developed from monetary analysis, and is similar to the specie-flow approach.

L Yeager, *International Monetary Relations: Theory, History and Policy* (New York, 1976)

PH

species concept *Biology See* SPECIES, THEORY OF.

species essentialism (*c.*400 BC) *Biology/Philosophy* Also called the natural state model of species, this was based on the ideas of Greek philosopher Aristotle (384–322 BC), and applied by Swedish botanist Carolus Linnaeus (1707–78) and others in their search for the perfect 'type specimen' for each species.

It is the concept that all members of a species share a common natural state that serves to define and separate them from other species, with observed variations in individuals of a species being caused by forces that interfere with the organism's attainment of its natural state. This pre-evolution perspective holds that species are fixed entities, a view no longer accepted by biologists. *See also* SPECIATION, THEORY OF, SPECIES, THEORY OF. *Compare with* POPULATION THINKING.

KH

species taxa *Biology See* SPECIES CONCEPT.

species, theory of (18th century) *Biology/Philosophy* (Also referred to as the biological species concept, the isolation species concept, the species concept, and the species taxa.) Most often associated with Ernst Mayr (1904–), an evolutionary biologist at Harvard University (but many other biologists have theorized on the subject before and since). This is the idea that animals and plants can be considered groups of interbreeding natural populations that are reproductively isolated from other such groups.

While the definition of species may appear obvious to a layperson, biologists and philosophers continue to debate its parameters. Some suggest that the wide variety of species concepts being advocated by biologists and philosophers merely proves that there is no unique factor common to all species, and that the idea itself should be abandoned. *See also* ALLOPATRIC SPECIATION, COHESION SPECIES CONCEPT, EFFECT HYPOTHESIS, GENETIC REVOLUTION, PARAPATRIC SPECIATION, RECOGNITION CONCEPT OF SPECIES, SALTATION SPECIATION, SPECIATION THEORY OF, SPECIES ESSENTIALISM and SYMPATRIC SPECIATION.

M Ereshefsky, *The Units of Evolution: Essays on the Nature of Species* (Cambridge, Mass., 1992)

KH

speciesism (1970s) *Biology/Philosophy* A term attributed to British psychologist

Richard Ryder, author of *Victims of Science* (1975), it was popularized by Australian philosopher Peter Singer in *Animal Liberation* (1975). Speciesism is the doctrine that certain species are innately superior to others; and is used especially to describe the exploitation of lower species by humans. *See also* CHAIN OF BEING.

T Regan and P Singer, *Animal Rights and Human Obligations* (Englewood Cliffs, NJ, 1976)

KH

specificity theory *Psychology* Not attributable to any one person, this is a generic term denoting any psychological theory holding that specific constellations of factors produce specific psychosomatic disorders.

Specificity theories fall into four general groups: (1) personality specificity theory which claims definite personality traits lead to specific physical symptoms; (2) conflict specificity theory which claims that specific conflicts are associated with specific diseases; (3) emotion specificity theory which claims that specific emotions lead to definite somatic changes and eventually to particular somatic disturbances; (4) response pattern specificity, which rests on individual differences in stress-response pattern and places symptom choice at the physiological level.

Z J Lipowski, D R Lipsitt and P C Whybrow, eds., *Psychosomatic Medicine: Current Trends and Clinical Application*, (New York: 1977)

NS

specious present *Philosophy* An idea to deal with the problem that we can apparently only be aware of what is present, and what is present must be momentary (otherwise it would include the future or past and not be all present), yet anything real must exist for at least some time: so how can we be aware of anything real, and of things as continuing in time? Introduced by E R Clay and quoted by William James (1842–1910) in *The Principles of Psychology* (1901).

The specious present is a short period (various claims have been made by psychologists as to its length), allegedly presented to consciousness as all present at once, though in reality never more than one moment is present at once (hence the 'specious'). Our awareness of it, however, does take place all at once, being a single momentary act, or an element of a continuous series of such acts.

Objections to this include the view that the problem of generating our awareness of the passing of time from momentary experiences (whatever they are of) is a false one since such momentary experiences are a myth. *See also* PROCESS PHILOSOPHY and the writings of Henri Bergson (1859–1941) on what he calls 'the intuition of duration'.

W James, *The Principle of Psychology*, 1 (1902), 609; C W K Mundle, 'How Specious is the "Specious Present"?', *Mind* (1954)

ARL

spectral theorem (20th century) *Mathematics* A bounded linear operator T can be expressed in terms of the complex numbers z such that $zI - T$ is not a bounded linear operator; that is, such that z is in the spectrum of T. *See also* OPERATOR THEORY.

JB

spectroscopic displacement law *Chemistry* A law which states that the spectrum of an un-ionized atom resembles that of a singly-ionized atom of the element one place higher in the periodic table, and that of a doubly-ionized atom two places higher in the table, and so forth.

GD

speech act theory (1930s–1960s) *Linguistics/Philosophy* Also 'illocutionary act theory'. Originally formulated by the British philosopher John Langshaw Austin (1911–60), and developed by the American John Rogers Searle (1932–), it is a branch of PRAGMATICS.

When saying something, one is simultaneously *doing* something. An 'utterance act' is performed in voicing words and sentences; a 'propositional act' is carried out by referring to entities and predicating states and actions. The interpersonal act performed in speaking is an 'illocutionary act' (the central concept): 'I promise to pay you 5.00' counts as an act of promising if certain SINCERITY CONDITIONS or FELICITY CONDITIONS are fulfilled. The intended effect on the addressee is a 'perlocutionary act'. *See also* NEGATION, EMOTIVISM, PRESCRIPTIVISM, PERFORMATIVE THEORY OF TRUTH (one form of) and SUBJECTIVIST THEORIES OF PROBABILITY.

J L Austin, *How to Do Things with Words* (Oxford, 1962); J R Searle, *Speech Acts* (Cambridge, 1969)

RF

speech community (1930s–) *Linguistics* Always implicit in linguistic theory, particularly DIALECTOLOGY, and much discussed in modern SOCIOLINGUISTICS since the 1960s.

A set of people with a common language, or who share a repertoire of varieties (accents, styles, even languages in multilingualism); people who live together and interact through language; people with shared social attributes (young people, lawyers, women); people in the same social system. The term is most relevant to small, well-defined, stable communities.

J J Gumperz, 'The Speech Community' (1968), repr. in P P Giglioli, ed., *Language and Social Context* (London, 1972), 219–31

RF

Sperner's lemma (20th century) *Mathematics* Named after its discoverer E Sperner.

Suppose that an *n*-dimensional simplex (the *n*-dimensional analogue of the triangle) is decomposed into simplices of various dimensions, but meeting face to face; and suppose that the vertices of these are numbered onto the set of numbers 0, 1, 2, . . ., *n* so that if any subset of these numbers define a face with vertex *v* on it then the number for *v* is in the subset. Then one of the simplices in the set needs all the numbers to count it, so it has the same dimension as the original simplex.

JB

sphere packing problem (20th century) *Mathematics* One of the *n*-dimensional analogues of the problem of packing skittle balls in a cuboidal box in order to utilize the space in the box as efficiently as possible.

JB

spin conservation rule *Chemistry* In both radiative and radiation-less transitions the principle applies that transitions between terms of the same multiplicity are spin-allowed whereas transitions between terms of different multiplicity are spin-forbidden.

Compendium of Chemical Terminology: IUPAC Recommendations (Oxford, 1987)

MF

spin-statistics theorem *Physics* A relativistic quantum field theorem concerning the quantization of spins. Particles with half integer spins must obey Fermi-Dirac statistics to be quantized consistently. Similarly, particles with integer spins must obey BOSE-EINSTEIN statistics. This theorem is the basis for the PAULI EXCLUSION PRINCIPLE.

GD

spiritus animi (early 17th century) *Biology/Medicine* This term was coined in the 1600s, but its meaning was based on a theory of Greek physician Erasistratus of Julius (*c*.330 BC–*c*.250 BC). *Spiritus animi* refers to the hypothetical substance that swells a muscle, causing its contraction.

Erasistratus thought this substance was *pneuma* (Greek for breath), which was translated into Latin as *spiritus* and was believed to flow from the mind (Latin, *animi*). Towards the end of the 17th century it was shown that muscles fail to swell, or even decrease slightly in volume, during contraction. *See also* 'SUCCUS NERVEUS'.

J Fulton, *Selected Readings in the History of Physiology* (Springfield, Ill., 1966)

KH

Spoerer's law (1859) *Astronomy* Named after its originator, the German astronomer Gustav Friedrich Wilhelm Spoerer (1822–95), this law states that the mean latitude of sunspots varies systematically through a complete cycle. The law was also stated by the English astronomer Richard Christopher Carrington (1826–75).

S Mitton, ed., *The Cambridge Encyclopaedia of Astronomy* (London, 1973)

MS

spontaneity (1800) *Literary Theory* Theory developed by English poet William Wordsworth (1770–1850).

A tenet of ROMANTICISM. Poetry arises naturally in the poet and is not the product of existing ideas or labour with a conventional form. Wordsworth: 'Poetry is the spontaneous overflow of powerful feelings. . . .'

W Wordsworth, 'Preface to the Second Edition of the *Lyrical Ballads*' [1800], *Literary Theory since Plato*, H Adams, ed. (New York, 1971), 433-43

RF

spontaneous generation (17th century) *Biology* This is an ancient concept, but the term was coined by J Duncan in *Beetles* (1835). Also called abiogenesis, it is the (now discredited) idea that living organisms can arise spontaneously from inanimate matter.

This theory was supported by such observations as maggots appearing on rotten meat left outdoors, which led to the mistaken conclusion that the meat itself – rather than adult flies laying eggs in the meat – created the maggots. *Compare with* BIOGENESIS.

J Farley, *The Spontaneous Generation Controversy* (Baltimore, 1977)

KH

spontaneous order (20th century) *Politics/Sociology* Theory of Austrian social scientist F A Hayek (1899–1992).

Social order is not deliberately created but arises out of the natural selection of those institutions and values and practices which are effective, so that a traditional framework develops over time within which individuals may operate securely.

F A Hayek, *Law, Legislation and Liberty* (London, 1982)

RB

spot theory *Psychology* Proposed by the German physiologist Max von Frey (1852–1932).

The skin is tested spot by spot to determine whether different spots respond to different stimuli and thereby yield different sensory qualities. The theory cannot account for a variety of sensory phenomena but was important in discriminating different sensory sub-systems for touch, warmth, cold and pain. *See* CONCENTRATION THEORY OF COLD.

R S Woodworth and H Schlosberg, *Experimental Psychology* (London, 1966)

NS

Sprachbund *Linguistics See* AREAL LINGUISTICS.

spread of effect hypothesis *Psychology* Proposed by the American psychologist Edward Lee Thorndike (1874–1949).

To account for the phenomenon of stimulus generalization, Thorndike proposed that the effect of 'satisfiers' or 'annoyers' spread to other stimuli present at the time of the response or to stimuli similar in nature to the originally reinforced stimulus.

E L Thorndike, *Animal Intelligence* (New York, 1911)

NS

square-cube law (17th century) *Physics* Discovered by the Italian scientist Galileo Galilei (1564–1642), this law states that the strength of any structure which is likely to fail because the material fractures cannot be predicted from models or from scaling up from previous experience.

The reason for this is that, in scaling up, the weight of a structure will increase as the cube of the dimensions, but the cross-sectional area of the members that have to carry this load increases only as the square of the dimensions. Consequently, the stress in the members of the scaled-up structure goes up linearly with the dimensions.

J Thewlis, ed., *Encyclopaedic Dictionary of Physics* (New York, Oxford and London, 1962)

MS

squeeze law *Mathematics See* SANDWICH THEOREM.

stable and unstable manifold theorem (1964) *Mathematics* The theorem that if a system of ordinary DIFFERENTIAL EQUATIONS has a hyperbolic fixed point, then there exist local stable and unstable manifolds of the same dimensions tangent to the corresponding eigenspaces of the linearized system. There are generalizations of this theorem to maps on Banach spaces.

R L Devaney, *An Introduction to Chaotic Dynamical Systems* (New York, 1989)

ML

stadium paradox (5th century BC) *Mathematics* A paradox ascribed to Zeno of Elea (*c.*490–435 BC).

To put the paradox in a modern form, let us assume that both time and space are quantized; that is, have minimal quantities. If two objects one quantum long travel in opposite directions on parallel tracks at the speed of one quantum of space in one quantum of time, then they pass at a relative speed of two quanta of space in one quantum of time. They therefore take half a quantum of time to pass and cover half a quantum of relative space in doing it. Therefore time and space cannot both be quantized.

JB

stage theory *Psychology* Associated with the work of the American psychologist Arnold Lucius Gesell (1880–1961).

This term describes any developmental theory which characterizes growth as a progression through a succession of stages. Stage theories tend to be either maturational or interactionist and usually have at least four criterial properties: (1) they predict qualitative differences in behaviour over time and experience; (2) they assume invariance of stages; (3) they assume structural cohesiveness of a stage; (4) there is some hierarchical integration of structures from stage to stage.

S A Rathus, *Understanding Child Development* (New York, 1988)

NS

Stalinism (20th century) *Politics* Doctrines and policies associated with the Soviet politician Joseph Stalin (1879–1953).

A concentration of repressive power in the hands of a single leader and the ruthless suppression of opposition and dissent, combined with the advocacy of SOCIALISM IN ONE COUNTRY. Less a doctrine espoused by Stalin than a style of politics attributed to him and to other despotic Communists.

David McLellan, *Marxism After Marx* (London, 1980)

RB

standard theory (1965) *Linguistics* First major revision of his TRANSFORMATIONAL-GENERATIVE GRAMMAR by the American linguist Avram Noam Chomsky (1928–).

Transformations do not change meaning. The base (syntactic) component of the grammar now contains the phrase structure rules and information about which transformations (for example, Question, Negative, and so on) are to apply; the base is responsible for a 'deep stucture' which is interpreted by the semantic component of the grammar. Syntactic transformations relate the deep structure to a 'surface structure' which is interpreted by the rules for pronunciation, the phonological component of the grammar. The 'standard theory' or '*Aspects* grammar' was much modified in the late 1960s and early 1970s – *see* EXTENDED STANDARD THEORY – and challenged by other models, for example GENERALIZED PHRASE-STRUCTURE GRAMMAR.

N Chomsky, *Aspects of the Theory of Syntax* (Cambridge, Mass., 1965)

RF

standardization (process in English from the 14th century; theory, 1960s) *Linguistics* The choice of one dialect, or one language, as the official language of a nation: to facilitate communication and education, and to express nationhood. The chosen variety is codified through dictionaries and grammars, and receives prestige through use by privileged groups (therefore, in Great Britain, divisive rather than unifying). Definition by Milroy and Milroy is: 'Intolerance of optional variability in language.'

J Milroy and L Milroy, *Authority in Language* (London, 1985)

RF

starch-statolith hypothesis *Biology See* STATOLITH HYPOTHESIS.

Stark-Einstein law of photochemical equivalence *Physics See* EINSTEIN THEORY OF PHOTOCHEMICAL EQUIVALENCE.

stasipatric speciation *Biology* Proposed by M J D White, this theory postulates that a new chromosomal arrangement can arise anywhere in the range of a species, becoming increasingly common over time and spreading geographically, until individuals carrying this new trait as heterozygotes become so common that they breed, causing homozygotes to appear. *See also* SPECIATION, THEORY OF, and SPECIES, THEORY OF. *Compare with* ALLOPARAPATRIC SPECIATION, ALLOPATRIC SPECIATION, PARAPATRIC SPECIATION and SYMPATRIC SPECIATION.

C Barigozzi, *Mechanisms of Speciation* (New York, 1981)

KH

state *Politics* Central conception of political science.

There is no single theory of the state, but the concept of the state either as the institution exercising ultimate legitimate power in a territory, or as the highest expression of the will of the people, has been at the heart of political science and political theory.

David Miller *et al.*, eds, *The Blackwell Encyclopaedia of Political Thought* (Oxford, 1987)

RB

state capitalism (20th century) *Politics* Theory of state under COMMUNISM.

In Russia after the revolution of 1917, what developed was not communism but the assumption of the powers and functions of capitalism by the state. In an amended version (as 'state monopoly capitalism' or 'STAMOCAP'), it is argued that in western capitalist society state and capitalism have joined in alliance to create a new form of capitalism.

David Miller, ed., *The Blackwell Encyclopaedia of Political Thought* (Oxford, 1987)

RB

state of nature *Politics* Theory frequently employed in conjunction with that of a SOCIAL CONTRACT.

Before the formation of society or of states, people lived in a state of nature which has been variously characterized as utopian or brutal. Government is then a damaging limitation of primal innocence, or a necessary check on original depravity.

David Miller *et al.*, eds, *The Blackwell Encyclopaedia of Political Thought* (Oxford, 1987)

RB

stationary state (1776) *Economics* Referred to by Scottish economist Adam Smith (1723–90), this is a situation of zero growth in which the stock of goods is always the same (that is, quantity consumed equals quantity supplied in the same time period) and rewards to factors of production are at a minimum. *See* SECULAR STAGNATION THEORY and MALTHUSIAN POPULATION THEORY.

A Smith, *An Inquiry into the Nature and Causes of the Wealth of Nations* (London, 1776)

PH

statistical learning theory *Psychology* Proposed by the American psychologist William K Estes (1919–).

Introduced in the early 1950s to denote a variety of attempts to formulate and quantify basic principles of learning theory, statistical learning theory introduced a range of statistical and mathematical techniques which proved useful in the identification of patterns of learning not previously noticed, and in achieving a greater integration of the diverse empirical work in the field. The theory is sometimes criticized for replacing psychological concepts and processes with mathematical and statistical processes. *See* MATHEMATICAL LAW(S), and STIMULUS SAMPLING THEORY.

W K Estes, 'Toward a Statistical Theory of Learning', *Psychological Review*, vol. LVII (1950), 94–107

NS

statistical mechanics *Physics* A theory by which the properties of macroscopic systems are predicted from the statistical behaviour of their constituent particles. In a large collection of molecules, for example, the total energy is the sum of the vibrational, rotational, translational and electronic energies of the individual molecules. The system as a whole can have any number of possible energy levels with energies given by the MAXWELL-BOLTZMANN DISTRIBUTION LAW and weighted by the statistical factor for each level. The sum over all energy levels is called the partition function. In principle, statistical mechanics can be used to obtain thermodynamic properties of a system from a knowledge of the energy levels of its components. However, in practice it is difficult to evaluate the partition functions because of interactions between the particles.

Robert M. Besançon, ed., *The Encyclopedia of Physics* (New York, 1985)

GD

statistics (18th century) *Mathematics* Early contributors to the theory of statistics include the French mathematician Adrien-Marie Legendre (1752–1833), who developed the method of least squares, as well as Gauss, Laplace and Fermat, Pascal, Huygens, Leibniz, Jakob and Johann Bernoulli and Arbuthnot in probability.

Between 1700 and 1900 the study of statistics spread from astronomy and geodesy to psychology, biology and to the social sciences. Also the use of probability in statistics advanced from games of chance to probability models for measurements and finally to inverse probability and statistical inference. The British statistician Karl Pearson (1857–1936) is considered to be the founder of 20th century statistics.

The mathematical study of methods of summarizing or describing data and then drawing inferences from the summary measures. *See also* HYPOTHESIS TESTING and PROBABILITY THEORY.

M Hazewinkel, ed., *Encyclopaedia of Mathematics* (Dordrecht, 1988)

ML

statolith hypothesis of gravity perception (1901–02) *Biology* Also called the starch-statolith hypothesis, this idea was originally proposed by Hungarian botanist Gottlieb Haberlandt (1854–1945) in Austria and Bohumil Nemec in Czechoslovakia, and was refined in the 1970s. The theory that plants perceive gravity because specialized starchy grains in 'amyloplasts' settle to the lower part of gravity-sensitive organs owing to gravitational pull.

Research in the 1970s supported the idea that amyloplasts are the 'statoliths' (from the Greek word *líthos*, meaning stone) of many botanical tissues. However, studies from the mid-1980s demonstrate that starch is not essential for gravity perception in all plants.

KH

steady-state theory of the universe (20th century) *Astronomy* A theory devised by the British astronomer Sir Fred Hoyle (1915–), William H McCrea, the Austrian-born astronomer Thomas Gold (1920–) and the British cosmologist Sir Hermann Bondi (1919–), this holds that the large-scale features of the universe do not change with time.

Because of the observed expansion of the universe, it is necessary to compensate for the separating effect of expansion by the continuous formation of new galaxies and clusters of galaxies. This rate of formation just compensates the effect of expansion and so gives a stable situation.

Bondi and Gold adopt the 'perfect cosmological principle' that the large-scale features of the universe are the same not only from every location in space but for every instant in time. This leads immediately to a steady-state universe and it is immaterial whether observers compare their pictures 'at the same time' or not. McCrea and Hoyle start with a mathematical definition of continuous creation within the framework of general relativity theory and then derive the steady-state solution as a consequence of the field equations. On this model, the large-scale properties of distant parts of the universe should be the same as those of nearer parts. The theory is no longer believed to hold.

S Mitton, ed., *The Cambridge Encyclopaedia of Astronomy* (London, 1973)

MS

Stefan's law (Stefan–Boltzmann law) *Physics See* BLACK BODY LAW.

Steinitz's exchange theorem (20th century) *Mathematics* Named after its discoverer Ernst Steinitz (1871–1928).

Let k and m be positive integers such that $k < m$, let $\mathbf{u}_1, \mathbf{u}_2, \ldots, \mathbf{u}_k$ and $\mathbf{v}_1, \mathbf{v}_2, \ldots, \mathbf{v}_m$ be linearly independent sets in a vector space. Then

$$\mathbf{v}_1, \mathbf{v}_2, \ldots, \mathbf{v}_m$$

can be written in a different order as

$$\mathbf{w}_1, \mathbf{w}_2, \ldots, \mathbf{w}_m$$

such that

$$\mathbf{u}_1, \mathbf{u}_2, \ldots, \mathbf{u}_k, \mathbf{w}_1, \mathbf{w}_2, \ldots, \mathbf{w}_{m-k}$$

is linearly independent. From this result it can be deduced that if the dimension of a vector space V is finite, then it is uniquely defined by V and any subspace of V of the same dimension as V must be equal to V.

JB

Steinitz's theorem (19th century) *Mathematics* Named after Ernst Steinitz (1871–1928) who proved this extension of a result by Bernhard Riemann (1826–66).

Let the infinite series

$$a_1 + a_2 + a_3 + \ldots + a_n + \ldots$$

have a sum to infinity. Then the sums of the various series that can be derived from the equation by changing the order of the terms are either all the same or range over all the real numbers as well as including series of which the sum tends to infinity or to minus infinity. This has generalizations when the terms are vectors in n-dimensional space.

JB

Steinmetz's law (1916) *Physics* Named after its discoverer, the American electrical engineer Charles Proteus Steinmetz (1865–1923). This is a useful rule for the approximate calculation of the work W needed to take a ferromagnetic material around its hysteresis loop.

$$W = \eta B_m^{1.68}$$

where B_m is the maximum value of the induction in the cycle and η is a coefficient that depends on the material, known as the Steinmetz coefficient.

J Thewlis, ed., *Encyclopaedic Dictionary of Physics* (New York, Oxford and London, 1962)

MS

Steno's laws (17th century–) *Geology* Formulated by the Danish naturalist Niels Steinsen (1638–86) (better known as Nicolas Steno), these recognized the organic origin of fossils.

Steno also implicitly established the laws of SUPERPOSITION OF STRATA and CROSS-CUTTING RELATIONSHIPS.

DT

stepwise phenomenon *Psychology* Identified by the German-American psychologist Max Wertheimer (1880–1943).

The principle that a sequence of separate steps along a continuum is normally perceived as an organized, smooth progression. The stepwise phenomenon occurs with a prothetic continuum but not with a metathetic one; for example, a sequence of lights of increasing brightness is seen as a stepwise progression, but a sequence of increasing wavelengths is seen as a discontinuous progression moving through various hues. *See* GESTALT THEORY.

A D Ellis, ed., *A Sourcebook of Gestalt Psychology* (London, 1938)

NS

Stevinus's theorem *Physics* Named after its discoverer, the Flemish mathematician and physicist Simon Stevinus (1548–1620), this states that the pressure upon the base of a vessel containing a given liquid is independent of the shape of the vessel, being determined only by the vertical height of the free surface of the liquid above the base.

J Thewlis, ed., *Encyclopaedic Dictionary of Physics* (New York, Oxford and London, 1962)

MS

Stewart-Kirchhoff law (1858) *Chemistry* Named after the Scottish physicist Balfour Stewart (1828–87) and the German physicist Gustav Robert Kirchhoff (1824–87).

The law states that for any substance, the ratio of emissive power and absorption coefficient depends only on the temperature and the frequency and plane of polarization of the radiation, and not on the nature of the substance.

GD

Stickelberger's theorem (19th century) *Mathematics* Named after its discoverer Ludwig Stickelberger (1850–1936).

We calculate modulo an odd prime number p; that is, we replace each integer by its remainder on division by p. Let $f(x)$ be a polynomial of degree d which is greater than 1 with all coefficients integers and the coefficient of xd equal to 1. Suppose that $f(x)$ has no repeated factor, which implies that its discriminant $D(f)$ is not 0. The number r of factors of $f(x)$ which do not factorize further has the same parity as d if and only if $D(f)$ is a square modulo p.

JB

stimulus-response model *Biology/Philosophy See* BEHAVIOURISM.

stimulus-response (S-R) theory *Psychology* Particularly associated with the work of the American psychologist John Broadus Watson (1878–1958).

This is a general term referring to any of the associationistic learning theories that have the forming of a bond between a stimulus and a response as their theoretical basis. *See* ASSOCIATIONISM, BEHAVIOURISM and CONNECTIONISM.

B F Skinner, *About Behaviourism* (New York, 1974)

NS

stimulus sampling theory *Psychology* Proposed by the American psychologist William K Estes (1919–).

This mathematical theory of learning can produce accurate predictions about behaviours in a variety of complex learning experiments. On each trial of a learning experiment the subject samples a portion of the elements that form the stimulus, and makes a response. These elements become conditioned to it when that response is reinforced. Therefore, the probability of that response on any given trial may be specified by the proportion of all the stimulus elements conditioned to that response, although the exact proportion cannot be determined. *See* MATHEMATICAL LAW(S) and STATISTICAL LEARNING THEORY.

W K Estes, 'Toward a Statistical Theory Of Learning', *Psychological Review*, vol. LXVII (1957), 94–107

NS

stochasticism (1950s) *Music* Associated with the music of the Greek composer Iannis Xenakis (1922–) and some of his contemporaries, this term refers to compositional processes governed by laws of probability or chance.

It applies to music that progresses as a chain of events with each new event somehow determined by the cumulative content of what has gone before, leading to an inevitable goal. 'Stochastic' has latterly come to connote music with a fixed overall plan that accommodates random small-scale events within some or all sections.

MLD

stock-flow analysis *Economics* Analysis of the behaviour of individuals or markets at a fixed point in time, and over periods of time.

The technique is increasingly used in the study of employment. *See* UNEMPLOYMENT.

D W Bunshaw and R W Glower, *Introduction to Mathematical Economics* (Homewood, Ill., 1957)

AV, PH

Stockholm syndrome *Psychology* Not attributable to any one originator, this is a popular term referring to the emotional bond which sometimes develops between hostages and their captors when they are held for long periods of time under emotionally straining circumstances. The name derives from the instance it was publicly noted in Stockholm after a bank robbery turned into a five-day hostage siege.

J T Tedeschi, S Lindskole and P Rosenfeld, *Introduction to Social Psychology* (St Paul, 1985)

NS

Stoicism *Philosophy* Philosophy named from the Stoa, or portico, in Athens where its adherents gathered. It was founded by Zeno of Citium (*c*.336–*c*.264 BC) – different from Zeno the Eleatic – but considerably developed by his successors, notably: Chrysippus (*c*.280–*c*.206 BC), Posidonius (*c*.135–*c*.51 BC), Seneca 'the Younger' (*c*.4 BC–AD 65), Epictetus (*c*.AD 50–138). The emperor Marcus Aurelius (AD 121–80) was its last famous adherent, and Cicero (106–43 BC) is one of our main sources. It rivalled EPICUREANISM and ancient SCEPTICISM through much of its history, and eventually gave way to NEO-PLATONISM and Christianity, both of which it heavily influenced.

The Stoics divided philosophy into three branches, logic, physics, and ethics. In logic they went substantially beyond Aristotle (384–322 BC), inventing the propositional calculus. In physics (which included metaphysics) they developed a pantheistic but materialist and determinist system contrasting with Epicurean atomism. In ethics they aimed at self-sufficiency and acceptance of fate, treating 'virtue' as the only real value, though among the remaining things ('indifferents') some were 'preferred'.

A A Long and D N Sedley, *The Hellenistic Philosophers* (1987); translations with commentary

ARL

Stokes's law of phosphorescence *Physics* Named after its originator, the British mathematician and physicist Sir George Gabriel Stokes (1819–1903), this states that if a solid is capable of absorbing light it may re-emit light.

The process is called fluorescence if this emission persists only during irradiation, and phosphorescence if the emission continues after irradiation ceases. According to Stokes's law, the wavelength of the emitted light is always longer than that of the absorbed light; a result that is not always true.

J Thewlis, ed., *Encyclopaedic Dictionary of Physics* (New York, Oxford and London, 1962)

MS

Stokes's law of terminal speed *Physics* Named after its discoverer, the British mathematician and physicist Sir George Gabriel Stokes (1819–1903), this states that when a spherical body of radius r is moving under streamline conditions with a speed v through a liquid of viscosity η, the viscous drag force F experienced by the sphere is given by

$$F = 6\pi\eta r v\ .$$

J Thewlis, ed., *Encyclopaedic Dictionary of Physics* (New York, Oxford and London, 1962)

MS

Stokes's theorem (19th century) *Mathematics* Named after Irish analyst and physicist George Gabriel Stokes (1819–1903), this states that for a smooth vector field **F**, the integral with respect to arc length of the tangential component of **F** around the boundary of a piecewise smooth oriented surface is equal to the surface integral of the component of **F** in the direction of the outer normal to the surface **F**. That is,

$$\int_{\partial S} \mathbf{F}\cdot\mathbf{T}\,ds = \int_S (\mathrm{curl}\mathbf{F})\cdot\mathbf{n}\,dA.$$

See also GREEN'S THEOREM and DIVERGENCE THEOREM.

H Anton, *Calculus with Analytic Geometry* (New York, 1980)

ML

Stone-Čech compactification (1937, 1939) *Mathematics* Named after the mathematicians Marshall Harvey Stone (1903–) and Eduard Čech (1893–1960), this is the compactification of a completely regular Hausdorff space X which is characterized (up to homeomorphism) as the unique compactification W with the property that any continuous mapping from X into a compact space Y extends uniquely to a continuous mapping from W to Y. In a certain sense, it can be considered the maximal compactification of X whereas the ALEXANDROFF COMPACTIFICATION is the minimal compactification. *See also* TOPOLOGY.

J R Munkres, *Topology: A First Course* (Prentice-Hall, 1975)

MB

Stone-Weierstrass theorem *Mathematics* Named after American mathematician Marshall Harvey Stone (1903–) and German mathematician Karl Theodor Wilhelm Weierstrass (1815–97).

This is the result whereby an algebra of continuous complex-valued functions on a compact set is uniformly dense if it separates points, vanishes nowhere and is self-adjoint. This result generalizes an earlier result of Weierstrass, known as the WEIERSTRASS APPROXIMATION THEOREM, which states that the polynomials are uniformly dense in the continuous functions on a closed bounded interval.

W Rudin, *Principles of Mathematical Analysis* (McGraw-Hill, 1976)

MB

story and plot (1927) *Literary Theory* Distinction drawn by the English novelist Edward Morgan Forster (1879–1970).

Story is a narrative of events linked by time-sequence; plot links a sequence of events by causation. *See also* 'FABULA' AND 'SJUZHET', 'HISTOIRE' AND 'DISCOURS'.

E M Forster, *Aspects of the Novel* (Harmondsworth, 1963 [1927])

RF

strain theory (1885) *Chemistry See* BAEYER STRAIN THEORY.

strange attractor (1963) *Mathematics* The first strange attractor was discovered by American meteorologist Edward Norton Lorenz (1917–).

A chaotic solution to a system of nonlinear DIFFERENTIAL EQUATIONS characterized by its occurring in a bounded set of zero volume called the non-wandering set.

I Peterson, *The Mathematical Tourist* (New York, 1988)

ML

stratificational grammar (1957) Originated in the work of the American linguist Sydney Macdonald Lamb (1929–), developed in collaboration with H A Gleason (jnr).

Theory of linguistic structure which describes language on four levels or 'stratal systems': sememic, lexemic, morphemic and phonemic. The lexemic stratal system or 'lexology' deals with much of the structuring traditionally covered by SYNTAX.

D G Lockwood, *Introduction to Stratificational Linguistics* (New York, 1972)

RF

stream of consciousness (1890) *Stylistics* Term used by the American psychologist William James (1842–1910) for associatively linked thought. Appropriated for literary criticism in the early 20th century.

A range of linguistic techniques for representing the thought-processes of CHARACTERS in FICTION, aiming to imitate the loose, fragmentary, associative character of thought. It became prominent from the 1920s with Modernist psychological novelists such as James Joyce, Virginia Woolf, Dorothy Richardson, Marcel Proust, and William Faulkner. Includes INTERIOR MONOLOGUE.

R Humphrey, *Stream of Consciousness in the Modern Novel* (Berkeley and Los Angeles, 1954)

RF

string theory (*c.*1969) *Physics* This is so named since it developed from the theory of vibrating strings and was an attempt to create a unified FIELD THEORY in four-dimensional space-time. Dealing with elementary particles, the theory was designed to overcome the singularity problems of general relativity which arise when a quantum treatment of gravity is attempted. The theory proposes that instead of point-like particles, the basic components are finite lines (strings) or closed loops formed by these strings. String theory explains why, in weak interactions, particles violate parity conservation.

It was upgraded to an emerging superstring theory by M B Green and J H Schwarz in 1984. The latter is more successful but requires ten-dimensional space to ensure uniqueness. *See also* SUPERSTRING THEORY and SUPERSYMMETRY.

M B Green, J H Schwarz and E Witten, *Superstring Theory* (Cambridge, 1987)

JB, GD

strong law of large numbers (1915) *Mathematics* A precise form of the law of large numbers proved by Emile Borel (1871–1956). *See also* LARGE NUMBERS, LAW OF and WEAK LAW OF LARGE NUMBERS.

JB

Stroop effect *Psychology* Proposed by the American psychologist J R Stroop (*fl.*1930s).

The Stroop interference test consists of a series of colour name words (blue, red, green) printed in non-matching colours; for example, the word 'green' may be printed in red ink. Most people find it extremely difficult to attend to the ink colour alone when asked to name the colour in which each word is printed, because of an automatic tendency to read the words producing interfering information.

A Baddeley, *Human Memory: Theory and Practice* (London, 1990)

NS

structural anthropology (1945) *Anthropology* Introduced into this field by French anthropologist Claude Lévi-Strauss (1908–), with whom the theory is strongly associated. Broadly, any approach which emphasizes relations between things rather than things themselves.

Structuralists hold that meaning exists only within systems, especially those made up of sets of oppositions (nature versus culture, good versus evil, high status versus low status, and so on). Structuralist ideas developed within linguistics in the early part of the 20th century and were introduced into anthropology in the 1940s as part of kinship studies. In the 1950s and 1960s structuralist theory proved useful in the analyses of symbolism and mythology. Structuralism is now thought of as passé in some circles, but it remains implicit in much anthropological work today. *See also* ALLIANCE THEORY.

E Leach, *Lévi-Strauss* (London, 1970)

AB

structural functionalism (20th century) *Politics/Sociology* A more systematic exposition of FUNCTIONALISM, particularly employed in comparative politics.

Societies and political systems may be compared, whatever their formal or institutional differences, in terms of the various functions which contribute to the operation of their overall structures.

Geoffrey Roberts and Alistair Edwards, *A New Dictionary of Political Analysis* (London, 1991)

RB

structural linguistics (1916; 1933) (1) Set of theoretical principles adumbrated by the linguist Ferdinand de Saussure (1857–1913); (2) American school of descriptive linguistics, most active during the 1940s and 1950s, based on the work of Leonard Bloomfield (1887–1949) and of the ANTHROPOLOGICAL LINGUISTS before him.

(1) For the main tenets of Saussurean theory, *see* DIACHRONIC AND SYNCHRONIC, LANGUAGE AND PAROLE, PARADIGM AND SYNTAGM, SIGN, VALUE. Saussure's ideas also form the basis of literary STRUCTURALISM and European SEMIOTICS.

(2) Empiricist practice of linguistic analysis which begins with the observable data of speech, segments it into PHONEMES which are grouped into MORPHEMES and so 'upwards' by CONSTITUENT STRUCTURE analysis. *See also* EMPIRICISM IN LINGUISTICS.

F de Saussure, *Course in General Linguistics*, W Baskin, trans. (Glasgow, 1974 [1916]); L Bloomfield, *Language*, revised edn (London, 1935 [1933])

RF

structuralism (1960s) *Literary Theory* Regarded as a French movement, much indebted to the French linguist Ferdinand de Saussure (1857–1913), but with Russian and Czech sources, and international implications. Principals: Roland Barthes (1915–80), Gérard Genette (1931–), Algirdas Julien Greimas, Roman Jakobson (1896–1982), Claude Lévi-Strauss (1908–), Tzvetan Todorov (1939–).

Literary texts, and other cultural products such as myth, are analyzed as systems of signifying units, like language. Preoccupation with formal symmetries, parallelisms, contrasts; with features of poetic language which differ from those of 'ordinary language'; with the general properties of literature (POETICS). Jakobson's POETIC PRINCIPLE and the NARRATIVE GRAMMAR of Vladimir Propp (1895–1970) are important foci and models.

T Hawkes, *Structuralism and Semiotics* (London, 1977)

RF

structure-conduct-performance theory (20th century) *Economics* This is a model used to link elements of market structure to business conduct and performance in industrial economics.

PH

structure-dependency (1960s–). Central principle of the GENERATIVE GRAMMAR of the American linguist Avram Noam Chomsky (1928–).

Universal 'design feature' of language. Linguistic rules must refer to structural information in order to operate. For example, question-formation in English (for example, 'Is Chomsky famous?') moves the structurally identified element 'auxiliary' to the front of the sentence, not the second or any *n*th word.

V Cook, *Chomsky's Universal Grammar* (Oxford, 1988)

RF

structurism (1952) *Art* Theory developed by the American artist Charles Bieder, and so named to distinguish it from an earlier concept of 'constructionism'. With the use of abstract reliefs he attempted to synthesize qualities of painting, sculpture and architecture which he viewed as structural processes of nature. Some confusion has arisen about the application of this term to MINIMALISM.

G Kepes, ed., *Structure in Art and Science* (1965)

AB

Stuffer law *Chemistry* The law states that sulphones which have two $=SO_2$ groups on adjacent carbon atoms are easily saponified.

GD

Sturm und Drang (1776) *Literary Theory* Meaning 'Storm and Stress', this phrase originated as the title of a play by the German dramatist Friedrich Maximilian von Klinger (1752–1831). It was adopted as the general label for a German literary movement of the 1770s.

Short-lived but important as a precursor to ROMANTICISM, the movement stressed the expressive rendering of personal feelings in highly charged language. *See* BILDUNGSROMAN and ORIGINALITY.

B Kieffer, *The Storm and Stress of Language* (Philadelphia, 1986)

RF

Sturm's theorem (19th century) *Mathematics* Named after its discoverer Jacques Sturm (1803–55).

Let $f(x)$ be a polynomial with real coefficients. Write

$$p_0(x) = f(x) \text{ and } p_1(x) = f'(x)$$

then put

$$p_i(x) = -r_i(x) \text{ for i} = 2, 3, \ldots, k$$

where

$$r_2(x), r_3(x), \ldots, r_k(x)$$

are successive remainders computed from $f(x)$ and $f'(x)$ by the Euclidean algorithm. Let a and b be real numbers such that neither $f(a)$ nor $f(b)$ is 0. Then the number of roots of $f(x)$ between a and b is the difference between the number of sign changes for

$$p_0(x), p_1(x), p_2(x), \ldots, p_k(x)$$

at a and at b. *See also* EUCLIDEAN ALGORITHM and DESCARTES'S RULE OF SIGNS.

JB

style, theory of (1900) *Stylistics* A problematic concept of traditional origin, much discussed since STYLISTICS became popular in the 1960s. Also some usage in SOCIOLINGUISTICS.

A distinctive manner of expression, appropriate to a type of communicative context ('formal style', 'tabloid style'); in literary studies, characteristic of a period ('Augustan style'), a set of conventions ('Gothic style'), and importantly, an author ('Hemingway's style'). A style may be identified as a range of linguistic choices (chiefly lexical and syntactic) where options are available.

There is a theoretical problem whether such choices can be regarded as different ways of expressing the same meaning; different linguistic theories (for example, deep and surface structure of Noam Chomsky's STANDARD GRAMMAR) have been applied to this problem without agreed resolution.

N E Enkvist, 'On Defining Style', in N E Enkvist and J Spencer and M J Gregory, *Linguistics and Style* (London, 1964), 3–57

RF

stylistics (1. *c*.1900; 2. 1960s) *Stylistics* (1) Traditional European literary-critical practice, for example Charles Bally (1865–1947); (2) principally American and British application of linguistics, with various pioneers.

(1) Study of the characteristic language of an author, period, movement, or genre; may be impressionistic, or (in authorship attribution studies, for example) quantitative.

(2) Study of the language of literature using the technical methods of linguistics. Stylistics in this sense does not necessarily focus on STYLE as such, but serves diverse critical approaches, often PRACTICAL CRITICISM.

(1) S Ullman, *Language and Style* (London, 1964); (2) K Wales, *A Dictionary of Stylistics* (London, 1989)

RF

subjacency (1973) *Linguistics* Proposed by the American linguist Avram Noam Chomsky (1928–).

Restriction on how far transformations may move constituents. In GOVERNMENT AND BINDING, items may not be moved across more than one 'bounding node' (S, S', NP); thus sentences like 'Which car did she believe Tom's story that I drove?' are ungrammatical.

N Chomsky, 'Conditions on Transformations', *A Festschrift for Morris Halle*, S R Anderson and P Kiparsky, eds (New York, 1973)

RF

subjective idealism *Philosophy* Form of idealism represented primarily by George Berkeley (1685–1753), though his own name for it was 'immaterialism'. Berkeley distinguished minds or spirits (including both God and finite spirits like us), which are active, from ideas which are their contents and are passive. To be is to perceive, in the case of spirits, or to be perceived, in the case of ideas; 'perceive' here really means 'have as content', and 'be perceived' means 'be had as content'.

Berkeley was mainly concerned to reject the notion of matter, which he regarded as unknowable and the source of paradoxes, and itself stemming from the doctrine of 'abstract ideas', which he made his first target. The term 'subjective idealism', used of Berkeley and also of Immanuel Kant (1724–1804) (*see* TRANSCENDENTAL IDEALISM) by objective idealists, perhaps depends on emphasizing only one side of Berkeley's view, that to be is to be perceived; and in the case of Kant, his treatment of ideas as dependent on our minds.

G Berkeley, *A Treatise concerning the Principles of Human Knowledge* (1710)

ARL

subjectivism *Philosophy* Any theory treating a given subject matter as dependent on human beliefs and attitudes, whether those of an individual, a social group, or humanity generally.

A subjectivist theory of ethics, for example, might analyze an utterance like 'Abortion is wrong' as meaning that the speaker, or his society, or people in general, disapproves of abortion; alternatively it might offer a SPEECH ACT analysis of it. Subjectivism is akin to, but not the same as, RELATIVISM. The view that 'Abortion is wrong' means that most people disapprove of it is subjectivist but not relationist. The view that jailing an innocent man to prevent a riot is (objectively) right on UTILITARIANISM but wrong on Kantianism and that there is no 'correct' answer, is relativist but not subjectivist. The contrast term is OBJECTIVISM.

ARL

subjectivist theories of probability *Philosophy* Theories which analyze probability in terms of beliefs or attitudes rather than anything in the world itself.

For one theory, associated mainly with Bruno De Finetti (1906–85), the degree of probability of something is the degree of the speaker's belief, measured by his betting behaviour, but subject to the constraint that his bets must be 'coherent'; that is, he must not bet in such a way as to lose whatever happens (sometimes called 'having a Dutch book made against one'). This constraint still leaves probabilities dependent on the vagaries of individual attitudes, unless we substitute those of 'the rational man' – but that takes us away from subjectivism. Others, notably S E Toulmin (1922–), offer a SPEECH ACT THEORY whereby to call something probable is to assert it, though only tentatively. This may well apply to some uses of 'probably', but hardly to all, and shares the objections to other speech act theories.

H E Kyburg, *Probability and Inductive Logic* (1970), ch. 6

ARL

sublimity (18th century) *Art/Literary Theory* This emerged as an aesthetic term in 18th-century literature and art, denoting the extraordinary and the marvellous, and was initially based on the concept of 'elevation' as propounded by Longinus (*c.* AD 213–73) in his thesis *Peri Hupsous*.

The 1674 translation of Longinus by French critic Boileau (1636–1711) made this concept more widely known in Europe, but the most significant contribution to its understanding was made by English statesman and philosopher Edmund Burke (1729–97) in his *Philosophical Enquiry*. . . . He proposed the power of suggestiveness as a stimulus for invention, and the importance of the emotion of terror. English painter and critic Jonathan Richardson (1665–1745) expounded on the nature of the sublime in art, singling out excellence, the greatest of ideas and invention as important features of the theory.

W Burke, *Philosophical Enquiry into the Origin of our Ideas of the Sublime and Beautiful* (1757); J Richardson, *An Essay on the Theory of Painting* (1725)

AB

subsidiarity (20th century) *Politics* Theory of allocation of public or governmental functions associated with the European Community.

Tasks should never be allocated to a body higher up in a political hierarchy if they can be effectively carried out by a body lower down.

RB

substitution hypothesis *Psychology* Proposed by the Austrian founder of psychoanalysis Sigmund Freud (1856–1939).

This states that if psychoneurotic symptoms are superficially treated without dealing with the underlying causes of the disorder, symptom substitution will occur. This effect is frequently used by psychodynamically-oriented clinicians as a criticism of various forms of behaviour therapy and cognitive behavioural therapy, which often tackle the symptoms rather than the underlying cause.

G C Davison and J M Neale, *Abnormal Psychology* (Chichester, 1982)

NS

substitution theorem *Mathematics* The theorem of logic that a universally quantified statement implies any instance of it, and that an existentially quantified statement is implied by an instance of it.

ML

substitution theory (20th century) *Economics* A fundamental part of modern economic theory developed by, among others, Soviet economist Evgeny Slutsky (1880–1948) and English economist John Hicks (1904–89); this is the analysis of the manner in which consumers, faced with a constant level of real income, change purchasing decisions in the wake of price changes. *See* SLUTSKY THEOREM.

J R Hicks and R G D Allen, 'A Reconsideration of the Theory of Value', *Econometrica NS* (1934), 52–76, 196–219

AV, PH

succus nerveus (1664) *Biology* Proposed by English physiologist William Croone (1633–84), this was the hypothetical substance that interacts with the blood in muscles to produce contraction.

Succus nerveus ('juice of the nerves') was an elaboration of the earlier idea of animal spirits (*spiritus animi*), and it persisted until German physiologist Hermann Ludwig Ferdinand von Helmholtz (1821–94) clarified the role of electricity in nerve impulses.

W Croone, *De ratione motus musculorum* (1664); J Fulton, *Selected Readings in the History of Physiology* (Springfield, Ill., 1966)

KH

sudden life *Biology* A reference to the set of theories that hold that life, usually in the form of a single cell, appeared suddenly on the Earth, without the gradual appearance of pre-biotic forms.

These include theories that life came from outer space or was placed on earth by God. *See also* ORIGINS OF LIFE. *Compare with* PRE-CELLULAR EVOLUTION.

M Barbieri, *The Semantic Theory of Evolution* (Chur, Switzerland, 1985)

KH

sufficient reason, principle of *Philosophy* Principle that there must be a sufficient reason – causal or otherwise – for why whatever exists or occurs does so, and does so in the place, time and manner that it does.

The principle goes back to at least the early 5th century BC – being used by Parmenides (*see* ELEATICISM) in his Fragment 8, lines 9–10 – but it is most famously associated with Gottfried Wilhelm Leibniz (1646–1716), who used it to exclude all arbitrariness, and to account for 'truths of fact' (while the law of CONTRADICTION accounted for 'truths of reason'). He also derived the IDENTITY OF INDISCERNIBLES from it. *See also* principle of PERFECTION and CAUSAL PRINCIPLE.

H G Alexander, ed., *The Leibniz-Clark Correspondence* (1956)

ARL

suffix effect (20th century) *Psychology* Identified by the American psychologists D Salter and J G Colley, this is the phenomenon whereby an extraneous stimulus (the 'suffix') presented immediately after the full list of to-be-recalled materials depresses the recall of that material. This occurs even when the subject is informed in advance that the 'suffix' stimulus is irrelevant to the task and should be ignored.

D Salter and J G Colley, 'The Stimulus Suffix: A Paradoxical Effect', *Memory and Cognition*, vol. v (1976), 257–62

NS

sum of squares theorem *Mathematics* The theorem that if the matrix of the quadratic form of a sum of squares of normal random variables is independent of rank r, then the sum of squares is distributed proportionally to a chi-square distribution with r degrees of freedom.

ML

summability theory *Mathematics* The study of methods used to sum quantities, especially of methods of assigning values to divergent series and integrals. *See also* ABEL SUMMATION, CESARO SUMMATION and TAUBERIAN THEOREMS.

K Knopp, *Theory and Application of Infinite Series* (New York, 1990)

ML

sunspot theory (1884) *Economics* Advanced by English economist William Jevons (1835–82), this is a TRADE CYCLE THEORY stating that trade is linked to the regular occurrence of solar flares, or spots, which affect the earth's climate and agricultural output.

Critics, while acknowledging the cyclical nature of the sunspot activity and that of agricultural production, view the theory as unduly simplistic. Consequently, it has little validity today. *See* HECKSCHER-OHLIN TRADE THEORY and TECHNOLOGICAL GAP THEORY.

W S Jevons, *Investigations in Currency and Finance* (London, 1884)

AV, PH

superconductivity, theory of *Physics See* BCS THEORY.

superdense theory (20th century) *Astronomy* The cosmological theory that the universe has evolved from a 'superdense' agglomeration of matter that suffered an enormous explosion. *See* BIG BANG THEORY.

S Mitton, ed., *The Cambridge Encyclopaedia of Astronomy* (London, 1973)

MS

superfluous man (19th century) *Literary Theory* In Russian *lishni chelovek*, this term was applied by the poet Alexander Pushkin (1799–1837) to the hero of his verse fiction *Eugene Onegin* (1823–31). The term came into general usage after publication of *Diary of a Superfluous Man* in 1850, a story by Ivan Turgenev (1818–83).

Type of character in 19th century Russian fiction: an intelligent, sensitive and critical idealist who fails to act on his beliefs because of personal diffidence and/or external constraints. The archetypal superfluous man is Oblomov in the 1857 novel of that name by Russian writer Ivan Alexandrovich Goncharov (1812–91). *See* ANTI-HERO and OUTSIDER.

J V Clardy, *The Superfluous Man in Russian Letters* (Washington, D.C., 1980)

RF

superposition of strata, law of (17th century–) *Geology* Proposed by the Danish naturalist Nicolas Steno (1638–86), this states that in any undisturbed succession of rocks, the lowest strata are older than the upper beds.

This law remains a fundamental principle in the relative dating of rocks. In very disturbed beds, the entire sequence may be inverted and requires the identification of 'way up' criteria, such as mud cracks, to determine the original orientation. *See also* STENO'S LAWS.

DT

superposition principle *Mathematics* The principle that any linear combination of solutions to a homogeneous linear DIFFERENTIAL EQUATION is also a solution to that equation.

W E Boyle and R C Di Prima, *Elementary Differential Equations and Boundary Value Problems* (New York, 1977)

ML

superposition principle *Physics* When a physical system is acted on by a number of independent influences, the resultant effect is the appropriate sum of the individual influences, provided that the behaviour of the systems can be expressed by linear differential equations.

In particular, when several particles interact, the force of interaction between each pair is independent of the presence of the other particles. The net force on any particle due to the presence of the other particles is found by calculating its interaction with each of the particles in turn and then adding the separate effects vectorially.

J Thewlis, ed., *Encyclopaedic Dictionary of Physics* (New York, Oxford and London, 1962)

MS

superstrings (1980s) *Physics* A theory combining ideas from string theory with those of supersymmetry which is thought to be a more useful route to a unified theory of fundamental interactions than quantum field theory. Superstring theory avoids the problems of infinities by introducing particles of spin 2, known as gravitons. Superstring theories also contain higher dimensional spaces than the normal four of space-time. There is no direct experimental evidence for superstrings, although they have estimated lengths of about 10^{-35} m and energies of 10^{19} GeV. A further refinement of the theory postulates two-dimensional entities, that is 'supermembranes'. *See also* STRING THEORY and SUPERSYMMETRY.

Paul Davies, ed., *The New Physics* (Cambridge, 1989)

GD

superstructuralism (1987) *Literary Theory* Coined by literary theorist Richard Harland, this term designates an inclusive intellectual field encompassing not only STRUCTURALISM, POSTSTRUCTURALISM and SEMIOTICS, but also related work in adjacent disciplines such as Foucault's philosophy of knowledge and Althusser's political theory.

R F R Harland, *Superstructuralism: The Philosophy of Structuralism and Post-Structuralism* (London, 1987)

RF

supersymmetry (1970s) *Physics* An invariance symmetry principle which treats bosons and fermions equally. Theories using supersymmetry assign boson partners to fermions and vice versa. For example, the boson partners of fermion electrons, quarks and leptons are called selectrons, squarks and sleptons. Correspondingly, the fermion partners of boson photons, gluons, W and Z-particles are called photinos, gluinos, winos and zinos. Although supersymmetries have not been observed experimentally, they are thought to be important in the search for a grand unified theory. *See also* GRAND UNIFIED THEORIES, STRING THEORY and SUPERSTRINGS.

Paul Davies, ed., *The New Physics* (Cambridge, 1989)

GD

supply-side economics (20th century) *Economics* With its emphasis on aggregate supply, rather than aggregate demand (as in KEYNESIAN ECONOMICS), this theory is concerned with the productive capacity of the economy.

Free market supply-side economics emerged in the 1980s as the complement to MONETARISM. Government measures included: tax cuts; measures to facilitate the mobility of labour; reduction in public expenditure; and deregulation. *See* THATCHERISM, REAGANOMICS, SAY'S LAW and NEW CLASSICAL MACROECONOMICS.

P Minford, *The Supply Side Revolution in Britain* (Aldershot, 1991)

PH

support theorem *Mathematics* One of the key geometric ideas in FUNCTIONAL ANALYSIS. If a convex subset C of a real Banach space has a non-empty interior then any point x on the boundary is a *support point* (*see also* BISHOP-PHELPS THEOREM). In other words, there is a supporting hyperplane (geometrically a tangent) to C at x.

This result is fundamental in OPTIMIZATION THEORY for example, since it leads to the existence of subgradients of convex functions, which in turn correspond to LAGRANGE MULTIPLIERS in such results as the KARUSH-KUHN-TUCKER THEOREM or to dual solutions in the FENCHAL DUALITY THEOREM. *See also* HAHN-BANACH THEOREM and SEPARATION THEOREM OF MAZUR.

AL

suprasegmental phonology (1940s–1950s) *Linguistics* American STRUCTURAL LINGUISTICS in the tradition of Leonard Bloomfield (1887–1949). Best-known formulation by George L Trager and H L Smith.

Segmental PHONEMES cut up the stream of speech linearly: /b//a//g/. Trager and Smith proposed additional phonemes not tied to single segments: four different degrees of stress (loudness), four levels of pitch (tone) and three types of juncture between syllables and between words. The theory is now out of favour.

G L Trager and H L Smith, *An Outline of English Structure* (Norman, Ok., 1951)

RF

suprematism (1915) *Art* A movement in ABSTRACT ART launched by Russian Kasimir

Malevich (1878–1935). In painting, Suprematism aims for pure art using pristine geometrical shapes (particularly the square) devoid of personal feeling, but expressing 'non-objective sensation'. Art should be non-utilitarian.

In the (former) Soviet Union, such ideas greatly influenced Liubov Popova (1889–1924), Alexander Rodchenko (1891–1956), El Lissitzky (1890–1941) and Naum Gabo (1890–1977), although some moved towards industrial design and an art of social utility. Malevich's theories were also of major importance to European CONSTRUCTIVISM.

K Malevich, *The Non-Objective World* (Chicago, 1959)

LL

surface structure *Linguistics See* STANDARD THEORY.

surplus value (19th century) *Economics* Cited by English economist David Ricardo (1772–1823) and German political economist Karl Marx (1818–1883), this term describes the surplus derived from the use of a factor of production over its cost.

Marx noted that an employee works more hours than is necessary to provide basic subsistence for himself and his family, thereby creating a surplus, or profit. *See* THEORIES OF PROFITS.

K Marx, *Capital*, vols I–III (Harmondsworth, 1976–81)

AV, PH

Surrealism (1917) *Art/Literary Theory* A term coined by the French poet Guillaume Apollinaire (1880–1918) in 1917, and adopted as a name in the first Surrealist manifesto (1924), written by poet and critic André Breton (1896–1966) in Paris.

It included the following declaration: 'Surrealism rests in the belief in the superior reality of certain forms of association neglected heretofore; in the omnipotence of the dream and in the disinterested play of thought.' It arose out of DADA, but was more ambitious and internationally influential. Like Dada, it declared the importance of the absurd, the irrational and involvement in political anarchy as a means of effecting social change. A literary as well as artistic movement, by the end of World War II it had largely disbanded as a coherent movement.

A Breton, *Manifestoes of Surrealism* (Michigan, 1969)

AB

survival of the fittest (1864) *Biology/Sociology* A phrase coined by British philosopher Herbert Spencer (1820–1903) in his book *Principles of Biology* (1864), but often associated with the evolutionary theorist Charles Darwin (1809–82).

This is the idea that those individuals best adapted to their environment are those most likely to live to reproduce. The phrase was adopted by Darwin as another way of conveying natural selection, and the idea was used by proponents of Social Darwinism to justify unregulated, 'LAISSEZ-FAIRE' capitalism. *Compare with* ALTRUISM and DARWINIAN FITNESS.

KH

survival of the fittest (19th century) *Politics/Sociology* Theory of social progress of English social scientist Herbert Spencer (1820–1903), wrongly attributed to Charles Darwin (1809–82).

If government is kept to its minimum functions of the defence of persons and property, the enforcement of contracts, and the defence of the frontiers, individuals will flourish according to their ability or fitness to adapt to changing circumstances. Although Spencer believed in the inheritance of superior and inferior skills, and in the passing on of skills acquired in one generation to the next, he did not share the fatalism of Darwin's theory of natural selection. Thus desirable human characteristics were naturally rewarded, and undesirable ones naturally discouraged.

Rodney Barker, *Political Ideas in Modern Britain* (London, 1978)

RB

suspension of disbelief *Literary Theory See* WILLING SUSPENSION OF DISBELIEF.

Sutherland's law of gas viscosity (1893) *Physics* Named after its discoverer, the Scottish theoretical physicist William Sutherland (1859–1911).

If η_T is the viscosity of the gas at a temperature T and η_{273} is the viscosity at a temperature of 273 K,

$$\frac{\eta_T}{\eta_{273}} = \frac{273 + C}{T + C}\left(\frac{T}{273}\right)^{2/3}.$$

J Thewlis, ed., *Encyclopaedic Dictionary of Physics* (New York, Oxford and London, 1962)

MS

Sutton's law *Psychology* Not attributable to any one originator, this is a rarely used term named after the notorious bank robber, Willie Sutton.

This principle of diagnosis states that one should look for a disorder where or in whom the disorder is most likely to be found. It supports the hypothesis that all diseases and disorders have groups of predisposing factors.

G C Davison and J M Neale, *Abnormal Psychology* (Chichester, 1982)

NS

sweeping theory *Meteorology* As rain drops of different mass move at different velocities through a cloud, the larger droplets sweep up smaller droplets in their path. *See also* COALESCENCE THEORY.

DT

Sylow's theorems (1872) *Mathematics* Named after Norwegian mathematician Peter Ludwig Sylow (1832–1918), these are three theorems in GROUP THEORY which assert the existence of subgroups of a certain prescribed order in an arbitrary finite group and describe the properties of such subgroups.

(1) Sylow's first theorem asserts that if p is a prime number and α is a positive integer such that p^α divides the order $|G|$ of a finite group G, then G has a subgroup of order p^α. A subgroup of order p^m where p^m divides $|G|$ but p^{m+1} does not divide $|G|$ is called a *Sylow p-subgroup.*

(2) Sylow's second theorem asserts that all Sylow p-subgroups are conjugate and hence isomorphic.

(3) Sylow's third theorem asserts that the number of Sylow p-subgroups for a given p is congruent to 1 mod p.

I N Herstein, *Topics in Algebra* (Wiley, 1975)

MB

Sylvester's law of inertia (1852) *Mathematics* Stated by Joseph Sylvester (1814–97) and proved by Carl Jacobi (1804–51).

For a homogeneous quadratic polynomial Q in a finite number of indeterminates, the number of positive coefficients and the number of non-zero coefficients in any polynomial with square terms only into which Q can be transformed is constant for Q. This has geometrical significance, for example in the classification of conic sections or quadric surfaces.

JB

Sylvester's theorem (1893) *Mathematics* Proposed as a problem by Joseph Sylvester (1814–97).

Show that it is impossible to arrange any finite number of points in the Euclidean plane so that a line through any two of them also passes through a third unless they all lie on one line. The problem was solved by T Gallai in 1933. *See also* MOTZKIN'S THEOREM.

JB

symbiotic theory of cellular evolution (1880s) *Biology* Also called endosymbiosis, and hereditary symbiosis, this theory was originated by German botanist Andreas Franz Wilhelm Schimper (1856–1901) and Altmann in the 1880s and refined by Lynn Margulis (as Lynn Sagan) (1938–) in 1967. The theory that mitochondria, chloroplasts, and possibly other 'organelles' in animal and plant cells were once free-living cells which evolved by living in symbiosis with early cells before being engulfed as permanent components.

This theory assumes that prokaryotes (bacteria) are the primitive cells which gave rise to eukaryotes (the nucleated cells of animals and plants). Originally viewed with scepticism, the symbiotic theory gained adherents in the 1960s, when some organelles were found to carry prokaryotic molecular systems. *See also* AKARYOTIC THEORY, EUKARYOTIC THEORY, PROKARYOTIC THEORY and SERIAL SYMBIOSIS THEORY.

KH

symbol (traditional) *Literary Theory/ Linguistics* This concept has been particularly important in the 20th century: in literary theory since SYMBOLISM; and in SEMIOTICS.

An expression which designates an object which stands for something else (for example, 'The Stars and Stripes', and 'The Cross'); secondary signification, less diffuse than connotation. The relationship of symbolism is determined by cultural conventions, not by resemblance or analogy. A literature tends to possess a number of very powerful recurrent symbolisms; *see also* ARCHETYPE and MYTH.

Recent semiotics instructively demonstrates the prevalence of symbolism in everyday communication.

R Barthes, *Mythologies* [1957], A Lavers, trans. (London, 1972)

RF

Symbolism (later 19th century) *Art* In art, the movement's aims were set out by the French critic Albert Aurier in an article in the *Mercure de France* (March, 1891). He asserted that a work of art must be 'ideaed'; that is, the expression of the idea. Symbolist, since it expresses idea through form; synthetic as its method of representation is general, and subjective and decorative. Adherents' interest in the occult, poetry and religion often imbued their work with an ethereal and mystical character.

AB

Symbolism (later 19th century) *Literary Theory* A movement originating in France and influencing early 20th-century English poetry, it is characterized by uses of language that demand an unusual degree of symbolic interpretation of words and objects, offering novel systems of SYMBOLS. The effect is anti-mimetic and obscure; hence the objections of IMAGISM.

C Chadwick, *Symbolism* (London, 1971)

RF

Symbolism (later 19th century) *Music* Direct musical themes may be used as symbols (for example, Richard Wagner's *leitmotifs*), but more often the musical language of a passage is designed to reveal the emotional state implicit in the narrative; for example, *Pelléas et Mélisande*, music by Claude Debussy (1862–1918), text by Maurice Maeterlinck (1862–1949).

MAD

sympatric speciation (mid-1900s) *Biology* Advocated by Bush, this is the concept that a new species can evolve within the range of the parent population, implying that reproductive isolation can occur biologically (for example, by feeding on different plant species) without the need for geographic isolation.

Polyploidy in plants, which causes 'instant' speciation, is accepted as sympatric speciation by most biologists, but otherwise the possibility of sympatric speciation remains somewhat controversial. *Compare with* ALLOPARAPATRIC SPECIATION, ALLOPATRIC SPECIATION, PARAPATRIC SPECIATION, STASIPATRIC SPECIATION. *See also* SPECIATION, THEORY OF and SPECIES, THEORY OF.

C Barigozzi, *Mechanisms of Speciation* (New York, 1982)

KH

synaesthesia (1892) *Literary Theory* Relevant sense of the term appears in the thesis *'Audition coloré'* by J Millet.

Description of a sense-impression in terms of another sense, for example 'loud perfume', 'sparkling noyse' (Donne, Crashaw, early 17th century). A device much used in French Symbolist poetry, but found throughout Western literature.

J Downey, *Creative Imagination* (1929)

RF

synchromism (1913) *Art* Colour theory proposed by the American artists Stanton Macdonald-Wright (1890–1973) and Morgan Russell (1886–1953) at joint exhibitions held in Munich and Paris. Related to the colour theories of NEO-IMPRESSIONISM and ORPHISM (2), synchromism asserts that colour alone provides the form and subject of a painting.

Synchromism and Related Color Principles in American Painting, 1910–1930, exh. cat., Knoedler Gallery (New York, 1965)

AB

synchronic *Linguistics See* DIACHRONIC AND SYNCHRONIC.

syndicalism (19th century–) *Politics/Economics* Theory of direct working class or trade union power, founded in the theories of French social philosopher Georges Sorel (1847–1922).

Since political POWER arises from economic power, capitalism is most effectively replaced by workers organizing on the basis of their occupation or workplace, and then using their economic power either through strikes or occupations. A popular movement in Italy, particularly under Benito Mussolini (1883–1945). *See* INDUSTRIAL DEMOCRACY, COLLECTIVE BARGAINING THEORY.

G Sorel, *Reflections on Violence* (New York, 1941); David Miller *et al.*, eds, *The Blackwell Encyclopaedia of Political Thought* (Oxford, 1987)

RB, PH, AV

syntagm *Linguistics See* PARADIGM AND SYNTAGM.

syntax (traditional) *Linguistics* The major preoccupation of TRANSFORMATIONAL-GENERATIVE GRAMMAR since the 1950s.

The central component of a GRAMMAR, concerned with patterns of arrangement of words and phrases, not covering patterns of sound or meaning.

K Brown and J Miller, *Syntax*, 2nd edn (London, 1990)

RF

synthetic theory of evolution *Biology See* NEO-DARWINISM.

synthetism (1889) *Art* Term used by the Pont Aven artists (in particular, Emile Bernard (1868–1941)) for their exhibition at the Exposition Universelle of 1889; and by the *Groupe Synthétiste*, formed in 1891, which included French painter Paul Gauguin (1848–1903) among its members.

The term refers to an emphasis on the simplification of drawing and pattern, and the expressive purity of colours in contrast to the scientific developments of NEO-IMPRESSIONISM.

H R Rookmaker, *Synthetist Art Theories* (Amsterdam, 1959)

AB

systemic grammar (1960s) *Linguistics* Developed by the British linguist Michael Alexander Kirkwood Halliday (1925–) from SCALE AND CATEGORY GRAMMAR.

A grammar is seen as a network of systems from which options are made to generate particular structures: *see* for example TRANSITIVITY. *See also* FUNCTIONAL GRAMMAR.

C S Butler, *Systemic Linguistics* (London, 1985)

RF

systems music *Music See* MINIMALISM.

systems theory (20th century) *Mathematics* The mathematical analysis of models of physical systems in engineering. *See also* CONTROL THEORY.

JB

systems theory (20th century) *Politics/Sociology* Also termed 'systems analysis'. Theory of politics and government as a system associated with American political scientist David Easton (1917–).

Government and politics constitute a system of inputs and outputs, with 'gatekeepers' who filter demands upon the system in order to avoid overload. Systems analysis like STRUCTURAL FUNCTIONALISM is contrasted with CONFLICT THEORY.

Geoffrey Roberts and Alistair Edwards, *A New Dictionary of Political Analysis* (London, 1991)

RB

Szent-Györgyi hypothesis *Biology/Chemistry* Named after the Hungarian-born American biochemist Albert von Nagyrapolt Szent-Györgyi (1893–1986) famous for his discovery of actin and myosin, and the isolation of Vitamin C. He won the Nobel prize in 1937.

The hypothesis states that a mechanism exists which allows the energy absorbed from light or from a chemical reaction in one part of a living system to be available for reaction in another part of the system without degradation.

GD

T

tachism (1952) *Art* Derived from the French word *tache* meaning 'patch', this term was coined by the French critic Michel Tapie in relation to post-war European ABSTRACT ART, and related to 'ART INFORMEL' and ABSTRACT EXPRESSIONISM. The patches and blots of colour on a canvas assume their own significance, as if applied at random, spontaneously, and act as projections of the artist's emotional state.

M Tapie, *Un Art Autre* (Paris, 1952)

AB

tacit knowledge *Philosophy* Primarily an idea developed by the Hungarian social philosopher Michael Polanyi (1891–1976).

Starting from such facts as our ability to recognize faces without knowing how we do so, and to be trained in a psychological laboratory to respond to certain perceived stimuli without knowing just what it is we are responding to, Polanyi claims that we transfer our attention from (for example) the specific features of a face to the face as a whole (we attend *from* the former *to* the latter, as he puts it), and thereby have tacit knowledge of the former. The idea is then developed to account for the foreshadowing by which a scientist sees first a problem and then a possible solution to it in his data. The important point, however, is not whether the knowledge in question is unconscious, but its function in being that which we 'attend from'. 'Tacit knowledge' has also been used for our knowledge of what gives meaning to the words and sentences of our own language.

M Polanyi, *The Tacit Dimension* (1966), ch. 1

ARL

tagmemics (1950s–) *Linguistics* Term 'tagmeme' coined by American linguist Leonard Bloomfield (1887–1949) in 1933. Theory developed by American linguist Kenneth Lee Pike (1912–) and others; also called 'slot and filler grammar'.

A tagmeme is the smallest unit of grammar, analogous to PHONEME in phonology. For Pike, it is not simply a type of CONSTITUENT like 'word', or a class, like 'noun', but simultaneously a class and a function; for example noun and head, occurring in a slot. Central to tagmemics is the integration of sentences into texts and ultimately the culture. Tagmemics has been applied to non-linguistic behaviour within a culture.

K L Pike, *Linguistic Concepts: An Introduction to Tagmemics* (Lincoln, 1982)

RF

Talbot-Plateau law *Psychology* Formulated by the English scientist William Henry Fox Talbot (1800–77) and the Belgian physicist Joseph Antoine Ferdinand Plateau (1801–83).

This principle states that when the rate of a flickering light is sufficiently high it comes to be seen as a continuous steady stream of light with a perceived brightness that is the mean of the periodic impressions. For instance, if the flickering is composed of equally long on and off periods, the steady state will have one-half the brightness of the on-phase. This fused brightness is called 'Talbot brightness'.

L Kaufman, *Perception* (New York, 1979)

NS

Talbot's law *Physics* When observation is made of a periodic light source, the image fuses when the frequency exceeds a certain critical value. When fusion has occurred the intensity of a periodic source of light is that of a steady source which emits the same amount of luminous flux over a complete period.

J Thewlis, ed., *Encyclopaedic Dictionary of Physics* (New York, Oxford and London, 1962)

MS

tangent rule (1579) *Mathematics* Trigonometry was first studied by Hipparchus and Ptolomy (Claudius Ptolomaeus) in about AD 150, but before 1500 the trigonometry of spherical surfaces was favoured. The tangent rule was proved by François Viète (1540–1603).

Let *A, B, C* be the angles of a plane triangle opposite the sides of length *a, b, c*. Then

$$\text{tab}\,\frac{\mathrm{B}-\mathrm{C}}{2}=\frac{b-c}{b+c}\cot\frac{A}{2}$$

JB

tangled bank (1859) *Biology* A phrase coined by eminent British evolutionist Charles Robert Darwin (1809–82) in the last paragraph of his book, *On the Origin of Species by Means of Natural Selection, or the Preservation of Favoured Races in the Struggle for Life*; this is a metaphor for the interconnectedness of animals and plants on the earth.

KH

target theory of radiation processes *Physics* A theory explaining the biological effects of ionizing radiation.

The theory assumes the presence of a small sensitive region (the 'target area') within each cell. To bring about an effect one or more 'hits' (ionizing events) may be necessary on each target area.

J Thewlis, ed., *Encyclopaedic Dictionary of Physics* (New York, Oxford and London, 1962)

MS

Tarski's fixed point theorem (20th century) *Mathematics* Named after the Polish-born American logician, algebraist and analyst Alfred Tarski (1902–1983).

The theorem that an isotone mapping on a complete lattice to itself has a fixed point.

M Hazewinkel, ed., *Encyclopaedia of Mathematics* (Dordrecht, 1988)

ML

Tauberian theorems *Mathematics* (19th century) Named after Austrian analyst Alfred Tauber (1866–1933), these are SUMMABILITY THEORY results which show that a series summed by a method converges to the value produced by that method. Tauber's original result states that if $\{na_n\}$ tends to zero and Σa_n is summable to S by the ABEL SUMMATION method, then Σa_n is actually equal to S. Correspondingly, theorems which assert that a given method is *regular* (giving the correct sum to a convergent series or sequence) are ABELIAN THEOREMS.

K Knopp, *Theory and Application of Infinite Series* (New York, 1990)

ML

tax incidence *Economics* First discussed by the Physiocrats in France, this is the analysis of the effect of a particular tax on the distribution of economic welfare.

The term also refers to the ultimate payer of a tax. If a government increases tax on petrol, oil companies may absorb it if competition is intense or they may pass it on to private motorists. Similarly, a taxi driver may pass on the tax increase to his passenger and a food distributor may pass it on to a supermarket, which in turn passes it on to its customer. *See* EQUAL SACRIFICE THEORY and ABILITY TO PAY PRINCIPLE.

J A Pechman, *Who Paid the Taxes, 1966–85?* (Washington, D.C., 1985)

AV, PH

Taylor's theorem (18th century) *Mathematics* Named after British mathematician Brook Taylor (1685–1731), much of whose work was so terse and difficult to understand that he never received credit for many of his innovations. The importance of Taylor's theorem was recognized, approximately 60 years after his death, by Lagrange. There is also evidence that Gregory used Taylor's theorem 44 years before Taylor.

The theorem of analysis that if a function f can be differentiated $(n + 1)$ times in an interval containing the point a and if

$$P_n(x) = f(a) + f'(a)(x-a) + \frac{f''(a)}{2!}(x-a)^2 + \ldots + \frac{f^{(n)}(a)}{n!}(x-a)^n$$

is the nth Taylor polynomial about $x = a$ for f, then for each x in the interval, there is at least one point c between a and x such that the remainder

$$R_n(x) \equiv f(x) - P_n(x) = \frac{f^{(n+1)}(c)}{(n+1)!}(x-a)^{n+1}.$$

H Anton, *Calculus with Analytic Geometry* (New York, 1980)

ML

technological gap theory *Economics* This proposes that changes in international trade are dictated by the relative technological sophistication of countries.

Some nations, such as the US or Japan, have a competitive trade advantage because of their ability to innovate. Over time, other countries will bridge a particular gap although the really innovative will have opened others. *See* PRODUCT LIFECYCLE THEORY.

A Heertje, *Economics and Technical Change* (London, 1977)

AV, PH

telegony (1890–95) *Biology* The theory that inheritance of an individual is influenced not only by the father, but also by all the previous mates of the mother. Experiments in modern genetics have disproved this notion. *See also* MENDELISM.

KH

telegraphic speech (1963) *Linguistics* A term coined by the American psycholinguist R Brown and colleagues.

The early speech of children omits function words (*the*, *is*, *will*, and so on) and inflections (*-s*, *-ing*, and so on), resembling adult telegrams where such words, carrying low information, are omitted to save cost.

R Brown, *A First Language* (London, 1973)

RF

teleology (1730–40) *Biology* The doctrine that purpose and design are revealed in the beauty and perfection of nature. *See also* NATURAL THEOLOGY and PALEY'S WATCHMAKER.

G Blandino, *Theories on the Nature of Life* (New York, 1969)

KH

teleology *Philosophy* In general, belief in or appeal to explanation in terms of ends or purposes. As an ethical doctrine teleology claims that our duties are specifiable in terms of the production of some value. Teleology is perhaps rather wider than CONSEQUENTIALISM as it includes such views as that an act is our duty if doing it will promote our own virtue. *See also* DEONTOLOGY.

ARL

telephone theory *Psychology See* FREQUENCY THEORY OF HEARING.

tellability (late 1960s) *Stylistics* Term coined by the American sociolinguist William Labov.

A story is tellable if it has a point for its audience, by being relevant, exemplary, or unexpected. *See also* NATURAL NARRATIVE.

M L Pratt, *Toward a Speech Act Theory of Literary Discourse* (Bloomington, 1977)

RF

Teller-Redlich rule *Chemistry* Named after Redlich and the Hungarian-born American physicist Edward Teller (1908–).

The rule states that the product of the frequency ratio values of all vibrations of a given symmetry-type of two isotopic molecules depends on the atomic masses and the geometric structures of the molecules and is independent of the potential constants.

GD

template theory *Biology See* INSTRUCTIVE THEORY.

tenor *Linguistics See* REGISTER.

tenor, vehicle and ground (1936) *Literary Theory* Components of the 'comparison' theory of METAPHOR (Ivor Armstrong Richards, 1893–1979).

The tenor of a metaphor is the topic referred to: 'love' in 'love is a rose'; the vehicle is the second term 'rose' to which love is said to be analogous; the ground is the basis or justification for the comparison, here sweetness, naturalness, sensuality, and so on.

I A Richards, *The Philosophy of Rhetoric* (London, 1936)

RF

term structure of interest rates *Economics* This refers to the relationship between the fixed amount of interest paid on a financial security (such as a government or corporate bond) and the amount of time before the bond reaches its maturity date. Early work on this theory of expectations was carried out by US economist Irving Fisher (1867–1947) and English economist John Hicks (1904–89). *See* ADAPTIVE EXPECTATIONS. *See also* RANDOM WALK HYPOTHESIS, RATIONAL EXPECTATIONS and TIME PREFERENCE THEORY OF INTEREST.

AV, PH

Terzaghi's principle *Geology* Proposed by Czechoslovakian-born American Karl Terzaghi (1883–1963) founder of soil mechanics and civil engineer. When a rock is subjected to a stress, this is opposed by the fluid pressure of fluids in pores within the rock. Thus, the effective stress equals the total stress less the pore fluid pressure.

Karl Terzaghi *Soil mechanics in engineering practice* (1948)

DT

testicular extract theory (1889) *Biology/Medicine* Espoused by Charles Edouard Brown-Séquard (1817–94), a French-American neurophysiologist, endocrinologist and physician; this is the belief that injecting liquefied dog or guinea pig testicle under the skin could rejuvenate the sexual prowess of an aged man.

Although his theory was immediately controversial and soon discredited, Brown-Séquard's advocacy of the technique encouraged serious study of sex hormones and the isolation of testosterone by Carl R Moore in 1929. *See also* AGEING, THEORIES OF and HUMORAL ACTION, CONCEPT OF.

J Fulton, *Selected Readings in the History of Physiology* (Springfield, Ill., 1966)

KH

testing effect *Psychology* Identified by the Anglo-American psychologist Raymond Bernard Cattell (1904–).

This refers to the effect that taking a test actually has upon what the test is designed to evaluate. It is a source of error in psychological testing which is particularly likely to occur where the use of pre-tests may modify the phenomenon which is subsequently measured or tested.

A Anastasi, *Psychological Testing* (London, 1982)

NS

tetranucleotide theory (early 1900s) *Biology* Based on work by Ascoli and Levene. This term describes the idea that the four bases of deoxyribonucleic acid (DNA) and ribonucleic acid (RNA) – adenine, cytosine, guanine, and thymine in DNA (or uracil in RNA) – are present in these compounds in equimolar amounts, and that an essential unit of these molecules is one containing each of the four bases. This theory was replaced by CHARGAFF'S RULES in 1950.

A Sturtevant, *A History of Genetics* (New York, 1965)

KH

text (traditional, 1970s) *Stylistics/Literary Theory/Linguistics* Numerous definitions attest to the importance of this term.

In literary work: authoritative version; written as opposed to spoken language; unit 'above' the sentence; communicative interaction. In POSTSTRUCTURALISM: a 'weaving' (etymology of this word) of codes. In SEMIOTICS: anything which signifies, in any medium. *See also* TEXT LINGUISTICS.

R de Beaugrande and W Dressler, *Introduction to Text Linguistics* (London, 1981)

RF

text linguistics (1970s) *Linguistics* Various proposals.

The sentence was traditionally the 'upper limit' of linguistic description; TRANSFORMATIONAL-GENERATIVE GRAMMAR confirmed this. Text linguistics provides a variety of models for the structure of whole texts and of their parts and their relationships. *See also* COHERENCE, COHESION, and MACROSTRUCTURE.

R de Beaugrande and W Dressler, *Introduction to Text Linguistics* (London, 1981)

RF

textual function *Linguistics See* FUNCTIONAL GRAMMAR.

textual world *Stylistics See* COHERENCE.

thalamic theory of emotion *Psychology* Proposed by the American physiologist Walter Bradford Cannon (1871–1945).

This theory claims that the integration of external expressions of emotion is controlled by the thalamus sending relevant excitations to the cortex of the brian at the same time that the hypothalamus controls the behaviour. This theory was put forward as a critique of the JAMES-LANGE THEORY which argued that the sensory feedback controlled the emotional expression. *See* ACTIVATION THEORY and CANNON-BARD THEORY.

H Gleitman, *Psychology* (New York, 1991)

NS

Thales' theorem (5th century BC) *Mathematics* Named after the Greek philosopher Thales of Miletus (*c.* 624–547 BC) who was the teacher of Pythagoras and is known as the 'father of philosophy'.

The theorem states that the locus of the vertices of all right angles whose arms pass through two points A and B is a circle with diameter AB.

W Gellert, S Gottwald, M Hellwich, H Kästner and H Küstner, eds, *The VNR Concise Encyclopedia of Mathematics* (New York, 1975)

ML

Thatcherism *Economics* This is the name given to the 'LAISSEZ-FAIRE' policy of Margaret Thatcher (1925–), British prime minister from 1979 to 1990.

An advocate of SUPPLY-SIDE ECONOMICS, among Thatcher's policies were: reductions in tax; manipulation of the money supply to reduce inflation; PRIVATIZATION of public industry; reduction of trade union power; reduction of government's role in the economy; and encouragement of people to save, work and buy property. In 1983, Thatcher's anti-inflationary policies resulted in the worst unemployment figures since 1923. Her critics argue that she sacrificed Britain's social well-being in the pursuit of her economic policy. *See* MONETARISM and REAGANOMICS.

Robert Studelsky, ed., *Thatcherism* (1989); Christopher Johnson, *The Economy under Mrs Thatcher* (London, 1991)

PH

Thatcherism (20th century) *Politics* Theory of politics associated with British Conservative Party Leader and Prime Minister Margaret Thatcher (1925–).

Term more frequently used by critics than by supporters of the cluster of NEW RIGHT policies pursued or affirmed by the government of Margaret Thatcher between 1979 and 1990. 'Thatcherism' was presented as a mixture of liberal free market economics, and conservative paternalist cultural and moral attitudes; together with a vigorous use of STATE power in military, police and 'security' areas.

Ruth Levitas, ed., *The Ideology of the New Right* (Cambridge, 1986)

RB

theatre of cruelty (1938) *Literary Theory* Type of drama called for by the French dramatist Antonin Artaud (1896–1948). By exposing audiences to emotional violence, plays and productions should shock them into self-realization and the shedding of repressions.

Theatricality, spectacle, special effects and mime are foregrounded over and above the text. Artaud's own experiments were unsuccessful. A prototype for this mode was the notorious *Ubu Roi* (1896) by French writer Alfred Jarry (1873–1907); a well known modern example is *Marat/Sade* (1964) by Peter Weiss.

A Artaud, *The Theatre and its Double* [1938], V Corti, trans. (London, 1970)

RF

theism *Philosophy* The religious belief that God is the creator of and supreme authority in the universe.

In most major religions God is a beneficent being (or beings) with a particular sympathy for mankind, which owes him an allegiance of obedience and worship. Philosophical objections to the idea include: the conflict inherent between an omnipotent

God and the expression of human freedom; the presence of evil in the world; and the ill-defined character of the supreme being. *See also* ATHEISM, DEISM and PANTHEISM.

DP

theme and rheme (1930s) *Linguistics* Proposed by PRAGUE SCHOOL founder V Mathesius (1882–1945) and developed after World War II, principally by F Danes and J Firbas. Adopted in British linguist Michael Alexander Kirkwood Halliday's FUNCTIONAL GRAMMAR.

The theme of a sentence is the topic about which it is communicating; the rheme is something said about that topic. In the work of Firbas, a third term 'transition' is used for the linking elements between theme and rheme.

J Firbas, 'On the Dynamics of Written Communication in the Light of the Theory of Functional Sentence Perspective', *Studying Writing: Linguistic Approaches*, C R Cooper and S Greenbaum, eds (Beverly Hills, 1986), 40–71

RF

theorem of parallel axes *Physics* The moment of inertia I of a body about any axis is related to the moment of inertia I_{cm} about a parallel axis passing through the centre of mass by the equation

$$I = I_{cm} + Mh^2$$

where M is the mass of the body and h is the distance between the two axes.

J Thewlis, ed., *Encyclopaedic Dictionary of Physics* (New York, Oxford and London, 1962)

MS

theorema egregium (19th century) *Mathematics* From the Latin phrase meaning 'remarkable theorem', this was proved by German mathematician and astronomer Carl Friedrich Gauss (1777–1855).

The theorem asserts that the Gaussian curvature K remains invariant under isometric mappings; that is, under motions, parameter transformations and bending.

W Gellert, S Gottwald, M Hellwich, H Kästner and H Küstner, eds, *The VNR Concise Encyclopedia of Mathematics* (New York, 1975)

ML

theory (17th century) *Natural sciences* Although it dates from Antiquity, Sir T Herbert first applied this term in its modern sense (to geology) in 1638. A theory is a general principle supported by a substantial body of scientific evidence which explains observed facts. As a probable explanation for observations, a theory offers an intellectual framework for future discussion, investigation and refinement.

R Giere, *Understanding Scientific Reasoning* (New York, 1979)

KH

theory of everything (20th century) *Physics* This is the theorist's dream and would incorporate the four fundamental interactions: gravitational, electromagnetic, strong nuclear and weak nuclear. The development of such a theory is at present based on grand unified theories which imply that there are in the universe only two kinds of particles: particles of matter (quarks and leptons) and force-carrying particles (gauge bosons such as gluons).

At present no theory of everything exists, but there is promise of a unique theory in one in which the universe began in ten dimensions, but only four of which expanded, to give space and time. In this theory, particles are entities extended in space with lengths of the order of 10^{-36}m, known as strings. *See* QUARK THEORY and GRAND UNIFIED THEORIES.

J Thewlis, ed., *Encyclopaedic Dictionary of Physics* (New York, Oxford and London, 1962)

MS

thermodynamic scale of temperature *Physics* *See* KELVIN TEMPERATURE SCALE.

thermodynamics, laws of *Physics* *See* ZEROTH LAW OF THERMODYNAMICS, FIRST LAW OF THERMODYNAMICS, SECOND LAW OF THERMODYNAMICS and NERNST'S HEAT THEOREM.

theta theory (1980s) *Linguistics* Devised by the American linguist Avram Noam Chomsky (1928–).

θ theory is the component of GOVERNMENT AND BINDING THEORY which assigns thematic roles like Agent, Patient, Goal to the arguments (roughly, nouns) of verbs. *See also* CASE GRAMMAR and TRANSITIVITY.

N Chomsky, *Lectures on Government and Binding* (Dordrecht, 1981)

RF

Thevenin's theorem *Physics* Named after its originator, this states that any linear circuit (that is, one in which the current flowing is proportional to the potential difference, having two terminals) can be represented by a source of e.m.f. and a resistor in series.

J Thewlis, ed., *Encyclopaedic Dictionary of Physics* (New York, Oxford and London, 1962)

MS

third law of thermodynamics (1906) *Physics* *See* NERNST'S HEAT THEOREM.

third man argument *Philosophy* One of a group of arguments presented by Plato (*c*.427–*c*.347 BC) in his dialogue *Parmenides* (§§131e–3a) in apparent criticism of PLATO'S THEORY OF FORMS.

Briefly, the argument might be put as follows. If a man is made to be what he is by participating in a Platonic Form (though Greek did not distinguish small and capital letters), then another Form will be needed to explain how both the man and the Form can be called 'man'. This Form will be a 'third man', and yet another Form (a 'fourth man') will be needed to explain how these three items can all be called 'man', and so on to an infinite regress. As Plato presents it, the argument requires Forms themselves to have the properties they are Forms of (*see* PARADIGMATISM), but the argument can be restated to avoid this, and in fact raises fundamental questions about objects and their properties, and predication.

The theory has been copiously discussed in the 20th century, and by Aristotle (384–322 BC) – to whom we owe the name, and also the example 'man'; Plato's own example being 'large' – for whom it played a major part in determining his reactions to Plato's metaphysics.

J A Passmore, *Philosophical Reasoning* (1961), ch. 2; modern treatment of Plato's problem

ARL

Thomas-Reiche-Kuhn sum rule *Chemistry* *See* F-SUM RULE.

Thomson effect (1854) *Physics* Named after Scottish mathematician and physicist William Thomson, 1st Baron Kelvin (1824–1907) who predicted the effect.

When heat is conducted along the wires of a thermocouple that is carrying no current, there is a uniform temperature gradient in each wire. When a current flows there is a change in the temperature distribution that is not entirely caused by Joule heating. Allowing for Joule heating, the heat that must be extracted laterally at all places along the wire to restore the initial temperature distribution is called the Thomson heat. *See* JOULE'S LAW OF ELECTRIC HEATING.

J Thewlis, ed., *Encyclopaedic Dictionary of Physics* (New York, Oxford and London, 1962)

MS

Thomson's theory of the atom (1907) *Physics* Named after its originator, the British mathematical physicist Joseph John Thomson (1856–1940). After his discovery of the electron, Thomson developed a model of the atom in which electrons were embedded in a sphere of positive charge of uniform density.

Initially, this model had most of the mass associated with the electrons so that even the lightest atom contained a large number of electrons. However, when X-ray scattering measurements showed that the number of electrons in an atom was of the order of the relative atomic mass, Thomson modified his model so that most of the mass was associated with the smeared-out positive charge. He pictured the electrons as lying in coplanar rings and rotating to give stability. This model is sometimes referred to as the 'currant bun' or 'plum duff' model. Despite some early successes, it was superseded when it could not explain the large-angle alpha particle scattering observed by Ernest Rutherford, 1st Baron Rutherford of Nelson (1871–1937).

J Thewlis, ed., *Encyclopaedic Dictionary of Physics* (New York, Oxford and London, 1962)

MS

Thomsonianism (19th century) *Medicine* Named after its originator, the American critic of academic medicine Samuel A Thomson (1769–1843), this is the theory that special vegetable regimens can cure disease. The popularity of Thomsonianism peaked before the mid-19th century and had disappeared by the end of the century.

W Hand, *Magical Medicine: The Folklore Component of Medicine in the Folk Belief, Custom, and Ritual of the Peoples of Europe and America* (Berkeley, Calif., 1980)

KH

Thorndike's law of effect *Psychology* *See* LAW OF EFFECT.

thought, the three laws of *Philosophy* Traditional name for the laws of identity, contradiction and excluded middle, regarded as being particularly basic to thinking.

The three laws are no longer singled out in quite this way. The law of excluded middle is subject to dispute (and also to a variant form, the law of BIVALENCE), and even the law of contradiction has received limited criticism (*see* DIALETHEISM, PARACONSISTENCY). Only the law of identity remains undiscussed, but nowadays attention tends to focus more on the laws of logic as a whole and on what sort of justification can be given for them, or for those of them which are accepted.

ARL

three age system (19th century) *History* Theory of historical periods. Prehistory can be divided into Stone, Bronze, and Iron Ages. The terms have survived longer than the historical theory on which they were based.

Alan Bullock, Oliver Stallybrass, and Stephen Trombley, eds, *The Fontana Dictionary of Modern Thought*, 2nd edn (London, 1988)

RB

three-cell model *Climatology* A simplified model of atmospheric circulation in which a net transport of energy from low to high latitudes is caused by the action of three adjoining cells of vertically-rotating air masses.

The low and high latitude cells are primarily Hadley cells, separated by a Ferrel cell in intermediate latitudes. The polar cell is separated from the Ferrel cell along the polar front. While conceptually sound on a very broad scale, the process is in fact vastly more complicated, with the influence of upper atmospheric circulation (particularly the jet streams) needing to be taken into account. *See also* HADLEY CELL MODEL and ROSSBY MODEL.

J M Wallace and P V Hobbs, *Atmospheric Science: An Introductory Survey* (New York, 1977)

DT

three circle theorem (1896) *Mathematics* So called because it refers to three concentric circles, this theorem was proved by Jacques Hadamard (1865–1963).

For a function $f(z)$ of a complex variable z which has a differential coefficient at all points between two concentric circles, let $m(r)$ be the maximum value of the modulus of $f(z)$ on concentric circle of radius r between the other two. Then the logarithm

$$\ln [m(r)]$$

is a function of $\ln r$ such that chords of the graph always lie above the graph. The theorem can be used to measure the rate of increase of the modulus of $f(x)$.

JB

3n + 1 problem *Mathematics* *See* COLLATZ SEQUENCE.

three series theorem (Kolmogorov) (20th century) *Mathematics* Named after Russian probabilist, topologist and analyst Andrei Nikolaevich Kolmogorov (1903–87).

This theorem of PROBABILITY THEORY states that a series of independent random variables ΣX_i converges with probability 1 if and only if three particular series converge and with probability zero otherwise. When the random variables are uniformly bounded, it is only necessary to verify the convergence of the sums of the expectations and the variances.

W Feller, *An Introduction to Probability Theory and Its Applications* (New York, 1966)

three worlds theory (20th century) *Politics* With the end of conventional European imperialism, there were three groups of states, or three worlds; the industrial, developed, capatalist world; the socialist world of Eastern Europe and China; and the third world of newly liberated and underdeveloped states. The theory, which with the abdication of communist despotisms in Eastern Europe became out of date, was more generally criticized for over-simplification; and for being applicable only to states, not to peoples or to political movements.

Aijaz Ahmad, *In Theory: Classes, Nations, Literatures* (London, 1992)

RB

threshold effect *Biology See* TOLERANCE, LAW OF.

Thue-Siegel-Roth theorem (1909, 1921, 1955) *Mathematics* Named after Norwegian mathematician Axel Thue (1863–1922), German mathematician Carl Ludwig Siegel (1896–1981) and British mathematician Klaus Friedrich Roth (1925–).

This is the result in NUMBER THEORY whereby for any algebraic irrational number α and for any $k > 2$, the inequality

$$|\alpha - \frac{p}{q}| > \frac{1}{q^k}$$

is true for all but a finite number of rational numbers p/q with $q > 0$.

This theorem, which is extremely difficult to prove, has important consequences in the study of DIOPHANTINE EQUATIONS. In particular, if for some $k > 0$ the inequality is violated infinitely often then α is transcendental. For his substantial improvement on the earlier results of Thue and Siegel, Roth was awarded a Fields Medal. *See also* DIOPHANTINE ANALYSIS, DIRICHLET'S THEOREM ON DIOPHANTINE APPROXIMATION and HURWITZ'S THEOREM.

A Baker, *A Concise Introduction to the Theory of Numbers* (Cambridge, 1984)

MB

Thurstone's law of comparative judgment (1927) *Psychology* Formulated by the American psychologist Louis Leon Thurstone (1887–1955).

This law states that the effect of each stimulus on the subject can be summarized by a single number which varies from trial to trial but has normal distribution, and that the subject selects the stimuli associated with the larger number for each trial when presented with two or more stimuli and asked to choose between them with respect to some specified attribute. *See also* LIKERT SCALE.

R S Woodworth and H Schlosberg, *Experimental Psychology* (London, 1966)

NS

tidal theories of planetary formation (19th century–) *Astronomy* Such theories assume that the material to form the planets was pulled out of the Sun by the tidal forces exerted by some passing object.

An early theory of this type (1898) was that of W F Sedgwick, who posited the tidal effect of a passing star. This was suggested independently by the English mathematical physicist Sir James Hopwood Jeans (1877–1946) in 1901, and a similar hypothesis was proposed by the American geologist Thomas Chrowder Chamberlin (1843–1928) and Forest Ray Moulton (1872–1952) in 1905 (*see* PLANETISIMAL THEORY OF PLANETARY FORMATION).

J H Jeans and the English geophysicist Sir Harold Jeffreys (1891–) showed that on the close approach to the Sun by a second star, an immense tidal effect would be produced and a large tongue of gas drawn from both the Sun and the second star, with each tongue given some angular momentum. Condensation would then occur through gravitational instability, the largest planets being formed near the middle of the tongue. These planets would travel in elongated orbits, but their motions would be damped by the residual material of the tongues and they would then take up nearly circular orbits. The difficulties with theories of this type are: such a collision is unlikely; there are angular momentum problems; and much of the material drawn out of the star would probably fall back in again.

S Mitton, ed., *The Cambridge Encyclopaedia of Astronomy* (London, 1973)

MS

Tiebout hypothesis (1956) *Economics* This theory is named after American economist Charles Tiebout (1924–68), who proposed that if public goods/services were provided by a large number of local governments, consumers would have a greater diversity of choice. *See* SOCIAL WELFARE FUNCTION.

C Tiebout, 'A Pure Theory of Local Government Expenditure', *Journal of Political Economy*, vol. LXIV (1956), 416–24

PH

Tietze's extension theorem *Mathematics* Named after Austrian mathematician Heinrich Franz Friedrich Tietze (1880–1964), this is the result in TOPOLOGY whereby a topological space is normal if and only if every continuous mapping of a closed subset of the space into the unit interval may be extended to a continuous mapping of the entire space into the unit interval. *See also* URYSOHN'S LEMMA.

J R Munkres, *Topology: A First Course* (Prentice-Hall, 1975)

MB

time *Physics* A fundamental quantity usually indicating duration or a precise moment. The second is the SI unit of time, defined as the duration of 9,192,631,770 periods of the radiation corresponding to the transition between the two hyperfine levels of the ground state of caesium-133. Time in an absolute or Newtonian sense is that which enables a date to be uniquely associated with any given event. Operational time allows time to be treated as an independent variable infinitely divisible and extending to an infinite distance in both positive or negative directions. The concept of quantized time arises in quantum mechanics, with a quantum particle, or chronon, corresponding to the time taken for a photon to cross the diameter of an electron ($4.5 = 10^{-24}$ s). Solar time relates to successive intervals between transits of the Sun across the meridian, while sidereal time, used in astronomy, relates to successive intervals between transits of a point (related to the vernal equinox) across the meridian. *See also* SPACE-TIME.

GD

time preference theory of interest (1871) *Economics* Developed first by Austrian economist Carl Menger (1840–1921), this is the analysis of how individuals or firms will sacrifice present utility in the hope of greater future returns. The expected rate of return is highly subjective. *See* TERM STRUCTURE OF INTEREST RATES, RANDOM WALK HYPOTHESIS.

C Menger, *Principles of Economics*, J Dingwall and B Hoselitz, eds (Glencoe, Ill., 1950)

AV, PH

Tinbergen's hierarchical model (1951) *Biology* Proposed by Nikolaas Tinbergen (1907–88), a British zoologist born in the Netherlands who shared the 1973 Nobel prize for Physiology or Medicine. This is a model for animal behaviour that attempts to explain how behaviour is organized over extended periods of time.

Tinbergen theorized that instincts are arranged in a hierarchy, with high-level behavioural centres (such as those for feeding or reproduction) in the central nervous system ultimately controlling specific actions. Motivational energy flows from the centres down through various blocks (called innate releasing mechanisms, or IRMs) which inhibit the behaviour until an appropriate stimulus allows release. *Compare with* LORENZ'S HYDRAULIC MODEL.

C Barnard, *Animal Behaviour: Ecology and Evolution* (New York, 1983)

KH

token identity theory *Psychology See* IDENTITY THEORY.

tolerance, law of (1913) *Biology* (Also called the dose-response curve, and the saturation curve; and including the notion of threshold effect.) Named by Victor E Shelford, this is the theory that the response of an organism to some essential environmental parameter (such as water or sunlight) follows a predictable pattern.

At first the organism fails to respond to the parameter until a certain threshold is reached. Then, so long as supplying more of the factor generates an increased response in the organism, the factor is said to be deficient. If providing still more fails to change the organism's response, then the factor is present in the zone of tolerance (also called saturation). When even more of the factor is provided, such that the organism begins to exhibit less response than at the saturation stage, then inhibition or toxicity has been reached. *Compare with* BLACKMAN CURVES and LIEBIG'S LAW.

KH

tonality (antiquity) *Music* Sometimes attributed (as *tonalité*) to the French music critic Castil-Blaze (1784–1857), this term is now used to cover a wider range of meanings than he postulated.

A theoretical system in which selected pitches (those belonging to a 'key') are ordered according to a hierarchic pattern, dependent upon a central pitch, or 'tonic'; the pattern is deemed to repeat at intervals of one octave. 'Chromatic' notes (those not belonging to the key) may be integrated into the music provided that the tonic does not lose its supremacy. The inherent hierarchy of chords within the tonal system creates a feeling of harmonic movement and the use of goal-directed progressions generates long-term structure.

R Reti, *Tonality, Atonality, Pantonality: A Study of Some Trends in Twentieth Century Music* (London, 1958)

MLD

tone *Literary theory See* FOUR KINDS OF MEANING.

Tonelli's theorem *Mathematics* The theorem of MEASURE THEORY which gives conditions for the order of integration of an iterated integral to be reversed. If $(X, \mathcal{M}, \mu)$ and $(Y, \mathcal{N}, \nu)$ are sigma finite measure spaces and f is a non-negative measurable function with respect to $\mu \times \nu$, then

$$\int fd(\mu\times\nu) = \int \left[\int f(x,y)d\nu(y)\right] d\mu(x) = \int \left[\int f(x,y)d\mu(y)\right] d\nu(x).$$

See also FUBINI'S THEOREM.

G B Folland, *Real Analysis* (New York, 1984)

ML

tonic sol-fa (1840s) *Music* A SOLMIZATION system based on existing models, it was refined by the English clergyman and educationalist John Curwen (1816–80).

Curwen postulated that each scale degree has its own character ('soh' = firm, 'ray' = expectant) and that by personal discovery of the 'feel' of each degree, a person may be prompted to sing a note by recalling its quality, rather than by counting its distance from 'doh'. Individual scale degrees may be indicated by hand signs or by notated sol-fa symbols (rhythm can be notated too). The system's simplicity and suitability for teaching massed choirs make it an important alternative to orthodox methods, though its application is limited to the teaching/learning of sung melody.

J Curwen, *Tonic Sol-fa* (London, 1878)

MLD

topology *Mathematics* The branch of GEOMETRY which is concerned with the properties of a geometric figure that remain invariant when the figure is bent, stretched, shrunk, or deformed in any way that does not create new points or fuse existing ones.

In the 20th century, the subject consists of two somewhat separate areas: namely, point set topology, where the geometrical figures in question are collections of points; and *combinatorial* or ALGEBRAIC TOPOLOGY (formerly known as *analysis situs*), in which the geometric figures are considered as aggregates of smaller building blocks. Point set topology may also be considered as a generalization of concepts such as continuity and limits to sets other than the real and complex numbers.

J R Munkres, *Topology: A First Course* (Prentice-Hall, 1975)

MB

Torricelli's law of efflux (1644) *Physics* Named after its discoverer, the Italian mathematician and physicist Evangelista Torricelli (1608–47), this states that the speed of efflux of a liquid through an orifice is equal to that which a body would gain in falling freely from the free surface of the liquid to the orifice.

For example, for a can with a hole a distance h below the free liquid surface, the speed of efflux v is given by

$$v^2 = 2gh$$

where g is the acceleration of free fall. The law assumes that there are no viscous effects.

J Thewlis, ed., *Encyclopaedic Dictionary of Physics* (New York, Oxford and London, 1962)

MS

toryism *Politics* Theory of paternalist CONSERVATISM.

Societies are characterized both by traditional values and institutions; and by division into a skilled and perceptive minority, and a less capable and less wise majority. It is the responsibility of the minority to further the INTERESTS of both TRADITION and the masses.

S H Beer, *Modern British Politics* (London, 1965)
RB

total history (20th century) *History* View of history associated with the French '*annales*' group of historians.

History should attempt to give as encompassing an account of the social, economic, and cultural structures of the past as possible, rather than narrate events.

P Burke, ed., *Economy and Society in Early Modern Europe* (London, 1972)
RB

total probability theorem (20th century) *Mathematics* The probability of an event A is given by the sum of the products of the probabilities that A occurs given that B_i occurs for $i = 1, 2, \ldots, n$.
JB

total serialism (1950s) Developed by various composers, most notably Karlheinz Stockhausen (1928–) and Pierre Boulez (1925–); this is a theoretical system, used as a compositional technique, in which all parameters (for example pitch, rhythm, timbre) are divided into a series of 12 different components and then organized according to the rules of TWELVE-NOTE TECHNIQUE. *See also* SERIALISM.
MLD

total theatre (1920s) *Literary Theory* In German *Totaltheater*, this term is associated with the work of the German director Erwin Piscator (1893–1966), for whom the architect Walter Gropius (1883–1969) planned in 1926 to build an appropriate theatre.

It is a form of theatrical production in which the text is less important than special and spectacular technical effects such as film projection, lighting, striking sets and costumes. *See* THEATRE OF CRUELTY.

J Willett, *The Theatre of Erwin Piscator: Half a Century of Politics in the Theatre* (London, 1978)
RF

total war (20th century) *Politics* Theory of modern war.

Total mobilization of economies and populations for war is a unique feature of the modern world and of the 19th and 20th centuries.

Graham Evans and Jeffrey Newnham, *The Dictionary of World Politics* (Hemel Hempstead, 1990)
RB

totalitarianism (20th century) *Politics* Theory of despotic government.

Totalitarian regimes are characterized by an ambition for permanence; attempted total control over all aspects of the lives of their subjects; concentration of power in a single leader; and use of mass propaganda and public ritual. The theory of totalitarianism attempts to generalize from the characteristics of the German Nazi and Soviet Stalinist regimes, and to create a more widely applicable descriptive category. When it began to be applied to right-wing military regimes, conservative theorists devised the notion of AUTHORITARIANISM to distinguish such regimes from the totalitarian pure type.

David Miller *et al.*, eds, *The Blackwell Encyclopaedia of Political Thought* (Oxford, 1987)
RB

trace theory (1970s) *Linguistics* Proposed by the American linguist Avram Noam Chomsky (1928–).

Condition on movement transformations developed in EXTENDED STANDARD THEORY. When a transformation moves a constituent out of its original place, for example in the formation of a question like 'Which book did Alice borrow?', a phonetically empty but syntactically active and indexically informative trace is left in the original position: 'Which book did Alice borrow *t*?'. Trace ensures grammaticality of sentences like 'Which men did Alice say are to be fired?'.

N Chomsky, *Reflections on Language* (London, 1975)
RF

trace theory of memory *Philosophy* Theory that if we are correctly said to remember some fact or event (as against relearning it, guessing it, and so on) there must be some physiologically identifiable trace in the brain which carried the information in question right through from the time when we first learnt it. The trace need not be a physical object; it could be an electrical circuit or such like.

H A Bursen, *Dismantling the Memory Machine* (1978); critical

ARL

trace theory *Psychology See* DECAY THEORY.

trade cycle *Economics* First observed by the English economist Sir William Petty (1623–87), this phenomenon is defined as the existence of fluctuations in NATIONAL INCOME over a variable timespan. Government policy is used to dampen the magnitude of the fluctuations in order to maintain stability in the economy.

Petty's findings were later developed by English economists Thomas Malthus (1766–1834) and John Stuart Mill (1806–73), the German political economist Karl Marx (1818–83) and the Norwegian economist Ragnar Frisch (1895–1973).

There are several explanations for these cycles. (1) In the 1940s and 1950s, Samuelson, Hicks, Goodwin, Phillips and Kalecki combined the multiplier with the accelerator theory of investment (multiplier-accelerator). (2) Friedman asserts that business cycles are a monetary phenomenon. (3) They may be the effect of changes in technology and taste changes. (4) Frisch found that a dynamic system with certain mathematic properties produced a 'damped' cycle with wavelengths of four to eight years.

See BUSINESS CYCLE, KONDRATIEFF CYCLES, SUNSPOT THEORY, PRODUCT LIFE-CYCLE THEORY, MULTIPLIER-ACCELERATOR, FINE-TUNING and POLITICAL BUSINESS CYCLE.

R C O Matthews, *The Trade Cycle* (Cambridge, 1959); R F Harrod, *The Trade Cycle* (Oxford, 1936); E D Domar, 'Capital Expansion, Rate of Growth, and Unemployment', *Econometrica*, vol. XIV (April, 1946), 137–47

PH

tradition *Politics* A central element in the political theory of CONSERVATISM.

Societies and constitutions are ordered by slowly established and time-tested values and institutions, rather than by rational schemes or plans. The task of government is to cultivate and work with these traditions, amending them where necessary, rather than to devise utopian blueprints.

David Miller *et al.*, eds, *The Blackwell Encyclopaedia of Political Thought* (Oxford, 1987)

RB

traditional grammar (Classical Greece and Rome) *Linguistics* 'School grammar': European system of grammatical concepts and terminology (parts of speech, subject and predicate, and so on) which, though criticized by both STRUCTURAL LINGUISTICS and GENERATIVE GRAMMAR, underpins Western linguistics. Originally delineated by Greek grammarians in the 5th–4th centuries BC, adapted to Latin by the Romans, and developed by medieval grammarians, it became firmly embedded in the education systems of Europe, and highly prescriptive from the 18th century onwards (*see also* RULE).

F Palmer, *Grammar* (Harmondsworth, 1971)

RF

tragedy (*c*.330 BC) *Literary Theory* The Greek philosopher Aristotle (384–322 BC) definitively formulated the theory of tragedy in his analysis of Greek drama of the 5th century BC.

A serious play presenting the struggle and downfall of a noble, courageous, central character with whom the audience empathizes and whose fate promotes pity and fear in the audience (*see* CATHARSIS). The hero's downfall stems from a TRAGIC FLAW in his character.

Plays of William Shakespeare and Christopher Marlowe around 1600 are regarded as the modern expression of classic tragedy. There is another tradition of tragedy, more melodramatic and violent, derived from the Roman dramatist Seneca (*c*.4 BC–AD 65), and a moralistic medieval tradition of narratives concerning the falls of the great.

Aristotle, *Poetics*, S H Butcher, trans., *Critical Theory since Plato*, H Adams, ed. (New York, 1971), 48–66; A Poole, *Tragedy: Shakespeare and the Greek Example* (Oxford, 1987)

RF

tragedy of the commons (20th century) *Politics* An illustration of GAME THEORY.

The tragedy of the commons is an illustration of how the rational pursuit of individual advantage appears to lead to solutions which are in the best INTERESTS of neither individuals nor the community as a whole. A group of peasants is assumed to have grazing rights, but if each peasant puts his cattle on the common seven days a week the pasture will be exhausted and the cattle will starve. If everyone would voluntarily limit themselves to four days a week, there would be enough. But if everyone limits their use, then the self-interested individual has no reason to limit his. If everyone does not limit their use, he has even less reason. So the commons are destroyed. *See also* CHICKEN, PRISONER'S DILEMMA, and ZERO-SUM.

Christopher Hood, *Administrative Analysis* (Brighton, 1986)

RB

tragic flaw (*c*. 330 BC) *Literary Theory* Usual translation of the word *hamartia* used by Aristotle (384–322 BC) in his theory of TRAGEDY.

For Aristotle, the tragic flaw which causes the hero's downfall is an error of judgement on his part, inconsistent with his nobility. Critics have extrapolated other kinds of tragic flaw, including moral weakness (for example Greek *hubris*, meaning 'pride') and psychological defects, as in some modern discussions of Shakespeare's tragic heroes.

Aristotle, *Poetics*, S H Butcher, trans., *Critical Theory since Plato*, H Adams, ed. (New York, 1971), 48–66

RF

tragicomedy (classical) *Literary Theory* The term originates with the Roman poet and playwright Plautus (*c*.254–184 BC), but the genre was first theorized by the Italian playwright and critic Giambattista Guarini (1538–1612) in his *Compendio della Poesia Tragicomica* (1601).

Tragicomedy mixes the 'high' and 'low' comedy: a tragic plot with a happy ending, a social mixture of nobles and servants, shifts and clashes of tone and language. A popular form in 16th-century Italy and 17th-century England and France, it persisting in the 20th century through the work of such playwrights as Samuel Beckett.

D L Hirst, *Tragicomedy* (London, 1984)

RF

trahison des clercs (1927–) *Politics* Term originally used by the French philosopher and novelist Julien Benda (1867–1956) to describe the betrayal of intellectual values by the right wing.

More generally, intellectuals by allying themselves too closely with government, states, or political parties betray the independence which is essential if they are to contribute to public discussion.

Roger Scruton, *A Dictionary of Political Thought* (London, 1982)

RB

transactional analysis (TA) *Psychology* Proposed by the American psychologist Eric Lennard Berne (1910–).

This theory of personality and social behaviour is used as a vehicle for psychotherapy, and is practised in a group setting where the primary goal is to have the client achieve an adaptive, mature and realistic attitude towards life. A central concept of TA is that of 'stroking': the process of stimulating and giving recognition to fellow human beings. Stroking patterns form a common theme in the main subsections of TA: (1) personality structure; (2) communication; (3) games; (4) feelings analysis; and (5) script analysis.

E Berne, *Intuition and Egostates* (San Francisco, 1977)

NS

transactional theory *Psychology* Proposed by the German psychologist O Kulpe (1862–1915).

This theory assumes that perception results from acquired but unconscious assumptions about the environment (represented as probabilities of transactions occurring within it) and that what is perceived is dependent on knowledge gathered from interactive experiences with the environment. The real world and its perceptual properties are created in the transaction.

E Berne, *Intuition and Egostates* (San Francisco, 1977)

NS

transcendental idealism *Philosophy* Form of IDEALISM espoused by Immanuel Kant (1724–1804), who called himself a transcendental idealist but an empirical realist. He meant, roughly, that what we experience can only be representations, not things in themselves, of which we can know nothing except that they must exist in order to ground the representations.

The idealism is 'transcendental' because we are forced into it by considering that our knowledge has necessary limitations and that we could not know things as they are, totally independent of us. But there is nothing to stop us knowing the appearances as *they* are, presented to us as from outside and not invented by us, for we could have no way of inventing them if they were not really presented to us in this way. (Perhaps compare, though Kant does not say this, the way we cannot invent new colours in imagination.)

I Kant, *Critique of Pure Reason*, trans. by N K Smith (1953, German originals 1781 and 1787); especially A366–80 (pp. 344–52 in Smith's translation)

ARL

transfer pricing (1950s) *Accountancy* This is the fixing of internal prices charged for transactions within departments in a firm, or within semi-autonomous divisions. It can be used to minimize the payment of tariffs/taxes or to transfer profits from a high-taxation country to a low-taxation one.

J Hirshleifer, 'On the Economics of Transfer Pricing', *Journal of Business*, vol. XXIX (July, 1956), 172–84

PH

transference principle *Psychology* Proposed by the Austrian founder of psychoanalysis Sigmund Freud (1856–1939).

This psychoanalytic term refers to the displacement of feelings and attitudes (having roots in an earlier crucial relationship, usually with a parent or sibling) onto the analyst. Transference may be either positive or negative depending on whether the person develops pleasant or hostile attitudes towards the analyst. Teasing out the transference patterns enables the patient to achieve important insight into the nature and origins of his problem.

S Freud, 'Psychopathology of Everyday Life' (vol. XVI) and 'Introductory Lectures' (vol. XV), *Standard Edition of the Complete Works of Sigmund Freud* (London and New York, 1901 and 1916)

NS

transformation (1957) *Linguistics* Developed by the American linguist Avram Noam Chomsky (1928–), adapting the concept from Zellig Sabbetai Harris's 1952 use of 'transformation'.

A PHRASE STRUCTURE GRAMMAR generates a set of 'underlying strings' which roughly correspond to the simple sentences of a language. Transformational rules operate on these strings, rearranging their constituents, deleting constituents or adding new ones, to account for more complex sentences and for other grammatical facts like sentence-relatedness (for example, Active and Passive).

The 1957 version is now of historical interest. Modern generative grammars either have no transformations, or a general instruction to 'move' some unit.

N Chomsky, *Syntactic Structures* (The Hague, 1957)

RF

transformation theory *Chemistry* In the field of quantum mechanics, the term refers to the study of co-ordinate and other transformations, especially those which leave some properties of the system invariant. It also signifies that radiation can display particle properties (for example, the Compton effect); and that particles can display wave properties (for example, electron diffraction).

P W Atkins, *Physical Chemistry* (Oxford, 1990)

GD

transformational cycle (1962) *Linguistics* Proposed by the American linguist Avram Noam Chomsky (1928–).

Originally applied to PHONOLOGY, then to SYNTAX: a set order of application of transformations, starting with the lowest embedded sentence and working 'bottom-up' from smaller to larger units until the highest sentence is reached. Of historical interest.

A Radford, *Transformational Syntax* (Cambridge, 1981)

RF

transformational-generative grammar (1957) *Linguistics* Formulated by the American linguist Avram Noam Chomsky (1928–).

A GENERATIVE GRAMMAR need not in principle employ transformations. In 1957 Chomsky both defined the goals of generative grammar and also proposed a particular kind of generative grammar based on PHRASE STRUCTURE and TRANSFORMATIONAL rules.

J Lyons, *Chomsky*, revised edn (Glasgow, 1977)

RF

transition state theory *Chemistry See* ABSOLUTE REACTION RATE THEORY.

transitivity (1960s) *Linguistics* Traditional term reinterpreted by the British linguist Michael Alexander Kirkwood Halliday (1925–).

Important system in SYSTEMIC GRAMMAR, serving the 'ideational function' (*see* FUNCTIONAL GRAMMAR): within the clause, the choices of processes, participants and circumstances which are made to represent the structure of an idea or experience.

M A K Halliday, *An Introduction to Functional Linguistics* (London, 1985)

RF

transpiration-cohesion-tension theory *Biology See* COHESION THEORY.

trapezoidal rule *Mathematics* The method of approximation of a definite integral

$$\int_a^b f(x)dx \approx \frac{\delta}{2}[f(a)+f(a+\delta)+f(a+2\delta)+f(a+3\delta)+\ldots+f(b)]$$

where

$$\delta = (b-a)/n$$

The error is

$$(b-a)^3M/12n^2$$

where M is the maximum of the second derivative of f on the interval. The formula is exact for linear functions. When $f(x)$ is positive this is equivalent to approximating the area under a curve by a sequence of trapezoids formed by connecting the nodes. *See also* SIMPSON'S RULE.

H Anton, *Calculus with Analytic Geometry* (New York, 1980)

Traube's rule *Chemistry* Named after the chemist Isidor Traube (1860–1943), this rule states that for a homologous series of fatty acids, the concentration in dilute aqueous solution at which a given lowering of surface tension is observed decreases threefold for each additional methylene group in the series.

GD

tree diagram (1957) *Linguistics* Notation for CONSTITUENT STRUCTURE favoured in GENERATIVE GRAMMAR.

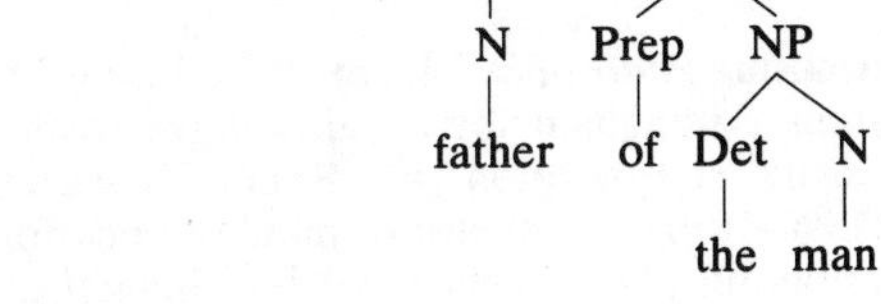

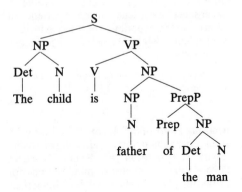

K Brown and J Miller, *Syntax*, 2nd edn (London, 1991)

RF

triads, Döbereiner's law of (1817 and 1829) *Chemistry* Named after the German chemist Johann Wolfgang Döbereiner (1780–1849).

Döbereiner pointed out that chemical elements which resembled each other often occurred in groups of three, and that there was often a close correspondence between the arithmetic mean of the atomic numbers of the lightest and heaviest elements and the third member of the group. In this way, he grouped together lithium, sodium and potassium; calcium, strontium and barium; sulphur, selenium and tellurium; and chlorine, bromine and iodine. This quantitative relationship prompted others to speculate that atoms might not be the ultimate building blocks and that the differences in weight between successive members of a triad represented weights of some more fundamental unit.

M Freemantle, *Chemistry in Action* (Basingstoke, 1987)

GD

trial and error theory of learning *Psychology* Proposed by the Englishman, Conwy Lloyd Morgan (1852–1936).

This theory proposes that learning consists of the process of succeeding in an attempt, by trying repeatedly and subsequently gaining knowledge from one's failures. It is sometimes referred to as the theory of 'trial, error and accident success'. This theory is now outdated, being unable to explain, among other things, successful learning in the absence of evidence of the experience of failure and the experience of 'sudden insight' in problem solving. *See* IRRADIATION THEORY OF LEARNING.

G A Kimble, *Hilgard and Marquis Conditioning and Learning* (New York, 1961)

NS

trialism *Philosophy* Term introduced by John Cottingham for an alternative to the usual interpretation of René Descartes (1596–1650) as a dualist of mind of body for whom all phenomena involving thought or consciousness belong to mind and all those involving extension belong to body.

The trialist interpretation keeps the two *substances* of mind and body, but introduces a third *attribute*, sensation, alongside thought and extension and belonging to the union of mind and body. Among other things this allows animals, which do not have thought, to be regarded as having sensation and not, as in the traditional dualist interpretation, as being mere automata. Cottingham does not claim, however, that Descartes developed this trialism with complete consistency.

J Cottingham, 'Cartesian Trialism', *Mind* (1985)

ARL

trichromatic theory of colour vision (19th century) *Physics/Psychology* This theory assumes that the colour response of the human eye is such that the effects of all colours can be produced by mixing, in various proportions, three so-called primary colours, and no others.

The colours usually chosen are red, green and blue. It is a simple step from this theory to the hypothesis that three different types of receptor in the eye must be responsible for colour vision. *See* YOUNG-HELMHOLTZ THEORY OF COLOUR VISION.

J Thewlis, ed., *Encyclopaedic Dictionary of Physics* (New York, Oxford and London, 1962)

MS

trickle-down theory *Economics* A theory of economic development that claims higher standards of living for the poor will develop gradually and not at the overt expense of the more affluent. *See* DUAL ECONOMY THEORY.

AV, PH

tridimensional theory of feeling *Psychology* Proposed by the German experimental psychologist Wilhelm Wundt (1832–1920).

This theory postulates that all feelings can be represented as being composed of three dimensions: pleasure-pain, excitement; depression; and tension-relaxation. It has been superseded by more sophisticated accounts of feelings and emotions.

W Wundt, *Butrage zur Theories der Sinneswahrnehmung* (Leipzig and Heidelberg, 1862)

NS

Tristram Shandy paradox (1759) *Philosophy* Named after a fictional character created by English author Lawrence Sterne (1713–68).

Shandy finds that in two years of writing he has covered two days of his autobiography and doubts whether he will ever complete the work. However, even at that poor rate he could finish the work provided that he had an infinite amount of time in which to do so.

JB

trompe l'oeil *Art* A French term meaning 'deception of the eye', used to describe a highly illusionistic painting in which objects are depicted with photographic realism or have extremely realistic perspective. *See* ILLUSIONISM.

AB

trope (traditional) *Literary Theory* Part of RHETORIC.

Division (with SCHEME) of FIGURES OF SPEECH. Tropes are figurative devices based on deviations of meaning rather than on formal pattern: notably METAPHOR.

G N Leech, *A Linguistic Guide to English Poetry* (London, 1969)

RF

tropism (1939) *Literary Theory* Theory of French novelist Nathalie Sarraute (1902–), this is an atomistic conception of experience as a multitude of individual responses to a stream of minute stimuli.

Stylistically, tropism leads in the NOUVEAU ROMAN to a technique of representation in fragmentary details. *See* 'CHOSISME'.

N Sarraute, *Tropisms* [1939], M Jolas, trans. (London, 1963)

RF

tropisms, theory of (*c*.1912) *Biology/ Philosophy* Proposed by Jacques Loeb (1859–1924), a physiologist and physician who was associated with the Rockefeller Institute in New York. The concept that all the activities of animals and humans are determined by tropisms, just as plant movements are determined by tropisms.

Loeb believed that matters of the mind and inner life will ultimately be explained by physio-chemical mechanisms that remain elusive today.

G Blandino, *Theories on the Nature of Life* (New York, 1969)

KH

Trotskyism (20th century) *Politics* Views originally developed by Russian revolutionary Leon Trotsky (1879–1940).

Revolution, once begun, must continue until workers' power is established worldwide. PERMANENT REVOLUTION is thus preferred to SOCIALISM IN ONE COUNTRY.

David Miller *et al.*, eds, *The Blackwell Encyclopaedia of Politcal Thought* (Oxford, 1987)

RB

Trouton's rule (1884) *Chemistry* Named after the Irish-born physicist Frederick Thomas Trouton (1863–1922) who was Professor of Physics at University College London from 1902.

This empirical rule states that many liquids have approximately the same value for the molar entropy of vaporization of 85 J K^{-1} mol^{-1}. Highly structured liquids with appreciable intermolecular forces (for example, water which is hydrogen-bonded) show marked deviations from this value.

P W Atkins, *Physical Chemistry* (Oxford, 1990)

GD

truth-conditional semantics *Linguistics/ Philosophy See* SEMANTICS, TRUTH-CONDITIONAL.

truth theory (1935) *Philosophy* Semantic concept formalized by the Polish-American mathematician and logician Alfred Tarski (1902–83), although other thinkers had previously discussed the idea.

Truth theory concerns the truth-values of sentence structures in various formal logical languages. Tarski suggested a table by which these values could be determined (although he was less sure about whether the same rules could be applied to natural languages). *See also* CORRESPONDENCE THEORY OF TRUTH.

A Tarski, 'The Concept of Truth in Formalized Languages', *Logic, Semantics and Metamathematics* (1956)

DP

tunica-corpus theory (1924) *Biology* Proposed by Schmidt, this is the theory of plant development holding that embryonic tissues differentiate into outer layers, called tunica, and an inner cell mass, called corpus.

The corpus ultimately becomes the interior part of the plant, while the tunica differentiates the outer layers, including the epidermis. Since its proposal, this theory has been modified several times, with botanists finding it increasingly difficult to distinguish the two parts. *Compare with* HISTOGEN THEORY.

KH

Turing machine (1937) *Mathematics* Invented by English mathematician Alan Mathison Turing (1912–54), this is a notional computing machine which helped in the invention of the electronic computer (which it resembles in outline). However, the Turing machine has a potentially infinite input tape but a finite control system. The object of the machine is to provide an abstract model of a machine which can perform all recursive computations. *See also* RECURSIVE FUNCTION THEORY.

JB

turn-taking *Linguistics See* CONVERSATIONAL ANALYSIS.

turnpike theory *Economics* (1958) Named by the American economists Robert Dorfman (1911–), Paul Samuelson (1915–) and Robert Solow (1924–); this theory asserts that it is sometimes better to adopt a maximum or near maximum possibility balanced growth-path to allow an economy to move to a more satisfactory state quickly, even if consumption is lower in the interim than at the beginning or end.

An optimal growth theory, turnpike theory was named after the American term for motorway. *See* HARROD-DOMAR GROWTH MODEL and SOLOW ECONOMIC GROWTH MODEL.

R Dorfman, P A Samuelson, and R Solow, *Linear Programming and Economic Analysis* (New York, 1958)

PH

twelve-bar blues (early 20th century) *Music* The black American blues song emerged at the turn of the century as a lament, representing, perhaps, an attempt to drive away the 'blues'. A theoretical structure of three four-bar phrases, forming the harmonic sequence I-I-I-I, IV-IV-I-I, V- IV-I-I, that underlies a typical blues song.

The harmonic framework remains implicitly present throughout, although the melody and accompaniment may at times depart so far from a given chord that no note of it is actually present.

P Oliver, *Early Blues Songbook* (London, 1982)

MLD

twelve-note technique (1920s) *Music* Developed independently by the Austrian-born composers Arnold Schoenberg (1874–1951) and Josef Hauer (1883–1959), this is a theoretical system, used as a compositional method, in which all 12 chromatic pitches within an octave are treated as equal.

The 12 notes are placed in a fixed order (called a note row, or tone row). In strict technique, no note may be repeated before the other 11 are sounded. The row may be presented in its original ('prime') form, in retrograde, inversion or retrograde inversion, and may be transposed so as to start on any pitch. In freer technique, segments of the row may be repeated, superimposed or sounded as a single vertical chord. *See also* SERIALISM.

G Perle, *Twelve-tone Tonality* (Berkeley, Calif., 1978)

MLD

twin law (19th century) *Physics* This is the name given to the statement that defines the twin orientation in a particular example of crystal twinning.

For example, in the case of spinel, the twin face (or plane) is an octahedral plane.

J Thewlis, ed., *Encyclopaedic Dictionary of Physics* (New York, Oxford and London, 1962)

MS

twin paradox (1905) *Physics* A paradox arising out of relativity theory. A pair of identical twins are initially together in an inertial system. The first twin stays behind while the second twin is accelerated to a high velocity on a long space journey and then returns to rest beside the first twin. The special theory of relativity predicts that the second twin will have aged less than the first. From the second twin's point of view, however, the first twin should be younger when they meet up again. The paradox is resolved by realising that the experiences of the twins have not been equivalent. The second twin has undergone two accelerations and two decelerations before finally returning to the first twin's inertial frame of reference. Therefore, since he has not travelled at constant velocity, the theory of special relativity cannot be applied to his experience of the passage of time. *See also* RELATIVITY, THEORY OF SPECIAL.

GD

twin primes *Mathematics* A pair of prime numbers which differ by two. It is an outstanding conjecture that the number of such pairs is infinite. In 1974, Jing-Run Chen (1933–) showed that the conjecture is valid if one of the primes is replaced by a number with at most two prime factors.

A Baker, *A Concise Introduction to the Theory of Numbers* (Cambridge, 1984)

MB

two-body theories of planetary formation *Astronomy See* COLLISION THEORIES OF PLANETARY FORMATION and TIDAL THEORIES OF PLANETARY FORMATION.

two-factor learning theory *Psychology* Proposed by the American psychologist Orval Hobart Mowrer (1907–82), and also known as 'two press theory'.

Two-factor theorists assume that there are fundamental distinctions between classical and instrumental conditioning, and that it is impossible to reduce one type of conditioning to the other even though both processes occur together in any investigation of learning. These theorists have also assumed that classical conditioning to stimuli occurring at the same time as a reinforcer on an instrumental experiment endow these stimuli with classically-conditioned motivational states which influence the performance of the instrumental response.

G A Kimble, *Hilgard and Marquis Conditioning and Learning* (New York, 1961)

NS

two-gene hypothesis (1965) *Biology* Proposed by Dreyer and Bennett, this is the theory that two genes are required to encode for one antibody chain; specifically, that the 'constant' (C) region of peptides is produced by a single gene, while the diverse V region is produced by another.

Much experimental evidence now exists to support this theory, which represents an exception to the one gene-one polypeptide chain hypothesis. *See also* ANTIBODY, THEORIES OF.

T Kindt and J Capra, *The Antibody Enigma* (New York, 1984)

KH

two-waves theory (20th century) *Feminism* Conception of the history of feminism, attributed to writers rather than claimed by them.

Feminism is seen as having two waves. The first was concerned principally with political, legal, and constitutional rights, and with gaining the vote, up until the third decade of the 20th century. The second wave, from the 1960s on, was concerned with attacking sexual, economic, social and domestic oppression. Critics of this view say that it denies the great variety of early feminism.

HR

Tychonoff conditions (1935) Defined by Russian mathematician and physicist Andrei Nikolaevich Tychonoff (1906–), these are axioms denoted by T_0, T_1, T_2, T_3, T_4, T_5 which progressively determine increasing segregation of pairs of points in a topological space. *See also* TOPOLOGY.

JB

Tychonoff's theorem (20th century) *Mathematics* Named after Russian mathematician and physicist Andrei Nikolaevitch Tychonoff (1906–), this is the result, of fundamental importance in TOPOLOGY, whereby the infinite Cartesian product of any compact topological spaces is compact.

J R Munkres, *Topology: A First Course* (Prentice-Hall, 1975)

MB

types, ramified theory of *Philosophy See* TYPES, SIMPLE THEORY OF. For the simple theory, which Frank Plumpton Ramsey (1903–30) separated out from the ramified theory, properties of objects are of type one, properties of type one properties are of type two, and so on.

The ramified theory further classifies properties of each type into orders. A first-order type $n+1$ property is a property of things of type n. A second-order type $n+1$ property is still of things of type n, but it involves a reference to first-order type $n+1$ properties. *Red* is a first-order type one property applying to objects. *Applying to objects* is a first-order type two property applying to type one properties. *Having some first-order type one property* is a second-order type one property; it still applies to objects, but makes reference to first-order properties. Thus each type has its own ramification (or branching) of orders. (But when accuracy is not needed, 'type' and 'order' are often used more loosely or even interchangeably.)

Because of certain unwanted technical results, the ramified theory led Bertrand Russell (1872–1970) to introduce the axiom of REDUCIBILITY. *See also* VICIOUS CIRCLE PRINCIPLE.

F P Ramsey, *Foundations of Mathematics* (1931), ch. 1

ARL

types, simple theory of *Philosophy* Theory developed by Bertrand Russell (1872–1970) to deal with paradoxes like his *paradox of classes*: is the class of all classes that are not members of themselves a member of itself? If yes, no; if no, yes.

Russell said there is no such class. Classes (and also properties) cannot all be lumped together, but form a hierarchy. Objects are of type zero. Classes of objects are of type one. Classes of type one classes are of type two, and so on. The class of all type *n* classes is of type *n*+1, but there is no class of *all* classes (or property which applies to all properties). This is the simple theory, which Russell developed only as part of the ramified theory (*see* ramified theory of TYPES). This latter (which also applies to propositions) resolves the various paradoxes, but lets us speak only of classes and so on of a given level; notions like class-membership, property, and truth (in the case of propositions) become systematically ambiguous over the hierarchy of levels of classes, properties, or propositions. Other solutions have therefore been sought, but Russell's paradox and similar ones revolutionized logic early in the 20th century.

I M Copi, *The Theory of Logical Types* (1971)

ARL

tyranny of the majority (19th century) *Politics* Reservation about democracy expressed by French historian Alexis de Tocqueville (1805–59).

An avoidable danger of democracy was a tyranny of majority opinion which would suppress the drive to individual judgment inherent in democratic theories. The term has also been used (as it was not by de Tocqueville) to suggest that democratic states would use coercive and oppressive measures even more rigorously than their aristocratic and monarchic predecessors.

David Miller, ed., *The Blackwell Encyclopaedia of Political Thought* (Oxford, 1987)

RB

U

uncertainty *Economics* The element of risk that is unpredictable and has no measurable probability. Profit is usually regarded as the reward a company earns for enduring uncertainty in business activity. *See* BOUNDED RATIONALITY.

K H Borch, *The Economics of Uncertainty* (Princeton, N.J., 1968)

AV, PH

uncertainty principle *Physics See* HEISENBERG'S UNCERTAINTY PRINCIPLE.

unconscious memory *Psychology* Proposed by the Austrian founder of psychoanalysis Sigmund Freud (1856–1939).

This is the term given to the memory of prior events, feelings and emotions encountered by the individual which may or may not be available for conscious retrieval. Psychodynamic approaches suggest that unconscious memories are memories that have been 'repressed'. Information theorists suggest that failure to access stored information can be accounted for in terms of information retrieval failure.

A J Parkin, *Memory and Amnesia: An Introduction* (Oxford, 1987)

NS

underconsumption theories (20th century) *Economics* Analysis of how total output fails to be sold at the cost of production plus normal profit. Such insufficient consumption within a depressed economy aggravates the economic decline of the state.

A further theory suggests that when there is inadequate buying power in an economy, the government should give periodic injections of money to consumers. *See* SECULAR STAGNATION THEORY.

G Harberler, *Prosperity and Depression* (Geneva, 1937)

AV, PH

underlexicalization (1980s) *Linguistics* A term coined by English linguist Richard Fowler (1938–) on the basis of Michael Alexander Kirkwood Halliday's OVERLEXICALIZATION, of which it is the converse.

Lack, in a speaker or a text, of words to designate specific concepts, leading to a use of more general, or tangential, terms; used in LINGUISTIC CRITICISM.

R Fowler, *Linguistic Criticism* (Oxford, 1986)

RF

unemployment *Economics* The under-utilization of labour in the creation of wealth.

Unemployment can take many forms (such as voluntary, involuntary, frictional, structural or demand deficient) and it can be measured both as a stock and a flow. Classical economists saw unemployment as a temporary phenomenon until price flexibility restored an economy to full employment. Later theorists, notably English economist John Maynard Keynes (1883–1946), challenged this view. *See* EQUILIBRIUM THEORY.

AV, PH

unexpected examination paradox (1948) *Mathematics* Also called the hangman paradox when the event is an execution, this was found by Lennard Ekbom.

One day next week there will be an examination but the students will not know the day in advance. As Friday is the last day, students would expect it then, so it cannot be later than Thursday. Similarly Thursday and all other days are impossible. This makes occurrence of the examination on any day a surprise.

JB

unified field theories *Physics* Theories which attempt to incorporate the gravitational, electromagnetic, weak and strong forces within a unified framework so that a consistent mathematical formalism can be used to predict all their characteristics. It is not yet certain whether such theories are intrinsically possible, but there is considerable optimism that a single unified theory will emerge out of the use of extended objects such as superstrings and supersymmetries. *See also* GRAND UNIFIED THEORIES.

Paul Davies, ed., *The New Physics* (Cambridge, 1989)

GD

uniformitarianism (18th century) *Geology* Formulated by the Scottish geologist James Hutton (1726–97), this principle was later summarized by the Scottish geologist Sir Charles Lyell (1797–1875) as 'the present is the key to the past'. This is now taken to mean that all past forces that operated on the Earth's surface can be recognized in the forces currently operating.

Well established by 1820, the principle's main features were the invocation of natural rather than supernatural past processes (unlike CATASTROPHISM and NEPTUNISM), and the proposal of a uniform progression of events (*see* PLUTONISM). In Europe, this principle is usually known as ACTUALISM.

S Gould, *Time's Arrow, Time's Cycle: Myth and Metaphor in the Discovery of Geological Time* (Cambridge, Mass., 1987); A Hallam, *Great Geological Controversies* (Oxford, 1983)

KH, DT

uniformity of nature, principle of the *Philosophy* A claim that may be offered as a grounding for the INDUCTIVE PRINCIPLE, though it is not always distinguished from the principle itself. It may be crudely formulated as 'Nature is uniform', or 'The future will resemble the past', or – in a more refined version like that given under inductive principle – with 'will be found' for 'can be assumed'.

The main problem is to specify the respects in which the resemblance holds, since the future will obviously not resemble the past in all respects, whereas it can be shown that *whatever* happens (bar perhaps the total annihilation of the universe) the future will resemble the past in some respects (indeed in infinitely many). Even the more refined versions are not immune to this danger, if the class and the property concerned are left quite unrestricted. A more hopeful version of the principle may say that mere position in space or time cannot by themselves be relevant to whether some phenomenon occurs or not. *See also* INDUCTIVISM.

N Goodman, 'Seven Strictures on Similarity', *Experience and Theory*, L Foster and J W Swanson, eds (1970)

ARL

unilateralism (20th century) *Politics* Theory of (nuclear) disarmament.

The competition between nuclear states can be broken by one side reducing or disposing of its nuclear weapons, thus removing the threat which had induced the other side to arm in the first place. *See also* MULTILATERALISM.

David Robertson, *The Penguin Dictionary of Politics* (London, 1986)

RB

unit membrane hypothesis or model (1950s–1960s) *Biology* Also called the *Danielli-Davson-Robertson model*, this was proposed by American anatomist, physician and biochemist James David Robertson (1922–) who used electron microscopy to elaborate on the DANIELLI-DAVSON MODEL.

The theory asserts that all cell membranes consist of a biomolecular lipid layer sandwiched between two protein layers, resulting in a membrane about 65–85 angstroms (65–85 ten-millionths of a millimetre) thick. Although Robertson allowed for chemical differences among membranes, he hypothesized that all membranes exhibit this same pattern of organization, an idea that has since been abandoned. *Compare with* FLUID MOSAIC MODEL.

P Sheller, *Cell and Molecular Biology* (New York, 1987)

KH

unity, the Unities (4th century BC) *Literary Theory* The basic idea was articulated by the Greek philosophers Plato (*c*.427–*c*.347 BC) and Aristotle (384–322 BC).

Literary theory has long stipulated as a fundamental requirement that an artistic text should be an integrated whole, its parts co-operating to a single purpose (though the NOVEL is a permissively loose form). Unity is a requirement both of the CHICAGO CRITICS and of NEW CRITICISM.

Aristotle specified 'unity of action' for TRAGEDY, a simple coherent plot. In 1570 the Italian Lodovico Castelvetro (1505–71) added unities of place (single location) and time (12 hours), making up the 'Three Unities', a highly restrictive set of constraints on neo-classical theatre.

L Castelvetro, 'The *Poetics* of Aristotle Translated and Explained' [1570], R L Montgomery, trans., *Critical Theory since Plato*, H Adams, ed. (New York, 1971), 144–53

RF

universal code theory (20th century) *Biology* The theory that all forms of life use the genetic code, which consists of systematic strings of nucleotides appearing in groups of three in deoxyribonucleic acid (DNA) and ribonucleic acid (RNA). The code governs protein synthesis. Relatively few exceptions exist to this theory.

I Asimov, *The Genetic Code* (Orion, 1962)

KH

universal grammar *Linguistics See* LINGUISTIC UNIVERSALS.

universal gravitation, law of *Physics See* gravitation, law of universal.

universal law of learning *Biology See* LEARNING THEORY.

universalism *Philosophy* Term used – usually in its adjectival forms: *universalist(ic)* – as a contrast term to EGOISM and ALTRUISM when referring to UTILITARIANISM and similar topics. It is summed up in the slogan of Jeremy Bentham (1748–1832), 'Everyone to count for one and no-one for more than one'.

ARL

universalizability *Philosophy* A complex and controversial notion which has been used both to distinguish the moral from the non-moral and to distinguish the moral from the immoral – two jobs which tend to get in each other's way.

'What if everyone did that?' is often a relevant question in moral contexts; but 'did what exactly?'. The same action can be described in many ways. And what counts as 'everyone'? Presumably everyone with certain characteristics or in a certain situation, but which characteristics or situations? As for distinguishing the moral from the non-moral, no doubt if I act in arbitrary ways I cannot claim to be acting on a moral principle – but will I be acting on any other kind of principle either? And what counts as 'arbitrary'?

These questions of course give only the general flavour of discussions in this area; and though the terms are often confused, one should distinguish the universal (as against particular or individual) from the general (as against specific): 'Always help the blind' is universal but specific. The appeal to universalizability in ethics dates from Immanuel Kant (1724–1804).

D Locke, 'The Trivializability of Universalizability', *Philosophical Review* (1968)

ARL

unreadiness, law of *Psychology* Proposed by the American psychologist Edward Lee Thorndike (1874–1949).

This law forms part of Thorndike's theory of learning and postulates that an unpleasant effect is produced from the functioning of behavioural 'condition units' which were not ready to function. There is also a corresponding law of readiness. These laws are now mainly of historical importance.

G A Kimble, *Hilgard and Marquis Conditioning and Learning* (New York, 1961)

NS

urban dialectology *Linguistics See* CORRELATIONAL SOCIOLINGUISTICS.

Ursatz (1920s) *Music* A German term meaning 'fundamental structure' coined by the music theorist Heinrich Schenker (1868–1935). He proposed that, through successive reductions of the contrapuntal fabric of a piece of tonal music, a basic background structure can be identified.

This structure comprises a fundamental melodic line (*Urlinie* – a stepwise descent to the tonic from the third, fifth or octave) and a bass arpeggiation (*Bassbrechung* – a leap upwards from tonic to dominant, returning to tonic). Together, the melody and arpeggiation represent an unfolding of the tonic harmony, implying that the music is a 'composing out' in temporal terms of a single chord. The theory is controversial as it is not based on empirical evidence; however, it is important in that it stresses the significance of musical layers and their interrelationships.

A Forte and S Gilbert, *Introduction to Schenkerian Analysis* (New York, 1982)

MLD

Urysohn's lemma (20th century) *Mathematics* Named after Russian mathematician Paul Samuilovich Urysohn (1898–1924), this is the result in TOPOLOGY whereby a topological space S is normal if and only if any two disjoint closed sets A and B can be separated in the sense that there is a continuous function $f: S \rightarrow [0,1]$ with $f(A) = 0$ and $f(B) = 1$. *See also* SEPARATION AXIOMS and TIETZE'S EXTENSION THEOREM.

J R Munkres, *Topology: A First Course* (Prentice-Hall, 1975)

MB

Urysohn's metrization theorem (20th century) *Mathematics* Named after Russian mathematician Paul Samuilovich Urysohn (1898–1924), this is the result in TOPOLOGY whereby every regular topological space with a countable basis is metrizable; that is, the topological space may be defined by a metric.

J R Munkres, *Topology: A First Course* (Prentice-Hall, 1975)

MB

use inheritance *Biology See* ACQUIRED CHARACTERISTICS, INHERITANCE OF.

use, law of *Psychology* Proposed by the American psychologist Edward Lee Thorndike (1874–1949).

Behavioural responses, functions, associations and so on which are sufficiently practised, exercised or rehearsed are strengthened relative to those which go unused. The law has wide application in teaching, training and education.

E L Thorndike, *The Fundamentals of Learning* (New York, 1932)

NS

use theories of meaning *Philosophy* Theories springing mainly from Ludwig Wittgenstein (1889–1951), that the meaning of a word or sentence is to be sought in its use, not in its correspondence to some entity (as NAMING and CORRESPONDENCE THEORIES OF MEANING in general imply).

The use in question normally means actual usage, but may also refer to an alleged correct usage; or the meaning of a word may be explained in terms of *rules* for its use (*see also* 'DE FACTO' AND 'DE JURE' THEORIES OF MEANING). Sometimes such theories were regarded not so much as theories *of* meaning but as replacing theories of meaning, as in the slogan 'Don't ask for the meaning, ask for the use' (common in the heyday of LINGUISTIC PHILOSOPHY, with which use theories were closely associated). It may be objected, however, that use is indeed different from meaning, but cannot simply replace it since it presupposes it.

L Wittgenstein, *Philosophical Investigations* (1952), especially part 1, §§1–43

ARL

ut pictura poesis (15th century–)*Art* A Latin phrase meaning 'explaining painting and poetry', this argument's origins lie in the comparisons made between the two disciplines in Aristotle's *Poetics* and Horace's *Ars Poetica*. These formed the bases during the Renaissance and Baroque periods for several treatises on similar theories.

The fundamental assumption is that painting, like poetry, is the ideal imitation of human action. The principal aims of the artist were imitation, invention, expression, instruction and decorum. Through the comparison with poetry, painting was elevated to the status of a liberal art.

R E Lee, *Ut Pictura Poesis – The Humanistic Theory of Painting* (New York, 1967)

AB

utilitarianism *Philosophy* Any of a variety of views all of which are consequentialist or teleological, being distinguished from other forms of CONSEQUENTIALISM (if any) by saying that the consequence to be pursued is the maximization of good. This maximization may refer to the greatest total good or the greatest average good, but the slogan 'greatest good of the greatest number' is ambiguous. (*See* BENTHAM'S THEORY OF UTILITARIANISM.) Utilitarianism is always universalistic, not egoistic or altruistic.

Utilitarianism may be divided into ACT and RULE UTILITARIANISM, into HEDONISTIC, IDEAL, and PREFERENCE UTILITARIANISM, and also into ordinary and NEGATIVE UTILITARIANISM; these divisions cutting across each other. Utilitarians distinguish the rightness of an action, which depends on its actual consequences (or those of the relevant rule, for rule utilitarians), from the moral goodness of the agent, which depends on his motives and intentions. Objections have usually centred on the practicality of utilitarianism (especially act utilitarianism) in view of our limited predictive powers; and on whether it can cater for our intuition about justice and 'backward-looking' obligations like that of promise-keeping. Our ability to control population numbers also raises issues.

J J C Smart and B Williams, *Utilitarianism For and Against* (1973); debate with annotated bibliography

ARL

utilitarianism, Bentham's theory of (19th century–) *Philosophy/Politics* A theory originally developed by English political philosopher Jeremy Bentham (1748-1832), this asserts that actions and institutions should be judged by their contribution to utility, which is measured by calculating the relative contribution to happiness or pleasure, as opposed to pain. The aim of government should thus be 'the greatest happiness of the greatest number'.

It has been pointed out that not only is pleasure difficult to measure, but that utilitarianism provides neither any guarantee of individual rights against majority interests, nor any means of weighing high levels of pleasure for a few against lower levels of pleasure for greater numbers.

David Miller, ed., *The Blackwell Encyclopaedia of Political Thought* (Oxford, 1987)

RB

Utopianism *Politics/Philosophy* Label usually applied in a hostile sense to those who advocate – or are wildly optimistic in thinking they can achieve – a state of affairs perfect in some or all respects. One charge is that the excessive and unrealistic pursuit of some good can lead to gross neglect of other goods and even of elementary justice. The term derives from an essay written in 1516 by English statesman and humanist Sir Thomas More (1478–1535), who constructed it from the Greek for, roughly, 'no place' (or, if he had in mind another etymology, 'good place').

Utopias are frequent in literature, the earliest serious one being Plato's *Republic* (*c.*380 BC). Recently utopias have tended to be replaced by 'dystopias' (on the analogy of 'dyspepsia', 'dyslexic' and so on), such as Aldous Huxley's *Brave New World* (1931) and George Orwell's *1984* (1948). *See also* NEGATIVE UTILITARIANISM.

Krishnan Kumar, *Utopia and Anti-Utopia in Modern Times* (Oxford, 1987)

ARL

utterance act *Linguistics* *See* SPEECH ACT THEORY.

V

vacuum (quantum theory of fields) *Physics* A theory due to the English mathematical physicist Paul Adrien Maurice Dirac (1902–84).

Dirac applied a quantum treatment to Maxwell's electromagnetic field and described it in terms of a large number of oscillators with discrete energy levels. According to Heisenberg's uncertainty principle, there is an intrinsic limitation to the precision in specifying both time and energy, which permits the law of conservation of energy to be suspended for very short time intervals. Each oscillator has a fixed minimum energy therefore, with the result that even a vacuum is filled with virtual activity and virtual particles. The theory is sullied by the occurrence of infinities in the mathematical framework which string theory has recently attempted to surmount. *See also* STRING THEORY and HEISENBERG'S UNCERTAINTY PRINCIPLE.

Paul Davies, ed., *The New Physics* (Cambridge, 1989)

GD

valence bond theory *Chemistry* A theory of the American chemist Linus Pauling (*see* RESONANCE, THEORY OF). Valence bond theory is the alternative approach to that offered by MOLECULAR ORBITAL THEORY to the question of chemical bonding.

A covalent bond requires a pair of electrons and suitably oriented electron orbitals on the atoms to be bonded. Although both theories regard the strength of the chemical bond as a result of the accumulation of electron density in the internuclear region of two atoms, valence bond theory does not allow ionic terms to be included in the electron distribution wave function. Valence bond theory has also been less extensively developed since it is less adaptable to computer methods.

P W Atkins, *Physical Chemistry* (Oxford, 1990)

GD

valence shell electron pair repulsion theory (VSEPR) *Chemistry* A theory which attributes the shapes of molecules to the minimization of energy achieved when the electron pairs forming chemical bonds repel each other. For example, methane (CH_4) with four equivalent carbon-hydrogen bonds has perfect tetrahedral symmetry. Ammonia (NH_3), with three equivalent nitrogen-hydrogen bonds, adopts a pyramidal shape; but the bond angles are less than the tetrahedral angle of $109°28'$ because of greater repulsion by the non-bonded lone pair of electrons on the nitrogen atom. In water (H_2O), the tetrahedral angle is further compressed by the repulsion of the two lone pairs on the oxygen atom. The theory does not take account of delocalization of electron density. *See also* VALENCE BOND THEORY.

M Freemantle, *Chemistry in Action* (Basingstoke, 1987)

GD

value (1916) *Linguistics* Foundational principle of linguistics and of STRUCTURALISM formulated by the French linguist Ferdinand de Saussure (1857–1913).

The meaning of a SIGN is not what it refers to, but its place in a system of oppositions with other signs: for example, *tree*, *bush*, *shrub*, and so on.

F de Saussure, *Course in General Linguistics*, W Baskin, trans. (Glasgow, 1974 [1916])

RF

van Deemter rate theory *Chemistry* The theory relates to gas chromatography and considers the sample phase to flow continuously rather than stepwise.

GD

van der Waerden's conjecture (1926) *Mathematics* Named after its proposer, Bertil van der Waerden.

Let A be a matrix with n rows and columns of non-negative real numbers such that the sum of elements in each row or column is 1. If the sum of all products with one factor taken from every row and one taken from every column has the lowest possible value, then A is the matrix in which every element is $1/n$. This has now been proved.

JB

van Niel hypothesis (1930s) *Biology* Named after C B van Niel, this is the theory that the oxygen (O_2) released by plants is derived from water, not from carbon dioxide in the atmosphere, even though carbon dioxide is necessary for photosynthesis.

Although some experimental support for Niel's hypothesis appeared a few years after his proposal, strong evidence for the theory was lacking until 1941, with convincing evidence being reported in 1975.

F Salisbury, *Plant Physiology* (Belmont, Calif., 1992)

KH

van Valen's law *Biology See* RED QUEEN HYPOTHESIS.

van't Hoff principle (1884) *Chemistry* Named after the Dutch chemist Jacobus Henricus van't Hoff (1852–1911), this principle states that for substances at equilibrium with each other, an increase in temperature shifts the equilibrium in the direction that absorbs heat (endothermic), and a decrease in temperature shifts the equilibrium in the direction that evolves heat (exothermic). *See also* LE CHER PRINCIPLE.

D M Considine, *Van Nostrand's Scientific Encyclopedia* (New York, 1989)

GD

van't Hoff law (1885) *Chemistry* Named after the Dutch chemist Jacobus Henricus van't Hoff (1852–1911) who was Professor of Chemistry at Amsterdam, Leipzig and Berlin and won the Nobel prize for Chemistry in 1901. The law (a generalization of AVOGADRO'S LAW) states that the osmotic pressure of a dissolved substance at a given temperature is the same as the gas pressure it would exert if it were an ideal gas occupying the same volume as that of the solution.

Many aqueous solutions were found to deviate from this behaviour, and the Swedish chemist Svante Arrhenius (1859–1927) pointed out this was because solutes dissociated into ions. Van't Hoff took account of these deviations with an empirical factor (i) in his famous equation

$$\pi V = iRT$$

where π is the osmotic pressure, V is the volume, R is the gas constant and T the absolute temperature.

GD

van't Hoff theory *Chemistry See* VAN'T HOFF LAW.

vanguard party (20th century) *Politics* Originally a theory justifying Communist Party leadership in tsarist Russia.

Because the working class (allegedly) suffer from FALSE CONSCIOUSNESS, it was argued by Vladimir Ilyich Lenin (1870–1924) that the Communist Party should take the lead in mobilizing industrial workers and organizing them into a revolutionary body, rather than waiting for workers to arrive themselves at revolutionary opinions.

L Kolakowski, *Main Currents of Marxism*, 3 vols (Oxford, 1978)

RB

variable proportions, law of *Economics See* DIMINISHING RETURNS, LAW OF.

Vegard's law (1921) *Physics* Named after its discoverer, this states that if two salts have the same crystal structure and form solid solutions, the lattice spacings of the latter are a linear function of the composition; that is, the lattice spacing-composition curves are straight lines joining the values for the two pure salts. Vegard's law is often valid for salts but seldom for metallic solid solutions.

J Thewlis, ed., *Encyclopaedic Dictionary of Physics* (New York, Oxford and London, 1962)

MS

veil of ignorance (1971) *Politics* Element in the theory of justice of American political theorist John Rawls (1921–).

Conditions for just social life can be sketched if people are imagined in an 'original position' where they decide upon social rules whilst behind a 'veil of ignorance' which prevents their knowing anything about their own situation in the hypothesized society.

David Miller *et al.*, eds, *The Blackwell Encyclopaedia of Political Thought* (Oxford, 1987)

RB

Venn diagram (19th century) *Mathematics* Named after its inventor, John Venn (1834–1923), this is a diagram in which mathematical sets or the terms of categorical statements are represented by overlapping circles (perhaps with some similar figures) within a boundary to represent the universal set.

vent for surplus (18th, 20th centuries) *Economics* Originally formulated by Scottish economist Adam Smith (1723–90), this is an explanation of why nations export goods, thereby creating international trade.

Domestic economies often are too small to absorb all the output of their markets, thus generating a surplus. The theory has been developed by Burmese-born British economist Hla Myint (1920–) into a study of economic development. *See* DEMOGRAPHIC TRANSITION.

H Myint, *The Economies of the Developing Countries* (London, 1963)

AV, PH

Verdoorn's law (1948) *Economics* Named after Dutch economist P J Verdoorn, this relates to the long-term dynamic relationship between the rate of growth in output and the growth of productivity due to increasing returns.

P J Verdoorn, 'Verdoorn's Law in Retrospect: A Comment', *Economic Journal*, vol. XC (June, 1980), 382–85

AV, PH

verifiability (or verification) principle *Philosophy* Principle that to be meaningful a sentence or proposition must be either verifiable by means of the five senses or a tautology of logic.

The verifiability might be required in practice or (more usually) in principle, and might need to be conclusive (*strong verifiability*) or could be merely partial (*weak verifiability*). Mathematical sentences are treated as tautologies. All others (of metaphysics, ethics, religion, and so on) have meaning, if at all, only in some secondary way (*see also* SPEECH ACT THEORIES). Sometimes the principle says that the meaning is the method of verification, and then the principle can be called the *verifiability (or verification) theory of meaning*, though this title sometimes refers simply to the claim that the principle, however formulated, should be accepted.

Among objections to the principle are that it cannot apply to itself, and is in danger of excluding too much (propositions in science and history, and so on); but a derivative of it has recently appeared as ANTIREALISM. *See also* LOGICAL POSITIVISM.

A J Ayer, *Language, Truth and Logic* (1936); 2nd edn with important new 'Introduction', 1946; main exposition in English

ARL

verismo (late 19th century) *Art* A movement in late 19th-century Italian art and literature, based in Naples. It emphasized the importance of popular culture and a realistic portrayal of contemporary life, breaking both with academicism and ROMANTICISM. Typical were the paintings of A Mancini (1852–1930) and the sculptures of V Gemito (1852–1929).

The term is also used to refer to the most extreme form of REALISM in 20th-century art (for example, in some aspects of SURREALISM and PHOTOREALISM). Occasionally it is applied to the harsh realism of German Expressionists such as Max Beckmann (1884–1950).

C Maltese, *Realismo e Verismo nella pittura Italiana* (Milan, 1967)

LL

verismo (1890s) *Music* Italian for REALISM (and taking its name from the literary movement), this term became musical currency following the first performance of the opera *Cavalleria Rusticana* by Italian composer Pietro Mascagni (1863–1945). It is a style of operatic composition, inspired by the 'naturalistic' literature of French and Italian novelists Émile Zola (1840–1902) and Giovanni Verga (1840–1922).

In an attempt to convey the passions of ordinary folk, composers and librettists wrote in conversational rather than poetic language. They set their operas in contemporary, often rural surroundings, deliberately avoiding the artifice of operas concerning nobility and royalty. However, in exploring the violence and sordidness of human nature there is a tendency to sensationalize, thus diverging from the intended realism.

J Budden, *The Operas of Verdi*, vol. III (London, 1981), 278–81

MLD

Verner's law (1875) *Linguistics* Named after Danish historical linguist Karl Verner (1846–96).

Verner gave a systematic explanation of exceptions to GRIMM'S LAW. Together these two 'laws' provide an excellent illustration of the workings of 19th century historical linguistics.

R Anttila, *Introduction to Historical and Comparative Linguistics* (New York, 1972)

RF

vibrational sum rule *Chemistry* The rule that the sums of the band strengths of all emission/absorption bands with the same upper/lower state, respectively, is proportional to the number of molecules in the upper/lower state, where the band strength is the emission/absorption intensity divided by the fourth power of the frequency.

GD

vicariance biogeography (1974) *Biology* A term coined by Leon Croizat in 1974 based on his theory of biogeography published in *Panbiogeography* (1958).

The distribution of organisms depends entirely on their normal means of dispersion; they do not overcome existing barriers. Disjunct distributions are explained by the removal of barriers in the past, such as the disappearance of land bridges over bodies of water. This is largely accepted, although there are exceptions.

P Vincent, *The Biogeography of the British Isles* (London, 1990)

RA

vicious circle principle *Philosophy* Principle introduced by Bertrand Russell (1872–1970) as a basis for the ramified theory of TYPES. It reads: 'Whatever involves all of a collection must not be one of that collection'; or 'If, provided a certain collection had a total, it would have members only definable in terms of that total, then the said collection has no total.'

A N Whitehead and B A W Russell, *Principia Mathematica*, 1 (1910), Introduction, ch. 2, §1

ARL

Vienna Circle *Philosophy* Properly *Der Wiener Kreis*, a group of philosophers working in Vienna in the 1920s who originated LOGICAL POSITIVISM.

Its leading members included: Moritz Schlick (1882–1936), Rudolf Carnap (1891–1970), Otto Neurath (1882–1945), Herbert Feigl (1902–), Kurt Gödel (1906–78), Friedrich Waismann (1896–1959); with Carl G Hempel (1905–) and H Reichenbach (1891–1953) as associates in Berlin and Alfred Jules Ayer (1910–89) in England, and Karl Raimund Popper (1902–) and Ludwig Wittgenstein (1889–1951) on its fringes.

The Circle was formally constituted as such in 1929, and dispersed after the German invasion of Austria in 1938, mainly to England and the USA. It published a journal called *Erkenntnis*.

V Kraft, *The Vienna Circle* (1969)

ARL

Vierdot's law *Psychology* Formulated by the German, K Vierdot (1818–84).

This principle states that the two-point threshold of a mobile part of the body is directly related to: (1) the mobility of that part of the body; (2) the distance of the body part from the central axis.

A S Reber, *Dictionary of Psychology* (Harmondsworth, 1986)

NS

viral theory (1929) *Biology* Attributed to the British biologist John Burdon Sanderson Haldane (1892–1964), this is the idea that viruses were the precursors to the first living cells.

Most proponents of this theory believe that viruses based on ribonucleic acid (RNA) preceded those based on deoxyribonucleic acid (DNA) because RNA is the more reactive molecule. The mechanism by which viruses – which depend on living cells to reproduce – could have evolved into cells has yet to be described. *Compare with* GENOTYPE THEORY, PHENOTYPE THEORY, and PRECELLULAR EVOLUTION.

M Barbieri, *The Semantic Theory of Evolution* (Chur, Switzerland, 1985)

KH

virial theorem *Physics See* CLAUSIUS VIRIAL THEOREM.

virogene-oncogene hypothesis *Biology See* ONCOGENE HYPOTHESIS.

virtual evolution *Biology See* ANAGENESIS.

virtual reality (1980s–) *Art* A term borrowed from science fiction. Computer technology is used to simulate multiple audio, visual and tactile experiences. As the technology continues to be improved, 'spectators' are increasingly able to interact with more and more convincing virtual realities.

LL

virtual representation (18th century) *Politics* Justification of pre-democratic politics.

People are represented, even though they do not choose their representatives, if those representatives have concern themselves with their INTERESTS. Thus voting is unnecessary since those who wield power (or choose those who do) 'virtually represent' the disenfranchized be they the poor, or women, or non-whites.

A H Birch, *Representative and Responsible Government* (London, 1964)

RB

virtual work principle *Physics* The principle states that under applied forces, a system with workless constraints is in equilibrium if, and only if, zero (virtual) work is done by the applied forces in an arbitrary infinitesimal displacement satisfying the constraints.

GD

visual cliff phenomenon *Psychology* Identified by the Americans J J Gibson (1904–80), and R D Walk (1920–).

An experimental apparatus is used consisting of a large box with a heavy glass top and a narrow board across the centre of the glass. This board separates a shallow 'safe' side from an apparent drop (the 'visual cliff'). The subject (usually a neonate) is placed on the board: consistent movement toward the shallow side is presumed to indicate the ability to perceive depth. Most species which can locomote at birth show an immediate avoidance of the 'cliff'.

C H Graham, ed. *Vision and Visual Perception* (New York, 1965)

NS

visual perspective hypothesis *Psychology* Proposed by the German psychologist F Heider (1896–).

Part of ATTRIBUTION THEORY, this hypothesis indicates that people tend to attribute the cause of their behaviours to external factors rather than to their own personality; and that this varies as a function of perspective. As an individual generally views his environment more often than he critically views himself, there emerges a judgmental bias in explaining behavioural events in the world.

R A Baron and D Byrne, *Social Psychology: Understanding Human Interaction* 4th edn (Boston, 1984)

NS

Vitali covering theorem (20th century) *Mathematics* Named after Italian mathematician Giuseppe Vitali (1875–1932) this is the result whereby if a set E in Euclidean n-space has a Vitali covering (that is, a covering by hypercubes such that for every element e of E, there is a member of the cover which contains e and has arbitrarily small positive measure), then there is a countable sequence of pairwise disjoint members of the covering whose union has Lebesgue outer measure equal to that of *E*.

H L Royden, *Real Analysis* (1968)

MB

vitalism (17th–19th centuries) *Biology/Chemistry/Philosophy* Any of various views insisting, in contrast to MECHANISM, that life involves a special principle and cannot be explained in terms of physical and chemical properties alone.

The theory has its origins in the classification of compounds in 1675 by the French chemist Nicolas Lemery (1645–1715). He considered them as animal, vegetable or mineral according to how they originated. Another French chemist, Antoine Laurent Lavoisier (1743–94), grouped the animal and vegetable compounds together but still retained the original classification. By the start of the 19th century, it was believed that definite and fundamental differences existed between 'organic' and 'inorganic' compounds, as substances had come to be known. In 1815, the Swedish chemist Johan Jakob Berzelius (1779–1848) proposed that the two classes of compounds were produced from their elements by entirely different laws. Organic compounds were produced under the influence of a vital force and so were incapable of being prepared artificially. This distinction was ended in 1828 when German chemist Friedrich Wöhler (1800–82) synthesized the organic compound urea from the purely inorganic ammonium cyanate.

In philosophical terms, the life principle involved may take the form of *entelechies* within living things, which are responsible for their growth and development (according to Hans Driesch (1867–1941)); or of a general life force like the *élan vital* of Henri Bergson (1859–1941), who rejected the kind of vitalism that postulated individual entelechies. *See also* EMERGENCE THEORIES, HOLISM and HUMORALISM, CONCEPT OF. *Compare with* REDUCTIONISM.

E Sinnott, *The Bridge of Life: From Matter to Spirit* (New York, 1966); H Driesch, *The History and Theory of Vitalism* (1914); I L Finar, *Organic Chemistry* (London, 1973)

GD, ARL, KH

volley theory of hearing (20th century) *Psychology* Proposed by the American psychologists E G Wever and C W Bray.

This theory claims that single nerve fibres in the basilar membrane of the ear need not respond to every succesive wave of the sound energy stimulus but to every second, third or fourth wave. Each wave is thought to excite a group of nerve fibres, the next wave to fire another group, and so on. The pattern of neural impulses reaching the brain by the successive volley represents the frequency of the sound.

R S Woodworth and H Schlosberg, *Experimental Psychology* (London, 1966)

NS

volonté générale *Politics See* GENERAL WILL.

voluntarism *Philosophy* Any of a number of doctrines emphasizing the existence, nature, or role of the will; whether a cosmic will, as with Artur Schopenhauer (1788–1860), or (more commonly, and also including Schopenhauer) the human will. Such doctrines may emphasize the role of the will in our thinking or acquisition of knowledge, or the reality of free decisions as against universal DETERMINISM, or the power of individuals' choices in shaping the course of history.

A Schopenhauer, *The Will to Live*, R Taylor, ed. (1962); selected writings

ARL

von Baer's laws (*c.*1837) *Biology* Also referred to as the law of corresponding stages, von Baer's laws are named after Karl Ernest Ritter von Baer (1792–1876), an Estonian embryologist known for discovering the mammalian ovum. His findings were based on experiments by his friend Christian Pander (1794–1865), who described the three layers of the vertebrate embryo in 1817.

Von Baer postulated that the goal of embryonic development is the formation of three germ layers (the ectoderm, endoderm and mesoderm) from which all other structures are uniquely formed; and that the younger the embryo of a species, the stronger the resemblance between the embryo of that species and the embryo of some other vertebrate species. This comparison supported the concept of epigenesis. *Compare with* PREFORMATION.

D Abbott, *Biologists* (New York, 1983)

KH

von Neumann problem (20th century) *Mathematics* Named after Hungarian-born American mathematician John von Neumann (1903–57), this is similar to the DIRICHLET PROBLEM except that the normal derivative, $\partial u/\partial n$, of u is specified at each point of the boundary of a region rather than the values of u itself.

J E Marsden, *Basic Complex Analysis* (Freeman, 1973)

MB

von Restorff effect *Psychology* Identified by the German psychologist von Restorff.

Also called the isolation effect, this is the principle that when given a series of stimuli to be learned (for example, a list of words), if one of them is made physically distinctive from the others (for example, by using a different colour or font) it will be easier to learn and recall.

A D Baddeley, *The Psychology of Human Memory* (New York, 1976)

NS

von Weizsäcker's theory of planetary formation (20th century) *Astronomy* Named after German physicist Baron Carl Friedrich von Weizsäcker (1912–) who introduced it, this is essentially the same as the NEBULAR HYPOTHESIS OF PLANETARY FORMATION.

It is distinct in its recognition of a difference in the behaviour of the gaseous part of the nebulous cloud and the dust portion, composed of small solid particles of 'terrestrial materials'. The outer parts of the gas would escape into space while the inner parts would condense into the Sun. The denser dust particles would execute elliptical orbits around the Sun and aggregate, first as a result of collisions and, later, through gravitational attraction. This process continues until only a few large chunks (planets) remain that are too far apart to interfere with one another. This process of dust aggregation runs concurrently with the dissipation of the gaseous part of the original cloud.

S Mitton, ed., *The Cambridge Encyclopaedia of Astronomy* (London, 1973)

MS

Vorticism (1914) *Art/Literary Theory* An avant-garde movement in painting, sculpture and literature, launched by the English painter and writer Wyndham Lewis (1882–1957) and the American poet Ezra Pound (1885–1972) who coined the term.

The image of the vortex (whirlpool, spinning cone) signifies a point of stillness and concentration at the centre of energetic motion. At the centre is, for Pound, the 'primary pigment', different for each of the arts; in poetry, it is the image (*see* IMAGISM). Pound's early 'vorticist' poems are based on concentrated, concrete, nominal images.

Vorticism was primarily an artistic movement, whose adherents admired, but remained separate from, Italian FUTURISM. Painters Lewis, Edward Wadsworth (1889–1949), Christopher Richard Wynne Nevinson (1889–1946), David Bomberg (1890–1957), William Roberts (1895–1980) and sculptors Jacob Epstein (1880–1959) and Henri Gaudier-Brzeska (1891–1915) were the main exhibitors and exponents of the group. Two issues of *Blast*, their magazine/manifesto were published (1913–14). Internal quarrels and, especially, World War I led to the group's dissolution and a change in attitude.

R Cork, *Vorticism and Abstract Art in the First Machine Age* (London, 1976); R W Dasenbrock, *The Literary Vorticism of Ezra Pound and Wyndham Lewis* (Baltimore, 1985)

RF, LL

vulcanism (18th–19th centuries) *Geology* Whilst essentially the same as PLUTONISM, this term is customarily used when considering the main contention between adherents of PLUTONISM and NEPTUNISM: the origin of basalts.

The dispute mostly centred around the problem of whether basalts in the Auvergne (France) were deposited from aqueous solutions, as proposed by the neptunists, or were the result of volcanic eruptions. By about 1820, it was resolved that these and, by implication, all other basalts were associated with vulcanism.

A Hallam, *Great Geological Controversies* (Oxford, 1983)

W

wages fund doctrine *Economics* A precept held by the classical school of economists, it states that an employer uses a fund of money, or fixed amount of capital, to pay workers; so, if employees demand higher wages, unemployment of some workers will be necessary.

The doctrine was also extended to a macroeconomic scale, but was later criticized on the grounds that labour rates were determined by supply and demand. *See* CLASSICAL MACROECONOMIC MODEL.

F W Taussig, *Wages and Capital* (London, 1896)

AV, PH

Wagner's law *Economics*. Named after German economist Adolph Wagner (1835–1917), this theory states that the development of an industrial economy will be accompanied by an increased share of public expenditure in GNP.

PH

Walden's rule (1906) *Chemistry* Named after the Latvian-born Russian chemist Paul Walden (Pavel Ivanovich Valden) (1863–1957). This empirical rule states that the product of the molar conductivity (at infinite dilution of an ionic solute and the viscosity of the solvent in which it is dissolved) is approximately constant for the same ion in different solvents at constant temperature.

The rule is based on the assumption that ions behave like Stokes spheres in a viscous continuum. In practice, ions are solvated to different degrees in different solvents, making the rule only approximately correct.

P W Atkins, *Physical Chemistry* (Oxford, 1990)

GD

Wallace's line (1869) *Biology* Named after British naturalist Alfred Russel Wallace (1823–1913) who described the line in *The Malay Archipelago* (1869) and *The Geographical Distribution of Animals* (1876). It is an hypothetical line dividing animals derived from Asian species on the west from those derived from Australian species on the east.

The line runs through a strait dividing Borneo from the Celebes, southeast of the Philippines and east of Bali. Wallace proposed the line based on his observations of the similarities and differences in the species living on either side. In the mid-20th century, studies of the ocean floor and PLATE TECTONICS proved that Wallace's line lies on the perimeter of the Indo–Australian plate, an area of relatively recent crustal activity.

R Milner, *The Encyclopedia of Evolution: Humanity's Search for Its Origins* (New York, 1990)

KH

Walras's stability *Economics* Named after French- born economist Léon Walras (1834–1910), who laid the groundwork for a unified model which included theories of exchange, production, formation and theory.

Walras asserted that if there are *n* markets and $n-1$ markets are in equilibrium, then the last market must be in equilibrium as well. *See* GENERAL EQUILIBRIUM, PARTIAL EQUILIBRIUM THEORY.

L Walras, *Eléments d'économie politique pure* (Lausanne, 1874–77); M Morishima, *Equilibrium, Stability and Growth* (Oxford, 1964)

PH

Walther's law *Geology* Proposed by German geologist Johannes Walther (1860–1937) who studied the deserts of North Africa and the Middle East. Changes in the nature of sediments in a succession of rocks indicate changes in the depositional environments that can also be observed laterally.

This model is generally applicable to lowland, desert areas where rivers and sands migrate laterally and give rise to both vertical and lateral environmental sequences. It is a dangerous concept to apply in areas of tectonism.

DT

Wang's problem (1974) *Mathematics* Named after Chinese-born American mathematician Hao Wang (1921–).

If a proposition is undecidable in some system, it would still be true in the system because a counter example would be elementarily provable.

JB

Warburg's theory *Biology/Medicine* The theory that cancer results from a change in metabolism in the cell caused by an irreversible injury to the cell's respiration. The main difference between a normal cell and a tumour cell is that the tumour cell shifts toward anaerobic (without free oxygen) metabolism, which emphasizes glycolysis and fermentative-type metabolism for energy production. This theory is one of many competing explanations for cancer. *See also* CANCER, THEORIES OF.

KH

Waring's problem (1770) *Mathematics* Named after British mathematician Edward Waring (1736–98) who made the assertion without proof in his *Meditationes Algebraicae*.

It is the problem of showing that for every natural number $k \geqslant 2$, there is a number s such that every natural number is the sum of at most s kth powers of natural numbers; the least such s, if it exists, is denoted by $g(k)$. For example, LAGRANGE'S THEOREM states that $g(2) = 4$ while $g(3) = 9$ and $g(4) = 19$. In 1909, German mathematician David Hilbert (1862–1943) proved the existence of $g(k)$ for every k. The exact value of $g(k)$ is now known for every k except $k = 4$. Much of this numerical evaluation is based on the HARDY-LITTLEWOOD METHOD.

T M Apostol, *Introduction to Analytic Number Theory* (Springer, 1976)

MB

water-potential concept (1960) *Biology* Developed by Australian biologist Ralph O Slatyer and the American Sterling A Taylor, this concept proposes that plants respond physiologically to a value called water potential.

Water potential is usually defined as the chemical potential of water in a system, compared with the chemical potential of pure water at the same temperature and height (to control for the effects of gravity) and at atmospheric pressure. The chemical potential of the reference water is set at zero. Although widely accepted by both soil and plant scientists, in the 1980s the water-potential concept was questioned by some researchers, who suggested that gravity or the plant's relative water content may be the most important determinants of wilting and other physiological responses.

F Salisbury, *Plant Physiology* (Belmont, Calif., 1992)

KH

Watson-Crick hypothesis or model (1953) *Biology* Named after American scientist James Dewey Watson (1928–) and British scientist Francis Harry Compton Crick (1916–), who, with British scientist Maurice Hugh Frederick Wilkins (1916–), shared the 1962 Nobel prize for Physiology or Medicine.

The double helix, right-handed model for the structure of deoxyribonucleic acid (DNA) in which sugar phosphate backbones form the basic, twisting shape while providing a rigid framework for pairs of base compounds. Other DNA models have since been discovered, such as the left-handed Z form, but the Watson-Crick model remains well substantiated and has exerted enormous influence on molecular genetics.

H Judson, *The Eighth Day of Creation: The Makers of the Revolution in Biology* (London, 1979)

KH

wave theory of light (19th–20th centuries) *Physics* A quantitative wave theory of light was devised by the French physicist Augustin Jean Fresnel (1788–1827) who assumed that light is an elastic vibration in a material medium called the luminiferous AETHER. He derived expressions for the fractions of light reflected and refracted at the interface between two transparent media.

Other important contributions were made by: (1) the English physicist Thomas Young (1773–1829), who used a wave model to explain the observations of the interference of light; (2) the Danish physicist Erasmus Bartholinus (1625–98), whose discovery of the polarization of light showed that the light waves must be transverse; (3) the Scottish physicist James Clerk-Maxwell (1831–79), whose 1865 theory of the electromagnetic field showed that light was essentially the periodic fluctuation in the strengths of electric and magnetic fields at right angles to the direction of propagation of the light.

A wave model of light is satisfactory for discussing the propagation of light but not its interaction with matter, for which a particle model is essential.

J Thewlis, ed., *Encyclopaedic Dictionary of Physics* (New York, Oxford and London, 1962)

MS

weak law of large numbers (1915) *Mathematics* A precise form of the law of large numbers proved by Emile Borel (1871–1956). *See also* LARGE NUMBERS, LAW OF and STRONG LAW OF LARGE NUMBERS.

JB

Weber-Fechner law *Physics/Psychology See* FECHNER'S LAW and WEBER'S LAW.

Weber's law (*c.*1834) *Psychology* Formulated by the German physiologist Ernst Heinrich Weber (1795–1878). The just noticeable difference (JND) between two stimuli is proportional to the magnitude of the original stimulus.

This law is usually presented as $\Delta I/I = K$ *where* ΔI is the change required for a JND in stimulation, I is the stimulus magnitude, and K is a constant for the particular sense. The value of K is called the Weber ratio. The law tends to break down when very low or high intensity stimuli are used but holds reasonably well for the mid-range of most stimulus dimensions. *See also* FECHNER'S LAW and FULLERTON-CATTELL LAW.

L Kaufman, *Perception* (New York, 1978)

NS

Weber's theory of the location of the firm (1909) *Economics* Formulated by German economist and sociologist Alfred Weber (1868–1958), this states that the location of firms is determined by attempts to minimize costs.

Thus, if production costs are the same everywhere, transport costs will govern the choice of location. *See* CENTRAL PLACE THEORY, LOCATION THEORY, GRAVITY MODELS and LEAST COST LOCATION THEORY.

A W Weber, *Theory of the Location of Industries* (Chicago, 1929)

AV, PH

Wedderburn-Artin theorem (20th century) *Mathematics* Named after Scottish mathematician Joseph Henry Maclagan Wedderburn (1882–1948) and German-born American mathematician Emil Artin (1898–1962).

This is the fundamental structure theorem in ring theory that every semi-simple right *Artinian ring* is the direct sum of a finite number of simple right Artinian rings each of which is isomorphic to the ring of $n \times n$ matrices over some division ring, for some positive integer n.

T W Hungerford, *Algebra* (Springer, 1974)

MB

Wegener-Bergeron-Findeism theory *Meteorology See* BERGERON THEORY.

Weierstrass approximation theorem (19th century) *Mathematics* Named after German mathematician Karl Wilhelm Theodor Weierstrass (1815–97), this states that if X is a compact subset of $\Re^n$, the restriction of the polynomial functions on X are dense in the continuous functions on X.

G B Folland, *Real Analysis* (New York, 1984)

ML

Weierstrass M test (19th century) *Mathematics* Proved but not published by German mathematician Karl Weierstrass (1815–97).

An infinite sum in which each term varies with a complex variable z

$$a_0(z) + a_1(z) + \ldots + a_n(z) + \ldots$$

is uniformly convergent to a sum $f(z)$ if, for all values of z under consideration, the modulus of $a_n(z)$ is less than or equal to the non-negative number M_n, which does not vary with z, such that

$$M_0 + M_1 + \ldots + M_n + \ldots$$

has a sum to infinity. This is the most common method of proving that an infinite series is uniformly convergent, which is a valuable property because it allows a limit as z tends to d of $f(z)$ to be evaluated in terms of

$$b_0 + b_1 + \ldots + b_n + \ldots$$

where b_n is the limit of $a_n(z)$ as z tends to d.

W L Ferrar, *A Textbook of Convergence* (Oxford, 1938)

JB

Weismannism *Biology See* GERM PLASM THEORY.

Weiss theory of ferromagnetism (1905) *Physics* Named after the French physicist Pierre Weiss (1865–1940).

A theory based on the hypotheses that below the Curie point, a ferromagnetic substance is composed of an ensemble of small, independent magnetized regions called domains, and that each domain is spontaneously magnetized under the influence of an applied magnetic field which tends to align the individual magnetic moments within the domain. Also known as molecular field theory. *See also* MAGNETISM, THEORIES OF

Robert M. Besançon, ed., *The Encyclopedia of Physics* (New York, 1985)

GD

Weiss zone law (1804) *Physics* Named after its discoverer, the German mineralogist Christian Samuel Weiss (1780–1856), this states that if the crystal face with Miller indices (hkl) is a member of the zone of faces whose axis has the direction with indices [UVW], then

$$hU + kV + lW = 0.$$

J Thewlis, ed., *Encyclopaedic Dictionary of Physics* (New York, Oxford and London, 1962)

MS

Weissenberg effect *Physics* Named after its discoverer, this is the phenomenon whereby when a vertical rod is rotated about its axis in a visco-elastic liquid, the liquid climbs up the rod.

The reason for this behaviour is that, in a visco-elastic liquid, the stresses in the direction of the streamlines and in a direction normal to the streamlines are not equal. The difference between these stresses causes the visco-elastic liquid to climb the rotating rod.

J Thewlis, ed., *Encyclopaedic Dictionary of Physics* (New York, Oxford and London, 1962)

MS

welfarism (20th century) *Politics* A pejorative term indicating preference for collective social services.

The social provision of goods and services is to be preferred to their purchase or procurement by individual effort and choice. The 'theory' is a construct of those who use the term critically, and does not occur in the arguments of those to whom it is attributed.

RB

well-ordering principle (20th century) *Mathematics* The theorem or axiom that every set A can be well ordered; that is, there is an order relation on A such that every non-empty subset of A has a least element with respect to this ordering. This assertion is equivalent to ZORN'S LEMMA and the AXIOM OF CHOICE. When the set in question is the positive integers, the assertion is equivalent to the principle of INDUCTION.

P R Halmos, *Naive Set Theory* (Springer, 1974)

MB

Welter's rule *Chemistry* This rule states that an approximate value for the heat of combustion of an organic compound may be calculated by subtracting the oxygen and hydrogen in the proportion of H_2O and then adding the heats of combustion of the remaining carbon and hydrogen atoms.

GD

Wenzel's law *Chemistry* This law states that the rate of dissolution of a solid in a liquid is proportional to the surface area of the solid exposed to the liquid.

GD

Werner theory of co-ordination compounds (1893) *Chemistry* Named after the German-born Swiss chemist Alfred Werner (1866–1919) who was Professor of Chemistry in Zürich and winner of the Nobel prize in 1913. Werner proposed a theory to account for compounds of the type represented by $CoCl_3.6NH_3$ whose formation had greatly puzzled chemists up to then.

Werner postulated that the six ammonia molecules were symmetrically co-ordinated to the central cobalt ion through 'subsidiary valencies' of the metal, while the 'principal valencies' were satisfied by the chloride ions. Over the next 20 years he went on to elaborate and perfect his theory, and it has needed relatively little modification since although the terms principal 'valencies' and 'subsidiary valencies' are no longer used. *See also* LIGAND FIELD THEORY.

M Freemantle, *Chemistry in Action* (Basingstoke, 1987)

GD

Wernicke's area (1874) *Linguistics* Named after 19th century German neurologist Carl Wernicke (1848–1905).

Area of the angular gyrus of the left cerebral hemisphere, damage to which may result in 'Wernicke's aphasia': defective comprehension of speech and writing, and apparently fluent speech which is ungrammatical in details and contains non-words and semantically empty words. *See also* BROCA'S AREA.

M Garman, *Psycholinguistics* (Cambridge, 1990)

RF

Wever-Bray effect *Psychology See* VOLLEY THEORY OF HEARING.

Whig interpretation of history (1931) *History* Term coined by the English historian Sir Herbert Butterfield (1900–79) to describe an optimistic account of history in terms of its ends.

History can be seen as the successive victories of PROGRESS over reaction, and in terms of movement towards the goal eventually reached.

Sir Herbert Butterfield, *The Whig Interpretation of History* (London, 1931)

RB

white man's burden (1898) *Politics* Phrase coined by English writer Rudyard Kipling (1865–1936).

The government and development of empire is a burden, dutifully accepted by white Europeans, but unappreciated by those over whom they exercise their paternal power. Kipling's poetic expression neatly encapsulates a theory of imperialism which was often more assumed than articulated.

A P Thornton, *Doctrines of Imperialism* (London, 1965)

RB

Wicksell's theory of capital (1893) *Economics* Named after Swedish economist Knut Wicksell (1851–1926), this theory examines factor prices as derived from the value of the marginal product.

Wicksell pointed out that in an equilibrium situation, the interest rate would exceed the value of the marginal product of capital because the aggregate stock of capital would be revalued due to changes in the interest rate.

K Wicksell, *Uber Wert, Kapital und Rente* (Jena, 1893)

PH

Wiedemann-Franz-Lorenz law (1854) *Physics* Named after its discoverers, the German physicist Gustav Heinrich Wiedemann (1826–99), Rudolph Franz and the Danish physicist Ludwig Valentin Lorenz (1829–91). Wiedemann and Franz stated the law (known as the Wiedemann-Franz law) that, at a given temperature, the ratio of the thermal conductivity k to the electrical conductivity σ has a constant value for all metals. Lorenz (1872) extended the law by adding that the ratio of the two conductivities is proportional to the temperature T; that is

$$\frac{k}{\sigma T} = \text{constant}$$

where the constant has the same value for all metals. The law is only approximately valid for a limited range of temperature.

J Thewlis, ed., *Encyclopaedic Dictionary of Physics* (New York, Oxford and London, 1962)

MS

Wiedemann's additivity law (1863) *Chemistry* An empirical law named after the German chemist Gustav Heinrich Wiedemann (1826–99) who became professor of Germany's first chair of physical chemistry at Leipzig in 1871 and later Professor of Physics at the same university. It states that the molar magnetic susceptibility of a mixture or solution of components is the sum of the susceptibilities of the components in the proportion of their weight fractions.

GD

Wien's displacement law (1893) *Physics* Named after its discoverer, the German physicist Wilhelm Wien (1864–1928), this states that if curves of the spectral emissive power of a black body against wavelength λ are compared for different temperatures T, it is found that the maximum spectral emissive powers occur at wavelengths λ_m that satisfy the condition

$$\lambda_m T = \text{constant} .$$

J Thewlis, ed., *Encyclopaedic Dictionary of Physics* (New York, Oxford and London, 1962)

MS

Wien's distribution law (1896) *Physics* Named after its discoverer, the German physicist Wilhelm Wien (1864–1928).

Let $u_\lambda d\lambda$ be the equilibrium energy per unit volume, in the wavelength range λ to $\lambda + d\lambda$, in an evacuated cavity with opaque walls at a temperature T. Then,

$$u_\lambda = C\lambda^{-5}F(\lambda T)$$

where C is a constant and f is a universal function of λ and T. For long wavelengths, an approximate form of Wien's distribution law, known as Wien's formula, is

$$u\lambda = A\lambda^{-5}\exp(-B/\lambda T)$$

where A and B are constants.

J Thewlis, ed., *Encyclopaedic Dictionary of Physics* (New York, Oxford and London, 1962)

MS

Wilcoxen test (20th century) *Mathematics* Discovered by Frank Wilcoxen (1892–1965), this test examines whether a population has a given median by examining the ranking orders of one sample.

JB

willing suspension of disbelief (1815) *Literary Theory* Proposed by the English poet and critic Samuel Taylor Coleridge (1772–1834).

The reader's agreement to accept the provisional truth of fiction in poetry by granting to 'shadows of imagination' a temporary credulity: 'that willing suspension of disbelief for the moment, which constitutes poetic faith.' *See* 'ALS OB'.

S T Coleridge, *Biographia Literaria* [1815] (London, 1952), ch. 14

RF

Wilson cycle *Geology* The Canadian oceanographer J Tuzo Wilson suggested that oceans may have a cycle of opening and closing, with the join marking the approximate location of the subsequent opening.

While this model has some validity for the North and Central Atlantic, there is no evidence for such a cycle in other oceans. However, the timescales involved (some 400 million years) may well relate to those of convective motions within the Earth's mantle. *See also* SEA-FLOOR SPREADING.

J T Wilson, *The Way the Earth Works – Readings from Scientific American* (San Francisco, 1970)

DT

Wilson's theorem *Mathematics* Named after British mathematician John Wilson (1741–93), but the statement was apparently first published by fellow British mathematician Edward Waring (1736–98) in his *Meditationes algebraicae* of 1770 and was later proved by French mathematician Joseph Louis Lagrange (1736–1813).

This is the result whereby a natural number n is prime if and only if n divides $1 + (n - 1)!$.

A Baker, *A Concise Introduction to the Theory of Numbers* (Cambridge, 1984)

MB

wing rule *Biology See* RENSCH'S LAWS.

wit (17th century) *Literary Theory* In ordinary usage, verbal cleverness and quickness designed to produce a comic or elegant effect. But in literary theory, 'wit' is a more serious faculty: the poetic linking of apparently divergent ideas, exemplified in the

METAPHYSICAL POETS' use of CONCEITS. *See also* IMAGINATION and PARADOX.

G Williamson, *The Proper Wit of Poetry* (London, 1961)

RF

withering away of the state (19th century) *Politics* Term used intially by the German collaborator of Karl Marx, Friederich Engels (1820–95).

After the seizure of power by the working class, the DICTATORSHIP OF THE PROLETARIAT will be used to abolish capitalism and, hence, classes. Since states only exist to regulate class conflict, the STATE will thereafter be redundant and will wither away.

David McLellan, *Marxism after Marx* (London, 1980)

RB

Witt colour theory (1876) *Chemistry* Witt showed that dyestuffs and other coloured organic compounds (chromogens) contained groups with multiple bonds (for example, NO_2, NO, or N=N) which he called chromophores. Other (non-chromophoric) groups (for example, OH, NH_2, Cl, and COOH) intensified the colour, and he called these auxochromes. Witt also coined the terms bathochromic shift and hypsochromic shift to signify, respectively, the intensification and lightening of colour caused by chromophoric groups.

I L Finar, *Organic Chemistry* (London, 1973)

GD

wobble hypothesis (1966) *Biology* Proposed by the British biophysicist Francis Harry Compton Crick (1916–).

DNA codes for amino acids in a code of three bases. There are many more code combinations in existence than amino acids, therefore some of the codes are considered degenerate. However, some amino acids have more than one code, the combination differing only in the third base. Crick postulated that the first two bases formed the necessary bonding for the code to be read, whilst the third was bound less tightly, with a certain degree of movement or wobble; possibly such that in some cases a similar base would do equally well.

RA, KH

Woese's dogma (1977) *Biology* Named after American evolutionist and biophysicist Carl R Woese (1928–), this is the belief that the evolution of heavy-molecular-weight, ribosomal ribonucleic acids (RNAs) was a necessary precursor to the evolution of complex life-forms.

Because heavy ribosomes are essential for the accurate transmission of specific hereditary characteristics (that is, the translation of genes into specific proteins), this dogma implies that the existence of heavy ribosomes (and not merely the coexistence of proteins and genes) was an essential precondition for the appearance of biological specificity. *See also* PROGENOTE THEORY.

M Barbieri, *The Semantic Theory of Evolution* (Chur, Switzerland, 1985)

KH

Wonderful Life theory (1989) *Biology* Also called contingency theory. Named after the book *Wonderful Life* by Stephen Jay Gould (1942–), a noted biologist and evolutionist at Harvard University and the theory's chief proponent. This controversial idea asserts that during the Cambrian period (500–600 million years ago), many hundreds of distinct body types evolved that have since become extinct, leaving the relatively few phyla that are in existence today.

Before Gould put forward his hypothesis, most biologists assumed that evolution since the Cambrian has expanded species diversity, not contracted it.

S Gould, *Wonderful Life: the Burgess Shale and the Nature of History* (New York, 1989)

KH

Woodward-Hoffmann rules (20th century) *Chemistry* Named after the American chemist Robert Burns Woodward (1917–79), Professor of Science at Harvard University from 1950 and winner of the Nobel prize for Chemistry in 1965; and the Polish-born American chemist Roald Hoffmann (1937–), who shared the Nobel prize for Chemistry with Kenichi Fukui in 1981. The rules govern the paths and stereochemistry of both intramolecular cyclization and concerted intermolecular addition reactions.

The rules are based on the conservation of orbital symmetry, according to which a combination of p_X-orbitals to form a sigma-bond occurs by overlap of 'like-sign' orbitals. The way in which this overlap is achieved, however, depends on the number of π-electrons involved in the reaction and also on whether the reaction is under thermal or photochemical control.

S B Parker, *McGraw-Hill Encyclopedia of Science and Technology* (New York, 1987)

GD

work vs text (1971) *Stylistics* Distinction proposed by the French structuralist Roland Barthes (1915–80).

Two attitudes to a piece of literature. The classical approach sees it as a 'work': an objective, fixed entity, property of the author, interpretable as having a single authoritative meaning. By contrast, the 'text' is multiple in meaning, a tissue of references to other texts, no longer owned by an author but open to a reader's exploration. *See also* 'LISIBLE' AND 'SCRIPTIBLE'.

R Barthes, 'From Work to Text', *Textual Strategies*, J Harari, ed. (Ithaca, 1979), 73–81

RF

workers' control (20th century) *Politics* Theory of industrial and economic power.

Production should be controlled by those directly engaged in it, rather than by non-productive owners. Workers' control – of which GUILD SOCIALISM is an example – is to be distinguished from revolutionary doctrines such as SYNDICALISM or ANARCHO-SYNDICALISM in that it does not seek to use economic power in order to sieze political power.

Rodney Barker, *Political Ideas in Modern Britain* (London, 1978)

RB

world society (20th century) *Politics* Theory of international relations propounded by Australian political scientist John W Burton (1915–). A more ambitious version of the theory of international society.

Graham Evans and Jeffrey Newnham, *The Dictionary of World Politics* (Hemel Hempstead, 1990)

RB

writerly (text) *Stylistics See* 'LISIBLE' AND 'SCRIPTIBLE'.

Wulff's theorem *Physics* Named after its originator, the Ukrainian physicist Georg (Yuri Viktorovich) Wulff (1863–1925), this theorem answers the question of what shape minimizes the surface energy of a crystal in equilibrium.

A plane is drawn perpendicularly through the end of each radius vector in the polar diagram of surface energy for the crystal considered. The body formed by all points that can be reached from the origin without crossing any of these planes has the shape of minimum surface energy.

J Thewlis, ed., *Encyclopaedic Dictionary of Physics* (New York, Oxford and London, 1962)

MS

Wüllner's law *Chemistry* The law states that modification of osmotic and other properties of water by a dissolved substance is proportional to the concentration of the solute. *See also* RAOULT'S LAW.

GD

X

X-bar theory (1970) *Linguistics* Proposed by the American linguist Avram Noam Chomsky (1928–).

(Notations X′, X) Account of the structure of phrases which ensures that a phrase is identified in terms of its 'head'. For example, in 'The President's speech to the alumni' the head is the N[oun] 'speech', and the phrases 'speech to the alumni' and 'The President's speech to the alumni' are 'projections' to the head which must receive the same category symbol with different 'bars' or 'primes', N′ and N″, respectively.

N Chomsky, 'Remarks on Nominalization', *Readings in English Transformational Grammar*, R Jacobs and P S Rosenbaum, eds (Waltham, Mass., 1970), 184–221; A Radford, *Transformational Syntax* (Cambridge, 1981)

RF

x-efficiency (1966) *Economics* Formulated by American economist Harvey Leibenstein (1922–), this describes the general efficiency of a firm (judged on managerial and technological criteria) in transforming inputs at minimum cost into maximum profits. *See* THEORY OF THE FIRM, MANAGERIAL THEORY OF THE FIRM, SATISFICING, AGENCY THEORY, SCALAR PRINCIPLE and PARKINSON'S LAW.

H Leibenstein, 'Allocative Efficiency vs "X-efficiency"', *American Economic Review*, vol. LVI (June, 1966), 392–415

AV, PH

Y

Yang-Mills theory (1954) *Physics* Named after the Chinese-born American physicist Chen Ning Yang (1922–), joint winner of the 1957 Nobel prize for Physics, and the English mathematician Robert Laurence Mills (1927–).

Yang and Mills produced a nonlinear version of Maxwell's equations incorporating a non-Abelian (non-commutative) group. Originally formulated for Lorentz space-time, it commonly uses Euclidean space-time when applied in quantum theory. The theory has been incorporated into nearly every model of particle physics since the mid-1970s based on the hypothesis that the nucleon-nucleon forces can be derived by imposing local isospin invariance. This implies that the interaction must occur through the exchange of three massless vector bosons.

Douglas M. Considine, ed., *Van Nostrand's Scientific Encyclopedia* (New York, 1989)

GD

Yerkes-Dodson law *Psychology* Formulated by the American Robert Mearns Yerkes (1876–1956) and J D Dodson.

Yerkes and Dodson discovered that mice were better able to learn a simple task in order to avoid a painful shock, and better able to learn a complex task when doing so to avoid a mild shock. Later researchers interpreted this to mean that there is an optimum level of arousal associated with the difficulty of each task in which an animal might engage. The finding is wrongly labelled a 'law' in that it is really based on little more than supposition.

R M Yerkes and J D Dodson, 'The Relation of Strength of Stimulus to Rapidity of Habit Formation', *Journal of Comparative Neurology and Psychology*, vol. XVIII (1908) 459–82

NS

Young-Helmholtz trichromatic theory (1802) Biology/Medicine/Psychology Named after its originator, the British physician Thomas Young (1773–1829) and German physician and physiologist Herman von Helmholtz (1821–94) who refined it. This theory asserts that colour vision is produced by three different kinds of cones in the retina, each containing a different pigment that is most sensitive to light in the red, green or violet parts of the spectrum.

The perception of other colours results from the stimulation of combinations of these pigments. According to this theory, colour blindness arises from a deficiency in reception of one of the cones. Studies of goldfish, monkey and human vision support this theory, but competing theories are also supported by experimental results.

M Ali, *Vision in Vertebrates* (New York, 1985)

KH

Z

Zaitsev's rule *Chemistry* The rule states that dehydrohalogenation of an alkyl halide yields the more substituted of the two possible alkenes as the predominant product. Alternatively, hydrogen is preferentially eliminated from the carbon atom joined to the least number of hydrogen atoms. *See also* MARKOWNIKOFF'S RULE.

I L Finar, *Organic Chemistry* (London, 1973)

GD

Zeeman effect (1896) *Physics* Named after its discoverer, Dutch physicist Pieter Zeeman (1865–1943), this is the splitting of optical lines into several closely-spaced components when the source of light is placed in a strong magnetic field. These components are polarized in a way that depends on the direction from which the source is viewed relative to the direction of the magnetic field.

J Thewlis, ed., *Encyclopaedic Dictionary of Physics* (New York, Oxford and London, 1962)

MS

Zeigarnik effect *Psychology* Identified by the Russian psychologist B Zeigarnik (1900–).

This refers to the phenomenon whereby the recall ratio for tasks interrupted at the middle or latter end of task completion is higher than for tasks interrupted at or near the beginning. Other research has indicated that: (1) the Zeigarnik effect is likely to appear if the subject is ego-involved in the task to some extent; (2) the effect is more likely to appear if the task doesn't seem to be part of the experimental game-plan; (3) the effect is most likely to appear if the subject has set a genuine level of aspiration in the interrupted task.

A D Baddeley, *The Psychology of Memory* (New York, 1976)

NS

Zeno's paradox (5th century BC) *Mathematics* Found by Zeno of Elea (*c.*490–435 BC).

Examples of this include the Achilles paradox, in which Achilles cannot catch a tortoise since it is always a fraction ahead. To obtain a paradox it must be assumed that a sum of an infinity of terms must be infinite, but both time and distance are given by the finite numbers which are the sums of geometric series. *See also* DISTANCE PARADOX and STADIUM PARADOX.

JB

Zermelo-Fraenkel set theory (1908, 1922) *Mathematics* Named after German mathematicians Ernst Friedrich Ferdinand Zermelo (1871–1953) and Abraham A Fraenkel (1891–1965).

Often given the abbreviation ZF, this is now the most widely recognized axiomatization of SET THEORY. When Zermelo's axiom of choice is included, the resulting system is denoted ZFC. It may be argued that ZFC is sufficient for the formulation of every mathematical result. *See also* AXIOMATIC SET THEORY and NAIVE SET THEORY.

Encyclopedic Dictionary of Mathematics (MIT Press, 1987)

MB

Zermelo's theorem (1904) *Mathematics* Named after Ernst Zermelo (1871–1953) who proved the existence of the ordering.

The AXIOM OF CHOICE is equivalent to the statement that an order relation can be defined for any set so that there is a least element in every subset. *See also* ZORN'S LEMMA and WELL-ORDERING PRINCIPLE.

JB

zero-base budgeting (1969) *Accountancy* First used by the American economist Peter Pyhrr, this is a budgeting method which starts with an evaluation of projects on their merits to produce a total.

Zero-base budgeting contrasts with traditional budgeting, which starts with a fixed sum of money on the basis of which spending decisions are then made. In zero-base budgeting, managers justify their expenditure anew each year as if for the first time. The method, which is particularly suited to rapidly-changing industries, was introduced in the US by the Carter administration during the late 1970s.

P A Pyhrr, *Zero-base Budgeting* (New York, 1973); P C Sarant, *Zero-base Budgeting in the Public Sector: A Pragmatic Approach* (Reading, Me, 1978)

AV, PH

zero-one law (1933) *Mathematics* Discovered by Andrei Nikolaevich Kolmogorov (1903–87), this is the result of PROBABILITY THEORY whereby an event which does not depend on any finite subsequence of previous successive terms of a sequence of random variables has probability zero or one.

JB

zero-sum (20th century) *Politics* Concept in GAME THEORY.

There are circumstances in which one person can only win at the expense of another, or vice-versa. Such an assumption underlay MERCANTALISM and WAGES FUND THEORY.

Roger Scruton, *A Dictionary of Political Thought* (London, 1982)

RB

zeroth law of thermodynamics *Physics* So named by English physicist Ralph Howard Fowler (1889–1944), this law states that two closed systems in thermal equilibrium with a third are in thermal equilibrium with each other. This is not an independent law and can be deduced from the SECOND LAW OF THERMODYNAMICS.

J Thewlis, ed., *Encyclopaedic Dictionary of Physics* (New York, Oxford and London, 1962)

MS

zeta function *Mathematics* Also known as the Riemann zeta function, after German mathematician Georg Friedrich Bernhard Riemann (1826–66) who in 1860 demonstrated its close relationship to questions concerning the distribution of prime numbers.

This is the function defined by the series

$$\zeta(s) = \sum_{n=1}^{\infty} n^{-s}$$

or the infinite product

$$\zeta(s) = \prod_{p} (1 - p^{-s})^{-1}$$

where the product is taken over all primes p. The latter representation was given by Swiss mathematician Leonhard Euler (1707–83) in 1749. *See also* APERY'S THEOREM, RIEMANN HYPOTHESIS and PRIME NUMBER THEOREM.

A Baker, *A Concise Introduction to the Theory of Numbers* (Cambridge, 1984)

Zipf's law *Psychology* Formulated by the American philologist George Kingsley Zipf (1902–50).

Zipf proposed that there is an inverse relationship between the length of a word and the frequency with which it will occur; that is, short words will occur frequently and long words infrequently. The law does not explain why this should occur, nor does it hold with equal effect for different languages.

G K Zipf, *Human Behavior and the Principle of the Least Effort* (Cambridge, Mass., 1949)

NS

zonal theory *Climatology* This term refers to the way in which study of the average pattern of frequently-changing atmospheric pressure zones and wind circulation enables the zonal circulation to be measured in terms of the mean pressure differences between lines of latitude.

Strong differences (a high zonal index) generally evolve through four main stages: (1) strong sea-level westerly winds and long wave upper atmospheric patterns; (2) a reduction in the zonal index and shorter wavelength upper atmospheric circulation; (3) minimum zonal index and breakdown in the upper atmospheric circulation pattern; (4) a gradual increase in sea-level zonal index and the development of open upper atmospheric circulation patterns. Such an evolution is commonly blocked during periods of low zonal index and can also be strongly influenced by factors affecting the upper atmospheric jet stream circulation patterns.

J G Lockwood, *Causes of Climate*, (New York, 1979)

DT

zoo hypothesis *Astronomy* This is the untestable hypothesis that the Earth has been visited by extraterrestrial civilizations whose members have left no signs of their visits, presumably because they did not wish to disturb the development of the primitive life-forms they found. Such visitors might even be observing the activities on Earth much as humans observe animal behaviour in a zoo.

S Mitton, *The Cambridge Encyclopaedia of Astronomy* (London, 1973)

MS

Zorn's lemma *Mathematics* Named after German-born American mathematician Max August Zorn (1906–), this is the result in SET THEORY whereby an ordered set in which every chain has an upper bound must contain a maximal element.

A powerful tool in modern mathematics, Zorn's lemma can be used to prove, among other things, that every vector space has a basis. *See also* WELL-ORDERING PRINCIPLE.

P R Halmos, *Naive Set Theory* (Springer, 1974)

MB

Bibliography

The titles shown below provide a basic reading list for those interested in the subject area.

ANTHROPOLOGY

R Fox, *Kinship and Marriage* (Harmondsworth, 1967)
A Kuper, *Anthropology and Anthropologists: The Modern British School* (London, 1983)
A Kuper, *The Invention of Primitive Society: Transformations of an Illusion* (London, 1988)
E Leach, *Lévi-Strauss* (London, 1970)
C Lévi-Strauss, *The Elementary Stuctures of Kinship* (London, 1969 [French edn, 1949])
O Lewis, *The Children of Sanchez* (New York, 1961)
R H Lowie, *The History of Ethnological Theory* (New York, 1937)
H W Scheffler and F G Lounsbury, *A Study in Structural Semantics: The Siriono Kinship System* (Englewood Cliffs, NJ, 1971)
J Tennekes, *Anthropology, Relativism and Method* (Assen, Holland, 1971)
A van Gennep, *The Rites of Passage* (London, 1960 [French edn, 1908])
C Wissler, *Man and Culture* (New York, 1923)

BIOLOGY

D Abbott, *Biologists* (New York, 1983)
I Asimov, *The Genetic Code* (Orion, 1962)
W Beck, K Liem and G Simpson, *Life: An Introduction to Biology* (New York, 1991)
P Bowler, *Evolution: The History of an Idea* (Berkeley, Calif., 1984)
W F Bynum, E J Browne and R Porter, *Dictionary of the History of Science* (Princeton, NJ, 1981)
C Darwin, *On the Origin of Species by Means of Natural Selection, or the Preservation of Favoured Races in the Struggle for Life* (London, 1859)
P Davis, *Historical and Current Aspects of Plant Physiology* (Ithaca, New York, 1975)
R Dawkins, *The Selfish Gene* (Oxford, 1976)
R Dawkins, *The Blind Watchmaker: Why the Evidence of Evolution Reveals a Universe Without Design* (New York, 1986)
N Eldredge, *Life Pulse: Episodes from the Story of the Fossil Record* (New York, 1987)
J Fulton, *Selected Readings in the History of Physiology* (Springfield, Ill., 1966).
S J Gould, *Ontogeny and Phylogeny* (Cambridge, Mass., 1977)
F Hitching, *The Neck of the Giraffe: Where Darwin Went Wrong* (New Haven, Conn., 1987)
R King, *Handbook of Genetics* (New York, 1976)
H Kruger, *Other Healers, Other Cures: A Guide to Alternative Medicine* (1974)
J Lamarck, Zoological Philosophy (1809; Chicago, 1984)
R McGrew, *Encyclopedia of Medical History* (New York, 1985)
F Magill, *Magill's Survey of Science: Life Science Series* (Englewood Cliffs, NJ, 1991)
E Mayr, *The Growth of Biological Thought* (Cambridge, Mass., 1982)

Pauline M H Mazumdar, *Eugenics, Human Genetics and Human Failings* (London, 1991)
P B Medawar and J S Medawar, *Aristotle to Zoos: A Philosophical Dictionary of Biology* (Oxford, 1983)
R Milner, *The Encyclopedia of Evolution: Humanity's Search for Its Origins* (New York, 1990)
R Olby, *The Origins of Mendelism* (London and Chicago, 1985)
R Owen, *Camouflage and Mimicry* (Chicago, 1982)
H Plotkin, *The Role of Behavior in Evolution* (Cambridge, Mass., 1988)
J Rowley, *Chromosomes and Cancer: From Molecules to Man* (New York, 1983)
A Roy and B Chatterjee, *Molecular Basis of Aging* (Orlando, Fla, 1984)
F Salisbury, *Plant Physiology* (Belmont, Calif., 1992)
N Scott-Ram, *Transformed Cladistics, Taxonomy and Evolution* (Cambridge, 1990)
B F Skinner, *About Behaviourism* (New York, 1974)
G Stine, *The New Human Genetics* (Dubuque, Ia, 1989)
J Watson, *The Double Helix: A Personal Account of the Discovery of the Structure of DNA* (London, 1981)

CHEMISTRY
P W Atkins, *Physical Chemistry* (Oxford, 1990)
Compendium of Chemical Terminology: IUPAC Recommendations (Oxford, 1987)
D M Considine, *Van Nostrand's Scientific Encyclopedia* (New York, 1989)
I L Finar, *Organic Chemistry* (London, 1973)
M Freemantle, *Chemistry in Action* (Basingstoke, 1987)
C C Gillespie, *Dictionary of Scientific Biography* (New York, 1981)
J Grant, *Hackh's Chemical Dictionary* 5edn (New York, 1987)
S B Parker, *McGraw-Hill Encyclopedia of Science and Technology* (New York, 1987)
R C Weast, *CRC Handbook of Chemistry and Physics* (Boca Raton, 1992)

EARTH SCIENCES
M Allaby, *Air: The Nature of Atmosphere and the Climate* (Oxford, 1992)
M Begon, J L Harper and C R Townsend, *Ecology: Individuals, Populations and Communities* (London, 1987)
N Eldredge, *Unfinished Synthesis* (Oxford 1985)
N Eldredge and S J Gould, 'Punctuated Equilibrium: An Alternative to Phyletic Gradualism', *Models in Palaeobiology*, T T M Schopf, ed. (San Francisco, 1972)
J B S Haldane, *Causes of Evolution* (London, 1932)
A G Loewy and P Siekevitz, *Cell Structure and Function* (London, 1970)
E Mayr, 'Change in genetic environment and evolution', *Evolution as a Process*, J Huxley, A G Harvey and E B Ford, eds (London, 1954), 157–80
D M Moore, *Green Planet: The Story of Plant Life on Earth* (Cambridge 1982)
P Vincent, *The Biogeography of the British Isles* (London, 1990)

ECONOMICS
Y Balasko, *Foundations of the Theory of General Equilibrium* (New York, 1986)
M Blaug, *The Methodology of Economics, or How Economists Explain* (Cambridge and New York, 1980)
M Boskin, *Reagan and the Economy* (San Francisco, 1989)
J E Cairns, *Some Leading Principles of Political Economy* (London, 1874)
A D Chandler, *The Visible Hand: The Managerial Revolution in American Business* (Cambridge, Mass., 1977)
M Friedman, ed., *Studies in the Quantity of Money* (Chicago, 1956)
M Friedman, 'A Theory of the Consumption Function', *National Bureau of Economic Research* (Princeton, N.J., 1957)

H G Harcourt, *Some Cambridge Controversies in the Theory of Capital* (Cambridge, 1972)
R F Harrod, *The Trade Cycle* (Oxford, 1936)
J F Henry, *The Making of Neoclassical Economics* (London, 1990)
J Hicks, *The Crisis in Keynsian Economics* (Oxford, 1974)
K D Hoover, *The New Classical Macroeconomics* (Oxford, 1988)
Christopher Johnson, *The Economy under Mrs Thatcher* (London, 1991)
J M Keynes, *General Theory of Employment, Interest and Money* (New York, 1936)
T R Malthus, *An Essay on the Principle of Population* (London, 1798)
K Marx, *Capital*, (Harmondsworth, 1976–81)
R L Meek, *Studies in the Labour Theory of Value* (London, 1973)
P Minford, *The Supply Side Revolution in Britain* (Aldershot, 1991)
A M Okun, *The Political Economy of Prosperity* (Washington, 1970)
C N Parkinson, *Parkinson's Law or The Pursuit of Progress* (London, 1958)
E T Penrose, *The Theory of the Growth of the Firm* (Oxford, 1959)
L Peter and R Hull, *The Peter Principle: Why Things Always Go Wrong* (London, 1969)
J Rawls, *A Theory of Justice* (Oxford, 1971)
D Ricardo, *On the Principles of Political Economy and Taxation* (London, 1817)
P A Samuelson and W D Nordhaus, *Economics* (Mcgraw-Hill, 1992), 234
E F Schumacher, *Small is Beautiful: A Study of Economics as if People Mattered* (London, 1973)
Joseph A Schumpeter, *History of Economic Analysis* (New York, 1957)
A Smith, *An Inquiry into the Nature and Causes of the Wealth of Nations* (London, 1776)
J von Neumann and O Morgenstern, *The Theory of Games and Economic Behaviour* (Princeton, N.J., 1944)
E R Weintraub, *General Equilibrium Theory* (London, 1974)
K Wicksell, *Lectures on Political Economy* (London, 1911)

LINGUISTICS

J Aitchison, *The Articulate Mammal*, 3rd ed. (London, 1989)
R Anttila, *Introduction to Historical and Comparative Linguistics* (New York, 1972)
R E Asher, ed., *The Encyclopedia of Language and Linguistics* (Oxford, 1993)
R de Beaugrande and W Dressler, *Introduction to Text Linguistics* (London, 1981)
W Bright, ed., *International Encyclopedia of Linguistics* (New York, 1992)
G Brown and G Yule, *Discourse Analysis* (Cambridge, 1983)
V Cook, *Chomsky's Universal Grammar: An Introduction* (Oxford, 1988)
D Crystal, *A Dictionary of Linguistics and Phonetics* (Oxford, 1985)
D Crystal, *The Cambridge Encyclopedia of Language* (Cambridge, 1987)
W Downes, *Language and Society* (London, 1984; revised ed. forthcoming)
W N Francis, *Dialectology* (London, 1983)
M Garman, *Psycholinguistics* (Cambridge, 1990)
M A K Halliday, *Introduction to Function Grammar* (London, 1985)
W Labov, *Sociolinguistic Patterns* (Philadelphia, 1972)
E H Lenneberg, *Biological Foundations of Language* (New York, 1967)
S C Levinson, *Pragmatics* (Cambridge, 1983)
J Lyons, *Semantics* (Cambridge, 1977)
K Malmkjr, ed., *The Linguistics Encyclopedia* (London, 1991)
J Milroy and L Milroy, *Authority in Language* (London, 1985)
F J Newmeyer, *Linguistic Theory in America* (New York, 1980)
F J Newmeyer, ed., *Linguistics: The Cambridge Survey*, 4 vols (Cambridge, 1988)
S Pit Corder, *Introducing Applied Linguistics* (Harmondsworth, 1973)
A Radford, *Transformational Grammar* (Cambridge, 1988)
R H Robins, *A Short History of Linguistics* (London, 1967)

LITERARY THEORY AND STYLISTICS

M H Abrams, *The Mirror and the Lamp* (New York, 1953)
M H Abrams, *A Glossary of Literary Terms*, 6th ed. (New York, 1993)
H Adams, ed., *Critical Theory Since Plato* (New York, 1971)
H Adams and L Searle, eds., *Critical Theory Since 1965* (Tallahassee, 1986)
C Belsey, *Critical Practice* (London, 1980)
D Birch, *Language, Literature and Critical Practice* (London, 1989)
M Bradbury and J McFarlane, eds., *Modernism* (Harmondsworth, 1976)
T V F Brogan and A Preminger, eds., *Princeton Encyclopedia of Poetry and Poetics*, 3rd ed. (Princeton, 1993)
R Carter, ed., *Language and Literature: An Introductory Reader in Stylistics* (London, 1982)
M Coyle, P Garside, M Kelsall and J Peck, eds., *Encyclopedia of Literature and Criticism* (London, 1990)
J A Cuddon, *A Dictionary of Literary Terms and Literary Theory* (Oxford, 1991)
J Culler, *Structuralist Poetics* (London, 1975)
T Eagleton, *Literary Theory* (Oxford, 1983)
U Eco, *A Theory of Semiotics* (London, 1977)
V Erlich, *Russian Formalism: History, Doctrine* (The Hague, 1965)
A Fowler, *Kinds of Literature* (Oxford, 1982)
R Fowler, *Linguistic Criticism* (Oxford, 1986)
R Fowler, 'Literature', *Encyclopedia of Literature and Criticism*, M Coyle *et al.*, eds. (London, 1990), 3–26
J V Harari, ed., *Textual Strategies* (London, 1979)
T Hawkes, *Structuralism and Semiotics* (London, 1977)
G N Leech, *A Linguistic Guide to English Poetry* (London, 1969)
D Lodge, *Twentieth Century Literary Criticism: A Reader* (London, 1972)
D Lodge, *Modes of Modern Writing* (London, 1977)
D Lodge, *Modern Criticism and Theory: A Reader* (London, 1988)
W Nash, *Rhetoric*, (Oxford, 1989)
A Ortony, ed., *Metaphor and Thought* (Cambridge, 1989)
S Rimmon-Kenan, *Narrative Fiction: Contemporary Poetics* (London, 1983)
S J Schmidt, *Foundations of the Empirical Study of Literature* (Hamburg, 1982)
M J Toolan, *Narrative: A Critical Linguistic Introduction* (London, 1988)
K Wales, *A Dictionary of Stylistics* (London, 1989)
R Wellek, *Concepts of Criticism*, S G Nicholls, ed. (New Haven, 1962)
R Wellek and A Warren, *Theory of Literature* (London, 1963 [1949])
R Williams, *Keywords* (Glasgow, 1976)
W K Wimsatt (jnr) and C Brooks, *Literary Criticism: A Short History* (New York and London, 1957)

MATHEMATICS

A V Aho, J E Hopcroft and J D Ullman, *Data Structures and Algorithms* (Reading, Mass., 1983)
H Anton, *Calculus with Analytic Geometry* (New York, 1980)
T M Apostol, *Introduction to Analytic Number Theory* (Springer, 1976)
T M Apostol, *Mathematical Analysis* (Addison-Wesley, 1974)
A Baker, *A Concise Introduction to the Theory of Numbers* (Cambridge, 1984)
A Baker, *Transcendental Number Theory* (Cambridge, 1979)
W E Boyle and R C Di Prima, *Elementary Differential Equations and Boundary Value Problems* (New York, 1977)
E W Cheney *Introduction to Approximation Theory* (New York, 1982)
H S M Coxeter, *Introduction to Geometry* (New York, 1961)
R L Devaney, *An Introduction to Chaotic Dynamical Systems* (New York, 1989)
Encyclopedic Dictionary of Mathematics (MIT Press, 1987)

W Feller, *An Introduction to Probability Theory and Its Applications* (New York, 1966)
W Gellert, S Gottwald, M Hellwich, H Kästner and H Küstner, eds, *The VNR Concise Encyclopedia of Mathemat- ics* (New York, 1975)
M B Green, J H Schwarz and E Witten, *Superstring Theory* (Cambridge, 1987)
P R Halmos, *Naive Set Theory* (Springer 1974)
G H Hardy and E M Wright, *An Introduction to the Theory of Numbers*, 5th edn (Oxford, 1979)
P Hartman, *Ordinary Differential Equations* (New York, 1964)
M Hazewinkel, ed., *Encyclopaedia of Mathematics* (Dordecht, 1988)
I N Herstein, *Topics in Algebra* (Wiley, 1975)
M Kline, 'Non-Euclidean Geometry', *Mathematical Thought from Ancient to Modern Times* (New York, 1972)
K Knopp, *Theory and Application of Infinite Series* (New York, 1990)
T W Körner, *Fourier Analysis* (Cambridge, 1988)
J E Marsden, *Basic Complex Analysis* (Freeman, 1973)
J R Munkres, *Topology: A First Course* (Prentice-Hall, 1975)
M Reid, *Undergraduate Algebraic Geometry* (Cambridge, 1990)
W Rudin, *Principles of Mathematical Analysis* (McGraw-Hill, 1976)
D Ruelle, *Chance and Chaos* (Princeton, N.J., 1991)
J L von Neumann and O Morgenstern, *Theory of Games and Economic Behaviour* (Princeton, 1947)

MUSIC

C G Allaire, *The Theory of Hexachords, Solmization and the Modal System* (Rome, 1972)
R Bullivant, *Fugue* (London, 1971)
J Curwen, *Tonic Sol-fa* (London, 1878)
A Forte, *The Structure of Atonal Music* (New Haven and London, 1973)
A Forte and S Gilbert, *Introduction to Schenkerian Analysis* (New York, 1982)
K Meyer-Baer, *Music of the Spheres and the Dance of Death* (Princeton, N.J., 1970)
E Narmour, *Beyond Schenkerism* (Chicago, 1977)
M Nyman, *Experimental Music: Cage and Beyond* (London, 1974)
C Palmer, *Impressionism in Music* (London, 1973)
G Perle, *Serial Composition and Atonality* (Berkeley, Calif., 1962)
G Perle, *Twelve-tone Tonality* (Berkeley, Calif., 1978)
R Réti, *The Thematic Process in Music* (New York, 1951)
R Reti, *Tonality, Atonality, Pantonality: A Study of Some Trends in Twentieth Century Music* (London, 1958)
F W Robinson and S G Nichols, eds, *The Meaning of Mannerism* (Hanover, N.H., 1972)
E Salzman, *Twentieth-Century Music: An Introduction* (Englewood Cliffs, N.J., 1967), 43–67
A Whittall, *Romantic Music* (London, 1987)
J Willett, *Expressionism* (London, 1970)

PHILOSOPHY

Aquinas, St Thomas, *Summa Theologiae*
Aristotle, *De Anima*
Aristotle, *De Generatione et Corruptione*
Aristotle, *De Interpretatione*
Aristotle, *Metaphysics*
Aristotle, *Nicomachean Ethics*
Aristotle, *Physics*
J L Austin, *Collected Papers* (1961)
J L Austin, *How to Do Things with Words* (1962)
H L Bergson, *L'évolution creatrice* (1907), trans. as *Creative Evolution* (1911)

G Berkeley, *A Treatise concerning the Principles of Human Knowledge* (1710)
R Carnap, *Logical Foundations of Probability* (1950)
R Descartes, *Meditations on the First Philosophy* (1641)
G Frege, 'Negation', *Translations from the Philosophical Writings of Gottlob Frege*, P T Geach and M Black, eds (1952)
G Hegel, *The Philosophy of Right* (1821)
M Heidegger, *Sein und Zeit* (1927)
T Hobbes, *Leviathan* (1651)
David Hume, *A Treatise of Human Nature* (1739–40)
I Kant, *Critique of Pure Reason*, trans. by N K Smith (1953, German originals 1781 and 1787); especially A366–80 (pp. 344–52 in Smith's translation)
J Locke, *An Essay Concerning Human Understanding* (1690)
J S Mill, *Utilitarianism* (1861), ch. 2 (reprinted in J Plamenatz, *The English Utilitarians* (1949), 137)
David Miller *et al*, eds *The Blackwell Encyclopaedia of Political Thought* (Oxford, 1987)
G E Moore, *Principia Ethica* (1903)
I Newton, *Philosophiae Naturalis Principia Mathematica* (1687)
Plato, *Dialogues*
K R Popper, *The Logic of Scientific Discovery* (1959)
W V O Quine, *Word and Object* (1960)
B Russell, *An Inquiry into Meaning and Truth* (1940)
G Ryle, *The Concept of Mind* (1949)
B de Spinoza, *Ethics* (1677)
A N Whitehead and B A W Russell, *Principia Mathematica*, 1 (1910)
L Wittgenstein, *Philosophical Investigations* (1952), especially part 1, §§1–43

PHYSICS

Robert M Besançon, ed., *The Encyclopedia of Physics* (New York, 1985)
Douglas M Considine, ed., *Van Nostrand's Scientific Encyclopedia* (New York, 1989)
Peter Coveney and Roger Highfield, *The Arrow of Time* (London, 1991)
John Daintith, ed., *Dictionary of Physics* (New York, 1982)
Paul Davies, ed., *The New Physics* (Cambridge, 1989)
Charles C Gillispie, ed., *Dictionary of Scientific Biography* (New York, 1981)
Valerie Illingworth, ed., *Macmillan Dictionary of Astronomy* (London, 1985)
Valerie Illingworth, ed., *Penguin Dictionary of Physics* (London, 1990)
S Mitton, ed., *The Cambridge Encyclopaedia of Astronomy* (London, 1973)
Jayant V Narlikar, *Introduction to Cosmology* (Cambridge, 1993)
Sybil B Parker, ed., *McGraw-Hill Encyclopedia of Science and Technology* (New York, 1987)
J Thewlis, ed., *Encyclopaedic Dictionary of Physics* (New York, Oxford and London, 1962)
R C Weast, ed., *CRC Handbook of Chemistry and Physics* (Boa Raton, 1992)

POLITICS

Robert Abrams, *Foundations of Political Analysis* (New York, 1979)
Geoff Andrews, ed., *Citizenship* (London, 1991)
Rodney Barker, *Political Ideas in Modern Britain* (London, 1978)
Rodney Barker, *Political Legitimacy and the State* (Oxford, 1990)
Tom Bottomore, *A Dictionary of Marxist Thought*, 2nd edn (Oxford, 1991)
Valerie Bryson, *Feminist Political Theory: an Introduction* (London, 1992)
Alan Bullock, Oliver Stallybrass and Stephen Trombley, eds, *The Fontana Dictionary of Modern Thought*, 2nd edn (London, 1988)
Alan Bullock and R B Woodings, eds, *The Fontana Dictionary of Modern Thinkers*, (London, 1983)
Diana Coole, *Women in Political Theory*, 2nd edn (Brighton, 1992)

Patrick Dunleavy, *Democracy, Bureaucracy and Public Choice* (London, 1991)
Roger Eatwell and Anthony Wright, eds, *Modern Political Ideologies* (London,1993)
Graham Evans and Jeffrey Newnham, *The Dictionary of World Politics* (Hemel Hempstead, 1990)
David Held, *Models of Democracy* (Oxford, 1987)
Christopher Hood, *Administrative Analysis* (Brighton, 1986)
Maggie Humm, *The Dictionary of Feminist Theory* (London, 1989)
Maggie Humm, *Feminism: a Reader* (Hemel Hempstead, 1992)
Desmond King, *The New Right: Politics, Markets and Citizenship* (London, 1987)
L Kolakowski, *Main Currents of Marxism*, 3 vols (Oxford, 1978)
Ruth Levitas, ed, *The Ideology of the New Right* (Cambridge, 1986)
Terry Lovell, ed, *British Feminist Thought: a Reader* (Oxford, 1990)
David McLellan, *Marxism After Marx* (London, 1980)
David Miller *et al.*, eds, *The Blackwell Encyclopaedia of Political Thought* (Oxford, 1987)
Juliet Mitchell and Ann Oakley, eds, *What Is Feminism?* (Oxford, 1986)
Carol Pateman, *The Disorder of Women* (Cambridge, 1989)
Anne Phillips, *Engendering Democracy* (Cambridge, 1991)
Geoffrey Roberts and Alistair Edwards, *A New Dictionary of Political Analysis* (London, 1991)
Helen Roberts, ed., *Doing Feminist Research*, 2nd edn (London, 1990)
David Robertson, *The Penguin Dictionary of Politics* (London, 1986)
Roger Scruton, *A Dictionary of Political Thought* (London, 1982)
Quentin Skinner, ed., *The Return of Grand Theory in the Human Sciences* (Cambridge, 1985)

PSYCHOLOGY

A Anastasi, *Fields of Applied Psychology* (New York, 1979)
A D Baddeley, *The Psychology of Human Memory* (New York, 1976)
A Bandura, *Social Learning Theory* (Englewood Cliffs, N.J., 1977)
R A Baron and D Byrne, *Social Psychology: Understanding Human Interaction* (Boston, 1984)
M Billig, *Social Psychology and Intergroup Relations* (London, 1976)
G H Bower and E R Hilgard, *Theories of Learning* (Englewood Cliffs, N.J., 1981)
J S Bruner, *Beyond the Information Given* (New York, 1973)
N R Carlson, *Physiology of Behavior* (London, 1986)
R B Cattell, *Personality and Motivation Structure and Measurement* (New York, 1957)
F Denmark, 'Styles of Leadership', *Psychology of Women Quarterly*, vol. II (1977), 99–113.
H J Eysenck, *Fact and Fiction in Psychology* (Harmondsworth, 1974)
S Freud, 'Psychopathology of Everyday Life' (vol. XVI) and 'Introductory Lectures' (vol. XV), *Standard Edition of the Complete Works of Sigmund Freud* (London and New York, 1901 and 1916)
E Fromm, *To Have or to Be?* (New York, 1979)
E Goffman, *The Presentation of Self in Everyday Life* (New York, 1959)
C G Jung, *Analytic Psychology: Its Theory and Practice* (New York, 1968)
L Kaufman, *Perception* (New York, 1979)
B Kolb and I Q Whishaw, *Fundamentals of Human Neuropsychology*, 3rd edn. (San Francisco, 1990)
D A Norman, *Memory and Attention: An Introduction to Human Information Processsing* (New York, 1977)
K N Walsh, *Neuropsychology* (London, 1978)

Index of People

The names listed below are of those people who are mentioned in entries throughout the book. The entries in which they appear are listed below their names.

Boole, George (1815–64)
English mathematician
logic and foundations

Booth, Wayne C (1921–)
American literary critic
Chicago critics
implied author

Bopp, Franz (1791–1867)
German philologist
Indo-European

Borel, Félix Edouard Émile (1871–1956)
French mathematician and politician
alternation theorem
Borel-Cantelli lemma
strong law of large numbers
weak law of large numbers

Borelli, Giovanni Alfonso (1608–1679)
Italian mathematician and physician
iatrophysics

Borges, Jorge Luis (1899–1986)
Argentinian writer
fantastic
magic realism

Born, Max (1882–1970)
German physicist
Born-Haber cycle
Born's theory of melting

Borsuk, Karol
Mathematician
Borsuk-Ulam theorem

Bose, Satyendra Nath (1894–1974)
Indian physicist
Bose-Einstein statistical and distributive law

Botschiver, Theo
Artist
eventstructure

Bouguer, Pierre (1698–1758)
French physicist
Lambert's law of absorption

Boulez, Pierre (1925–)
French conductor and composer
total serialism

Boveri, Theodor Heinrich (1862–1915)
German biologist
chromosome theory of cancer
somatic mutation hypothesis or theory

Bowditch, Henry Pickering (1840–1911)
American physiologist
all-or-none principle

Bowen, Norman Levi (1887–1956)
Canadian/American geologist
Bowen's reaction principle

Bower, Gordon Howard (1932–)
Psychologist
associative learning
encoding specificity principle
encoding variability principle

Bowman, Sir William (1816–92)
English physician and surgeon
Bowman-Heidenhain hypothesis

Boyle, The Hon Robert (1627–91)
Irish physicist and chemist
Boyle's law

Bradley, Francis Herbert (1846–1924)
Welsh philosopher
absolute
associationism
identity theory of truth
objective idealism
psychologism

Bragg, William Lawrence (1890–1971)
British physicist
Bragg's law

Brahe, Tycho (1545–1601)
Danish astronomer
Kepler's laws of planetary motion

Braid, James
Scottish surgeon and hypnotist
animal magnetism
Braid's theory of hypnosis

Braine, M D S (1795–1860)
American psycholinguist
pivot grammar

Brans, Carl Henry (1935–)
American physicist
Brans-Dicke theory

Braque, Georges (1882–1963)
French painter
Cubism
Fauvism

Bray, C W
American psychologist
volley theory of hearing

Bréal, Michel (1832–1915)
French philologist
polysemy

Brecht, Bertolt Eugen Friedrich (1898–1956)
German playwright and poet
alienation effect
epic theatre

Bredt, Konrad Julius (1855–1937)
German organic chemist
Bredt's rule

Brehm, Jack Williams
American psychologist
reactance theory

Breit, Gregory (1899–1981)
Russian/American physicist
Breit-Wigner equation
compound nucleus theory of Bohr and Breit and Wigner

Brentano, Franz (1838–1917)
German psychologist and philosopher
phenomenology

Bresnan, J W
American linguist
lexical-functional grammar

Breton, André (1896–1966)
French writer and critic
Surrealism

Brewster, Sir David (1781–1868)
Scottish physicist
Brewster's law

Brianchon, Charles-Julien (1785–1864)
French mathematican
Brianchon's theorem

Bridgman, Percy Williams (1882–1961)
American physicist
operationalism

Britten, Roy John (1919–)
American geneticist
Britten-Davidson model

Broadbent, Donald Eric (1926–)
English psychologist
filter theory

Broca, Paul Pierre (1824–80)
French surgeon and anthropologist
Broca's area

Brønsted, Johannes Nicolaus (1879–1947)
Danish physical chemist
Brønsted-Lowry theory
Brønsted relation

Brooks, Cleanth (1906–)
American literary critic
heresy of paraphrase
new criticism
paradox

Brouwer, Luitzen Egbertus Jan (1881–1966)
Dutch mathematician
Brouwer's theorem
fixed point theorems
intuitionism
invariance of domain theorem

Brower, Lincoln Pierson and Jane van Zandt
American ecologists and zoologists
automimicry

Brown, 'Capability' (Lancelot Brown) (1716–83)
English landscape gardener
picturesque

Brown, John (1735–88)
Scottish physician
Brownian system

Brown, Penelope (1944–)
Linguist
politeness

Brown, R
American linguist
power and solidarity
telegraphic speech

Brown, Thomas (1778–1820)
Scottish philosopher
association, laws of
contrast, law of
frequency, law of

chiaroscuro
naturalism

Cardano, Girolamo (1501–76)
Italian scientist and philosopher
Cardano's formula
propensity theory of probability

Carey, (Samuel) Warren (1914–)
Australian geologist
expanding Earth hypothesis

Carleson, Lennart Axel Edvard (1928–)
Danish mathematican
Carleson's theorem

Carlyle, Thomas (1795–1881)
Scottish historian
great man theory

Carnan, R (1891–1970)
Psychologist
identity theory

Carnap, Rudolf (1891–1970)
German/American philosopher
physicalism
Vienna Circle

Carneades, (*c.*214–*c.*129 BC)
Greek philosopher
scepticism

Carnot, Nicholas Léonard Sadi (1796–1832)
French physicist
Carnot cycle
Carnot's theorem

Carra, Carlo (1881–1966)
Italian painter
arte mat
Futurism

Carracci, Annibale (1540–1609)
Italian artist
Caravaggism

Carrington, Richard Christopher (1826–75)
English astronomer
Spoerer's law

Carroll, Lewis *see* Dodgson, Charles Lutwidge

Carter, R A
British stylistician
re-registration
sovereignty

Cary, John (*died c.*1720)
English merchant and writer
machinery question

Casorati, Felice (1835–90)
Italian mathematican
Casorati-Weierstrass theorem

Cassel, (Karl) Gustav (1866–1945)
Swedish economist
purchasing power parity

Castelvetro, Lodovico (1505–71)
Italian literary theorist
unity, the Unities

Castil-Blaze (1784–1857)
French music critic
tonality

Castle, William Ernest (1867–1962)
American biologist
Hardy-Weinberg equilibrium, law or principle

cat, Schrödinger's, (short-lived/eternal)
Schrödinger's cat

Catalan, Eugène Charles (1814–94)
Belgian mathematican
Catalan's constant

Cattell, Raymond Bernard (1905–)
English psychologist
cultural bias hypotheses
dynamic effect law
Fullerton-Cattell law
testing effect

Cauchy, Augustin Louis, Baron (1789–1857)
French mathematician
Cauchy condensation test
Cauchy criterion
Cauchy integral theorem
Cauchy lemma
Cauchy mean value theorem
Cauchy-Hadamard theorem
Cauchy-Kowalewska theorem
Cauchy-Schwartz inequality
Cauchy's stress theorem
comparison test
Padé approximates
ratio test
residue theorem of Cauchy
root test

Caudwell, Christopher (1907–37)
Critic
Marxist criticism

Cavalieri, Francesco Bonaventura (1598–1647)
Italian mathematician
Cavalieri's principle

Caxton, William (*c.*1422–*c.*1491)
First English printer
fabulation

Cayley, Arthur (1821–95)
English mathematician
Cayley-Hamilton theorem
Cayley representation theorem

Cazamian, Louis
French literary historian
social novel

Čech, Eduard (1893–1960)
Mathematician
homology theory
Stone-Čech compactification

Celant, Germano
Italian art critic
arte povera

Celsius, Anders (1701–44)
Swedish astronomer
Celsius temperature

Ceroli, Mario
Italian artist
arte povera

Cervantes Saavedra, Miguel de (1547–1616)
Spanish novelist
fabulation

Cesaro, Ernesto (1859–1906)
Italian mathematician
Cesaro summation

Ceva, Giovanni (1647–1734)
Italian mathematician
Ceva's theorem

Cézanne, Paul (1839–1906)
French painter
Cubism
post-Impressionism

Chailakhyan, Mikhail
Russian botanist
florigen

Chamberlin, Edward (1899–1967)
American economist
monopolistic competition

Chamberlin, Thomas Chrowder (1843–1928)
American geologist
catastrophic event theories of the solar system
planetesimal theory of planetary formation
tidal theories of planetary formation

Chandrasekhar, Subrahmanyan (1910–)
Indian/American astrophysicist
Chandrasekhar's limit

Chapman, Sydney (1888–1970)
English mathematician and geophysicist
Enskog and Chapman theory of thermal diffusion

Chargaff, Erwin (1905–)
Czech/American biochemist
Chargaff's rules

Charles x, (1757–1836)
King of France
king's touch

Charles, Jacques Alexandre César (1746–1823)
French physicist
Charles's law

Charpentier, P M A (1852–1916)
French psychologist
Charpentier's law

Chaucer, Geoffrey (*c.*1340–1400)
English poet
courtly love

Chebyshev, Pafnutii Lvovich (1821–94)
Russian mathematician
prime number theorem

Chen, Jing-Run (1933–)
Chinese mathematician
twin primes

Chevreul, Michel Eugène (1786–1889)
French chemist
impressionism

Chicherin, A N (1889–1960)
Russian writer
constructivism

Child, Clement Dexter
American scientist
Child's law

Chirico, Giorgio de (1888–1978)
Italian artist
arte mat

Cholodny, N
American biologist
Cholodny-Went model

Chomsky, Avram Noam (1928–)
American linguist and political activist
adequacy, levels of
ambiguity
behaviourism in linguistics
case grammar
communicative competence
constituent, constituent structure
core grammar
corpus
correlational sociolinguistics
discovery procedure
embedding
evaluation measure or procedure
extended standard theory
generative phonology
generative grammar
government and binding
grammaticality
innate ideas
innateness hypothesis
intuition
kernel sentence
linguistic competence
mentalism in linguistics
modularity
phrase structure grammar
rationalism in linguistics
recursion
rewriting rule
standard theory
structure-dependency
style, theory of
subjacency
theta theory
trace theory
transformational cycle
transformational-generative grammar
transformation
X-bar theory

Chrétien de Troyes, (*died c.*1183)
French poet
courtly love

Christaller, Walter (1894–1975)
German geographer
central place theory

Chrysippus, (*c.*280–*c.*206 BC)
Greek philosopher
Stoicism

Cicero, Marcus Tullio (106–43BC)
Roman orator and statesman
Stoicism

Clark, John Bates (1847–1938)
American economist
marginal productivity theory of distribution

Clark, John M (1884–1963)
American economist
acceleration principle
neo-classical theory

Clausius, Rudolf Julius Emanuel (1822–88)
German physicist
Clausius' theorem and inequality
Clausius' virial theorem
entropy, law of increase of
kinetic theory of gases

Clements, Frederic Edward (1874–1945)
American ecologist
climax theory
organismal model

Clerk-Maxwell, James (1831–79)
Scottish physicist
kinetic theory of gases
wave theory of light

Clower, Robert Wayne (1926–)
American economist
dual decision hypothesis

Coase, Ronald Harry (1910–)
English/American economist
Coase theorem

Coates, Robert Myron (1897–1973)
American art critic
abstract expressionism

Coates, J
Linguist
feminist linguistics

Cobb, Charles
American mathematician
Cobb-Douglas production function

Cobden-Sanderson, Thomas James (1840–1922)
English printer and bookbinder
Arts and Crafts Movement

Cole, George Douglas Howard (1889–1959)
English economist and historian
encroaching control

Coleridge, Samuel Taylor (1772–1834)
English poet
clerisy
fancy
harmonization of impulses
imagination
organic form
Romanticism
willing suspension of disbelief

Collatz, Lothar (1910–91)
Mathematician
Collatz sequence

Colley, J G
American psychologist
suffix effect

Collingwood, Robin George (1889–1943)
English philosopher and historian
empathy, historical

Comfort, Alex (1920–)
British author
epiphenomenalist theory of ageing

Compton, Arthur Holly (1892–1962)
American physicist
Compton's rule

Comte, Auguste (1798–1857)
French philosopher
altruism
empiriocriticism
positivism

Conconi, L
Italian painter
scapigliatura

Condon, Edward Uhler (1902–74)
American physicist
Franck-Condon principle
Gamow-Gurney-Condon theory of alpha particle decay

Condorcet, Marie Jean Antoine Nicolas Caritat, Marquis de (1743–94)
French mathematician
Condorcet's principle

Confucius (prop. K'ung Fu-tzu, the Master K'ung) (*c.*551–479 BC)
Chinese philosopher
Confucianism
golden rule

Conybeare, William Daniel (1785–1873)
English clergyman
diluvianism

Coolidge, Calvin (1872–1933)
30th president of the USA
Coolidge effect

Cooper, Anthony Ashley, 3rd Earl of Shaftesbury (1671–1713)
English philosopher and politician
moral sense theories

Cooper, Leon Neil (1930–)
American physicist
BCS theory

Cope, Edward Drinker (1840–97)
American naturalist and palaeontologist
Cope's law or rule

Copernicus, Nicolaus (1473–1543)
Polish astronomer
heliocentric theory of the solar system

Copper, A
Psychologist
indoleamine hypothesis of depression

Cottingham, John
Philosopher
trialism

de Coriolis, Gaspard Gustave (1792–1843)
French physicist
Coriolis's theorem

De Finetti, Bruno (1906–85)
Philosopher and logician
subjectivist theories of probability

de Kooning, Willem (1904–)
Dutch/American artist
action painting

De Morgan, Augustus (1806–71)
English mathematician
De Morgan's laws

De Quincey, Thomas (1785–1859)
English critic and essayist
falling rate of profit

de Saussure, Horace Benédict (1740–99)
Swiss physicist and geologist
glacial theory

de Secondat Montesquieu, Charles-Louis (1689–1755)
French political theorist
separation of powers

de Sitter, Willem (1872–1934)
Dutch astronomer
Einstein-de Sitter universe

de Tocqueville, Alexis (1805–59)
French political theorist
tyranny of the majority

de Tracy, Destutt (1754–1836)
French philosopher
ideology

de Vries, Hugo (Marie) (1848–1935)
Dutch botanist and geneticist
mutation theory

de Wilde, E
Art critic
fundamental art (painting)

Debreu, Gerard (1921–)
French/American econmist
Arrow-Debreu model
demand theory

Debussy, Claude (1862–1918)
French painter
impressionism
Symbolism

Debye, Peter Joseph Wilhelm (1884–1966)
Dutch/American physicist
Debye-Hückel theory
Debye's T^3 law
Debye's theory of heat capacity

Dedekind, Julius Wilhelm Richard (1831–1916)
German mathematician
Dedekind cut
Peano's axioms

Delaunay, Robert (1885–1941)
French painter
Blaue Reiter
Orphism
simultanéisme

Delbouf, J (1831–96)
French psychologist
progression, law of

DeLorean, John (1925–)
American entrepreneur
DeLorean theory

Dement, W
American psychologist
rebound effect

Democritus (*c.*460–370 BC)
Greek philosopher
atomism
Epicureanism

Denckla, W D
American biologist
ageing hormone

Denmark, Florence (1932–)
American psychologist
great woman theory

Derain, André (1880–1954)
French artist
expressionism
Fauvism

Derrida, Jacques (1930–)
French philosopher
deconstruction
différance
grammatology
logocentrism
phonocentrism
poststructuralism

Deryagin, B V (*fl.*1940s)
Chemist
DLVO theory

Desargues, Girard (1591–1661)
French mathematician
Desargues's theorem

Descartes, René (1596–1650)
French philosopher
affections, doctrine of the
algebraic geometry
Continental rationalists
contracting Earth hypothesis
Descartes's rule of signs
Eleaticism
foundationalism
immaterialism
interactionism
materialism
occasionalism
rationalism in linguistics
trialism

Deutsch, J Anthony (1927–)
Biologist
Deutsch's feedback model

Dewey, John (1859–1952)
American philosopher
context effect
holism
instrumentalism
pragmatism

Dicey, Albert Venn (1835–1922)
English jurist
rule of law

Dicke, Robert Henry (1916–)
American physicist
Brans-Dicke theory

Dickens, Charles (1812–70)
English author
Bildungsroman
social novel

Dieterici, Conrad
Chemist
Dieterici's rule

Dietzgen, Josef (1820–88)
German philosopher
dialectical materialism

Dillon, James (1950–)
English composer
new complexity

Dilthey, Wilhelm (1833–1911)
German philosopher
hermeneutics
historicism

Dini, Ulisse (1845–1918)
Italian geometer
Dini's theorem

Diogenes of Sinope (*c.*410–*c.*320 BC)
Greek philosopher
cynicism

Diophantus of Alexandria (*fl.*250 AD)
Greek mathematician
Diophantine analysis
Diophantine equations
Lagrange's theorem

Dirac, Paul Adrien Maurice (1902–84)
English mathematical physicist
antimatter
Dirac's quantum theory
Fermi-Dirac statistical and distributive law
hole theory of electrons
vacuum (quantum theory of fields)

Dirichlet, Peter Gustav Lejeune (1805–59)
German mathematician
Dirichlet's problem
Dirichlet's theorem
Dirichlet's theorem on arithmetical progressions
Dirichlet's theorem on Diophantine approximation
pigeon-hole principle

Dix, Otto (1891–1969)
German painter
Neue Sachlichkeit

Dixon, Henry H (1869–1953)
Irish plant physiologist
cohesion theory

Djilas, Milovan (1911–)
Yugoslavian politician
new class

Döbereiner, Johann Wolfgang (1780–1849)
German chemist
triads, Döbereiner's law of

Fautrier, Jean (1898–1964)
French artist
art informel

Fechner, Gustav Theodor (1801–87)
German physicist, psychologist and philosopher
comparative judgment, law of
Fechner's law
Fechner's paradox
mathematical law
monism
parallel law

Feigenbaum, Mitchell Jay (s) (1944–)
American physicist
Feigenbaum's constant
Feigenbaum's period-doubling cascade

Feigl, Herbert (1902–)
Austrian/American philosopher of science
Vienna Circle

Feit, Walter (1930–)
Mathematician
Feit-Thompson theorem

Fejer, Lipót (1880–1959)
Hungarian mathematician
Fejer's theorem

Fénéon, Félix (1861–1944)
French art critic
neo-Impressionism

Fenichle, O
Psychologist
multiple factor theory

Ferenczi, Sándor (1873–1933)
Hungarian psychologist
relaxation principle

Ferguson, Charles Albert (1921–)
American linguist
diglossia

Fermat, Pierre de (1601–65)
French mathematician
algebraic geometry
derivative tests
Fermat primes
Fermat's last theorem
Fermat's little theorem
Fermat's principle of least time
Fermat's (Steiner's) problem
integral test
integration by parts
Lagrange's theorem
Pell's equation
probability theory
product rule
quotient rule

Fermi, Enrico (1901–54)
Italian/American nuclear physicist
Fermi-Dirac statistical and distributive law
Fermi's ageing theory
Fermi's paradox
Fermi's theory of beta decay

Ferneyhough, Brian (1943–)
English composer
new complexity

Ferry, E S (1868–1956)
American psychologist
Ferry-Porter law

Festinger, Leon (1919–)
American psychologist
cognitive consistency theory
dissonance theory
social comparison theory

Feynman, Richard (1918–88)
American physicist
Hellmann-Feynman theorem

Fibonacci, Leonardo (*c.*1170–1250)
Italian mathematician
Fibonacci numbers
recursion principle

Ficino, Marsilio (1433–99)
Italian philosopher
humanism

Fick, Adolf Eugen (1829–1901)
German physicist
diffusion theory

Fiedler, Fred Edward (1922–)
Austrian psychologist
contingency theory of leadership
Fiedler's contingency model of leadership

Fielding, Henry (1707–54)
English novelist
Bildungsroman

Fillmore, Charles J (1929–)
American linguist
case grammar

Findeism, Walter
German meteorologist
Bergeron's theory

Firbas, J
Linguist
functional sentence perspective
theme and rheme

Firth, John Rupert (1890–1960)
English linguist
collocation
context of situation
scale-and-category grammar

Fischer, Emil Hermann (1852–1919)
German chemist
lock and key model

Fischer, Ernst (1875–1959)
Mathematician
large numbers, law of

Fisher, Irving (1867-1947)
American economist
adaptive expectations
life-cycle hypothesis
neo-classical theory
quantity theory of money
term structure of interest rates

Fisher, Osmond
British scientist
fission hypothesis

Fisher, Sir Ronald Aylmer (1890–1962)
English statistician and geneticist
evolutionary determinism
Fisher-Behrens problem
Fisher's fundamental theorem
overspecialization
Riesz-Fischer theorem

Fitts, Paul Morris (1912–65)
American psychologist
Fitts's law

Fitzgerald, George Francis (1851–1901)
Irish physicist
Fitzgerald-Lorentz contraction

Fizeau, Armand Hippolyte Louis (1819–96)
French physicist
Doppler effect

Flaubert, Gustave (1821–80)
French novelist
realism

Fleming, Sir John Ambrose (1849–1945)
English physicist and electrical engineer
Fleming's rules

Floquet, G
French mathematician
Floquet theory

Flourens, Pierre Jean Marie (1794–1867)
French physiologist
localization theory

Fock, Vladimir Alexandrovich (1898–)
Russian theoretical physicist
self-consistent field theory

Fodor, Jerry Alan
American psycholinguist
componential analysis
interpretive semantics
language of thought
sausage machine

Fontana, Lucio (1899–1968)
Italian artist
spatialism

Forge, L de la (1632–66)
Philosopher
occasionalism

Forster, Edward Morgan (1879–1970)
English author
flat and round characters
story and plot

Forster, K I
Psycholinguist
search model

Forte, Allen (1926–)
American music theorist
pitch-class set theory

Foucault, Michel (1926–84)
French philosopher
author, death of the
cultural materialism
poststructuralism

Fourier, Jean Baptiste Joseph, Baron de (1768–1830)
French mathematician
Fourier principle
Fourier series
Fourier transform
Fourier heat conduction equation
harmonic analysis

Fowler, Ralph Howard (1889–1944)
English physicist
zeroth law of thermodynamics

Fowler, Richard (1938–)
English linguist
linguistic criticism
literature as discourse
mind-style
underlexicalization

Fracastoro, Girolamo (1483–1553)
Italian scholar and physician
germ theory of disease

Fraenkel, Abraham A (1891–1965)
German mathematician
Zermelo-Fraenkel set theory

Franck, James (1882–1964)
German/American physicist
Franck-Condon principle

Franz, Rudolph (*fl.* 1850s)
German physicist
Wiedemann-Franz-Lorenz law

Fraser-Darling, Sir Frank (1903–1979)
British ecologist and geneticist
Fraser-Darling effect

Frasier, L
American psycholinguist
sausage machine

Fredholm, Erik Ivar (1866–1927)
Swedish mathematician and physicist
alternative theorem
Fredholm alternative

Frege, Friedrich Ludwig Gottlob (1848–1925)
German mathematician and philosopher
identity theory of truth
logicism
negation, performative (or speech act) theory of
psychologism
sense and reference

Frenet, Jean-Frederic (1816–1900)
fundamental theorem of space curves

Frenkel, Yakov Ilyich (1894–1954)
Russian physicist
Frenkel-Andrade theory of liquid viscosity
hole theory of liquids

Fresnel, Augustin Jean (1788–1827)
French physicist
Fresnel-Arago laws
wave theory of light

Freud, Sigmund (1856–1939)
Austrian neurologist and founder of psychoanalysis
abstinence rule
cloaca theory
constancy, law/principle of
depth theory
dream theory
dynamic theory
ego-alter theory
frustration aggression hypothesis
hydraulic theory
infantile birth theories
Jungian theory
modernism
motivated forgetting
object relations theory
omnipotency theory
pain principle
parapraxis
penis envy
pleasure principle
psychic energy
psychoanalytic criticism
psychoanalytical theory
reality principle
substitution hypothesis
transference principle
unconscious memory

Frey, Max von (1852–1932)
German physiologist
paradoxical cold
spot theory

Freytag, Gustav (1816–95)
German writer
Freytag's pyramid

Friedel, Georges (1865–1933)
French physicist
Friedel's law

Friedmann, Alexander (1888–1925)
Russian physicist
Friedmann universes

Friedrich, Caspar David (1774–1840)
German painter
Romanticism

Friedman, Milton (1912–92)
American economist
monetarism
NAIRU (non-accelerating inflation rate of unemployment)
NARU (natural rate of unemployment)
negative income tax
permanent income hypothesis
Reaganomics

Friesz, (Émile) Othon (1874–1949)
French painter
Fauvism

Frisch, Ragnar Anton Kittil (1895–1973)
Norwegian economist
trade cycle

Frobenius, Ferdinand Georg (1849–1917)
German mathematician
Frobenius theorem

Fromm, Erich (1900–80)
German/American psychoanalyst and philosopher
activism
Fromm's theory of personality

Fry, Roger Eliot (1866-1934)
English art critic
formalism
post-Impressionism

Frye, Northrop (1912–)
English literary critic
genre
myth criticism

Fubini, Guido (1879–1943)
Italian mathematician
Fubini's theorem

Fukui, Kenichi (1919–)
Japanese chemist
Woodward-Hoffmann rules

Fukuyama, Francis
Japanese historian
end of history

Fullerton, G S (1859–1925)
American psychologist
Fullerton-Cattell law

Gabo, Naum (Naum Neevia Pevsner) (1890–1977)
Russian/American sculptor
constructivism
kinetic art
suprematism

Gadamer, Hans-Georg (1900–)
German philosopher
hermeneutics

Galbraith, John Kenneth (1908–)
Canadian/American economist
countervailing power

Galen (Claudius Galenus) (*c.* 129–200 AD)
Greek physician
animal heat
constitutional theory
Erasistratus' theory of circulation
four humours
healing power of nature
humoralism

Galilei, Galileo (1564–1642)
Italian astronomer, mathematician and philosopher
square-cube law

Gall, Franz Josef (1758–1828)
German anatomist
faculty theory
mental faculty theory
phrenology

Gally, J A
American biologist
somatic recombination model

Galois, Évariste (1811–32)
French mathematician
Galois theory

Galton, Sir Francis (1822–1911)
English scientist
defect theory
eugenics
factor theory
filial regression, law of
geneticism
nature and nurture

Galvani, Luigi (1737–98)
Italian physiologist
animal electricity

Gamow, George (1904–68)
Russian/American physicist
Gamow-Gurney-Condon theory of alpha particle decay

Gandhi, Mahatma (Mohandâs Karamchand) (1869–1948)
Indian political leader
civil disobedience
satyagraha

Garcia, J (1917–86)
American psychologist
Garcia effect/toxicosis

Garner, Wightman W (*fl.* 1920s)
American agricultural scientist
photoperiodism

Gascoyne, David (1916–)
British poet
anti-literature

Gaskell, Mrs Elizabeth Cleghorn, né Stevenson (1810–65)
English novelist
social novel

Gass, William Howard (1924–)
American novelist
metafiction

Gaudier-Brzeska, Henri (1891–1915)
French sculptor
Vorticism

Gauguin, Eugène Henri Paul (1848–1903)
French painter
post-Impressionism
synthetism

Gause, George Francis
American zoologist
competitive-exclusion principle

Gauss, Carl Friedrich (1777–1855)
German mathematician, astronomer and physicist
dynamics of an asteroid
divergence theorem
fundamental theorem of algebra
Gauss's lemma
Gauss's theorem in electrostatics
parallelism
prime number theorem
theorema egregium

Gay-Lussac, Joseph Louis (1778–1850)
French chemist and physicist
Charles's law
Gay-Lussac's law of combining volumes

Gazdar, Gerald
British linguist
generalized phrase-structure grammar

Geiger, Hans Wilhelm (1882–1945)
German physicist
Geiger and Nuthall rule
Geiger's law

Gelb, I J
Linguist
grammatology

Gelfond, Alexander Osipovich (1906–68)
Russian mathematician
Gelfond-Schneider theorem

Gell-Mann, Murray (1929–)
American theoretical physicist
quark theory

Gelting, J
British economist
balanced budget multiplier

Gemito, V (1852–1929)
Sculptor
verismo

Genette, Gérard (1931–)
French stylistician
focalization

Gentileschi, Orazio (1563–1647)
Italian painter
Caravaggism

Gesell, Arnold Lucius (1880–1961)
American psychologist
stage theory

Geulincx, Arnold (1625–69)
Belgian philosopher
Continental rationalists
pre-established harmony, doctrine of

Gibbs, Josiah Willard (1839–1903)
American mathematical physicist
equilibrium law
Gibbs' phenomenon
Gibbs' theorem
phase rule

Gibrat, Robert (1904–80)
French economist
Gibrat's rule of proportionate growth

Gibson, James Jerome (1904–80)
American psychologist
Gibson effect
visual cliff hypothesis

Gide, André Paul Guillaume (1869–1951)
French writer
mise en abyme

Giffen, Sir Robert (1837–1910)
Scottish economist and statistician
Giffen paradox

Giles, Howard
British social psychologist
accommodation

Gilliéron, J (1854–1926)
Swiss linguist
dialectology

Gilman, A
American linguist
power and solidarity

Gladstone, John Hall (1827–1902)
Physicist
Gladstone and Dale law

Glashow, Sheldon Lee (1932–)
American physicist
electroweak theory

Glass, Philip (1937–)
American composer
minimalism

Gleason, Henry Allan
American botanist
individualistic model
stratificational grammar

Gloger, C L
Biologist
Gloger's law or rule

Gödel, Kurt (1906–78)
Czech/American mathematician
computability theory
continuum hypothesis
Gödel-Rosser theorem
Gödel's completeness theorem
Gödel's incompleteness theorem
logicism
Vienna Circle

Goebbels, Paul Joseph (1897–1945)
German Nazi politician
art of the Third Reich

Goethe, Johann Wolfgang von (1749–1832)
German writer
Bildungsroman
classicism/Romanticism

Goffman, Erving (1922–82)
Canadian/American sociologist
impression management theory
politeness
role-role theory

Gold, Thomas (1920–)
Austrian/American astronomer
cosmological principle
steady-state theory of the universe

Goldbach, Christian (1690–1764)
Prussian mathematician
Goldbach conjecture

Goldschmidt, Richard Benedikt (1878–1958)
German biologist
hopeful monsters
macroevolution
microevolution
saltation speciation

Goldschmidt, Victor Moritz (1888–1947)
Swiss/Norwegian chemist
Goldschmidt's law/rule

Halberg, Franz
American biologist
circadian rhythms

Haldane, John Burdon Sanderson (1892–1964)
Anglo-Indian biologist
Haldane's law or rule
Oparin-Haldane hypothesis
overspecialization
viral theory

Hall, Edwin H (1855–1938)
Physicist
Hall effect

Hall, Granville Stanley (1844–1924)
American psychologist and educationalist
recapitulation theory

Hall, Robert Ernest (1943–)
Economist
average cost pricing

Halle, Morris
American linguist
analysis-by-synthesis
generative phonology
metrical theory

Halliday, Michael Alexander Kirkwood (1925–)
British linguist
antilanguage
cohesion
context of situation
functional grammar
linguistic criticism
overlexicalization
paradigm and syntagm
register
relexicalization
scale-and-category grammar
social semiotic, language as
systemic grammar
theme and rheme
transitivity
underlexicalization

Hallwachs, Wilhelm Ludwig Franz (1859–1922)
German physicist
Hallwachs's effect
photoelectric effect

Hamilton, Richard (1922–)
English artist
Pop art

Hamilton, William Donald (1936–)
British theoretical biologist
haplodiploid hypothesis
inclusive fitness

Hamilton, Sir William Rowan (1805–65)
Irish mathematician
Cayley-Hamilton theorem
Hamilton's principle

Hansen, Alvin (1887–1975)
American economist
IS-LM model

Hardy, Sir Alister Clavering (1896–1985)
English marine biologist
aquatic theory of human evolution

Hardy, Godfrey Harold (1877–1947)
English mathematician
Hardy-Littlewood method
Hardy-Weinberg equilibrium, law or principle

Harland, Richard
Literary theorist
superstructuralism

Harlow, Harry Frederick (1905–81)
American psychologist
resonance theory of learning

Harman, Denham (1916–)
American physician and biochenist
free radical theory of ageing

Harris, Zellig Sabbetai
Linguist
transformation

Harrod, (Sir Henry) Roy Forbes (1900-78)
English economist
Harrod-Domar growth model
life-cycle hypothesis
multiplier-accelerator
natural and warranted rates of growth

Hart, Albert (1909–)
American economist
automatic stabilization

Hartley, David (1705–57)
English psychologist and philosopher
associationism

Hartree, Douglas Rayner (1897–1958)
English mathematician and physicist
self-consistent field theory

Hartung, Hans (1904–89)
German/French artist
art informel

Harvey, William (1578–1657)
English physician
epigenesis

Hasan, Ruqaiya
British linguist
cohesion

Hauer, Josef (1883–1959)
Austrian composer
twelve-note technique

Hausdorff, Felix (1868–1942)
German/Polish mathematician
Hausdorff maximality theorem

Haushofer, Karl (1869–1946)
German political geographer
geopolitics

Haüy, René Just (1743–1822)
French mineralogist
rational indices, law of

Havighurst, Robert J
American psychologist
activity theory of ageing

Haxel, O
Physicist
shell model of the atomic nucleus

Hayek, Friedrich August von (1899–1992)
Austrian/British political economist
catallaxy
spontaneous order

Hayflick, Leonard (1928–)
American biologist
Hayflick limit

Hebb, Donald Olding (1904–85)
Canadian psychologist
consolidation theory of learning
decay theory
Hebb's theory of perception learning
phase sequence hypothesis
short-circuiting law

Heckel, Erich (1883–1970)
German painter
Die Brücke

Heckscher, Eli Filip (1879–1952)
Swedish economist and economic historian
Heckscher-Ohlin trade theory

Hegel, Georg Wilhelm Friedrich (1770–1831)
German philosopher
absolute
absolutism
alienation
civil society
dialectic
end of history
monism
neo-Platonism
objective idealism

Heidegger, Martin (1889–1976)
German philosopher
deconstruction
existentialism

Heidenhain, Rudolf Peter Heinrich (1834–97)
German physician
Bowman-Heidenhain hypothesis

Heider, F (1896–)
Austrian/American psychologist
assimilation-contrast theory
attribution theory
balance theory
leniency effect
psychologist's fallacy
reciprocity principle
visual perspective hypothesis

Heine, Heinrich (1821–81)
German mathematician
Heine-Borel covering theorem
Heine's theorem

Heisenberg, Werner Karl (1901–76)
German theoretical physicist
Copenhagen interpretation
Heisenberg's theory of ferromagnetism
Heisenberg's uncertainty principle

camera lucida
dynamical theory of heat
Hooke's law

Hopf, Heinz (1894–1971)
German mathematician
Hopf bifurcation theorem

Hopper, Edward (1882–67)
American painter
regionalism

Horace, Quintus Horatius Flaccus (65–8 BC)
Roman poet and satirist
decorum
ut pictura poesis

Horner, Johann Friedrich (1831–86)
Swiss opthalmologist
Horner's law

Horney, Karen (1885–1952)
American psychoanalyst
psychoanalytical theory

Horstmann, August Friedrich (1842–1929)
German physical chemist
equilibrium law

Hotelling, Harold (1895–1973)
American economist
Hotelling's law
public utility pricing

House, R J
American psychologist
path-goal theory of leadership

Hovland, Carl Iver (1912–61)
American psychologist
sleeper effect

Hoyle, Fred (1915–)
British astronomer
steady-state theory of the universe

Hubble, Edwin Powell (1889–1953)
American astronomer
Hubble's law

Hückel, Erich (*fl.*1920s)
German physicist
Debye-Hückel theory
Hückel ($4n + 2$) rule
Hückel molecular orbital theory

Huebner, Robert Joseph (1923–)
American scientist
oncogene hypothesis

Hull, Clark Leonard (1884–1952)
American psychologist
continuity theory/hypothesis
drive reduction theory
drive stimulus theory
factor theory of learning
frustration fixation hypothesis
Hullian theory
multiple response principle
reinforcement

Hulme, Thomas Ernest (1883–1917)
English critic, poet and philosopher
Imagism

Hume, David (1711–76)
Scottish philosopher and historian
associationism
associative learning
British empiricists
Hume's law
logical positivism
naturalized epistemology
Pyrrhonism
regularity theory of causation
scepticism
specie-flow mechanism

Hume-Rothery, William (1899–1968)
English chemist
Hume-Rothery's rule

Hund, Friedrich (*fl.*1920s)
Chemist
Hund's rule

Hunt, William Holman (1827–1910)
English painter and art historian
Pre-Raphaelite

Hunter, John (1728–93)
Scottish physiologist and surgeon
blastema theory

Hurwitz, Adolf (1859–1919)
German mathematician
Hurwitz's theorem

Husserl, Edmund Gustav Albrecht (1859–1938)
German philosopher
phenomenology
psychologism

Hutcheson, Francis (1694–1746)
English philosopher
moral sense theories

Hutton, James (1726–97)
Scottish geologist
actualism
glacial theory
neptunism
plutonism
uniformitarianism

Huxley, Aldous Leonard (1894-1963)
English writer
Utopianism

Huxley, Sir Julian Sorell (1887–1975)
English biologist and humanist
cladogenesis

Huxley, Thomas Henry (1825–95)
English biologist
agnosticism
biogenesis

Huygens, Christiaan (1629–93)
Dutch physicist
Huygens's principle

Hyatt, Alpheus (1838–1902)
American evolutionist
anagenesis

Hyman, H (1928–)
American psychologist
Hick-Hyman law
Hyman's law

Hymes, Dell H
American ethnographer of communication
communicative competence
ethnography of communication
literary competence

Ingarden, Roman Witold (1893–1970)
Polish philosopher
reception theory

Ionesco, Eugène (1912–)
Romanian/French playwright
absurd, theatre of the

Ipsen, G
Linguist
semantic field

Iser, Wolfgang (1926–)
German aesthetician and critic
implied reader
indeterminacy
reception theory

Isou, Isidore (1925–)
Romanian/French poet
lettrism

Ives, Charles Edward (1874–1954)
American composer
bitonality

Jackson, John Hughlings (1835–1911)
English neurologist
Jackson's law
mixed cerebral dominance theory

Jacob, François (1920–)
French biochemist
Jacob-Monod model

Jacobi, Karl Gustav Jacob (1804–51)
German mathematician
Jacobi's theorem
Sylvester's law of inertia

Jahn, H A
Physicist
Jahn-Teller theorem

Jakobson, Roman Osipovich (1896–1982)
Russian/American linguist
aphasia
binary opposition
clinical linguistics
coupling
distinctive feature
dominant
formalism, Russian
literariness
literary language
metalanguage
metaphor and metonymy
parallelism
poetic principle
Prague School
shifters
structuralism

Langford, C H
Logician
relevance logics

Langland, William (*c.*1332–1400)
English poet
allegory

Langmuir, Irving (1881–1957)
American chemist
Child's law
Langmuir isotherm
Langmuir's theory
octet rule

Lankester, Sir Edwin Ray (1847–1929)
English zoologist
degeneration theory

Lapicque, Louis (1866–1952)
French physiologist and anthropologist
chronaxie

Lapicque, Marcelle
French physiologist and anthropologist
chronaxie

Laplace, Pierre-Simon, Marquis de (1749–1827)
French mathematician and astronomer
central limit theorem
classical theory of probability
determinism
nebular hypothesis of planetary formation
potential theory
range theories of probability

Laposky, Ben F
American artist
electronic art

Larionov, Mikhail Fyodorovich (1881–1964)
Russian painter
rayonism

Larmor, Sir Joseph (1857–1942)
Irish mathematician
dynamo theory
Larmor's theorem

Lashley, Karl Spencer (1890–1958)
American psychologist
equipotential theory
multiple control principle

Lassalle, Ferdinand (1825–64)
German social democrat
iron law of wages

Latham, John
British artist
eventstructure

Lavoisier, Antoine Laurent (1743–94)
French chenist
phlogiston theory
vitalism

Le Châtelier, Henry (1850–1936)
French chemist
Le Châtelier's principle

Le Parc, Julio (1928–)
Argentinian artist
minimalism

Leavis, Frank Raymond (1895–1978)
English literary critic
canon
practical criticism

Leavitt, Henrietta Swan (1868–1921)
American astronomer
period-luminosity law for Cepheid variables

Lebesgue, Henri Léon (1875–1941)
French mathematician
dominated convergence theorem
measure theory
monotone convergence theorem
Riemann-Lebesgue lemma

Lederberg, Geoffrey N (1925–),
American biologist and geneticist
somatic mutation theory of antibody diversity

Legendre, Adrien-Marie (1752–1833)
French mathematician
prime number theorem
statistics

Leibenstein, Harvey (1922–)
American economist
x-efficiency

Leibniz, Gottfried Wilhelm (1646–1716)
German philosopher and mathematician
alternating series test, Leibniz's
chain of being
Continental rationalists
continuity, law or principle of
contracting Earth hypothesis
derivative tests
differential equations
Eleaticism
exotic theory of fossil origins
fundamental theorem of calculus
hylozoism
indiscernibility of identicals
integral test
integration by parts
Leibniz's law
Leibniz's theorem
modal realism
panpsychism
parallelism
perfection, principle of
pre-established harmony, doctrine of
product rule
quotient rule
sufficient reason, principle of

Lemery, Nicolas (1645–1715)
French chemist
vitalism

Lenin, Vladimir Ilyich (1870–1924)
Russian revolutionary
bolshevism
dictatorship of the proletariat
empiriocriticism
imperialism
Leninism
vanguard party

Lenneberg, Eric Heinz
American neurolinguist
critical period

Lenz, Heinrich Friedrich Emil (1804–65)
Russian/German physicist
Lenz's law

Leonardo da Vinci (1452–1519)
Italian artist
aerial perspective
Amonton's laws of friction
proportion

Leontief, Wassily (1906–)
Russian/American economist
Heckscher-Ohlin trade theory
input-output analysis
Leontief paradox

LePage, R B
British linguist
act of identity

Lerner, Abba (1905–82)
Romanian economist
Marshall-Lerner principle

Leśniewski, Stanislaw (1886–1936)
Polish logician
categorial grammar
mereology
ontology

Leuba, C
Psychologist
optimal stimulation, principle of

Leucippus (5th century BC)
Greek philosopher
atomism
Epicureanism

Levene, Phoebus Aaron Theodor (orig. Fishel Aaronovich Lenin) (1869–1940)
Russian/American biochemist
tetranucleotide theory

Lévi-Strauss, Claude (1908–)
French social anthropologist
alliance theory
bricolage
narrative grammar
structural anthropology
structuralism

Levinson, Steven C
Linguist
politeness

Levin, Samuel R
American stylistician
coupling
deviation

Lewes, George Henry (1817–78)
English writer
anthropomorphism

Lewin, Kurt (1890–1947)
German/American psychologist

brain field theory
cognitive personality theory
contemporaneity principle
field theory
interaction principle
needs-press theory of personality

Lewis, Clarence Irving (1883–1964)
American philosopher
relevance logics

Lewis, David Kellogg
Philosopher
modal realism
mutual knowledge

Lewis, Gilbert Newton (1875–1946)
American physical chemist
electronic theory of valency
Lewis's colour theory
Lewis's theory of acids and bases
octet rule

Lewis, Oscar (1914–70)
American anthropologist
culture of poverty

Lewis, W C McC
Chemist
collision theory

Lewis, (Percy) Wyndham (1882–1957)
English novelist, painter and critic
Vorticism

Lewitt, Sol (1928–)
American artist
cold art
conceptual art

l'Hôpital, Guillaume François Antoine de, Marquis de St Mesmé (1661–1704)
French mathematician
l'Hôpital's rule

Liberman, A M (1917–)
Psychologist
motor theory of speech perception

Lie, Marius Sophus (1842–99)
Norwegian mathematician
group theory

Liebig, Justus, Freiherr von (1803–73)
German chemist
Blackman curve

Likert, R A (1903–81)
American psychologist
Likert scale

Lindahl, Erik (1891–1960)
Swedish economist
benefit approach principle
Lindahl equilibrium

Lindemann, Frederick Alexander, 1st Viscount Cherwell (1886–1957)
English physicist
Lindemann's theory
Lindemann's theory of melting
Nernst-Lindemann theory of heat capacities

Linnaeus, Carolus (1707–78)
Swedish naturalist and physician
binomial nomenclature
essentialism
species essentialism

Liouville, Joseph (1809–82)
French mathematician
Liouville's theorem

Lipsey, Richard (1928–)
Canadian economist
second best, theory of

Lissitzky, El (Eliezer Markowich) (1890–1941)
Russian painter and designer
suprematism

Listing, Johann Benedict (1808–82)
German mathematician
Möbius strip

Littlewood, John Edensor (1885–1977)
English mathematician
dynamics of an asteroid
Hardy-Littlewood method

Lloyd, L
Biologist
master-slave hypothesis

Lobachevsky, Nicholai Ivanovich (1793–1856)
Russian mathematician
parallel postulate

Locke, John (1632–1704)
English philosopher
absolute zero
associationism
associative learning
benefit approach principle
British empiricists
contractualism
empiricism
environmentalism
human capital theory
ideational theories of meaning
immaterialism
innate ideas
mercantalism
nature and nurture
psychologism
Rawls' theory of justice

Loeb, Jacques (1859–1924)
German/American biologist
forced movement theory
mechanistic psychology
tropisms, theory of

Loftus, Elizabeth F (1944–)
American psychologist
pseudo-forgetting

Logan, Frank Anderson (1924–)
American psychologist
correspondence theory/law
micromolar theory

Lombroso, Cesare (1836–1909)
Italian physician and criminologist
atavism
Lombrosian theory

London, Fritz Wolfgang (1900–54)
German/American physicist
Heitler-London covalence theory

Long, Richard (1945–)
English artist
land art

Longinus (*c*.213–73 AD)
Greek literary critic
sublimity

Lord, Albert Bates
American literary theorist
oral-formulaic poetry

Lorentz, Hendrik Antoon (1853–1928)
Dutch physicist
Fitzgerald-Lorentz contraction
Lorentz force law
Lorentz-Lorenz law

Lorenz, Edward Norton (1917–)
American meteorologist
Butterfly effect
strange attractor

Lorenz, Konrad Zacharias (1903–89)
Austrian zoologist and ethologist
analogy
instinct
Lorenz's hydraulic model
homology

Lorenz, Ludwig Valentin (1829–91)
Danish physicist
Lorentz-Lorenz law
Wiedemann-Franz-Lorenz law

Lorraine, Claude (1600–82)
French painter
picturesque

Lotze, Rudolph Hermann (1817–81)
German philosopher
local sign theory

Lounsbury, Floyd G (1914–)
American linguist and anthropologist
extension of kinship terms

Lovelock, James Ephraim (1919–)
English scientist
Gaia hypothesis

Löwenheim, Leopold (1878–1940)
Mathematician
Löwenheim-Skolem theorem

Lowry, Thomas Martin
English chemist
Brønsted-Lowry theory)

Lubbock, Sir John, 1st Baron Avebury (1834–1913)
English politician
animal language hypothesis

Lubbock, Percy (1879–1965)
English critic and biographer
point of view

Lucas, Howard Johnson (1885–)
American chemist
Lucas's theory

Lucas, Robert (1937–)
American economist
new classical macroeconomics

Luce, R Duncan (1925–)
American psychologist
probability theory of learning

Lucretius, Titus Lucretius Carus (*c*.99–55 BC)
Roman poet and philosopher
effluxes (or effluences), theory of
Epicureanism

Lüders, G
Physicist
CPT theorem

Ludwig, Karl Friedrich Wilhelm (1816–95)
German physiologist
Cushny's theory
Ludwig's theory
reductionism

Lukács, Georg Szegedy von (1885–1971)
Hungarian philosopher and critic
Marxist criticism

Lusin, Nikolai Nikolaevitch (1883–1950)
Russian mathematician
Lusin's theorem

Luther, Martin (1483–1546)
German religious reformer
extermination theory of fossil origins

Lydekker, R (1849–1915)
British naturalist
Lydekker's line

Lyell, Sir Charles (1797–1875)
Scottish geologist
actualism
drift theory
extermination theory of fossil origins
uniformitarianism

Lyons, John (1932–)
British linguist
semantics, structural
sense relations

Lypapanov, A A
Russian mathematician
Lypapanov's theorem

Lysenko, Trofim Denisovich (1898–1976)
Russian biologist
Lysenkoism

Lysippus (*fl.c.*360–316 BC)
Greek sculptor
Lysippan proportions
proportion

Macdonald-Wright, Stanton (1890–1973)
American artist
synchromism

Mach, Ernst (1838–1916)
Austrian physicist and philosopher
empiriocriticism
Mach's principle
neutral monism
sensationalism

Macherey, Pierre (1938–)
French literary critic
absence
cultural materialism
literary mode of production
Marxist criticism

Machiavelli, Niccolò di Bernardo dei (1469–1527)
Italian statesman, writer and political philosopher
leadership theories
Machiavellianism

Macke, August (1887–1914)
German Expressionist painter
Blaue Reiter

MacKenzie, Henry (1745–1831)
Scottish writer
sensibility

Mackinder, Sir Halford John (1861–1947)
English geographer and politician
heartland theory

MacLane, Saunders (1909–)
American mathematician
category theory

Maclaurin, Colin (1698–1746)
Scottish mathematician
Maclaurin's theorem

MacLean, P D (1913–)
American psychologist
MacLean's theory of emotion

Maeterlinck, Count Maurice (1862–1949)
Belgian dramatist
Symbolism

Magendie, François (1783–1855)
French physiologist
Bell-Magendie law

Magoun (jnr), Francis Peabody
Poet
oral-formulaic poetry

Magoun, Horace Winchell (1907–)
American neuroscientist
reticular activating system

Malebranche, Nicolas (1638–1715)
French philosopher
Continental rationalists
occasionalism

Malevich, Kasimir Severinovich (1878–1935)
Russian painter and designer
suprematism

Malinowski, Bronislaw (1884–1942)
Polish/British anthropologist
anthropological linguistics
context of situation
functionalist theory
phatic communion

Malory, Sir Thomas (*d.*1471)
English writer
romance

Malpighi, Marcello (1628–94)
Italian anatomist
preformation

Malthus, Reverend Thomas Robert (1766–1834)
English economist and clergyman
Malthusian population theory
trade cycle

Malus, Étienne Louis (1775–1812)
French physicist
Malus's law
Malus's theorem

Malynes, Gerald (1586–1641)
Belgian/English merchant
mercantalism

Mancini, A (1852–1930)
Italian painter
verismo

Mandelbrot, Benoit (1924–)
Polish/American mathematician
fractal
Mandelbrot set

Manfredi, Bartolommeo (*c.*1587–1620/1)
Italian painter
Caravaggism

Manguin, Henri Charles (1874–1949)
French painter
Fauvism

Mani of Persia (*c.*215–76 AD)
Persian philosopher
Manicheism

Mantegna, Andrea (1431–1506)
Italian painter
illusionism

Marbe, C
German psychologist
Marbe's law

Marc, Franz (1880–1916)
German Expressionist painter
Blaue Reiter

Marcel, Gabriel Honoré (1889–1973)
French philosopher and dramatist
existentialism

Margulis, Lynn (1938–)
English scientist
Gaia hypothesis
symbiotic theory of cellular evolution

Marinetti, Emilio Filippo Tommaso (1876–1944)
Italian writer
Futurism

Mariotte, Edmé (1620–84)
French physicist and priest
Boyle's law
Mariotte's law

Markov, Andrei Andreevich (1856–1922)
Russian mathematician
algebraic topology

Menelaus of Alexandria
Roman mathematician
Ceva's theorem
Menelaus' theorem

Menger, Carl von (1840–1921)
Austrian economist
time preference theory of interest

Mercalli, Giuseppe (1850–1914)
Italian seismologist
Mercalli scale

Merkel, Friedrich Siegismund (1845–1919)
German anatomist
Merkel's law

Merkle, Ralph
Mathematician
knapsack cryptosystem

Mersenne, Marin (1588–1648)
French mathematician and scientist
Mersenne primes
Mersenne's laws

Merz, Mario (1925–)
Italian artist
arte povera

Mesmer, Friedrich Anton, or Franz (1743–1815)
Austrian physician
animal magnetism

Messiaen, Olivier Eugène Prosper Charles (1908–92)
French composer and organist
serialism

Meyer, Leonard B (1918–)
American musicologist
implication-realization
information theory

Meyer, Viktor (1848–97)
German chemist
Meyer's law of esterification

Michener, Charles Duncan (1918–)
American biologist
mutualistic hypothesis

Milankovitch, M (1879–1958)
Yugoslavian geophysicist
Milankovitch hypothesis

Milhaud, Darius (1892–1974)
French composer
bitonality

Mill, James (1773–1836)
Scottish philosopher, historian and economist
associationism

Mill, John Stuart (1806–73)
English philosopher and social reformer
ability-to-pay principle
crowding hypothesis
denotation and connotation
economic methodology
hedonism
human capital theory
machinery question
Mill's canons
non-competing groups
phenomenalism
psychologism
trade cycle

Millais, Sir John Everett (1829–96)
English painter
Pre-Raphaelite

Miller, G A (1920–)
American psychologist
derivational theory of complexity
magical number seven

Miller, Merton (1923–)
American economist
Modigliani-Miller theory of the cost of capital

Miller, William Hallowes (1801–80)
Welsh chemist
rational intercepts, law of

Millet, J
literary theorist
synaesthesia

Mills, Robert Laurence (1927–)
English mathematician
Yang-Mills theory

Mil'man, David Pinkhusovich (1913–)
Russian mathematician
Kreĭn–Milman theorem

Milne, Edward Arthur (1896–1950)
English astrophysicist
cosmological principle

Milroy, Lesley
Linguist
social networks and language

Milton, John (1608–74)
English poet
dissociation of sensibility
epic

Minardi, T (1787–1871)
Italian painter
purism

Minford, Patrick (1943–)
British economist
new classical macroeconomics

Minkowski, Hermann (1864–1909)
Russian/German mathematician
geometry of numbers
Minkowski's inequality
Minkowski's theorem on convex bodies
space-time

Minsky, Marvin (1927–)
American philosopher
frame

Mises, Ludwig von (1881–1973)
Austrian economist
forced saving

Mises, Richard von (1883–1953)
Austrian/American mathematician
frequency theory of probability
impossibility of a gambling system, principle of the

Mitchell, Peter Dennis (1920–)
English biochemist
chemiosmotic theory

Mitscherlich, Eilhard (1794–1863)
German chemist
Mitscherlich's law of isomorphism

Möbius, August Ferdinand (1790–1868)
German mathematician
Möbius strip

Modigliani, Franco (1918–)
Italian/American economist
life-cycle hypothesis
Modigliani-Miller theory of the cost of capital

Mohr, Christian Otto (1835–1918)
German engineer
Mohr-Coulomb theory of failure

Mohs, Friedrich (1773–1839)
German mineralogist
Mohs' scale

Moivre, Abraham de (1667–1754)
French mathematician
central limit theorem
de Moivre's formula

Molyneux, William (1656–1738)
Anglo-Irish philosopher
nativism

Moment, Gairdner B
American biologist
Shangri-La phenomenon

Mondrian, Piet (Pieter Cornelius Mondriaan) (1872–1944)
Dutch artist
De Stijl

Monet, Claude (1840–1926)
French Impressionist painter
impressionism

Monod, Jacques (1910–76)
French biochemist
Jacob-Monod model

Montague, R (1930–70)
American philosopher
Montague grammar

Montel, Paul (1876–1975)
French mathematician
Montel's theorem

Moore, Carl R
Biologist
testicular extract theory

Moore, Eliakim Hastings (1862–1932)
Mathematician
Moore-Osgood theorem

Moore, George Edward (1873–1958)
English philosopher
fact/value distinction
ideal utilitarianism
identity, law of
identity theory of truth
naturalism
organic unities, principle of

Newton's method
potential theory
product rule
quotient rule

Neymann, Jerzy
Mathematician
Neyman-Pearson lemma

Nicholson, G
Biologist
fluid-mosaic model

Nicod, Jean (1893–1924)
French philosopher
Nicod's criterion

Nielsen, Niels (1865–1931)
Mathematician
Schreier-Nielsen theorem

Nietzsche, Friedrich Wilhelm (1844–1900)
German philosopher, scholar and writer
Apollonian/Dionysian
existentialism
perspectivism

Nikodym, Otton
Mathematician
Radon-Nikodym theorem

Noguera, R
Biochemist
indoleamine hypothesis of depression

Nolde, Emil (pseud. of Emil Hansen) (1867–1956)
German painter and printmaker
Die Brücke
expressionism

Nuthall, John Mitchell (1890–1958)
British physicist
Geiger and Nuthall rule

Nyquist, Harry (1889–1976)
Swedish/American communications engineer
Nyquist's noise theorem

Ockham, William of (*c.*1285–1349)
English philosopher, theologian and political writer
language of thought

Oesterlé, J
Mathematician
abc conjecture

Ogden, Charles Kay (1889–1957)
English linguistic reformer
emotive/referential language

Ohlin, Bertil Gottard (1899–1979)
Swedish economist and politician
Heckscher-Ohlin trade theory

Ohm, Georg Simon (1787–1854)
German physicist
Ohm's acoustic law
Ohm's law of electricity
Ohm's law of hearing

Okun, Arthur (1926–80)
American economist
Okun's law

Olbers, Heinrich Wilhelm Matthäus (1758–1840)
German physician and astronomer
Olbers' paradox

Oldenburg, Claes Thure (1929–)
Swedish/American sculptor
minimalism

Olson, Elder (James) (1909–)
American literary critic
Chicago critics

Olson, Mancur (1932–)
American political scientist
collective action, logic of

Onsager, Lars (1903–76)
Norwegian/American physical chemist
Onsager reciprocal theorem

Oparin, Aleksandr Ivanovich (1894–1980)
Russian biochemist
Oparin-Haldane hypothesis
phenotype theory

Orgel, Leslie K (1927–)
American biologist
error catastrophe theory of ageing

Ortega y Gasset, José (1883–1955)
Spanish writer and philosopher
perspectivism

Orwell, George (pseud. of Eric Arthur Blair) (1903–50)
English writer
allegory
Utopianism

Osborn, Henry Fairfield (1857–1935)
American zoologist and palaeontologist
adaptive radiation

Osgood, Charles Egerton (1916–)
American psychologist
congruity theory/principle

Osgood, William Fogg (1864–1943)
Mathematician
Moore-Osgood theorem

Osthoff, H (1847–1909)
German linguist
neo-grammarians

Ostwald, Friedrich Wilhelm (1853–1932)
Latvian chemist
Ostwald's dilution law

Oswald, Ian
American psychologist
restoration theory

Overbeck, Johann Friedrich (1789–1869)
German painter
Nazarene Brotherhood

Overbeek, J Th G (*fl.*1940s)
Chemist
DLVO theory

Owen, Robert (1771–1858)
Welsh social reformer
Owenism

Ozenfant, Amédée (1886–1966)
French artist
purism

Pacioli, Luca (*c.*1445–1517)
Italian mathematician
double entry accounting

Paik, Nam June (1932–)
Canadian artist
electronic art

Paivio, Allan Uhro (1925–)
Canadian psychologist
associative learning
dual-coding theory

Paley, Reverend William (1743–1805)
English theologian
adaptation
natural theology
Paley's watchmaker

Pander, Christian Heinrich (1794–1865)
Russian/German anatomist
von Baer's laws
germ layer theory

Paneth, Friedrich Adolf (1887–1958)
Austrian chemist
Paneth's adsorption rule

Pane, Gina (1939–)
French artist
body art

Paolozzi, Eduardo Luigi (1924–)
Scottish/Italian sculptor and printmaker
Pop art

Papez, J W (1883–1958)
American psychologist
Papez's theory of emotion

Pareto, Vilfredo Federigo Damaso, Marquis (1848–1923)
Italian economist and sociologist
circulation of élites
compensation principle
demand theory
élitism
neo-classical theory
Pareto efficiency
Pareto optimality

Parker, Robert L
American earth scientist
plate tectonics

Parkinson, Cyril Northcote (1909–93)
English political scientist
Parkinson's law

Parmenides of Elea (early 5th century BC)
Greek philosopher
absolute

Plateau's problem
Talbot-Plateau law

Plato (*c.*427–347 BC)
Greek philosopher
chain of being
contractualism
dialectic
diegesis
eclecticism and syncretism
epic
essentialism
hylomorphism
idealization
identity theory of predication
imitation
interactionism
materialism
mimesis
narrator
neo-Platonism
neo-Pythagoreanism
one over many principle
Orphism
Platonic mathematics
Platonism
relativism
third man argument
unity, the Unities
Utopianism

Plautus (Titus Maccius, or Maccus) (*c.*254–184 BC)
Roman playwright
tragicomedy

Playfair, John (1748–1819)
Scottish mathematician and physicist
actualism
parallel postulate

Plotinus (205–70 AD)
Roman philosopher
neo-Platonism

Poe, Edgar Allan (1809–49)
American poet and writer
fantastic

Poetzl, O
Austrian psychologist
Poetzl effect

Poincaré, Jules Henri (1854–1912)
French mathematician
algebraic topology
conventionalism
galactic rotation theory
Poincaré's conjecture
Poincaré's lemma
Poincaré's theorem

Poisseuille, Jean Léonard Marie (1797–1869)
French physicist
Poiseuille's law

Poisson, Siméon Denis (1781–1840)
French mathematical physicist
Poisson's ratio

Polanyi, Michael (1891–1976)
Hungarian/British physicial chemist and philosopher
absolute reaction rate, theory of
tacit knowledge

Pollock, (Paul) Jackson (1912–56)
American painter
abstract expressionism
action painting
all-over painting
Jungian aesthetics

Polya, George (1887–1985)
Mathematician
Polya's theory of enumeration

Polyclitus (*fl.c.*232 BC)
Greek sculptor
Lysippan proportions

Polyclitus of Samos (*fl.*5th century BC)
Greek sculptor
Polyclitan school
proportion

Pomeranchuk, Isaak Yakovlevich (1913–66)
Russian theoretical physicist
Pomeranchuk's theorem

Poncelet, Jean Victor (1788–1867)
French mathematician and engineer
fundamental theorem of projective geometry

Pope, Alexander (1688–1744)
English poet
bathos

Pope, William Jackson (1870–1939)
English chemist
Barlow's rule

Popova, Liubov Sergeevna, née Eding (1889–1924)
Russian painter and stage director
suprematism

Popper, Sir Karl Raimund (1902–)
Austrian/British philosopher
deductivism
falsificationism
historicism
hypothetico-deductive method
improbabilism
negative utilitarianism
open society
piecemeal social engineering
propensity theory of probability
Vienna Circle

Porter, T C (*fl.*1890s–1900s)
English psychologist
Ferry-Porter law

Posidonius (*c.*135–51 BC)
Greek philosopher and scientist
Stoicism

Potter, Van Rensselaer (1911–)
American scientist
deletion hypothesis
feedback deletion hypothesis

Poulantzas, Nicos (1936–79)
Greek political theorist
relative autonomy

Pound, Ezra Loomis (1885–1972)
American poet and critic
Imagism
modernism
Vorticism

Poussin, Nicolas (1594–1665)
French painter
idealization
picturesque
Poussinism

Poynting, John Henry (1852–1914)
English physicist
Poynting's theorem

Prange (jnr), A J
Biochemist
indoleamine hypothesis of depression

Pratt, Frederick Haven (1873–1958)
American physiologist
all-or-none principle

Pratt, John Henry (1809–71)
English clergyman and geophysicist
Pratt hypothesis

Pratt, Mary Louise (1948–)
Stylistician
natural narrative

Prebisch, Raul (1901–86)
Argentinian economist
Prebisch-Singer thesis

Premack, David (1935–)
American psychologist
Premack principle

Previati, L
Italian painter
scapigliatura

Prévost, Pierre (1751–1839)
Swiss physicist and philosopher
Prévost's theory of exchanges

Priessnitz, Vincent (1799–1851)
German farmer
hydropathy

Prince, Gerald
Literary critic
narratee

Propp, Vladimir (1895–1970)
Russian linguist
actantial theory
formalism, Russian
kernel, catalyzer, index
narrative grammar
structuralism

Protagoras (*c.*490–420 BC)
Greek sophist and teacher
lawyer paradox

Proudhon, Pierre-Joseph (1809–65)
French journalist and socialist
property is theft

Proust, Louis Joseph (1755–1826)
French chemist
constant composition, law of

Prout, William (1785–1850)
English chemist and physiologist
Prout's hypothesis

Schleyer, J M (1832–1912)
German priest and linguist
artificial languages

Schlick, Moritz (1882–1936)
German philosopher
consistent empiricism
Vienna Circle

Schmidt-Rottluff, Karl (1884–1976)
German painter and printmaker
Die Brücke

Schneider, Theodor (1911–)
Mathematician
Gelfond-Schneider theorem

Schnirel'man, Lev Genrikhovich (1905–38)
Russian mathematican
Schnirelmann's density theorem

Schoenberg, Arnold Franz Walter (1874–1951)
Austro-Hungarian Jewish composer
Blaue Reiter
expressionism
serialism
twelve-note technique

Scholes, Robert (1929–)
American literary critic
fabulation

Schopenhauer, Artur (1788–1860)
German philosopher
voluntarism

Schreier, Otto (1901–29)
Mathematician
Schreier-Nielsen theorem

Schrieffer, John Robert (1931–)
American physicist
BCS theory

Schröder, Friedrich Wilhelm Karl Ernst (1841–1902)
German mathematician
Schroeder-Bernstein theorem

Schrödinger, Erwin (1887–1961)
Austrian physicist
Schrödinger's cat

Schuchardt, Hugo (1842–1927)
German philologist
creoles and pidgins

Schultz, Theodore William (1902–)
American economist
human capital theory

Schumacher, Ernst (1911–77)
German/British economist and businessman
small is beautiful principle

Schumpeter, Joseph (1883–1946)
American economist and political scientist
democratic élitism

Schur, Issai (1875–1941)
Russian/German mathematician
Schur's lemma

Schuster, P
Biologist
hypercycle theory

Schwann, Theodor (1810–82)
German physiologist
cell theory

Schwartz, Hermann Amandus (1843–1921)
German mathematician
Cauchy-Schwartz inequality
reflection principle of Schwartz
Schwartz's lemma

Schwitters, Kurt (1887–1948)
German artist
environment art

Scitovsky, Tibor (1910–)
Hungarian/American economist
Scitovsky paradox

Searle, John Rogers (1932–)
American philosopher
speech act theory

Sechenov, Ivan Mikhaylovich (1829–1905)
Russian psychologist and physiologist
nervism

Sedgwick, Adam (1787–1857)
English geologist
diluvianism

Sedgwick, W F
Astronomical physicist
tidal theories of planetary formation

Seebeck, Thomas Johann (1770–1831)
Estonian/German physicist
Seebeck effect

Segonzac, André Denoyer de (1884–1974)
French painter and engraver
expressionism

Selberg, Atle (1917–)
Mathematician
prime number theorem

Seligman, Martin E P (1942–)
American psychologist
behaviour constraint theory
learned helplessness

Seliker, L
Linguist
interlanguage

Sell, Roger D (1942–)
Linguist
literary pragmatics

Sellars, Roy Wood (1880–1973)
American philosopher
critical realism

Selvinski, I (1899–1968)
Russian writer
constructivism

Selye, Hans Hugo Bruno (1907–82)
Austrian/Canadian physicist
epiphenomenalist theory of ageing

Seneca Lucius Annaeus 'the Younger' (*c.*4 BC–65 AD)
Roman philosopher and writer
Stoicism
tragedy

Serret, Joseph Alfred (1819–85)
Mathematician
fundamental theorem of space curves

Seurat, Georges Pierre (1859–91)
French artist
neo-Impressionism
pointillism
post-Impressionism

Severini, Gino (1883–1966)
Italian artist
Futurism

Sextus Empiricus (2nd century AD)
Greek philosopher and physician
Pyrrhonism
scepticism

Shakespeare, William (1564–1616)
English playwright and poet
nature and nurture
tragedy

Shannon, Claude Elwood (1916–)
American mathematician
communication theory
mathematical law
Shannon's theorem

Shaw, George Bernard (1856–1950)
Irish playwright and writer
Pygmalion effect

Shaw, Jeffrey (1944–)
Artist
eventstructure

Sheldon, William Herbert (1898–1970)
American psychologist
constitutional theory
Sheldon's constitutional theory

Shelford, Victor Ernest (1877–1968)
American zoologist
tolerance, law of

Shelley, Percy Bysshe (1792–1822)
English poet
poet as legislator
Romanticism

Sherif, M (1906–)
Turkish psychologist
assimilation theory

Shklovsky, Viktor (1893–1984)
Russian literary theorist
art as device
deautomatization
fabula and *sjuzhet*
formalism, Russian

Sidgwick, Henry (1838–1900)
English philosopher
externalities

Siegel, Carl Ludwig (1896–1981)
German mathematician
Thue-Siegel-Roth theorem

Signac, Paul (1863–1935)
French painter
divisionism

Simon, Sir Franz Eugen (1893–1956)
German physicist
Nernst's heat theorem (third law of thermodynamics)

Wedgwood, Thomas
British early photographer
camera obscura

Wegener, Alfred Lothar (1880–1930)
German geophysicist
Bergeron's theory
continental drift theory

Weierstrass, Karl Theodor Wilhelm (1815–97)
German mathematician
Bolzano-Weierstrass theorem
Casorati-Weierstrass theorem
dynamics of an asteroid
Lindemann's theorem
maximum value theorem
Stone-Weierstrass theorem
Weierstrass approximation theorem
Weierstrass M test

Weinberg, Steven (1933–)
American physicist
electroweak theory

Weinberg, Wilhelm (1862–1937)
German geneticist
Hardy-Weinberg equilibrium, law or principle

Weismann, August Friedrich Leopold (1834–1914)
German biologist
germ plasm theory
programmed ageing
selfish DNA
senescence gene hypothesis

Weiss, Christian Samuel (1780–1856)
German chemist
rational intercepts, law of
Weiss zone law

Weiss, Peter Ulrich (1916–82)
German dramatist
theatre of cruelty

Weiss, Pierre (1865–1940)
French physicist
Curie-Weiss law
Weiss theory of ferromagnetism

Weizsäcker, Carl Friedrich, Baron von (1912–)
German philosopher and physicist
liquid drop model of the atomic nucleus

Wellek, René (1903–)
Literary theorist
new criticism

Wellesley-Miller, Sean
Artist
eventstructure

Went, Frits Warmolt (1903–90)
Dutch/American botanist
Cholodny-Went model

Werner, Abraham Gottlob (1749–1817)
German geologist
neptunism

Werner, Alfred (1866–1919)
German/Swiss chemist
Werner theory of co-ordination compounds

Wernicke, Carl (1848–1905)
German neurologist
Wernicke's area

Wertheimer, Max (1880–1943)
German psychologist and philosopher
equality, law of
gestalt theory
good continuation, law of
good shape, law of
phi phenomenon
pragnanz
reorganization theory
simplest path law
stepwise phenomenon

Wever, E G
American psychologist
volley theory of hearing

Wexler, P J
British linguist
grammetrics

Wharf, B L (1897–1941)
Philosopher
perspectivism

Wheeler, John Archibald (1911–)
American theoretical physicist
black hole
Bohr-Wheeler theory of nuclear fission
many worlds hypothesis

Whewell, William (1794–1866)
English scholar
catastrophism
hypothetico-deductive method

Whistler, James Abbott McNeill (1834–1903)
American artist
aestheticism

White, Michael James Denham (1910–)
Australian cytogeneticist
stasipatric speciation

Whitehead, Alfred North (1861–1947)
English mathematician and philosopher
panpsychism
process philosophy

Whitehead, John Henry Constantine (1904–60)
Mathematician
K-theory

Whorf, Benjamin Lee (1897–1941)
American linguist
Sapir-Whorf hypothesis

Whybrow, Peter C
Biochemist
indoleamine hypothesis of depression

Wicksell, Knut (1851–1926)
Swedish economist
loanable funds theory of the rate of interest
neo-classical theory
Wicksell's theory of capital

Wicksteed, Philip (1844–1927)
English economist
adding-up problem

Wiedemann, Gustav Heinrich (1826–99)
German physicist
Wiedemann-Franz-Lorenz law
Wiedemann's additivity law

Wien, Wilhelm (1864–1928)
German physicist
Wien's displacement law
Wien's distribution law

Wiener, Norbert (1894–1964)
American mathematician
cybernetics

Wigner, Eugene Paul (1902–)
Hungarian/American physicist
absolute reaction rate, theory of
Breit-Wigner equation
compound nucleus theory of Bohr and Breit and Wigner

Wilcoxen, Frank (1892–1965)
mathematician
Wilcoxen test

Wilde, Oscar Fingal O'Flahertie Wills (1854–1900)
Irish writer and wit
aestheticism

Wilkins, Maurice Hugh Frederick (1916–)
British physicist
Watson-Crick hypothesis or model

William (snr), Nassau
Economist
economic methodology

Williams, Raymond (1921–88)
Welsh critic and novelist
cultural materialism
Marxist criticism

Williams, William Carlos (1883–1963)
American poet and novelist
objectivism

Willis, Thomas (1621–73)
English physician
animal heat

Wilson, Allan
Biologist
molecular clock

Wilson, Colin Henry (1931–)
English writer
outsider

Wilson, Deirdre
Linguist
relevance

Wilson, Edward Osborne (1929–)
American biologist
sociobiology

Subject Area Index

Entries are arranged alphabetically under their relevant subject area(s).

Climatology

Computing

Economics

Literary theory

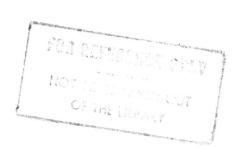